Lecture Notes in Computer Science 702
Edited by G. Goos and J. Hartmanis

E. Börger G. Jäger H. Kleine Büning
S. Martini M. M. Richter (Eds.)

Computer Science Logic

6th Workshop, CSL '92
San Miniato, Italy
September 28 - October 2, 1992
Selected Papers

Springer-Verlag
Berlin Heidelberg New York
London Paris Tokyo
Hong Kong Barcelona
Budapest

Series Editors

Gerhard Goos
Universität Karlsruhe
Postfach 69 80
Vincenz-Priessnitz-Straße 1
D-76131 Karlsruhe, Germany

Juris Hartmanis
Cornell University
Department of Computer Science
4130 Upson Hall
Ithaca, NY 14853, USA

Volume Editors

Egon Börger
Simone Martini
Dipartimento di Informatica, Università di Pisa
Corso Italia, 40, I-56125 Pisa, Italy

Gerhard Jäger
Universität Bern, Institut für Informatik und angewandte Mathematik
Länggasstraße 51, CH-3012 Bern, Switzerland

Hans Kleine Büning
FB 17, Mathematik/Informatik, Universität - GH Paderborn
Postfach 1621, D-33095 Paderborn, Germany

Michael M. Richter
FB Informatik, Universität Kaiserslautern
Postfach 30 49, D-67653 Kaiserslautern, Germany

CR Subject Classification (1991):F, I.2.3-4, G.2, D.3

ISBN 3-540-56992-8 Springer-Verlag Berlin Heidelberg New York
ISBN 0-387-56992-8 Springer-Verlag New York Berlin Heidelberg

Printed in Germany

Typesetting: Camera-ready by authors
Printing and binding: Druckhaus Beltz, Hemsbach/Bergstr.
45/3140-543210 - Printed on acid-free paper

Preface

The *Computer Science Logic* Workshop CSL'92 was held in San Miniato (Pisa) from September 28 to October 2, 1992. It took place in the charming environment of the Centro Studi *I Cappuccini*, a nicely restored monastery made available by the Cassa di Risparmio of San Miniato. CSL'92 was the sixth of the series and the first one which was held as Annual Conference of the *European Association for Computer Science Logic*, founded in Schloß Dagstuhl in July 1992 by computer scientists and logicians from 14 countries.

The workshop was attended by 78 participants from 15 countries; 8 invited lectures and 25 talks, selected from 72 submissions, were presented. Following the traditional procedure for CSL volumes, full versions of the original contributions have been collected after their presentation at the workshop and a regular reviewing procedure has been started. On the basis of 58 reviews, 26 papers were selected for publication. They appear here in revised final form.

We thank the referees, without whose help we would not have been able to accomplish the difficult task of selecting among the many valuable contributions.

We also gratefully acknowledge the generous sponsorship by the following institutions:

Consiglio Nazionale delle Ricerche (CNR)
Cassa di Risparmio di San Miniato
Università degli Studi di Pisa
Dipartimento di Informatica dell'Università di Pisa
Cassa di Risparmio di Pisa
Hewlett-Packard Italiana S.p.A., Pisa Science Center

Finally, we would like to thank the following persons who generously helped in various ways in the organization of the conference: Antonella D'Alessandro, Paola Glavan, Stefania Gnesi, Elvinia Riccobene.

March 1993

E. Börger G. Jäger H. Kleine Büning S. Martini M.M. Richter

List of Referees

K. Ambos-Spies, Heidelberg
H. Barendregt, Nijmegen
M. Bezem, Utrecht
E. Börger, Pisa
S. Buss, San Diego
T. Coquand, Göteborg
E. Dahlhaus, Sydney
R. De Nicola, Roma
M. Dezani-Ciancaglini, Torino
A. Goerdt, Paderborn
E. Grädel, Aachen
E. Grandjean, Caen
Y. Gurevich, Ann Arbor
M. Hanus, Saarbrücken
R. Hasegawa, Paris
F. Honsell, Udine
M. Hyland, Cambridge
N. Immerman, Amherst
G. Jäger, Berne
H. Jervell, Oslo
H. Kleine Büning, Paderborn
P. Kolaitis, Santa Cruz
P. Lincoln, Stanford
J. Makowsky, Haifa
S. Martini, Pisa
W. McCune, Argonne
E. Moggi, Genova
F. Montagna, Siena
L. Pacholski, Wroclaw
M. Parigot, Paris
C. Paulin-Mohring, Lyon
L. Priese, Koblenz
W. Reisig, München
M.M. Richter, Kaiserslautern
D. Rosenzweig, Zagreb
L. Roversi, Pisa
D. Sangiorgi, Edinburgh
A. Scedrov, Philadelphia
W. Schönfeld, Heidelberg
H. Schwichtenberg, München
D. Seese, Karlsruhe
J. Shepherdson, Bristol
E. Speckenmeyer, Düsseldorf
D. Spreen, Siegen
R. Stärk, Münich
W. Thomas, Kiel
K. Wagner, Würzburg

Table of Contents

A Universal Turing Machine

Stål Aanderaa

University of Oslo
staal.aanderaa@math.uio.no

Abstract. The aim of this paper is to give an example of a universal Turing machine, which is somewhat small. To get a small universal Turing machine a common constructions would go through simulating tag system (see Minsky 1967). The universal machine here simulate two-symbol Turing machines directly.

The Turing machine is defined by the Figure 1 or the Table 1. Suppose the universal Turing machine should simulate the Turing machine defined by the Figure 2 or by the Table 2, starting by the instantaneous description

$$010p_0100 \tag{1}$$

Then the universal Turing machine UTm should start by the instantaneous description

$$01AABq_0baac^3d^{599}cd^{14}c^3d^{708}cd^{15}c^3d^{599}cd^{10}c^3d^{644}cd^{11}c. \tag{2}$$

Here the first three symbols in (2): $01A$, code the first three symbols in (1): 010. The next two symbols in (2): AB code the state symbol p_0 in (1). q_0 in (2) denote the state of the universal Turing machine. The symbols baa in (2) code the last three symbols 100 in (1). The last part of (2):

$$c^3d^{599}cd^{14}c^3d^{708}cd^{15}c^3d^{599}cd^{10}c^3d^{644}cd^{11}c. \tag{3}$$

codes the Turing machine Tme defined in Table 2. The exponents are calculated as follows:

	A	B	C	D	E	F
(i)	$0p_00 \vdash 0H0$	$AABA$	$(0010)_2 = 2$	$RAAA0$,	$(21113)_4$	$= 599$
(ii)	$0p_01 \vdash_L p_101$	$AABB$	$(0011)_2 = 3$	$LBB01$	$(00032)_4$	$= 14$
(iii)	$0p_10 \vdash_R 01p_0$	$ABBA$	$(0110)_2 = 6$	$R0BAB$	$(23010)_4$	$= 708$
(iv)	$0p_11 \vdash_L p_100$	$ABBB$	$(0111)_2 = 7$	$LBB00$	$(00033)_4$	$= 15$
(v)	$1p_00 \vdash 1H0$	$BABA$	$(1010)_2 = 10$	$RAAA0$,	$(21113)_4$	$= 599$
(vi)	$1p_01 \vdash_L p_111$	$BABB$	$(1011)_2 = 11$	$LBB11$	$(00022)_4$	$= 10$
(vii)	$1p_10 \vdash_R 11p_0$	$BBBA$	$(1110)_2 = 14$	$R1BAB$	$(22010)_4$	$= 644$
$(viii)$	$1p_11 \vdash_L p_110$	$BBBB$	$(1111)_2 = 15$	$LBB10$	$(00023)_4$	$= 11$

In row (ii) the exponents of d is calculated to be 14, in order to simulate the move stated in column A. First we have to calculate where to put the information.

This is done coding $0p_01$ in the following way. Replace 0, p_0 and 1 by A, AB and B, respectively. Then we get the word $AABB$ in column B. This word is interpreted as a binary number which is calculated in column C to be 3. This means that the information about the move is to be located between the c number 4 and 5. The L in the word $LBB01$ means that the head moves to the left in this move. The rest $BB01$ of the word $LBB01$ codes the word p_101, where p_1 is replaced by BB and the rest of the word is kept unchanged. Then $LBB01$ is interpreted as a base 4 number in the following way: L, B, 0 and 1 are interpreted as the digits 0, 0, 3 and 2, respectively. The result is $(00032)_4$ which is the decimal number 14.

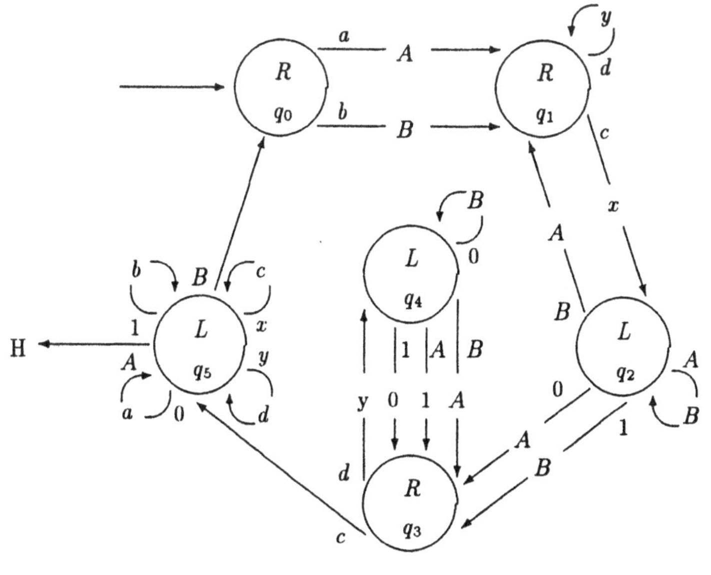

Fig. 1. Universal Turing machine: UTm

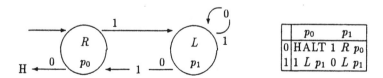

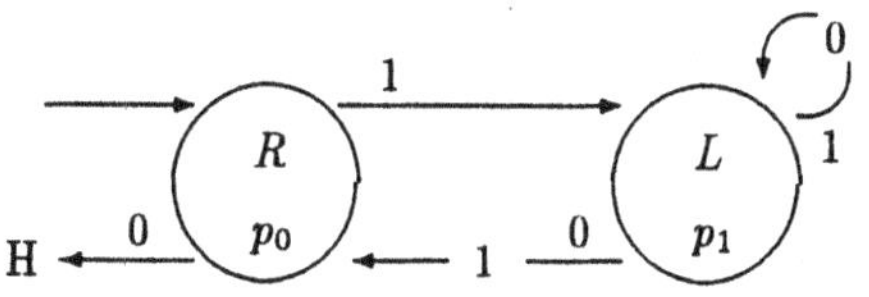

	p_0	p_1
0	HALT	1 R p_0
1	1 L p_1	0 L p_1

Fig. 2. Turing machine example: Tme

In row (*iii*) the columns A, B and C are made in the same way as in row (*ii*). In column D the word $R0BAB$ represents the subword $01p_0$ of column A in the following way. R means that the move is a right move. 0 is kept unchanged. 1 is

	q_0	q_1	q_2	q_3	q_4	q_5
a	$A\ R\ q_1$	$a\ R\ q_1$	$a\ L\ q_2$	$a\ R\ q_3$	$a\ L\ q_4$	$a\ L\ q_5$
b	$B\ R\ q_1$	$b\ R\ q_1$	$b\ L\ q_2$	$b\ R\ q_3$	$b\ L\ q_4$	$b\ L\ q_5$
c	$c\ R\ q_0$	$x\ L\ q_2$	$c\ L\ q_2$	$c\ L\ q_5$	$c\ L\ q_4$	$c\ L\ q_5$
d	$d\ R\ q_0$	$y\ R\ q_1$	$d\ L\ q_2$	$y\ L\ q_4$	$d\ L\ q_4$	$d\ L\ q_5$
x	$x\ R\ q_0$	$x\ R\ q_1$	$x\ L\ q_2$	$x\ R\ q_3$	$x\ L\ q_4$	$c\ L\ q_5$
y	$y\ R\ q_0$	$y\ R\ q_1$	$y\ L\ q_2$	$y\ R\ q_3$	$y\ L\ q_4$	$d\ L\ q_5$
A	$A\ R\ q_0$	$A\ R\ q_1$	$B\ L\ q_2$	$A\ R\ q_3$	$1\ R\ q_3$	HALT
B	$B\ R\ q_0$	$B\ R\ q_1$	$A\ R\ q_1$	$B\ R\ q_3$	$A\ R\ q_3$	$B\ R\ q_0$
0	$0\ R\ q_0$	$0\ R\ q_1$	$A\ R\ q_3$	$0\ R\ q_3$	$B\ L\ q_4$	$a\ L\ q_5$
1	$1\ R\ q_0$	$1\ R\ q_1$	$B\ R\ q_3$	$1\ R\ q_3$	$0\ R\ q_3$	$b\ L\ q_5$

Table 1. Universal Turing machine: UTm

replaced by B and p_0 is replaced by AB. Then R, 0, B and A are interpreted as the digits 2, 3, 0 and 1,respectively, in the number system of the base four. The result is $(23010)_4$ which is calculated to be 708 in decimal.

To simulate the move $010p_0100 \vdash 01p_10100$ of the Tme machine, the UTm machine will have to use about 20 000 steps. Among the instantaneous descriptions which will occur are the following instantaneous descriptions (Compare row (ii) above):

$$
\begin{aligned}
&01AABq_0baac^3d^{599}cd^{14}c^3d^{708}cd^{15}c^3d^{599}cd^{10}c^3d^{644}cd^{11}c.\\
&01AABBq_1aac^3d^{599}cd^{14}c^3d^{708}cd^{15}c^3d^{599}cd^{10}c^3d^{644}cd^{11}c.\\
&01AABBaq_2axc^2d^{599}cd^{14}c^3d^{708}cd^{15}c^3d^{599}cd^{10}c^3d^{644}cd^{11}c.\\
&0Bq_3BBBBaax^3y^{599}xd^{14}c^3d^{708}cd^{15}c^3d^{599}cd^{10}c^3d^{644}cd^{11}c.\\
&0BBBBBaax^3y^{599}q_4xyd^{13}c^3d^{708}cd^{15}c^3d^{599}cd^{10}c^3d^{644}cd^{11}c.\\
&0BBB01aax^3y^{599}xy^{13}q_5yc^3d^{708}cd^{15}c^3d^{599}cd^{10}c^3d^{644}cd^{11}c.\\
&0BBBq_0abaac^3d^{599}cd^{14}c^3d^{708}cd^{15}c^3d^{599}cd^{10}c^3d^{644}cd^{11}c.
\end{aligned}
$$

To simulate the next move $01p_10100 \vdash 011p_0100$ of the Tme machine, the following instantaneous descriptions will occur (Compare row (vii) above):

$$
\begin{aligned}
&0BBBq_0abaac^3d^{599}cd^{14}c^3d^{708}cd^{15}c^3d^{599}cd^{10}c^3d^{644}cd^{11}c.\\
&0BBBAq_1baac^3d^{599}cd^{14}c^3d^{708}cd^{15}c^3d^{599}cd^{10}c^3d^{644}cd^{11}c.\\
&0BBBAbaq_2axc^2d^{599}cd^{14}c^3d^{708}cd^{15}c^3d^{599}cd^{10}c^3d^{644}cd^{11}c.\\
&Aq_3BBBBbaax^3y^{599}xy^{14}x^3y^{708}xy^{15}x^3y^{599}xy^{10}x^3d^{644}cd^{11}c.\\
&ABBBBbaax^3y^{599}xy^{14}x^3y^{708}xy^{15}x^3y^{599}xy^{10}x^2q_4xyd^{643}cd^{11}c.\\
&01BABbaax^3y^{599}xy^{14}x^3y^{708}xy^{15}x^3y^{599}xy^{10}x^3y^{643}q_5ycd^{11}c.\\
&01BABq_0baac^3d^{599}cd^{14}c^3d^{708}cd^{15}c^3d^{599}cd^{10}c^3d^{644}cd^{11}c.
\end{aligned}
$$

The loop q_1 - q_2 converts from binary to unary in order to find the information about the next move.

Before the information is evaluated the UTm machine guesses that the next move will be a left move. Hence to be prepared for a left move, the UTm machine changes the content of a square when moving from state q_2 to the state q_3. (a 0 is changed to A and a 1 is canged to B). If the move turn out to be a right move, the guess was wrong, and in order to change back each R in colemn D above is replaced by 2 in colemn E. If the move was a left move, the guess was correct, and nothing has to be changed and aech L in colemn D above is replaced by 0 in colemn E.

The loop q_3 - q_4 converts from unary to a number in base four in order to code the next state and to code the new position of the head.

Reference

Minsky, Marvin *Computation : Finite and Infinite Machines.* Prentice-Hall 1967.

Recursive Inseparability in Linear Logic

Stål Aanderaa[1] and Herman Ruge Jervell[2]

[1] University of Oslo
staal.aanderaa@math.uio.no
[2] Universities of Oslo and Tromsø
herman.jervell@ilf.uio.no

Abstract. We first give our version of the register machines to be simulated by proofs in propositional linear logic. Then we look further into the structure of the computations and show how to extract "finite counter models" from this structure. In that way we get a version of Trakhtenbrots theorem without going through a completeness theorem for propositional linear logic. Lastly we show that the interpolant I in propositional linear logic of a provable formula $A \multimap B$ cannot be totally recursive in A and B.

We use results and notations about linear logic as given by Troelstra in his lectures [T].

1 ANDOR machines

We consider register machines with a finite number of registers **a b c d** .... The machines have a number of states say **p q r s** ... **t**. The computations are controlled by instructions. There are four forms of instruction

$p : a + (q)$ add 1 to register a and proceed to state q
$p : a - (q)$ subtract 1 from register a if possible and proceed to state q
$p : and(q, r)$ spawn off two processes and proceed with states q and r
$p : or(q, r)$ proceed nondeterministically to one of the states q and r

There are the following important differences from ordinary register machines

1. In the ordinary register machine the subtraction instruction is combined with a branch. If the register is empty we proceed in one direction, else we subtract 1 and proceed in another direction. This is not the case with andor machines. If the subtraction is not possible, we go over into a waiting state from which the computation does not proceed.
2. In the and-branching we spawn off two new processes from a given process.
3. A process halts if it comes to a halting state and all registers are empty.
4. To get termination of the machine we demand that all spawned processes halt.

Let us see that for register machine computations with empty halting states we get the same computations with andor machines. This is done by showing that register machine instructions can be translated directly into andor machine instructions. To do this we need a simple trick (see [LMSS]). Assume we have one halting state **h** and two registers **a b** and have the following register machine instruction

$$p : \text{if } a = 0 \text{ then goto } q \text{ else } a - (r)$$

To translate this instruction we need some new auxiliary states **k l m n** . The following andor machine instructions do the job for the case of two registers **a b** and one halting state **h**. The general case with more registers and more halting states is done in a similar way. The instruction above is replaced by

p	$: or(k, l)$
k	$: and(m, q)$
m	$: or(n, h)$
n	$: b - (m)$
l	$: a - (r)$

The trick is to allow a number of garbage computations which does not matter. After having come to state **k** , you can only come to state **q** if it is possible to get to a halting state **h** after emptying all other registers than **a** (in our case register **b**).

Any computation on a register machine ending in a halting state with empty registers can then be transferred to a terminating computation on an andor machine.

2 The structure of computations

A computation in a register machine can be thought of as a transition between storage states. In an andor machine we must also take into account all the processes that are spawned. This is done by the following syntax

register :: a | b | c | d | ...
state :: p | q | r | s | ...
storagestate :: **empty** | **register** | **state** | **storagestate** · **storagestate**
configuration :: **setof storagestate**

The product of storage states given by · is assumed to be commutative. We are only interested in storage states which have at most one state present - even if we have notation for more. The storage state **aaabbq** indicates that we are in state **q** with 3 in register **a** and 2 in register **b**. The following is a configuration {**aabq**, **aar**, **abs**} . Configurations are interpreted conjunctively so that the configuration above can be thought of as three processes with storage states **aabq aar abs** . Configurations are sets and not multisets. A typical halting state is a configuration { **h** } with empty registers. The computations in an andor machine can be thought of as transitions between configurations in an obvious way.

Recursive inseparability 1 *There is an instruction set I with two halting configurations C and D such that no state is transformed into both C and D, and the configurations transformed into C and those into D are recursively inseparable.*

To get to the algebraic structures we first remark that the storage state together with the concatenation $\cdot$ and **empty** makes a commutative monoid with unit. We use this to make an IL-algebra (see [T]) in the standard way

$$X * Y := \{x \cdot y : x \in X, y \in Y\}$$
$$X \multimap Y := \{z : \forall x \in X (z \cdot x \in Y)\}$$

Then $(\textbf{configurations}, \cap, \cup, \emptyset, \multimap, *, \{\textbf{empty}\},)$ is an IL-algebra. (See [T] Proposition 8.9). On the top of this algebraic structure we introduce the transition made by the andor machine. The transition $\rightarrow$ gives a relation between configurations. For the transition we have the following laws

T1 : If $C \rightarrow D$, then also $A * C \rightarrow A * D$
T2 : $C \rightarrow C$
T3 : If $C \rightarrow D$ and $D \rightarrow E$ then $C \rightarrow E$

Note T1. This is the crucial property in the construction of andor machines. It is not true for register machines.

An instruction set I is given by a finite number of relations of the following forms

I1 : $A * C \rightarrow D$ (from $a+$)
I2 : $C \rightarrow A * D$ (from $a-$)
I3 : $C \rightarrow D \cup E$ (from *and*)
I4 : $C \rightarrow D$ and $C \rightarrow E$ (from *or*)

Given an andor machine we can transform it faithfully into a configuration space with transitions and a finite number of equations of form I1-4. A configuration space with transition is a very simple structure. The laws are just sufficient to simulate computations in an andor machine given the instructions.

3 Linear logic

Now consider propositional linear logic. In most of our discussions we use the fragment MALLA — multiplicative additive propositional linear logic with assumptions. We write

$$A, B, C, \ldots \vdash F, G, H, \ldots$$

Here the assumptions are to the left of $\vdash$. The formulas A, B, C, F, G, H are from the multiplicative additive fragment of linear logic. The intended meaning is that assumptions can be brought into derivations any number of times – or possibly not at all. Using exponentials we can interpret this as

$$!A \star !B \star !C \multimap F + G + H$$

In MALLA there is an obvious interpretation of instruction sets for andor machines

- the letters in the instruction set is taken as atomic formulas
- $*$ into multiplicative conjunction $\star$
- $\cap$ into additive conjunction $\sqcap$
- $\cup$ into additive disjunction $\sqcup$
- $\rightarrow$ into linear implication $\multimap$
- {**empty**} into the multiplicative unit 1

Given an andor machine with the instruction set I. Let $I^{\#}$ be the corresponding formula in propositional linear logic. A configuration C can be transformed into a propositional linear formula $C^{\#}$ in the same way. Then

Theorem 1. *Let I be an instruction set and C and D configurations in an andor machine. The following are equivalent*

- *I transfers C into D*
- *from the equations I we can derive C→D using T1-3 and the laws of IL-algebra*
- $I^{\#} \vdash C^{\#} \multimap D^{\#}$ *is derivable in propositional linear logic*

The only slightly difficult part here is to prove that there are not more formulas of the form $I^{\#} \vdash C^{\#} \multimap D^{\#}$ provable in propositional linear logic than those given by the andor machine. This is proved in [LMSS] by a cut elimination argument.

Recursive inseparability 2 *There is an instruction set I and halting configuration D such that for configurations C we cannot recursively separate whether*

- *we can prove* $I^{\#} \vdash C^{\#} \multimap D^{\#}$ *in propositional linear logic*
- *a configuration structure with transition gives a counter model that C→D using instructions I*

4 Getting a phase structure

The configuration space is of course not much more than a convenient language for our syntactical transformations. The information about the particular andor machine we use is contained in the transitions. We now bring the transitions into the IL-algebra. A transition is a binary relation between configurations. We can interpret this in an IL-algebra by repeating the usual subset construction on the configuration space to get an IL-algebra with

Domain : Sets of configurations
Product : $X \star Y = \{x * y : x \in X, y \in Y\}$
Implication : $X \multimap Y = \{z : \forall x \in X(z * x \in Y)\}$
Lattice : Union and intersection of sets

On this IL-algebra we can interpret transitions as operators

$$C(X) = \{y : y \to x \wedge x \in X\}$$

taking sets of configurations into sets of configurations.
A closure operation in an IL-algebra satisfies

1. $X \subseteq C(X)$
2. $X \subseteq Y \Rightarrow C(X) \subseteq C(Y)$
3. $CC(X) \subseteq C(X)$
4. $C(X) * C(Y) \subseteq C(X * Y)$

We observe that the crucial property T1 for transitions is exactly what is needed to get property 4 above.

Lemma 2. *C is a closure operation.*

Let M be an andor machine. The phase structure corresponding to M is the IL-algebra given by the transition closure C above. We denote this algebra by ΦM.

5 Finite parts of machines

Each particular computation uses only finite information. This is the extra trick used to get the Trakhtenbrot theorem. Let us introduce machines with bounded registers. Such a machine is given by a number MAX. Assume that register **a** contains the number N. If N < MAX the instructions **a+** and **a-** behaves as usual. If N = MAX , then **a+** gives MAX while **a-** gives a nondeterministic choice between MAX and MAX-1. If M is the original machine, we let M/MAX be the machine with registers bounded by MAX. This construction works both for register machines and andor machines.

Lemma 3. *For register machines or andor machines. Let C and D be configurations with all registers $\leq$ MAX . Then if M transfers C into D, so does also M/MAX.*

Lemma 4. *For a deterministic register machine M. Let D and E be halting configurations with empty registers. Assume that M transfer C into D, but not E. Then neither does M/MAX for large enough MAX.*

We would like to get lemma 3 for arbitrary machines. We do not need to do this in general. It is only necessary to do this for andor machines derived from deterministic register machines. The andor machines are going to be nondeterministic in general with nondeterminism introduced by the branching instructions in the deterministic register machine. The nondeterminism dervived from the branching

$$p : \text{if } a = 0 \text{ then goto } q \text{ else } a-(r)$$

is

p	$: or(k,l)$
k	$: and(m,q)$
m	$: or(n,h)$
n	$: b-(m)$
l	$: a-(r)$

The branch going down from state k will only be used to get down to state q exactly when register a is empty. The same construction works equally well for an andor machine with bounded registers. Hence

Lemma 5. *Let M be an andor machine derived from a deterministic register machine and let D and E be halting configurations with empty registers. Assume that M transfers C into D, but not E. Then neither does M/MAX for large enough MAX.*

Recursive inseparability 3 *There is an instruction set I and halting configuration D such that for configuration C we cannot recursively separate whether*

- *we can prove in propositional linear logic that C gives D in I*
- *a finite configuration structure gives a countermodel that C⟶D in I*

The construction of the phase structure ΦM from andor machine M can also be repeated using bounded registers giving ΦM_{MAX} for registers bounded by MAX.

6 Exponentials

To interpret exponentials in an IL-algebra we need what Troelstra calls a modality; we then get an IL-algebra with storage ([T] Definition 8.16). A modality is a subset F of the domain X of the IL-algebra satisfying certain conditions (idempotency and ≤ 1) and

$$\{\bot, 1\} \subseteq F \subseteq \{x : x \star x = x \land x \leq 1\}$$

In all such modalities our transitions are going to be interpreted as 1. Because given an interpetation of !, a transition $C \to D$ and the phase structure ΦM (or ΦM_{MAX}), $C \to D$ is translated into a formula F in the multiplicative additive fragment of linear logic. In the phase structure we have $1 \leq F$, and hence $1 = !F$ for any modality.

Recursive inseparability 4 *It is recursively inseparable whether for formulas F in linear logic*

- *F is derivable in intuitionistic linear logic*
- *A finite phase structure with modality gives a counter example to F*

7 Introducing duality

So far the recursive inseparability is in intuitionistic linear logic. The outline of the argument is as follows. We started with an andor machine with two halting states where we cannot recursively separate whether a particular starting configuration ends in one or the other of the halting states. The computations can be represented as proofs in linear logic or by interpretations in a configuration space with transitions. The configuration space is later taken over to a phase structure. To get recursive inseparability in classical linear logic it is sufficient to show that an IL-algebra can be embedded into a CL-algebra.

We start with an IL-algebra L. Let $L^\perp$ be the dual lattice. We then construct a lattice $LL^\perp$ by using the sum of L and $L^\perp$ - put L and $L^\perp$ between the top and bottom elements $\top$ and $\perp$.The lattice operations in $LL^\perp$ are defined in the obvious way. We have also defined a duality $\perp$. We need to define products in the extended lattice. To do that we divide up into cases depending on whether the elements come from $\top$, L, $L^\perp$ or $\perp$. We define products by the following multiplication table

$\star$	$\top$	A_1	$B_1^\perp$	$\perp$
$\top$	$\top$	$\top$	$\top$	$\perp$
A_0	$\top$	$A_0 \star A_1$	$(A_0 \multimap B_1)^\perp$	$\perp$
$B_0^\perp$	$\top$	$(A_1 \multimap B_0)^\perp$	$\top$	$\perp$
$\perp$	$\perp$	$\perp$	$\perp$	$\perp$

To prove that $LL^\perp$ is a CL-algebra we note

- $\star$ is commutative
- 1 is a multiplicative unit
- $\star$ is associative
- $X \multimap Y = (X \star Y^\perp)^\perp$
- $X^\perp = X \multimap 0$

Having constructed the embedding of an intuitionistic linear algebra into a classical linear algebra we can conclude

Recursive inseparability 5 *For formulas F in propositional linear logic we cannot recursively separate whether*

- *we can prove F in classical propositional linear logic*
- *a finite CL-algebra with modality gives a counter-model to F*

8 Interpolation

The interpolation theorem is true for classical and intuitionistic linear logic (see Roorda's thesis [R]), and we can show many of the usual properties of the interpolant. In particular, we can conclude that if we have two formulas A, B without any other common propositional symbols than 0 and 1 and if we can prove

$$\vdash A \multimap B$$

then there is an interpolant I built up from the propositional constants alone such that

$$\vdash A \multimap I$$

and

$$\vdash I \multimap B.$$

But the following argument shows that there is no total recursive function giving the interpolant as a function of the two formulas A and B. As in classical predicate logic we need to know the derivation of the formula $A \multimap B$.

Let A and B be the formulas expressing that an andor machine terminates in a halting state a respectively b. We know that there is no recursive function which can decide whether A or B is derivable. Consider now the following implication

$$C = !(!A \multimap \bot) \multimap !B$$

Observe

Lemma 6. *For A B and C as above we have (in both intuitionistic and classical logic) :*

- $\vdash A \Rightarrow \vdash C$
- $\vdash B \Rightarrow \vdash C$

Now let I be an interpolating formula for the implication C. I is built up from the propositional constants alone. We have

$$\vdash !(!A \multimap \bot) \multimap I$$

$$\vdash I \multimap !B$$

We have then

$$\vdash !(!A \multimap \bot) \multimap !I$$

$$\vdash !I \multimap !B$$

Observe that in our models of linear logic with the trivial modality – $\bot$ and 1 – we have

The Basic Logic of Proofs

Sergei Artëmov*
Steklov Mathematical Institute,
Vavilov str. 42,
117966 Moscow, Russia.
e-mail: art@log.mian.su

Tyko Straßen†
University of Berne, IAM,
Länggassstr. 51,
CH- 3012 Berne.
e-mail: strassen@iam.unibe.ch

Abstract

Propositional Provability Logic was axiomatized in [5]. This logic describes the behaviour of the arithmetical operator "y is provable". The aim of the current paper is to provide propositional axiomatizations of the predicate "x is a proof of y" by means of modal logic, with the intention of meeting some of the needs of computer science.

1 Introduction

The Propositional Provability Logic GL was axiomatized in [5]. This logic describes the behaviour of the arithmetical operator "y is provable" by means of modal logic. Although some properties of this logic are relevant for computer science (e.g. various forms of Gödel's incompleteness theorem for consistency proofs in databases), GL is rather a mathematical domain. One reason is that in computer science not only the *provability* of a statement is of interest, but also in many cases the *proofs themselves*, respectively informations about the time or memory expenditure for a proof are known. These considerations lead to a different situation. For example it is well-known that a powerful machine cannot prove its own consistency, but it is very well possible for such a machine to demonstrate that a given proof does not derive $0 = 1$, or that no computation within a fixed time comes to that answer. The studies on the Logic of Proofs have been initiated by a series of questions by G. Jäger related to this topic. One was to give an arithmetically complete propositional axiomatization of the predicate "x is a proof of y". The modal systems $\mathcal{P}$ and $\mathcal{PF}$ introduced below solve this problem.

Most definitions in this introduction are in accordance with those of the classical Provability Logic [5]. Nevertheless, the Basic Logic of Proofs is entirely different from the Provability Logic, and the arithmetical completeness proof for it does not use the Solovay argument.

*Supported by the Swiss Nationalfonds (project 21-27878.89) during a stay at the University of Berne in January 1992.

†Financed by the Union Bank of Switzerland (UBS/SBG) and by the Swiss Nationalfonds (projects 21-27878.89 and 20-32705.91).

$$\vdash K \Rightarrow \models !K = 1$$

$$\vdash K \multimap L \Rightarrow \models K \leq L$$

Our model can separate between derivations terminating in a and terminating in b. We have

$$\nvdash A \Rightarrow \models !A = \bot$$

$$\nvdash B \Rightarrow \models !B = \bot$$

Applying this we get

$$\vdash A \Rightarrow \nvdash B \rightarrow \models !B = \bot \Rightarrow !I = \bot$$

$$\vdash B \Rightarrow \nvdash A \Rightarrow \models !A = \bot \Rightarrow !(!A \multimap \bot) = 1 \Rightarrow !I = 1$$

The interpolation formula $!I$ is built up from the propositional constants. The interpretation of $!I$ is given as $\bot$ or 1 depending on whether I as a classical propositional formula has truth value **false** or **true**. So the interpretation of $!I$ is recursively given by I.

Theorem 7. *In (classical or intuitionistic) propositional linear logic there is no totally recursive interpolation function.*

9 References

B : E Börger. *Berechenbarkeit, Komplexität, Logik.* Vieweg Verlag 1985. English translation published by North-Holland.

LMSS : P D Lincoln, J Mitchell, A Scedrov, N Shankar. *Decision problems for propositional linear logic.* In *Proc. 31st Annual IEEE Symposium on Foundations of Computer Science, St Louis, Missouri, October 1990.*

R : D Roorda *Resource Logics: Proof-theoretical Investigations.* Thesis. Amsterdam 1991.

T : A S Troelstra. *Lectures on linear logic.* CSLI Lecture Notes No 29. 1991

1.1 Definition The modal language contains two sorts of variables,

$$p_0, p_1, \ldots \text{ (called } \textit{proof variables}\text{)},$$
$$S_0, S_1, \ldots \text{ (called } \textit{sentence variables}\text{)},$$

the usual boolean connectives, truth values $\top$ (for truth) and $\bot$ (for absurdity), and the labeled modality symbol $\Box_{p_i}$ for each proof variable p_i. The modal language is generated from the atoms $\top, \bot, S_0, S_1, \ldots$ by the boolean connective $\rightarrow$ as usual, and by the unary modal operators $\Box_{p_0}(\cdot), \Box_{p_1}(\cdot), \ldots$ as follows: if p is a proof variable and A a modal formula then $\Box_p(A)$ ($\Box_p A$ for short) is a modal formula.

Parentheses are avoided whenever possible by the usual conventions on precedence along with the modal convention that $\Box_{p_i}(\cdot)$ is given the minimal scope. Small letters $p, q, r, \ldots$ are used for proof variables, capital letters $S, T, \ldots$ for sentence variables and $A, B, C, \ldots$ for modal formulas.

The clear intention is to interpret $\Box_p A$ as "p is a proof of A". In order to allow iterations of modalities, which is an essential principle of the Logic of Proofs, the modal language must be interpreted in theories, which are able to link theorems and proofs after some natural coding. These considerations lead to the notion of the *arithmetical interpretation* of the modal language.

1.2 Definition Let the theory T be a recursive extension of $I\Sigma_1$ which is valid in the standard model of arithmetic, for example let T be Peano Arithmetic **PA**. Greek letters $\varphi, \psi, \ldots$ denote arithmetical formulas. In this paper it will not be distinguished between the number n and its numeral $\overline{n}$.

1.3 Definition An arithmetical formula $Prf(\cdot,\cdot)$ is called a *proof predicate* in T iff

- $Prf(x, y)$ is (provably-in-T equivalent to) a recursive formula in x and y.
- $\mathrm{T} \vdash \varphi \iff \exists n \in \mathbf{N}: \ Prf(n, \ulcorner\varphi\urcorner)$ for all arithmetical formulas φ.

A proof predicate is thus nothing but a recursive enumeration of the theorems of T.

1.4 Example

1. A *standard Gödel proof predicate* for the theory T is a recursive formula $\widetilde{Prf}(\cdot,\cdot)$, such that for every n, m: $\widetilde{Prf}(n, m)$ is true iff n codes a proof in T of the formula coded by m. Examples of such formulas can be found e.g. in [3] or [4]. Note that, according to $\widetilde{Prf}(\cdot,\cdot)$, (a) each proof codes exactly one theorem but each theorem has infinitely many proofs, and, (b) each proof is longer than the theorem proved by it, provided the usual Gödel numbering is used.

2. A modification $Prf_1(\cdot,\cdot)$ of $\widetilde{Prf}(\cdot,\cdot)$ is obtained if one allows not only proofs as first argument, but also programs. Note that property (b) of the first example fails for $Prf_1(\cdot,\cdot)$: $Prf_1(x, y)$ does not imply in general $x \geq y$ because short programs can compute long theorems.

3. In the context of ***resource bounded reasoning*** one can construct the proof predicate $Prf_2(x,y) := \exists p \leq x : Prf_1(p,y)$, i.e. which is true iff there is a a program of size not greater than x which computes y. Note that also property (a) of the first example fails for $Prf_2(\cdot,\cdot)$: for some x there can be several formulas φ such that $Prf_2(x,\ulcorner\varphi\urcorner)$ holds.

1.5 Definition A proof predicate is called *functional* iff for all $n, k_1, k_2 \in \mathbf{N}$:

$$\text{If } Prf(n,k_1) \text{ and } Prf(n,k_2) \text{ then } k_1 = k_2.$$

The standard Gödel proof predicate $\widetilde{Prf}(\cdot,\cdot)$ and $Prf_1(\cdot,\cdot)$ from example 1.4 are examples of such functional proof predicates.

1.6 Definition Let $Prf(\cdot,\cdot)$ be a proof predicate in **T**, and let ϕ be a function that assigns to each proof variable p_i some $n \in \mathbf{N}$ and to each sentence variable S_i a sentence of **T**. An *arithmetical interpretation* $(\cdot)^*$ is a pair $(Prf(\cdot,\cdot),\phi)$ of such $Prf(\cdot,\cdot)$ and ϕ. The arithmetical interpretation $(A)^*$ (A^* for short) of a modal formula A is the extension of ϕ to all modal formulas by:

- $\top^* := (0=0) \quad \bot^* := (0=1) \quad p_i{}^* := \phi(p_i) \quad S_i{}^* := \phi(S_i)$
- $(A \to B)^* := A^* \to B^*$
- $(\Box_p A)^* := Prf(p^*, \ulcorner A^* \urcorner)$

1.7 Definition Let $\equiv$ denote the syntactical identity of formulas (e.g. $\varphi \equiv \varphi$, but $\varphi \wedge \varphi \not\equiv \varphi$, $S_0 \not\equiv S_1$ and $\Box_{p_0}\top \not\equiv \Box_{p_1}\top$). An arithmetical interpretation $(\cdot)^* = (Prf(\cdot,\cdot),\phi)$ is called *functional* iff $Prf(\cdot,\cdot)$ is functional and $(\cdot)^*$ is injective, which means that $A^* \equiv B^*$ implies $A \equiv B$. The requirement of injectivity can be eliminated as it is demonstrated in [2].

The Basic Logic of Proofs is not concerned with occasional details about the coding of proofs in **T** by means of one fixed $Prf(\cdot,\cdot)$. Rather the Basic Logic of Proofs describes those basic principles which are true for *all* proof predicates of a given class.

In this paper the decidable modal logics $\mathcal{P}$ and $\mathcal{PF}$ are introduced, with axioms and rules of inference as follows:

$$\left.\begin{array}{l}\left.\begin{array}{ll}\textbf{(A1)} & \text{All (boolean) tautologies}\\ \textbf{(A2)} & \Box_p A \longrightarrow A\\ \textbf{(R1)} & \dfrac{A \quad A \to B}{B}\end{array}\right\}\mathcal{P}\\ \begin{array}{ll}\textbf{(A3)} & \Box_p A \longrightarrow \neg\Box_p B \quad (A \not\equiv B)\end{array}\end{array}\right\}\mathcal{PF}$$

where A and B are modal formulas and p is a proof variable.
(A2) is the *Reflexivity Axiom* and (A3) the *Functionality Axiom*.

The main result of the paper claims that for each modal formula A:

$$\mathcal{P} \vdash A \iff A^* \text{ is true for every interpretation } (\cdot)^*$$
$$\mathcal{PF} \vdash A \iff A^* \text{ is true for every functional interpretation } (\cdot)^*$$

Moreover, for each of these completeness theorems a proof predicate $Prf(\cdot,\cdot)$ can uniformly be chosen.

It is easy to see that neither $\mathcal{P}$ nor $\mathcal{PF}$ is closed under the necessitation $A \vdash \Box_p A$, or under the substitution rule $A \leftrightarrow B \vdash \Box_p A \leftrightarrow \Box_p B$.

The standard Gödel proof predicate $\widetilde{Prf}(\cdot,\cdot)$ enjoys an additional property, namely to be monotonic:

1.8 Definition A proof predicate is called *monotone* iff $Prf(n,k)$ implies $n \geq k$ for all $n, k \in \mathbb{N}$. An arithmetical interpretation $(\cdot)^* = (Prf(\cdot,\cdot), \phi)$ is called *monotone* iff $Prf(\cdot,\cdot)$ is a monotonic proof predicate.

A logic $\mathcal{PFM}$ containing all those formulas which are true under all interpretations based on any fixed functional and monotonic proof predicate has been established in [1], too. Thereby, $\mathcal{PFM}$ is the logic of the standard Gödel proof predicate $\widetilde{Prf}(\cdot,\cdot)$. $\mathcal{PFM}$ will not be presented in this paper.

In section 2 the soundness and completeness for $\mathcal{P}$ is proved, in section 3 the same is done for $\mathcal{PF}$ and section 4 is devoted to uniform proof predicates.

2 Arithmetical soundness and completeness of $\mathcal{P}$

The aim of this section is to prove soundness and completeness of the modal system $\mathcal{P}$ with respect to arithmetical interpretations.

2.1 Theorem Let A be a modal formula. Then

$$\begin{aligned} \mathcal{P} \vdash A \quad & \iff \quad \forall^*:\ \mathrm{T} \vdash A^* \\ & \iff \quad \forall^*:\ A^* \text{ is true} \end{aligned}$$

Proof of the soundness of $\mathcal{P}$: Let $(\cdot)^*$ be some arithmetical interpretation. One has to show that $\mathrm{T} \vdash A^*$. Induction on the complexity of the proof of A:

(A1) and (R1) straightforward.

(A2) 1st case: $\mathrm{T} \vdash Prf(p^*, \ulcorner A^* \urcorner)$. It follows that $\mathrm{T} \vdash A^*$, hence $\mathrm{T} \vdash Prf(p^*, \ulcorner A^* \urcorner) \to A^*$.
2nd case: $\mathrm{T} \nvdash Prf(p^*, \ulcorner A^* \urcorner)$. As $Prf(\cdot,\cdot)$ is recursive $\mathrm{T} \vdash \neg Prf(p^*, \ulcorner A^* \urcorner)$, hence $\mathrm{T} \vdash Prf(p^*, \ulcorner A^* \urcorner) \to A^*$.

■

The next task is to prove the arithmetical completeness of $\mathcal{P}$.

2.1 Gentzen style system for $\mathcal{P}$

In the following, a *sequent* is a formal expression $\Gamma \supset \Delta$, where Γ and Δ are finite sets of modal formulas. If $\Gamma = A_1, \ldots, A_k$ then $\bigwedge \Gamma := A_1 \wedge \ldots \wedge A_k$; analogous definition for $\bigvee \Delta$.

2.2 Definition $\mathcal{P_G}$ is the sequent calculus with axioms and rules of inference as follows:

- Sequent calculus for classical propositional logic including the cut-rule.
- $\dfrac{A, \Gamma \supset \Delta}{\Box_p A, \Gamma \supset \Delta}\text{refl}$

$\mathcal{P}_\mathcal{G}^-$ is the system $\mathcal{P_G}$ without the cut rule.

2.3 Soundness of $\mathcal{P_G}$ w.r.t. $\mathcal{P}$ For each sequent $\Gamma \supset \Delta$:

$$\mathcal{P_G} \vdash \Gamma \supset \Delta \quad \Longrightarrow \quad \mathcal{P} \vdash \bigwedge \Gamma \to \bigvee \Delta$$

Proof Straightforward induction on the complexity of the $\mathcal{P_G}$-proof of $\Gamma \supset \Delta$.
■

2.4 Definition The sequent $\Gamma \supset \Delta$ is called *saturated*, if the following statements hold for all modal formulas A, B and for all proof variables p:

1. $A \to B \in \Delta$ implies $A \in \Gamma$ and $B \in \Delta$,
2. $A \to B \in \Gamma$ implies $B \in \Gamma$ or $A \in \Delta$,
3. $\Box_p A \in \Gamma$ implies $A \in \Gamma$.

2.5 Saturation Lemma Let $\Gamma \supset \Delta$ be a sequent such that $\mathcal{P}_\mathcal{G}^- \nvdash \Gamma \supset \Delta$. Then there exists a saturated sequent $\Gamma' \supset \Delta'$ such that

(i) $\Gamma \subset \Gamma'$, $\Delta \subset \Delta'$,

(ii) $\Gamma \cup \Delta$ and $\Gamma' \cup \Delta'$ have the same subformulas,

(iii) $\mathcal{P}_\mathcal{G}^- \nvdash \Gamma' \supset \Delta'$.

Furthermore $\Gamma' \supset \Delta'$ is effectively computable from $\Gamma \supset \Delta$.
Such a $\Gamma' \supset \Delta'$ is called a *saturation* of $\Gamma \supset \Delta$.

Proof This is a fairly standard lemma. Here just a recursive algorithm is given, which accepts a sequent as input and which saturates this sequent provided it is not $\mathcal{P}_\mathcal{G}^-$-provable, otherwise the algorithm fails.

2.6 Saturation Algorithm Given $\Gamma \supset \Delta$, for each subformula S of $\Gamma \cup \Delta$ nondeterministically try to perform one of the following steps:

- if $S = A \to B \in \Delta$ then $\Gamma := \Gamma \cup \{A\}$ and $\Delta := \Delta \cup \{B\}$.
- if $S = A \to B \in \Gamma$ then either $\Gamma := \Gamma \cup \{B\}$ and branch, or $\Delta := \Delta \cup \{A\}$ and branch.
- if $S = \Box_p A \in \Gamma$ then $\Gamma := \Gamma \cup \{A\}$.
- if $\perp \in \Gamma$ or $\top \in \Delta$ or $\Gamma \cap \Delta \neq \emptyset$ (i.e. $\Gamma \supset \Delta$ is an axiom) then backtrack.

Properties of the Saturation Algorithm:

- Termination: there are only finitely many subformulas in $\Gamma \cup \Delta$ and at most two branches in each step.
- (i) and (ii) clearly hold.
- If the algorithm fails then each branch in the computation contains an axiom, and one can readily construct a $\mathcal{P}_{\mathcal{G}}^{-}$-proof of $\Gamma \supset \Delta$.
- If the algorithm succeeds then the resulting sequent $\Gamma' \supset \Delta'$ is saturated and not $\mathcal{P}_{\mathcal{G}}^{-}$-provable. Otherwise, assume that $\Gamma' \supset \Delta'$ is $\mathcal{P}_{\mathcal{G}}^{-}$-provable and hence, as it is saturated, an axiom. Starting with $\Gamma' \supset \Delta'$ and according to the saturation process one can construct a $\mathcal{P}_{\mathcal{G}}^{-}$-proof of $\Gamma \supset \Delta$.

■

2.2 Arithmetical completeness

2.7 Main Lemma for $\mathcal{P}$ Let $\Gamma' \supset \Delta'$ be a saturated sequent which is not $\mathcal{P}_{\mathcal{G}}^{-}$-provable. Then there exists an arithmetical interpretation $(\cdot)^*$ which makes all formulas in Γ' true and all formulas in Δ' false, i.e.

$$(\bigwedge \Gamma' \to \bigvee \Delta')^* \quad \text{is false.}$$

Proof For the proof and sentence variables let $(\cdot)^*$ be defined as:

$$p_i{}^* := 2i, \qquad S_i{}^* := \begin{cases} \forall x_i(x_i = x_i) & \text{if } S_i \in \Gamma', \\ \forall x_i(x_i \neq x_i) & \text{else.} \end{cases}$$

By induction on the complexity of a modal formula it follows that $(\cdot)^*$ is an injective arithmetical interpretation of the modal language.

Write $\bigwedge \Gamma'$ as

$$\bigwedge_{i=0}^{m} \bigwedge_{j=0}^{J_i} \Box_{p_i} A_{i,j} \;\wedge \Gamma''$$

where Γ'' contains no formula of the form $\Box_p A$.

By some variant of the arithmetical fixed point argument (Diagonalization Lemma) one can find a predicate $Prf(\cdot,\cdot)$ – and this $Prf(\cdot,\cdot)$ completes the interpretation $(\cdot)^*$ – which solves the following fixed point equation: **T** proves

$$\begin{array}{lll} Prf(u,v) & \longleftrightarrow & \forall r \le u \Big[\; u = 2r+1 \to \widetilde{Prf}(r,v) \quad \wedge \\ & & \quad u = 2r \to \Big[\; \bigwedge_{k=0}^{m} \big(r = k \to \bigvee_{j=0}^{J_k}(v = \ulcorner A_{k,j}{}^{*}\urcorner)\big) \wedge \\ & & \qquad r > m \to v = \ulcorner \forall x_0 \forall x_0 (x_0 = x_0)\urcorner \; \Big] \; \Big] \end{array}$$

Note that $Prf(\cdot,\cdot)$ may occur in each $A_{i,j}{}^*$ and remind that $\widetilde{Prf}(\cdot,\cdot)$ is the Gödel proof predicate for **T**.

The first task is to show that $Prf(\cdot,\cdot)$ can be constructed in this way.

Let the formula F be defined as

$$\begin{array}{lll} F(x,y,z_{0,0},z_{0,1},\ldots,z_{m,J_m}) & \longleftrightarrow & \forall r \le x \Big[\; x = 2r+1 \to \widetilde{Prf}(r,y) \quad \wedge \\ & & \quad x = 2r \to \Big[\; \bigwedge_{k=0}^{m} \big(r = k \to \bigvee_{j=0}^{J_k}(y = z_{k,j})\big) \wedge \\ & & \qquad r > m \to y = \ulcorner \forall x_0 \forall x_0 (x_0 = x_0)\urcorner \; \Big] \; \Big] \end{array}$$

and let $Sb_{i,j}(a,b)$ be a standard term corresponding to the primitive recursive function, determined by the following specification:

> If a is the Gödel number of a formula $B(x,y,z)$ then consider the interpretation $(\cdot)^{**} := (B(x,y,b),\phi)$ where ϕ is defined on proof and sentence variables identical to $(\cdot)^*$ and put
>
> $$Sb_{i,j}(a,b) := \ulcorner A_{i,j}{}^{**}\urcorner$$

Finally let

$$\begin{array}{rcl} B(x,y,z) & := & F(x,y,Sb_{0,0}(z,z),Sb_{0,1}(z,z),\ldots,Sb_{m,J_m}(z,z)) \\ g & := & \ulcorner B(x,y,z)\urcorner \\ Prf(x,y) & := & B(x,y,g) \end{array}$$

Now observe that $Sb_{i,j}(g,g) = \ulcorner A_{i,j}{}^{*}\urcorner$ as if $a = b = g$ then

$$\begin{array}{rcl} (\cdot)^{**} & \equiv & (B(x,y,g),\phi) \\ & \equiv & (Prf(x,y),\phi) \\ & \equiv & (\cdot)^* \end{array}$$

So it follows that

$$Prf(x,y) \equiv B(x,y,g) \equiv F(x,y,Sb_{0,0}(g,g),Sb_{0,1}(g,g),\ldots,Sb_{m,J_m}(g,g))$$

which is provably equivalent to

$$F(x, y, \ulcorner A_{0,0}{}^{*}\urcorner, \ulcorner A_{0,1}{}^{*}\urcorner, \ldots, \ulcorner A_{m,J_m}{}^{*}\urcorner)$$

Hence this so-defined $Prf(\cdot,\cdot)$ is a solution of the fixed point equation above.

The next task is to show that $(\cdot)^*$ has the desired properties, i.e. that $(\cdot)^*$ makes all formulas in Γ' true and all formulas in Δ' false, and afterwards the proof will be completed by the observation that $Prf(\cdot,\cdot)$ really is a proof predicate.

Let D be some modal formula from $\Gamma' \cup \Delta'$. Note that $Prf(\cdot,\cdot)$ is recursive. Therefore D^* is a closed, recursive arithmetical formula and

$$\begin{array}{lcl} D^* \text{ is true} & \Longleftrightarrow & \mathrm{T} \vdash D^* \\ D^* \text{ is false} & \Longleftrightarrow & \mathrm{T} \vdash \neg D^* \end{array}$$

By induction on the complexity of D it follows that:

$$\begin{array}{lcl} D \in \Gamma' & \Longrightarrow & D^* \text{ is true} \\ D \in \Delta' & \Longrightarrow & D^* \text{ is false} \end{array}$$

- The basis of the induction and the boolean cases are straightforward.
- $D = \Box_{p_i} A_{i,j} \in \Gamma'$: $(\Box_{p_i} A_{i,j})^* = Prf(p_i{}^*, \ulcorner A_{i,j}{}^{*}\urcorner) = Prf(2i, \ulcorner A_{i,j}{}^{*}\urcorner)$ is true by the fixed point equation.
- $D = \Box_{p_i} B \in \Delta'$ for some $i \leq m$: $(\Box_{p_i} B)^* = Prf(2i, \ulcorner B^{*}\urcorner)$ is false by the fixed point equation as $\ulcorner B^{*}\urcorner \neq \ulcorner A_{i,j}{}^{*}\urcorner$ for any $j \leq J_i$. Here is made use of the fact that Γ' and Δ' are disjoint.
- $D = \Box_{p_i} C \in \Delta'$ for some $i > m$: $(\Box_{p_i} C)^* = Prf(2i, \ulcorner C^{*}\urcorner)$ is false by the fixed point equation as there exists no modal formula C such that $\ulcorner C^{*}\urcorner = \ulcorner \forall x_0 \forall x_0 (x_0 = x_0)\urcorner$.

So it remains to show that $Prf(\cdot,\cdot)$ can be used as a proof predicate in T:

$$\mathrm{T} \vdash \varphi \qquad \Longleftrightarrow \qquad \exists n \in \mathbf{N}: \; Prf(n, \ulcorner \varphi \urcorner) \text{ is true}$$

Let $\mathrm{T} \vdash \varphi$. By the definition of the Gödel proof predicate there exists an $n_0 \in \mathbf{N}$ such that $\widetilde{Prf}(n_0, \ulcorner \varphi \urcorner)$ holds, hence by the fixed point equation $Prf(2n_0 + 1, \ulcorner \varphi \urcorner)$. Conversely if $Prf(n_0, \ulcorner \varphi \urcorner)$ is true for some $n_0 \in \mathbf{N}$, consider the three cases:

1st case: $n_0 = 2k+1$. If $Prf(2k+1, \ulcorner \varphi \urcorner)$ then by the fixed point equation $\widetilde{Prf}(k, \ulcorner \varphi \urcorner)$, hence $\mathrm{T} \vdash \varphi$.

2nd case: $n_0 = 2k$ and $k \leq m$. By the fixed point equation $\ulcorner \varphi \urcorner = \ulcorner A_{k,j}{}^{*}\urcorner$ for some $\Box_{p_k} A_{k,j} \in \Gamma'$. Then, by the injectivity of the Gödel numbering $\varphi \equiv A_{k,j}{}^*$. But $A_{k,j}$ is in Γ' as $\Gamma' \supset \Delta'$ is saturated, and thus φ is true and provable in T.

3rd case: $n_0 = 2k$ and $k > m$. It follows $\varphi \equiv \forall x_0 \forall x_0 (x_0 = x_0)$ from the fixed point equation. Trivially $\mathbf{T} \vdash \varphi$.

So $Prf(\cdot,\cdot)$ is a proof predicate for T and the Main Lemma for $\mathcal{P}$ is proved. Even something more has been proved, namely that there exists an arithmetical interpretation which makes all formulas in Γ' provable and all formulas in Δ' refutable, i.e.

$$\mathbf{T} \vdash \neg(\bigwedge \Gamma \to \bigvee \Delta)^*$$

■

2.8 Corollary Let $\Gamma \supset \Delta$ be a sequent. Then

$$\left[\forall^* : \ (\bigwedge \Gamma \to \bigvee \Delta)^* \text{ is true}\right] \implies \mathcal{P}_{\mathcal{G}}^- \vdash \Gamma \supset \Delta$$

Proof Assume that $\mathcal{P}_{\mathcal{G}}^- \nvdash \Gamma \supset \Delta$. By the Saturation Lemma there exists a saturated sequent $\Gamma' \supset \Delta'$ which is not $\mathcal{P}_{\mathcal{G}}^-$-provable. Hence by the Main Lemma there exists an arithmetical interpretation $(\cdot)^*$ which makes $\bigwedge \Gamma' \to \bigvee \Delta'$ false. But as $\Gamma \subset \Gamma'$ and $\Delta \subset \Delta'$, this interpretation falsifies also $\bigwedge \Gamma \to \bigvee \Delta$.
■

The proof of theorem 2.1 can be completed now as follows: If $\mathcal{P} \nvdash \bigwedge \Gamma \to \bigvee \Delta$, then $\mathcal{P}_{\mathcal{G}}^- \nvdash \Gamma \supset \Delta$ and, by corollary 2.8, $(\bigwedge \Gamma \to \bigvee \Delta)^*$ is false for some interpretation $(\cdot)^*$.

2.9 Corollary

- $\mathcal{P}_{\mathcal{G}}$ is equivalent to $\mathcal{P}$ and admits cut elimination.
- Both $\mathcal{P}$ and $\mathcal{P}_{\mathcal{G}}$ are decidable.

Proof Apply the saturation algorithm 2.6 to the sequent $\Gamma \supset \Delta$. If the algorithm finds a saturation then $\Gamma \supset \Delta$ is not provable, otherwise $\Gamma \supset \Delta$ is provable.
■

3 Functionality

This section is devoted to the proof of arithmetical soundness and completeness of the modal system $\mathcal{PF}$ with respect to functional interpretations. The proofs go along the lines of the previous section on the basic system $\mathcal{P}$.

3.1 Theorem Let A be a modal formula. Then

$$\begin{aligned} \mathcal{PF} \vdash A \quad &\iff \quad \forall^*(\text{functional}): \ \mathbf{T} \vdash A^* \\ &\iff \quad \forall^*(\text{functional}): \ A^* \text{ is true} \end{aligned}$$

Proof of the soundness of $\mathcal{PF}$: Using theorem 2.1 it is sufficient to check the soundness of the Functionality Axiom which is straightforward.
■

3.1 Gentzen style system for $\mathcal{PF}$

3.2 Definition $\mathcal{PF_G}$ **is a sequent calculus with axioms and rules of inference as follows:**

- Sequent calculus for classical propositional logic including the cut-rule.
- $\dfrac{A, \Gamma \supset \Delta}{\Box_p A, \Gamma \supset \Delta}\text{refl} \qquad \dfrac{\Gamma \supset \Delta, \Box_p A}{\Box_p B, \Gamma \supset \Delta}\text{func} \quad (A \not\equiv B)$

$\mathcal{PF}_\mathcal{G}^-$ is the system $\mathcal{PF_G}$ without the cut rule.

3.3 Soundness of $\mathcal{PF_G}$ w.r.t. $\mathcal{PF}$ For each sequent $\Gamma \supset \Delta$:

$$\mathcal{PF_G} \vdash \Gamma \supset \Delta \qquad \Longrightarrow \qquad \mathcal{PF} \vdash \bigwedge \Gamma \to \bigvee \Delta$$

■

3.4 Definition The sequent $\Gamma \supset \Delta$ is called *functionally saturated*, if the following statements hold for all modal formulas A, B and for all proof variables p:

1. $A \to B \in \Delta$ implies $A \in \Gamma$ and $B \in \Delta$,
2. $A \to B \in \Gamma$ implies $B \in \Gamma$ or $A \in \Delta$,
3. $\Box_p A \in \Gamma$ implies $A \in \Gamma$ and $\Box_p B \in \Delta$ for each subformula $\Box_p B$ of $\Gamma \cup \Delta$ such that $A \not\equiv B$.

3.5 Saturation Lemma for $\mathcal{PF_G}$ Let $\Gamma \supset \Delta$ be a sequent such that $\mathcal{PF}_\mathcal{G}^- \nvdash \Gamma \supset \Delta$. Then there exists a functionally saturated sequent $\Gamma' \supset \Delta'$ such that

(i) $\Gamma \subset \Gamma'$, $\Delta \subset \Delta'$,

(ii) $\Gamma \cup \Delta$ and $\Gamma' \cup \Delta'$ have the same subformulas,

(iii) $\mathcal{PF}_\mathcal{G}^- \nvdash \Gamma' \supset \Delta'$.

Furthermore $\Gamma' \supset \Delta'$ is effectively computable from $\Gamma \supset \Delta$.
Such a $\Gamma' \supset \Delta'$ is called a *functional saturation* of $\Gamma \supset \Delta$.

Proof The Saturation Algorithm for $\mathcal{PF_G}$ works similarly to that for $\mathcal{P_G}$.
■

3.2 Arithmetical completeness

3.6 Main Lemma for $\mathcal{PF}$ Let $\Gamma' \supset \Delta'$ be a functionally saturated sequent which is not $\mathcal{PF}_{\mathcal{G}}^{-}$-provable. Then there exists a functional interpretation $(\cdot)^*$ which makes all formulas in Γ' true and all formulas in Δ' false, i.e.

$$(\bigwedge \Gamma' \to \bigvee \Delta')^* \quad \text{is false.}$$

Proof For the proof and sentence variables let $(\cdot)^*$ be defined as in case of $\mathcal{P}$, namely:

$$p_i^* := 2i, \qquad S_i^* := \begin{cases} \forall x_i(x_i = x_i) & \text{if } S_i \in \Gamma', \\ \forall x_i(x_i \neq x_i) & \text{else.} \end{cases}$$

Again it follows that $(\cdot)^*$ is an injective arithmetical interpretation of the modal language.

Now write $\bigwedge \Gamma'$ as

$$\bigwedge_{i=0}^{m} \Box_{p_i} A_i \ \wedge \Gamma''$$

where Γ'' contains no formula of the form $\Box_p A$. This notion is possible, as, due to the functional saturation and non-provability of $\Gamma' \supset \Delta'$, there is no proof variable p such that both $\Box_p A$ and $\Box_p B$ may be contained in Γ' for distinct A and B.

Again let $\widetilde{Prf}(\cdot,\cdot)$ be a standard proof predicate for $\mathbf{T}$. The *fixed point equation* for $Prf(\cdot,\cdot)$ is simply a special case of that for $\mathcal{P}$: $\mathbf{T}$ proves

$$\begin{aligned} Prf(u,v) \longleftrightarrow \forall r \le u \Big[\; & u = 2r+1 \to \widetilde{Prf}(r,v) \quad \wedge \\ & u = 2r \to \Big[\bigwedge_{k=0}^{m} \big(r = k \to v = \ulcorner A_k^* \urcorner\big) \quad \wedge \\ & \qquad r > m \to v = \ulcorner \forall x_0 \forall x_0 (x_0 = x_0) \urcorner \; \Big] \; \Big] \end{aligned}$$

Clearly $Prf(\cdot,\cdot)$ is functional.

The remaining part of the proof is exactly as in the case for $\mathcal{P}$, and again it is clear that this interpretation $(\cdot)^*$ even has the property that

$$\mathbf{T} \vdash \neg(\bigwedge \Gamma \to \bigvee \Delta)^*$$

■

3.7 Corollary Let $\Gamma \supset \Delta$ be a sequent. Then

$$\Big[\forall^*(\text{functional}): \ (\bigwedge \Gamma \to \bigvee \Delta)^* \ \text{is true}\Big] \implies \mathcal{PF}_{\mathcal{G}}^{-} \vdash \Gamma \supset \Delta$$

In order to complete the proof of theorem 3.1 one has to notice that $\mathcal{PF} \nvdash \bigwedge \Gamma \to \bigvee \Delta$ implies $\mathcal{PF}_{\mathcal{G}} \nvdash \Gamma \supset \Delta$ and, by corollary 3.7, $(\bigwedge \Gamma \to \bigvee \Delta)^*$ is false for some functional interpretation $(\cdot)^*$.

3.8 Corollary $\mathcal{PF}_{\mathcal{G}}$ is equivalent to $\mathcal{PF}$ and admits cut elimination; both $\mathcal{PF}$ and $\mathcal{PF}_{\mathcal{G}}$ are decidable.

4 Uniform proof predicates

A natural question is whether there exists a fixed (i.e. uniform) proof predicate $Prf(\cdot,\cdot)$ under which for every modal formula A

$$\mathcal{P} \vdash A \iff \forall^*: \ \mathrm{T} \vdash A^*$$

respectively, if there exists a fixed functional proof predicate under which

$$\mathcal{PF} \vdash A \iff \forall^*: \ \mathrm{T} \vdash A^*$$

So in this case $\forall^*$ quantifies only proof and sentence variables. Uniform proof predicates are in a certain sense proof predicates ***without any special properties.*** For example, a uniform proof predicate for $\mathcal{P}$ may not be functional for obvious reasons.

There exist uniform proof predicates for $\mathcal{P}$ and $\mathcal{PF}$. The construction of a uniform proof predicate for $\mathcal{PF}$ will be described in the following; the case of $\mathcal{P}$ can be treated similarly.

4.1 Theorem There exists a functional proof predicate $\widehat{Prf}(\cdot,\cdot)$ such that for every modal formula A:

$$\mathcal{PF} \vdash A \iff \mathrm{T} \vdash A^* \text{ for each } (\cdot)^* \text{ based on } \widehat{Prf}(\cdot,\cdot)$$

Proof It follows from the proof of Lemma 3.5 that the saturation procedure for $\mathcal{PF}$ is primitive recursive, i.e. that $\mathcal{PF}$ is primitive recursive. Let $A_0, A_1, \ldots$ be a primitive recursive list of all modal formulas not provable in $\mathcal{PF}$, and let $\Gamma_0 \supset \Delta_0, \Gamma_1 \supset \Delta_1, \ldots$ be a primitive recursive list of sequents such that for every i, $\Gamma_i \supset \Delta_i$ is a saturation of $\supset A_i$. Let $<\cdot,\cdot>$ be a primitive recursive pairing function and let $(\cdot)_1$, $(\cdot)_2$ be the corresponding projection functions. Fix a primitive recursive Gödel numbering $\ulcorner\cdot\urcorner$ such that, if E is a proper subformula of D then $\ulcorner E\urcorner < \ulcorner D\urcorner$. Let $C(x)$ be a natural formalization of

"There exists a modal formula B such that $\Box_{p_{(x)_2}} B \in \Gamma_{(x)_1}$".

Note that $C(x)$ is primitive recursive, since the existential quantifier occurring in it can be bounded primitive recursively in x. The construction of Lemma 3.6 gives primitive recursively, for each formula A_n, a proof predicate and an interpretation of the sentence and proof variables such that $A_n{}^*$ is false.

For each n let the interpretation ϕ_n of the proof and sentence variables be defined as:

$$\phi_n(p_i) := 2\cdot<n,i>, \qquad \phi_n(S_i) := \begin{cases} \forall x_i(x_i = x_i) & \text{if } S_i \in \Gamma_n, \\ \forall x_i(x_i \neq x_i) & \text{else.} \end{cases}$$

Notice that the interpretation of both proof and sentence variables depends from n.

The predicate $\widehat{Prf}(\cdot,\cdot)$ can now be defined by the following *fixed point equation*:

$$\widehat{Prf}(u,v) \longleftrightarrow \forall r \leq u \Bigg[$$

$$\begin{array}{lll} u = 2r+1 & \rightarrow & \widetilde{Prf}(r,v) \qquad \wedge \\ u = 2r & \rightarrow & \Big[\; C(r) \rightarrow v = \ulcorner B^{*}\urcorner \text{ for a modal formula } B \text{ such that } \Box_{p_{(r)_2}} B \in \Gamma_{(r)_1} \text{ (such a formula } B \text{ is unique as } \Gamma_{(r)_1} \supset \Delta_{(r)_1} \text{ is functionally saturated) and the interpretation } (\cdot)^{*} = (\widehat{Prf}(\cdot,\cdot), \phi_{(r)_1}). \\ & & \;\;\neg C(r) \rightarrow v = \ulcorner \forall x_0 \forall x_0 (x_0 = x_0)\urcorner \qquad \Big] \;\Bigg] \end{array}$$

4.2 Lemma Let D be a modal formula contained in $\Gamma_n \cup \Delta_n$. Then:

$$\begin{array}{lll} D \in \Gamma_n & \Longrightarrow & D^{*} \text{ is true} \\ D \in \Delta_n & \Longrightarrow & D^{*} \text{ is false} \end{array}$$

Proof Induction on the complexity of D:

- D is atomic: by the definition of $(\cdot)^{*}$.
- The case of boolean connectives is straightforward.
- $D = \Box_{p_i} B \in \Gamma_n$. Then $(\Box_{p_i} B)^{*} = \widehat{Prf}(2\cdot\langle n,i\rangle, \ulcorner B^{*}\urcorner)$ is true according to the fixed point equation.
- $D = \Box_{p_i} B \in \Delta_n$. Then $C(\langle n,i\rangle)$ is violated, and $\ulcorner B^{*}\urcorner = \ulcorner \forall x_0 \forall x_0 (x_0 = x_0)\urcorner$ is also false as there exists no modal formula B such that $B^{*} \equiv \forall x_0 \forall x_0 (x_0 = x_0)$. Therefore $\widehat{Prf}(2\cdot\langle n,i\rangle, \ulcorner B^{*}\urcorner)$ is false, too.

■

4.3 Lemma

(a) $\widehat{Prf}(\cdot,\cdot)$ is primitive recursive and functional.

(b) $\mathrm{T} \vdash \varphi \Longleftrightarrow \widehat{Prf}(n, \ulcorner \varphi \urcorner)$ for some n.

Proof

(a) It is easy to see that the right side of the fixed point equation is provably equivalent to a primitive recursive formula because all the quantifiers in the descriptions of functions and predicates are bounded by the corresponding primitive recursive functions. Thus $\widehat{Prf}(\cdot,\cdot)$ is primitive recursive. Obviously, $\widehat{Prf}(\cdot,\cdot)$ is functional, too.

(b) Let $\mathrm{T} \vdash \varphi$ and m be the Gödel number of the proof of φ in T. Then $\widetilde{Prf}(m, \ulcorner \varphi \urcorner)$ holds and thus $\widehat{Prf}(2m+1, \ulcorner \varphi \urcorner)$. Let now $\widehat{Prf}(k, \ulcorner \varphi \urcorner)$ for some k.

If $k = 2m+1$ then $\widetilde{Prf}(m, \ulcorner \varphi \urcorner)$ holds, so m is the Gödel number of a proof of φ, hence $\mathrm{T} \vdash \varphi$.

If $k = 2m$ and $C(m)$ then $\varphi \equiv D^*$ for some modal formula D such that $D \in \Gamma_{(m)_1}$ and the interpretation $(\cdot)^*$ corresponding to $\Gamma_{(m)_1} \supset \Delta_{(m)_1}$. By lemma 4.2, D^* is a true primitive recursive formula; again $\mathrm{T} \vdash \varphi$.

If $k = 2m$ and not $C(m)$ then $\varphi \equiv \forall x_0 \forall x_0 (x_0 = x_0)$ and so trivially $\mathrm{T} \vdash \varphi$.

By the formalization of (b) one can also prove that $\mathrm{T} \vdash \forall y\, (\widehat{Pr}(y) \leftrightarrow \widetilde{Pr}(y))$.
■

Thus theorem 4.1 is proved.
■

After the uniformization of the proof predicate, a natural question is of course whether it is also possible to choose a fixed interpretation for the sentence or proof variables. This question will be answered up to the end of this section.

To recall the definition, each interpretation $(\cdot)^*$ consists of three natural parts:

(i) a (functional) proof predicate $Prf(\cdot,\cdot)$ for T,
(ii) an evaluation α of proof variables as natural numbers,
(iii) an evaluation β of sentence variables as sentences of T.

So the completeness parts of theorems 2.1 and 3.1 state that

$$\forall A:\ \exists \alpha, \beta, Prf(\cdot,\cdot):\ (\mathrm{T} \vdash A^* \ \Rightarrow\ \mathcal{P}/\mathcal{PF} \vdash A)$$

Theorem 4.1 shows that also

$$\exists Prf(\cdot,\cdot):\ \forall A:\ \exists \alpha, \beta:\ (\mathrm{T} \vdash A^* \ \Rightarrow\ \mathcal{P}/\mathcal{PF} \vdash A)$$

As the proofs of theorems 2.7 and 3.6 demonstrate, the interpretation α of the proof variables alone is uniformizable (e.g. $p_i{}^* = 2i$). The completeness theorems can therefore be formulated as

$$\exists \alpha:\ \forall A:\ \exists \beta, Prf(\cdot,\cdot):\ (\mathrm{T} \vdash A^* \ \Rightarrow\ \mathcal{P}/\mathcal{PF} \vdash A)$$

It is not possible to use a uniform proof predicate in addition to α. Assume that

$$\exists \alpha, Prf(\cdot,\cdot):\ \forall A:\ \exists \beta:\ (\mathrm{T} \vdash A^* \ \Rightarrow\ \mathcal{P}/\mathcal{PF} \vdash A)$$

and let $\hat{\alpha}$ be such an α and $\widehat{Prf}(\cdot,\cdot)$ be such a $Prf(\cdot,\cdot)$. As $\mathcal{P}/\mathcal{PF} \nvdash \Box_{p_0} \top$ it follows that $\mathrm{T} \nvdash \widehat{Prf}(p_0{}^{\hat{\alpha}}, \ulcorner 0 = 0 \urcorner)$ and then $\mathrm{T} \vdash \neg \widehat{Prf}(p_0{}^{\hat{\alpha}}, \ulcorner 0 = 0 \urcorner)$. But this is equivalent to $\mathcal{P}/\mathcal{PF} \vdash \neg \Box_{p_0} \top$, which is known to be false.

The interpretation β of the sentence variables is not uniformizable at all. Assume that

$$\exists \beta:\ \forall A:\ \exists \alpha, Prf(\cdot,\cdot):\ (\mathrm{T} \vdash A^* \ \Rightarrow\ \mathcal{P}/\mathcal{PF} \vdash A)$$

and let $\hat{\beta}$ be such a fixed β. As $\mathcal{P}/\mathcal{PF} \nvdash \neg \Box_{p_0} S_0$, this implies that

$$\exists \alpha, Prf(\cdot,\cdot):\ \mathrm{T} \nvdash \neg Prf(p_0{}^{\alpha}, \ulcorner S_0{}^{\hat{\beta}} \urcorner)$$

which is equivalent to

$$\exists \alpha, Prf(\cdot,\cdot) : \ \mathrm{T} \vdash Prf({p_0}^{\alpha}, \ulcorner {S_0}^{\widehat{\beta}} \urcorner)$$

from which follows that $\mathrm{T} \vdash {S_0}^{\widehat{\beta}}$. As a consequence,

$$\forall \alpha, Prf(\cdot,\cdot) : \ \mathrm{T} \nvdash Prf({p_0}^{\alpha}, \ulcorner \neg {S_0}^{\widehat{\beta}} \urcorner)$$

hence

$$\forall \alpha, Prf(\cdot,\cdot) : \ \mathrm{T} \vdash \neg Prf({p_0}^{\alpha}, \ulcorner \neg {S_0}^{\widehat{\beta}} \urcorner)$$

which implies $\mathcal{P}/\mathcal{PF} \vdash \neg \Box_{p_0} \neg S_0$, but again this is known to be false.

The situation can therefore be summarized as:

> One can either choose a uniform proof predicate as $\widehat{Prf}(\cdot,\cdot)$ in this section, or one can choose a uniform interpretation of the proof variables as in sections 2 and 3; all other combinations of uniformization are impossible. So it is not possible to have a uniform proof predicate and a uniform interpretation of the proof variables simultaneously or a uniform interpretation of the sentence variables.

The authors wish to thank the anonymous referee for his valuable suggestions.

References

[1] S. Artëmov and T. Straßen, "The Basic Logic of Proofs," Tech. Rep. IAM 92-018, Department for computer science, University of Berne, Switzerland, September 1992.

[2] S. Artëmov and T. Straßen, "Functionality in the Basic Logic of Proofs," Tech. Rep. IAM 93-004, Department for computer science, University of Berne, Switzerland, January 1993.

[3] G. Boolos, *The unprovability of consistency: an essay in modal logic.* Cambridge: Cambridge University Press, 1979.

[4] C. Smoryński, "The incompleteness theorems," in *Handbook of Mathematical Logic* (J. Barwise, ed.), ch. D.1, S3, pp. 821–865, North-Holland, Amsterdam, 1977.

[5] R. M. Solovay, "Provability interpretations of modal logic," *Israel Journal of Mathematics*, vol. 25, pp. 287–304, 1976.

Algorithmic Structuring of Cut-free Proofs

Matthias Baaz* Richard Zach**

Technische Universität Wien, Austria

Abstract. The problem of algorithmic structuring of proofs in the sequent calculi **LK** and $\mathbf{LK_B}$ (**LK** where blocks of quantifiers can be introduced in one step) is investigated, where a distinction is made between linear proofs and proofs in tree form. In this framework, structuring coincides with the introduction of cuts into a proof. The algorithmic solvability of this problem can be reduced to the question of *k/l-compressibility:*

"Given a proof of $\Pi \rightarrow \Lambda$ of length k, and $l \leq k$: Is there is a proof of $\Pi \rightarrow \Lambda$ of length $\leq l$?"

When restricted to proofs with universal or existential cuts, this problem is shown to be (1) undecidable for linear or tree-like **LK**-proofs (corresponds to the undecidability of second order unification), (2) undecidable for linear $\mathbf{LK_B}$-proofs (corresponds to the undecidability of semi-unification), and (3) decidable for tree-like $\mathbf{LK_B}$-proofs (corresponds to a decidable subproblem of semi-unification).

1 Introduction

Most classical algorithms in proof theory eliminate the structure of given proofs to extract information, e.g., Herbrand disjunctions (as obtained via cut-elimination or the ε-theorem), or normal forms of functional interpretations. The problem of *structuring of proofs* is inverse to these procedures: How to structure a proof by decomposition and introduction of propositions?

In sequent calculi, structuring of proofs can be identified with the insertion of cuts into a proof. This provides us with a general basis for formal approaches to the problem above. All usual cut-elimination procedures for first order logic found in the literature (such as those of Gentzen [1934] and Tait [1968], where substitution is the only operation on terms) produce cut-free proofs of increased term complexity relative to the original proof. If we view the structuring problem as the inverse problem to cut-elimination and restrict ourselves to such procedures, we can of course find a simpler proof with cuts that yields the given proof after cut-elimination if such a proof exists. Such procedures, however, depend on *specific* methods for cut-elimination, and the view of proofs as literal objects.

Since we would actually like to disregard *term* structure in favour of *proof* structure (i.e., we would like to consider proofs as schemata of a certain form, and as equivalent up to substitutions), we take a more general approach here: given a proof

* Technische Universität Wien, Institut für Algebra und Diskrete Mathematik E118.2, Wiedner Hauptstraße 8–10, A-1040 Wien, Austria, baaz@logic.tuwien.ac.at

** Technische Universität Wien, Institut für Computersprachen E185.2, Resselgasse 3/1, A-1040 Wien, Austria, zach@logic.tuwien.ac.at

and end sequent, we ask for a shorter proof with possibly increased structure. In sequent calculus this corresponds to the introduction of stronger cuts (if the proof cannot be abbreviated trivially, of course). We will be able to solve this problem if we can construct a procedure that solves the following central question:

1.1. k/l-COMPRESSIBILITY Given a proof of $\Pi \rightarrow \Lambda$ of length k, and $l \leq k$: Is there is a proof of $\Pi \rightarrow \Lambda$ of length $\leq l$?

In what follows, we study proofs in **LK** and $\mathbf{LK_B}$ (**LK** where blocks of quantifiers can be introduced in one step) considered as acyclic graphs (not only tree-like proofs). We restrict ourselves to the fragments with only universal or existential cuts (the cut formulas are pure universal or existential formulas), denoted $\mathbf{LK}^{\Pi\Sigma}$ and $\mathbf{LK}_\mathbf{B}^{\Pi\Sigma}$, respectively. We show that k/l-compressibility is

(1) undecidable for $\mathbf{LK}^{\Pi\Sigma}$-proofs,
(2) undecidable for linear $\mathbf{LK}_\mathbf{B}^{\Pi\Sigma}$-proofs, but is
(3) decidable for tree-like $\mathbf{LK}_\mathbf{B}^{\Pi\Sigma}$-proofs.

Since we consider k/l-compressibility as central, and since bounds on cut elimination do only depend on the length of the given proof, it makes no difference whether the given proof is cut-free or not. However, structuring of cut-free proofs is important to computer science, where deduction systems are usually quantifier-free.

In the following, we assume familiarity with BUSS [1991] and KRAJÍČEK and PUDLÁK [1988]

2 Basic definitions

We follow BUSS [1991] in the definition of sequent calculus **LK**, with the exception that axioms and weakenings are restricted to atomic formulas.
The calculus $\mathbf{LK_B}$ is **LK** with the rules ($\forall$:left) and ($\forall$:right) replaced by

$$\frac{A(t_1,\ldots,t_r), \Gamma \rightarrow \Delta}{(\forall x_1)\ldots(\forall x_r)A(x_1,\ldots,x_r), \Gamma \rightarrow \Delta}\ \forall_B\text{: left}$$

and

$$\frac{\Gamma \rightarrow \Delta, A(b_1,\ldots,b_r)}{\Gamma \rightarrow \Delta, (\forall x_1)\ldots(\forall x_r)A(x_1,\ldots,x_r)}\ \forall_B\text{: right}$$

respectively ($b_1, \ldots, b_r$ must not occur in the lower sequent). ($\exists$-left) and ($\exists$-right) are analogously replaced by ($\exists_B$-left) and ($\exists_B$-right).

2.1. DEFINITION A *(linear) proof* is a directed acyclic graph s.t.

(1) every node is labeled with a sequent and the name of a rule of inference,
(2) every node with indegree 0 is labeled by an axiom sequent,
(3) exactly one node has outdegree 0 (labeled by the *end sequent*),
(4) all other nodes have outdegree ≥ 1, and

(5) if an edge connects a node labeled by sequent R to a node labeled by S, then R is a premise to the inference associated with S, and the edge is labeled by L or R according to whether R is the left or right premise of the rule, and unlabeled if the rule has only one premise.

A proof is called *tree-like* if it is a tree, i.e., if every node has outdegree 1. The *length* of a proof is the number of its nodes. For simplicity, we identify nodes with the sequents they are labeled with.

2.2. DEFINITION A *proof analysis* is like a proof except that nodes are only labeled with names of inference rules, and nodes corresponding to axioms and weakenings additionally carry the corresponding predicate symbol.

A proof *realizes* a proof analysis P with end sequent $\Pi \rightarrow \Lambda$, if there is a bijection between the nodes and edges in the proof and the proof analysis s.t. corresponding nodes are labeled by the same rule names, axioms and weakening formulas have the predicate symbol determined by the corresponding label in P, corresponding edges have the same labels, and the end sequent of the proof is $\Pi \rightarrow \Lambda$. If there is such a proof, P is called *realizable* with end sequent $\Pi \rightarrow \Lambda$.

The decision problem of whether a given proof analysis with end sequent can be realized by a proof is called the *realizability problem*.

The decision problem of whether there is a proof of a given sequent of length $\leq k$ is called the *k-provability problem*.

2.3. *Remark* It is easily seen that the decidability of realizability implies decidability of k-provability (enumerate all proof analyses up to length k), which in turn implies the decidability of k/l-compressibility, but the converse is not immediately obvious. Consider the class of proof analyses with undecidable realizability problem given in KRAJÍČEK and PUDLÁK [1988], §5: The end sequents $A \rightarrow A, P(s^n 0)$ are trivially derivable by one weakening, and hence k-provability is decidable. To see that the undecidability of k-provability need not imply the undecidability of k/l-compressibility, consider a system of first order logic with all true formulas as axioms and with sound rules: k-provability is undecidable, but k/l-compressibility is decidable.

2.4. *Remark* The restriction to atomic axioms and weakenings makes the use of proof analyses easier, since we can do without a number of case distinctions: In the cut-free case, the end sequent determines the logical form of all formulas, but in the presence of cuts and non-atomic axioms and weakenings, we only have a bound on the logical complexity of the cut-formulas (by PARIKH [1973], Theorem 2). Consequently we have to add information on the logical form of cut-formulas to the proof analyses.

3 k/l-Compressibility is undecidable for $\mathbf{LK}^{\Pi\Sigma}$

We derive the undecidability of k/l-compressibility for $\mathbf{LK}^{\Pi\Sigma}$ from the undecidabilty of k-provability: To establish the undecidability of k-provability, we associate with a non-recursive r.e. set $X \subseteq \omega$ a sequence of proof analyses P_i and end sequents $\Pi_i \rightarrow \Lambda_i$, $i \in \omega$, s.t.

$$n \in X \iff P_n \text{ is realizable with end sequent } \Pi_n \rightarrow \Lambda_n,$$

and, furthermore, that all proofs of $\Pi_n \rightarrow \Lambda_n$ for $n \in \omega \setminus X$ are longer than P_n.

In fact, there is a recursive superset X^* of X such that $\Pi_n \rightarrow \Lambda_n$ is *provable* for all $n \in X^*$, since k-provability for *cut-free* proofs is decidable (cf. KRAJÍČEK and PUDLÁK [1988], Theorem 6.1). If $\Pi_n \rightarrow \Lambda_n$ is of the form $\Pi \rightarrow \Lambda, A(s^n(0))$, then X^* is even co-finite.

To show that k/l-compressibility is undecidable, it suffices to bound the length of the proofs of $\Pi_n \rightarrow \Gamma_n$. This is the statement of the following theorem, which can be gathered from BUSS [1991]:

3.1. THEOREM *For every r.e. set $X \neq \emptyset$ there is a formula $A_X(c)$ and $k \in \omega$ s.t. $n \in X$ iff $\rightarrow A_X(s^n(0))$ has an* **LK**- *(by construction* $\mathbf{LK}^{\Pi\Sigma}$-*) proof of length k and $\rightarrow A(s^n(0))$ has an* **LK**- *(by construction* $\mathbf{LK}^{\Pi\Sigma}$-*) proof of length $k+1$ for all $n \in \omega$.*

Proof. Every r.e. set $X \subseteq \omega$ can be represented by a set Ω of partial substitution equations obeying the special restriction s.t., $n \in X$ iff $\Omega \cup \{\beta_1 = s^n(0)\}$ has a solution (BUSS [1991], Theorem 3). The proof of this fact is via Matijacevič's Theorem by encoding diophantine equations as partial substitution equations. Let $\Omega \cup \{\beta_1 = s^n(0)\}$ be the set of equations characterizing the r.e. set X.

In the proof of the Main Theorem of BUSS [1991] a formula $A_X(s^n(0))$ and an integer N are constructed s.t. $\rightarrow A_X(s^n(0))$ has an **LK**-proof of $\leq N$ steps iff the above equations have a solution, and is provable in $N+1$ steps, if all but one of the equations have a solution (Section 4, see in particular p. 93, first paragraph). The first part of the theorem now follows from the fact that the system encodes X and hence is solvable iff $n \in X$. For the second part, we replace β_1 by $s^r(0)$ for some $r \in X$, $r \neq n$. Then $s^r(0) = s^n(0)$ is the only equation not satisfied (regardless of whether $n \in X$ or not).

The proofs constructed are all tree-like, use only existential cuts, atomic axioms and atomic weakenings. The central Propositon 8 of BUSS [1991] (as noted there) can be adapted to the non-tree-like case. Hence the arguments extend to the case of linear $\mathbf{LK}^{\Pi\Sigma}$-proofs. □

3.2. COROLLARY *k/l-Compressibility is undecidable for* $\mathbf{LK}^{\Pi\Sigma}$-*proofs (whether linear or tree-like).*

3.3. *Remark* If the end sequent contains only unary function symbols, k/l-compressibility is decidable: cf. PARIKH [1973], Theorem 1 for the case of one and FARMER [1991], Corollary 5.20 for several unary function symbols. It is also decidable if we are looking for shorter proofs with *quantifier-free* cuts (cf. KRAJÍČEK and PUDLÁK [1988], Section 2).

3.4. *Remark* The theorem shows that, in the worst case, we have to pay for introduced structure by a significant—in fact non-recursive—increase in the term structure, *even in decidable subcases*. This situation could be alleviated by taking into account known properties of the function symbols, such as associativity and commutativity.

4 k/l-Compressibility is undecidable for linear $\mathbf{LK}_{\mathrm{B}}^{\Pi\Sigma}$-proofs

To be able to deal with block inferences of quantifiers, we introduce the concept of *semi-unification*:

4.1. DEFINITION (cf. BAAZ [1993], KFOURY *et al.* [1990], PUDLÁK [1988]) A substitution δ is called a *semi-unifier* of the semi-unification problem $\{(s_1, t_1), \ldots, (s_p, t_p)\}$ iff there exist $\sigma_1, \ldots, \sigma_p$ such that $s_1\delta = t_1\delta\sigma_1, \ldots, s_p\delta = t_p\delta\sigma_p$. In other words, a semi-unifier makes the s_i substitution instances of the corresponding t_i.

4.2. EXAMPLE $\delta = \{f(x, f(x,x))/z\}$ is a semi-unifier of $(f(x,z), f(x, f(x,y)))$ because

$$f(x,z)\{f(x,f(x,x))/z\} = f(x,f(x,y))\{f(x,f(x,x))/z\}\{f(x,x)/y\}.$$

There is no semi-unifier of $(f(x,y), f(x, f(x,y)))$, since no simultaneous substitution will make the left side a substitution instance of the right side.

4.3. THEOREM *Realizability is undecidable for linear* $\mathbf{LK}_{\mathrm{B}}^{\Pi\Sigma}$*-analyses.*

This follows immediately from the undecidability of semi-unification (KFOURY *et al.* [1990]) and the following proposition:

4.4. PROPOSITION *Let the language contain a binary function symbol f. For every semi-unification problem $\Omega = \{(s_1,t_1), \ldots, (s_p,t_p)\}$, there is a proof analysis P_Ω and a sequent $\Pi_\Omega \to \Lambda_\Omega$, s.t. there is an* $\mathbf{LK}_{\mathrm{B}}^{\Pi\Sigma}$*-proof realizing P_Ω with end sequent $\Pi_\Omega \to \Lambda_\Omega$ iff Ω is solvable.*

Proof. First note that the semi-unification problem can be reduced to a semi-unification problem $\{(s_1^*, t), \ldots, (s_p^*, t)\}$ with $s_i^* = f(\cdots f(a_{i_1}, a_{i_2}) \ldots s_i) \ldots a_{i_p})$ and $t = f(\cdots f(t_1, t_2), \ldots t_p)$, where a_{i_j} are new free variables.

Let $A_\Omega(a_1, \ldots, a_n) \equiv P(t) \wedge ((P(s_1^*) \wedge \ldots \wedge P(s_p^*)) \supset Q)$, where all free variables are among $a_1, \ldots, a_n$ and do not occur in Q. We sketch the construction of a proof analysis as follows:

$$\left.\begin{array}{l}\text{propositional inferences}\end{array}\right.$$

$$
\begin{array}{ll}
(a) & A_\Omega(a_1, \ldots, a_n)\delta \to A_\Omega(a_1, \ldots, a_n)\delta \\
(a+1) & \underline{(\forall x_1) \ldots (\forall x_n) A_\Omega(x_1, \ldots, x_n) \to A_\Omega(a_1, \ldots, a_n)\delta}
\end{array}
$$

$$\left.\begin{array}{l}\text{propositional inferences including} \\ \text{propositional cuts from } (a+1)\end{array}\right.$$

$$
\begin{array}{ll}
(b) & (\forall x_1) \ldots (\forall x_n) A_\Omega(x_1, \ldots, x_n) \to P(t)\delta \\
(b+1) & \underline{(\forall x_1) \ldots (\forall x_n) A_\Omega(x_1, \ldots, x_n) \to (\forall y_1) \ldots (\forall y_m) R(y_1, \ldots, y_m)}
\end{array}
$$

$$\left\{\begin{array}{l}\text{propositional inferences including}\\ \text{propositional cuts from } (a+1)\end{array}\right.$$

$$(c) \qquad P(s_1^*)\delta, \ldots, P(s_p^*)\delta, (\forall x_1)\ldots(\forall x_n)A_\Omega(x_1,\ldots,x_n) \rightarrow Q$$

$$\left\{\begin{array}{l}p\ (\forall_B\text{-left})\text{-inferences, exchanges}\\ \text{and contractions from } (c)\end{array}\right.$$

$$\begin{array}{ll}(d) & (\forall z_1)\ldots(\forall z_s)R'(z_1,\ldots,z_s), (\forall x_1)\ldots(\forall x_n)A_\Omega(x_1,\ldots,x_n) \rightarrow Q\\ (e) & (\forall x_1)\ldots(\forall x_n)A_\Omega(x_1,\ldots,x_n), (\forall x_1)\ldots(\forall x_n)A_\Omega(x_1,\ldots,x_n) \rightarrow Q\\ (e+1) & (\forall x_1)\ldots(\forall x_n)A_\Omega(x_1,\ldots,x_n) \rightarrow Q\end{array}$$

Here, $(a+1)$ is obtained from (a) by ($\forall_B$:left), $(b+1)$ from (b) by ($\forall_B$:right), (e) from $(b+1)$ and (d) by cut, and $(e+1)$ from (e) by contraction. Note that $(\forall y_1)\ldots(\forall y_m)R(y_1,\ldots,y_m) \equiv (\forall z_1)\ldots(\forall z_s)R'(z_1,\ldots,z_s)$ by the cut rule and hence δ is forced to be a semi-unifier. The label $(a+1)$ is ancestor of *both* sides of the cut, the skeleton is therefore *not* in tree form. (The length of the skeleton is linear in n.) □

4.5. *Remark* If $p = 1$, then the realizability of this analysis is decidable (cf. PUDLÁK [1988], Theorem (i)).

4.6. *Remark* Note that we do not, and indeed cannot, have a result like this:

For every r.e. set $X \subseteq \omega$ there is a proof analysis P_X and a sequent $\Pi_X \rightarrow \Lambda_X, A_X(a)$ s.t. there is an $\mathbf{LK}_\mathrm{B}^{\Pi\Sigma}$-proof realizing P_X with end sequent $\Pi_X \rightarrow \Lambda_X, A_X(s^n(0))$ iff $n \in X$.

This follows from the fact that for every proof analysis P and every sequent $\Pi \rightarrow \Lambda$ with free variable a, there is a semi-unification problem

$$\Omega = \{(s_1(a), t_1(a)), \ldots, (s_p(a), t_p(a))\}$$

s.t. P is realizable by an $\mathbf{LK}^{\Pi\Sigma}$-proof with end sequent $(\Pi \rightarrow \Lambda)\{s^n(0)/a\}$ iff $\Omega\{s^n(0)/a\}$ has a solution.

But $\Omega\{s^n(0)/a\}$ is either solvable for all $n \geq m$ and unsolvable for $n < m$, or for only one n. To see this, calculate the most general semi-unifier δ of

$$\{(f(s_1, a), f(t_1, a)), \ldots, (f(s_p, a), f(t_p, a))\}$$

(see below, Proposition 5.4). δ assigns to a either a term of the form $s^m(0)$ (one solution for $n = m$) or one of the form $s^m(b)$ (a solution for every $n \geq m$) (cf. BAAZ [1993]).

For $\mathbf{LK}^{\Pi\Sigma}$, such an undecidable proof analysis exists, cf. KRAJÍČEK and PUDLÁK [1988], Section 5.

4.7. THEOREM *k/l-Compressibility is undecidable for linear $\mathbf{LK}_\mathrm{B}^{\Pi\Sigma}$-proofs.*

Proof. We exhibit a class C of semi-unification problems whose solvability is undecidable and then show that for $\Omega \in C$ there is a sequent $\Pi_\Omega \rightarrow \Lambda_\Omega$ s.t.

(1) $\Pi_\Omega \to \Lambda_\Omega$ has a proof (with cut) of length l iff Ω has a solution, and
(2) $\Pi_\Omega \to \Lambda_\Omega$ has a proof of length $l + C$.

Let $\mathcal{C}$ consist of $\Omega = \{(s_1, t), (s_2, t)\}$ where

(1) $(\forall x_1, \ldots, x_n) A_\Omega(x_1, \ldots, x_n) \to Q$ is valid for $A_\Omega \equiv P(t) \wedge \big(P(s_1) \wedge P(s_2) \supset Q\big)$,
(2) s, t_1, t_2 are pairwise not unifiable.

We have to prove that $\mathcal{C}$ has the desired property that the proof analysis in Proposition 4.4 describes an optimal proof of $(\forall \bar{x}) A_\Omega(\bar{x}) \to Q$ if Ω is solvable, and that proofs are longer if Ω has no solution. Then we construct a longer proof analysis that is realizable by an $\mathbf{LK}_{\mathrm{B}}^{\Pi\Sigma}$-proof with the same end sequent for all $\Omega \in \mathcal{C}$.

First of all, $\mathcal{C}$ is undecidable because of the following: (a) By Theorem (ii) of PUDLÁK [1988], every semi-unification problem can be translated into a problem of the form $\Psi = \{(s_1', t'), (s_2', t')\}$. Every such problem can in turn be rewritten as $\Psi' = \big\{\big(f(g(a), s_1), f(a, t)\big), \big(f(h(a), s_2), f(a, t)\big)\big\}$, where a is a new variable. Ψ' obviously has the same solutions as Ψ, but the components of the two equations are pairwise not unifiable.

(b) Validity of $(\forall \bar{x}) A_\Omega(\bar{x}) \to Q$ is decidable. This follows from the fact that the following resolution proof exists iff $(\forall \bar{x}) A_\Omega(\bar{x}) \to Q$ is valid:

$$\cfrac{\cfrac{\{P(t)\sigma_2\} \qquad \cfrac{\{P(t)\sigma_1\} \quad \{\neg P(s_1), \neg P(s_2), Q\}}{\{\neg P(s_2)\delta_1, Q\}}\,\delta_1}{\{Q\}}\,\delta_2 \qquad \{Q\}}{\Box}\,id$$

where σ_1, σ_2 are renamings of variables. Consequently the following equations hold:

$$\begin{aligned} P(t)\sigma_1\delta_1\delta_2 &= P(s_1)\delta_1\delta_2 \\ (\text{since } P(t)\sigma_1\delta_1 &= P(s_1)\delta_1) \\ P(t)\sigma_2\delta_2 &= P(s_2)\delta_1\delta_2 \end{aligned}$$

The crucial point for the encoding of semi-unification problems by the proof analysis and end sequent $(\forall \bar{x}) A_\Omega(\bar{x}) \to Q$ is that $(\forall \bar{x}) A_\Omega(\bar{x})$ is "produced" only once, i.e., that $(a + 1)$ is ancestor to both premises of the cut (d). We can force this to be the case by replacing $A_\Omega(\bar{a})$ by $\neg^{2r} A_\Omega(\bar{a})$, where r is sufficiently large to make a separate deduction—by copying the part of the analysis above $(a + 1)$—too costly.

Let $(\forall \bar{x})\neg^{2r} A_\Omega(\bar{x}) \to A'$ be the sequent at $(a+1)$. We have (1) $\neg^{2r} A_\Omega(\bar{x})\delta \to A'$ for some δ and (2) Q has to be derived from A'. Take the shortest derivations of (1) and (2). The shortest derivation of Q must contain a quantified cut, since s_1, s_2, t are pairwise not unifiable. If $\big\{(s_1, t), (s_2, t)\big\}$ is not semi-unifiable, one universal or existential cut is not sufficient. The universal cut in the analysis given in the proof of Proposition 4.4 is the simplest possible one(This is intuitively clear, a rigorous proof would use analoga to Propositions 4–9 of BUSS [1991]).

Now we show that there *is* a uniform way of deriving valid sequents

$$(\forall \bar{x})\neg^{2r} A_\Omega(\bar{x}) \to Q$$

(which of course is longer than the one using the solution to the semi-unification problem Ω). Given σ_1, σ_2, δ_1, δ_2 from the above resolution deduction, the following gives a proof:

$$\Big\{ \text{propositional inferences}$$

$$\begin{array}{ll} (a) & \neg^{2r} A(a_1, \ldots, a_n) \rightarrow A(a_1, \ldots, a_n) \\ (a+1) & (\forall x_1) \ldots (\forall x_n) \neg^{2r} A(x_1, \ldots, x_n) \rightarrow A(a_1, \ldots, a_n) \\ \hline \end{array}$$

$$\Big\{ \begin{array}{l} \text{propositional inferences including} \\ \text{propositional cuts from } (a+1) \end{array}$$

$$\begin{array}{ll} (b) & (\forall x_1) \ldots (\forall x_n) \neg^{2r} A(x_1, \ldots, x_n) \rightarrow P(t) \\ (b+1) & (\forall x_1) \ldots (\forall x_n) \neg^{2r} A(x_1, \ldots, x_n) \rightarrow (\forall y_1) \ldots (\forall y_m) P(t) \\ \hline \end{array}$$

$$\begin{array}{ll} (\alpha) & P(t)\sigma_1\delta_1\delta_2 \rightarrow P(t)\sigma_1\delta_1\delta_2 \\ (\alpha+1) & (\forall x_1) \ldots (\forall x_n) P(t) \rightarrow P(t)\sigma_1\delta_1\delta_2 \\ (\beta) & P(t)\sigma_2\delta_2 \rightarrow P(t)\sigma_2\delta_2 \\ (\beta+1) & (\forall x_1) \ldots (\forall x_n) P(t) \rightarrow P(t)\sigma_2\delta_2 \\ (\gamma) & (\forall x_1) \ldots (\forall x_n) P(t) \rightarrow P(t)\sigma_1\delta_1\delta_2 \wedge P(t)\sigma_2\delta_2 \\ \hline \end{array}$$

$$\Big\{ \begin{array}{l} \text{propositional inferences including} \\ \text{propositional cuts from } (a+1) \end{array}$$

$$\begin{array}{ll} (c) & P(s_1) \wedge P(s_2), (\forall x_1) \ldots (\forall x_n) \neg^{2r} A_\Omega(x_1, \ldots, x_n) \rightarrow Q \\ (c+1) & (\exists x_1) \ldots (\exists x_n)(P(s_1) \wedge P(s_2)), (\forall x_1) \ldots (\forall x_n) \neg^{2r} A_\Omega(x_1, \ldots, x_n) \rightarrow Q \\ \hline \end{array}$$

$$\Big\{ \text{propositional inferences}$$

$$\begin{array}{ll} (\delta) & P(s_1)\delta_1\delta_2 \wedge P(s_2)\delta_1\delta_2 \rightarrow P(s_1)\delta_1\delta_2 \wedge P(s_2)\delta_1\delta_2 \\ (\delta+1) & P(s_1)\delta_1\delta_2 \wedge P(s_2)\delta_1\delta_2 \rightarrow (\exists x_1) \ldots (\exists x_n)(P(s_1) \wedge P(s_2)) \end{array}$$

$$\Big\{ \text{cut from } (\gamma) \text{ and } (\delta+1)$$

$$\begin{array}{ll} (\varepsilon) & (\forall x_1) \ldots (\forall x_n) P(t) \rightarrow (\exists x_1) \ldots (\exists x_n)(P(s_1) \wedge P(s_2)) \end{array}$$

$$\Big\{ \text{two cuts from } (b+1), (c+1), (\varepsilon)$$

$$\begin{array}{ll} (e) & (\forall x_1) \ldots (\forall x_n) \neg^{2r} A_\Omega(x_1, \ldots, x_n), (\forall x_1) \ldots (\forall x_n) \neg^{2r} A_\Omega(x_1, \ldots, x_n) \rightarrow Q \\ (e+1) & (\forall x_1) \ldots (\forall x_n) \neg^{2r} A_\Omega(x_1, \ldots, x_n) \rightarrow Q \end{array}$$

For the cut resulting in (ε), recall that $P(t)\sigma_1\delta_1\delta_2 = P(s_1)\delta_1\delta_2$ and $P(t)\sigma_2\delta_2 = P(s_2)\delta_1\delta_2$. $\square$

5 k/l-Compressibility is decidable for tree-like $\mathbf{LK}_{\mathrm{B}}^{\Pi\Sigma}$-proofs

For tree-like $\mathbf{LK}_{\mathrm{B}}^{\Pi\Sigma}$-analyses there is a procedure to decide realizability, given the analysis and end sequent. This procedure uses special semi-unification problems to determine the term structure of the proof. These problems are decidable, and furthermore a most general solution can be found, which guarantees term-minimal proofs.

5.1. DEFINITION A semi-unifier σ of a semi-unification problem Ω is called *most general semi-unifier*, if every semi-unifier σ' of Ω can be written as $\sigma\delta$, for some substitution δ. The most general semi-unifier is unique up to renaming of variables.

In contrast to second order unification, semi-unification has the property that most general semi-unifiers exist, if any exist at all:

5.2. PROPOSITION *There is an algorithm computing the most general semi-unifier of a given semi-unification problem Ω if any semi-unifier for Ω exists.*

See BAAZ [1993] or KFOURY *et al.* [1990] for details. The algorithm works roughly as follows: Let $\{(s_1,t_1),\ldots,(s_n,t_n)\}$ be the given semi-unification problem, and let α_i be disjoint canonical renamings of the variables in t_i. Unify $t_i\alpha_i$ with s_i. Apply the resulting unifier to the problem and repeat the process, until the unifier is only a renaming of variables or until unification fails, in which case there is no semi-unifier. The procedure will not always terminate, since semi-unification is undecidable, but will produce a most general semi-unifier if there is any semi-unifier. In what follows we will only use a decidable class of semi-unification problems for which the algorithm terminates after one step:

5.3. DEFINITION Let t be a term and $a_1, \ldots, a_n$ be a sequence of variables.

$$t * \langle a_1,\ldots,a_n\rangle := f(\ldots f(f(t,a_1),a_2)\ldots a_n)$$

5.4. PROPOSITION *Let Ω be a semi-unification problem of the form*

$$\{(s_1 * \langle a_1,\ldots,a_n\rangle, t_1 * \langle a_1,\ldots,a_n\rangle),\ldots,(s_r * \langle a_1,\ldots,a_n\rangle, t_r * \langle a_1,\ldots,a_n\rangle)\},$$

where the variables in $s_1, \ldots, s_r$ are among $a_1, \ldots, a_n$, and let α_i be disjoint canonical renamings of the variables in t_i. Let σ be the most general unifier of

$$\{(s_1 * \langle a_1,\ldots,a_n\rangle, t_1 * \langle a_1,\ldots,a_n\rangle\alpha_1),\ldots,(s_r * \langle a_1,\ldots,a_n\rangle, t_r * \langle a_1,\ldots,a_n\rangle\alpha_r)\},$$

If σ exists, then σ is also a most general semi-unifier of Ω, otherwise Ω is unsolvable.

Proof. σ is also a most general unifier of $\{(s_1,t_1\alpha_1'),\ldots,(s_r,t_r\alpha_r')\}$, where α_i' is a renaming of the variables occuring in t_i *other than* $a_1, \ldots, a_n$. Let $t_i \equiv t_i(a_1,\ldots,a_n,b_1,\ldots,b_m)$. Then

$$t_i\sigma = t_i(a_1\sigma,\ldots,a_n\sigma,b_1,\ldots,b_m)$$

($b_1, \ldots, b_m$ do not occur in $s_1, \ldots, s_r$!) and

$$s_i\sigma = t_i(a_1\sigma,\ldots,a_n\sigma,b_1\alpha_i'\sigma,\ldots,b_m\alpha_i'\sigma). \quad \Box$$

5.5. PROPOSITION *Let P be a tree-like proof analysis with given end sequent. If there is an $\mathbf{LK}_B$-proof D realizing P, then there also is a proof D' with the following properties:*

(1) D' is regular (no two strong quantifier inferences have the same eigenvariable and eigenvariables do not occur in the end sequent).
(2) If P contains a sequence of applications of ($\forall_B$:left) to the same formula, then D' introduces all quantifiers in the first of these applications, and all following ($\forall_B$:left) inferences in the sequence are empty introductions. Similarly for ($\exists_B$:right)
(3) If P contains a sequence of applications of ($\forall_B$:right) to the same formula, then D' introduces all quantifiers in the last of these applications, and all preceding ($\forall_B$:right) inferences in the sequence are empty introductions. Similarly for ($\exists_B$:left)

Proof. (1) In a tree-like proof, eigenvariables can be renamed to ensure regularity. (2), (3) If strong quantifier inferences are moved downwards and weak quantifier inferences are moved upwards in a regular proof tree, the eigenvariable conditions can be protected by renaming. □

5.6. THEOREM *Realizability is decidable for tree-like $\mathbf{LK}_B^{\Pi\Sigma}$-proof analyses.*

Proof. Given a tree-like proof analysis P and an end sequent $\Pi \to \Lambda$, we construct a *preproof* $\Psi(P, \Pi \to \Lambda)$. A preproof is an assignment of formulas to the nodes of the analysis P such that all inferences except quantifier inferences introducing cut-formulas are in correct form (i.e., valid applications of the rules), and a substitution for free variables will "correct" the cuts as well. Ψ is term-minimal, i.e., if D is a proof realizing P, then D can be written as $\Psi\sigma$, for some substitution σ. The construction is similar to the construction of cut-free term-minimal tree-like proofs in KRAJÍČEK and PUDLÁK [1988], Section 2.

Constructing a preproof Since $\mathbf{LK}_B^{\Pi\Sigma}$-analyses contain the names of predicates in axioms and weakenings, the logical structure of a proof is uniquely determined (cf. Proposition 5.5) except for the quantifier prefix of the cut formulas in universal and existential cuts. We index the universal and existential cuts by α_1, $\alpha_2, \ldots$

(1) Determine the propositional structure of Ψ from P. Use different free variables for *every* term position in the predicates. For quantifier prefixes use special *quantifier prefix variables* ($\forall_B$-α_i), ($\exists_B$-α_i).
(2) Unify the end sequent of Ψ with $\Pi \to \Lambda$, and proceed upwards in the proof tree as follows:
    (a) Unify conclusions of propositional inferences, exchanges, contractions, and weakenings with the respective premises.
    (b) In strong quantifier inferences not introducing cut formulas, e.g.,

$$\frac{\Gamma' \to \Delta', A'}{\Gamma \to \Delta, (\forall x_1)\ldots(\forall x_n)A(x_1,\ldots,x_n)}\ (\forall_B\text{:right})$$

unify Γ, Δ with Γ',Δ', and $A(c_1,\ldots,c_n)$ with A', where c_1, ..., c_n are new free variables *which are handled as constants* to avoid substitution into eigenvariables, similarly for ($\exists_B$:left).

(c) In weak quantifier inferences not introducing cut formulas, e.g.,

$$\frac{A',\Gamma' \to \Delta'}{(\forall x_1)\ldots(\forall x_n)A(x_1,\ldots,x_n),\Gamma \to \Delta}\;(\forall_B\text{:left})$$

unify Γ, Δ with Γ',Δ', and $A(z_1,\ldots,z_n)$ with A', where z_1, ..., z_n are new variables, similarly for ($\exists_B$:right).

(d) Unify A and A' in axioms $A \to A'$ and unify the cut formulas in the premises of a cut.

As can easily be seen, the steps in the construction are all as general as possible and the restrictions imposed by the unifications are all necessary. If the procedure fails to find a preproof (i.e., one of the unifications fails or eigenvariable conditions are violated), P is not realizable with end sequent $\Pi \to \Lambda$.

To complete the preproof to a proof we now have to determine the quantifier prefixes and the term structure of the universal and existential cut formulas. We first illustrate this:

5.7. EXAMPLE

$$\dfrac{\dfrac{\dfrac{\begin{array}{c}\vdots\\ \dfrac{\begin{array}{c}\vdots\\ \Gamma_1 \to \Delta_1, A_1\end{array}}{\Gamma_1 \to \Delta_1, (\forall_{\mathrm{B}}\text{-}\alpha)A}\,*\\ \vdots\\ \Gamma_2 \to \Delta_2, (\forall_{\mathrm{B}}\text{-}\alpha)A, A_2\end{array}}{\Gamma_2 \to \Delta_2, (\forall_{\mathrm{B}}\text{-}\alpha)A, (\forall_{\mathrm{B}}\text{-}\alpha)A}\,**}{\Gamma_2 \to \Delta_2, (\forall_{\mathrm{B}}\text{-}\alpha)A} \qquad \begin{array}{c} \dfrac{\begin{array}{c}\vdots\\ A_3,\Gamma_3 \to \Delta_3\end{array}}{(\forall_{\mathrm{B}}\text{-}\alpha)A,\Gamma_3 \to \Delta_3} \quad \dfrac{\begin{array}{c}\vdots\\ A_4,\Gamma_4 \to \Delta_4\end{array}}{(\forall_{\mathrm{B}}\text{-}\alpha)A,\Gamma_4 \to \Delta_4}\\ \vdots\\ (\forall_{\mathrm{B}}\text{-}\alpha)A,\Gamma_5 \to \Delta_5\end{array}}{\Gamma_2,\Gamma_5 \to \Delta_2,\Delta_5}$$

Let P_1 (P_2) denote the part of the preproof above the end sequent and below $*$ ($**$). If σ is an extension of the preproof to a proof, then (a) the eigenvariables of ($*$) do not occur in $P_1\sigma$ and (b) the eigenvariables of ($**$) do not occur in $P_2\sigma$. This leads to the semiunification problem

$$\{(A_2 * \langle\bar{a}\rangle, A_1 * \langle\bar{a}\rangle), (A_3 * \langle\bar{a}\rangle, A_1 * \langle\bar{a}\rangle), (A_4 * \langle\bar{a}\rangle, A_1 * \langle\bar{a}\rangle)\},$$

where $\bar{a}$ are the free variables in P_1. Let δ_1 be the most general semi-unifier. Next, determine the most general semi-unifier δ_2 of

$$\{(A_3\delta_1 * \langle\bar{b}\rangle, A_1\delta_1 * \langle\bar{b}\rangle), (A_4\delta_1 * \langle\bar{b}\rangle, A_1\delta_1 * \langle\bar{b}\rangle)\},$$

where $\bar{b}$ are the free variables in $P_2\delta_1$. We obtain:

$$\begin{aligned}
A_1\delta_1\delta_2 &= A(c_1,\ldots,c_r)\\
A_2\delta_1\delta_2 &= A(g_1(d_1,\ldots,d_s),\ldots,g_r(d_1,\ldots,d_s))\\
A_3\delta_1\delta_2 &= A(g_1(t_1,\ldots,t_s),\ldots,g_r(t_1,\ldots,t_s))\\
A_4\delta_1\delta_2 &= A(g_1(t'_1,\ldots,t'_s),\ldots,g_r(t'_1,\ldots,t'_s))
\end{aligned}$$

Since the c_i do not occur in $P_2\delta_1\delta_2$, c_i can be replaced by $g_i(d'_1, \ldots, d'_s)$. Finally, replace $(\forall_B\text{-}\alpha)A$ by $(\forall z_1 \ldots z_s)A(g_1(z_1, \ldots, z_s), \ldots, g_r(z_1, \ldots, z_s))$. (Any permutation of $z_1 \ldots z_s$ can be chosen.)

Correction of cuts To correct cuts in the general case, we associate with each universal or existential cut α_i the set of strong propositional premises $\mathrm{Prm}_s(\alpha_i)$ and weak propositional premises $\mathrm{Prm}_w(\alpha_i)$. Recall that $\forall$ ($\exists$)-introduction is strong (weak) on the right side and weak (strong) on the left side. Thus, $\mathrm{Prm}_s(\alpha_i)$ is the set of those formulas A_j that are ancestors to the cut formula on the strong side of the cut (A_1, A_2 in Example 5.7), and $\mathrm{Prm}_w(\alpha_i)$ is the set of those formulas A_j that are ancestors to the cut formula on the weak side of the cut (A_4, A_5). Let $\mathcal{D} = \bigcup_{\alpha_i} \mathrm{Prm}_s(\alpha_i)$

Define a partial order $\leq$ on $\mathcal{D}$, according to where in the proof A_j is quantified to yield the cut formula $(\forall_B\text{-}\alpha_i)A$: $A_j \leq A_k$ if A_j is quantified below A_k ($A_2 \leq A_1$). The *exclusion area* $D(A_j)$ of the proof corresponding to A_j is the part above the end sequent and below the premise of this quantifier inference ($D(A_1) = P_1$, $D(A_2) = P_2$).

Balancing cuts Select a maximal element $A_j \in \mathcal{D}$ and compute the most general semi-unifier σ_j of the problem

$$\begin{aligned}&\big\{(A_k * \langle a_1, \ldots, a_n\rangle, A_j * \langle a_1, \ldots, a_n\rangle) \bigm| A_k \in \mathrm{Prm}_w(\alpha_i)\big\} \cup \\ &\cup \big\{(A_l * \langle a_1, \ldots, a_n\rangle, A_j * \langle a_1, \ldots, a_n\rangle) \bigm| A_l \in \mathrm{Prm}_s(\alpha_i), A_j \not\leq A_l\big\}\end{aligned}$$

where $A_j \in \mathrm{Prm}_s(\alpha_i)$ and $a_1, \ldots, a_n$ are the free variables in $D(A_j)$. Apply σ_j to the preproof and repeat this process for $\mathcal{D} := (\mathcal{D} \setminus A_j)\sigma_j$ until $\mathcal{D} = \emptyset$.

Call a free variable in A_j *critical for* A_j if it does not occur in $D(A_j)$ and let $\mathrm{crit}(A_j)$ be the set of all free variables critical for A_j. A variable is *critical for the cut* α_i if it is critical for one of its strong premises.

The critical variables of a strong premise A_j of α_j are the potential eigenvariables for the introduction of quantifiers on A_j: The above semi-unifications make all strong and weak premises A' in $D(A_j)$ corresponding to the same cut as A_j substitution instances of A_j (A_2 is a substitution instance of A_1, and A_3, A_4 are substitution instances of both A_1, A_2). By the *-construction in the semi-unification problems above, if $A' = A_j\delta$ for some substitution δ, then δ only acts on $\mathrm{crit}(A_j)$. Note that the critical variables fulfill the eigenvariable condition.

If A_j and A_k are two premises of the cut α_i, then the critical variables of A_j do not occur in A_k and vice versa (If $c \in \mathrm{crit}(A_j)$ and $A_j \leq A_k$ then c occurs in $D(A_k)$ and hence cannot be critical for A_k. If c would occur in A_k, then it would also occur in a weak premise of the cut (by the above semi-unifications), but this premise is in $D(A_j)$. If, on the other hand, A_k is in $D(A_j)$, then c does not occur in A_k by definition). Critical variables for one cut premise are not critical variables for any other cut (For any two premises A_j and A_k, either A_j is in $D(A_k)$ or vice versa).

Unifying premises Now let $A_j(c_1, \ldots, c_s)$ be one of the possibly several $\leq$-minimal strong premises of the cut α_i, where $c_1, \ldots, c_s$ are the critical variables of A_j. A_j is the least general of the strong premises and therefore determines the term structure of the cut formula (A_2 in the example). Unify every other premise A' of α_i with $A_j\delta$, where δ is a a disjoint canonical renaming of the critical variables of A_j. The unifier acts only on *critical* variables of *this cut*. This makes all strong premises of α_i equal up to renaming of critical variables. Recall that the weak premises are substitution instances of A_j and hence, are now of the form $A_j(t_1,\ldots,t_s)$. Replace the cut formula $(\forall_B\text{-}\alpha_i)A$ by $(\forall v_1)\ldots(\forall v_s)A(v_1,\ldots,v_s)$ ($(\exists_B\text{-}\alpha_i)A$ by $(\exists v_1)\ldots(\exists v_s)A(v_1,\ldots,v_s)$). Repeat this step for every cut in the preproof. The resulting proof is uniquely determined up to the order of the quantifiers in cut formulas.

The unifying of premises may influence other cuts, but since critical variables are disjoint for different cuts, this has no effect on other cuts being balanced or unified. All correction steps with exception of the last one are most general and forced by the information provided by the proof analysis and end sequent. Hence, if the correction fails at any step, or if eigenvariable conditions on variables introduced in the construction of the preproof are violated, there is no proof extending the preproof. □

5.8. COROLLARY *k-Provability is decidable for* $\mathbf{LK}_{\mathrm{B}}^{\Pi\Sigma}$.

5.9. COROLLARY *k/l-Compressibility is decidable for* $\mathbf{LK}_{\mathrm{B}}^{\Pi\Sigma}$.

5.10. *Remark* The term depth d' of the constructed proof can be very roughly bounded by $d' \leq d \cdot 2^{m \cdot l^l}$, where d is the maximal term depth and m the number of quantified variables in the given end sequent $\Pi \rightarrow \Lambda$, and l is the length of the proof analysis.

The construction of the preproof for $\Pi \rightarrow \Lambda$ introduces at most m new variables in each step, and at most ml overall. The correction of a strong premise introduces at most $(l-1)v$ variables, where v is the current number of variables. The disappearance of a variable in a unification step increses the term depth at most by a factor of 2. If every bound variable occurs only once in the end sequent, then $d' = d$.

6 Conclusion

Two fundamental distinctions have been made in this paper:

(a) The distinction between systems that introduce one (or any fixed number) quantifier and systems that introduce blocks (an unknown number) of quantifiers of the same type in one introduction. Our results show that a committment on the *form* of these blocks of quantifiers, while irrelevant for cut elimination, is disadvantageous for the algorithmic introduction of cuts into a given proof. This is generally the case with constructions that depend on operations on the term structure, e.g. when generalizing proofs, and is essentially due to the fact that second order unification problems (that correspond to single introduction of quantifiers) do not have most general solutions, in contrast to semi-unification problems (that correspond to block introduction of quantifiers, cf. BAAZ [1993])

(b) The distinctions between linear and tree-like ways to write proofs. Until the 1950s, linear notation of proofs was commonplace in logic, but since then has almost disappeared. In computer science, linear proofs have been reintroduced, cf. resolution deductions where one and the same clause is used several times. The more space efficient linear notation, however, has serious drawbacks when the relationship between quantifiers in a given proof is investigated.

The problem of structuring of proofs itself will be of importance to computer science, since it is closely related to structuring of *programs.* If we conceive of proof complexity as the degree of entanglement (e.g., as the topological genus of the proof analysis, cf. STATMAN [1974]), then structuring means algorithmic simplification.

For proof theory, the significance of the problem is that it enables us to separate model-theoretically indistinguishable systems according to their structural properties (cf. (a), (b) above). For a detailed discussion of this aspect, cf. G. Kreisel's postscript to BAAZ and PUDLÁK [1993].

References

BAAZ, M.

[1993] Note on the existence of most general semi-unifiers. In *Arithmetic, Proof Theory and Computational Complexity*, P. Clote and J. Krajíček, editors, pp. 19–28. Oxford University Press.

BAAZ, M. and P. PUDLÁK.

[1993] Kreisel's conjecture for $L\exists_1$. In *Arithmetic, Proof Theory and Computational Complexity*, P. Clote and J. Krajíček, editors, pp. 29–59. Oxford University Press.

BUSS, S. R.

[1991] The undecidability of k-provability. *Ann. Pure Appl. Logic*, **53**, 75–102.

FARMER, W. M.

[1991] A unification-theoretic method for investigating the k-provability problem. *Ann. Pure Appl. Logic*, **51**, 173–214.

GENTZEN, G.

[1934] Untersuchungen über das logische Schließen I–II. *Math. Z.*, **39**, 176–210, 405–431.

KFOURY, A. J., J. TIURYN, and P. URZYCZYN.

[1990] The undecidability of the semi-unification problem. In *Proc. 22nd ACM STOC*, pp. 468–476. journal version to appear in *Inf. Comp.*

KRAJÍČEK, J. and P. PUDLÁK.

[1988] The number of proof lines and the size of proofs in first order logic. *Arch. Math. Logic*, **27**, 69–84.

PARIKH, R. J.

[1973] Some results on the length of proofs. *Trans. Am. Math. Soc.*, **177**, 29–36.

PUDLÁK, P.

[1988] On a unification problem related to Kreisel's conjecture. *Comm. Math. Univ. Carol.*, **29**(3), 551–556.

STATMAN, R.

[1974] *Structural Complexity of Proofs.* PhD thesis, Stanford University.

TAIT, W. W.

[1968] Normal derivability in classical logic. In *The Syntax and Semantics of Infinitary Languages*, J. Barwise, editor, pp. 204–236. Springer, Berlin.

Optimization Problems: Expressibility, Approximation Properties and Expected Asymptotic Growth of Optimal Solutions

Thomas Behrendt* Kevin Compton† Erich Grädel‡

1 Introduction

Although the notion of NP-completeness was defined in terms of *decision problems*, the prime motivation for its study and development was the apparent intractability of a large family of combinatorial *optimization problems*. NP-completeness of a decision problem rules out the possibility of finding an optimal solution of the corresponding optimization problem in polynomial time unless P = NP. It does not exclude, however, the possibility that there are efficient algorithms which produce *approximate* solutions. In fact, for many optimization problems with NP-complete decision problems, there are simple and efficient algorithms that produce solutions differing from optimal solutions by at most a constant factor. For some problems, there even exist so-called *polynomial-time approximation schemes* (PTAS), which produce approximate solutions to any desired degree of accuracy. For other problems, notably the Traveling Salesperson Problem, there do not exist efficient approximations unless P = NP (see [7]). Until now the "structural" reasons for the different approximation properties of NP optimization problems have not been sufficiently understood.

Papadimitriou and Yannakakis [17] provided a new perspective by relating the approximation properties of optimization problems to their logical representation. Exploiting Fagin's characterization of NP by existential second order logic [6], they defined two classes of optimization problems, MAX SNP and MAX NP, and showed that all problems in these classes are approximable in polynomial time up to a constant factor. They also identified a host of problems which are complete for MAX SNP with respect to reductions that preserve polynomial-time approximation schemes. Very recently, the classes MAX SNP and MAX NP have received a lot of attention due to results by Arora et al. [2] showing that problems which are hard for MAX SNP cannot have a PTAS, unless P = NP.

We present the syntactic criterion of Papadimitriou and Yannakakis in the more general form and notation provided by Kolaitis and Thakur [11].

Definition 1.1 Recall that Σ_n (respectively Π_n) are prefix classes in first order logic, consisting of formulae in prefix normal form with n alternating blocks of quantifiers

*Mathematisches Institut, Universität Basel, Rheinsprung 21, CH-4051 Basel
†EECS Department, University of Michigan, Ann Arbor MI 48109-2122, kjc@eecs.umich.edu
‡Lehrgebiet Mathematische Grundlagen der Informatik, RWTH Aachen, Ahornstr. 55, 5100 Aachen, graedel@mjoli.informatik.rwth-aachen.de

beginning with $\exists$ (respectively $\forall$). The classes MAX Σ_n (respectively MAX Π_n) consist of maximization problems Q whose input instances are finite structures A of a fixed signature σ, such that the cost of an optimal solution of Q on input A is definable by an expression

$$opt_Q(A) = \max_{\bar{S}} |\{\bar{x} : A \models \psi(\bar{x}, \bar{S})\}|$$

where $\psi(\bar{x}, \bar{S})$ is a Σ_n-formula (respectively a Π_n-formula) and where $\bar{S}$ are predicate variables not contained in σ.

Examples.

- MAX CUT (MC) is the problem of decomposing the vertex set of a given graph G into two subsets such that the number of edges between them is maximal. It is in MAX Σ_0:

$$opt_{MC}(G) = \max_{U} |\{(x, y) : G \models Exy \wedge (Ux \leftrightarrow \neg Uy)\}|.$$

- MAX SAT is the problem of finding an assignment that satisfies the maximal number of clauses in a given propositional formula in CNF. Such a formula can be represented by a structure $F = (U; P, N)$ with universe U consisting of the clauses and the variables, and with binary predicates P and N where Pxy and Nxy say that variable y occurs positively, respectively negatively in clause x. MAX SAT is in MAX Σ_1 with the defining expression

$$opt_Q(F) = \max_{S} |\{x : F \models (\exists y)((Sy \wedge Pxy) \vee (\neg Sy \wedge Nxy))\}|.$$

  Kolaitis and Thakur proved that MAX SAT $\notin$ MAX Σ_0.

- MAX CLIQUE is the problem of finding a clique of maximal size in a graph. The size of such a clique in G is usually denoted by $\omega(G)$. MAX CLIQUE is in MAX Π_1 because

$$\omega(G) = \max_{C} |\{x : G \models Cx \wedge (\forall y)(\forall z)(Cy \wedge Cz) \rightarrow (y = z \vee Eyz)\}|$$

  Simple monotonicity arguments [16] show that MAX CLIQUE is not in MAX Σ_1. By very recent results in [2], there exists an $\varepsilon > 0$ such that MAX CLIQUE cannot be approximated in polynomial-time within a factor of n^{ε}.

The syntactic criteria for MAX SNP and MAX NP used by Papadimitriou and Yannakakis are those for MAX Σ_0 (where $\psi(\bar{x}, \bar{S})$ is quantifier-free) and for MAX Σ_1 (where $\psi(\bar{x}, \bar{S})$ is existential). However, two remarks about the definitions of these classes should be made. First, the definition of MAX Σ_n as given above is not really sufficient to establish that all problems in MAX Σ_1 are approximable up to a constant factor, at least if approximability means — as usually understood — that we can actually find in polynomial time a nearly optimal solution. The criterion as given by Definition 1.1 only allows us to determine the *cost* of an optimal solution up to a constant factor. We will therefore propose a modified notion for the logical representation of an optimization problem which requires that the formula models (in some sense to be made precise later) *all feasible solutions* of the problem, and not just the cost of an optimal one.

Second, it should be noted that in most papers the definitions of the classes MAX SNP and MAX NP have been interpreted differently than what was originally intended in [17]. While most authors (see [8, 12, 15, 16]) understood MAX SNP, respectively MAX NP to be *precisely* MAX Σ_0 and MAX Σ_1, Papadimitriou and Yannakakis actually had in mind their closures under the appropriate reductions (although they did not really make this clear; but see the remark in [18]). In particular, these extended versions of MAX SNP and MAX NP can also contain minimization problems. Kann [10] defines yet another, intermediate version of MAX SNP. We think that these different classes all have their merits, but it is important not to confuse them. The "pure" syntactic classes are interesting because they provide a logical criterion for approximability, and provide an opportunity to prove results about optimization problems using tools from logic (or, more precisely, finite model theory). In logic we have lower bound techniques that have no counterpart in computational complexity theory. In many cases these techniques (e.g. monotonicity arguments, Ehrenfeucht-Fraïssé games and limit laws) show that a problem does not satisfy a certain syntactic criterion, and thus establish separation and hierarchy results among the syntactic classes (without referring to unproved hypotheses from complexity theory). On the other hand, the closure classes may be appropriate if one is interested in pure complexity results. But closing syntactic classes under a class of reductions that are defined in terms of computational complexity, rather than logical definability, precludes the use of the logical techniques.

This paper is about syntactic classes. One of our goals was to find a more general syntactic criterion for approximability than the one provided by Papadimitriou and Yannakakis. This is achieved using other results from finite model theory than just Fagin's theorem, in particular the close connection between *fixpoint logic* and polynomial-time computability. To avoid confusion, we use the names MAX Σ_0 and MAX Σ_1, introduced by Kolaitis and Thakur, rather than MAX SNP and MAX NP.

Kolaitis and Thakur [11, 12] systematically investigated the logical expressibility of optimization problems. They proved that the class MAX $\mathcal{PB}$, consisting of all polynomially bounded maximization problems, coincides with MAX Π_2 and that there is proper hierarchy of four levels.

Theorem 1.2 MAX $\Sigma_0 \subsetneq$ MAX $\Sigma_1 \subsetneq$ MAX $\Pi_1 \subsetneq$ MAX $\Pi_2 =$ MAX $\mathcal{PB}$.

It is interesting that the classes MAX Π_2 and MAX Π_1 are separated by MAXIMUM CONNECTED COMPONENT (MCC), the problem of finding a connected component of maximal cardinality in a graph. This optimization problem is clearly solvable in polynomial time.

New expressibility classes of optimization problems. In this paper we extend the classes MAX Σ_i and MAX Π_i in several ways. Most importantly we allow the formulae to contain relations definable in least fixpoint logic. In addition, we maximize not only over relations but also over constants. We call the extended classes MAX Σ_i^{FP} and MAX Π_i^{FP}. The proof of [17] can be extended to MAX Σ_1^{FP} to show that all problems in this class are approximable. Some problems, such as MCC, descend from the highest level MAX Π_2 in the original hierarchy to the lowest level MAX Σ_0^{FP} in the new hierarchy. *Thus our extended class* MAX Σ_1^{FP} *provides a more powerful sufficient criterion for approximability than the original class* MAX Σ_1 *of* [17]. However, we also prove that even MAX Σ_1^{FP} does not contain all approximable problems; in fact, it does not even contain all polynomial time optimization problems. We discuss the question of how far the class MAX Σ_1 can be extended while preserving approximability.

Separation of the extended classes by the probabilistic method. We separate also the extended classes; e.g. we prove that

$$\text{MAX } \Sigma_0^{\text{FP}} \subsetneq \text{MAX } \Sigma_1^{\text{FP}} \subsetneq \text{MAX } \Pi_1^{\text{FP}} \subsetneq \text{MAX } \Pi_2^{\text{FP}} = \text{MAX } \mathcal{PB}.$$

Also, we prove that a number of important problems do not belong to MAX Σ_1^{FP}. These include MAX CLIQUE, MAX INDEPENDENT SET, V-C DIMENSION and MAX COMMON INDUCED SUBGRAPH. To do this we have to use more sophisticated methods than the techniques of [16, 11] which break down in the presence of fixpoint definitions. We use two alternative methods.

The first method, introduced in the present paper, characterizes rates of growth of average optimal solution sizes. For instance, it is known [4] that the expected size of a maximal clique in a random graph of cardinality n grows asymptotically like $2 \log n$. We show that no problem in MAX Σ_1^{FP} can have this property, thus proving that MAX CLIQUE is not in MAX Σ_1^{FP}. This technique is related to *limit laws for various logics* [5, 13, 14] and to the *probabilistic method* from combinatorics [1]. We believe that this method may be of independent interest.

The second method uses special classes of structures where fixpoint logic has no more expressive power than quantifier-free formulae. On such classes we can apply monotonicity arguments that break down on arbitrary finite structures. With this technique we give an alternative proof that MAX CLIQUE is not in MAX Σ_1^{FP}. We also show that MAX MATCHING is not expressible by existential sentences with fixpoint definitions.

In contrast to the recent results on the non-approximability of many maximization problems, among them MAX CLIQUE, our results do not depend on any unproved hypothesis from complexity theory, such as $\text{P} \neq \text{NP}$.

2 Preliminaries

Definition 2.1 An NP *optimization problem* is a quadruple $Q = (I_Q, \mathcal{F}_Q, f_Q, opt)$ such that

- I_Q is the set of *input instances* for Q.
- $\mathcal{F}_Q(I)$ is the set of *feasible solutions* for input I. Here, "feasible" means that the size of the elements $S \in \mathcal{F}_Q(I)$ is polynomially bounded in the size of I and that the set $\{(I, S) : S \in \mathcal{F}_Q(I)\}$ is recognizable in polynomial time.
- $f_Q : \{(I, S) : S \in \mathcal{F}_Q(I)\} \to \mathbb{N}$ is a polynomial-time computable function, called the *cost function*.
- $\text{opt} \in \{\max, \min\}$.

For every NP optimization problem Q, the following decision problem is in NP: given an instance I of Q and a natural number k, is there a solution $S \in \mathcal{F}_Q(I)$ such that $f_Q(I, S) \geq k$ when $\text{opt} = \max$, (or $f_Q(I, S) \leq k$, when $\text{opt} = \min$).

Let $opt_Q(I) := \text{opt}_{S \in \mathcal{F}_Q(I)} f_Q(I, S)$. An NP optimization problem is said to be *polynomially bounded* if there exists a polynomial p such that $opt_Q(I) \leq p(|I|)$ for all instances I. We denote by MAX $\mathcal{PB}$ (MIN $\mathcal{PB}$) the set of all polynomially bounded maximization (minimization) problems.

Approximation. The *performance ratio* of a feasible solution S for an instance I of Q is defined as $R(I, S) := opt_Q(I)/f_Q(I, S)$ if Q is a maximization problem and as

$R(I,S) := f_Q(I,S)/opt_Q(I)$ if Q is a minimization problem.

Definition 2.2 We say that an NP optimization problem Q is *approximable up to a constant factor* if there exists a constant $c > 0$ and a polynomial-time algorithm Π which produces, for every instance I of Q, a feasible solution $\Pi(I)$ with performance ratio $R(I, \Pi(I)) \leq c$. APX is the class of all NP optimization problems that are approximable up to a constant factor.

A weaker notion of approximability that is sometimes used requires only that the *cost* of an optimal solution can be approximated; for instance, in the case of MAX CLIQUE it would only be required that the algorithms approximates the clique number $\omega(G)$, not that it actually finds a nearly optimal clique.

Logical representation of optimization problems. Let Q be an optimization problem whose input instances are finite structures of fixed vocabulary σ. The definition of the classes MAX Σ_n and MAX Π_n as given by Kolaitis and Thakur requires only that there is an appropriate logical definition of $opt_Q(A)$, the cost of an optimal solution. However, optimization problems can be modelled by logical formulae in a much closer way.

Definition 2.3 A formula $\psi(\bar{x}, \bar{S})$ of vocabulary $\sigma \cup \{S_1, \ldots, S_r\}$ *represents* Q if and only if the following holds.

(i) For every instance A and every feasible solution $S_0 \in \mathcal{F}_Q(A)$, there exists an expansion $B = (A, S_1, \ldots, S_r)$ of A such that $f_Q(A, S_0) = |\{\bar{x} : B \models \psi(\bar{x}, \bar{S})\}|$.

(ii) Conversely, every expansion $B = (A, S_1, \ldots, S_r)$, for which the set $L = \{\bar{x} : B \models \psi(\bar{x}, \bar{S})\}$ is non-empty defines a feasible solution S_0 for A with $f_Q(A, S_0) = |L|$; moreover, this solution can be computed in polynomial time from B.

In particular, $opt_Q(A) = \max_{\bar{S}} |\{\bar{x} : A \models \psi(\bar{x}, \bar{S})\}|$.

In all examples that we consider, the feasible solution defined by $(A, S_1, \ldots, S_r)$ will either be one of the S_i or the set $\{\bar{x} : A \models \psi(\bar{x}, \bar{S})\}$ itself. Lautemann [15] has independently considered this more detailed logical representation of optimization problems. A more constructive alternative to Definition 1.1 might then be the following.

Definition 2.4 MAX Σ_n is the class of all maximization problems that can be represented by Σ_n-formulae. The classes MAX Π_n, MIN Σ_n and MIN Π_n are defined analogously.

Clearly this definition is more restrictive than the one used by Kolaitis and Thakur. We think that it is justified by the following observations. First, the more restrictive definition is necessary to establish the result of Papadimitriou and Yannakakis that MAX $\Sigma_1 \subseteq$ APX. Second, on all natural examples in the literature, the two definitions make no difference. Third, the results of Kolaitis and Thakur, in particular the fact that

$$\text{MAX } \Sigma_0 \subsetneq \text{MAX } \Sigma_1 \subsetneq \text{MAX } \Pi_1 \subsetneq \text{MAX } \Pi_2 = \text{MAX } \mathcal{PB}$$

remain true with the more restrictive definition. (However, the proof that MAX Π_2 = MAX $\mathcal{PB}$ needs some modification.) All results presented in this paper are true for both possible choices of logical representation of optimization problems. However, if the more liberal one (modelling only the cost of the optimal solution) is chosen, then the more liberal definition of approximability must also be adopted.

Fixpoint logic. It is well-known that the expressive power of first-order logic is limited by the lack of a mechanism for unbounded iteration or recursion. The most notable example of a query that is not first-order expressible is the transitive closure (TC). This has motivated the study of more powerful languages that add recursion in one way or another to first-order logic. The most prominent of these are the various forms of *fixpoint logics.*

Let σ be a signature, P an r-ary predicate not in σ and $\psi(\bar{x})$ be a formula of the signature $\sigma \cup \{P\}$ with only positive occurrences of P and with free variables $\bar{x} = x_1, \ldots, x_r$. Then ψ defines for every finite σ-structure A with universe $|A|$ an operator ψ^A on the class of r-ary relations over $|A|$ by

$$\psi^A : P \longmapsto \{\bar{a} : (A, P) \models \psi(\bar{a})\}.$$

Since P occurs only positively in ψ, this operator is monotone, i.e. $Q \subseteq P$ implies that $\psi^A(Q) \subseteq \psi^A(P)$. Therefore this operator has a *least fixed point* which may be constructed inductively beginning with the empty relation. Set $\Psi^0 := \emptyset$ and $\Psi^{j+1} := \psi^A(\Psi^j)$. At some stage i, this process reaches a stable predicate $\Psi^i = \Psi^{i+1}$, which is the *least fixed point* of ψ on A, and denoted by Ψ^∞. Since $\Psi^i \subseteq \Psi^{i+1}$, the least fixed point is reached in a polynomial number of iterations, with respect to the cardinality of A.

The fixed point logic (FO + LFP) is defined by adding to the syntax of first order logic the *least fixed point formation rule:* if $\psi(\bar{x})$ is a formula of the signature $\sigma \cup \{P\}$ with the properties stated above and $\bar{u}$ is an r-tuple of terms, then

$$[\mathrm{LFP}_{P,\bar{x}}\ \psi](\bar{u})$$

is a formula of vocabulary σ (to be interpreted as $\Psi^\infty(\bar{u})$).

Example. Here is a fixpoint formula that defines the transitive closure of the binary predicate E:

$$[\mathrm{TC}\ E](u, v) \equiv [\mathrm{LFP}_{T,x,y}\ (x = y) \vee (\exists z)(Exz \wedge Tzy)](u, v).$$

On the class of all finite structures, (FO + LFP) has strictly more expressive power than first-order logic — it can express the transitive closure — but is strictly weaker than PTIME-computability. However, Immerman [9] and Vardi [19] proved that on *ordered structures* the situation is different. There (FO + LFP) characterizes precisely the queries that are computable in polynomial time. On the other hand, on very simple classes of structures, such as structures with empty signatures (i.e. sets), (FO + LFP) collapses to first-order logic.

3 Optimization problems definable by fixpoint logic

The fact that the problem MAXIMUM CONNECTED COMPONENT (MCC) appears only in the highest level MAX Π_2 of the expressibility hierarchy suggests that we do not yet have the "right" definitions. After all, MCC is computationally a very simple problem, and it appears high in the expressibility hierarchy just because first-order logic cannot express the transitive closure.

It is possible that there will always remain a certain "mismatch" between computational complexity and logical expressibility. But this mismatch is certainly not as big as the difference between first-order logic and PTIME. If we base our definitions on fixpoint logic (or other logical systems that allow recursion) rather than first-order logic, we obtain a closer relationship between logical and computational complexity.

Definition 3.1 Let Q be a maximization problem whose instances are finite structures over a fixed vocabulary σ. We say that Q belongs to the class MAX Σ_i^{FP} if there exists a Σ_i-formula $\psi(\bar{x}, \bar{c}, \bar{S}, \bar{P})$ of vocabulary $\sigma \cup \{\bar{S}, \bar{c}\}$ (where $\bar{S}$ and $\bar{c}$ are tuples of predicate symbols and constants that do not occur in σ) such that

- $\bar{P} = P_1, \ldots, P_r$ are global predicates on σ-structures that are definable in fixpoint logic.
- the formula $\psi(\bar{x}, \bar{c}, \bar{S}, \bar{P})$ represents Q (in the sense of Definition 2.3, with the obvious modifications). In particular, $opt_Q(A) = \max_{\bar{S},\bar{c}} |\{\bar{x} : A \models \psi(\bar{x}, \bar{c}, \bar{S}, \bar{P})\}|$.

The classes MAX Π_i^{FP}, MIN Σ_i^{FP} and MIN Π_i^{FP} are defined in an analogous way.

We insist that the fixpoint-predicates must not depend on the relations $\bar{S}$ over which we maximize; we therefore call them *predefined fixpoint predicates.* (We will discuss this condition below).

The results of Kolaitis and Thakur [11] translate to the extended classes and prove that

$$\text{MAX } \Sigma_0^{\mathrm{FP}} \subsetneq \text{MAX } \Sigma_1^{\mathrm{FP}} \subsetneq \text{MAX } \Pi_1^{\mathrm{FP}} \subsetneq \text{MAX } \Pi_2^{\mathrm{FP}} = \text{MAX } \mathcal{PB}.$$

The increased expressive power provided by the fixpoint predicates has the effect that some problems occur in lower levels in the new hierarchy than they did in the original one.

Example. The problem MAXIMUM CONNECTED COMPONENT belongs to MAX Σ_0^{FP}. Its optimum on a graph $G = (V, E)$ is definable by

$$opt_{\mathrm{MCC}}(G) = \max_c |\{x : G \models [\mathrm{TC}\ E](c, x)\}|.$$

Thus MCC descends from the highest level (MAX Π_2) of the original hierarchy to the lowest level (MAX Σ_0^{FP}) of the new hierarchy. This is interesting because, as we will see, also the extended class MAX Σ_1^{FP} contains only problems in APX. Thus our approach provides a more powerful syntactic criterion for approximability than the original class MAX Σ_1. Let us consider to what extent our definitions are adequate and discuss some alternatives.

Maximization over constants. Does maximization over constants really give more power? We prove that it does up to the level MAX Σ_1^{FP}.

Proposition 3.2 MAX CONNECTED COMPONENT *is not expressible in* MAX Σ_1^{FP} *without maximization over constants.*

PROOF. Let G_n be the graph with n vertices and no edges. Obviously, $opt_{\mathrm{MCC}}(G_n) = 1$ for all n. Moreover, for every fixpoint-definable predicate P, there exists a natural number n_0 such that P is in fact Σ_0-definable on $\{G_n : n > n_0\}$. Therefore, if MCC is expressible in MAX Σ_1^{FP} without constants, then there is an existential formula $\psi(\bar{x}, \bar{S})$ (without fixpoint predicates), such that for all $n > n_0$

$$opt_{\mathrm{MCC}}(G_n) = \max_{\bar{S}} |\{\bar{x} : G_n \models \psi(\bar{x}, \bar{S})\}| = 1.$$

Choose a tuple $\bar{u} \in G_n$ and predicates $\bar{S}$ such that $G_n \models \psi(\bar{u}, \bar{S})$. Note that G_{2n} consists of two disjoint copies of G_n; let $\bar{S}^*$ be the union of the two copies of $\bar{S}$. Existential sentences are preserved by extensions, so there exist at least two tuples $\bar{u}$, satisfying $G_{2n} \models \psi(\bar{u}, \bar{S}^*)$. This contradicts the fact that $opt_{\mathrm{MCC}}(G_{2n}) = 1$. ∎

Another simple problem that requires maximization over constants is the *maximal degree* $\Delta(G)$ of a graph, defined by $\Delta(G) = \max_c |\{x : G \models Ecx\}|$. Similar monotonicity arguments as above show that $\Delta(G)$ is not definable in MAX Σ_1^{FP} without maximization over constants.

However, constants can be replaced by monadic predicates at the expense of a universal subformula.

Proposition 3.3 *Every problem in* MAX Π_1^{FP} *or* MAX Π_2^{FP} *can be expressed without maximization over constants.*

Fixpoint definitions over the new predicates. To strengthen our classes we could modify the definition of MAX Σ_i^{FP} so that the fixpoint predicates might depend also on the predicates $\bar{S}$ over which we maximize. In fact, one could propose classes of all maximization problems Q such that

$$opt_Q(A) = \max_{\bar{S}} |\{\bar{x} : A \models \psi(\bar{x}, \bar{S})\}|$$

where $\psi(\bar{x}, \bar{S})$ is a formula of (FO + LFP), possibly with restrictions on the quantifier structure. If we stipulate that $\psi(\bar{x}, \bar{S})$ has the form $[\mathrm{LFP}_{R,\bar{z}}\ \varphi](\bar{u})$ with φ quantifier-free then we will remain inside MAX Σ_0 because fixpoints over quantifier-free formulae are again Σ_0-definable. The next stronger possible class, motivated by the existential nature of the class MAX Σ_1 is the class MAX EFP defined as above with the condition that $\psi(\bar{x}, \bar{S}) \equiv [\mathrm{LFP}_{R,\bar{z}}\ \varphi](\bar{u})$ where φ is existential. In particular $\psi(\bar{x}, \bar{S})$ is a formula in *existential fixpoint logic* [3]. However, this class is already too expressive.

Proposition 3.4 *If* P $\neq$ NP, *then* MAX EFP *contains non-approximable problems.*

PROOF. We consider the following variant of circuit satisfiability. A circuit is described by a finite structure $C = (V, E, I, out)$ where (V, E) is a directed acyclic graph, $I \subseteq V$ is the set of sources (vertices with no incoming edges) describing the input nodes, every node in $V - I$ has fan-in two, and out is a sink (no outgoing edges). Every non-input node is considered as a NAND-gate and out is the output node. Every subset $S \subset I$ defines an assignment to the input nodes, and therefore a value $f_C(S) \in \{0, 1\}$, the value computed by C for input S.

Now the circuit-satisfiability problem is

$$\text{CIRCUIT-SAT} := \{C : (\exists S \subseteq I) f_C(S) = 1\}.$$

Since a Boolean formula is a special case of a circuit, it is clear that CIRCUIT-SAT generalizes SAT and is therefore NP-complete.

On the other hand, it is not difficult to construct a formula $\psi(S)$ in existential fixpoint logic such that

$$C \models \psi(S) \iff f_C(S) = 1.$$

We now can define a problem $Q \in$ MAX EFP by

$$opt_Q(C) = \max_S |\{x : C \models \psi(S)\}|.$$

Note that x does not occur freely in ψ, so $opt_Q(C) = |V|$ if $C \in$ CIRCUIT-SAT and $opt_Q(C) = 0$ otherwise. Therefore, if Q were approximable up to any constant $\varepsilon > 0$, then the corresponding approximation algorithm would solve the CIRCUIT-SAT problem, and it would follow that P = NP. ∎

Maximization over total orderings. Papadimitriou and Yannakakis [17] proposed another direction for generalization: to maximize over total orderings. A natural problem expressible in this way is MAX SUBDAG: given a digraph G, find an acyclic subgraph of G with maximal number of edges. The expression defining the optimum for this problem is

$$opt_Q(G) = \max_{<} |\{(x,y) : G \models Exy \wedge x < y\}|.$$

This suggests the following definition.

Definition 3.5 For every class M, as defined in Definitions 2.3 or 3.1, let $M(<)$ defined in the same way as M, except that some of relations over which we optimize are binary predicates $<_1, <_2, \ldots$, which do not run over all binary predicates, but only over total orderings of the given structure. We write the defining expression of an optimization problem in $M(<)$ on an input structure A as

$$opt_Q(A) = \operatorname*{opt}_{\bar{<},\bar{S},\bar{c}} |\{\bar{x} : A \models \psi(\bar{x}, \bar{c}, \bar{<}, \bar{S}, \bar{P})\}|.$$

As remarked in [17] this feature does not destroy approximability: MAX $\Sigma_1^{\mathrm{FP}} \subseteq$ APX. (We prove a more general fact below.)

Note that for classes MAX Π_1 and above, maximization over orderings does not increase the expressive power, because total orderings are axiomatizable by Π_1-formulae. Thus MAX $\Pi_1(<) =$ MAX Π_1 and MAX $\Pi_1^{\mathrm{FP}}(<) =$ MAX Π_1^{FP}, etc.

Maximization over general classes of relations. How far can we generalize the idea of the previous paragraph? Instead of just maximizing over orderings, we could maximize over any specified class of predicates. Let $\bar{C} = C_1, \ldots, C_q$ where each C_i is a class of relations of some fixed arity r_i. We now maximize over tuples $\bar{S} = S_1, \ldots, S_q$ of relations, subject to the condition that $S_i \in C_i$. The expression defining the cost of an optimal solution then has the form

$$opt_Q(A) = \operatorname*{opt}_{\bar{S}\in\bar{C},\bar{c}} |\{\bar{x} : A \models \psi(\bar{x}, \bar{c}, \bar{S}, \bar{P})\}|.$$

For any class M of Definitions 2.3 or 3.1, and any $\bar{C}$, this defines a new class $M(\bar{C})$. The classes $M(<)$ are special cases.

In view of our goal to find a good criterion for approximability, we ask what condition $\bar{C}$ must satisfy so that problems in MAX $\Sigma_1^{\mathrm{FP}}(\bar{C})$ be approximable. We give such a condition here, and prove a general form of the Theorem of Papadimitriou and Yannakakis.

Definition 3.6 For every $n \in \mathbb{N}$, let $C(n)$ be a class of relations of fixed arity r over n and let $C = \bigcup_n C(n)$. We say that C is *well-behaved* if for every $k \in \mathbb{N}$ there exists a number $\varepsilon(k) > 0$ such that for all n and for any set of conditions $\alpha_1, \ldots, \alpha_k$ of the form $S(\bar{u})$ or $\neg(S\bar{u})$, (where $\bar{u}$ are r-tuples over n) the probability that $\alpha_1 \wedge \cdots \wedge \alpha_k$ is satisfied by a randomly chosen relation $S \in C(n)$, is computable in polynomial time, and is either 0 or at least $\varepsilon(k)$.

Examples.

- The class of all r-ary relations is well-behaved with $\varepsilon(k) = 2^{-k}$.
- If we fix, in a consistent manner, an ordering on k pairs of elements then there are at least $n!/(2k)!$ ways to extend this to a total ordering. Thus, the class of all total orderings is also well-behaved with $\varepsilon(k) = ((2k)!)^{-1}$.

- The class of unary functions, represented by binary relations, is not well-behaved. If we fix one value $f(u) = v$, then the probability that a function on n satisfies this condition is $1/n$.

A tuple $\bar{C}$ of well-behaved classes is again well-behaved, in the sense that the probability that any k atomic formulae $S_i(\bar{u})$ or $\neg S_i(\bar{u})$ are satisfied has the properties required by Definition 3.6.

Theorem 3.7 *If $\bar{C}$ is well-behaved then every maximization problem in* MAX $\Sigma_1^{\mathrm{FP}}(\bar{C})$ *is approximable up to a constant factor.*

PROOF. Let Q be a maximization problem, such that for all input structures $A \in I_Q$

$$opt_Q(A) = \max_{\bar{S}\in\bar{C},\bar{c}} |\{\bar{x} : A \models (\exists\bar{y})\psi(\bar{x},\bar{y},\bar{c},\bar{S},\bar{P})\}|$$

where ψ is a quantifier-free formula with predefined fixpoint predicates $\bar{P}$. Fix an input structure A of cardinality n. The fixpoint predicates can be evaluated in polynomial time, so we may assume that they are part of the input, and not worry about them anymore. Moreover, there are only polynomially many possible tuples $\bar{c}$, so we can compute the optimum for each of them separately; thus we can assume that the value of $\bar{c}$ is fixed. To enhance readability we drop $\bar{P}$ and $\bar{c}$ and write $\psi(\bar{x},\bar{y},\bar{S})$ in the sequel.

We consider the class $\bar{C}(n)$ as a probability space Ω with uniform distribution. Define the random variable X^A on Ω by

$$X^A(\bar{S}) = |\{\bar{u} : A \models (\exists\bar{y})\psi(\bar{u},\bar{y},\bar{S})\}|.$$

Obviously $opt_Q(A) = \max(X^A) \geq E(X^A)$ where $E(X^A)$ is the expected value of X^A. Note that X^A can be written as the sum of the indicator random variables

$$X^A_{\bar{u}}(\bar{S}) = \begin{cases} 1 & \text{if } A \models (\exists\bar{y})\psi(\bar{u},\bar{y},\bar{S}) \\ 0 & \text{otherwise} \end{cases}$$

By linearity of expectation $E(X^A) = \sum_{\bar{u}} E(X^A_{\bar{u}})$. In this sum we can discard those $X^A_{\bar{u}}$ which are identically 0, so let

$$B^A := \{\bar{u} : X^A_{\bar{u}} \neq 0\} = \{\bar{u} : A \models (\exists\bar{y})\psi(\bar{u},\bar{y},\bar{S}) \text{ for some } \bar{S} \in \bar{C}(n)\}$$

Now

$$E(X^A) = \sum_{\bar{u}\in B^A} E(X^A_{\bar{u}}) \leq |B^A|.$$

Fix $\bar{u} \in B^A$. There exist predicates $\bar{S}^* \in \Omega$ and a tuple $\bar{v}$ such that $A \models \psi(\bar{u},\bar{v},\bar{S}^*)$. This formula depends only a fixed number $\alpha_1,\ldots,\alpha_k$ of $\bar{S}$-atoms, and every tuple $\bar{S} \in \Omega$ respecting the values of $\alpha_1,\ldots,\alpha_k$ on $\bar{S}^*$ will also satisfy $\psi(\bar{u},\bar{v},\bar{S})$. Since $\bar{C}$ is well-behaved, the probability that a randomly chosen $\bar{S}$ has this property is at least $\varepsilon(k)$. Thus, $E(X^A_{\bar{u}}) \geq \varepsilon(k)$ for every $\bar{u} \in B^A$. This implies

$$|B^A| \geq \max(X^A) \geq E(X^A) \geq \varepsilon(k)|B^A|$$

and, in particular, $E(X^A) \geq \varepsilon(k)\,\mathrm{opt}_Q(A)$.

It remains to prove that an assignment $\bar{S}^* \in \Omega$ with $X^A(S^*) \geq E(X^A)$ can be found in polynomial time. Enumerate all atoms $S_i(\bar{u})$ as $\alpha_1,\ldots,\alpha_m$. Clearly m is a polynomial in n. We determine truth-values for $\alpha_1,\ldots,\alpha_m$ as follows.

Suppose values for $\alpha_1,\ldots,\alpha_i$ are already computed, and let β_i be the conjunction of those α_j and $\neg\alpha_j$ (for $j \leq i$) that have been set to TRUE. Now we define α_{i+1} to be TRUE if

$$E(X^A|(\beta_i \wedge \alpha_{i+1})) \geq E(X^A|(\beta_i \wedge \neg\alpha_{i+1}))$$

and FALSE otherwise. Note that these conditional expectations can be computed in polynomial time, by the same arguments as in the first part of this proof and because $\bar{C}$ is well-behaved. At the end, β_m determines relations $\bar{S}^*$, so $E(X^A|\beta_m) = X^A(S^*)$. Note that

$$E(X^A|\beta_i) = P(\alpha_{i+1}|\beta_i)E(X^A|(\beta_i \wedge \alpha_{i+1})) + P(\neg\alpha_{i+1}|\beta_i)E(X^A|(\beta_i \wedge \neg\alpha_{i+1}))$$

$$\leq E(X^A|\beta_{i+1})$$

where $P(\alpha_{i+1}|\beta_i)$ is the conditional probability that α_{i+1} holds, given that β_i is TRUE. In particular

$$E(X^A) = E(X^A|\beta_0) \leq E(X^A|\beta_m) = X^A(\bar{S}^*).$$

This technique for "derandomizing" a probabilistic argument is well known. Alon and Spencer [1] call it *the method of conditional expectations.* ∎

4 Probabilistic methods

Let Q be an optimization problem whose input instances are finite structures over a fixed vocabulary σ. We now consider the behaviour of $opt_Q(A)$ on a *randomly chosen* σ-structures A. Fix n and let Ω be the probability space of all σ-structures over universe n, with uniform distribution; then opt_Q is a random variable on Ω whose expected value is denoted by $E(opt_Q)$.

We will establish a probabilistic (necessary) criterion for membership in MAX Σ_1^{FP}. In fact it holds for any class MAX $\Sigma_1^{FP}(\bar{C})$ provided that $\bar{C}$ is Σ_2-axiomatizable. In particular, our criterion applies to MAX $\Sigma_1^{FP}(<)$, since linear orderings are in fact Π_1-axiomatizable.

Theorem 4.1 (Probabilistic criterion for MAX Σ_1^{FP}) *Suppose that Q is a problem in MAX $\Sigma_1^{FP}(\bar{C})$ where $\bar{C}$ is Σ_2-axiomatizable. Then there exists a polynomial $p(n)$ and a constant $\varepsilon > 0$ such that*

$$p(n) \geq E(opt_Q) \geq \varepsilon p(n)$$

or $E(opt_Q)$ decreases to 0 exponentially fast as n goes to infinity.

Before we prove Theorem 4.1, we assemble some results from the theory of asymptotic probabilities that we need. As usual, we denote by $\mu_n(\psi)$ the probability that the sentence ψ is true in a random structure with universe n.

Fact 4.2 *For every formula $\varphi(\bar{x})$ in fixpoint logic, there exists a quantifier-free first-order formula $\alpha(\bar{x})$ and a constant $c > 0$ such that*

$$\mu_n((\forall\bar{x})[\varphi(\bar{x}) \leftrightarrow \alpha(\bar{x})]) > 1 - c^n$$

for large enough n.

In fact this result is true even for stronger logics than fixpoint logic, e.g. the infinitary logic $L^{\omega}_{\infty\omega}$. It is essentially Theorem 3.13 in [14]. The second fact that we need is a generalization of the 0-1 law for strict Σ_1^1-sentences, due to Kolaitis and Vardi [13]. Strict Σ_1^1-formulae have the form $(\exists\bar{S})(\exists\bar{y})(\forall\bar{z})\varphi$ where φ is quantifier-free. A dyadic rational is a rational number whose denominator is a power of two.

Fact 4.3 *Let $\psi(\bar{x})$ be a strict Σ_1^1-formula with free variables $\bar{x} = x_1, \ldots, x_k$. For every k-tuple $\bar{u} \in \mathbb{N}^k$, there exists a dyadic rational $p_{\bar{u}}$, such that $\mu_n(\psi(\bar{u}))$ tends to $p_{\bar{u}}$ exponentially fast. Moreover, $p_{\bar{u}}$ only depends on the equality type of $u_1, \ldots, u_k$ (not on $\bar{u}$ itself).*

Finally we will need a Lemma about binomial distributions $b(n,k,p) := \binom{n}{k} p^k q^{n-k}$ where $0 \leq p \leq 1$ and $q = 1 - p$. For a proof, see [4, p. 10] or [1, Appendix A].

Lemma 4.4 *If $\varepsilon > 0$ and $k \geq (1+\varepsilon)pn$, then $b(n,k,p)$ tends to 0 exponentially fast as n goes to infinity.*

PROOF OF THEOREM 4.1. Let $Q \in \text{MAX}\ \Sigma_1^{\text{FP}}(\bar{C})$. We first assume that opt_Q can be expressed without maximization over constants, i.e. that there exists a Σ_1-formula $\psi(\bar{x}, \bar{S})$ (with predefined fixpoint predicates) such that

$$opt_Q(A) = \max_{\bar{S} \in \bar{C}} |\{\bar{x} : A \models \psi(\bar{x}, \bar{S})\}|.$$

The proof of Theorem 3.7 shows that for some constant $\varepsilon > 0$, $|B^A| \geq opt_Q(A) \geq \varepsilon |B^A|$ where

$$B^A = \{\bar{u} : A \models (\exists \bar{S} \in \bar{C})\psi(\bar{u}, \bar{S})\}.$$

On Ω, we define the random variable $X(A) := |B^A|$. It follows that $E(X) \geq E(opt_Q) \geq \varepsilon E(X)$. It suffices to prove that $E(X)$ converges to a polynomial $F(n)$. We write X as the sum of the indicator random variables

$$X_{\bar{u}}(A) := \begin{cases} 1 & \text{if } \bar{u} \in B^A \\ 0 & \text{otherwise} \end{cases}$$

By linearity of expectation, $E(X) = \sum_{\bar{u}} E(X_{\bar{u}})$. Let $\alpha(\bar{S})$ be a Σ_2-axiom for $\bar{C}$. Then $E(X_{\bar{u}})$ is the probability that the formula

$$\eta(\bar{u}) \equiv (\exists \bar{S})(\alpha(\bar{S}) \wedge \psi(\bar{u}, \bar{S}))$$

holds on a random structure with universe n. Fact 4.2 tells us that except on a exponentially decreasing fraction of structures, the predefined fixpoint predicates are definable by quantifier-free formulae. If we substitute them into $\eta(\bar{u})$, then we obtain a strict Σ_1^1-formula $\varphi(\bar{u})$ such that, for some constant $c > 0$,

$$|E(X_{\bar{u}}) - \mu_n(\varphi(\bar{u}))| < c^n.$$

Now, by Fact 4.3, the probability $\mu_n(\varphi(\bar{u}))$ converges exponentially fast to a dyadic rational $p_{\bar{u}}$ which only depends on the equality type of $\bar{u}$. If k is fixed then the number of equality types of k-tuples is also fixed; moreover, the number of k-tuples of equality type e over n is a polynomial $f_e(n)$. Let p_e be the asymptotic probability of $\varphi(\bar{u})$ for tuples of equality type e. It follows that $E(X)$ converges exponentially fast to the polynomial

$$F(n) = \sum_e p_e f_e(n).$$

With maximization over constants, the situation becomes more complicated. We now have

$$opt_Q(A) = \max_{\bar{c}, \bar{S}} |\{\bar{x} : A \models \psi(\bar{x}, \bar{c}, \bar{S})\}|.$$

To establish Theorem 3.7 we fixed for every input structure A an optimal tuple $\bar{c}$ which then was considered as part of the input. Since $\bar{c}$ depends on A this no longer works when A is a random input. Therefore, let

$$B^A(\bar{c}) := \{\bar{x} : A \models (\exists \bar{S} \in \bar{C})\psi(\bar{u}, \bar{c}, \bar{S})\}.$$

On every input structure A we then have

$$\max_{\bar{c}} |B^A(\bar{c})| \geq opt_Q(A) \geq \varepsilon \max_{\bar{c}} |B^A(\bar{c})|$$

for a fixed constant $\varepsilon > 0$. Let $X := \max_{\bar{c}} |B^A(\bar{c})|$; it suffices to prove that there exists a polynomial $F(n)$ such that $E(X) \sim F(n)$.

As above, we find a strict Σ_1^1-formula $\varphi(\bar{c}, \bar{x})$ such that, for any fixed $(\bar{c}, \bar{u})$, the expectation that $\bar{u} \in B^A(\bar{c})$ is exponentially close to the asymptotic probability of $\varphi(\bar{c}, \bar{u})$. Again, the asymptotic probability of $\varphi(\bar{c}, \bar{u})$ is a dyadic rational that depends only on the equality type of $(\bar{c}, \bar{u})$.

Let D be the set of equality types of $\bar{c}$ in n; clearly, the size of D is bounded (independently of n) and the cardinality of every $d \in D$ is a polynomial $f_d(n)$. Each equality type e of tuples $(\bar{c}, \bar{u})$ is an extension of an equality type $d \in D$; we write $d \prec e$ when this occurs. If $\bar{c} \in d \prec e$, let $U_e(\bar{c}) = \{\bar{u} : (\bar{c}, \bar{u}) \in e\}$. The cardinality of $U_e(\bar{c})$ is described by a polynomial $g_e(n)$ (which depends only on e). We denote the asymptotic probability of $\varphi(\bar{c}, \bar{u})$ (for $(\bar{c}, \bar{u}) \in e$) by p_e. If $\bar{c} \in d$ is fixed, then the arguments in the first part of this proof show that $E(|B^A(\bar{c})|) = \sum_{d \prec e} E(|B^A(\bar{c}) \cap U_e(\bar{c})|)$ converges exponentially fast to $G_d(n) := \sum_{d \prec e} p_e g_e(n)$ which is a polynomial. Eventually one of the $G_d(n)$ will dominate all the other ones, so asymptotically $F(n) := \max_{d \in D} G_d(n)$ is a polynomial. This implies that

$$E(X) = E(\max_{\bar{c}} |B^A(\bar{c})|) \geq \max_{\bar{c}} E(|B^A(\bar{c})|) \sim \max_{d \in D} \sum_{d \prec e} p_e g_e(n) = F(n).$$

It remains to prove that asymptotically $E(X)/F(n) < 1 + \varepsilon$ for every $\varepsilon > 0$. We first prove a Lemma.

Lemma 4.5 *Let $d \in D$ and $d \prec e$. Then, for every $\varepsilon > 0$, the probability that there exists a tuple $\bar{c} \in D$ such that*

$$|B^A(\bar{c}) \cap U_e(\bar{c})| \geq (1 + \varepsilon) p_e g_e(n)$$

tends to 0 exponentially fast.

PROOF. Fix $k = k(n)$ and define the random variable $Y(A)$ to be the number of tuples $\bar{c} \in d$ such that $B^A(\bar{c}) \cap U_e(\bar{c})$ has cardinality k. We can write Y as the sum of the indicator random variables

$$Y_{\bar{c},U}(A) = \begin{cases} 1 & \text{if } U = B^A(\bar{c}) \cap U_e(\bar{c}) \\ 0 & \text{otherwise} \end{cases}$$

where $\bar{c}$ has equality type d and U is a subset of $U_e(\bar{c})$ of cardinality k. Let $m := g_e(n)$, $p := p_e$ and $q := 1 - p$. Markov's inequality and linearity of expectation give

$$P(Y \geq 1) \leq E(Y) = \sum_{\bar{c},U} E(Y_{\bar{c},U}) = f_d(n) \binom{m}{k} p^k q^{m-k} = f_d(n) b(m, k, p).$$

By Lemma 4.4, if $k \geq (1 + \varepsilon)pm = (1 + \varepsilon)p_e g_e(n)$ then $b(m, k, p)$ converges to 0 exponentially fast. Thus the same holds for the probability that there exists a tuple $\bar{c}$ for which $|B^A(\bar{c}) \cap U_e(\bar{c})|$ exceeds $(1 + \varepsilon)p_e g_e(n)$. ∎

Suppose that $E(X) \geq (1 + \varepsilon)F(n)$. Then there is a constant $\varepsilon > 0$ such that there exists with non-negligible probability at least one tuple $\bar{c}$ (of equality type, say, d) with $|B^A(\bar{c})| \geq (1 + \varepsilon)G_d(n)$. But then there must exist an extension e of d such that with non-negligible probability there is a $\bar{c}$ with $|B^A(\bar{c}) \cap U_e(\bar{c})| \geq (1 + \varepsilon)p_e g_e(n)$.

The Lemma just proved shows that this is not the case. This proves the theorem. ∎

Applications. As usual in graph theory, let $\omega(G)$, $\alpha(G)$ and $\chi(G)$ denote the size of a maximum clique, the size of a maximum independent set and the chromatic number of a graph G. We use the following results from the theory of random graphs (see [1, 4]).

Fact 4.6 *(i)* $E(\omega) = E(\alpha) \sim 2\log n$,

(ii) $E(\chi) \sim n/(2\log n)$.

Together with our probabilistic criterion, this implies that MAX CLIQUE and MAX INDEPENDENT SET are not in MAX $\Sigma_1^{\mathrm{FP}}(<)$.

There are other important maximization problems Q for which $E(opt_Q)$ does not grow like $\Theta(p(n))$ for any polynomial $p(n)$, and which therefore are not in MAX $\Sigma_1^{\mathrm{FP}}(<)$. Examples include the following.

V-C DIMENSION. Given a collection $S_1, \ldots, S_m$ of subsets of a finite set M, find a set $T \subseteq M$ of maximal cardinality which is shattered by $S_1, \ldots, S_m$ (this means that every subset of T occurs as $T \cap S_i$ for some i). The cardinality of T is called the *Vapnik-Chervonenkis dimension* of $S_1, \ldots, S_m$; it plays an important rôle e.g. in learning theory. We can represent a collection $S_1, \ldots, S_m$ of subsets of n by a binary predicate S over $\max(n, m)$ such that $S_i = \{j : (i, j) \in S\}$.

MAX COMMON INDUCED SUBGRAPH (MCIS). Given two graphs G and H, find a graph of maximal cardinality which is an induced subgraph of both G and H.

LONGEST CHORDLESS PATH (LCP). Given a graph G, find a set V of nodes, as large as possible, such that $G|_V$ is a simple path.

Note that to apply the probabilistic criterion to these (and other) problems, it is not necessary to determine $E(opt_Q)$ explicitly. Theorem 4.1 implies the following proposition.

Proposition 4.7 *Let Q be a maximization problem. If, over the probability space Ω of all σ-structures with universe n,*

$$1 \leq E(opt_Q) = o(n)$$

then $Q \notin$ MAX $\Sigma_1^{\mathrm{FP}}(<)$.

In fact, if we can show that for all $\varepsilon > 0$, the probability that $opt_Q(A) \geq \varepsilon|A|$ tends to 0 exponentially fast then $E(opt_Q) = o(n)$ follows because opt_Q is polynomially bounded.

There is a large class of graph problems, for which this can be established by the following method. For any property P of graphs, let $c_n(P)$ be the number of graphs with vertex set n that satisfy P. Let MAX INDUCED SUBGRAPH WITH PROPERTY P (MIS(P)) be the problem of maximizing the cardinality of a set of nodes V in a given graph G, such that the induced subgraph $G|_V$ has property P. The problems MAX CLIQUE, MAX INDEPENDENT SET, LONGEST CHORDLESS PATH are special cases of MIS(P); another example is MAX INDUCED k-COLOURABLE SUBGRAPH.

Theorem 4.8 *If* $0 \leq \log c_n(P) \leq n^2/2 - n^{1+\delta}$ *for some* $\delta > 0$, *then* MIS(P) $\notin$ MAX $\Sigma_1^{\mathrm{FP}}(<)$.

PROOF. For $r = r(n)$, let $Y_r(G)$ be the random variable whose value is the number of induced subgraphs in G of cardinality r with property P. Clearly $\mathrm{opt}_{\mathrm{MIS}(P)}(G) = \max\{r : Y_r(G) > 0\}$. By linearity of expectation

$$E(Y_r) = \binom{n}{r} c_r(P) 2^{-\binom{r}{2}}.$$

Stirling's formula implies that $\binom{n}{r} = 2^{O(r)}$ for $n = O(r)$. It follows that

$$P(Y_r \geq 1) \leq E(Y_r) = 2^{O(r)+\log c_r(P)-\binom{r}{2}} \leq 2^{-r^{1+\delta}+O(r)}$$

which tends to 0 exponentially fast as n, and hence r, goes to infinity. Hence, $1 \leq E(\mathrm{opt}_{\mathrm{MIS}(P)}) = o(n)$ and by Proposition 4.7 it follows that $\mathrm{MIS}(P) \notin$ MAX $\Sigma_1^{\mathrm{FP}}(<)$. ■

The arguments showing that MCIS and V-C-DIMENSION are not in MAX $\Sigma_1^{\mathrm{FP}}(<)$ are very similar.

We also obtain results for minimization problems.

Theorem 4.9 MIN COLOURING *is not in* MIN Σ_1^{FP}.

PROOF. Recall that MIN Σ_1^{FP} = MIN Σ_0^{FP}. Suppose that the chromatic number could be defined by

$$\chi(G) = \min_{\bar{S},\bar{c}} |\{\bar{x} : G \models \psi(\bar{x}, \bar{c}, \bar{S})\}|$$

where ψ is quantifier-free (with predefined fixpoint predicates) and $\bar{x} = x_1, \ldots, x_k$. It follows that $\chi(G)$ can be defined as

$$\chi(G) = n^k - \max_{\bar{S},\bar{c}} |\{\bar{x} : G \models \neg\psi(\bar{x}, \bar{c}, \bar{S})\}| = n^k - opt_Q(G)$$

with a maximization problem $Q \in$ MAX Σ_0^{FP}. But this implies that $E(\chi) = n^k - \Theta(p(n))$ for some polynomial p, which is the not the case, since $E(\chi) \sim n/(2\log n)$. ■

There also is a probabilistic criterion for membership in MAX Π_1^{FP}, which will allow us to separate MAX Π_2^{FP} from MAX Π_1^{FP}.

Theorem 4.10 (Probabilistic criterion for MAX Π_1^{FP}**)** *For every optimization problem $Q \in$ MAX Π_1^{FP} and every natural number $k \in \mathbb{N}$, the property that $opt_Q(A) > k$ satisfies a 0-1 law.*

PROOF. We again use the fact that except on a exponentially decreasing fraction of structures, fixpoint predicates are definable by quantifier-free formulae. Thus, the optimum for $Q \in$ MAX Π_1^{FP}, is defined on almost all structures by an expression

$$opt_Q(A) = \max_{\bar{S},\bar{c}} |\{\bar{x} : A \models \psi(\bar{x}, \bar{c}, \bar{S})\}|$$

where $\psi(\bar{x}, \bar{c}, \bar{S})$ is a Π_1-formula.

Then, the property that $opt_Q(A) > k$ is expressed by the strict Σ_1^1-formula

$$(\exists \bar{S})(\exists \bar{c})(\exists \bar{x}_0) \cdots (\exists \bar{x}_k) \bigwedge_{0 \leq i < j \leq k} (\bar{x}_i \neq \bar{x}_j) \wedge \bigwedge_{0 \leq i \leq k} \psi(\bar{x}_i, \bar{c}, \bar{S}).$$

The theorem now follows from the 0-1 law for strict Σ_1^1-formulae. ■

We did not find natural optimization problems that do not satisfy this criterion. However, we can cook up artificial ones such as MAXIMUM CONNECTED COMPONENT WITH PERFECT MATCHING (MCCPM) which, given a graph, asks for a maximum connected component that admits a perfect matching. Note that MCCPM is solvable in polynomial time.

Proposition 4.11 MCCPM $\notin$ MAX Π_1^{FP}.

PROOF. With probability tending to 1, a random graph is connected. Obviously, a graph of odd cardinality cannot have a perfect matching, but almost all graphs of even cardinality admit a perfect matching [4]. Thus, $opt_{\text{MCCPM}}(G)$ tends to 0 on graphs with odd cardinality, and to $|G|$ on graphs of even cardinality. ∎

Corollary 4.12 MAX $\Pi_1^{\text{FP}} \subsetneq$ MAX $\Pi_2^{\text{FP}} =$ MAX $\mathcal{PB}$.

5 Monotonicity properties

The simple monotonicity properties that were used in [16, 11] to separate expressibility classes of optimization problems do not survive in the presence of fixpoint definitions. Nevertheless we can make use of them to prove inexpressibility results for MAX Σ_1^{FP} and MIN Σ_1^{FP} by looking at classes of structures where fixpoint logic collapses to Σ_0-formulae. Although such classes cannot contain very interesting structures, they somewhat surprisingly suffice to show that prominent problems such as MAX CLIQUE, MAX MATCHING and MIN COLOURING are not in MAX Σ_1^{FP} and MIN Σ_1^{FP} respectively.

As usual, we identify n with the set $\{0, \ldots, n-1\}$ and denote the group of all permutations on n by S_n.

Definition 5.1 For all natural numbers n, m, we define the graphs $K_{n;m} = (n \times m, E_{n,m})$ where $E_{n,m} = \{((x,y)(x',y')) : x \neq x'\}$. $K_{n;m}$ is the complete n-partite graph where each partition class contains exactly m vertices.

Let $\sigma \in S_n$ and $\pi_1, \ldots, \pi_n \in S_m$. Then the bijection $(i,j) \longmapsto (\sigma i, \pi_i j)$ on $n \times m$ is an automorphism of $K_{n;m}$, and every automorphism of $K_{n;m}$ can be described in this way.

Definition 5.2 For k-tuples $\bar{u}, \bar{v}$ in $K_{n;m}$ we write $\bar{u} \equiv_k \bar{v}$ if there is an automorphism f of $K_{n;m}$ with $fu_i = v_i$ for $i = 1, \ldots, k$. The equivalence classes with respect to $\equiv_k$ are called $\equiv_k$-types.

The following lemma is easy to prove.

Lemma 5.3 *There exists a function $f : \mathbb{N} \rightarrow \mathbb{N}$ such that the number of $\equiv_k$-types in $K_{n;m}$ is bounded by $f(k)$ (independently of n and m). Furthermore every $\equiv_k$-type is uniformly definable by a quantifier-free formula. This means that there exist quantifier-free formulae $e_1(\bar{x}), \ldots, e_{f(k)}(\bar{x})$ such that every $\equiv_k$-type in any $K_{n;m}$ is defined by precisely one formula $e_i(\bar{x})$.*

Note that for small n, m, some of the $\equiv_k$-types may not occur in $K_{n;m}$. However, for every k, there exists a n_0 such that for $n, m > n_0$, the number of $\equiv_k$-types in $K_{n;m}$ is precisely $f(k)$.

Proposition 5.4 *For any formula $\varphi(\bar{x}) \in L^{\omega}_{\infty\omega}$ there exists a Σ_0-formula $\alpha(\bar{x})$ and a number n_0 such that for all $n, m > n_0$*

$$K_{n;m} \models (\forall \bar{x})(\varphi(\bar{x}) \leftrightarrow \alpha(\bar{x})).$$

PROOF. Take n, m large enough such that every $\equiv_k$-type e_i is realized by some k-tuple $\bar{u}_i$ in $K_{n;m}$. Let $I(\varphi) = \{i \leq f(k) : K_{n;m} \models \varphi(\bar{u}_i)\}$ and set

$$\alpha(\bar{x}) \equiv \bigvee_{i \in I(\varphi)} e_i(\bar{x}).$$

∎

By proposition 5.4 there exists for every optimization problem Q in MAX $\Sigma_1^{FP}(<)$ or MIN $\Sigma_1^{FP}(<)$ an existential first-order formula $\psi(\bar{x}, \bar{c}, \bar{<}, \bar{S})$ (without fixpoint predicates) such that

$$opt_Q(K_{n;m}) = \underset{\bar{S},\bar{<},\bar{c}}{\text{opt}} |\{\bar{x} : K_{n;m} \models \psi(\bar{x}, \bar{c}, \bar{<}, \bar{S})\}|$$

for all large enough n, m. Using the fact that existential formulae are closed under extensions, we can then establish a useful monotonicity criterion to prove inexpressibility results even for MAX $\Sigma_1^{FP}(<)$ and MIN $\Sigma_1^{FP}(<)$:

Theorem 5.5 *Let Q be an optimization problem on graphs in* MAX $\Sigma_1^{FP}(<)$ *or in* MIN $\Sigma_1^{FP}(<)$. *Then either $opt_Q(K_{n;m}) = O(1)$ or there exists a constant n_0 such that $opt_Q(K_{n;m}) < opt_Q(K_{n;m+1})$ for all all $n, m > n_0$.*

Obviously, $\omega(K_{n;m}) = \chi(K_{n;m}) = n$, for all m. Thus Theorem 5.5 implies that MAX CLIQUE $\not\in$ MAX $\Sigma_1^{FP}(<)$ and MIN COLOURING $\not\in$ MIN $\Sigma_1^{FP}(<)$.

A very similar criterion applies to the structures $K_n^d = (n, R^d)$ with the d-ary predicate $R^d = \{(a_1, \dots, a_d) : a_i \neq a_j \text{ for all } i \neq j\}$.

Theorem 5.6 *Let Q be an optimization problem on d-ary relations in* MAX $\Sigma_1^{FP}(<)$ *or* MIN $\Sigma_1^{FP}(<)$. *Then either $opt_Q(K_n^d) = O(1)$ or there exists a constant n_0 such that $opt_Q(K_n^d) < opt_Q(K_{n+1}^d)$ for all $n > n_0$.*

As applications, we present the problems to find a maximum matching in a graph and to find a maximum (disjoint) covering in a d-dimensional predicate.

MAX MATCHING(MM) is the problem of finding a set of independent edges of maximal size in a given graph. It is well-known that this is solvable in polynomial-time. A *covering* of a d-dimensional predicate R is a subset $M \subseteq R$ of mutually disjoint d-tuples (i.e. if $\bar{u} \in M$ and $\bar{v} \in M$ then $u_i \neq v_j$ for all $i, j \leq d$). MAX d-COVER (*MdC*) is the problem of finding a maximum covering of a given d-dimensional predicate. Note that MAX 2-COVER is MAX MATCHING. For all d, MAX d-COVER is in APX. Panconesi and Ranjan [16] proved that MAX d-COVER is not in MAX Σ_1. Since $opt_{MdC}(K_{dn}^d) = opt_{MdC}(K_{dn+1}^d) = n$ we can extend this to the following result.

Theorem 5.7 MAX MATCHING *and* MAX d-COVER *are not in* MAX $\Sigma_1^{FP}(<)$.

To complete the picture we note that the problem MAX SAT separates MAX Σ_1 from MAX Σ_0^{FP}.

Theorem 5.8 MAX SAT $\not\in$ MAX Σ_0^{FP}.

Corollary 5.9 MAX $\Sigma_0^{FP} \subsetneq$ MAX $\Sigma_1^{FP} \subsetneq$ MAX $\Pi_1^{FP} \subsetneq$ MAX $\Pi_2^{FP} =$ MAX $\mathcal{PB}$.

References

[1] N. Alon and J. Spencer, *The Probabilistic Method*, Wiley, New York (1991).

[2] S. Arora, C. Lund, R. Motwani, M. Sudan and M. Szegedy, *Proof verification and intractability of approximation problems*, Proceedings of Annual IEEE Symposium on Foundations of Computer Science (1992), 14–23.

[3] A. Blass and Y. Gurevich, *Existential fixed-point logic*, in: "Computation Theory and Logic" (E. Börger, Ed.), Lecture Notes in Computer Science Nr. 270, Springer 1987, 20–36.

[4] B. Bollobás, *Random graphs*, Academic Press, London (1985).

[5] K. Compton, *0-1 laws in logic and combinatorics*, in: "NATO Advanced Study Institute on algorithms and Order", (I. Rival, Ed), pp 353–383, Reidel, Dordrecht (1988).

[6] R. Fagin, *Generalized first-order spectra and polynomial time recognizable sets*, Complexity of Computations, SIAM-AMS Proc. 7 (1974), Richard Karp, ed., American Math. Soc., Providence, RI, 43-73.

[7] M. R. Garey and D. S. Johnson, *Computers and Intractibility: A Guide the to the Theory of NP-Completeness*, Freeman, New York (1979).

[8] T. Hirst and D. Harel, *Taking it to the Limit: On Infinite Variants of NP-Complete Problems*, unpublished manuscript (1992).

[9] N. Immerman, *Relational Queries Computable in Polynomial Time*, Information and Control **68** (1986), 86–104.

[10] V. Kann, *On the Approximability of NP-complete Optimization Problems*, Dissertation (1992), Royal Institute of Technology, Stockholm.

[11] Ph. Kolaitis and M. Thakur, *Logical definability of NP optimization problems*, to appear in Information and Computation.

[12] Ph. Kolaitis and M. Thakur, *Approximation properties of NP minimization classes*, Proceedings of 6th IEEE Conference on Structure in Complexity Theory (1991), 353–366, to appear in Journal of Computer and System Sciences.

[13] Ph. Kolaitis and M. Vardi, *The decision problem for the probabilities of higher-order properties*, Proceedings of 19th Annual ACM Symposium on Theory of Computing (1987), 425–435.

[14] Ph. Kolaitis and M. Vardi, *Infinitary logic and 0-1 laws*, Information and Computation **98** (1992), 258–294.

[15] C. Lautemann, Logical definability of NP-optimization problems with monadic auxiliary predicates, Informatik-Bericht Nr. 1/92, Institut für Informatik, Johannes Gutenberg-Universität Mainz (1992).

[16] A. Panconesi and D. Ranjan, *Quantifiers and approximation*, Proceedings of 22nd Annual ACM Symposium on Theory of Computing (1990), 446–456.

[17] Ch. Papadimitriou and M. Yannakakis, *Optimization, approximation and complexity*, Journal of Computer and System Sciences **43** (1991), 425–440.

[18] Ch. Papadimitriou and M. Yannakakis, *On the complexity of computing the V-C dimension*, Extended Abstract (1992).

[19] M. Vardi, *Complexity of Relational Query Languages*, Proc. of 14th Annual ACM Symposium on Theory of Computing (1982), 137–146.

Linear λ-Calculus and Categorical Models Revisited

Nick Benton[1], Gavin Bierman[1], Valeria de Paiva[1] and Martin Hyland[2]

[1] Computer Laboratory, University of Cambridge, UK
[2] Department of Pure Mathematics and Mathematical Statistics, University of Cambridge, UK

1 Intuitionistic Linear Logic

Girard's Intuitionistic Linear Logic [7] is a refinement of Intuitionistic Logic, where formulae must be used exactly once. In other words, the familiar Weakening and Contraction rules of Gentzen's sequent calculus [17] are removed. To regain the expressive power of Intuitionistic Logic, these rules are returned, but in a controlled manner. A logical operator, '!', is introduced which allows a formula to be used as many times as required (including zero).

In this paper we shall consider *multiplicative exponential* linear logic (MELL), i.e. the fragment which has multiplicative conjunction or tensor, $\otimes$, linear implication, $\multimap$, and the logical operator "exponential", !. We recall the rules for MELL in a sequent calculus system in Fig. 1. We use capital Greek letters Γ, Δ for sequences of formulae and A, B for single formulae. The *Exchange* rule simply allows the permutation of assumptions.

The '! rules' have been given names by other authors. $!_{\mathcal{L}-1}$ is called *Weakening*, $!_{\mathcal{L}-2}$ *Contraction*, $!_{\mathcal{L}-3}$ *Dereliction* and $(!_{\mathcal{R}})$ *Promotion*[1]. (We shall use these terms throughout this paper.) In the *Promotion* rule, $!\Gamma$ means that every formula in the set Γ is modal, in other words, if Γ is the set $\{A_1, A_2, \ldots A_n\}$, then $!\Gamma$ denotes the set $\{!A_1, !A_2, \ldots !A_n\}$.

2 Categorical considerations and term assignment

The sequent calculus is best thought of as providing not proofs themselves, but a meta-theory concerning proofs. Hence a formulation in these terms does not always provide clear clues as to how it should be enriched to a term assignment system. Fortunately we can use the general form of a categorical model (of the proof theory) of the logic to derive an appropriate term assignment system for the sequent calculus formulation of this logic.

The fundamental idea of the categorical treatment of proof theory is that propositions should be interpreted as the objects of a category (or multicategory, or polycategory) and proofs should be interpreted as maps; operations transforming proofs into proofs then correspond (if possible) to natural transformations (between appropriate hom-functors) in the categorical sense. The maps modelling proofs are built up using these categorical operations and so the problem of a term assignment is

[1] Girard, Scedrov and Scott [8] prefer to call this rule *Storage*.

$$\frac{}{A \vdash A}\ \mathit{Identity}$$

$$\frac{\Gamma, A, B, \Delta \vdash C}{\Gamma, B, A, \Delta \vdash C}\ \mathit{Exchange}$$

$$\frac{\Gamma \vdash B \qquad B, \Delta \vdash C}{\Gamma, \Delta \vdash C}\ \mathit{Cut}$$

$$\frac{\Gamma \vdash A}{\Gamma, I \vdash A}\ (I_{\mathcal{L}}) \qquad \frac{}{\vdash I}\ (I_{\mathcal{R}})$$

$$\frac{\Gamma, A, B \vdash C}{\Gamma, A \otimes B \vdash C}\ (\otimes_{\mathcal{L}}) \qquad \frac{\Gamma \vdash A \qquad \Delta \vdash B}{\Gamma, \Delta \vdash A \otimes B}\ (\otimes_{\mathcal{R}})$$

$$\frac{\Gamma \vdash A \qquad \Delta, B \vdash C}{\Gamma, \Delta, A \multimap B \vdash C}\ (\multimap_{\mathcal{L}}) \qquad \frac{\Gamma, A \vdash B}{\Gamma \vdash A \multimap B}\ (\multimap_{\mathcal{R}})$$

$$\frac{\Gamma \vdash B}{\Gamma, !A \vdash B}\ (!_{\mathcal{L}-1}) \qquad \frac{\Gamma, !A, !A \vdash B}{\Gamma, !A \vdash B}\ (!_{\mathcal{L}-2})$$

$$\frac{\Gamma, A \vdash B}{\Gamma, !A \vdash B}\ (!_{\mathcal{L}-3}) \qquad \frac{!\Gamma \vdash A}{!\Gamma \vdash !A}\ (!_{\mathcal{R}})$$

Fig. 1. Multiplicative Exponential Linear Logic

essentially the problem of providing a syntax expressing these operations. Here we carry out this programme for MELL.

Deriving the term formation rules

Since we are dealing with sequents $\Gamma \vdash A$, in principle we should deal with multicategories. However it simplifies things to assume at once that the multicategorical structure is represented by a tensor product $\bullet$, so that we are dealing with a monoidal category [13]. We shall write $\langle\rangle$ for the unit of this tensor product. To simplify the presentation we use the same symbols both for propositions of linear logic and for their denotations in our monoidal category. The idea then is that a sequent of form

$$C_1, C_2, \ldots, C_n \vdash A$$

will be interpreted as a map $C_1 \bullet C_2 \bullet \ldots \bullet C_n \to A$ from the tensor product of the C_i to A. (Thus a coherence result is assumed [11].) When Γ is the sequence $C_1, C_2, \ldots, C_n$, we write $\Gamma \rightarrow A$ for this map. We seek to enrich the sequent judgement to a term assignment judgement of the form

$$x_1 : C_1, x_2 : C_2, \ldots, x_n : C_n \vdash e : A$$

where the x_i are (distinct) variables and e is a term; usually we suppress (irrelevant) variables and write $\Gamma \vdash e : A$ for this term assignment.

The whole process is based upon some simple assumptions about the interpretation of the basic structural rules, and a simple procedure for dealing with the logical rules, which we describe in turn.

2.1 Structural Rules

The sequent representing the *Identity* rule is interpreted as the (canonical) identity arrow $A \xrightarrow{1} A$ from A to A. The corresponding rule of term formation is $x : A \vdash x : A$. The rule of *Exchange* we interpret by assuming that we have a symmetry for the tensor product $\bullet$ (making our model a *symmetric* monoidal category). We henceforth suppress *Exchange* and the corresponding symmetry; thus we really consider multisets of formulae, and as a result no term forming operations result from this rule. The *Cut* rule

$$\frac{\Gamma \vdash A \qquad A, \Delta \vdash B}{\Gamma, \Delta \vdash B} \; Cut$$

is then interpreted as a generalized form of composition: if the maps $\Gamma \xrightarrow{f} A$ and $A \bullet \Delta \xrightarrow{g} B$ are the interpretations of hypotheses of the rule, then the composite

$$\Gamma \bullet \Delta \xrightarrow{f \bullet 1_\Delta} A \bullet \Delta \xrightarrow{g} B$$

is the interpretation of the conclusion. We take as the corresponding rule of term formation a textual substitution:

$$\frac{\Gamma \vdash f : A \qquad x : A, \Delta \vdash g : B}{\Gamma, \Delta \vdash g[f/x] : B} \; Cut$$

One should note that the contexts Γ and Δ are disjoint; namely the variables which occur in Γ do not occur in Δ. This restriction holds for all the binary multiplicative rules.

2.2 Logical rules for Multiplicatives

We shall make the assumption that any logical rule corresponds to an operation on maps of the category which is *natural* in (the interpretations of) the components of the sequents which remain unchanged during the application of a rule. Composition corresponds to *Cut* so clearly the logical significance is that we are assuming that our operations commute (where appropriate) with *Cut*.

We start by considering the connective $\otimes$. The $(\otimes_{\mathcal{L}})$ rule

$$\frac{\Gamma, A, B \vdash C}{\Gamma, A \otimes B \vdash C}\,(\otimes_{\mathcal{L}})$$

gives an operation taking maps $\Gamma \bullet A \bullet B \to C$ to maps $\Gamma \bullet (A \otimes B) \to C$. An appropriate syntax is

$$\frac{\Gamma, x : A, y : B \vdash f : C}{\Gamma, z : A \otimes B \vdash \mathsf{let}\ z\ \mathsf{be}\ x \otimes y\ \mathsf{in}\ f : C}\,(\otimes_{\mathcal{L}})$$

where we understand that the variables x and y are bound in the term $\mathsf{let}\ z\ \mathsf{be}\ x \otimes y\ \mathsf{in}\ f$. Naturality in Γ is clear since we may substitute for the corresponding variables, whilst naturality in C gives rise to an equation

$$g[\mathsf{let}\ z\ \mathsf{be}\ x \otimes y\ \mathsf{in}\ f/w] = \mathsf{let}\ z\ \mathsf{be}\ x \otimes y\ \mathsf{in}\ g[f/w] \tag{1}$$

The $(\otimes_{\mathcal{R}})$ rule

$$\frac{\Gamma \vdash A \qquad \Delta \vdash B}{\Gamma, \Delta \vdash A \otimes B}\,(\otimes_{\mathcal{R}})$$

gives an operation taking arrows $\Gamma \to A$ and $\Delta \to B$ to an arrow $\Gamma \bullet \Delta \to A \otimes B$. This might suggest a quite complex syntax, but fortunately our naturality assumptions imply that this operation is completely determined by a map $A \bullet B \to A \otimes B$. It follows that an appropriate syntax is

$$\frac{\Gamma \vdash e : A \qquad \Delta \vdash f : B}{\Gamma, \Delta \vdash e \otimes f : A \otimes B}\,(\otimes_{\mathcal{R}})$$

The $(I_{\mathcal{L}})$ rule

$$\frac{\Gamma \vdash A}{\Gamma, I \vdash A}\,(I_{\mathcal{L}})$$

gives an operation taking maps $\Gamma \to A$ to maps $\Gamma \bullet I \to A$. An appropriate syntax is

$$\frac{\Gamma \vdash e : A}{\Gamma, x : I \vdash \mathsf{let}\ x\ \mathsf{be}\ *\ \mathsf{in}\ e : A}\,(I_{\mathcal{L}})$$

so that in effect we simply introduce a dummy free variable for the assumption I. Naturality in Γ is clear since we may substitute for the corresponding (free) variables. However naturality in A gives rise to an equation

$$f[\mathsf{let}\ x\ \mathsf{be}\ *\ \mathsf{in}\ e/y] = \mathsf{let}\ x\ \mathsf{be}\ *\ \mathsf{in}\ f[e/y] \tag{2}$$

The $(I_{\mathcal{R}})$ rule

$$\frac{}{\vdash I}\,(I_{\mathcal{R}})$$

gives simply a map $\langle\rangle \to I$. An appropriate syntax is

$$\frac{}{\vdash * : I}\,(I_{\mathcal{R}})$$

Our treatment of the $(\multimap_{\mathcal{L}})$ rule

$$\frac{\Gamma \vdash A \qquad \Delta, B \vdash C}{\Gamma, A \multimap B, \Delta \vdash C}\,(\multimap_{\mathcal{L}})$$

follows traditional treatments of the left implication rule in sequent systems (which all involve a Yoneda Lemma argument). It follows from our naturality assumptions by a straightforward application of a Yoneda Lemma that an operation as above is determined by its action on a pair of identity arrows. Thus it is enough to give an operation of application $\mathsf{app}: A \bullet (A \multimap B) \longrightarrow B$. Then given arrows $e: \Gamma \to A$, $f: B \bullet \Delta \to C$ the required arrow $\Gamma \bullet (A \multimap B) \bullet \Delta \to C$ is the composite

$$\Gamma \bullet (A \multimap B) \bullet \Delta \xrightarrow{\quad e \bullet 1 \bullet 1 \quad} A \bullet (A \multimap B) \bullet \Delta \xrightarrow{\quad \mathsf{app} \bullet 1 \quad} B \bullet \Delta \xrightarrow{\quad f \quad} C$$

and an appropriate syntax is

$$\frac{\Gamma \vdash e : A \qquad \Delta, x : B \vdash f : C}{\Gamma, g : A \multimap B, \Delta \vdash f[(ge)/x] : C}\,(\multimap_{\mathcal{L}})$$

All the naturality assumptions are now dealt with by substitution. The $(\multimap_{\mathcal{R}})$ rule

$$\frac{\Gamma, A \vdash B}{\Gamma \vdash A \multimap B}\,(\multimap_{\mathcal{R}})$$

gives an operation taking an arrow $\Gamma \bullet A \to B$ to an arrow $\Gamma \to A \multimap B$. This is a form of abstraction and an appropriate syntax is

$$\frac{\Gamma, x : A \vdash e : B}{\Gamma \vdash \lambda x{:}\, A.e : A \multimap B}\,(\multimap_{\mathcal{R}})$$

2.3 Logical rules for the connective '!'

Next we consider the '!' connective. The left rules are reasonably straightforward, the right rule is a bit more involved. We consider the *Dereliction* and *Promotion* rules first.

Dereliction **and** *Promotion.* Consider the *Dereliction* rule

$$\frac{\Gamma, A \vdash B}{\Gamma, !A \vdash B}\,\textit{Dereliction}$$

Since it gives an operation taking an arrow $\Gamma \bullet A \to B$ to an arrow $\Gamma \bullet !A \to B$, an appropriate syntax is

$$\frac{\Gamma, x : A \vdash e : B}{\Gamma, z :!A \vdash \mathsf{let}\ z\ \mathsf{be}\ !x\ \mathsf{in}\ e : B}\,\textit{Dereliction}$$

and indeed this is the syntax given by Abramsky [1]. With this formulation naturality in B gives rise to an equation

$$f[\mathsf{let}\ z\ \mathsf{be}\ !x\ \mathsf{in}\ e/y] = \mathsf{let}\ z\ \mathsf{be}\ !x\ \mathsf{in}\ f[e/y]$$

However it is a consequence of naturality that our operation is determined by its effect on identity arrows, thus it is enough to give a map $!A \xrightarrow{\varepsilon} A$. Then given an arrow $e: \Gamma \bullet A \to B$, the required arrow $\Gamma \bullet !A \to B$ is the composite

$$\Gamma \bullet !A \xrightarrow{1 \bullet \varepsilon} \Gamma \bullet A \xrightarrow{e} B$$

so another appropriate syntax (and the one we shall use in what follows as it surpresses further naturality equations) is

$$\frac{\Gamma, x : A \vdash e : B}{\Gamma, z :!A \vdash e[\mathsf{derelict}(z)/x] : B}\ \textit{Dereliction}$$

Next consider the problematic *Promotion* rule

$$\frac{!\Gamma \vdash A}{!\Gamma \vdash !A}\ \textit{Promotion}$$

This gives an operation (of *Promotion*) taking an arrow $!\Gamma \to A$ to an arrow $!\Gamma \to !A$. Now it is not a priori clear what form of naturality should be assumed for this rule. If we assume that the operation should be natural in $!\Gamma$, then Abramsky's rule [1, Section 3],

$$\frac{\overline{x} :!\Gamma \vdash e : A}{\overline{x} :!\Gamma \vdash !e :!A}$$

would give an appropriate syntax[2]. However nothing in the idea of a categorical model suggests this assumption. (Note in passing that the categorically appealing assumption would be that ! is a functor and that we have naturality in Γ.) The important point to realize is that if the operation is not natural in $!\Gamma$, then the operation should not preserve substitution for the free variables implicitly declared in $!\Gamma$. Hence we are restricted to giving an operation on 'higher-order' terms, where the variables which appear initially must be bound and fresh variables introduced. These considerations lead to the term assignment rule

$$\frac{\overline{x} :!\Gamma \vdash e : A}{\overline{y} :!\Gamma \vdash \mathsf{promote}\ \overline{y}\ \mathsf{for}\ \overline{x}\ \mathsf{in}\ e :!A}\ \textit{Promotion}$$

We do not claim that there is a clear reason in terms of the category theory given so far to prefer one rule to the other, but we choose our rule simply so as to avoid any premature assumptions.

---

[2] This assumption has the effect that in the categorical model, which we shall consider later, the comonad is *idempotent*: a point noted by Wadler [18].

Weakening and *Contraction.* Finally we consider the *Weakening* and *Contraction* rules. The rule

$$\frac{\Gamma \vdash B}{\Gamma, !A \vdash B}\ \textit{Weakening}$$

gives an operation taking an arrow $\Gamma \to B$ to an arrow $\Gamma \bullet !A \to B$. An appropriate syntax is

$$\frac{\Gamma \vdash e : B}{\Gamma, z :!A \vdash \mathsf{discard}\ z\ \mathsf{in}\ e : B}\ \textit{Weakening}$$

where we have simply introduced a fresh dummy variable of type $!A$. Naturality in Γ is as before clear since we may substitute for the corresponding variables. Naturality in B gives rise to an equation

$$f[\mathsf{discard}\ z\ \mathsf{in}\ e/y] = \mathsf{discard}\ z\ \mathsf{in}\ f[e/y] \tag{3}$$

The *Contraction* rule

$$\frac{\Gamma, !A, !A \vdash B}{\Gamma, !A \vdash B}\ \textit{Contraction}$$

gives an operation taking an arrow $\Gamma \bullet !A \bullet !A \to B$ to an arrow $\Gamma \bullet !A \to B$. An appropriate syntax is

$$\frac{\Gamma, x :!A, y :!A \vdash e : B}{\Gamma, z :!A \vdash \mathsf{copy}\ z\ \mathsf{as}\ x, y\ \mathsf{in}\ e : B}\ \textit{Contraction}$$

where we understand that the variables x and y are bound in the term $\mathsf{copy}\ z \mathsf{as}\ x, y \mathsf{in}\ e$. Naturality in Γ is clear since we may substitute for the corresponding variables, while naturality in B gives rise to an equation

$$f[\mathsf{copy}\ z\ \mathsf{as}\ x, y\ \mathsf{in}\ e/w] = \mathsf{copy}\ z\ \mathsf{as}\ x, y\ \mathsf{in}\ f[e/w] \tag{4}$$

This concludes our derivation of a term assignment system for MELL from general considerations of the form of a categorical model. We display this system of term assignment in Fig. 2. We stress that rather elementary assumptions and unsophisticated categorical observations have been used in this analysis. However, our analysis has not only led us to a term assignment system, but has also uncovered a series of *naturality equations*, which are listed in Fig. 3.

3 Linear Natural Deduction

In the previous section we have provided a term assignment for a sequent calculus presentation of linear logic. Here we briefly consider a corresponding natural deduction formulation. In such a system a deduction is a derivation of a proposition from a finite set of assumption packets by means of inference rules. In intuitionistic logic these packets consist of (possibly empty) multisets of propositions. The restriction needed to make the derivations linear is that packets contain exactly one proposition, i.e. a packet is now equivalent to a proposition. Whereas before we typically had rules discharging many packets of an assumption we now only discharge the one. Thus we can label every proposition with a unique natural number.

$$x : A \vdash x : A$$

$$\frac{\Gamma \vdash e : A \qquad \Delta, x : A \vdash f : B}{\Gamma, \Delta \vdash f[e/x] : B}\ Cut$$

$$\frac{\Gamma \vdash e : A \qquad \Delta, x : B \vdash f : C}{\Gamma, g : A \multimap B, \Delta \vdash f[(ge)/x] : C}\ (\multimap_{\mathcal{L}}) \qquad \frac{\Gamma, x : A \vdash e : B}{\Gamma \vdash \lambda x{:}A.e : A \multimap B}\ (\multimap_{\mathcal{R}})$$

$$\frac{\Gamma \vdash e : A}{\Gamma, x : I \vdash \mathsf{let}\ x\ \mathsf{be}\ *\ \mathsf{in}\ e : A}\ (I_{\mathcal{L}}) \qquad \frac{}{\vdash * : I}\ (I_{\mathcal{R}})$$

$$\frac{\Delta, x : A, y : B \vdash f : C}{\Delta, z : A \otimes B \vdash \mathsf{let}\ z\ \mathsf{be}\ x \otimes y\ \mathsf{in}\ f : C}\ (\otimes_{\mathcal{L}}) \qquad \frac{\Gamma \vdash e : A \qquad \Delta \vdash f : B}{\Gamma, \Delta \vdash e \otimes f : A \otimes B}\ (\otimes_{\mathcal{R}})$$

$$\frac{\Gamma \vdash e : B}{\Gamma, z :!A \vdash \mathsf{discard}\ z\ \mathsf{in}\ e : B}\ Weakening$$

$$\frac{\Gamma, x :!A, y :!A \vdash e : B}{\Gamma, z :!A \vdash \mathsf{copy}\ z\ \mathsf{as}\ x, y\ \mathsf{in}\ e : B}\ Contraction$$

$$\frac{\Gamma, x : A \vdash e : B}{\Gamma, z :!A \vdash e[\mathsf{derelict}(z)/x] : B}\ Dereliction$$

$$\frac{\overline{x} :!\Gamma \vdash e : A}{\overline{y} :!\Gamma \vdash \mathsf{promote}\ \overline{y}\ \mathsf{for}\ \overline{x}\ \mathsf{in}\ e :!A}\ Promotion$$

Fig. 2. Term Assignment System for sequent calculus

Others have considered systems of natural deduction for linear logic [15, 18, 14]. Our main contribution is in our treatment of the *Promotion* rule. Previous authors formulated it as the following:

$$\begin{array}{c} !A_1 \cdots !A_n \\ \vdots \\ \dfrac{B}{!B}\ Promotion \end{array}$$

Clearly this rule is not closed under substitution. To ensure that the rule enjoys closure under substitution we use the following formulation:

$$f[\mathsf{let}\ x\ \mathsf{be}\ *\ \mathsf{in}\ e/y] = \mathsf{let}\ x\ \mathsf{be}\ *\ \mathsf{in}\ f[e/y]$$

$$f[\mathsf{let}\ z\ \mathsf{be}\ x{\otimes}y\ \mathsf{in}\ g/w] = \mathsf{let}\ z\ \mathsf{be}\ x{\otimes}y\ \mathsf{in}\ f[g/w]$$

$$f[\mathsf{discard}\ z\ \mathsf{in}\ e/y] = \mathsf{discard}\ z\ \mathsf{in}\ f[e/y]$$

$$f[\mathsf{copy}\ z\ \mathsf{as}\ x, y\ \mathsf{in}\ e/w] = \mathsf{copy}\ z\ \mathsf{as}\ x, y\ \mathsf{in}\ f[e/w]$$

Fig. 3. Naturality Equations

$$\dfrac{\begin{matrix} \vdots & & \vdots & [!A_1 \cdots !A_n] \\ & & & \vdots \\ !A_1 & \ldots & !A_n & B \end{matrix}}{!B}\ \textit{Promotion}$$

One should be aware that this rule carries an implicit side condition that not only must *all* assumptions be exponential, but that *all* are discharged (and re-introduced). Our subsequent term assignment is given in Fig. 4. We note at once a significant property of the term assignment system for linear natural deduction. Essentially the terms code the derivation trees so that any valid term assignment has a *unique* derivation.

Theorem 1 (Unique Derivation). *For any term t and proposition A, if there is a valid derivation of the form $\Gamma \vdash t : A$, then there is a unique derivation of $\Gamma \vdash t : A$.*

Proof. By induction on the structure of t. □

As mentioned above, our system enjoys closure under substitution.

Theorem 2 Substitution. *If $\Gamma \vdash a : A$ and $\Delta, x : A \vdash b : B$ then $\Gamma, \Delta \vdash b[a/x] : B$*

Proof. By induction on the derivation $\Delta, x : A \vdash b : B$. □

As one would expect there is an exact equivalence between the natural deduction and sequent calculus formulations (indeed the substitution property is essential for this). The details of this equivalence are given in [2].

4 Cut Elimination

In this section we consider cut elimination for the sequent calculus formulation of MELL, extended or decorated with terms. Suppose that a derivation in the term assignment system of Fig. 2 contains a cut:

$$x : A \vdash x : A$$

$$\frac{\Gamma, x : A \vdash e : B}{\Gamma \vdash \lambda x{:}A.e : A \multimap B}\,(\multimap_{\mathcal{I}}) \qquad \frac{\Gamma \vdash e : A \multimap B \qquad \Delta \vdash f : A}{\Gamma, \Delta \vdash ef : B}\,(\multimap_{\mathcal{E}})$$

$$\vdash * : I \qquad \frac{\Gamma \vdash e : A \qquad \Delta \vdash f : I}{\Gamma, \Delta \vdash \mathsf{let}\ f\ \mathsf{be}\ *\ \mathsf{in}\ e : A}\,(I_{\mathcal{E}})$$

$$\frac{\Gamma \vdash e : A \qquad \Delta \vdash f : B}{\Gamma, \Delta \vdash e \otimes f : A \otimes B}\,(\otimes_{\mathcal{I}}) \qquad \frac{\Gamma \vdash e : A \otimes B \qquad \Delta, x : A, y : B \vdash f : C}{\Gamma, \Delta \vdash \mathsf{let}\ e\ \mathsf{be}\ x \otimes y\ \mathsf{in}\ f : C}\,(\otimes_{\mathcal{E}})$$

$$\frac{\Delta_1 \vdash e_1 :!A_1 \quad \cdots \quad \Delta_n \vdash e_n :!A_n \qquad x_1 :!A_1, \ldots, x_n :!A_n \vdash f : B}{\Delta_1, \ldots, \Delta_n \vdash \mathsf{promote}\ e_1, \ldots, e_n\ \mathsf{for}\ x_1, \ldots, x_n\ \mathsf{in}\ f :!B}\,\textit{Promotion}$$

$$\frac{\Gamma \vdash e :!A \qquad \Delta \vdash f : B}{\Gamma, \Delta \vdash \mathsf{discard}\ e\ \mathsf{in}\ f : B}\,\textit{Weakening}$$

$$\frac{\Gamma \vdash e :!A \qquad \Delta, x :!A, y :!A \vdash f : B}{\Gamma, \Delta \vdash \mathsf{copy}\ e\ \mathsf{as}\ x, y\ \mathsf{in}\ f : B}\,\textit{Contraction}$$

$$\frac{\Gamma \vdash e :!A}{\Gamma \vdash \mathsf{derelict}(e) : A}\,\textit{Dereliction}$$

Fig. 4. Term Assignment System for Linear Natural Deduction

$$\frac{\dfrac{}{\Gamma \vdash e : A}\,D_1 \qquad \dfrac{}{\Delta, x : A \vdash f : B}\,D_2}{\Gamma, \Delta \vdash f[e/x] : B}\,\textit{Cut}$$

If $\Gamma \vdash e : A$ is the direct result of a rule D_1 and $\Delta, x : A \vdash f : B$ the result of a rule D_2, we say that the cut is a (D_1, D_2)-cut. A step in the process of eliminating cuts in the derivation tree will replace the subtree with root $\Gamma, \Delta \vdash f[e/x]: B$ with a tree with root of the form $\Gamma, \Delta \vdash t : B$. The terms in the remainder of the tree may be affected as a result.

Thus to ensure that the cut elimination process extends to derivations in the term assignment system, we must insist on an equality $f[e/x] = t$, which we can read from left to right as a term reduction. In fact we must insist on arbitrary substitution instances of the equality, as the formulae in Γ and Δ may be subject to cuts in the derivation tree below the cut in question. In this section we are mainly concerned to describe the equalities/reductions which result from the considerations just described. Note, however, that we cannot be entirely blithe about the process of eliminating cuts at the level of the propositional logic. As we shall see, not every apparent possibility for eliminating cuts should be realized in practice.

As things stand there are 11 rules of the sequent calculus aside from *Cut* (and *Exchange*) and hence 121 a priori possibilities for (D_1, D_2)-cuts. Fortunately most of these possibilities are not computationally meaningful in the sense that they have no effect on the terms. We say that a cut is *insignificant* if the equality $f[e/x] = t$ we derive from it as above is actually an identity (up to α-equivalence) on terms (so in executing the cut the term at the root of the tree does not change).

Note that any cut involving an axiom rule

$$\frac{}{x : A \vdash x : A}\ \mathit{Identity}$$

is insignificant; and the cut just disappears (hence instead of 121 we must now account for 100 cases). These 100 cases of cuts we will consider as follows: 40 cases of cuts the form (R, D) as we have 4 right rules and 10 others; 24 cases of cuts of the form (L, R) as we have 6 left-rules and 4 right ones and finally 36 cases of cuts of the form (L, L). Let us consider these three groups in turn.

Firstly we observe that there is a large class of insignificant cuts of the form (R, D) where R is a right rule: $(\otimes_{\mathcal{R}})$, $(I_{\mathcal{R}})$, $(\multimap_{\mathcal{R}})$, *Promotion*. Indeed all such cuts are insignificant with the following exceptions:

- *Principal cuts.* These are the cuts of the form $((\otimes_{\mathcal{R}}), (\otimes_{\mathcal{L}}))$, $((I_{\mathcal{R}}), (I_{\mathcal{L}}))$, $((\multimap_{\mathcal{R}}), (\multimap_{\mathcal{L}}))$, (*Promotion*, *Dereliction*), (*Promotion*, *Weakening*), (*Promotion*, *Contraction*) where the cut formula is introduced on the right and left of the two rules.
- Cases of the form $(R, \mathit{Promotion})$ where R is a right rule. Here we note that cuts of the form $((\otimes_{\mathcal{R}})), \mathit{Promotion})$, $((I_{\mathcal{R}})), \mathit{Promotion})$ and $((\multimap_{\mathcal{R}}), \mathit{Promotion})$ cannot occur; so the only possibility is (*Promotion*, *Promotion*).

Next any cut of the form (L, R) where L is one of the left rules $(\otimes_{\mathcal{L}})$, $(I_{\mathcal{L}})$, $(\multimap_{\mathcal{L}})$, *Weakening*, *Contraction*, *Dereliction* and R is one of the simple right rules $(\otimes_{\mathcal{R}})$, $(I_{\mathcal{R}})$, $(\multimap_{\mathcal{R}})$ is insignificant (18 cases). Also cuts of the form $((\multimap_{\mathcal{L}}), \mathit{Promotion})$ and (*Dereliction*, *Promotion*) are insignificant (2 cases). There remain four further cases of cuts of the form $(L, \mathit{Promotion})$ where L is a left rule.

Lastly we have to consider the 36 cuts of the form (L_1, L_2), where the L_i are both left rules. Again we derive some benefit from our rules for $(\multimap_{\mathcal{L}})$ and *Dereliction*: cuts of the form $((\multimap_{\mathcal{L}}), L)$ and $(\mathit{Dereliction}, L)$ are insignificant. There are 24 remaining cuts of interest.

We now summarize the cuts of which we need to take some note. They are:

- Principal cuts. There are six of these.
- Secondary cuts. The single (strange) form of cut (*Promotion*, *Promotion*) and the four remaining cuts of form $(L, \mathit{Promotion})$ where L is a left rule other than $(\multimap_{\mathcal{L}})$ or (*Dereliction*).

- Commutative cuts. The twenty-four remaining cuts of the form (L_1, L_2) just described. These correspond almost[3] case by case to the commutative conversions for natural deduction (considered in [3]) and are not considered further here.

4.1 Principal Cuts

We do not dwell on the cases of principal cuts involving tensor, the constant I and linear implication as they are standard. We shall consider in detail the principal cuts involving the *Promotion* rule.

- (*Promotion*, *Dereliction*)-cut. The derivation

$$\dfrac{\dfrac{!\Gamma \vdash B}{!\Gamma \vdash !B}\,\textit{Promotion} \qquad \dfrac{B, \Delta \vdash C}{!B, \Delta \vdash C}\,\textit{Dereliction}}{!\Gamma, \Delta \vdash C}\,\textit{Cut}$$

is reduced to

$$\dfrac{!\Gamma \vdash B \qquad B, \Delta \vdash C}{!\Gamma, \Delta \vdash C}\,\textit{Cut}$$

This reduction yields the following term reduction

$$(f[\mathsf{derelict}(q)/p])[\mathsf{promote}\ y_i\ \mathsf{for}\ x_i\ \mathsf{in}\ e/q] = f[e/p]$$

- (*Promotion*, *Weakening*)-cut. The derivation

$$\dfrac{\dfrac{!\Gamma \vdash B}{!\Gamma \vdash !B}\,\textit{Promotion} \qquad \dfrac{\Delta \vdash C}{!B, \Delta \vdash C}\,\textit{Weakening}}{!\Gamma, \Delta \vdash C}\,\textit{Cut}$$

is reduced to

$$\dfrac{\Delta \vdash C}{!\Gamma, \Delta \vdash C}\,\textit{Weakening}^*$$

where *Weakening** corresponds to many applications of the *Weakening* rule.

This gives the term reduction

$$\mathsf{discard}\ (\mathsf{promote}\ e_i\ \mathsf{for}\ x_i\ \mathsf{in}\ f)\ \mathsf{in}\ g = \mathsf{discard}\ e_i\ \mathsf{in}\ g$$

- (*Promotion*, *Contraction*)-cut. The derivation

$$\dfrac{\dfrac{!\Gamma \vdash B}{!\Gamma \vdash !B}\,\textit{Promotion} \qquad \dfrac{!B, !B, \Delta \vdash C}{!B, \Delta \vdash C}\,\textit{Contraction}}{!\Gamma, \Delta \vdash C}\,\textit{Cut}$$

is reduced to

[3] The exceptions are the cases where $(\multimap_{\mathcal{L}})$ is the (second) rule above the cut. In these cases we obtain slightly stronger rules.

$$\dfrac{\dfrac{!\Gamma \vdash B}{!\Gamma \vdash !B}\,\mathit{Promotion} \qquad \dfrac{\dfrac{!\Gamma \vdash B}{!\Gamma \vdash !B}\,\mathit{Promotion} \qquad !B, !B, \Delta \vdash C}{!\Gamma, !B, \Delta \vdash C}\,\mathit{Cut}}{\dfrac{!\Gamma, !\Gamma, \Delta \vdash C}{!\Gamma, \Delta \vdash C}\,\mathit{Contraction}^*}\,\mathit{Cut}$$

or to the symmetric one where we cut against the other $!B$ first. This gives the term reduction

$$\begin{aligned} &\mathsf{copy}\ (\mathsf{promote}\ e_i\ \mathsf{for}\ x_i\ \mathsf{in}\ f)\ \mathsf{as}\ y, y'\ \mathsf{in}\ g = \\ &\mathsf{copy}\ e_i\ \mathsf{as}\ z_i, z_i'\ \mathsf{in}\ g[\mathsf{promote}\ z_i\ \mathsf{for}\ x_i\ \mathsf{in}\ f/y, \mathsf{promote}\ z_i'\ \mathsf{for}\ x_i\ \mathsf{in}\ f/y'] \end{aligned}$$

As would be expected these principal cuts correspond to the β-reductions which can be derived from the natural deduction system outlined in Section 3 (and detailed in [3]).

$$\begin{aligned}
&\mathsf{let}\ f \otimes g\ \mathsf{be}\ x \otimes y\ \mathsf{in}\ h && = h[f/x, g/y] \\
&\mathsf{let}\ *\ \mathsf{be}\ *\ \mathsf{in}\ h && = h \\
&h[(\lambda x{:}\, A.f)g/y] && = h[f[g/x]/y] \\
&(f[\mathsf{derelict}(q)/p])[\mathsf{promote}\ y_i\ \mathsf{for}\ x_i\ \mathsf{in}\ e/q] && = f[e/p] \\
&\mathsf{discard}\ (\mathsf{promote}\ e_i\ \mathsf{for}\ x_i\ \mathsf{in}\ f)\ \mathsf{in}\ g && = \mathsf{discard}\ e_i\ \mathsf{in}\ g \\
&\mathsf{copy}\ (\mathsf{promote}\ e_i\ \mathsf{for}\ x_i\ \mathsf{in}\ f)\ \mathsf{as}\ y, y'\ \mathsf{in}\ g && = \mathsf{copy}\ e_i\ \mathsf{as}\ z_i, z_i'\ \mathsf{in} \\
& && \quad g[\mathsf{promote}\ z_i\ \mathsf{for}\ x_i\ \mathsf{in}\ f/y, \\
& && \quad\ \mathsf{promote}\ z_i'\ \mathsf{for}\ x_i\ \mathsf{in}\ f/y']
\end{aligned}$$

Fig. 5. Principal reduction rules

4.2 Secondary Cuts

We now consider the cases where the *Promotion* rule is on the right of a cut rule. The first case is the 'strange' case of cutting *Promotion* against *Promotion*, then we have the four cases $(\otimes_{\mathcal{L}})$, $(I_{\mathcal{L}})$, *Weakening* and *Contraction* against the rule *Promotion*. Here we discuss only the 'strange' case of

• (*Promotion*, *Promotion*)-**cut**. The derivation

$$\dfrac{\dfrac{!\Gamma \vdash B}{!\Gamma \vdash !B}\,\mathit{Promotion} \qquad \dfrac{!B, !\Delta \vdash C}{!B, !\Delta \vdash !C}\,\mathit{Promotion}}{!\Gamma, !\Delta \vdash !C}\,\mathit{Cut}$$

reduces to

$$\dfrac{\dfrac{\dfrac{!\Gamma \vdash B}{!\Gamma \vdash !B}\,\mathit{Promotion} \qquad !B, !\Delta \vdash C}{!\Gamma, !\Delta \vdash C}\,\mathit{Cut}}{!\Gamma, !\Delta \vdash !C}\,\mathit{Promotion}$$

Note that it is always possible to permute the cut upwards, as all the formulae in the antecedent are modal. This gives the term reduction

promote (promote z for x in f) for y in g = promote w for z in (g[promote z for x in f/y])

We present all the term equalities given by the secondary cuts in Fig. 6. Observe that the last four equations are particular instances of the naturality equations described in Section 2, while the first encapsulates the naturality of the Kleisli operation of *Promotion*. One is tempted to suggest that perhaps the reason why the rule *Promotion* gives us reductions with some sort of computational meaning is because this rule is not clearly either a left or a right rule. It introduces the connective on the right (so it is mainly a right rule), but it imposes conditions on the context on the left. Indeed there does not appear to be any analogous reductions in natural deduction.

promote (promote z for x in f) for y in g	= promote w for z in g[promote z for x in f/y]
promote (discard x in f) for y in g	= discard x in (promote f for y in g)
promote (copy x as y, z in f) for y in g	= copy x as y, z in (promote f for y in g)
promote (let z be $x \otimes y$ in f) for w in g	= let z be $x \otimes y$ in (promote f for w in g)
promote (let z be $*$ in f) for w in g	= let z be $*$ in (promote f for w in g)

Fig. 6. Secondary reduction rules

5 The Categorical Model

Much work has been done on providing such (categorical) models of Intuitionistic Linear Logic. Here we shall just mention the work of Seely [16] and de Paiva [4]. With a view to understanding what is involved here, let us consider the traditional analysis of the proof theory of some basic intuitionistic logic via the notion of a cartesian closed category. (Lambek and Scott [12] is a good source for this material.) In that case, the basic normalization process gives rise to β-equality on the terms of the typed λ-calculus. The β-equality rule is valid in any cartesian closed category, but the attractive categorical assumption of being cartesian closed amounts to requiring $\beta\eta$-equality, that is, to a further 'extensionality' assumption. Thus one way to understand what we do is that we make a minimal number of attractive simplifying assumptions about the basic categorical set up introduced in Section 2 which at least entail the (desired) equalities between proofs. In this section we simply discuss the categorical assumptions we make and give the resulting equations.

5.1 Categorical interpretation of the multiplicatives

We start by considering the connective $\otimes$. The categorical significance of the β-rule for $\otimes$ is that any map of the form $\Gamma \bullet A \bullet B \to C$ factors canonically through the map $A \bullet B \xrightarrow{\otimes} A \otimes B$ which results from the instance of the $(\otimes_{\mathcal{R}})$ rule

$$\frac{A \vdash A \qquad B \vdash B}{A, B \vdash A \otimes B} (\otimes_{\mathcal{R}})$$

The simplifying 'extensionality' assumption is then that this factorization is *unique*. This can be expressed by saying that (generalized) composition with $A \bullet B \to A \otimes B$ induces a natural isomorphism between maps

$$\frac{\Gamma \bullet (A \otimes B) \to C}{\overline{\Gamma \bullet A \bullet B \to C}}$$

In other words that the operation of composing with $A \bullet B \to A \otimes B$ provides an inverse to the $(\otimes_{\mathcal{L}})$-operation taking maps $\Gamma \bullet A \bullet B \to C$ to maps $\Gamma \bullet (A \otimes B) \to C$. Thus we may as well assume that the logical $\otimes$ coincides with $\bullet$. We get two equations expressing this isomorphism. One of these equations is, of course, the β-rule for tensor:

$$\mathsf{let}\ u \otimes v\ \mathsf{be}\ x \otimes y\ \mathsf{in}\ f = f[u/x, v/y] \tag{5}$$

The other can be regarded as an η-equality:

$$\mathsf{let}\ u\ \mathsf{be}\ x \otimes y\ \mathsf{in}\ f[x \otimes y/z] = f[u/z] \tag{6}$$

The case of I is like that for $\otimes$. Thus (generalized) composition with $\langle\rangle \to I$ induces a natural isomorphism between maps

$$\frac{\Gamma \bullet I \to C}{\overline{\Gamma \bullet \langle\rangle \to C}}$$

We identify $\langle\rangle$ and I, and use I both as a logical operator and to interpret the empty sequence on the left hand side of a sequent. As before we get two equations expressing the natural isomorphism. One is the β-rule and the other can again be regarded as an η-equality:

$$\mathsf{let}\ *\ \mathsf{be}\ *\ \mathsf{in}\ f = f \tag{7}$$

$$\mathsf{let}\ u\ \mathsf{be}\ *\ \mathsf{in}\ f[*/z] = f[u/z] \tag{8}$$

The β-rule for $\multimap$ has a slightly more complicated interpretation, it means that any map $f\colon A\otimes B \to C$ factors as

$$A\otimes B \xrightarrow{1\otimes\mathsf{cur}(f)} A\otimes(A\multimap C) \xrightarrow{\mathsf{app}} C$$

where $\mathsf{app}\colon A\otimes(A\multimap C) \to C$ is the map that results from an instance of the $(\multimap_{\mathcal{L}})$ rule

$$\frac{A \vdash A \qquad C \vdash C}{A, A\multimap C \vdash C}\,(\multimap_{\mathcal{L}})$$

Again the natural simplifying assumption is that the factorization is *unique*, which means that there exists a natural isomorphism between maps

$$\frac{A\otimes B \longrightarrow C}{\overline{A \longrightarrow B\multimap C}}$$

Thus $\multimap$ provides us with a closed structure on our category corresponding to the tensor $\otimes$. Again we have two equations to express our natural isomorphism. One is the β-rule and the other is the (linear form of the) traditional η-rule:

$$(\lambda x.f)e = f[e/x] \tag{9}$$

$$\lambda x.fx = f \tag{10}$$

5.2 Categorical interpretation of *Dereliction* and *Promotion*

Now we consider the meaning of the β-rule for ! involving *Dereliction*. The categorical import of this rule is that any map $!\Gamma \to A$ factors in a canonical way as a composite

$$!\Gamma \longrightarrow\ !A \xrightarrow{\varepsilon_A} A$$

where $!A \xrightarrow{\varepsilon_A} A$ is the canonical map obtained by *Dereliction* from the identity as described in Section 2. Given any proof $\Gamma \vdash B$ there is obviously a canonical two-step process that transforms it into a proof $!\Gamma \vdash !B$ by applying the *Dereliction* rule (several times) followed by *Promotion*.

$$\cfrac{\cfrac{\Gamma \vdash B}{!\Gamma \vdash B}\,\textit{Dereliction}^*}{!\Gamma \vdash !B}\,\textit{Promotion}$$

If $\Gamma \xrightarrow{f} B$ interprets the original proof, we write the resulting arrow as $!\Gamma \xrightarrow{!f} !B$. As a preliminary simplification, we assume that this definition gives the extension of ! to a multicategorical functor. This amounts to the assumption that ! is a *monoidal* functor [5]; that is, the functor ! comes equipped with a natural transformation

$$m_{A,B}: !A \otimes !B \rightarrow !(A \otimes B)$$

(natural in A and B) and a morphism $m_I: I \rightarrow !I$ making a standard collection of diagrams commute. Note that the β-rule for *Dereliction* certainly implies that for any $f: \Gamma \rightarrow A$, the equation $!f; \varepsilon_A = \varepsilon_\Gamma; f$ holds. Either composite gives the effect of *Dereliction* on f. This shows that $\varepsilon: ! \rightarrow 1$ will be a multicategorical natural transformation and so a monoidal natural transformation.

We need one further piece of structure. We apply the *Promotion* rule to the axiom $!A \vdash !A$ to obtain the derivation

$$\frac{!A \vdash !A}{!A \vdash !!A}\ \textit{Promotion}$$

In other words, from an identity arrow $!A \longrightarrow !A$ we can get a canonical arrow $\delta_A: !A \rightarrow !!A$. With the equations to hand we know rather little about δ. One can easily check that the composite

$$!A \xrightarrow{\delta_A} !!A \xrightarrow{\varepsilon_{!A}} !A$$

is the identity on $!A$, and that is one of the triangle identities for a comonad, but that is about it. However it is compelling to add to our preliminary assumption that ! is a monoidal functor, the assumption that δ (as well as ε) is a monoidal natural transformation and that $(!, \varepsilon, \delta)$ forms a comonad on our category. Note that given a monoidal comonad $(!, \varepsilon, \delta)$, the *Promotion* rule can be interpreted as follows: given a map $f: !C_1 \otimes \ldots \otimes !C_n \longrightarrow A$ we obtain the 'promoted' map as the composite

$$!C_1 \otimes \ldots \otimes !C_n \xrightarrow{\delta} !!C_1 \otimes \ldots \otimes !!C_n \xrightarrow{m} !(!C_1 \otimes \ldots \otimes !C_n) \xrightarrow{!f} !A$$

We can formulate the conditions that $(!, \varepsilon, \delta)$ be a monoidal comonad directly in terms of the basic operations given by MELL. In addition to the β-equality (equation (11) below) we obtain:

$$\mathsf{derelict}(\mathsf{promote}\ e_i\ \mathsf{for}\ x_i\ \mathsf{in}\ f) = f[e_i/x_i] \qquad (11)$$

$$\mathsf{promote}\ z\ \mathsf{for}\ x\ \mathsf{in}\ (\mathsf{derelict}(x)) = z \qquad (12)$$

$$\begin{aligned}&\mathsf{promote}\ (\mathsf{promote}\ z_i\ \mathsf{for}\ x_i\ \mathsf{in}\ f), w_j\ \mathsf{for}\ y, y_j\ \mathsf{in}\ g = \\ &\mathsf{promote}\ z_i, w_j\ \mathsf{for}\ z_i', y_j\ \mathsf{in}\ (g[\mathsf{promote}\ z_i'\ \mathsf{for}\ x_i\ \mathsf{in}\ f/y]).\end{aligned} \qquad (13)$$

Equation (12) can be thought of as an η-rule, as it provides a kind of uniqueness of the factorization mentioned above; equation (13) expresses an appropriate form of naturality of the operation of *Promotion* and it arises from a secondary cut elimination.

5.3 Categorical interpretation of *Weakening* and *Contraction*

We finally consider the categorical significance of the β-rules involving *Weakening* and *Contraction*. To do so let us first introduce a further canonical pair of maps. Using *Weakening* (and the right rule for I) we have a deduction

$$\frac{\vdash I}{!A \vdash I}\ \textit{Weakening}$$

which gives a canonical map $!A \xrightarrow{e_A} I$ (where e is used to remind the reader that this map corresponds to 'erasing' the assumption). From the rules ($\otimes_{\mathcal{R}}$) and *Contraction* we obtain

$$\cfrac{\cfrac{!A \vdash !A \qquad !A \vdash !A}{!A, !A \vdash !A \otimes !A}\ (\otimes_{\mathcal{R}})}{!A \vdash !A \otimes !A}\ \textit{Contraction}$$

which gives a canonical map $!A \xrightarrow{d_A} !A \otimes !A$ (again d is used to hint at 'duplication' of assumptions).

It follows from the β- and η-rules for $\otimes$ and I as well as from the naturality assumptions on *Contraction* and *Weakening* described in Section 2 that any map $\Gamma \otimes !A \xrightarrow{f} B$ arising from the use of the rule of *Weakening* is the composite

$$\Gamma \otimes !A \xrightarrow{1 \otimes e_A} \Gamma \otimes I \cong \Gamma \xrightarrow{\bar{f}} B$$

Similarly the effect of the rule of *Contraction* is that any map $!A \otimes \Gamma \xrightarrow{f} B$ arising from the use of *Contraction* is the composite

$$!A \otimes \Gamma \xrightarrow{d_A \otimes 1_\Gamma} !A \otimes !A \otimes \Gamma \xrightarrow{\bar{f}} B$$

The β-equalities for *Contraction* and *Weakening* namely,

$$\text{discard (promote } e_i \text{ for } x_i \text{ in } t) \text{ in } u = \text{discard } e_i \text{ in } u \tag{14}$$

$$\begin{aligned} &\text{copy (promote } e_i \text{ for } x_i \text{ in } t) \text{ as } y, z \text{ in } u = \\ &\text{copy } e_i \text{ as } x_i', x_i'' \text{ in } u[\text{promote } x_i' \text{ for } x_i \text{ in } t/y, \text{promote } x_i'' \text{ for } x_i \text{ in } t/z] \end{aligned} \tag{15}$$

say that maps obtained using the rule *Promotion* preserve the structure (on objects of the form $!A$) given by e and d. It follows at once that the canonical morphisms (e and d) are natural transformations. One might also expect that e and d give structure on the coalgebras, or (what amounts to the same thing) that they are themselves maps of coalgebras. This leads to the equations

$$\text{promote } e, e_i \text{ for } x, x_i \text{ in discard } x \text{ in } t = \text{discard } e \text{ in promote } e_i \text{ for } x_i \text{ in } t \tag{16}$$

$$\begin{aligned} &\text{promote } e, e_i \text{ for } z, z_i \text{ in copy } z \text{ as } x, y \text{ in } t = \\ &\text{copy } e \text{ as } x', y' \text{ in promote } x', y', e_i \text{ for } x, y, z_i \text{ in } t \end{aligned} \tag{17}$$

We believe that there is some computational sense to this interplay between *Promotion* on the one hand, and *Weakening* and *Contraction* on the other. Furthermore our intuitions about the processes of discarding and copying suggest strongly that the natural transformations e and d give rise to the structure of a (commutative) comonoid on the free !-coalgebras. (As a consequence all coalgebras have (and all maps of coalgebras preserve) the structure of a (commutative) comonoid.)

5.4 The categorical model of Intuitionistic Linear Logic

Much of the *categorical* analysis that we have just given is quite familiar, though the corresponding equational calculus seems new (if only because our syntax is new). We note however that (following Seely [16]) it has become standard to analyze the categorical meaning of *Weakening* and *Contraction* in terms of the relationship between the additives and the multiplicatives. Our analysis dispenses with additives and hence gives a more general account of the force of the exponentials. Even in the presence of the additives our formulation is not equivalent to Seely's and it certainly covers cases of interest not covered by his. To sum up the analysis in this section we give the following definition.

Definition 3. A categorical model for MELL[4] consists of:

1. a symmetric monoidal closed (multi)category (modelling tensor and linear implication);
2. together with a comonad $(!, \varepsilon, \delta)$ with the following properties:
   (a) the functor part '!' of the comonad is a monoidal functor and ε and δ are monoidal natural transformations,
   (b) every (free) !-coalgebra carries naturally the structure of a commutative comonoid[5] in such a way that coalgebra maps are comonoid maps.

We have indicated in the text above what are the equations in our term assignment system corresponding to this notion of categorical model. We display all these equations in Fig. 7.

The connection between these equations and the notion of a categorical model can be made precise along the following lines. First assume that we have a signature given by a collection of ground types and of typed function symbols. From this data, types and terms in context are defined inductively by the clauses of Fig. 2 giving rise to what we call a *term logic* for MELL.

Next assume that **C** is a categorical model for MELL. Then in particular **C** has the structure outlined in Sect. 2. Here $\langle\rangle$ and $\bullet$ are identified with the I and $\otimes$ of **C**. The required operations for I, $\otimes$, and $\multimap$ are given by standard operations in a symmetric monoidal closed category. As explained in Sect. 5.2 the map $\varepsilon\colon !A \to A$ used to interpret the *Dereliction* rule can be identified with the counit $\varepsilon_A\colon !A \to A$, while the operation of *Promotion* involves the comultiplication δ and the map which gives the monoidal structure of the functor !. Finally as explained in Sect. 5.3 the operations for *Weakening* and *Contraction* are given in terms of the comonoid structure on the (free) coalgebras.

[4] Note that in our formulation it is not necessary to consider the additives to model the exponential.

[5] This means not only that each !-coalgebra $(A, h_A\colon A \to !A)$ comes equipped with morphisms $e\colon A \to I$ and $d\colon A \to A \otimes A$ but also that e and d are coalgebra maps.

let $*$ be $*$ in e	$= e$
let u be $*$ in $f[*/z]$	$= f[u/z]$
let $e \otimes t$ be $x \otimes y$ in u	$= u[e/x, t/y]$
let u be $x \otimes y$ in $f[x \otimes y/z]$	$= f$
$(\lambda x{:}\, A.t)e$	$= t[e/x]$
$\lambda x{:}\, A.tx$	$= t$
derelict(promote e_i for x_i in t)	$= t[e_i/x_i]$
promote z for x in derelict(x)	$= z$
promote (promote z_i for x_i in f), w_j for y, y_j in g	$=$ promote z_i, w_j for z_i', y_j in (g[promote z_i' for x_i in f/y])
discard (promote e_i for x_i in t) in u	$=$ discard e_i in u
promote e, e_i for x, x_i in discard x in t	$=$ discard e in promote e_i for x_i in t
copy (promote e_i for x_i in t) as y, z in u	$=$ copy e_i as x_i', x_i'' in u[promote x_i' for x_i in t/y, promote x_i'' for x_i in t/z]
promote e, e_i for z, z_i in copy z as x, y in t	$=$ copy e as x', y' in promote x', y', e_i for x, y, z_i in t
copy e as x, y in discard x in t	$= t[e/y]$
copy e as x, y in discard y in t	$= t[e/x]$
copy e as x, y in t	$=$ copy e as y, x in t
copy e as x, w in copy w as y, z in t	$=$ copy e as w, z in copy w as x, y in t
f[let z be $*$ in e/w]	$=$ let z be $*$ in $f[e/w]$
f[let z be $x \otimes y$ in e/w]	$=$ let z be $x \otimes y$ in $f[e/w]$
f[discard z in e/w]	$=$ discard z in $f[e/w]$
f[copy z as x, y in e/w]	$=$ copy z as x, y in $f[e/w]$

Fig. 7. Categorical equalities

Given the structure outlined in Sect. 2, for any interpretation of a signature Σ in **C** there is a standard inductive definition of the interpretation of types and of terms in context of the term logic given by Σ in **C**. The steps in this inductive definition[6] were considered in Sect. 2 and for the convenience of the reader we present an indication of these steps in Fig. 8.

$$A \to A$$

$$\frac{\Gamma \to A \qquad A \bullet \Delta \to B}{\Gamma \bullet \Delta \to B}\ \mathit{Cut}$$

$$\frac{\Gamma \to A}{\Gamma \bullet I \to A}\ (I_{\mathcal{L}}) \qquad \frac{}{() \to I}\ (I_{\mathcal{L}})$$

$$\frac{\Gamma \bullet A \bullet B \to C}{\Gamma \bullet (A \otimes B) \to C}\ (\otimes_{\mathcal{L}}) \qquad \frac{\Gamma \to A \qquad \Delta \to B}{\Gamma \bullet \Delta \to A \otimes B}\ (\otimes_{\mathcal{R}})$$

$$\frac{\Gamma \to A \qquad \Delta \bullet B \to C}{\Gamma \bullet (A \multimap B) \bullet \Delta \to C}\ (\multimap_{\mathcal{L}}) \qquad \frac{\Gamma \bullet A \to B}{\Gamma \to A \multimap B}\ (\multimap_{\mathcal{R}})$$

$$\frac{\Gamma \to B}{\Gamma \bullet !A \to B}\ \mathit{Weakening} \qquad \frac{\Gamma \bullet !A \bullet !A \to B}{\Gamma \bullet !A \to B}\ \mathit{Contraction}$$

$$\frac{\Gamma \bullet A \to B}{\Gamma \bullet !A \to B}\ \mathit{Dereliction} \qquad \frac{!\Gamma \to A}{!\Gamma \to !A}\ \mathit{Promotion}$$

Fig. 8. (Outline of the) interpretation of Term Logic

The interpretation is sound and complete in the following sense.

Theorem 4.

1. *(Soundness) For any signature and interpretation of the corresponding system in a categorical model for Intuitionistic Linear Logic (all the equational consequences of) the equations in Fig. 7 hold in the sense that the interpretation of either term gives the same map in the category.*
2. *(Completeness) For any signature there is a categorical model for Intuitionistic Linear Logic and an interpretation of the system in it with the following property:*

[6] Note that strictly speaking the induction is on the *derivation* (in the sequent calculus) of $\Gamma \vdash e : A$. Hence one has to show that the interpretation in **C** is independent of the derivation. It is laborious to show this directly and the result also follows from a consideration of the equivalent natural deduction formulation sketched in Sect. 3.

- *If $\Gamma \vdash t{:}A$ and $\Gamma \vdash s{:}A$ are derivable in the system then t and s are interpreted as the same map $\Gamma \to A$ just when $t = s{:}A$ is provable from the equations in Fig. 7 (in typed equational logic).*

We can make some comments on the proof of soundness and completeness. We derived the equations of Fig. 7 from a consideration of the categorical model. So the proof of soundenss amounts to filling in the details of that derivation. As all too often the proof of completeness is given by a construction of a categorical term model. One has to check that the equations given are sufficient to establish all the properties of a categorical model as exhibited in Defn. 3.

Now we try to make clear the force of our definition in terms of a discussion of (the background to) Girard's translation of intuitionistic propositional logic into linear logic. We start by recalling some folklore results about the Eilenberg-Moore category of coalgebras.

Theorem 5.

1. *If a symmetric monoidal category* $\mathbf{C}$ *is equipped with a monoidal comonad* $(!, \varepsilon, \delta)$, *then the tensor product of* $\mathbf{C}$ *induces a symmetric monoidal structure on the category of coalgebras* $\mathbf{C}_!$.
2. - *If, furthermore,* $\mathbf{C}$ *is symmetric monoidal closed, then all free coalgebras are 'exponentiable' in* $\mathbf{C}_!$ *(in the sense appropriate to the monoidal structure); what is more any power of a free coalgebra is a free coalgebra. So the full subcategory of finite tensor products of free coalgebras forms a symmetric monoidal closed category containing the category of free coalgebras.*
   - *If, in addition, the (Kleisli) category of free coalgebras is closed under the tensor product in* $\mathbf{C}_!$, *then the category of free coalgebras is symmetric monoidal closed.*
3. *If on the other hand* $\mathbf{C}$ *is symmetric monoidal closed and* $\mathbf{C}_!$ *has equalizers of coreflexive pairs of arrows then* $\mathbf{C}_!$ *is symmetric monoidal closed.*

We make clear what is the force of our stipulation that every (free) !-coalgebra carries naturally the structure of a commutative comonoid in such a way that coalgebra maps are comonoid maps.

Theorem 6.

1. *If a symmetric monoidal category* $\mathbf{C}$ *is equipped with a comonad* $(!, \varepsilon, \delta)$ *satisfying part 2(b) of Definition 1, then the tensor product induced on the category* $\mathbf{C}_!$ *of coalgebras is a categorical product.*
2. *If, furthermore,* $\mathbf{C}$ *is symmetric monoidal closed, then all free coalgebras are exponentiable in* $\mathbf{C}_!$ *(in the standard sense); and so the full subcategory of exponentiable objects forms a cartesian closed category (containing the category of free coalgebras).*
3. *If, in addition, the (Kleisli) category of free coalgebras is closed under the product in* $\mathbf{C}_!$, *then the category of free coalgebras is cartesian closed. In particular this follows when* $\mathbf{C}$ *has finite products* $(1, \&)$ *and we have the natural isomorphisms*

$$I \cong !I$$
$$!A \otimes !B \cong !(A \& B)$$

4. If, on the other hand, $C_!$ has equalizers of coreflexive pairs of arrows then $C_!$ is cartesian closed.

This theorem, which in essence goes back to Fox [6], is the basis for the Girard translation of intuitionistic logic into Intuitionistic Linear Logic. In the usual formulation this translation is based on *3*, that is on the natural isomorphisms introduced by Seely [16], and so essentially takes place in the category of free coalgebras. (This option is still available in cases where the relevent natural isomorphisms do not hold.) However, the general theorem demonstrates that at the proof theoretic (computational) level a more subtle analysis (which involves the full category of coalgebras) is possible.

6 Conclusions and Future Work

We have described a new term assignment system for a sequent calculus version of MELL, based on a generic idea of a categorical model for MELL. This term assignment system, unlike its predecessors has an exact correspondence with a Linear Natural Deduction system which satisfies the essential property of closure under substitution (unlike all previous proposals). Using this term assignment system and an analysis of the process of cut-elimination we produced some β-equalities. Further analysis of the β-equalities as well as the judicious addition of some extra 'extensionality' assumptions (similar to the usual ones in Categorical Proof Theory) provided a precise notion of a categorical model, more general than the traditional one for Intuitionistic Linear Logic. For this general notion of categorical model we have soundness and completeness.

But we can identify a number of areas which need to be covered in the future. Clearly we need to consider the *additive* connectives. We should also like to consider quantifiers within this framework. Especially we should like to consider some of the many variants of Intuitionistic Linear Logic that have been proposed [10, 9, 8].

Acknowledgements

We should like to thank Andy Pitts and two anonymous referees for detailed comments on this work. This paper was prepared using Paul Taylor's TEX macros.

References

1. Samson Abramsky. Computational interpretations of linear logic. Technical Report 90/20, Department of Computing, Imperial College, London, October 1990.
2. Nick Benton, Gavin Bierman, Valeria de Paiva, and Martin Hyland. Term assignment for intuitionistic linear logic. Technical Report 262, Computer Laboratory, University of Cambridge, August 1992.
3. Nick Benton, Gavin Bierman, Valeria de Paiva, and Martin Hyland. A term calculus for intuitionistic linear logic. In *Proceedings of International Conference on Typed Lambda Calculi and Applications*, Lecture Notes in Computer Science, March 1993.

4. Valeria C.V. de Paiva. *The Dialetica Categories*. PhD thesis, Department of Pure Mathematics and Mathematical Statistics, University of Cambridge, 1988. Published as Computer Laboratory Technical Report 213, 1990.
5. S. Eilenberg and G.M. Kelly. Closed categories. In *Proceedings of Conference on Categorical Algebra, La Jolla*, 1966.
6. T. Fox. Coalgebras and cartesian categories. *Communications in Algebra*, 4(7):665–667, 1976.
7. Jean-Yves Girard. Linear logic. *Theoretical Computer Science*, 50:1–101, 1987.
8. Jean-Yves Girard, Andre Scedrov, and Philip Scott. Bounded linear logic: A modular approach to polynomial time computability. *Theoretical Computer Science*, 97:1–66, 1992.
9. Martin Hyland and Valeria de Paiva. Full intuitionistic linear logic. Unpublished manuscript, 1992.
10. Bart Jacobs. Semantics of weakening and contraction. Department of Pure Mathematics and Mathematical Statistics, University of Cambridge, unpublished manuscript, May 1992.
11. G.M. Kelly and S. Mac Lane. Coherence in closed categories. *Journal of Pure and Applied Algebra*, 1:97–140, 1971.
12. J. Lambek and P.J. Scott. *Introduction to higher order categorical logic*, volume 7 of *Cambridge studies in advanced mathematics*. Cambridge University Press, 1987.
13. Saunders Mac Lane. *Categories for the Working Mathematican*, volume 5 of *Graduate Texts in Mathematics*. Springer-Verlag, 1971.
14. Patrick Lincoln and John Mitchell. Operational aspects of linear lambda calculus. In *Proceedings of Symposium on Logic in Computer Science*, pages 235–246, June 1992.
15. Ian Mackie. Lilac: A functional programming language based on linear logic. Master's thesis, Department of Computing, Imperial College, London, September 1991.
16. R.A.G. Seely. Linear logic, *-autonomous categories and cofree algebras. In *Conference on Categories in Computer Science and Logic*, volume 92 of *AMS Contemporary Mathematics*, pages 371–382, June 1989.
17. M.E. Szabo, editor. *The Collected Papers of Gerhard Gentzen*. North-Holland, 1969.
18. Philip Wadler. There's no substitute for linear logic. Draft Paper, December 1991.

A self-interpreter of lambda calculus having a normal form

Alessandro Berarducci *
Università di Pisa, Dipartimento di Matematica
Via Buonarroti 2, 56100 Pisa, Italy.
Corrado Böhm †
Università di Roma, Dipartimento di Scienze dell'Informazione,
Via Salaria 113, 00198 Roma, Italy.

Abstract

We formalize a technique introduced by Böhm and Piperno to solve systems of recursive equations in lambda calculus without the use of the fixed point combinator and using only normal forms. To this aim we introduce the notion of a canonical algebraic term rewriting system, and we show that any such system can be interpreted in the lambda calculus by the Böhm - Piperno technique in such a way that strong normalization is preserved. This allows us to improve some recent results of Mogensen concerning efficient gödelizations $\lceil\ \rceil : \Lambda \to \Lambda$ of lambda calculus. In particular we prove that under a suitable gödelization there exist two lambda terms **E** (self-interpreter) and **R** (reductor), both having a normal form, such that for every (closed or open) lambda term M $\mathbf{E}\lceil M\rceil \to M$ and if M has a normal form N, then $\mathbf{R}\lceil M\rceil \to \lceil N\rceil$.

1 Introduction and summary

Λ is the set of all lambda terms. Λ_0 is the set of all closed lambda terms. The sign "$=$" between lambda terms denotes beta-convertibility (also written "$=_\beta$"), $\to$ is beta-reduction (possibly multistep), and "$\equiv$" is alpha convertibility, namely identity up to renamings of bound variables. $NF \subseteq \Lambda$ is the set of normal (i.e. irreducible) lambda terms, and $SN \subseteq \Lambda$ is the set of all strong normalizing lambda terms (i.e. all the reduction paths are finite).

In the first part of the paper we study the problem of interpreting an equational theory inside lambda calculus (section 2), and of interpreting a term

*Partially supported by MURST Research projects 40% "Modelli della computazione e dei linguaggi di programmazione", and 60% "Specifiche, Concorrenza e Logica computazionale", while the first author was holding a research position at the University of L'Aquila.

†Partially supported by MURST Research projects 40% "Fondamenti dei linguaggi funzionali e logici", and 60% "Progetto Ateneo".

rewriting system (TRS) inside lambda calculus (section 3). In the second part of the paper we apply our results to find a "self-interpreter" and a "reductor" of lambda calculus having a normal form. This improves some results of [1] and [9]. We consider equational theories and TRS's which are not necessarily first order, namely they involve some lambda abstractions and applications besides the usual rules for the formation of first order terms (on a given signature Σ). Interpreting an equational theory corresponds to solving a system of equations inside lambda calculus involving certain unknown lambda terms to be found (represented by the symbols of the given signature Σ). Statman proved that there is no algorithm which, given $F, G \in \Lambda$ decides whether there is $X \in \Lambda$ such that $FX = G$ (cf. [6]). So the general problem of solving equations in Λ is undecidable. In section 2 we define the notion of a *canonical* set (or "system") of equations and we prove (Theorem 2.5) that every canonical set of equations can be interpreted inside lambda calculus. This is essentially a formalization of a technique to solve recursive equations inside lambda calculus due to Böhm and Piperno, namely we prove that the Böhm-Piperno technique captures at least the canonical systems of equations. We are particularly interested in studying those systems of equations which admit a normal form solution (i.e. all the unknowns must be replaced by lambda terms in normal form). We show that if a canonical set of equations $\mathcal{E}$ is *algebraic* (i.e. first order), then the Böhm-Piperno technique yields a normal form solution of $\mathcal{E}$ (Theorem 2.5 part 2). Such canonical-algebraic systems are quite powerful since they can be used to find lambda representations of any partial recursive function (even on non-numerical data structures). Note that a normal form solution excludes the use of the fixed point combinator. Solvability problems of systems of equations have been extensively studied in [5,6,7]. Here we deal with a more restricted class of systems of equations suitable for functional programming. So our approach is more in the spirit of [4]. Our motivation is to represent programs and data-structures inside lambda calculus in such a way that the lambda-representation of every partial recursive function on such data-structures assumes a very simple form and can be automatically derived from an algebraic specification. In section 3 we strengthen the results of section 2 by considering TRS's instead of equational theories. The general problem of which TRS can be intepreted inside lambda calculus was raised in [8]. We show that every canonical TRS can be intepreted and every canonical algebraic TRS can be intepreted in a nice way, namely preserving strong normalization (Theorem 3.4). Starting with section 4 we begin the second part of the paper in which we study "efficient" gödelizations of lambda calculus. Given a lambda term M, let $\lceil M \rceil$ be the Church numeral of the Gödel number of M (with respect to a suitable Gödel numbering). Note that: 1. if $M \not\equiv N$, then $\lceil M \rceil \neq \lceil N \rceil$ (i.e. $\lceil M \rceil$ and $\lceil N \rceil$ are not beta-convertible). This seems to be an obvious requirement for any reasonable gödelization. From the representability of partial recursive functions in the lambda calculus, it follows that there is a lambda term $\mathbf{R}$ ("reductor"), such that: 2. if $M \in \Lambda_0$ has a normal form N, then $\mathbf{R}(\lceil M \rceil) = \lceil N \rceil$ (cf. [2]). Moreover there exists a lambda term $\mathbf{E}$ ("self-interpreter" or "enumerator") such that: 3. $\mathbf{E}(\lceil M \rceil) = M$ for all $M \in \Lambda_0$ (cf. [1]). We can think of $\mathbf{E}$ as the analogue of a universal Turing machine for the lambda calculus: $\mathbf{E}(\lceil M \rceil)x = Mx$. Any reasonable gödelization should satisfy the second fixed point theorem, namely the analogue of the recursion theorem in the theory of recursive functions: 4. $\forall F \in \Lambda\ \exists t \in \Lambda : F\lceil t \rceil = t$ (cf. [1]). We are not aware of any attempt to give an axiomatic definition of the notion "gödelization of lambda calculus", namely what properties should a map $\lceil\ \rceil : \Lambda \to \Lambda$ satisfy to be considered a gödelization? It seems to us that the fact that the range of $\lceil\ \rceil$ is a set of numerals is not to be included among the desired properties, however it is crucial that if M_1 and M_2 are not identical (i.e. not α-convertible), then $\lceil M_1 \rceil$ and $\lceil M_2 \rceil$ are not β-convertible. Mogensen [9] defined an innovative "gödelization" (there called "representation") $\lceil\ \rceil : \Lambda \to \Lambda$ in which $\lceil M \rceil$, far from beeing a numeral, is a lambda term with the same free variables as M. Mogensen's gödelization satisfies the above properties 1, 2, 3, 4 and has several advantages over the conventional gödelizations, for instance property 3, i.e. $\mathbf{E}(\lceil M \rceil) = M$, holds for any term $M \in \Lambda$, not necessarily closed. Mogensen's gödelization commutes both with substitutions of variables and α-conversion: $\lceil M \rceil[x := y] \equiv \lceil M[x := y] \rceil$ and α-convertible terms have α-convertible gödelizations. Even more strikingly, a lambda abstraction "λx" goes unchanged under the gödelization, i.e. $\lceil \lambda x.M \rceil \equiv U(\lambda x.\lceil M \rceil)$ where U is a suitable left-invertible lambda term. We push Mogensen's approach to its extreme consequences by letting the godelization of a bound variable to be the variable itself. This almost paradoxical move, together with an application of Theorem 2.5, will greatly simplify all the theorems concerning gödelizations, like the second fixed point theorem, the existence a reductor $\mathbf{R}$ and of an enumerator (or "self-interpreter") $\mathbf{E}$, etc. In particular we will define $\mathbf{E}$ and $\mathbf{R}$ as lambda terms having a normal form. This is quite surprising because it means that we can solve the necessarily involved recursive definitions needed to define $\mathbf{R}$ and $\mathbf{E}$ without using any fixed point combinator. In conclusion we have:

1. If $M_1 \not\equiv M_2$, then $\lceil M_1 \rceil$ and $\lceil M_2 \rceil$ are not beta-convertible.
2. If $M \in \Lambda$, then $\mathbf{E}(\lceil M \rceil) \twoheadrightarrow M$.
3. If $M \in \Lambda$ has a normal form N, then $\mathbf{R}(\lceil M \rceil) \twoheadrightarrow \lceil N \rceil$.
4. $\mathbf{E}$ and $\mathbf{R}$ have a normal form.
5. If M is strong normalizing, then $\mathbf{E}(\lceil M \rceil)$ and $\mathbf{R}(\lceil M \rceil)$ are strong normalizing.
6. The second fixed point theorem holds for $\lceil\ \rceil$.

It should be remarked that points 2 and 3 hold even when M is an open lambda term. Under Mogensen's approach, points 1, 2 and 6 hold, points 4 and 5 fail, and 3 holds restricted to closed lambda terms. We also consider a variant of our gödelization, closer to Mogensen's one, in which there is a self-interpreter given by $\mathbf{E} = \lambda x.x < \mathbf{K}, \mathbf{S}, \mathbf{C} >$ where $< \mathbf{K}, \mathbf{S}, \mathbf{C} >$ is the Church triple of the well known combinators $\mathbf{K} = \lambda xy.x$, $\mathbf{S} = \lambda xyz.xz(yz)$, $\mathbf{C} = \lambda xyz.xzy$. This is a quite simple solution for a lambda term which plays the role of a universal Turing machine.

2 Solving equations inside lambda calculus

Let Σ be a set of function symbols from a given signature. $\Lambda(\Sigma)$ denotes the set of *extended lambda terms* with symbols from the signature Σ. To be precise $\Lambda(\Sigma)$ can be defined by adding the following clause to the usual clauses for the formation of lambda terms: if $t_1, \ldots, t_n \in \Lambda(\Sigma)$ and $f \in \Sigma$ is an n-ary function symbols, then $f(t_1, \ldots, t_n) \in \Lambda(\Sigma)$. Note that $\mathrm{Ter}(\Sigma) \subseteq \Lambda(\Sigma)$ where $\mathrm{Ter}(\Sigma)$ is the set of first order terms with signature Σ. A *representation* of the signature Σ in Λ is a function

$$\phi : \Sigma \to \Lambda$$

Any such representation ϕ induces a map $\hat{\phi} : \Lambda(\Sigma) \to \Lambda$ in the obvious way, namely $\hat{\phi}(\lambda x.M) \equiv \lambda x.\hat{\phi}(M)$, $\hat{\phi}(MN) \equiv \hat{\phi}(M)\hat{\phi}(N)$ and for $f \in \Sigma$, $\hat{\phi}(f(M_1, \ldots, M_n)) \equiv \phi(f)\hat{\phi}(M_1)\ldots\hat{\phi}(M_n)$, abbreviated as $f^\phi M_1^\phi \ldots M_n^\phi$ (as usual in lambda calculus the omitted parenthesis are associated to the left). So if $\Sigma = \{0, f, g\}$, $(\lambda x.xf(x, g(0, x)))^\phi \equiv \lambda x.x(f^\phi x(g^\phi 0^\phi x))$. Note that $xf(x, g(0, x))$ denotes unambiguosly x applied to $f(x, g(0, x))$. Given a set of equations $\mathcal{E} = \{a_i = b_i | i \in J\}$ between extended lambda terms $a_i, b_i \in \Lambda(\Sigma)$, we say that a representation ϕ *satisfies* (or *solves*) $\mathcal{E}$ if for each equation $a_i = b_i$ in $\mathcal{E}$ we have $\hat{\phi}(a_i) =_\beta \hat{\phi}(b_i)$. If there exists a representation ϕ which satisfies $\mathcal{E}$ we say that $\mathcal{E}$ can be *interpreted* (or *represented* or *solved*) inside lambda calculus. We will only consider representations ϕ with range included in the set of closed lambda terms.

Definition 2.1 Let $\mathcal{E}$ be a set of equations in $\Lambda(\Sigma)$. We say that $\mathcal{E}$ is *canonical* if the function symbols in Σ can be partitioned in two disjoint subsets $\Sigma = \Sigma_0 \cup \Sigma_1$ so that, letting $\Sigma_0 = \{c_1, \ldots, c_r\}$ and $\Sigma_1 = \{f_1, \ldots, f_k\}$, each equation $t = t'$ of $\mathcal{E}$ has the form $f_i(c_j(x_1, \ldots, x_m), y_1, \ldots, y_n) = b_{i,j}$ where $f_i \in \Sigma_1$, $c_j \in \Sigma_0$, $b_{i,j} \in \Lambda(\Sigma)$ is a term depending on i and j, $n, m \geq 0$, the variables $x_1, \ldots, x_m$, $y_1, \ldots, y_n$ are all distinct and the free variables of $b_{i,j}$ are included among $x_1, \ldots, x_m$, $y_1, \ldots, y_n$. We call the elements of Σ_0 *constructors* and those of Σ_1 *programs*. We say that $\mathcal{E}$ is *complete* if for all $f_i \in \Sigma_1$ and $c_j \in \Sigma_0$, $\mathcal{E}$ contains an equation of the form $f_i(c_j(x_1, \ldots, x_m), y_1, \ldots, y_n) = b_{i,j}$.

Note that we allow some lambda abstractions and applications to appear on the right-hand-sides $b_{i,j}$ of a canonical system but not on the left-hand-sides.

Definition 2.2 A system of equations $\mathcal{E}$ is called *algebraic* if for each equation $a = b$ in $\mathcal{E}$, a and b belong to $\mathrm{Ter}(\Sigma)$ (note that for a canonical system we only have $a \in \mathrm{Ter}(\Sigma)$ and $b \in \Lambda(\Sigma)$). So an algebraic system of equations is simply an equational theory in the usual sense of first order logic, and a solution $\phi : \Sigma \to \Lambda$ of an algebraic system is an interpretation of $\mathcal{E}$ inside lambda calculus. A *canonical algebraic* system is a system which is both canonical and algebraic.

Example 2.3 The Ackermann's function can be defined by the following algebraic system (where 0 and s represent the constant zero and the successor function). 1. $Ack(0,y) = s(y)$, 2. $Ack(s(x),0) = Ack(x,s(0))$, 3. $Ack(s(x),s(y)) = Ack(x,Ack(s(x),y))$. The above system is not canonical, but it can be reduced to a canonical (and algebraic) system by enlarging the signature with a new function symbol f as follows. 1. $Ack(0,y) = s(y)$, 2. $Ack(s(x),y) = f(y,x)$, 3. $f(0,x) = Ack(x,s(0))$, 4. $f(s(z),x) = Ack(x,Ack(s(x),z))$. Similarly every partial recursive function can be defined by a canonical algebraic system (it is enough to verify closure under minimalization).

Remark 2.4 We can always enlarge a canonical system of equations by additional equations which force the constructors to by injective. For instance in the case of an n-ary constructor c we can add n program symbols $p_1,\ldots,p_n$ and the (canonical) equations $p_i(c(x_1,\ldots,x_n)) = x_i$. In all the applications we will give the constructors will always be interpreted as injective functions.

Theorem 2.5 *1. Any canonical system of equations $\mathcal{E}$ has a solution $\phi : \Sigma \to \Lambda$ inside lambda calculus.*

2. Any canonical algebraic system $\mathcal{E}$ has a solution $\phi : \Sigma \to \Lambda$ with range included in SN.

3. Moreover we can choose ϕ so that the restriction $\phi|\Sigma_0$ depends only on Σ_0 and not on $\mathcal{E}$, namely there is a fixed representation of the constructors.

The representation of the constructors $\phi|\Sigma_0$ is due to Böhm and Piperno and is defined as follows.

Definition 2.6 Suppose $\Sigma_0 = \{c_1, c_2, \ldots, c_r\}$. For $1 \le j \le r$ we define $\phi(c_j) \equiv \lambda x_1 \ldots x_n e.eU_j^r x_1 \ldots x_n e$ where n is the arity of c_j and $U_j^r \equiv \lambda x_1 \ldots x_r.x_j$. We call this representation the *canonical* representation of the constructors. Note that c_j^ϕ is injective.

It remains to define $\phi|\Sigma_1$, namely the representation of the programs. Without loss of generality we can assume that $\mathcal{E}$ is complete (otherwise adjoin more equations to make it complete).

Definition 2.7 Suppose $\Sigma_1 = \{f_1,\ldots,f_k\}$ and $\Sigma_0 = \{c_1,\ldots,c_r\}$. Consider $k \times r$ lambda terms $t_{i,j}$, $1 \le i \le k$, $1 \le j \le r$ to be defined later. Recall the definition of *Church r-tuple*: $< M_1,\ldots,M_r > \equiv \lambda x.xM_1\ldots M_r$. For $1 \le i \le k$ let

$$t_i \equiv < t_{i,1},\ldots,t_{i,r} >$$

and define:

$$\phi(f_1) \equiv < t_1, t_2, \ldots, t_k >$$

and for $i > 1$

$$\phi(f_i) \equiv < t_i, t_{i+1}, \ldots, t_{i-1} >$$

where $< t_i, t_{i+1}, \ldots, t_{i-1} >$ is the cyclic permutation of $< t_1, t_2, \ldots, t_k >$ beginning with t_i. By abuse of notation when $i = 1$ $< t_i, t_{i+1}, \ldots, t_{i-1} >$ denotes $< t_1, t_2, \ldots, t_k >$. Thus $\phi(f_i)$ is a Church k-tuple of Church r-tuples of lambda terms. The lambda terms $t_{i,j}$ are chosen in the only natural way which makes ϕ a solution of the canonical system of equations $\mathcal{E}$. More precisely consider a canonical equation $f_i(c_j(x_1, \ldots, x_m), y_1, \ldots, y_n) = b_{i,j}$ belonging to $\mathcal{E}$ ($b_{i,j} \in \Lambda(\Sigma)$). After applying ϕ the equation becomes:

$$< t_i, t_{i+1}, \ldots, t_{i-1} > (c_j^{\phi} \vec{x}) \vec{y} = b_{i,j}^{\phi}.$$

By definition of Church tuple, this simplifies to:

$$c_j^{\phi} \vec{x} t_i t_{i+1} \ldots t_{i-1} \vec{y} = b_{i,j}^{\phi}.$$

Recalling the definition of c_j^{ϕ} we have $c_j^{\phi} \vec{x} t_i = t_i U_j^r \vec{x} t_i = t_{i,j} \vec{x} t_i$. Hence the equation becomes:

$$t_{i,j} \vec{x} t_i t_{i+1} \ldots t_{i-1} \vec{y} = b_{i,j}^{\phi}.$$

We can now solve this equation for $t_{i,j}$ by replacing on both sides all the occurrences of $t_1, \ldots, t_k$ by fresh variables $v_1, \ldots, v_k$ and abstracting with respect to these variables. More precisely define:

$$t_{i,j} \equiv \lambda \vec{x} v_i v_{i+1} \ldots v_{i-1} \vec{y}. (b_{i,j}^{\psi})^{\phi}$$

where $\psi : \Sigma_1 \to \Lambda$ is defined by $\psi(f_s) = < v_s, v_{s+1}, \ldots, v_{s-1} >$ $(s = 1, \ldots, k)$. With this definition $t_{i,j} \vec{x} t_i t_{i+1} \ldots t_{i-1} \vec{y} \to b_{i,j}^{\phi}$ and all the equations will be satisfied.

For an illustration of the above definitions with $k = 1$ and $r = 3$ see the computations following Theorem 4.5. We have thus proved that every canonical system has a solution inside lambda calculus. To complete the proof of Theorem 2.5 it remains to prove that if $\mathcal{E}$ is canonical algebraic, then ϕ has range included in SN. A useful tool to prove strong normalization is the notion of perpetual reduction.

Definition 2.8 $\infty(M)$ means that M is not strong normalizing, namely $M \notin SN$. A *perpetual reduction* is a reduction $M \to N$, such that $\infty(M)$ implies $\infty(N)$.

In [1] we find the related notion of *perpetual redex*. A perpetual redex is a lambda term $\Delta = (\lambda x.P)Q$ such that any beta-reduction of the form $C[(\lambda x.P)Q] \to C[P[x := Q]]$ (for an arbitrary context $C[\,]$) is perpetual, namely $\infty(C[(\lambda x.P)Q]) \to \infty(C[P[x := Q]])$. It is known that I-redexes are always perpetual (see [1] Theorem 13.4.12). A theorem of Bergstra and Klop [3] characterizes the perpetual K-redexes (see also [1] Theorem 13.4.15). In particular the result of Bergstra and Klop implies the following:

Proposition 2.9 *Let $\Delta \equiv (\lambda x.P)Q$ where Q is a strong normalizing closed lambda term. Then Δ is a perpetual redex.*

Now we proceed to prove part 2 of Theorem 2.5.

Lemma 2.10 *Let $\mathcal{E}$ be a canonical set of equations and let ϕ and $t_{i,j}$ be as in the proof of Theorem 2.5. Then ϕ has range included in SN iff $t_{i,j} \in SN$ for all i's and j's.*

Proof. First note that for $c \in \Sigma_0$, c^ϕ is always a normal form. On the other hand for $f \in \Sigma_1$, f^ϕ is a Church k-tuple of Church r-tuples of the $t_{i,j}$'s. So if all the $t_{i,j}$'s are in SN, so are all the f^ϕ's (and conversely). QED

Lemma 2.11 *If $\mathcal{E}$ is a complete canonical algebraic system of equations and ϕ is the solution of $\mathcal{E}$ constructed in Theorem 2.5, then the terms $t_{i,j}$ defined in the proof of Theorem 2.5 are strong normalizing and closed.*

Proof. The fact that $t_{i,j}$ is a closed lambda term follows easily from the fact that in a canonical system of equations the free variables of the right-hand-sides are included among the free variables of the corresponding left-hand-sides. Recalling the definition of $t_{i,j}$, in order to prove $t_{i,j} \in SN$ it suffices to prove that $(b^\psi_{i,j})^\phi \in SN$ ($\psi : \Sigma_1 \to \Lambda$ and $\phi : \Sigma \to \Lambda$ are as in the proof of Theorem 2.5). We will prove the stronger claim that $(P^\psi)^\phi \in SN$ for every $P \in \text{Ter}(\Sigma)$. Write $P^{\psi\phi}$ for $(P^\psi)^\phi$. The proof is by induction on the length of P distinguishing the cases in which the outermost function symbol of P is in Σ_0 or in Σ_1. For the induction to work we need to prove a stronger result, namely we prove that: 1) $P^{\psi\phi} \in SN$ and 2) the normal form of $P^{\psi\phi}$ is either of the form $\lambda x.xM_1 \ldots M_u$ or of the form $xM_1 \ldots M_u$ and $P^{\psi\phi}$ can be reduced to its normal form by reducing I-redexes only (so perpetual redexes).

Case 1: P is $c(P_1, \ldots, P_n)$ where $c \in \Sigma_0$. Say $c = c_j$. Then $P^{\psi\phi} \equiv c_j^\phi P_1^{\psi\phi} \ldots P_n^{\psi\phi} \to \lambda e.eU_j^r P_1^{\psi\phi} \ldots P_n^{\psi\phi}$. Note that the displayed reduction contracts only I-redexes. By induction hypothesis all the $P_i^{\psi\phi}$'s belong to SN and satisfy (2). Thus clearly $P^{\psi\phi} \in SN$ and $P^{\psi\phi}$ satisfies (2).

Case 2. P is $f(P_1, \ldots, P_n)$ where $f \in \Sigma_1$. Say $f = f_1$ (the other cases are similar). Then $P^{\psi\phi} \equiv f^\psi P_1^{\psi\phi} \ldots P_n^{\psi\phi} \equiv < v_1, v_2, \ldots, v_k > P_1^{\psi\phi} \ldots P_n^{\psi\phi}$ $\to P_1^{\psi\phi} v_1 v_2 \ldots v_k P_2^{\psi\phi} \ldots P_n^{\psi\phi}$ where the displayed reduction contracts an I-redex. By induction hypothesis $P_1^{\psi\phi}$ belongs to SN and satisfies (2). Thus $P_1^{\psi\phi}$ reduces, by a sequence of I-reductions, either to a term of the form $xM_1 \ldots M_u$ or to a term of the form $\lambda x.xM_1 \ldots M_u$. In the first case $P^{\psi\phi} \to xM_1 \ldots M_u v_1 v_2 \ldots v_k P_2^{\psi\phi} \ldots P_n^{\psi\phi}$ by a sequence of I-reductions. In the second case $P^{\psi\phi} \to v_1 M_1[x := v_1] \ldots M_u[x := v_1] v_2 \ldots v_k P_2^{\psi\phi} \ldots P_n^{\psi\phi}$ by a sequence of I-reductions. So in either case $P^{\psi\phi}$ satisfies (1) and (2). QED

The proof of Theorem 2.5 is now complete.

Remark 2.12 In a "real life" situation the signature of constructors Σ_0 will in general contain the constructors of several data structures (numbers, lists, booleans, etc.). According to Definition 2.6 the lambda representation of a constructor depends not only on the data structure to which the given constructor belongs, but also on the cardinality of the whole set of constructors Σ_0. This might be a disadvantage if we later want to expand the signature of constructors by adjoining a new data structure, because we would have to change the lambda representation of the old data structures as well. This problem can be solved by applying Definition 2.6 to each data structure separately rather then to the whole set of constructors. If we do so it might happen that the constructors of two different data structures are represented by the same lambda term. However this will cause no harm if we assume that the programs specified by the equations $\mathcal{E}$ "respect the types" in the sense that if $\mathcal{E}$ contains two equations of the form $f_i(c_j(\vec{x}),\vec{y})$ and $f_i(c_{j'}(\vec{x}),\vec{y})$, then c_j and $c_{j'}$ both belong to the same data structure (this might be taken as a definition of "data structure"). Note that it is allowed that the codomain of f_i belong to a different data structure. In this setting the proof of Theorem 2.5 goes through with minor changes. For instance if the first argument of the program f is of type "integer" (with constructors $\{s,0\}$), then the lambda representation of f will involve a 2-tuple (since 2 is the cardinality of $\{s,0\}$) rather than a r-tuple (where $r=|\Sigma_0|$).

Remark 2.13 We say that $\mathcal{E}=\mathcal{E}_1\cup\mathcal{E}_2\cup\ldots\cup\mathcal{E}_n$ is *stratified* if $\mathcal{E}_i$ does not use any program symbol mentioned in $\mathcal{E}_{i+1}$ (although $\mathcal{E}_{i+1}$ may use program symbols in $\mathcal{E}_i$). In such a case instead of solving directly $\mathcal{E}$, it is simpler to solve the various $\mathcal{E}_i$'s consecutively. (Of course before solving $\mathcal{E}_{i+1}$ we need to substitute those program symbols of $\mathcal{E}_{i+1}$ which were mentioned in some previous $\mathcal{E}_j$ by the corresponding lambda terms.)

3 Interpreting a term rewriting system inside lambda calculus

In this section we consider the problem of interpreting a *term rewriting system* (TRS) inside lambda calculus. We strengthen the results of the previous section by taking into account reductions and not only equalities. First note that the notion of β-reduction "$\rightarrow$" makes sense also for extended lambda terms $t\in\Lambda(\Sigma)$. So for example $(\lambda x.f(x,y))z\rightarrow f(z,y)$. As usual α-convertible terms are considered identical.

Definition 3.1 Given a set of equations $\mathcal{E}$ on $\Lambda(\Sigma)$ we define a notion of reduction by orienting the equations of $\mathcal{E}$ from left to right. So $\mathcal{E}$ generates a binary relation $\rightarrow_{\mathcal{E}}$ on $\Lambda(\Sigma)$ defined as the least transitive relation containing $\rightarrow$ and the oriented equations of $\mathcal{E}$, and closed under substitutions and contexts. So if $\mathcal{E}=\{f(g(x),y)=x\}$, then $(\lambda x.f(x,y))g(\lambda v.v)\rightarrow_{\mathcal{E}} f(g(\lambda v.v),y)\rightarrow_{\mathcal{E}}\lambda v.v$.

Note that $(\mathcal{E},\rightarrow_{\mathcal{E}})$ can be considered as a generalized TRS in which the terms to be rewritten are in $\Lambda(\Sigma)$ rather than in $\mathrm{Ter}(\Sigma)$. If $\mathcal{E}$ is a set of algebraic equations and we restrict $\rightarrow_{\mathcal{E}}$ to Ter(Σ), we recover the usual notion of a TRS on Ter(Σ).

Definition 3.2 An *interpretation* of $(\mathcal{E}, \rightarrow_{\mathcal{E}})$ in Λ is a representation $\phi : \Sigma \rightarrow \Lambda$ which solves all the equations of $\mathcal{E}$ (in the sense of section 2) and preserves reduction, namely for all $M, N \in \Lambda(\Sigma)$, if $M \rightarrow_{\mathcal{E}} N$, then $M^{\phi} \rightarrow N^{\phi}$.

Lemma 3.3 *Let $M \in Ter(\Sigma_0)$, and $\phi : \Sigma_0 \rightarrow \Lambda$ be the canonical representation of the constructors (as in Definition 2.6). Then M^{ϕ} is strong normalizing.*

Proof. We reason by induction on the length of M. Suppose that $M \equiv c_j(M_1, \ldots, M_n)$ where $c_j \in \Sigma_0$. By inductive hypothesis $M_1^{\phi}, \ldots, M_n^{\phi}$ are strong normalizing. Recall that $c_j^{\phi} \equiv \lambda x_1 \ldots x_n e.eU_j^r x_1 \ldots x_n e$. Hence $M^{\phi} \equiv c_j^{\phi} M_1^{\phi} \ldots M_n^{\phi} \rightarrow \lambda e.eU_j^n M_1^{\phi} \ldots M_n^{\phi} e$. The latter term is strongly normalizing since each $M_i^{\phi} \in SN$. But then also M^{ϕ} is strongly normalizing since the above reduction is an I-reduction. QED

Theorem 3.4 *Let $\mathcal{E}$ be a canonical system and let $\phi : \Sigma \rightarrow \Lambda$ be the solution constructed in Theorem 2.5. Then:*

1. *ϕ is an interpretation of $(\mathcal{E}, \rightarrow_{\mathcal{E}})$ inside lambda calculus (i.e. if $M \rightarrow_{\mathcal{E}} N$, then $M^{\phi} \rightarrow N^{\phi}$).*
2. *Moreover if $\mathcal{E}$ is complete, canonical and algebraic, then for every closed term $M \in Ter(\Sigma)$, if M is strong normalizing relative to $\rightarrow_{\mathcal{E}}$, then M^{ϕ} is strong normalizing relative to $\rightarrow$ (i.e. ϕ preserves strong normalization of closed algebraic terms).*

Proof. To prove part 1 it is enough to follow the proof of Theorem 2.5 replacing everywhere equalities with reductions. To prove part 2 assume that $\mathcal{E}$ is complete, canonical and algebraic and $M \in$ Ter(Σ) is a closed term which is strong normalizing relative to $\rightarrow_{\mathcal{E}}$. Then there is an integer n such that all the $\mathcal{E}$-reduction paths starting from M have length $\leq n$. We prove that M^{ϕ} is strong normalizing (relative to $\rightarrow$) by induction on n.

If $n = 0$, then it easily follows that $M \in$ Ter(Σ_0) (here we use the assumption that M is closed and $\mathcal{E}$ is complete). Hence by Lemma 3.3 $M^{\phi} \in SN$.

If $n > 0$, then M contains at least one occurrence of a program symbol $f \in \Sigma_1$. By considering an innermost program symbol occurring in M we see M must contain an subterm of the form $f_i(c_j(\vec{X}), \vec{Y})$ where $f_i \in \Sigma_1$, $c_j \in \Sigma_0$ and $\vec{X} = X_1, \ldots, X_m$ and $\vec{Y} = Y_1, \ldots, Y_n$ are finite sequences of closed terms in Ter(Σ_0). Since $\mathcal{E}$ is complete, $\mathcal{E}$ contains an equation of the form $f_i(c_j(x_1, \ldots, x_m), y_1, \ldots, y_n) = b_{i,j}$, so the subterm $f_i(c_j(\vec{X}), \vec{Y})$ is an $(\rightarrow_{\mathcal{E}})$-redex. By contracting it we have $M \rightarrow_{\mathcal{E}} N$ where $N \in$ Ter(Σ) since $\mathcal{E}$ is algebraic. We write $\vec{X}^{\phi}$ for the sequence $X_1^{\phi}, \ldots, X_m^{\phi}$ and similarly for $\vec{Y}^{\phi}$. We then have the following reductions (using the notations of the proof of Theorem 2.5):

$$
\begin{aligned}
&(f_i(c_j(\vec{X}),\vec{Y}))^\phi\\
&\equiv < t_i, t_{i+1}, \dots, t_{i-1} > (c_j^\phi \vec{X}^\phi)\vec{Y}^\phi\\
&\to c_j^\phi \vec{X}^\phi t_i t_{i+1} \dots t_{i-1}\vec{Y}^\phi\\
&\to t_i U_j^r \vec{X}^\phi t_i t_{i+1} \dots t_{i-1}\vec{Y}^\phi\\
&\equiv < t_{i,1}, t_{i,2}, \dots, t_{i,r} > U_j^r \vec{X}^\phi t_i t_{i+1} \dots t_{i-1}\vec{Y}^\phi\\
&\to U_j^r t_{i,1} t_{i,2} \dots t_{i,r} \vec{X}^\phi t_i t_{i+1} \dots t_{i-1}\vec{Y}^\phi\\
&\to t_{i,j} \vec{X}^\phi t_i t_{i+1} \dots t_{i-1}\vec{Y}^\phi\\
&\equiv (\lambda \vec{x} v_i v_{i+1} \dots v_{i-1} \vec{y}. b_{i,j}^{\psi\phi})\vec{X}^\phi t_i t_{i+1} \dots t_{i-1}\vec{Y}^\phi\\
&\to b_{i,j}^\phi[\vec{x} := \vec{X}^\phi, \vec{y} := \vec{Y}^\phi]
\end{aligned}
$$

We claim that all the redexes reduced in the above reduction are perpetual. This is clear for the first three reductions since they reduce an I-redex. The last two reductions may reduce some K-redexes, but in any case they can only erase closed terms which are strong normalizing. In fact the terms X_i^ϕ and Y_j^ϕ are strong normalizing by Lemma 3.3 and all the various t_i's and $t_{i,j}$'s are strong normalizing by Lemma 2.11 because the system $\mathcal{E}$ is algebraic. Moreover all such terms are closed because M is closed and ϕ does not introduce free variables. Since N^ϕ is obtained from M^ϕ by the above sequence of reductions (inside a context), we can conclude that there is a perpetual reduction $M^\phi \to N^\phi$. By induction hypothesis N^ϕ is strong normalizing (since N certainly has a lower n). But since $M^\phi \to N^\phi$ is perpetual, also M^ϕ is strong normalizing. QED

Remark 3.5 The assumption $M \in \mathrm{Ter}(\Sigma)$ in Theorem 3.4 is necessary. Consider in fact the signature $\Sigma = \{s\}$ where s is a unary constructor. Take $\mathcal{E}$ to be the empty set of equations and $M \equiv s(\lambda x.xx)(s(\lambda x.xx)) \in \Lambda(\{s\})$. Then M is in normal form with respect to $\to_\mathcal{E}$ (for any $\mathcal{E}$), but M^ϕ is a lambda term without normal form (recall that according to Definition 2.6 $s^\phi \equiv \lambda xe.eU_1^1xe$). We believe that by changing Definition 2.6 it is possible to strengthen Theorem 3.4 by allowing M to be an arbitrary term in $\Lambda(\Sigma)$.

4 The self-interpreter

Let us fix a signature of constructors $\Sigma_0 \supseteq \{\mathrm{Var}, \mathrm{App}, \mathrm{Abs}\}$ (mnemonic for "variable", "application", "abstraction") where Abs, Var are unary, and App is binary.

Definition 4.1 Define a map $\lfloor\ \rfloor : \Lambda \to \Lambda(\Sigma_0)$ as follows:

1. $\lfloor x \rfloor \equiv x$ if x is a variable,
2. $\lfloor MN \rfloor \equiv \mathrm{App}(\lfloor M \rfloor, \lfloor N \rfloor)$,
3. $\lfloor \lambda x.M \rfloor \equiv \mathrm{Abs}(\lambda x.\lfloor M \rfloor[Var(x) := x])$ where $[Var(x) := x]$ is the operation of replacing all the occurrences of $Var(x)$ by x.

The above definition is similar to the one in Mogensen [9] except that Mogensen defines $\lfloor \lambda x.M \rfloor \equiv \mathrm{Abs}(\lambda x.\lfloor M \rfloor)$. The difference between the two approaches is that Mogensen puts Var around each variable, while we put Var only around the free variables (in fact equation 3 says to take off the Var in the case of a bound variable x). So with our definition $\lfloor \lambda x.xy \rfloor \equiv \mathrm{Abs}(\lambda x.\mathrm{App}(x, \mathrm{Var}(y)))$. We abbreviate $\mathrm{Abs}(\lambda x.t)$ by $\lfloor \lambda x \rfloor.t$. So equation 3 reads: $\lfloor \lambda x.M \rfloor \equiv \lfloor \lambda x \rfloor.\lfloor M \rfloor[\mathrm{Var}(x) := x]$.

Definition 4.2 The gödelization $\lceil M \rceil \in \Lambda$ of $M \in \Lambda$, is defined by $\lceil M \rceil = \hat{\phi}(\lfloor M \rfloor)$ where $\phi : \Sigma_0 \to \Lambda$ is given in Definition 2.6. So if the signature of construcotrs Σ_0 contains no other symbols other than Var, App, Abs we have: 1. $\mathrm{Var}^\phi = \lambda xe.eU_1^3 xe$, 2. $\mathrm{App}^\phi = \lambda xye.eU_2^3 xye$, and 3. $\mathrm{Abs}^\phi = \lambda xe.eU_3^3 xe$. (If other constructors are present we must replace the upper index 3 by the cardinality of Σ_0.)

Definition 4.3 Let $\Sigma = \Sigma_0 \cup \{E\}$. Consider the following set $\mathcal{E}_{\mathrm{E}}$ of equations in $\Lambda(\Sigma)$.

1. $E(\mathrm{Var}(x)) = x$,
2. $E(\mathrm{App}(x, y)) = E(x)E(y)$,
3. $E(\mathrm{Abs}(x)) = \lambda w.E(x\mathrm{Var}(w))$.

Lemma 4.4 *Let $\mathcal{E} = \mathcal{E}_{\mathrm{E}}$. Then $E(\lfloor t \rfloor) \to_{\mathcal{E}} t$ for every $t \in \Lambda$.*

Proof. By induction on the length of the term t. Equation 1 takes care of the case when t is a variable. By equation 2 we have $E(\lfloor MN \rfloor) \to_{\mathcal{E}} E(\lfloor M \rfloor)E(\lfloor N \rfloor)$ and we can apply the induction hypothesis. By equation 3, $E(\lfloor \lambda x.M \rfloor) \equiv E(\mathrm{Abs}(\lambda x.\lfloor M \rfloor[Var(x) := x])) \to_{\mathcal{E}} \lambda w.E((\lambda x.\lfloor M \rfloor[Var(x) := x])\mathrm{Var}(w)) \to_{\mathcal{E}} \lambda w.E(\lfloor M \rfloor[Var(x) := x][x := \mathrm{Var}(w)]) \to_{\mathcal{E}} \lambda w.E(\lfloor M \rfloor[Var(x) := \mathrm{Var}(w)]) \equiv \lambda x.E(\lfloor M \rfloor)$ (by α-conversion) and we can apply the induction hypothesis. QED

Note that $\mathcal{E}_{\mathrm{E}}$ is canonical (but not algebraic), with set of constructors Σ_0 and set of programs $\Sigma_1 = \{E\}$.

Theorem 4.5 *There exists a lambda term* $\mathbf{E}$ *(enumerator) such that* $\mathbf{E}(\lceil t \rceil) \to t$ *for every $t \in \Lambda_0$. Moreover if $t \in SN$ (strong normalizing), then* $\mathbf{E}(\lceil t \rceil) \in SN$.

Proof. Let ϕ be given by Theorem 2.5 applied to the set of equations $\mathcal{E} = \mathcal{E}_{\mathrm{E}}$. Let $\mathbf{E} = E^\phi$. By Lemma 4.4 $E(\lfloor t \rfloor) \to_{\mathcal{E}} t$. Thus $(E(\lfloor t \rfloor))^\phi \to t^\phi$, that is $\mathbf{E}(\lceil t \rceil) \to t$ ($t^\phi \equiv t$ since $t \in \Lambda$). An explicit computation of E^ϕ shows that E^ϕ has a normal form. We leave to the reader the verification that if $t \in SN$, then $\mathbf{E}\lceil t \rceil \in SN$, namely the reduction $\mathbf{E}\lceil t \rceil \to t$ is perpetual. Note that we cannot invoke directly Theorem 3.4 since $\mathcal{E}_{\mathrm{E}}$ is not algebraic. QED

To illustrate the method in detail we compute explicitly the lambda term $E^\phi \equiv \mathbf{E}$ in the case $\Sigma_0 = \{\mathrm{Var}, \mathrm{App}, \mathrm{Abs}\}$. Since $\Sigma_1 = \{E\}$ has cardinality

1, and Σ_0 has cardinality 3, according to Theorem 2.5 E^ϕ is a Church 1-tuple whose only element is a Church 3-tuple. Thus we have $E^\phi \equiv \lambda x.xE_0$ where $E_0 \equiv\ < E_{\mathrm{Var}}, E_{\mathrm{App}}, E_{\mathrm{Abs}} >$ for suitable choices of the lambda terms $E_{\mathrm{Var}}, E_{\mathrm{App}}, E_{\mathrm{Abs}}$. Applying ϕ to the equations $\mathcal{E}_{\mathrm{E}}$ we obtain:

1. $E^\phi(\mathrm{Var}^\phi x) = x$,
2. $E^\phi(\mathrm{App}^\phi xy) = E^\phi x(E^\phi y)$,
3. $E^\phi(\mathrm{Abs}^\phi x) = \lambda w.E^\phi(x(\mathrm{Var}^\phi w))$.

Since $E^\phi \equiv < E_0 > \equiv \lambda x.xE_0$ we obtain the equivalent set of equations:

1. $\mathrm{Var}^\phi xE_0 = x$,
2. $\mathrm{App}^\phi xyE_0 = < E_0 > x(< E_0 > y)$,
3. $\mathrm{Abs}^\phi xE_0 = \lambda w. < E_0 > (x(\mathrm{Var}^\phi w))$.

Recalling the definitions of Var^ϕ, App^ϕ, Abs^ϕ this simplifies as follows:

1. $E_0U_1^3xE_0 = x$,
2. $E_0U_2^3xyE_0 = < E_0 > x(< E_0 > y)$,
3. $E_0U_3^3xE_0 = \lambda w. < E_0 > (x(\mathrm{Var}^\phi w))$.

Since $E_0 \equiv < E_{\mathrm{Var}}, E_{\mathrm{App}}, E_{\mathrm{Abs}} >$ we obtain:

1. $E_{\mathrm{Var}}xE_0 = x$,
2. $E_{\mathrm{App}}xyE_0 = < E_0 > x(< E_0 > y)$,
3. $E_{\mathrm{Abs}}xE_0 = \lambda w. < E_0 > (x(\mathrm{Var}^\phi w))$.

We can now solve these equations in a natural way as follows.

1. $E_{\mathrm{Var}} \equiv \lambda xe.x$,
2. $E_{\mathrm{App}} \equiv \lambda xye. < e > x(< e > y) = \lambda xye.xe(ye)$,
3. $E_{\mathrm{Abs}} \equiv \lambda xew. < e > (x(\mathrm{Var}^\phi w)) = \lambda xew.x(\mathrm{Var}^\phi w)e$.

Thus:

$$\mathbf{E} = \lambda x.x < \mathbf{K}, \mathbf{S}, \lambda xew.x(\mathrm{Var}^\phi w)e >$$

where $\mathbf{K}, \mathbf{S}$ are the well known combinators and $\mathrm{Var}^\phi \equiv \lambda xe.eU_1^3xe$. This is a quite simple solution for a self-interpreter. Note in particular that $\mathbf{E}$ has a normal form.

Remark 4.6 If we modify equation 3 of Definition 4.1 by setting $\lfloor \lambda x.M \rfloor \equiv \mathrm{Abs}(\lambda x.\lfloor M \rfloor)$ we get the even simpler solution $\mathbf{E} = \lambda x.x < \mathbf{K}, \mathbf{S}, \mathbf{C} >$ where $\mathbf{C} \equiv \lambda xew.xwe$ is the well known combinator. However this alternative gödelization has the disadvantage that the reductor (see section 6) becomes much more complicated and does not work for closed terms. Moreover the proof of the second fixed point theorem (section 5) for this alternative gödelization would be quite difficult.

5 Second fixed point theorem

Any reasonable gödelization must satisfy the second fixed point theorem, namely for every $F \in \Lambda$ there exists $t \in \Lambda$ such that $F\ulcorner t \urcorner = t$. To prove this fact for our gödelization note that since Abs is a constructor, Abs^ϕ has a left inverse (for instance $\lambda x.xU_2^3$ is a left inverse).

Lemma 5.1 *Let* **Sub** *be a left inverse of* Abs^ϕ. *Then* $\mathbf{Sub}\ulcorner \lambda x.M \urcorner \ulcorner N \urcorner \rightarrow \ulcorner M[x := N] \urcorner$.

Proof. We have: $\mathbf{Sub}\ulcorner \lambda x.M \urcorner \ulcorner N \urcorner \equiv \mathbf{Sub}(\mathrm{Abs}^\phi(\lambda x.\ulcorner M \urcorner [Var^\phi x := x]))\ulcorner N \urcorner \rightarrow (\lambda x.\ulcorner M \urcorner [Var^\phi x := x])\ulcorner N \urcorner \rightarrow \ulcorner M \urcorner [Var^\phi x := x][x := \ulcorner N \urcorner] \equiv \ulcorner M \urcorner [Var^\phi x := \ulcorner N \urcorner] \rightarrow \ulcorner M[x := N] \urcorner$. QED

Theorem 5.2 *For every* $F \in \Lambda$ *there is* $t \in \Lambda$ *such that* $F\ulcorner t \urcorner = t$.

Proof. Let $A \equiv \lambda x.F(\mathbf{Sub}xx)$, and $t \equiv F(\mathbf{Sub}\ulcorner A \urcorner \ulcorner A \urcorner)$. Then we have: $t \equiv F(\mathbf{Sub}\ulcorner \lambda x.F(\mathbf{Sub}xx) \urcorner \ulcorner A \urcorner) \rightarrow F\ulcorner F(\mathbf{Sub}\ulcorner A \urcorner \ulcorner A \urcorner) \urcorner \equiv F\ulcorner t \urcorner$. QED

6 The reductor

In this section we consider a signature of contructors $\Sigma_0 \supseteq$ {Var, App, Abs, Bou, True, False} and the signature of programs $\Sigma_1 = \{H, \mathrm{Test}, R, R'\}$. The symbol Bou (mnemonic for "bounded") plays a role similar to Var and it is used in the definition of R. The program symbol H represents a "single-step head reductor" defined as follows.

Definition 6.1 Let $\mathcal{E}_H$ be the following set of equations.

1. $H(\mathrm{Var}(x), y) = H(\mathrm{Bou}(x), y) = \mathrm{Error}$,
2. $H(\mathrm{App}(u, v), y) = \mathrm{App}(H(u, v), y)$,
3. $H(\mathrm{Abs}(x), y) = xy$
   where "Error" is an arbitrary term.

It follows from equation 3 and Definition 4.1, that $H(\llcorner \lambda x.M \lrcorner, \llcorner N \lrcorner) \rightarrow_\mathcal{E} \llcorner M[x := N] \lrcorner$. Equations 2 and 3 imply: $H(\llcorner (\lambda x.M)Q_1 \ldots Q_k \lrcorner, \llcorner N \lrcorner) \rightarrow_\mathcal{E} \llcorner M[x := Q_1]Q_2 \ldots Q_k N \lrcorner$, namely H performs a single head-reduction. We can now define a reductor R by iterating H is a suitable way. We need a test for the presence of head redexes:

Definition 6.2 Let $\mathcal{E}_T$ be the following set of equations:

1. $\mathrm{Test}(\mathrm{Var}(x)) = \mathrm{Test}(\mathrm{Bou}(x)) = \mathrm{False}$.
2. $\mathrm{Test}(\mathrm{App}(x, y)) = \mathrm{Test}(x)$,
3. $\mathrm{Test}(\mathrm{Abs}(x)) = \mathrm{True}$,

Note that $\text{Test}(\lfloor xM_1 \dots M_k \rfloor) \to_{\mathcal{E}}$ False and $\text{Test}(\lfloor (\lambda x.M)M_1 \dots M_k \rfloor) \to_{\mathcal{E}}$ True, where $\mathcal{E} = \mathcal{E}_T$. We now give the canonical equations defining the reductor R.

Definition 6.3 Let $\mathcal{E}_R$ be the following set of equations.

1. $R(\text{Var}(x)) = \text{Var}(x)$,
2. $R(\text{Bou}(x)) = x$,
3. $R(\text{App}(x, y)) = R'(\text{Test}(x), x, y)$,
4. $R(\text{Abs}(x)) = \text{Abs}(\lambda w.R(x\text{Bou}(w))$,
5. $R'(\text{True}, x, y) = R(H(x, y))$,
6. $R'(\text{False}, x, y) = \text{App}(R(x), R(y))$.

Intuitively equations 3, 5 and 6 mean: $R(App(x, y))$ = if $\text{Test}(x)$, then $R(H(x, y))$, else $App(R(x), R(y))$. The purpose of Bou is to mark those variables which are bounded "externally" (namely the corresponding lambda abstractions are not in the scope of R).

Lemma 6.4 *Let $\mathcal{E} = \mathcal{E}_H \cup \mathcal{E}_T \cup \mathcal{E}_R$. If $M \in \Lambda$ has a normal form N, then $R(\lfloor M \rfloor) \to_{\mathcal{E}} \lfloor N \rfloor$.*

Proof. We must show that R finds the normal form of a term when it exists. Since the normal form of $\lambda x.t$ is λx followed by the normal form of t, R must somehow commute with $\lfloor \lambda x \rfloor$. By equation 4 we have: $R(\lfloor \lambda x.M \rfloor) \equiv R(\text{Abs}(\lambda x.\lfloor M \rfloor[\text{Var}(x) := x])) \to_{\mathcal{E}} \text{Abs}(\lambda w.R((\lambda x.\lfloor M \rfloor[\text{Var}(x) := x])\text{Bou}(w)))$ $\to_{\mathcal{E}} \text{Abs}(\lambda w.R(\lfloor M \rfloor[\text{Var}(x) := \text{Bou}(w)])) \equiv \text{Abs}(\lambda x.R(\lfloor M \rfloor[\text{Var}(x) := \text{Bou}(x)]))$ $\equiv \lfloor \lambda x \rfloor.R(\lfloor M \rfloor[\text{Var}(x) := \text{Bou}(x)])$. So R commutes indeed with $\lfloor \lambda x \rfloor$, but at the same time replaces x with $\text{Bou}(x)$ as a reminder that x has been bounded "externally". This will cause no harm since the protection will be eventually removed by equation 2. (Note that there is no danger that something will be substituted inside $\text{Bou}(x)$ because we never have two nested occurrences of R.) Equations 1 and 2 ensure that the process stops correctly when it has arrived at the level of the variables. Consider now $R(\text{App}(M, N))$. Equation 3 says that we must perform the test "Test" to verify the presence of head-redexes. If there is such an head redex, then equations 3 and 5 ensure that $R(\text{App}(M, N)) \to_{\mathcal{E}} R(H(M, N))$, namely first we contract the head redex by H, and then we search the normal form of the contracted term. If there is no head redex we must instead normalize the subterms coded by M and N. This is ensured by equations 3 and 6 which entail $R(\text{App}(M, N)) \to_{\mathcal{E}} \text{App}(R(M), R(N))$ as desired. QED

Applying Theorem 2.5 and the techniques of Section 3 to the above set of equations we easily obtain:

Theorem 6.5 *There is a lambda term* $\mathbf{R}$ *(reductor) such that* $\mathbf{R}$ *has a normal form and for every $M \in \Lambda_0$, if M has a normal form N, then $\mathbf{R}\lceil M \rceil \to \lceil N \rceil$. Moreover if M is strong normalizing $\mathbf{R}\lceil M \rceil$ is strong normalizing.*

Proof. Let $\mathbf{R} = R^{\phi}$ be the lambda representatio of R given by Theorem 2.5. $\mathbf{R}\lceil M\rceil \to \lceil N\rceil$ follows from $R(\lfloor M\rfloor) \to_{\mathcal{E}} \lfloor N\rfloor$. The fact that $\mathbf{R}$ has a normal form follows by a direct computation of R^{ϕ}. We leave to the reader the verification of the normalization properties. The idea is to reason as in the proof of Theorem 2.5 part 2. Note that we cannot invoke Theorem 2.5 directly since the third equation of the system defining H is not algebraic. QED

Notice that in general $\lceil N\rceil$ is not in normal form (although it is strongly normalizing). If we verify Theorem 6.5 on a computer using a standard lambda reducer, it may happen that $\lceil N\rceil$ be hardly recognizable from its normal form. Such a drawback will be eliminated in a forthcoming paper of Böhm and Piperno.

References

[1] H. P. Barendregt, The Lambda Calculus, Revised edition, North-Holland, Amsterdam 1984.

[2] H. P. Barendregt, Self interpretation in lambda calculus, Journal of functional programming 1 (2), 229 - 239, April 1991.

[3] J. A. Bergstra and J. W. Klop, Strong normalization and perpetual reductions in the lambda calculus, J. Inform. Process. Cybernet. 18 (718), 403 - 417, 1982.

[4] C. Böhm and A. Berarducci, Automatic synthesis of typed Λ-programs on term algebras, Theoretical Computer Science 39 (1985), 135 - 154.

[5] C. Böhm and A. Piperno, Characterizing X-separability and one-side-Invertibility in $\lambda - \beta - \Omega$-calculus, LICS 88, Edimburgh, Computer Soc. of the IEEE, 1988, 91-101.

[6] C. Böhm, A. Piperno and E. Tronci, Solving equations in Lambda-Calculus, in "Logic Colloquium '88, Proceedings of the Colloquium held in Padova, Italy, August 22-31, 1988", North-Holland 1989.

[7] C. Böhm and E. Tronci, About systems of equations in lambda calculus, Information and Computation, vol. 90, 1 (1991) 1 - 32.

[8] N. Deschowitz, J.-P. Jouannaud and J. W. Klop, Open problems in rewriting, in "Proceedings of RTA '91", Springer Lectures Notes in Computer Science, 488, 1991.

[9] Torben Æ. Mogensen, Efficient self interpretation in lambda calculus, to be published in J. of functional programming.

[10] A. Piperno and E. Tronci, Regular systems of equations in λ-calculus, International Journal of Foundations of Computer Science, Vol. 1, No. 3 (1990) 325-339.

An "Ehrenfeucht-Fraïssé Game" for Fixpoint Logic and Stratified Fixpoint Logic

Uwe Bosse

Institut für mathematische Logik der Albert-Ludwigs-Universität Freiburg i. Brsg., Abertstr. 23b, DW-7800 Freiburg.
email: bosse@sun8.ruf.uni-freiburg.de

1 Introduction

In recent years several extensions of first order logic have been investigated in the context of finite-model theory. With respect to computer science (database theory, complexity classes) fixpoint logic (FP) has turned out to be very important. FP is obtained from first order logic (FO) by adding least fixed points of positive formulas. For example, for finite structures which include a total ordering on the domain, fixpoint logic expresses exactly the class of polynomial time computable queries ([Imm86], [Va82]). But until now there is no algebraic characterization of the expressive power of fixpoint logic. In this context infinitary logic $L^{\omega}_{\infty\omega}$ has been applied (for example in [Imm82] or [dR87]); $L^{\omega}_{\infty\omega}$ is a proper extension of fixpoint logic and has a precise characterization in terms of pebble games (cf. [Ba77], [Imm82]). The relationship between $L^{\omega}_{\infty\omega}$ and FP is examined for example in [KV92]. In Sect. 3 we give a game theoretic characterization of fixpoint logic. As in the classical Ehrenfeucht-Fraïssé game each quantifier of a formula ($\exists$, $\forall$, k-ary LFP, k-ary $\neg$LFP) corresponds to a move in the game, hence a precise investigation of the expressive power of fixpoint logic should now be possible.

Stratified fixpoint logic (SFP) is a fragment of fixpoint logic and the logical equivalent of stratified DATALOG programs. (cf. [ChH85], [Ko91], [Co92]). Modifying the rules of our game we obtain in the 4th section an Ehrenfeucht-Fraïssé game for stratified fixpoint logic. In finite-model theory each fixpoint formula is equivalent to one in which the fixpoint construction is applied only once (cf. [Imm86] or [GS86]). This does not hold for SFP. The question arises when a new fixpoint construction (a new intensional variable in the corresponding stratified DATALOG-program) does in fact increase the expressive power. It is answered by a quantifier hierarchy within SFP established in the 5th section. This hierarchy is very similar to a hierarchy in transitive closure logic established in [GrC92] (there an Ehrenfeucht-Fraïssé game for TC is used) and motivated by some results in [Gro92]. In this context we also prove in a direct way that SFP has less expressive power than fixpoint logic.

On ordered structures FP is equivalent to alternate transitive closure logic ATC, introduced by Immerman ([Imm87]). Calò gives in [Cal90] an Ehrenfeucht-Fraïssé style characterization of ATC which has some similarities to the one for FP introduced in this paper.

2 Preliminaries

Fixpoint logic (FP) extends first order logic (FO) by closing FO under the least fixed points of positive operators. We choose the following way of defining FP-formulas of a signature σ, which for simplicity is assumed to be a finite set of predicate and constant symbols. Simultaneously we define $\text{rk}(\varphi)$, the ***rank*** of a FP-formula φ.

- Atomic and negated atomic first order formulas over σ and (positive) formulas of the form $Xt_1 \ldots t_n$ with an n-ary predicate variable X and terms (i.e. variables and constants) $t_1, \ldots, t_n$ are FP-formulas. Their rank is 0.
- If φ and ψ are FP-formulas then so are $(\varphi \wedge \psi)$ and $(\varphi \vee \psi)$. $\text{rk}(\varphi \wedge \psi) = \text{rk}(\varphi \vee \psi) = \max(\text{rk}(\varphi), \text{rk}(\psi))$.
- If φ is a FP-formula then so are $\exists x\varphi$ and $\forall x\varphi$. $\text{rk}(\exists x\varphi) = \text{rk}(\forall x\varphi) = \text{rk}(\varphi) + 1$.
- If $\varphi(x_1, \ldots, x_n, X)$ is a FP-formula with a free n-ary predicate variable X and free variables $x_1, \ldots, x_n$ (short: $\bar{x}$) [1] and $t_1, \ldots, t_n$ are terms then $[\text{LFP}_{\bar{x},X}\varphi(\bar{x}, X)]t_1, \ldots, t_n$ and $[\text{GFP}_{\bar{x},X}\varphi(\bar{x}, X)]t_1, \ldots, t_n$ are FP-formulas as well. ($\bar{x}, X$ need not to be the only free variables in φ. In $[\text{LFP}_{\bar{x},X}\varphi(\bar{x}, X)]\bar{t}$ and $[\text{GFP}_{\bar{x},X}\varphi(\bar{x}, X)]\bar{t}$ the variables $\bar{x}$ and X are bounded, the variables among $t_1, \ldots, t_n$ are free.) $\text{rk}([\text{LFP}_{\bar{x},X}\varphi(\bar{x}, X)]t_1, \ldots, t_n) = \text{rk}([\text{GFP}_{\bar{x},X}\varphi(\bar{x}, X)]t_1, \ldots, t_n) = \text{rk}(\varphi) + 1$.

$[\text{LFP}_{\bar{x},X}\varphi(\bar{x}, X)]$ is interpreted by the least and $[\text{GFP}_{\bar{x},X}\varphi(\bar{x}, X)]$ by the greatest fixed point of the n-ary monotone operator which is defined by $U \mapsto \{\bar{a} \mid \varphi(\bar{a}, U)\}$. Since negation is allowed only in front of atomic first order formulas we have introduced the greatest fixed point GFP. We are now able to negate least fixed points:

$$\neg[\text{LFP}_{\bar{x},X}\varphi(\bar{x}, X)]t_1, \ldots, t_n \longleftrightarrow [\text{GFP}_{\bar{x},X}\neg\varphi(\bar{x}, \neg X)]t_1, \ldots, t_n \ . \quad (1)$$

By $\neg\varphi(\bar{x}, \neg X)$ we mean not the formula itself but an equivalent one, where the negation $\neg$ appears only in front of the atomic subformulas. We notice that the predicate symbol X appears positively in $\neg\varphi(\bar{x}, \neg X)$. An example:
$\neg[\text{LFP}_{x,y,X}(x = y \vee \exists z(Rxz \wedge Xzy))]ab$ is equivalent to
$[\text{GFP}_{x,y,X}(\neg x = y \wedge \forall z(\neg Rxz \vee Xzy))]ab$.

It is known (cf. [Imm86] and [GS86]) that we can dispense with the GFP-quantifier if we restrict our attention to finite models: Each FP-formula is equivalent to one with only one occurence of the LFP- quantifier. Here, we consider infinite structures as well.

3 A Game Theoretic Characterization of FP

The least fixed point LFP(F) and the greatest fixed point GFP(F) of a monotone operator F satisfy the following equations:

$$\text{LFP(F)} = \cap\{X | \text{F}(X) \subset X\} \ ,$$

[1] $\bar{x}$ will always denote a tuple $(x_1, \ldots, x_n)$ or a sequence whose length n will be clear from the context.

$$\mathrm{GFP(F)} = \cup\{X | X \subset \mathrm{F}(X)\} \ .$$

These equations lead to the equivalences:

$$[\mathrm{LFP}_{\bar{x},X}\varphi(\bar{x},X)]\bar{t} \leftrightarrow \forall X(\neg X\bar{t} \rightarrow \exists\bar{x}(\neg X\bar{x} \wedge \varphi(\bar{x},X))) \ , \tag{2}$$
$$[\mathrm{GFP}_{\bar{x},X}\varphi(\bar{x},X)]\bar{t} \leftrightarrow \exists X(X\bar{t} \wedge \forall\bar{x}(X\bar{x} \rightarrow \varphi(\bar{x},X))) \tag{3}$$

which will be the keypoint of our Ehrenfeucht-Fraïssé game being played in the following way:

As in the classical Ehrenfeucht-Fraïssé game the game $\mathcal{G}_n^l(\mathfrak{A},\mathfrak{B})$ $(n,l \geq 1)$ is played by two players I and II on a pair $\mathfrak{A}$ and $\mathfrak{B}$ of structures. The game ends after n moves. There are two types of moves, point moves and fixpoint moves, corresponding to the point quantifiers $\exists,\forall$ and the fixpoint quantifiers LFP and GFP respectively. Each move extends an assignment $\bar{a} \mapsto \bar{b}$, $\bar{U} \mapsto \bar{V}$ with elements $a_1,\ldots,a_q \in A, b_1,\ldots,b_q \in B$ and sets $U_j \subset A^{s_j}, V_j \subset B^{s_j}$ $(1 \leq j \leq r$ and $1 \leq s_j \leq l$ for all $j \leq r)$. After each move player I chooses the kind of the next move to be played. If the game starts with the assignment $\bar{a} \mapsto \bar{b}$, $\bar{U} \mapsto \bar{V}$ we shall write $\mathcal{G}_n^l(\mathfrak{A},\bar{a},\bar{U},\mathfrak{B},\bar{b},\bar{V})$ instead of $\mathcal{G}_n^l(\mathfrak{A},\mathfrak{B})$. In each case (even when $n = 0$) the 'starting assignment' must first be accomplished by $c^A \mapsto c^B$ with any constant symbol $c \in \sigma$. We assume that (after a while or at the beginning of the game) the assignment $\bar{a} \mapsto \bar{b}, \bar{U} \mapsto \bar{V}$ has to be extended. Now the following moves are possible:

$\exists$-move: Player I chooses $a_{q+1} \in A$ and player II an element $b_{q+1} \in B$.
$\forall$-move: Player I chooses $b_{q+1} \in B$ and player II an element $a_{q+1} \in A$.

In each point move the assignment is extended by $a_{q+1} \mapsto b_{q+1}$.

LFP-move: Player I chooses $V_{r+1} \subset B^{s_{r+1}}$ with some $s_{r+1} \leq l$ and some already chosen $(b_{i_1},\ldots,b_{i_{s_{r+1}}}) \notin V_{r+1}$, $(1 \leq i_1,\ldots,i_{s_{r+1}} \leq q)$.
Player II then fixes $U_{r+1} \subset A^{s_{r+1}}$ with $U_{r+1} \neq A^{s_{r+1}}$.
Now player I chooses new elements $(a_{q+1},\ldots,a_{q+s_{r+1}}) \in A^{s_{r+1}} \setminus U_{r+1}$ and II answers with some $(b_{q+1},\ldots,b_{q+s_{r+1}}) \in B^{s_{r+1}} \setminus V_{r+1}$.
GFP-move: Player I chooses $U_{r+1} \subset A^{s_{r+1}}$ with some $s_{r+1} \leq l$ and some $(a_{i_1},\ldots,a_{i_{s_{r+1}}}) \in U_{r+1}$, $(1 \leq i_1,\ldots,i_{s_{r+1}} \leq q)$.
Player II then fixes $V_{r+1} \subset B^{s_{r+1}}$ with $V_{r+1} \neq \emptyset$.
Now player I chooses new elements $(b_{q+1},\ldots,b_{q+s_{r+1}}) \in V_{r+1}$ and II answers with some $(b_{q+1},\ldots,b_{q+s_{r+1}}) \in U_{r+1}$.

In each fixpoint move the assignment is extended by

$$U_{r+1} \mapsto V_{r+1},\ a_i \mapsto b_i \ \ (q < i \leq q + s_{r+1}) \ .$$

After n moves player II has won if the constructed element assignment $\bar{a} \mapsto \bar{b}$ is a partial isomorphism and for the subset assignment $\bar{U} \mapsto \bar{V}$ the following holds: for any $1 \leq j \leq r$ and $1 \leq i_1,\ldots,i_{s_j} \leq q$:

$$(a_{i_1},\ldots,a_{i_{s_j}}) \in U_j \quad \text{implies} \quad (b_{i_1},\ldots,b_{i_{s_j}}) \in V_j \ .$$

We call an assignment with these properties a *posimorphism*.

Remark. The fixpoint moves of the game $\mathcal{G}_n^l(\mathfrak{A},\mathfrak{B})$ seem to be asymmetrical with respect to $\mathfrak{A}$ and $\mathfrak{B}$. We yield a symmetric situation if, e.g. in the LFP-move, we pass over to the complements of the chosen subsets. Then the LFP-move looks like the GFP-move, only $\mathfrak{A}$ and $\mathfrak{B}$ have interchanged. The disadvantage of this symmetric version is that we have to distinguish between subsets fixed in an LFP-move and subsets chosen in a GFP-move and the definition of a posimorphism gets more complicated. On the other hand, in the formulas we have to distinguish between predicate variables which are bounded by an LFP-quantifier and those bounded by a GFP-quantifier. For our purposes the asymmetric version seems to be more appropriate, whereas for an application of the game the symmetric version would be better.

This game characterizes FP:

Theorem 1. *Let $\mathfrak{A}$ and $\mathfrak{B}$ be (finite or infinite) structures of the same finite signature. Equivalent are:*

(a) $\mathfrak{A}$ and $\mathfrak{B}$ satisfy the same sentences of FP .
(b) for all $n, l \geq 1$ player II has a winning strategy in the game $\mathcal{G}_n^l(\mathfrak{A},\mathfrak{B})$.

This theorem follows immediately from the more technical lemma below. But first we introduce a useful notation:
$\mathrm{FP}_n^l := \{\varphi \mid \varphi$ is a FP-formula with $\mathrm{rk}(\varphi) \leq n$ and the arity of the predicate variables in φ is $\leq l\}$.

Lemma 2 ($\mathrm{FP}_n^l \equiv \mathcal{G}_n^l(\mathfrak{A},\mathfrak{B})$). *For all (finite or infinite) structures $\mathfrak{A}$ and $\mathfrak{B}$ of the same finite signature σ, for all $n,l,q,r \geq 1$, $a_1,\ldots,a_q \in A$, $b_1,\ldots,b_q \in B$, $U_j \subset A^{s_j}$, $V_j \subset B^{s_j}$ $(1 \leq j \leq r$, all $s_j \leq l)$ the following are equivalent:*

(a) For all σ-FP-formulas $\varphi(x_1,\ldots,x_q,X_1,\ldots,X_r) \in \mathrm{FP}_n^l$ with free variables $x_1,\ldots,x_q$ and free predicate variables $X_1,\ldots,X_r$ of arity $s_1,\ldots,s_r$ respectively: $\mathfrak{A} \models \varphi(\bar{a},\bar{U})$ implies $\mathfrak{B} \models \varphi(\bar{b},\bar{V})$.
(b) Player II *has a winning strategy in the game $\mathcal{G}_n^l(\mathfrak{A},\bar{a},\bar{U},\mathfrak{B},\bar{b},\bar{V})$* .

Since negations of FP-formulas without free predicate variables are also FP-formulas (cf. (1)), Theorem 1 follows immediately from Lemma 2.

In the following proof we use the notation

$$T_q := \{x_1,\ldots,x_q\} \cup \{c \mid c \text{ is a constant symbol in } \sigma\},$$

and for a σ-structure $\mathfrak{A}$ with fixed elements $a_1,\ldots,a_q \in A$ and $t \in T_q$ we set

$$t^A := \begin{cases} a_i & \text{if } t = x_i \\ c^A & \text{if } t = c \end{cases} .$$

Proof. $(b) \Rightarrow (a)$: Let $n,l,q,r,\bar{a},\bar{b},\bar{U},\bar{V}$ be given as in Lemma 2. We now use an induction on the complexity of $\varphi \in \mathrm{FP}_n^l$.

If φ is of the form $Rt_1,\ldots,t_m$, $t_1 \equiv t_2$, $\neg Rt_1,\ldots,t_m$, $\neg t_1 \equiv t_2$ or $Xt_1,\ldots,t_m$ with m-ary predicate variable X and $t_j \in T_q$, then $\mathrm{rk}(\varphi) = 0$ and (b) say, that

$$\{t^A \mapsto t^B \mid t \in T_q\} \cup \bar{U} \mapsto \bar{V}$$

is a posimorphism; and this is what we need for the implication in (a).

If $\varphi = \varphi_1 \vee \varphi_2$ or $\varphi = \varphi_1 \wedge \varphi_2$ then we have $\mathrm{rk}(\varphi_1) \leq \mathrm{rk}(\varphi)$ and $\mathrm{rk}(\varphi_2) \leq \mathrm{rk}(\varphi)$ and induction hypothesis yields immediately: $\mathfrak{A} \models \varphi(\bar{a}, \bar{U}) \Rightarrow \mathfrak{B} \models \varphi(\bar{b}, \bar{V})$.

If $\varphi = \exists x \psi$ and $\mathfrak{A} \models \varphi(\bar{a}, \bar{U})$ e.g. $\mathfrak{A} \models \psi(\bar{a}, a, \bar{U})$ then we use the strategy of player II in (b) for an $\exists$-move to find a $b \in B$. As $\mathrm{rk}(\psi) \leq n-1$ the induction hypothesis yields $\mathfrak{B} \models \psi(\bar{b}, b, \bar{V})$ and hence $\mathfrak{B} \models \varphi(\bar{b}, \bar{V})$.

$\varphi = \forall x \psi$ is similar. Here we use the $\forall$-move.

If $\varphi = [\mathrm{LFP}_{\bar{x},X}\psi(x_1, \ldots, x_s, X)]\bar{t}$ $(s \leq l)$ and $\mathfrak{A} \models [\mathrm{LFP}_{\bar{x},X}\psi(\bar{x}, X)]\bar{t}^A$ we have to show $\mathfrak{B} \models [\mathrm{LFP}_{\bar{x},X}\psi(\bar{x}, X)]t_1^B, \ldots, t_s^B$.
We now remember the equivalence (2): for any $V \subset B^s \setminus \{(t_1^B, \ldots, t_s^B)\}$ we have to find some $(b_{q+1}, \ldots, b_{q+s}) \in B^s \setminus V$ with $\mathfrak{B} \models \psi(b_{q+1}, \ldots, b_{q+s}, V)$. To do this, we use the winning strategy in (b) of player II who finds the answer $U \subset A^s \setminus \{(t_1^A, \ldots, t_s^A)\}$ as an answer to V in an LFP-move. By (2) and since $\mathfrak{A} \models \varphi$ there is $(a_{q+1}, \ldots, a_{q+s}) \in A^s \setminus U$ with $\mathfrak{A} \models \psi(a_{q+1}, \ldots, a_{q+s}, U)$. Now II completes the LFP-move and finds $(b_{q+1}, \ldots, b_{q+s}) \in B^s \setminus V$ and has a winning strategy for $\mathcal{G}_{n-1}^l(\mathfrak{A}, a_1, \ldots, a_{q+s}, \bar{U}, U, \mathfrak{B}, b_1, \ldots, b_{q+s}, \bar{V}, V)$. $\psi \in \mathrm{FP}_{n-1}^l$ allows us to conclude with the induction hypothesis: $\mathfrak{B} \models \psi(b_{q+1}, \ldots, b_{q+s}, V)$.

$\varphi = [\mathrm{GFP}_{\bar{x},X}\psi(\bar{x}, X)]\bar{t}$ $(s \leq l)$ is similar. Instead of (2) we use (3) and a GFP-move.

$(a) \Rightarrow (b)$: To avoid struggling with subscripts we assume first $l = 1$. We fix a σ-structure $\mathfrak{A}$ and for any $n, q, r \in \omega$, $a_1, \ldots, a_q \in A$ and predicates $U_1, \ldots, U_r \subset A$ we define inductively FP-formulas $\varphi_{\bar{a},\bar{U}}^n(\bar{x}, \bar{X})$ ($\bar{x}, \bar{X}$ will be the only free variables in $\varphi_{\bar{a},\bar{U}}^n$) with the following properties $(\mathrm{i})_n$ - $(\mathrm{iii})_n$:

$(\mathrm{i})_n$ $\mathfrak{A} \models \varphi_{\bar{a},\bar{U}}^n(\bar{a}, \bar{U})$, $\varphi_{\bar{a},\bar{U}}^n \in \mathrm{FP}_n^l$.

$(\mathrm{ii})_n$ If $\mathfrak{B} \models \varphi_{\bar{a},\bar{U}}^n(\bar{b}, \bar{V})$ then player II has a winning strategy in the game $\mathcal{G}_n^l(\mathfrak{A}, \bar{a}, \bar{U}, \mathfrak{B}, \bar{b}, \bar{V})$.

$(\mathrm{iii})_n$ $\Phi_{q,r}^n := \{\varphi_{\bar{a},\bar{U}}^n \mid \mathfrak{A}$ is a σ-structure, $\bar{a} \in A$, $\bar{U} \subset A\}$ is a finite set.

(i) and (ii) and (a) of Lemma 2 then imply (b) of this lemma.

$n = 0$:

$$\varphi_{\bar{a},\bar{U}}^0 := \bigwedge\{\varphi \mid \varphi \text{ is of the form } t_1 \equiv t_2,\ Rt_1, \ldots, t_m \text{ or } X_i t \text{ with } t, \bar{t} \in T_q,\ 1 \leq i \leq r \text{ and } \mathfrak{A} \models \varphi(\bar{a}, \bar{U})\} \ \wedge$$
$$\bigwedge\{\varphi \mid \varphi \text{ is of the form } \neg t_1 \equiv t_2,\ \neg Rt_1, \ldots, t_m \text{ and } \mathfrak{A} \models \varphi(\bar{a}, \bar{U})\}.$$

$(\mathrm{i})_0$ and $(\mathrm{ii})_0$ hold, for $\mathfrak{B} \models \varphi_{\bar{a},\bar{U}}^0(\bar{b}, \bar{V})$ means that $\{t^A \mapsto t^B \mid t \in T_q\} \cup \bar{U} \mapsto \bar{V}$ is a posimorphism. $(\mathrm{iii})_0$ holds because for finite σ there are only finitely many possibilities to form formulas in the given way.

$n \mapsto n+1$:

$\varphi_{\bar{a},\bar{U}}^{n+1}$:= "II can carry out a $\forall$-move" $\wedge$ "II can carry out an $\exists$-move" $\wedge$ "II can carry out a GFP-move" $\wedge$ "II can carry out an LFP-move and then can continue for n more moves."

"II can carry out a $\forall$-move": $\forall x_{q+1} \bigvee_{a \in A} \varphi_{\bar{a}a\bar{U}}^n =: \alpha_{\bar{a}\bar{U}}^{n+1}$.

"II can carry out an $\exists$-move": $\bigwedge_{a \in A} \exists x_{q+1} \varphi_{\bar{a}a\bar{U}}^n =: \varepsilon_{\bar{a}\bar{U}}^{n+1}$.

Because of $(\mathrm{iii})_n$ the above disjunction $\bigvee_{a \in A}$ and conjunction $\bigwedge_{a \in A}$ are finite.

"II can carry out a GFP-move":

$$\bigwedge_{t\in T_q}\ \bigwedge_{\substack{U\subset A,\\ t^A\in U}} \exists X_{r+1}(X_{r+1}t \wedge \forall x_{q+1}(X_{r+1}x_{q+1} \rightarrow \bigvee_{a\in U} \varphi^n_{\bar{a},a,\bar{U},U}))\ . \tag{4}$$

The finiteness of $\Phi^n_{q+1,r+1}$ (induction hypothesis!) implies that the disjunction $\bigvee_{a\in U}$ and the conjunction $\bigwedge_{U\subset A,\ t^A\in U}$ are finite and we can transform (4) with (3) into:

$$\bigwedge_{t\in T_q}\ \bigwedge_{\substack{U\subset A,\\ t^A\in U}} [\mathrm{GFP}_{x_{q+1},X_{r+1}} \bigvee_{a\in U} \varphi^n_{\bar{a},a,\bar{U},U}]t\ =:\ \gamma^{n+1}_{\bar{a},\bar{U}}\ \in \mathrm{FP}^1_{n+1}\ .$$

"II can carry out an LFP-move:"

$$\bigwedge_{t\in T_q} \forall X_{r+1}(\neg X_{r+1}t \rightarrow \bigvee_{\substack{U\subset A\\ t^A\notin U}}\ \bigwedge_{a\in A\setminus U} \exists x_{q+1}(\neg X_{r+1}x_{q+1} \wedge \varphi^n_{\bar{a},a,\bar{U},U})). \tag{5}$$

Here again the disjunction and conjunction are finite. Now we transform the disjunctive normal form $\bigvee_{\dots}\bigwedge_{\dots}$ into a conjunctive one: with suitable (finite) $\mathcal{E}_t \subset \mathcal{P}(A)$ and $\Pi_t \subset {}^{\mathcal{E}_t}A$ we transform (5) into

$$\bigwedge_{t\in T_q} \forall X_{r+1}(\neg X_{r+1}t \rightarrow \bigwedge_{f\in \Pi_t}\ \bigvee_{U\in\mathcal{E}_t} \exists x_{q+1}(\neg X_{r+1}x_{q+1} \wedge \varphi^n_{\bar{a},f(U),\bar{U},U}))\ ,$$

which can easily be seen as equivalent to

$$\bigwedge_{t\in T_q}\ \bigwedge_{f\in\Pi_t} \forall X_{r+1}(\neg X_{r+1}t \rightarrow \exists x_{q+1}(\neg X_{r+1}x_{q+1} \wedge \bigvee_{U\in\mathcal{E}_t} \varphi^n_{\bar{a},f(U),\bar{U},U}))\ ,$$

and with (2) we get:

$$\bigwedge_{t\in T_q}\ \bigwedge_{f\in\Pi_t} [\mathrm{LFP}_{x_{q+1},X_{r+1}} \bigvee_{U\in\mathcal{E}_t} \varphi^n_{\bar{a},f(U),\bar{U},U}]t\ =:\ \lambda^{n+1}_{\bar{a},\bar{U}}\ \in\ \mathrm{FP}^1_{n+1}\ .$$

We now combine: $\varphi^{n+1}_{\bar{a},\bar{U}}\ :=\ \alpha^{n+1}_{\bar{a},\bar{U}}\ \wedge\ \varepsilon^{n+1}_{\bar{a},\bar{U}}\ \wedge\ \gamma^{n+1}_{\bar{a},\bar{U}}\ \wedge\ \lambda^{n+1}_{\bar{a},\bar{U}}$.
Easy combinatorics gives $(\mathrm{iii})_{n+1}$. $(\mathrm{i})_{n+1}$ and $(\mathrm{ii})_{n+1}$ are immediate (use (4) and (5)). This completes the proof of Lemma 2 for $l = 1$. □

The case $l > 1$ is handled similarly. Here, for any $a_1, \dots, a_q \in A$, $U_1 \subset A^{s_1}, \dots, U_r \subset A^{s_r}$ $(s_1, \dots, s_r \leq l)$ we define formulas $\varphi^n_{\bar{a},\bar{U}}(\bar{x}, \bar{X})$ with $(\mathrm{i})_n$ - $(\mathrm{iii})_n$ from above. The modifications in the definition of $\varphi^n_{\dots}$ should be clear.

One important consequence of Lemma 2 and its proof is the following

Corollary 3. *If $\varphi(\bar{x}, \bar{X}) \in \mathrm{FP}^l_n$ is a satisfiable FP-formula with free variables $x_1, \dots, x_q$ and free predicate variables $X_1, \dots, X_r$ of arity $s_1, \dots, s_r$ respectively then we have the equivalence*

$$\varphi \leftrightarrow \bigvee\{\varphi^n_{\bar{a},\bar{U}} | \mathfrak{A} \textit{ is a } \sigma-\textit{structure},\ \bar{a} \in A,\ U_j \subset A^{s_j}\ (j \leq r) \textit{ and } \mathfrak{A} \models \varphi(\bar{a}, \bar{U})\}.$$

4 An Ehrenfeucht-Fraïssé Game for Stratified Fixpoint Logic

Stratified fixpoint logic (SFP) is a fragment of FP with the expressive power of stratified DATALOG (S-DATALOG) – programs which are introduced in [ChH85] (cf. also [Ko91] and [Da87]. In [Ko91] EFP stands for SFP.) In the literature one finds different ways of defining SFP-formulas; most convenient for our purposes is the following: we keep the formation rules for FP-formulas but modify the rules for the LFP and GFP quantifier as follows:

- If $\varphi(\bar{x}, X)$ is an SFP-formula with n-ary predicate variable X and free variables $x_1, \ldots, x_n$ and X is in the scope neither of a $\forall$-quantifier nor of a fixpoint quantifier GFP or LFP then $[\mathrm{LFP}_{\bar{x},X}\varphi(\bar{x}, X)]\bar{t}$ is an SFP-formula.
- If $\varphi(\bar{x}, X)$ is an SFP-formula with n-ary predicate variable X and free variables $x_1, \ldots, x_n$ and X is in the scope neither of an $\exists$-quantifier nor of a fixpoint quantifier GFP or LFP then $[\mathrm{GFP}_{\bar{x},X}\varphi(\bar{x}, X)]\bar{t}$ is an SFP-formula.

It is not difficult to check, that this way of defining SFP-formulas leads to the same logic as the familiar ones. Corresponding to this modifications of the formation rules we have to modify the rules of the game: The kinds of moves remain the same. But according to the fact that in a subformula of an SFP-sentence at most one predicate variable occurs free, at each instance of the game at most one pair of subsets $U \subset A^s, V \subset B^s$ must be fixed. We therefore add the following rule: If the players carry out an LFP-move (GFP-move) and fix subsets $U \subset A^s$ and $V \subset B^s$, these subsets must be discarded as if they were never chosen as soon as a fixpoint move or a $\forall$-move ($\exists$-move) follows. This simplifies the game for player II – SFP is less powerfull than FP (as shown in [Ko91] or [Da87] or at the end of this paper). Instead of the notation $\mathcal{G}_n^l(\mathfrak{A}, \bar{a}, \bar{U}, \mathfrak{B}, \bar{b}, \bar{V})$ we use the notation S-$\mathcal{G}_n^l(\mathfrak{A}, \bar{a}, U, \mathfrak{B}, \bar{b}, V)_{\mathrm{L}}$ (S-$\mathcal{G}_n^l(\mathfrak{A}, \bar{a}, U, \mathfrak{B}, \bar{b}, V)_{\mathrm{G}}$) for the SFP-game starting with the indicated assignment where the subsets U and V are considered to be fixed in an LFP-move (GFP-move). U and V then must be discarded in a following $\forall$-move ($\exists$-move). Now Theorem 1 holds analogously:

Theorem 4. *Let $\mathfrak{A}$ and $\mathfrak{B}$ be (finite or infinite) structures of the same finite signature. Equivalent are:*

(a) $\mathfrak{A}$ and $\mathfrak{B}$ satisfy the same sentences of SFP.
(b) for all $n, l \geq 1$ player II has a winning strategy in the game S-$\mathcal{G}_n^l(\mathfrak{A}, \mathfrak{B})$.

Again this theorem follows from a more technical lemma:

Lemma 5. *For any two structures $\mathfrak{A}$ and $\mathfrak{B}$ of the same finite signature, and $n, l, q \geq 1$, $a_1, \ldots, a_q \in A$, $b_1, \ldots, b_q \in B$, $U \subset A^s$, $V \subset B^s$ $(s \leq l)$ the following are equivalent:*

(a) For all SFP-formulas $\varphi(x_1, \ldots, x_q, X) \in \mathrm{FP}_n^l$ with free variables $x_1, \ldots, x_q$ and free predicate variable X of arity s which appears only in the scope of $\exists$-quantifiers ($\forall$- quantifiers):

$$\mathfrak{A} \models \varphi(\bar{a}, U) \quad \textit{implies} \quad \mathfrak{B} \models \varphi(\bar{b}, V) .$$

(b) Player II *has a winning strategy in the game* $S\text{-}\mathcal{G}_n^l(\mathfrak{A}, \bar{a}, U, \mathfrak{B}, \bar{b}, V)_{\mathrm{L}}$ *(*$S\text{-}\mathcal{G}_n^l(\mathfrak{A}, \bar{a}, U, \mathfrak{B}, \bar{b}, V)_{\mathrm{G}}$*).*

The proof of $(b) \Rightarrow (a)$ remains the same, $(a) \Rightarrow (b)$ follows the same idea: We fix a structure $\mathfrak{A}$ and for any $n, l, q \geq 1$, $s \leq l$, $a_1, \ldots, a_q \in A$, $U \subset A^s$ we define inductively SFP-formulas $\psi^n_{\mathrm{L}\,\bar{a},U}$, $\psi^n_{\mathrm{G}\,\bar{a},U}$ and $\psi^n_{\bar{a}}$ with:

(i) $\mathfrak{A} \models \psi^n_{\mathrm{L}\,\bar{a},U}(\bar{a}, U) \wedge \psi^n_{\mathrm{G}\,\bar{a},U}(\bar{a}, U) \wedge \psi^n_{\bar{a}}(\bar{a})$,
the predicate variable in $\psi^n_{\mathrm{L}\,\bar{a},U}$ is in the scope neither of a fixpoint quantifier nor of a $\forall$-quantifier and the predicate variable in $\psi^n_{\mathrm{G}\,\bar{a},U}$ is in the scope neither of a fixpoint quantifier nor of an $\exists$-quantifier.

(ii) If $\mathfrak{B} \models \psi^n_{\mathrm{Q}\,\bar{a},U}(\bar{b}, V)$ then player II has a winning strategy in the game $\mathrm{S}\text{-}\mathcal{G}_n^l(\mathfrak{A}, \bar{a}, U, \mathfrak{B}, \bar{b}, V)_{\mathrm{Q}}$ ($\mathrm{Q} \in \{\mathrm{L}, \mathrm{G}\}$),
If $\mathfrak{B} \models \psi^n_{\bar{a}}(\bar{b})$ then player II has a winning strategy in the game $\mathrm{S}\text{-}\mathcal{G}_n^l(\mathfrak{A}, \bar{a}, \mathfrak{B}, \bar{b})$.

(iii) $\Psi^n_q := \{\psi^n_{\mathrm{Q}\,\bar{a},U}, \psi^n_{\bar{a}} \mid \mathfrak{A}$ is a σ-structure, $\bar{a} \in A$, $U \subset A$, $\mathrm{Q} \in \{\mathrm{L}, \mathrm{G}\}\}$ is a finite set.

The definition of $\psi^n_{\mathrm{Q}\bar{a},U}$ is immediate. For example we define (cf. proof of Lemma 2)

$$\psi^{n+1}_{\mathrm{L}\,\bar{a},U} := \bigwedge_{a \in A} \exists y \psi^n_{\mathrm{L}\,\bar{a}a,U} \wedge \forall y \bigvee_{a \in A} \psi^n_{\bar{a}a} \wedge \bigwedge_{t \in T_q} \bigwedge_{\substack{U' \subset A, \\ t^A \in U'}} [\mathrm{GFP}_{y,X} \bigvee_{a \in U'} \psi^n_{\mathrm{G}\,\bar{a},a,U'}]t$$

$$\wedge \bigwedge_{t \in T_q} \bigwedge_{f \in \Pi_t} [\mathrm{LFP}_{y,X} \bigvee_{U' \in \mathcal{E}_t} \psi^n_{\mathrm{L}\,\bar{a},f(U'),U'}]t \ .$$

□

5 A Hierarchy in Stratified Fixpoint Logic

As an application of the Ehrenfeucht-Fraïssé game for stratified fixpoint logic we establish a hierarchy theorem for quantifier classes in SFP. This theorem and the method of proving it is very similar to the hierarchy theorem for quantifier classes in infinitary logic in [GrC92]. As a byproduct we get a direct proof that SFP has less expressive power than FP.

Proviso: Throughout this section we assume that the signature σ contains at least two constant symbols c_1 and c_2. And we consider only those structures in which c_1 and c_2 are interpreted as different elements.

5.1 Quantifier Classes in SFP

Definition 6. With every SFP-formula φ we associate its *quantifier structure* $\mathrm{qs}(\varphi) \subset \{\exists, \forall, \mathrm{L}, \mathrm{G}\}^*$ as follows:

- If φ is quantifier-free then $\mathrm{qs}(\varphi) := \{\epsilon\}$ where ϵ is the empty word.
- If $\varphi = \varphi_1 \vee \varphi_2$ or $\varphi = \varphi_1 \wedge \varphi_2$ then $\mathrm{qs}(\varphi) := \mathrm{qs}(\varphi_1) \cup \mathrm{qs}(\varphi_2)$.

- If $\varphi = \exists x\psi$, then $\mathrm{qs}(\varphi) := \{\exists q \mid q \in \mathrm{qs}(\psi)\}$.
- If $\varphi = \forall x\psi$, then $\mathrm{qs}(\varphi) := \{\forall q \mid q \in \mathrm{qs}(\psi)\}$.
- If $\varphi = [\mathrm{LFP}_{\bar{x},X}\psi(\bar{x},X)]\bar{y}$, then $\mathrm{qs}(\varphi) := \{\mathrm{L}q \mid q \in \mathrm{qs}(\psi)\}$.
- If $\varphi = [\mathrm{GFP}_{\bar{x},X}\psi(\bar{x},X)]\bar{y}$, then $\mathrm{qs}(\varphi) := \{\mathrm{G}q \mid q \in \mathrm{qs}(\psi)\}$.

Definition 7. Let $p, q \in \{\exists, \forall, \mathrm{L}, \mathrm{G}\}^*$. We say p *appears in* q and write $p \subset q$ if p is obtained from q by deleting some characters of q.

Definition 8. For $p \in \{\exists, \forall, \mathrm{L}, \mathrm{G}\}^*$ we define the quantifier class

$$\mathrm{SFP}(p) := \{\varphi \mid \varphi \text{ is an SFP-formula, and for all } q \in \mathrm{qs}(\varphi) : q \subset p\}$$

Now it is our aim to distinguish the expressive power of the different quantifier classes. To the quantifier class SFP(p) corresponds the game S-$\mathcal{G}_p^l(\mathfrak{A}, \mathfrak{B})$: here the possibilities of player I are restricted; at the end of the game (the game ends not later than after $|p|$:=length(p) moves) the sequence of the played moves must appear in p in the sense of Definition 7. The following should not be very surprising and, following the idea of Lemma 5, is straightforward to prove:

Theorem 9 ($\mathrm{SFP}(p) \equiv \mathrm{S{-}}\mathcal{G}_p(\mathfrak{A}, \mathfrak{B})$). *For all σ-structures $\mathfrak{A}$ and $\mathfrak{B}$ the following are equivalent:*

(a) For each SFP(p)*-sentence φ holds: $\mathfrak{A} \models \varphi$ implies $\mathfrak{B} \models \varphi$.*
(b) For all $l \geq 1$ player II has a winning strategy for S-$\mathcal{G}_p^l(\mathfrak{A}, \mathfrak{B})$. □

Lemma 10 and 11 show which quantifier classes coincide:

Lemma 10. *If $p, q \in \{\exists, \forall, \mathrm{L}, \mathrm{G}\}^*$ then*

(a) if $\exists \not\subset q$ then $\mathrm{SFP}(p\mathrm{L}q) \equiv \mathrm{SFP}(pq)$, i.e. for all sentences $\varphi \in \mathrm{SFP}(p\mathrm{L}q)$ there is an equivalent one in $\mathrm{SFP}(pq)$,
(b) if $\forall \not\subset q$ then $\mathrm{SFP}(p\mathrm{G}q) \equiv \mathrm{SFP}(pq)$.

One possibility to prove this lemma is to show that whenever player II wins S-$\mathcal{G}_{pq}^l(\mathfrak{A}, \mathfrak{B})$ then he wins S-$\mathcal{G}_{p\mathrm{L}q}^l(\mathfrak{A}, \mathfrak{B})$: Since player I is not allowed to play an $\exists$-move after the LFP-move, player II can answer I's choice $V \subset B^s \setminus \{(b_{i_1}, \ldots, b_{i_s})\}$ by $U := A^s \setminus \{(a_{i_1}, \ldots, a_{i_s})\}$. Now player I is forced to complete the LFP-move with the choice of $(a_{i_1}, \ldots, a_{i_s})$ and II answers with $(b_{i_1}, \ldots, b_{i_s})$. Now U and V are discarded: the LFP-move had no consequences at all. □

Lemma 11. *Let $p, q \in \{\exists, \forall, \mathrm{L}, \mathrm{G}\}^*$. Then:*

(a) If $\exists \subset q$ then $\mathrm{SFP}(p\mathrm{LL}q) \equiv \mathrm{SFP}(p\mathrm{L}\exists q) \equiv \mathrm{SFP}(p\exists\mathrm{L}q) \equiv \mathrm{SFP}(p\mathrm{L}q)$.
(b) If $\forall \subset q$ then $\mathrm{SFP}(p\mathrm{GG}q) \equiv \mathrm{SFP}(p\mathrm{G}\forall q) \equiv \mathrm{SFP}(p\forall\mathrm{G}q) \equiv \mathrm{SFP}(p\mathrm{G}q)$.

Lemma 11 follows from the following more general lemma.

Lemma 12 (transitivity of LFP). *Let $\varphi(\bar{x}, X, S)$ be a* SFP-*formula where X (n-ary) and S (m-ary) are only in the scope of $\exists$-quantifiers. Let $\psi(\bar{y}, Y)$ be another* SFP-*formula, where Y (m-ary) is only in the scope of $\exists$-quantifiers. Then*
$[\mathrm{LFP}_{\bar{x},X}\varphi(\bar{x}, X, [\mathrm{LFP}_{\bar{y},Y}\psi(\bar{y}, Y)]\,\_)]\bar{t}$ *is equivalent to*
$[\mathrm{LFP}_{w,\bar{x},\bar{y},Z}\,\varrho(w, \bar{x}, \bar{y}, Z)](c_1, \bar{t}, \bar{c_1})$
with a new $n+m+1$-ary predicate variable Z and
$\varrho(w, \bar{x}, \bar{y}, Z) := ((w = c_1 \wedge \varphi(\bar{x}, Zc_1\,\_\,\bar{c_1}, Zc_2\bar{c_2}\,\_)) \vee (w = c_2 \wedge \psi(\bar{y}, Zc_2\bar{c_2}\,\_)))$

This lemma is a well known fact about positive induction. A proof can be found in [Mo74, Lemma 1C.2]. Of course this lemma holds analogously if we replace LFP by GFP and $\exists$ by $\forall$.

With the following equivalences we can deduce Lemma 11 from Lemma 12.

$$\exists x\varphi(x) \leftrightarrow [\mathrm{LFP}_{x,X}(\varphi(x) \vee \exists z\, Xz)]c_1$$
$$\forall x\varphi(x) \leftrightarrow [\mathrm{GFP}_{x,X}(\varphi(x) \wedge \forall z\, Xz)]c_1$$

□

Definition 13. We call a string $p \in \{\exists, \forall, \mathrm{L}, \mathrm{G}\}^*$ *reduced* if no reduction in the sense of Lemma 10 or Lemma 11 is possible. We call p *normal* if it is reduced and even the last L (G) is not followed by $\exists$ ($\forall$). For example: L∃LLG∃∀ is reducable to LG∃∀, which is normal.

Notation. Let p, q be reduced strings in $\{\exists, \forall, \mathrm{L}, \mathrm{G}\}^*$. We write $p \leq q$ if $p \subset q'$ where q' is obtained from q by changing some L to $\exists$ and some G to $\forall$.

5.2 The Hierarchy Theorem

Theorem 14 (Hierarchy theorem). *Let $q \in \{\exists, \forall, \mathrm{L}, \mathrm{G}\}^*$ be normal. Then there is a $\varphi_q \in \mathrm{SFP}(q)$, such that φ_q is equivalent to a formula $\psi \in \mathrm{SFP}(p)$ only if $q \leq p$. Thus*

$$\mathrm{SFP}(q) \not\subset \bigcup_{q \not\leq p} \mathrm{SFP}(p)$$

Remark. The claim in Definition 13, that L must not be followed by $\exists$, is necessary because L∃G∀ $\not\leq$ LG∀∃ but SFP(L∃G∀) $\subset$ SFP(L∃G∀∃) $\subset$ SFP(LG∀∃) (Lemma 11).

Proof. For normal q and $k \in \omega$ we construct structures $\mathfrak{A}_k(q)$ and $\mathfrak{B}_k(q)$ and give a formula $\varphi_q(x) \in \mathrm{SFP}(q)$ with $\mathfrak{A}_k(q) \models \varphi_q(c_1)$, $\mathfrak{B}_k(q) \models \neg\varphi_q(c_1)$ for all $k \geq 1$, and player II wins S-$\mathcal{G}_p^l(\mathfrak{A}_k(q), \mathfrak{B}_k(q))$ if $q \not\leq p$ and k sufficiently large ($k \geq l \cdot \mathrm{length}(p)$). Then we are finished.

$\mathfrak{A}_k(q)$ and $\mathfrak{B}_k(q)$ will be trees. They are inspired by the game trees in [ChH82], which are also used in [Ko91] and [Da87]. We use the binary predicate symbol E for the (directed) father-son relation. Moreover, we have some additional binary predicate symbols *Nephew*, *Grandson*, *Brothersgrandson* which are always interpreted in $\mathfrak{A}_k(q)$ and $\mathfrak{B}_k(q)$ as is suggested by their names. For example: *Nephew* $xy \leftrightarrow \exists b(\mathit{Brother}\, xb \wedge Eby)$ or *Brothersgrandson* $xy \leftrightarrow \exists b(\mathit{Brother}\, xb \wedge \mathit{Grandson}\, by)$. Then we use a unary predicate symbol *Black* for some of the leaves. Furthermore in the vocabulary of φ_q unary predicate symbols G_m ($m \leq |q| :=$ length(q)) appear. G_m will apply to an important generation within the trees $\mathfrak{A}_k(q)$ and $\mathfrak{B}_k(q)$. As we shall see later, we do not really need the predicates G_m, but they will simplify the definition of φ_q. Finally there are the constants *root* and *someoneelse*. *root* is always interpreted by the root and *someoneelse* e.g. by its leftmost son.

Now we present the inductive definition of $\mathfrak{A}_k(q)$ and $\mathfrak{B}_k(q)$ and $\varphi_q(x)$. The reader should check (*) from below in each case.

$\mathfrak{A}_k(\exists)$ 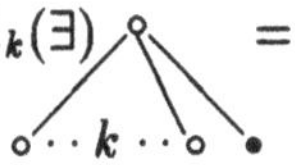

= A root with $k+1$ sons, exactly one of them black.

$\mathfrak{B}_k(\exists)$ 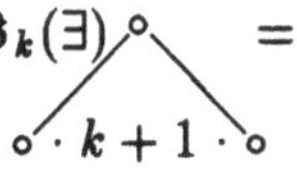

= A root with $k+1$ sons, none of them black.

For $\mathfrak{X} \in \{\mathfrak{A}_k(\exists), \mathfrak{B}_k(\exists)\}$ we define $G_1^{\mathfrak{X}} := \{\text{root of } \mathfrak{X}\}$ and

$$\varphi_\exists(x) := \exists z(Exz \wedge \mathit{Black}\, z) \ .$$

We also define $\varphi_q^{bro}(x)$, a formula which says that φ_q applies to a brother of x:

$$\varphi_\exists^{bro}(x) := \exists z(\mathit{Nephew}\, xz \wedge \mathit{Black}\, z) \ .$$

$$\mathfrak{A}_k(\forall) := \mathfrak{B}_k(\exists), \quad \mathfrak{B}_k(\forall) := \mathfrak{A}_k(\exists),$$

$$\varphi_\forall(x) := (G_1 x \wedge \forall z(Exz \rightarrow \neg \mathit{Black}\, z)) \ ,$$
$$\varphi_\forall^{bro}(x) := (G_1 x \wedge \forall z(\mathit{Nephew}\, xz \rightarrow \neg \mathit{Black}\, z)) \ .$$

Let $q \neq \epsilon$, $m := |q|$

<u>$\mathfrak{A}_k(\exists q)$</u>

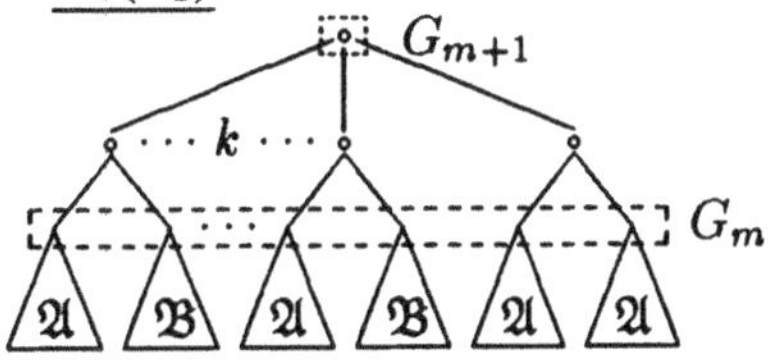

<u>$\mathfrak{B}_k(\exists q)$</u>

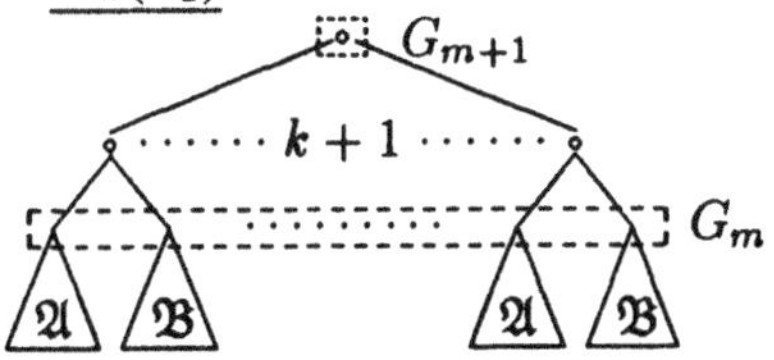

where $\mathfrak{A} := \mathfrak{A}_k(q)$ and $\mathfrak{B} := \mathfrak{B}_k(q)$.

$$\varphi_{\exists q}(x) := (G_{m+1}x \wedge \exists z(\mathit{Grandson}\, xz \wedge \varphi_q(z) \wedge \varphi_q^{bro}(z))) \ ,$$
$$\varphi_{\exists q}^{bro}(x) := (G_{m+1}x \wedge \exists z(\mathit{Brothersgrandson}\, xz \wedge \varphi_q(z) \wedge \varphi_q^{bro}(z))) \ .$$

$\mathfrak{A}_k(\forall q) := \mathfrak{B}_k(\exists\tilde{q})$, $\mathfrak{B}_k(\forall q) := \mathfrak{A}_k(\exists\tilde{q})$, where $\tilde{q}$ is obtained from q by changing $\exists$ to $\forall$ and vice versa and L to G and vice versa.

$$\varphi_{\forall q}(x) := (G_{m+1}x \wedge \forall z(\mathit{Grandson}\, xz \rightarrow (\varphi_q(z) \vee \varphi_q^{bro}(z)))) \ ,$$
$$\varphi_{\forall q}^{bro}(x) := (G_{m+1}x \wedge \forall z(\mathit{Brothersgrandson}\, xz \rightarrow (\varphi_q(z) \vee \varphi_q^{bro}(z)))) \ .$$

Let $\exists \subset q$, $m := |q|$. For $\mathfrak{X} \in \{\mathfrak{A}_k, \mathfrak{B}_k\}$ we define $\mathfrak{X}(\mathrm{L}q) := \mathfrak{X}(\overbrace{\exists \ldots \exists}^{k \text{ times}} q)$, but $G_{m+1}^{\mathfrak{X}} := \{\text{root of } \mathfrak{X}\}$.

$$\varphi_{\mathrm{L}q}(x) \mathrel{\widehat{:=}} (G_{m+1}x \wedge [\mathrm{LFP}_{x,X}(\varphi_q(x) \vee \exists z(Exz \wedge \exists^{\geq 2}s(Ezs \wedge Xs)))]x) \ ,$$
$$\varphi_{\mathrm{L}q}^{bro}(x) \mathrel{\widehat{:=}} (\exists z(Brother\, xz \wedge \varphi_{\mathrm{L}q}(z))) \ .$$

$\widehat{:=}$ indicates that we must use Lemma 11 to get an equivalent formula in SFP(Lq). If $\forall \subset q$ then $\mathfrak{A}_k(\mathrm{G}q) := \mathfrak{B}_k(\mathrm{L}\tilde{q})$, $\mathfrak{B}_k(\mathrm{G}q) := \mathfrak{A}_k(\mathrm{L}\tilde{q})$,

$$\varphi_{\mathrm{G}q}(x) \mathrel{\widehat{:=}} (G_{m+1}x \wedge \neg[\mathrm{LFP}_{x,X}(\varphi_{\tilde{q}}(x) \vee \exists z(Exz \wedge \exists^{\geq 2}s(Ezs \wedge Xs)))]x) \ ,$$
$$\varphi_{\mathrm{G}q}^{bro}(x) \mathrel{\widehat{:=}} (G_{m+1}x \wedge \forall z(Brother\, xz \rightarrow \varphi_{\mathrm{G}q}(z))) \ .$$

Here we use (1) and Lemma 11 to get a formula in SFP(Gq).

Now we see that for all $q \in \{\exists, \forall, \mathrm{L}, \mathrm{G}\}^*$ we have: $\varphi_q \in \mathrm{SFP}(q)$ and $\mathfrak{A}_k(q) \models \varphi_q(root)$ and $\mathfrak{B}_k(q) \not\models \varphi_q(root)$. More precisely:

(*) If $q = q'q''$ with $q', q'' \in \{\exists, \forall, \mathrm{L}, \mathrm{G}\}^*$ and $\mathfrak{X} = \mathfrak{A}_k(q)$ or $\mathfrak{X} = \mathfrak{B}_k(q)$ then $\mathfrak{X} \models \varphi_{q''}(x) \Leftrightarrow x$ is the root of a $\mathfrak{A}_k(q'')$-like tree.

It remains to show that player II has a winning strategy for S-$\mathcal{G}_p^l(\mathfrak{A}_k(q), \mathfrak{B}_k(q))$ if $q \not\leq p$ and k is sufficiently large. We proof it by induction on $|q|$.

$q = \exists$ or $q = \forall$: These cases are similar to $q = \exists q'$ and $q = \forall q'$ discussed below.

$q = \exists q'$: If $q \not\leq p$ then there exist $p'', p' \in \{\exists, \forall, \mathrm{L}, \mathrm{G}\}^*$ with $p = p''p'$, $q' \not\leq p'$ and $p'' \leq \mathrm{G}\forall\exists$. Hence it suffices to show that player II wins S-$\mathcal{G}_{\mathrm{G}\forall\exists p'}^{l}(\mathfrak{A}_k(\exists q'), \mathfrak{B}_k(\exists q'))$ if $k \geq l \cdot |\mathrm{G}\forall\exists p'|$.

$\mathfrak{A} := \mathfrak{A}_k(\exists q')$ $\qquad\qquad$ $\mathfrak{B} := \mathfrak{B}_k(\exists q')$

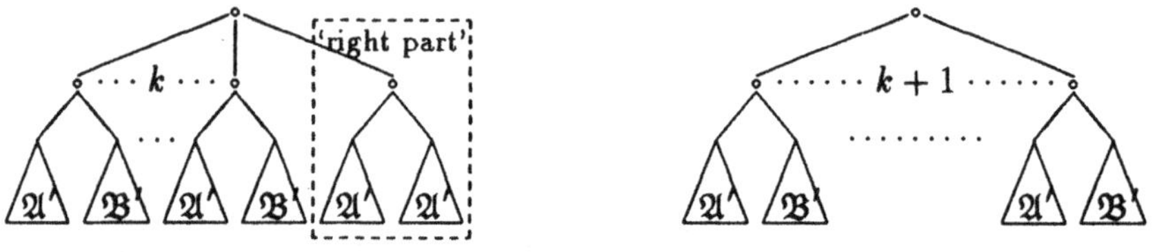

where $\mathfrak{A}' := \mathfrak{A}_k(q')$ and $\mathfrak{B}' := \mathfrak{B}_k(q')$.

We assume that player I has fixed $U \subset A^s$ in the GFP-move. Now player II uses the 'method of possible (pre)images': he defines for $n \leq k$, $(b_1, \ldots, b_n) \in B^n$ the set of possible preimages

$\mathrm{poss}(b_1, \ldots, b_n) := \{(a_1, \ldots, a_n) \mid$ there is a partial isomorphism $\pi : \mathfrak{A} \rightarrow \mathfrak{B}$ with $\{b_1, \ldots, b_n\} \subset \mathrm{Im}(\pi)$, $\mathrm{dom}(\pi) = A \setminus$ 'right part' and $a_i = \pi^{-1}(b_i)$ $(1 \leq i \leq n)\}$

and answers $V := \{(b_1, \ldots, b_s) \mid \mathrm{poss}(b_1, \ldots, b_s) \cap \mathrm{U} \neq \emptyset\}$.

If I chooses now $(b_1, \ldots, b_s) \in V$ then II answers $(a_1, \ldots, a_s) \in U \cap \mathrm{poss}(b_1, \ldots, b_s)$. In the following $\forall$-move II finds as an answer to $b_{s+1} \in B$ an element $a_{s+1} \in A$ such that $(a_1, \ldots, a_{s+1}) \in \mathrm{poss}(b_1, \ldots, b_{s+1})$. For every $i_1, \ldots, i_s \leq s+1$ we have $(a_{i_1}, \ldots, a_{i_s}) \in \mathrm{poss}(\mathrm{b}_{\mathrm{i}_1}, \ldots, \mathrm{b}_{\mathrm{i}_s})$, and hence $(a_{i_1}, \ldots, a_{i_s}) \in U$ implies $(b_{i_1}, \ldots, b_{i_s}) \in V$. The following $\exists$-move (U and V are discarded now) is interesting only if I chooses a_{s+2} from

the 'right part' 

. II then looks for a 'free' part 

in $\mathfrak{B}$

and there he finds $b_{s+2} \in \mathfrak{A}'$ corresponding to a_{s+2}. Now we have the following situation: There are subtrees $\mathfrak{A}' \subset \mathfrak{A}$ and $\mathfrak{B}' \subset \mathfrak{B}$ and an isomorphism $\pi : \mathfrak{A} \setminus \mathfrak{A}' \to \mathfrak{B} \setminus \mathfrak{B}'$ with $a_i \notin \mathfrak{A}'$, $b_i \notin \mathfrak{B}'$ and $b_i = \pi(a_i)$ $(1 \leq i \leq s+2)$. In the following game S-$\mathcal{G}^l_{p'}(\mathfrak{A}_k(\exists q'), \bar{a}, \mathfrak{B}_k(\exists q'), \bar{b})$ II answers the point moves within $\mathfrak{A}'$ or $\mathfrak{B}'$ according to his winning strategy for S-$\mathcal{G}^l_{p'}(\mathfrak{A}_k(q'), \mathfrak{B}_k(q'))$ (induction hypothesis!) and beyond $\mathfrak{A}'$ or $\mathfrak{B}'$ in accordance with π. For the fixpoint moves he uses the method of possible (pre)images: the possible (pre)images of elements beyond $\mathfrak{A}'$ or $\mathfrak{B}'$ are given by π and of elements within $\mathfrak{A}'$ or $\mathfrak{B}'$ by induction hypothesis. So we have proved that II wins S-$\mathcal{G}^l_{\mathrm{G}\forall\exists p'}(\mathfrak{A}_k(\exists q'), \mathfrak{B}_k(\exists q'))$.

$q = \forall q'$: This case is similar to the case $q = \exists q'$ with interchanged structures (but consider the remark after the definition of $\mathcal{G}^l_n(\mathfrak{A}, \mathfrak{B})$ in Sect. 3):

$\mathfrak{A} := \mathfrak{A}_k(\forall q')$ $\qquad$ $\mathfrak{B} := \mathfrak{B}_k(\forall q')$

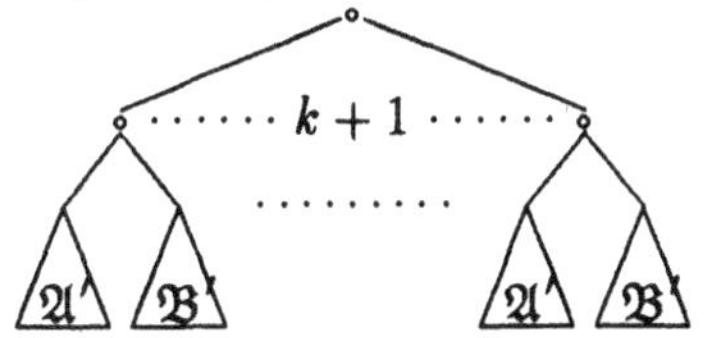

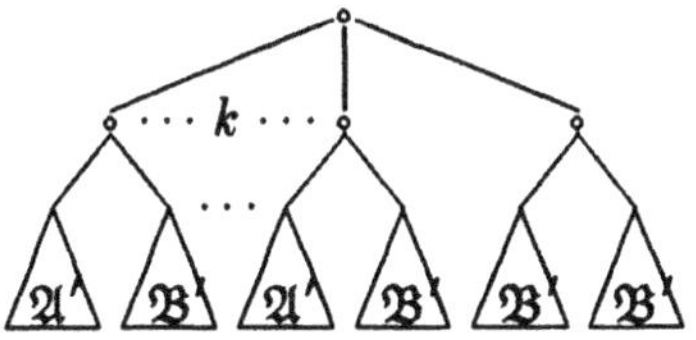

where $\mathfrak{A}' := \mathfrak{A}_k(q')$ and $\mathfrak{B}' := \mathfrak{B}_k(q')$.

If I fixes $V \subset B^s$ in an LFP-move then II's answer is
$U := \{(a_1, \ldots, a_s) \mid \mathrm{poss}(a_1, \ldots, a_s) \subset \mathrm{V}\}$, where poss is defined analogously to the above. This is the mentioned asymmetry between LFP- and GFP-moves.

$q = \mathrm{L}q'$: If $q \not\leq p$ then we find p' with $q' \not\leq p'$ and $p \leq (\mathrm{G}\forall\exists)^s p'$ with a $s \ll k$. (Notice that q is normal and hence $q' = \exists$ or q' does not begin with $\exists$).

$\mathfrak{A} := \mathfrak{A}_k(\mathrm{L}q') = \mathfrak{A}_k(\exists^k q')$ $\qquad$ $\mathfrak{B} := \mathfrak{B}_k(\mathrm{L}q') = \mathfrak{B}_k(\exists^k q')$

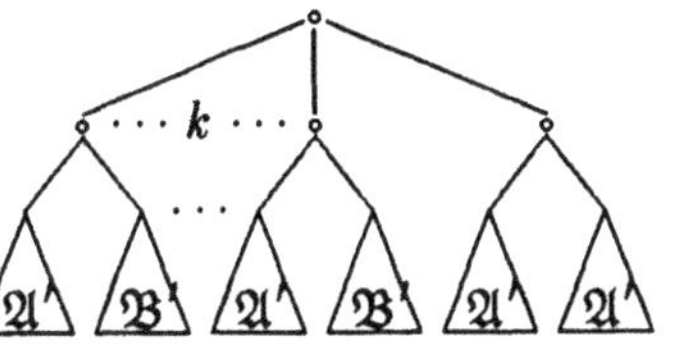

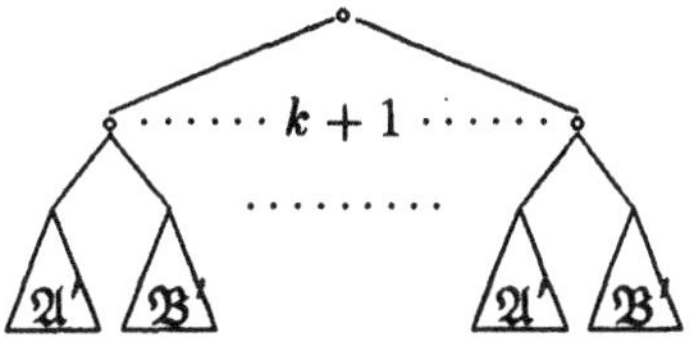

where $\mathfrak{A}' := \mathfrak{A}_k(\exists^{k-1} q')$ and $\mathfrak{B}' := \mathfrak{B}_k(\exists^{k-1} q')$.

We saw above that in this situation the sequence G$\forall\exists$ of moves leads two generations further down the tree. After s sequences of this kind there is a subtree $\mathfrak{A}_k(\exists^{k-s} q')$ in $\mathfrak{A}_k(\mathrm{L}q')$ and a subtree $\mathfrak{B}_k(\exists^{k-s} q')$ in $\mathfrak{B}_k(\mathrm{L}q')$ which do not contain the chosen elements, and beyond these subtrees we have an isomorphism π in accordance with the 'constructed' assignment. Of course we can get a 'free' $\mathfrak{A}_k(q')$ in $\mathfrak{A}$ and a free $\mathfrak{B}_k(q')$ in $\mathfrak{B}$ and extend π to an isomorphism between $\mathfrak{A} \setminus \mathfrak{A}_k(q')$ and $\mathfrak{B} \setminus \mathfrak{B}_k(q')$. From this situation player II can go on as above in the case $q = \exists q'$.

$q = \mathrm{G}q'$: Analogous to $q = \mathrm{L}\tilde{q}'$ with interchanged structures.

Now we have proved Theorem 14. □

How to fix the vocabulary. In order to get rid of the G_m in the definition of $\varphi_q(x)$ we find a formula $\gamma_q(x) \in \mathrm{SFP}(q)$ so that in our trees $\mathfrak{X}$ we have: $\mathfrak{X} \models \gamma_q(x) \leftrightarrow G_{|q|}x$. One possibility is to introduce a new unary predicate *Leaf* for the leaves of the trees, *Imp* for all memberships of important generations: $Imp^{\mathfrak{X}} := \bigcup_{m \geq 1} G_m^{\mathfrak{X}}$ and a new binary predicate *Sponsor* with $Sponsor^{\mathfrak{X}} := \bigcup_{m \geq 1} G_{m+1}^{\mathfrak{X}} \times G_m^{\mathfrak{X}}$. Then we define:

$$\gamma_\exists(x) := \exists z(Exz \wedge Leaf\, z) \ ,$$
$$\gamma_\forall(x) := \forall z(Exz \rightarrow Leaf\, z) \wedge \neg Leaf\, x \ ,$$
$$\gamma_{\exists q}(x) := \exists z(Sponsor\, xz \wedge \gamma_q(z)) \ ,$$
$$\gamma_{\forall q}(x) := \forall z(Sponsor\, xz \rightarrow \gamma_q(z)) \wedge Imp\, x \ ,$$
$$\gamma_{\mathrm{L}q}(x) :\widehat{\equiv} \gamma_{\exists q}(x) \ , \qquad \gamma_{\mathrm{G}q}(x) :\widehat{\equiv} \gamma_{\forall q}(x) \ .$$

In the definition of $\varphi_q(x)$ we now replace $G_{|q|}x$ by $\gamma_q(x)$.

5.3 SFP $\neq$ FP

We consider the structures $\mathfrak{A}_k := \mathfrak{A}_k((\forall\exists)^k)$ and $\mathfrak{B}_k := \mathfrak{B}_k((\forall\exists)^k)$. The inductive construction of $\mathfrak{A}_k$ and $\mathfrak{B}_k$ allows us to find a FP sentence that separates $\mathfrak{A}_k$ from $\mathfrak{B}_k$ (for all $k \geq 1$). We now assume the existence of a $\varphi \in \mathrm{SFP}(p) \cap \mathrm{FP}_n^l$ with $\mathfrak{A}_k \models \varphi$ and $\mathfrak{B}_k \models \neg\varphi$ for all $k \geq 1$. For $k \geq |p| \cdot l$ we have $(\forall\exists)^k \not\leq p$ and therefore a winning strategy of player II in $\mathcal{G}_p^l(\mathfrak{A}_k, \mathfrak{B}_k)$, a contradiction to Theorem 9

References

[Ba77] J. Barwise, *On Moschovakis' Closure Ordinals*, Journal of Symbolic Logic **42** (1977) 292-296.

[Cal90] Ariel Calò, *The Expressive Power of Transitive Closure*, M. Sc. Thesis, Technion-Israel Inst. of Technology, Haifa 1990.

[ChH82] A. Chandra and D. Harel, *Structure and Complexity of Relational Queries*, Journal of Computer and System Sciences **25** (1982) 99-128.

[ChH85] A. Chandra and D. Harel, *Horn Clause Queries and Generalizations*, J. Logic Programming 1 (1985) 1-15.

[Co92] K. J. Compton *Stratified Least Fixpoint Logic*, to appear.

[Da87] E. Dahlhaus, *Skolem Normal Forms Concerning the Least Fixed Point*, in: "Computation Theory and Logic" (Ed. E. Börger), Lecture Notes in Computer Science Nr. 270, Springer 1987, 101-106.

[dR87] M. de Rougemont, *Second Order and Inductive Definability on Finite Structures*, Zeitschrift f. mathem. Logik und Grundlagen der Mathematik **33** (1987) 47-63.

[GrC92] E. Grädel und G. L. McColm, *Hierarchies in Transitive Closure Logic, Stratified Datalog and Infinitary Logic*, to appear in Annals of Pure and Applied Logic.

[Gro92] M. Grohe, *Fixpunktlogiken in der endlichen Modelltheorie*, Diplomarbeit an der Universität in Freiburg, Institut für mathematische Logik.

[GS86] Y. Gurevich and S. Shelah, *Fixed Point Extensions of First Order Logic*, Annals of Pure and Applied Logic **32** (1986), 265-280.

[Imm82] N. Immerman, *Upper and Lower Bounds for First Order Expressibility*, Journal of Computer and System Sciences **25** (1982) 76-98.

[Imm86] N. Immerman, *Relational Queries Computable in Polynomial Time*, Information and Control **68** (1986), 86-104.

[Imm87] N. Immerman, *Languages that capture complexity classes*, SIAM Journal of computation, **16** (1987), 760-778.

[Ko91] Ph. Kolaitis, *The Expressive Power of Stratified Logic Programs*, Information and Computation **90** (1991) 50-66.

[KV92] Ph. Kolaitis and M. Vardi, *Fixpoint logic vs. infinitary logic in Finite-model Theory*, to appear.

[Mo74] Y. Moschovakis, *Elementary Induction on Abstract Structures*, North-Holland, Amsterdam 1974.

[Va82] M. Vardi, *Complexity of Relational Query Languages*, Proc. of 14th ACM Symposium on Theory of Computing (1982) 137-146.

The class of problems that are linearly equivalent to Satisfiability or a uniform method for proving NP-completeness.

Nadia CREIGNOU
Université de Caen, LAIAC
14032 CAEN Cedex FRANCE
email: Nadia.Creignou@univ-caen.fr

I. Introduction.

In this paper we discuss the class of problems that are linearly equivalent to Satisfiability and we present a uniform method for proving NP-completeness.

The study of linear classes is motivated by both theoretical and practical considerations. First, if two problems are mutually reducible each other in linear time then an algorithm for one of them can be transformed into an algorithm of the same complexity for the other. A second reason for studying low-order transformations lies in their simplicity and in their uniformity. Linear transformations are often easier to understand and to communicate than the previously published reductions, this aspect is reinforced by the fact that we present a uniform method based on a key problem, 3-colorability.

We are interested in extending the class of problems linearly equivalent to Satisfiability. As far as we know, A. K. Dewdney [5] was the first author who defined and exhibited linear time reductions between NP-complete problems, in particular he proved that 3-SAT and 3-colorability are mutually linear-time equivalent (on Turing machine). E. Grandjean [8] extended this class in introducing an other computational model, Turing machine with sorting; he proved that SAT$'$ is also linearly equivalent to Satisfiability. SAT$'$ is a problem of satisfiability of a conjunction of propositional clauses that are either usual clauses $(l_1 \vee \cdots \vee l_k)$ or special clauses 1-at-most$(l_1, \cdots, l_k)$ true if and only if at most one of the literals $l_1, \cdots, l_k$ is true, we often use it in our proofs of SAT-easiness by expressing our problems in SAT$'$.

Among natural NP-complete problems we consider the class of combinatorial problems that are roughly speaking characterized by following property: they are decided in a non deterministic way by choosing a fixed number of subsets such that the resulting instance must verify some property checkable in a deterministic linear time. For example the problem of 3-Colouring of a graph belongs to this class. Indeed, let us choose three subsets of the set of vertices. Then we have to check that they cover the set of vertices and we have to check that no edge has its two vertices in the same subset. These checkings can be done in a deterministic linear time. Inversely, the problem of k-Colouring, where k is not fixed, does not belong to this class. These problems correspond in fact to the problems of NSUBLINEAR [8] that do not contain a number in their instance. We show that many problems of this class are linear-time equivalent to Satisfiability. In particular, all but Hamiltonian Circuit and Disjoint Connecting Paths problems of this class that are listed in the well known book by Garey and Johnson are linearly equivalent to Satisfiability. More exactly let us say that a problem is SAT-easy if it is reducible to Satisfiability in linear time and that it is SAT-hard if it can be obtained by a linear-time transformation from Satisfiability. A problem both SAT-easy and SAT-hard is SAT-equivalent. We prove that the following problems are SAT-equivalent: Path With Forbidden Pairs, Kernel of Graph, Partition Into Triangles, Partition Into Hamiltonian Subgraphs, Cubic Subgraph, 3-Dimensional Matching, 3-Exact Cover, Exact Hitting Set, Restricted Perfect Matching, 3-Domatic Number, Partition into Paths of Length Two and 2-Partition into Perfect Matchings. The problem of k−Reduction of Incompletely Specified Automata, for k a fixed integer, is also SAT-equivalent, but the proof is not of the same nature than the others [3]. This shows that many combinatorial problems of widely different nature are linear-time equivalent. Moreover, lot of other problems that are not equivalent to Sat, like Vertex Cover, are however SAT-hard. This yields to the surprising result that Satisfiability, which is the first problem shown to be NP-complete [2] is, in some sense, a "minimal" NP-complete problem.

The reduction functions between different problems that we provide can always be computed by a Turing Machine which operates in linear time with in addition a fixed number of sortings. In particular, such functions have the property that the size of the value is bounded uniformly by a constant multiple of the size of the argument. Such transformations have also the property of being transitive. But in practice this property is not much used. Indeed, it appears that the SAT-easiness is essentially obtained by reduction to SAT′ [8] and that the SAT-hardness is almost always proved from 3-colorability.

This remark yields to the second important aspect of this paper. In studying SAT-hardness we present a new method for proving NP-completeness (with in addition linear-time reductions). The work presented here is similar in spirit to the paper by K. Compton and C. Henson [1]. In both cases a new and uniform method is presented, in one case for proving lower bounds on logical theories, in the other for proving NP-completeness. Moreover, both methods allow to improve the known results, i.e. to strengthen the reductions. Our method uses 3-colorability. All our proofs are based on the same model. They are of a "logical-combinatorial" nature in the sense that they are combinatorial since most of the problems concerned graphs and that they are logical since in fact the vertices and the edges of these graphs can be interpreted as predicates and logical connectives respectively. We are able to define a model of proof which applies to all the problems SAT-equivalent. So, we standardize many proofs of NP-completeness which existed before. So far as we know, no problem has appeared as a unique key problem for proving NP-completeness. M. Garey and D. Johnson just presented 6 problems which are among those that have been used most frequently [6] and T. Schaefer [14] proposed the CNF Satisfiability and its variants as key problems, these problems being used more widely in reductions that any other NP-complete problem. We hope that we will convince the reader that 3-colorability is a kind of canonical problem for proving NP-completeness in linear time.

One can ask: why is 3-colorability best suited for this task? The first reason which appears to us is that this problem is "user-friendly". Indeed, it is easy to represent and to visualize a graph. More seriously, we feel that the usefulness of 3-colorobality for linear time reductions is due to its symmetry. The roles of the three colors are completely symmetrical and the objects of the instance, vertices and edges, are homogenous. It is not the case for SAT for example where the negation often leads to some difficulties, in particular prevent to obtain linear reductions. In some sense, a logical sense, 3-colorability appears as a normalized form of the problem of satisfiability.

II. Preliminaries.

In this section we present the computational model, the definitions and notations used throughout the paper.

A *sort-lin* reduction is a mapping computable in linear time by a Turing Machine using in addition a fixed number of sortings (here, we need always at most three sortings). The Turing Machine used here is a Turing Machine with a fixed number of tapes: an input tape, a fixed number of work tapes and an one-way output tape. Let us notice that these reductions are transitive.

A problem A is *SAT-easy* if there is a sort-lin reduction from A to SAT. A problem B is *SAT-hard* if there is a sort-lin reduction from SAT to B. A problem both SAT-easy and SAT-hard is *SAT-equivalent* (or *linearly equivalent to SAT*).

It does not seem possible to have a more restricted computational model without making linear transformations impossible, in particular a fixed number of sortings is necessary to prove the SAT-hardness except for some problems. However, this computational model is satisfactory since a sort-lin function belongs to DLINEAR. Let us recall that DLINEAR is a robust and powerful notion of linear time given by E. Grandjean [8]. It is the class of functions that are computable within time $O(n/log(n))$, with uniform cost, by a RAM which can read its input in blocks and which only uses integers that are polynomially bounded into the input length n. A sort-lin function belongs to DLINEAR since $DTIME(n) \subset DLINEAR$ and since E. Grandjean has exhibited a sorting that belongs to DLINEAR [9].

Our sort-lin reductions concern mainly graph problems. One implicitly assume that the input of a graph is given by its set of vertices and its list of edges. Inputs of problems are assumed to be presented as strings of symbols from a fixed finite alphabet. The size of the input is then the length of the string.

III. The class of problems that are linearly equivalent to Satisfiability.

Listed in the diagram below are the problems presently known to be linearly equivalent to Satisfiability (SAT) under the computational model described in section II (except for Disjoint Connecting Paths and Hamiltonian Circuit which are only SAT-hard). The diagram also summarizes the linear transformations between various pairs of them. Most of these problems are described in the well known book by Garey and Johnson [6] (except for Exact Hitting Set which has been defined by R. M. Karp [11]) . Therefore, we only give the formal definitions of two of them which are either a restricted version of a more general problem or a new NP-complete problem.

SAT'

Instance: Set U of variables, a conjunction ϕ of usual clauses $(l_1 \vee \cdots \vee l_k)$ and special clauses 1-at-most$(l_1, \cdots, l_k)$ and 1-exactly$(l_1, \cdots, l_k)$ (these special clauses are true if and only if respectively at most one and exactly one of the literals $l_1, \cdots, l_k$ is true).

Question: Is there a satisfying truth assignment for ϕ?

3-Domatic Number

Instance: Graph $G = (V, E)$.

Question: Can V be partitioned into 3 disjoint sets V_1, V_2 and V_3 such that each V_i is a dominating set for G?, i.e., for each $u \in V \setminus V_i$ there exists $v \in V_i$ for which $\{u, v\} \in E$?

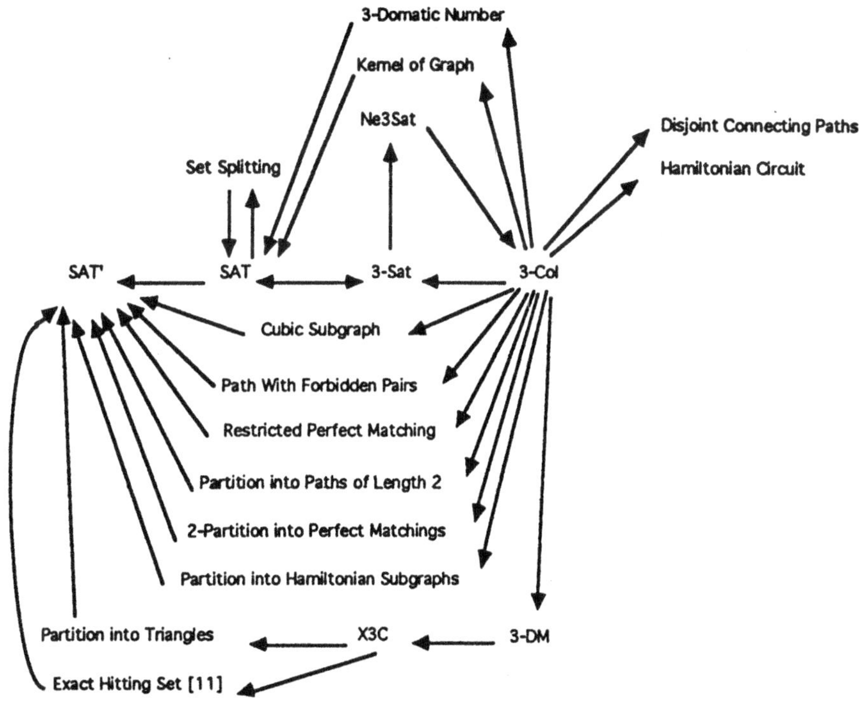

Diagram of linear reductions

The linear equivalence of Set Splitting and SAT (respectively 3-Sat and 3-col) has been shown by A. Dewdney [5] and the linear transformations between SAT, SAT′, and 3-SAT have been proved by E. Grandjean [8].

IV. Model of proof.

One of our goals is to define an effective and uniform method for proving NP-completeness. In this section, we present the methods that are used on the one hand for proving the SAT-easiness and on the other hand for proving the SAT-hardness (thus also the NP-completeness) of the quoted problems. In order to show that these problems are SAT-equivalent we mainly use the problems SAT′ and 3-colorability. Our results can then be deduced by transitivity of sort-lin functions.

The SAT-easiness of the considered problems is obtained by reduction to SAT or SAT′. The common method for finding a reduction to SAT is to give a logical encoding of the problem. Let us give an example and let us recall how one can prove that 3-colorability is SAT-easy . Suppose we are given an input for 3-colorability that is a graph $G = (V, E)$. Let us remind the reader that 3-colorability belongs to the class that we have described in the introduction. Indeed, it can be decided in a non deterministic way by choosing three subsets V_1, V_2, V_3 of V and by verifying firstly that these three subsets cover V, secondly that two vertices joined by an edge of E are not in the same subset. Let us point out that this informal characterization is always useful for proving SAT-easiness.

Let us now consider the following set of propositional variables:

$$U = \left\{p_x^1, p_x^2, p_x^3 \ : \ x \in V\right\}.$$

Intuitively, p_x^i will be true if and only if $x \in V_i$.

The formula F as input for SAT corresponding to the graph G is the conjunction of the two following formulae:

$$F_1 \equiv \bigwedge_{x \in V} \bigvee_{i=1}^{3} p_x^i$$

"V_1, V_2 and V_3 cover V."

$$F_2 \equiv \bigwedge_{\{x,y\} \in E} \bigwedge_{i=1}^{3} \neg(p_x^i \wedge p_y^i)$$

"for all edge $\{x, y\}$ of E, x and y are not in the same subset."

It is easy to see that F is satisfiable if and only if G is 3-colorable using the previous characterization.

In order to complete the proof of the SAT-easiness of 3-colorability it now suffices to verify that the described mapping is a sort-lin function: F_1 is obtained by a copy of V, in fact each x of V is rewrited three times as p_x^0, p_x^1, p_x^2, F_2 is obtained by a copy of E, in fact each edge $\{x, y\}$ of E is rewrited three times as $\neg(p_x^i \wedge p_y^i)$ for $i = 0, 1, 2$. Thus, a Turing machine can construct F in time $O(|V| + |E|)$.

All the proofs of SAT-easiness obey to this general scheme.

The SAT-hardness of the quoted problems is obtained in linear time from 3-colorability using "logical-combinatorial" proofs which are very uniform. In fact, most of our proofs are based on the same model and we describe below the method which we will use. Our examples concern mainly graph problems. The proofs are of a "logical-combinatorial" nature in the sense that the vertices and the edges of these graphs can be interpreted respectively as predicates and as logical connectives "$\vee$" , or "$\rightarrow$", or "$\leftrightarrow$".

Let $G = (V, E)$ be a graph given as input for 3-colorability. To prove the SAT-hardness of, say Cubic Subgraph or PHS (Partition into Hamiltonian Subgraphs), we need to construct in linear time a graph $G_1 = (V_1,\ E_1)$ such that:

for Cubic Subgraph, G_1 has a cubic subgraph if and only if G is 3-colorable;

for PHS, G_1 can be partitioned into cycles which contain at least 3 vertices if and only if G is 3-colorable.

We construct G_1 in the following manner. First, we number the occurrences of vertices of V in E. For each occurence x_i of a vertex of V in an edge of E (where x_i denotes the ith occurence of x in E) we define a *vertex-graph* G_{x_i}. For each edge $\{x_i, y_j\}$ we define an *edge-graph* $G_{x_i,\ y_j}$. Moreover we define a *control structure*, made up of a set of vertices and edges. All these graphs have a size fixed for all the inputs and then indepedent of G. The graph G_1 is then obtained by joining all these graphs with the control structure by appropriate edges whose the number is a linear function in the size of E.

Of course, the proposed subgraphs are defined according to the problem in consideration. For example, for Cubic Subgraph, the graphs will be constructed so that most vertices have degree 3. So, the problem of deciding which of these vertices and which edges belong to a possible cubic subgraph reduces to making a boolean decision for the vertices only. However, the different constructions of the vertex-graphs, edge-graphs and control structure have common characteristics and play a similar role in the different problems.

The hard part of the proof is then to use G_1 to define a 3-colouring for the vertices of G. Therefore, each vertex-graph G_{x_i} always contains the three vertices $c_0(x_i)$, $c_1(x_i)$, $c_2(x_i)$, which correspond intuitively to the predicates "the colour of x_i is 0, 1, 2". The control structure and the edges connecting it to the vertex-graphs are intended to define effectively a colouring of the occurrences of the vertices from V. The edges between vertex-graphs corresponding to occurrences of the same vertex insure that this 3-colouring is coherent, that is to say that all the occurrences of the vertex are coloured in the same way. Similarly, each edge-graph $G_{x_i,\ y_j}$ always contains the three vertices $c_0(x_i,\ y_j)$, $c_1(x_i,\ y_j)$, $c_2(x_i,\ y_j)$, which correspond intuitively to the predicates "at least one of the two vertices x_i and y_j has colour 0, 1, 2" respectively. The edges connecting $G_{x_i,\ y_j}$ with the vertex-graphs G_{x_i} and G_{y_j} will prevent the two adjacent vertices x and y from having the same colour.

We have to prove that such a reduction can be performed in linear time using the computational model that we have defined in section II.

Let us point out that the construction of all the subgraphs that we described (vertex-graphs, edge-graphs and control structure) can be performed in linear time by a Turing machine since each of them involve a fixed number of vertices. Moreover, the construction of the edges joining these subgraphs together can also be computed in linear time by a Turing machine since their number is a linear function in the size of E.

A technical problem that we encounter with Turing machine is the numbering of the occurrences in E of each of V. Indeed, this task cannot be done in linear time by a Turing machine. So, in order to overcome this difficulty we have introduced in addition a fixed number of sortings. Let us describe an algorithm which operates in linear time on Turing machine using in addition a fixed number of sortings and which from a list of couples (which in fact represents the set of the edges of a graph) construct the same list (perhaps in an other order) with a number associated to each occurence of variable (if v appears in two couples then v will be numbered v_i and v_j with $i \neq j$):

Instance: $L = (u_1, u'_1), (u_2, u'_2), \cdots, (u_p, u'_p)$ a list of edges

1. Erase the repeated couples (if $(u_i, u'_i) = (u_j, u'_j)$ for $i \neq j$ then erase one of these two couples). For that:
    1.1. Sort the couples in lexical order.
    1.2. Erase the repetitions (at this step, the repeated couples are consecutive).
2. Number the variables appearing as the first coordinate in a couple. For that:
    2.1. Sort the couples according to their first coordinate.
    2.2. Number the occurrences of each variable v which appears as first coordinate (these occurrences are grouped by 2.1). The ith occurence of v is denoted v_i.
    2.3. In the same time, on the second work tape, store the ordered list of distinct variables with the last number which has been used in 2.2 for each of them (if v appears m times as the first coordinate of a couple of L then store the couple (v, m)).
3. Number the variables appearing as the second coordinate in the same way using the list constructed in 2.3.

Let us examine the complexity of this algorithm. Let n be the size of the instance (considered as a word in a fixed alphabet). The number of couples p verifies $p = O(n/log\ n)$ since we have first erased all the repetitions. The additional length of the list that we have constructed (compared to the length of the instance) is owed to the length of the subscripts added to the variables. But, each added subscript is less or equal to $p = O(n/log\ n)$, thus its length is $O(log\ n)$. Moreover, the number of occurrences of variables is less or equal to $2p = O(n/log\ n)$. Therefore, the total length of the subscripts is: $O(n/log\ n) * O(log\ n)$. Thus, the size of the new list (also considered as a word in a fixed alphabet) is $O(n)$. Moreover, three sortings are necessary and the time required for the other computations is $O(n)$. Therefore, the construction that we propose can be performed in linear time by a Turing machine with an input tape, two work tapes and an one-way output tape, with in addition three sortings.

Without any change this model of proof is applied informally to reduce 3-colorability to the problems Cubic Subgraph, Partition Into Hamiltonian Subgraphs, Restricted Perfect Matching, Partition into Paths of Length 2 and Disjoint connecting paths. Moreover it gives a linear reduction from 3-colorability to Hamiltonian Circuit, although this is not proved here (a sort-lin reduction from Sat to Hamiltonian circuit is presented in [10]). For 3-Dimensionnal matching the idea is the same but the graphs are replaced by *vertex-triplets, edge-triplets* and *control triplets* since the input for 3-DM is a ternary relation.

The reductions from 3-colorability to the problems Path With Forbidden Pairs, Kernel of Graph, 3-Domatic Number and 2-Partition into Perfect Matchings are analogous but easier. Indeed, it is sufficient to introduce control vertices and vertex-graphs for each vertex of G independently from its number of occurrences in E.

We do not give here all the proofs (see [4]), we just give an example of a reduction using the simplified model of proof and an example of a reduction using exactly our model of proof.

VI. Linear transformations using a simplified model of proof.

The following problems are linearly equivalent to Satisfiability: Path with Forbidden pairs, Kernel of Graph, 3-Domatic Number and 2-Partition into Perfect Matchings. The reductions from 3-colorability to these problems are easier than the reductions from 3-colorability to the other problems that we will study in the next section. Indeed, let $G = (V, E)$ be a graph given as input for 3-colorability, it suffices to construct a vertex-graph for each vertex of G independently of its number of occurrences. For example, let us prove that 3-Domatic Number is SAT-equivalent.

Theorem 1. *3-Domatic Number is linearly equivalent to Satisfiability.*

Proof. First, we prove that 3-Domatic Number is SAT-easy by reducing it to SAT in linear time. Let $G = (V, E)$ be a graph given as input for 3-domatic number. We must construct a formula F which is satisfiable if and only if V can be partitioned in three subsets V_1, V_2 and V_3 such that each of them is a dominating set for G. The set of propositional variables of the corresponding input for SAT is:

$$U = \left\{p_x^1,\ p_x^2,\ p_x^3\ :\ x \in V\right\}.$$

Intuitively, the variable p_x^i will be true if and only if x is a vertex of V_i.

Let F be the following formula:

$$F \equiv \bigwedge_{x \in V} \left((p_x^1 \vee p_x^2 \vee p_x^3) \wedge \neg(p_x^1 \wedge p_x^2) \wedge \neg(p_x^1 \wedge p_x^3) \wedge \neg(p_x^2 \wedge p_x^3)\right)$$

" $V_1 \cup V_2 \cup V_3$ is a partition of V"

$$\wedge \bigwedge_{i=1}^{3} \bigwedge_{x \in V} (\neg p_x^i \rightarrow \bigvee_{\{x,y\} \in E} p_y^i)$$

"V_i is dominating set for $i = 1,\ 2$ and 3"

The reader will have no difficulty in verifying that F is satisfiable if and only if V is partitionable in three dominating sets. Moreover, F can clearly be constructed in linear time from G since $| F |= 9 | V | +3 * 2 | E |= 9 | V | +6 | E |$.

Now, in order to prove that 3-Domatic Number is SAT-hard we show that it can be obtained by reduction from 3-colorability in linear time. Let $G = (V, E)$ be an arbitrary instance of 3-colorability. We show how to construct a graph $G' = (V', E')$ such that V' can be partitioned in three dominating sets if and only if G is 3-colorable. For each x of V we define a vertex-graph, $G_x = (V_x, E_x)$, independently of its number of occurrences in E. The vertex-graph G_x is specified as follows.

The set of nodes is:

$$V_x = \left\{c_0(x),\ c_1(x),\ c_2(x),\ d_{01}(x),\ d_{12}(x),\ d_{20}(x)\right\}.$$

The set of edges is defined by:

$$E_x = \Big\{\{c_0(x),\ d_{01}(x)\},\ \{c_0(x),\ d_{20}(x)\}\Big\}$$
$$\bigcup\Big\{\{c_1(x),\ d_{01}(x)\},\ \{c_1(x),\ d_{12}(x)\}\Big\}$$
$$\bigcup\Big\{\{c_2(x),\ d_{20}(x)\},\ \{c_2(x),\ d_{12}(x)\}\Big\}.$$

(See Figure 1.1.) Let us notice that implicitely $d_{hk} = d_{kh}$.

Moreover, for each edge $\{x, y\}$ of E we define an edge-graph made up of the three vertices $c_0(x, y)$, $c_1(x, y)$ and $c_2(x, y)$.

Finally, these vertices are joined to the two corresponding vertex-graphs G_x and G_y by the following set of principal edges (See Figure 1.2.):

$$\Big\{\{c_k(x),\ c_k(x,y)\},\ \{c_k(y),\ c_k(x,y)\}\ :\ 0 \le k \le 2\Big\}.$$

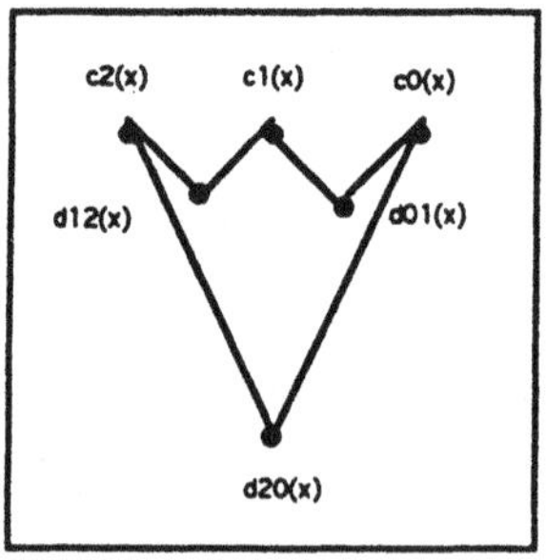

FIGURE 1.1

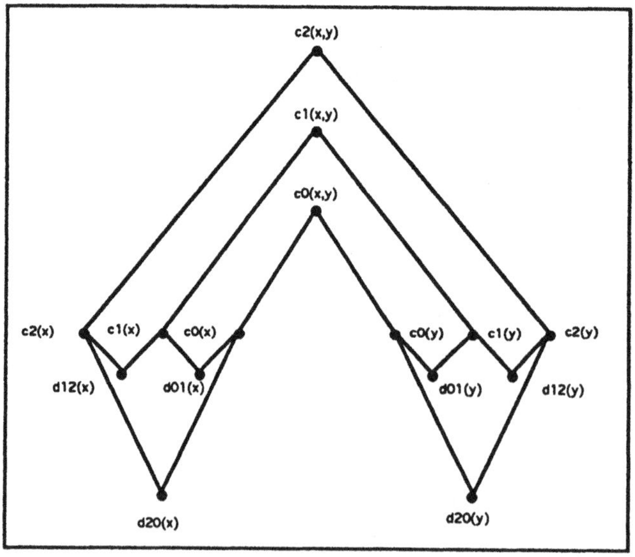

FIGURE 1.2

The graph $G' = (V', E')$ that we shall consider is formed by the union of all these sets of vertices and edges.

Proposition 1.1. *If G is 3-colorable then V' can be partitioned in three dominating sets.*

Proof. Let $col: V \longrightarrow \{0,1,2\}$ be a valid 3-colouring of G. Let us define a successor function S on $\{0,1,2\}$ by: $S(0)=1$, $S(1)=2$ and $S(2)=0$. Then, we can define a partition of V' in three dominating sets V_1', V_2' and V_3'. For each x in V such that $col(x)=k_1$, with $S(k_1)=k_2$ and $S(k_2)=k_3$, we have the following inclusions:
$\left\{c_{k_1}(x),\, d_{k_2k_3}(x)\right\} \subset V_1'$ and $\left\{c_{k_2}(x),\, d_{k_1k_3}(x)\right\} \subset V_2'$ and $\left\{c_{k_3}(x),\, d_{k_1k_2}(x)\right\} \subset V_3'$.

Thus, let us notice that for each $\{x,y\} \in E$ such that $col(x)=k_1$, with $S(k_1)=k_2$ and $S(k_2)=k_3$, and $col(y)=h_1$, with $S(h_1)=h_2$ and $S(h_2)=h_3$, we have $k_1 \neq h_1$ (since col is a valid 3-colouring of G) and then also $k_2 \neq h_2$ and $k_3 \neq h_3$. Therefore if $\{x,y\} \in E$ then for each l in $\{0,1,2\}$, the two vertices $c_l(x)$ and $c_l(y)$ belong to two different subsets: $c_l(x) \in V_i'$ and $c_l(y) \in V_j'$ for $i \neq j$. Hence, we set $c_l(x,y) \in V_k'$ where k is such that $\{i,j,k\}=\{0,1,2\}$. It is no hard to see that we have effectively defined a 3-partition of V' and that each subset V_i', for $i=1,2$ and 3, is a dominating set of G' •

Proposition 1.2. *If V' can be partitioned in three dominating subsets then G is 3-colorable.*

Proof. Suppose that V' can be partitioned in three subsets $V' = V_1' \bigcup V_2' \bigcup V_3'$ such that each V_i' is a dominating set for G'. Then, such a partition obeys some interesting properties, that we present in the two following lemmas.

Lemma 1.3. *Let x be a vertex of V'. If x has exactly two successors, y and z, then x, y and z belong to three different subsets: $x \in V_i'$, $y \in V_j'$ and $z \in V_k'$ with $\{i,j,k\}=\{0,1,2\}$.*

Proof. Otherwise one of the three subsets V_1', V_2' and V_3' would not be a dominating set of G'. •

Lemma 1.4. *For any x in V there exists i_0, i_1 and i_2 such that $\{i_0,i_1,i_2\} = \{0,1,2\}$ and $c_0(x) \in V_{i_0}'$ and $c_1(x) \in V_{i_1}'$ and $c_2(x) \in V_{i_2}'$.*

Proof. If $c_k(x)$ and $c_l(x)$, with $k \neq l$, both belong to the same subset V_i', then the two successors of $d_{kl}(x)$ belong to the same subset. This yields a contradiction, using Lemma 1.3. •

Let us now define a 3-colouring of G, $col: V \longrightarrow \{0,1,2\}$ by:

$$col(x) = k \text{ if and only if } c_k(x) \in V_1'.$$

This function is well defined by Lemma 1.4. We now show that it is a valid 3-colouring of G. Suppose that there exists an edge $\{x,y\}$ in E such that $col(x) = col(y) = k$. Then, by definition we have $c_k(x) \in V_1'$ and $c_k(y) \in V_1'$. Hence, the two successors of $c_k(x, y)$ belong to the same subset. This provides a contradiction, using Lemma 1.3 •

Thus, Theorem 1 is proved since it is easy to see that the construction of G' from G can be performed in linear time. •

VII. Linear transformations using exactly our model of proof.

The following problems are also linearly equivalent to Satisfiability: Cubic Subgraph, Partition into Hamiltonian Subgraphs, 3DM, X3C, Partition into Triangles, Restricted Perfect Matching and Partition into Paths of Length Two. The reductions from 3-colorability to these problems are harder than the previous transformations. Indeed, let $G = (V, E)$ be a graph given as input for 3-colorability, we must construct a vertex-graph for each occurence in E of a vertex of V. In particular we have to number the occurrences of vertices in E. Therefore, these reductions need a fixed number of sortings to be computable in linear time by a Turing machine.

For example, let us prove that Partition into Hamiltonian Subgraphs, PHS, is SAT-equivalent.

Theorem 2. *PHS is linearly equivalent to Satisfiability.*

Proof. First, we prove that this problem is SAT-easy by reduction to SAT'. Let $G = (V,\ A)$ be a directed graph as input for PHS. Notice that a partition of G into hamiltonian subgraphs can be given as a subset A' of A such that each vertex of the subgraph $G' = (V,\ A')$ is included in exactly one cycle and that there is no cycle of length two in G'. Hence, given the set of propositional variables

$$U = \{p_{x,y} \ / \ (x, y) \in A\ \},$$

Intuitively, the variable $p_{x,y}$ corresponds to the predicate "$(x, y) \in A'$ ".

Let us define the formula F :

$$F \equiv \bigwedge_{x \in V} \text{1-exactly}_{(x,y)\in A}\ (p_{x,y})\ \wedge \text{1-exactly}_{(z,x)\in A}\ (p_{z,x})$$

" any vertex x belongs to a cycle"

$$\wedge \bigwedge_{\substack{(x,y)\in A \\ (y,x)\in A}} p_{x,y} \longrightarrow \neg p_{y,x}$$

" there is no cycle of length two"

The reader will have no difficulty in verifying that there exists a partition of G into hamiltonian subgraphs if and only if F is satisfiable.

Moreover, the formula F can clearly be constructed in linear time from the graph G since $| F | \leq 6 \mid E \mid$, thus completing the proof of the SAT-easiness of PHS

Now we prove that PHS is SAT-hard by reduction from 3-colorability. Given a graph $G = (V,\ E)$, we show how to construct a directed graph $G_1 = (V_1,\ E_1)$ such that G_1 can be partitioned into hamiltonian subgraphs if and only if G is

3-colorable. The main difficulty of this reduction is due to the fact that we have two contradictory requirements. First we would like to construct a graph G_1 in which we have many choices of cycles. On the other hand we would like to have a "control" on these cycles, which in fact correspond to the valid 3-colouring of G. The main ideas of the proof are the following. First, we shall construct a graph with a "small" in-or-out degree; that is to say that each vertex has few predecessors and few successors, so that there are not two many cycles. Second we shall define control vertices which will prevent (or oblige) some vertices from being in the same hamiltonian subgraph.

First of all, let us number the occurrences of each vertex of V in E. In the present section x will denote a vertex of V, the variables $x_1, x_2, \ldots, x_n$ will denote its occurrences in E and $\{x_i, y_j\}$ an edge from E. In order to obtain cycles which contain at least three vertices we shall assume, without loss of generality, that we have $n \geq 2$.

For each vertex x of V we define two control vertices: $control_1(x)$, $control_2(x)$, and n vertex-graphs corresponding to the n occurrences of x in E : $G_{x_1}, \ldots, G_{x_n}$.

A vertex-graph G_{x_i} is given by the following sets of nodes and edges:

$$V_{x_i} = \Big\{ c_k(x_i),\ d_k(x_i),\ t_k^0(x_i),\ t_k^1(x_i),\ t_k^2(x_i) :\ 0 \leq k \leq 2 \Big\}$$

$$A_{x_i} = \Big\{ \big(c_k(x_i),\ d_k(x_i)\big) :\ 0 \leq k \leq 2 \Big\} \bigcup \Big\{ \big(c_k(x_i),\ t_k^0(x_i)\big),\ \big(t_k^2(x_i),\ c_k(x_i)\big) \Big\}$$
$$\bigcup \Big\{ \big(t_k^0(x_i),\ t_k^1(x_i)\big),\ \big(t_k^1(x_i),\ t_k^2(x_i)\big),\ \big(t_k^2(x_i),\ t_k^0(x_i)\big) \Big\}$$

(See figure 2.1.)

The n vertex-graphs, G_{x_i}, $i = 1, \ldots, n$ and the two control vertices $control_1(x)$ and $control_2(x)$, corresponding to the same vertex x from G are joined together by the following set of link-edges and control-edges:

$$\Big\{ \big(d_k(x_1),\ d_k(x_2)\big),\ \big(d_k(x_2),\ d_k(x_3)\big),\ \ldots,\ \big(d_k(x_{n-1}),\ d_k(x_n)\big) :\ 0 \leq k \leq 2 \Big\}$$
$$\bigcup \Big\{ \big(d_k(x_n),\ control_1(x)\big),\ \big(d_k(x_n),\ control_2(x)\big) :\ 0 \leq k \leq 2 \Big\}$$
$$\bigcup \Big\{ \big(control_1(x),\ d_k(x_1)\big),\ \big(control_2(x),\ d_k(x_1)\big) :\ 0 \leq k \leq 2 \Big\}$$

On the other hand, the corresponding edge-graph G_{x_i,y_j} for any edge (x_i, y_j) of E is defined as follows.

The set of vertices is given by

$$V_{x_i,y_j} = \Big\{ c_k(x_i, y_j),\ c_k(y_j, x_i),\ control_k(x_i, y_j)\ :\ 0 \leq k \leq 2 \Big\}$$

The set of edges is given by

$$A_{x_i,\, y_j} = \Big\{ \big(c_k(x_i, y_j),\ c_k(y_j, x_i)\big),\ \big(c_k(y_j, x_i),\ c_k(x_i, y_j)\big) :\ 0 \leq k \leq 2 \Big\}$$
$$\bigcup \Big\{ \big(c_k(x_i, y_j),\ control_k(x_i, y_j)\big)\big(control_k(x_i, y_j),\ c_k(x_i, y_j)\big) :\ 0 \leq k \leq 2 \Big\}$$
$$\bigcup \Big\{ \big(c_k(y_j, x_i),\ control_k(x_i, y_j)\big)\big(control_k(x_i, y_j),\ c_k(y_j, x_i)\big) :\ 0 \leq k \leq 2 \Big\}$$

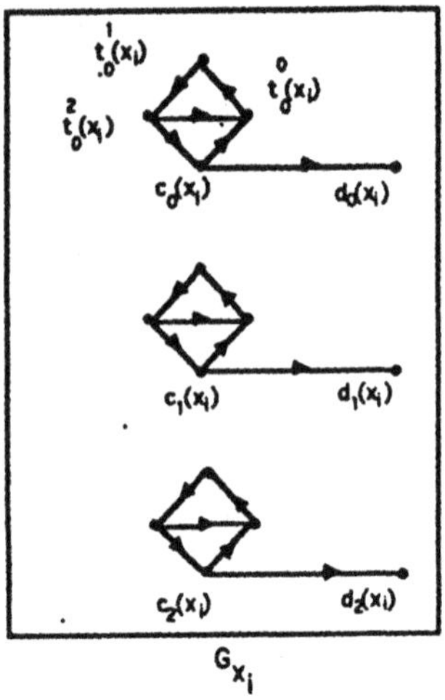

FIGURE 2.1

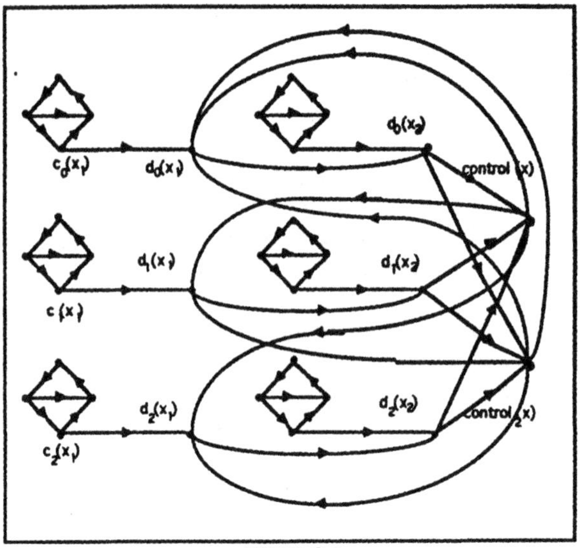

FIGURE 2.2

The set of principal edges which join any edge-graph $G_{x_i,\ y_j}$ to the two corresponding vertex-graphs G_{x_i} and G_{y_j} is specified as follows:

$$\Big\{ \big(d_k(x_i),\ c_k(x_i,y_j)\big),\ \big(c_k(y_j,x_i),\ c_k(x_i)\big) : \ 0 \le k \le 2 \Big\}$$

$$\Big\{ \big(d_k(y_j),\ c_k(y_j,x_i)\big),\ \big(c_k(x_i,y_j),\ c_k(y_j)\big) : \ 0 \le k \le 2 \Big\}$$

Roughly speaking the edges $\big(d_k(x_i),\ c_k(x_i,y_j)\big)$ $\big(c_k(y_j,x_i),\ c_k(x_i)\big)$ correspond to the fact: "if x_i is of colour k then one endpoint of the edge $\{x_i, y_j\}$ is of colour k". Let us remind the reader that, in order to define a 3-colouring of G from a partition into hamiltonian subgraphs of G_1, the edge-graphs are used to prevent two adjacent vertices from having the same colour. The link and control edges insure that all the occurrences x_i of a same vertex x of G are coloured in the same way.

Finally, the graph $G_1 = (V_1,\ A_1)$ that we shall consider is formed by the union of all the previous vertices and edges.

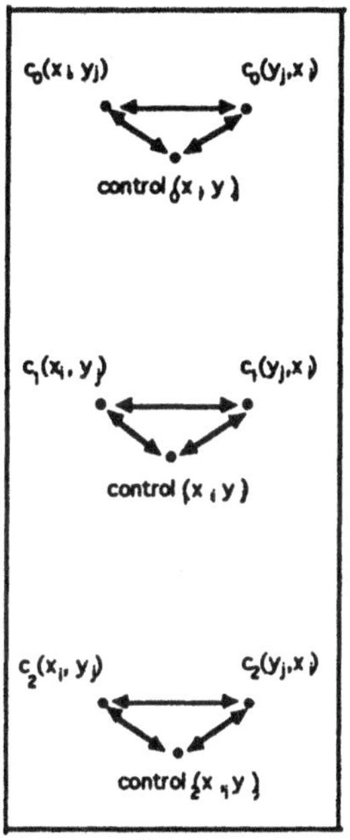

G_{x_i, y_j}

FIGURE 2.3

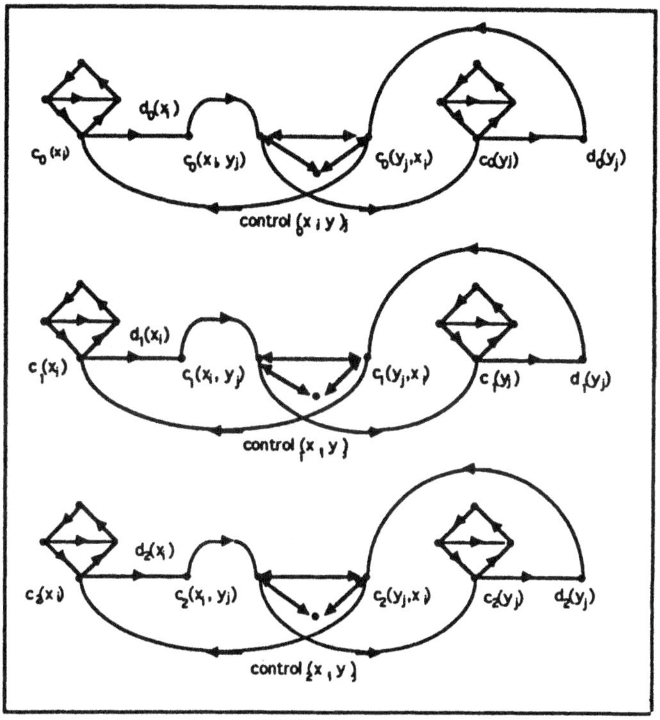

FIGURE 2.4

Let us recall that a partition into hamiltonian subgraphs of G_1 can be given by a subset of edges $A_1' \subseteq A_1$ such that any vertex of V_1 appears exactly once as an origine of an arc of A_1' and exactly once as endpoint of an arc of A_1'. All along this section we shall implicitly use this remark. Let us now introduce some notations.

$T_k(x_i)$ denotes the circuit (T as triangle): $\Big(t_k^0(x_i),\ t_k^1(x_i),\ t_k^2(x_i)\Big)$

$L_k(x_i)$ denotes the circuit (L as lozenge): $\Big(c_k(x_i),\ t_k^0(x_i),\ t_k^1(x_i),\ t_k^2(x_i)\Big)$

$CT_k(x_i,\ y_j)$ denotes the circuit: $\Big(c_k(x_i,\ y_j),\ control_k(x_i,\ y_j),\ c_k(y_j,\ x_i)\Big)$

Proposition 2.1 *If G is 3-colorable then G_1 can be partitioned into hamiltonian subgraphs.*

Proof. Let $col\colon V \longrightarrow \{0,\ 1,\ 2\}$ be a valid 3-colouring of G. Then, we can define a partition of G_1 in hamiltonian circuits by considering for each $\{x_i, y_j\} \in E$ with, say, $col(x) = 0$ and $col(y) = 1$ the following circuits:

$$T_0(x_i),\ L_0(y_j),\quad L_1(x_i),\ T_1(y_j),\quad L_2(x_i),\ L_2(y_j),$$

$$\Big(d_1(x_1),\dots,\ d_1(x_n),\ control_1(x)\Big),\quad \Big(d_2(x_1),\dots,\ d_2(x_n),\ control_2(x)\Big),$$

$$\Big(d_0(y_1),\dots,\ d_0(y_m),\ control_1(y)\Big),\quad \Big(d_2(y_1),\dots,\ d_2(y_m),\ control_2(y)\Big),$$

$$\Big(c_0(x_i),\ d_0(x_i),\ c_0(x_i,\ y_j),\ control_0(x_i,\ y_j),\ c_0(y_j,\ x_i),\ \Big)$$

$$\Big(c_1(y_j),\ d_1(y_j),\ c_1(y_j,\ x_i),\ control_1(x_i,\ y_j),\ c_1(x_i,\ y_j),\ \Big)$$

$$\text{and } CT_2(x_i,\ y_j).$$

Let us notice that all these circuits contain at least three vertices since we have assumed that each vertex appears at least in two edges of E.

Proposition 2.2. *If the graph G_1 can be partitioned into hamiltonian subgraphs, then the graph G is 3-colorable.*

Proof. Suppose $V_1 = \bigcup_{i=1}^{n} V_i'$ is a partition of V_1 with $|\ V_i'\ | \geq 3$. Moreover let A_1' be a subset of A_1 which induces an hamiltonian circuit on each of these subsets V_i'. The proof of the proposition 2.2 is not straightforward and requires the four following lemmas.

Lemma 2.3. *For any $k = 0$, 1 or 2 and for any edge $\{x_i,\ y_j\}$ in E, the three vertices*

$$c_k(x_i,\ y_j),\ control_k(x_i,\ y_j) \text{ and } c_k(y_j,\ x_i)$$

belong to the same circuit and consequently occur in this circuit either in that order or in the converse order.

Proof. The only predecessors of $control_k(x_i,\ y_j)$ are $c_k(x_i,\ y_j)$ and $c_k(y_j,\ x_i)$, and these two vertices are also its only successors. But, by hypothesis, every circuit contains at least three vertices. Thus these three vertices must belong to the same circuit. •

Lemma 2.4. *For any $k = 0$, 1, 2 and for any x_i, if $\big(c_k(x_i),\ t_k^0(x_i)\big) \in A_1'$ or $\big(t_k^2(x_i),\ c_k(x_i)\big) \in A_1'$ then $L_k(x_i)$ is the circuit of A_1' which contains them.*

Proof. See Figure 2.1. •

Lemma 2.5. *For any edge $\{x_i,\ y_j\}$ in E and for any $k \in \{0,\ 1,\ 2\}$, if $\big(d_k(x_i),\ c_k(x_i,y_j)\big) \in A_1'$ then the hamiltonian circuit which contains the vertex $c_k(x_i)$ is necessarily:*

$$\Big(d_k(x_i),\ c_k(x_i,y_j),\ control_k(x_i,y_j),\ c_k(y_j,x_i),\ c_k(x_i)\Big)$$

(this result also holds for y_j.)

Proof. Suppose $(d_k(x_i),\ c_k(x_i, y_j)) \in A'_1$. Then, from Lemma 2.3, this arc is followed on its circuit by the arcs

$$(c_k(x_i, y_j),\ control_k(x_i, y_j))$$

and

$$(control_k(x_i, y_j),\ c_k(y_j, x_i)).$$

Furthermore the only successors of $c_k(y_j,\ x_i)$ are $c_k(x_i, y_j)$, $control_k(x_i, y_j)$ (which have already been considered) and $c_k(x_i)$. Thus the arc $(c_k(y_j, x_i),\ c_k(x_i))$ belongs to A'_1.

On the other hand the vertex $c_k(x_i)$ has two successors $t^0_k(x_i)$ and $d_k(x_i)$ But if $(c_k(x_i),\ t^0_k(x_i)) \in A'_1$, then $L_k(x_i)$ is the circuit which contains $c_k(x_i)$, by Lemma 2.4, contradicting the fact that $(c_k(y_j,\ x_i),\ c_k(x_i))$ belongs to A'_1. So we have $(c_k(x_i),\ d_k(x_i)) \in A'_1$ and hence the circuit which contains $c_k(x_i)$ is

$$\Big(d_k(x_i),\ c_k(x_i,\ y_j),\ control_k(x_i,\ y_j),\ c_k(y_j,\ x_i),\ c_k(x_i)\Big). \bullet$$

Lemma 2.6. *Let x be a vertex of V. Let $x_1, ..., x_n$ be its occurrences in E and let $k = 0,\ 1,$ or 2.*

If $(d_k(x_1),\ d_k(x_2)) \in A'_1$ then $(d_k(x_i),\ d_k(x_{i+1})) \in A'_1$ for any $i = 1, ..., (n-1)$ and there exists a vertex s in $\{control_1(x),\ control_2(x)\}$ such that $(d_k(x_n),\ s) \in A'_1$.

Proof. Suppose $(d_k(x_1),\ d_k(x_2)) \in A'_1$. The successors of $d_k(x_2)$ are $d_k(x_3)$ and $c_k(x_2,\ y_{j_2})$. If $(d_k(x_2),\ c_k(x_2,\ y_{j_2})) \in A'_1$ then from Lemma 2.5 the hamiltonian circuit which contains $d_k(x_2)$ is:

$$\Big(d_k(x_2),\ c_k(x_2,\ y_{j_2}),\ control_k(x_2,\ y_{j_2}),\ c_k(y_{j_2},\ x_2)\ c_k(x_2)\Big),$$

contradicting the fact that $(d_k(x_1),\ d_k(x_2)) \in A'_1$. Thus $(d_k(x_2),\ d_k(x_3))$ is in A'_1. Similarly this would show that if $(d_k(x_{i-1}),\ d_k(x_i)) \in A'_1$ then $(d_k(x_i),\ d_k(x_{i+1})) \in A'_1$ for $i = 1, \ldots (n-1)$. Therefore the first statement of Lemma 2.6 follows by induction on i.

Suppose now that $(d_k(x_{n-1}),\ d_k(x_n)) \in A'_1$. The only successors of $d_k(x_n)$ are $control_1(x)$, $control_2(x)$ and $c_k(x_n, y_{j_n})$. Thus, using Lemma 2.5, we can conclude as above that A'_1 does not contain $(d_k(x_n),\ c_k(x_n, y_{j_n}))$ and then contains one of the two arcs $(d_k(x_n),\ control_1(x))$ and $(d_k(x_n),\ control_2(x))$. $\bullet$

Lemma 2.7. *Let x be a vertex of V and $x_1, \ldots, x_n$ its occurrences in E. Let us assume that $\{x_i,\ y_{j_i}\} \in E$ for $i = 1, \ldots, n$. Then, there exists a unique $k \in \{0,\ 1,\ 2\}$ such that $(d_k(x_i),\ c_k(x_i,\ y_{j_i})) \in A'_1$ for $i = 1, \ldots, n$.*

Proof. First we prove uniqueness. Suppose that there exist $k \neq l$ such that the two arcs $(d_k(x_n),\ c_k(x_n,\ y_{j_n}))$ and $(d_l(x_n),\ c_l(x_n,\ y_{j_n}))$ belong to A'_1. Then $d_h(x_n)$, where $\{h,\ k,\ l\} = \{0,\ 1,\ 2\}$, is the only valid predecessor of the two vertices $control_1(x)$ and $control_2(x)$ in A'_1, a contradiction. Therefore we have shown that k is unique.

We now prove the existence of such a k by induction on i. Let us first consider the case where $i = 1$. If $(d_k(x_1),\ c_k(x_1,\ y_{j_1})) \notin A'_1$ for $k \in \{0,\ 1,\ 2\}$. Then A'_1 contains the three arcs $(d_0(x_1),\ d_0(x_2))$, $(d_1(x_1),\ d_1(x_2))$, $(d_2(x_1),\ d_2(x_2))$ since the only successors of $d_k(x_1)$ are $c_k(x_1,\ y_{j_1})$ and $d_k(x_2)$. Then, using Lemma 2.6, we claim that there exist $s_0,\ s_1,\ s_2$ in $\{control_1(x),\ control_2(x)\}$ such that the three arcs $(d_0(x_n),\ s_0)$, $(d_1(x_n),\ s_1)$ and $(d_2(x_n),\ s_2)$ belong to A'_1. This provides a contradiction since that implies that one of the two vertices $control_1(x)$ and $control_2(x)$ appears as endpoint of at least two arcs of A'_1. Therefore there exists $k \in \{0,\ 1,\ 2\}$ such that $(d_k(x_1),\ c_k(x_1, y_{j_1})) \in A'_1$. Suppose now that $(i-1) \geq 1$ and $(d_k(x_{i-1}),\ c_k(x_{i-1},\ y_{j_{i-1}})) \in A'_1$. The only predecessors of $d_k(x_i)$ are $d_k(x_{i-1})$ and $c_k(x_i)$. But, by induction hypothesis, A'_1 contains $(d_k(x_{i-1}),\ c_k(x_{i-1},\ y_{j_{i-1}}))$. Thus necessarily $(c_k(x_i),\ d_k(x_i))$ belongs to A'_1. Moreover, the predecessors of $c_k(x_i)$ are $t_k^2(x_i)$ and $c_k(y_{j_i},\ x_i)$. From Lemma 2.4, if $(t_k^2(x_i),\ c_k(x_i)) \in A'_1$ then the hamiltonian circuit which contains $c_k(x_i)$ is $L_k(x_i)$, contradicting the fact that $(c_k(x_i),\ d_k(x_i))$ belongs to A'_1. Thus A'_1 contains $(c_k(y_{j_i},\ x_i),\ c_k(x_i))$. Furthermore, the three vertices $c_k(x_i,\ y_{j_i})$, $control_k(x_i,\ y_{j_i})$ and $c_k(y_{j_i},\ x_i)$ are consecutive in an hamiltonian circuit (Lemma 2.3). Thus A'_1 must contain the two arcs $(c_k(x_i, y_{j_i}),\ control_k(x_i, y_{j_i}))$ and $(control_k(x_i, y_{j_i}),\ c_k(y_{j_i}, x_i))$.

Finally, the only predecessors of $c_k(x_i, y_{j_i})$ are $control_k(x_i, y_j)$, $c_k(y_{j_i}, x_i)$ and $d_k(x_i)$. But the first and the second of them have already been used. Therefore, A'_1 must contain the arc $(d_k(x_i),\ c_k(x_i, y_{j_i}))$.

The existence of k such that $(d_k(x_i),\ c_k(x_i, y_{j_i})) \in A'_1$, for $i = 1, \ldots, n$, follows by induction on i, thus completing the proof of Lemma 2.7. •

We can now prove Proposition 2.2. Let us define a 3-colouring of G $col\colon V \longrightarrow \{0,1,2\}$ by:

$$col(x) = k \text{ if and only if } (d_k(x_i),\ c_k(x_i, y_{j_i})) \in A'_1 \text{ for } i = 1, \ldots, n.$$

This function is well defined by Lemma 2.7. We now show that it is a valid 3-colouring of G. If there exists some edge $\{x_i, y_j\}$ in E such that $col(x) = col(y) = k$, then the two arcs $(d_k(x_i),\ c_k(x_i, y_j))$ and $(d_k(y_j),\ c_k(y_j, x_i))$ belong to A'_1. So, using Lemma 2.5 the hamiltonian circuits which respectively contain $c_k(x_i)$ and $c_k(y_j)$ are:

$$\Big(d_k(x_i),\ c_k(x_i, y_j),\ control_k(x_i, y_j),\ c_k(y_j, x_i)\ c_k(x_i)\Big),$$

$$\text{and } \Big(d_k(y_j),\ c_k(y_j, x_i),\ control_k(x_i, y_j),\ c_k(x_i, y_j)\ c_k(y_j)\Big).$$

But these circuits are two distinct non disjoint circuits, a contradiction. •

Finally, the reader would easily verify that the construction of G_1 from G can be performed in linear time, thus completing the proof of Theorem 2. •

VIII. Conclusion.

We have widely extended the class of problems that are linearly equivalent to Satisfiability which has been introduced by A. K. Dewdney [5] and first expanded by E. Grandjean [9]. Moreover, we have developed a new method for proving NP-completeness and shown that 3-colorability is a key problem. So, we have unified many different proofs that existed before and defined a model of proof which could certainly be used in other contexts.

We have shown that many natural combinatorial problems are linearly equivalent to Satisfiability. In fact all the problems that we have studied belong to the class NSUBLINEAR [8], i.e. these problems can be decided in $O(n/logn)$ deterministic time with $O(n/log^2n)$ non deterministic instructions (typically: to guess a number and to write it in a register) by a Random Access Machine which only uses numbers that are polynomially bounded into the input length n. As far as we know, most of the problems in NSUBLINEAR which do not contain a number in their instance are linearly equivalent to Satisfiability. Two problems of this class, Hamiltonian Circuit and Disjoint Connecting Paths have been only proved to be SAT-hard. The question of whether they are SAT-easy is still open. The difficulty is due to the fact that even if we know how to express in linear time a local property in propositional logic , such as for example "x has exactly one successor", we do not know how to express non local property, such as connexity for example. The difficulty for the problems containing a number in their instance is of a different nature. It becomes from the predicate "be greater than K" that we do not know how to express in linear time in propositional logic. However, most of these problems such as Vertex Cover and Dominating Set (for example) are SAT-hard. This yields the surprising result that SAT is a kind of "minimal" NP-complete problem.

The relation of linear time equivalence splits the class of NP-complete problems into equivalence classes. The fact that the class of Satisfiability is so large is unexpected and yields the following question: Are there other natural linear classes (in particular among natural combinatorial problems containing a number in their instance) having a size comparable to the one of the class of Satisfiability? Two other classes, two candidates, are under our investigation, one containing Dominating Set and the other containing Max2Sat [4]. Moreover, we get the following linear-time reductions:

SAT $\longrightarrow$ Vertex Cover $\longrightarrow$ Dominating Set $\longrightarrow$ Max-2Sat.

As far as we know, these classes are not so large as the class of Satisfiability. Presently the proposed reductions are not very uniform and so it is not easy to increase these classes. But perhaps it is due to the fact that we have not yet found key problems like 3-colorability and SAT$'$ in the case of the class of Satisfiability.

Acknowledgments. I would like to take this opportunity to thank Etienne Grandjean for having introduced me to the interesting problems of linear time and NP-completeness and Pascal Weil for help with the english.

References.

[1] K. J. Compton and C. W. Henson, *A uniform method for proving lower bounds on the computational complexity of logical theories*, Annals of Pure and Applied Logic **48** (1990), 1-79.
[2] S. A. Cook, *The complexity of theorem-proving procedures*, Proc. 3rd Ann. ACM Symp. on Theory of Computing, Association for Computing Machinery, New York (1971), 151-158.
[3] N. Creignou, *Exact complexity of problems of incompletely specified automata*, T. R. Univ. de Caen, France, Cahiers du LIUC **92-9** (1992), , submitted to Annals of Mathematics and Artificial Intelligence, J. C. Baltzer Scientific Publishing Co. (Suisse), Special issue ed. M. Nivat and S. Grigorieff, (1993).
[4] N. Creignou, Temps linéaire et problèmes NP-complets , Thèse de doctorat, Univ. de Caen, France (1993).
[5] A. K. Dewdney, *Linear transformations between combinatorial problems*, Intern. J. Computer Math. **11** (1982), 91-110.
[6] M. R. Garey and G. S. Johnson, Computers and intractability: a guide to the theory of NP-completeness, W. H. Freeman and Co, San Franscisco (1979).
[7] E. Grandjean, *Nontrivial lower bound for a NP problem on automata*, SIAM J. Comput., **19**, (1990), 1-14.
[8] E. Grandjean, *Linear time algorithms and NP-complete problems* , T. R. Univ. de Caen, France, Cahiers du LIUC **91-9** (1991), to appear in Lect. Notes in Comput. Sci. CSL'92 (San Miniato) (1993).
[9] E. Grandjean, *Sorting, linear time and Satisfiability problem*, preprint (1992), submitted to Annals of Mathematics and Artificial Intelligence, J. C. Baltzer Scientific Publishing Co. (Suisse), Special issue ed. M. Nivat and S. Grigorieff, (1993).
[10] J. Hopcroft and J. D. Ullman, Introduction to Automata Theory, Languages and Computation, Addison-Wesley, MA (1979).
[11] R. M. Karp, *Reducibility among combinatorial problems*, in R. E. Miller and J. W. Thatcher (eds), Complexity of Computer Computations, Plenum Press, New York (1972), 85-103.
[12] D. G. Kirkpatrick and P. Hell, *On the complexity of a general matching problem*, Proc. 10th Ann. ACM Symp. of Theory of Computing, Association for Computing Machinery, New York (1978), 240-245.
[13] D. E. Knuth, *Axioms and Hulls*, Lect. Notes in Comput. Sci. **606**, Springer-Verlag.
[14] T. J. Schaefer, *The complexity of satisfiability problems*, Proc. 10th Ann. ACM Symp. of Theory of Computing, Association for Computing Machinery, New York, (1978), 216-226.

Model Building by Resolution

Christian G. Fermüller* Alexander Leitsch*

Abstract

Resolution calculi are best known as basis for algorithms testing the unsatisfiability of sets of clauses. Only recently more attention is paid to the fact that various resolution refinements may also benefitly be employed as decision procedures for a wide range of decidable classes of clause sets. In this proof theoretic approach to the decision problem one usually tries to test for satisfiability by termination of complete resolution procedures. Building on such types of decidability results we show that, for certain classes of clause sets, we can extract (representations of) models from the set of resolvents generated by hyperresolution. The process of model construction proceeds in two steps: First hyperresolution is employed to arrive at a finite set of atoms that represents a description of an Herbrand-model. In a second step we extract from this set of atoms a full representation of a model with finite domain. We emphasize that no backtracking is needed in our model constructing algorithm.

1 Introduction

Finding and investigating models of abstract structures is at the very heart of mathematical activity. In mathematical logic many results have been obtained about models of first order formulas. There are much less results, however, about algorithmical model building methods. In the area of first order prefix classes, most of the research was directed to prove the *existence* of models, without giving explicit *representations* of models. In particular, the existence of finite models for satisfiable formulas of a class Δ (such classes are called finitely controllable) implies the decidability of Δ. Thus showing finite controllability of classes is a key proof technique in the theory of decidable classes [DG79]. A characteristic of this technique is to compute a recursive bound $\alpha(F)$ for every F in Δ s.t. F is satisfiable iff there exists a model of domain size $\leq \alpha(F)$. The transformation of such a proof into an algorithmical method results in exhaustive search through all finite domain interpretations of size $\leq \alpha(F)$. But, even for small $\alpha(F)$, this is clearly inadequate for practical computing. In more recent time algorithmical finite model building based on theorem proving methods has been investigated by T. Tammet [Tam91],[Tam92], Manthey and Bry [MB88], and Caferra & Zabel [CZ91]. An earlier approach of S. Winker [Win82], although practically relevant and successful, did not define a general algorithmic method.

*Current address: Technische Universität Wien, Institut für Computersprachen, Resselgasse 3/1, A–1040 Vienna/Austria. E-mail: {chrisf,leitsch}@csdec1.tuwien.ac.at.

Tammet's approach, like ours, is based on resolution decision procedures. But his finite model building method applies to the monadic and Ackermann class only and is based on the termination of an ordering refinement. In the resulting model description the interpretation of the function symbols is given completely, but the interpretation of predicate symbols is only partial. Moreover, Tammet uses narrowing and works with equations on the object language level. In this paper the model building is based on termination sets for hyperresolution (which yield other decision classes). The finite model building method of section 5 is based on the transformation of Herbrand models; it does not use equality reasoning but filtration.

Caffera and Zabel define an (equational) extension of the resolution calculus, specifically designed for the purpose of finite model construction. They don't specify the model completely, but only give an interpretation of the predicate symbols. Their method is not based on "postprocessing" of termination set of resolution. Moreover, the decidable classes are completely different from ours.

Manthey and Bry describe a hyperresolution prover based on a model generation paradigm in [MB88]. Their method of model building, although similar in case of the class PVD_+, differs from ours in several aspects. They essentially use splitting of positive ground clauses and backtracking, features that are avoided in our approach. Moreover, we can handle cases, where no ground facts are produced. While Manthey and Bry construct Herbrand models only, the final products of our procedure are finite models.

We present a fully algorithmical method for the construction of models out of termination sets of positive hyperresolution provers. The basic idea is the following: If the (complete) theorem prover stops on a set of clauses $\mathcal{C}$ without derivation of $\Box$ (contradiction) then $\mathcal{C}$ must be satisfiable. Having obtained the (finite) set $\mathcal{D}$ of derivable clauses, we start to construct a model out of $\mathcal{D}$ (i.e., we use the information provided by positive hyperresolution). The first step of model building consists in transforming the set of derived positive clauses $\mathcal{D}_+$ into a finite set of unit clauses $\mathcal{U}$. Such a set $\mathcal{U}$ can be considered as a representation of an Herbrand model. In general, this model is infinite since its domain is the Herbrand universe. The second step is of a different flavor: We present a method which, for a wide range of classes decidable by hyperresolution, constructs finite models out of the set of unit clauses $\mathcal{U}$. Neither backtracking nor equality reasoning on the object level is required (the latter in contrast to [Tam91], [Tam92] and [CZ91], the former in contrast to [MB88]).

Although the finite models obtained in this way are not minimal (w.r.t. domain size) in general, the method is clearly superior to exhaustive search, since no kind of backtracking is involved. However, it is limited to syntax classes admitting filtration on the termination sets (i.e., Herbrand models can be mapped into finite models under preservation of truth values).

As the method presented in this paper is based on termination of positive hyperresolution with subsumption and condensing, the preprocessing step just consists of "ordinary" theorem proving. Only in case $\Box$ has not been derived, the proper model building algorithm is applied to the termination set. This is also a characteristic of Tammet's work, while Caferra & Zabel start with a model building procedure at once. Together with Tammet's results on finite model building, this paper can be considered as a starting point for model building methods based on resolution decision procedures for first order classes.

2 Basic notions

We assume the reader to be familiar with the concept of resolution:

Terms, *atoms*, and *literals* are defined as usual. By an *expression* we mean a term or a literal. A *clause* is a (finite) disjunction of literals. The *empty clause* (representing contradiction) is denoted by $\Box$. $V(E)$ denotes the set of variables occurring in an expression, clause or set of clauses E. An expression or clause E is called *ground* if $V(E) = \emptyset$.

To be able to argue concisely we need some additional notions:

Definition 2.1 *If C is a clause then C_+ is the set of positive (unsigned) literals in C, analogously C_- denotes the set of negative literals (negated atoms) in C. C is a called* Horn *iff it contains at most one positive literal, i.e. iff $|C_+| \leq 1$. C is* positive *iff $C_- = \emptyset$. For a set of clauses $\mathcal{C}$ $P(\mathcal{C})$ denotes the subset of positive clauses in $\mathcal{C}$ and $NP(\mathcal{C}) = \mathcal{C} - P(\mathcal{C})$, the subset of* non-positive *clauses.*

It is convenient to make use of some definitions concerning term structure that are well known from the term rewriting literature:

Definition 2.2 *By $\mathcal{P}(A)$ we denote the set of all* positions, *considered as finite sequences of integers, of an expression A (see eg. [Hul88]). $[A|p]$ denotes the subterm of A at position p. The* length of a position *$|p|$ is the number of integers in p.*

The number of occurrences *of a variable x in an expression A is defined as $OCC(x, A) = |\{p \mid p \in \mathcal{P}(A) \text{ and } [A|p] = x\}|$. The definition is extended to clauses and sets of expressions in an obvious manner.*

Definition 2.3 *The term depth of an expression A is defined as $\tau(A) = \max\{|p| \mid p \in \mathcal{P}(A)\}$.*

The maximal depth of occurrence *of a variable x in $V(A)$ is defined as $\tau_{\max}(x, A) = \max\{|p| \mid p \in \mathcal{P}(A) \text{ and } [A|p] = x\}$. Similarily, $\tau_{\min}(x, A) = \min\{|p| \mid p \in \mathcal{P}(A) \text{ and } [A|p] = x\}$. Again, these definitions are generalized to clauses and sets of expressions in the obvious way.*

Definition 2.4 *Let V be the set of variables and T be the set of terms. A* substitution *is a mapping $\sigma : V \rightarrow T$ s.t. $\sigma(x) = x$ almost everywhere. We call the set $dom(\sigma) = \{x | \sigma(x) \neq x\}$* domain *of σ and $rg(\sigma) = \{\sigma(x) | x \in dom(\sigma)\}$* range *of σ.*

A substitution σ is called ground *if all terms in $rg(\sigma)$ are ground. The set of all ground substitutions is denoted by GS.*

If E is an expression or clause and σ is a (ground) substitution then the result of applying σ to E is denoted by $E\sigma$. $E\sigma$ is called a *(ground) instance* of E.

Definition 2.5 *If T is a set of expressions then $G(T)$ denotes the* set of ground instances *of elements in T.*

Remark. We implicitly assume a fixed set of constant and function symbols when speaking of *the* set of ground substitutions or instances.

Most general unifiers (mgu) of sets of expressions and *(binary) resolvents* and *factors* of clauses are defined as usual (see, e.g., [CL73]).

Definition 2.6 *Let E_1 and E_2 be expressions, then $E_1 \leq_s E_2$—read: E_1* is more general than *E_2—iff there exists a substitution σ s.t. $E_1\sigma = E_2$.*

Similary, if C and D are clauses, $C \leq_s D$ iff there exists a substitution σ s.t. $C\sigma$ is a subclause of D. In this case we say that C subsumes *D.*

A smallest subclause C' of C s.t. $C \leq_s C'$ is called a condensation *of C.*

Definition 2.7 *For a set of clauses $\mathcal{C}$ $cond(\mathcal{C})$ denotes the set of condensations of all clauses in $\mathcal{C}$. (Remember that condensations are unique up to a renaming of variables, see [Joy76]).*

By $sub(\mathcal{C})$ we denote a subset of $\mathcal{C}$ s.t. for no pair of different clauses $C, D \in sub(\mathcal{C})$ it holds $C \leq_s D$. ($sub(\mathcal{C})$ is not unique in general, but can be made so by using specific selection strategies, see [Joy76]).

For any clause set $\mathcal{C}$, the Herbrand universe $\mathbf{HU}(\mathcal{C})$ and the Herbrand base $\mathbf{HB}(\mathcal{C})$, are defined as usual. An *interpretation* $\mathcal{I}$ is denoted as triple $\langle D, \varphi, d\rangle$, where D is the domain, φ the signature interpretation, and d the variable assignment of $\mathcal{I}$. Each interpretation $\mathcal{I} = \langle D, \varphi, d\rangle$ induces an evaluation function $v_\mathcal{I}$ in the usual way s.t. $v_\mathcal{I}(t) \in D$ for terms t and $v_\mathcal{I}(C) \in \{\mathbf{true}, \mathbf{false}\}$ for all literals and clauses C. A *model* for a clause $\mathcal{C}$ is an interpretation $\mathcal{I}$ s.t. $v_\mathcal{I}(C) = \mathbf{true}$ for all $C \in \mathcal{C}$.

Remark. A clause corresponds to the universal closure of the disjunction of its literals. This means that the truth value of a clause under an interpretation does not depend on the variable assignment. Since we are only interested in the truth values of clauses and ground atoms we may omit variable assigments from the definition of interpretations.

Definition 2.8 *An* H(erbrand)-interpretation *of a clause set $\mathcal{C}$ is an interpretation $\mathcal{I} = \langle D, \varphi\rangle$ s.t. $D = HU(\mathcal{C})$ and $\varphi(c) = c$ for all constants and $\varphi(f)(t_1, \ldots, t_n) = f(t_1, \ldots, t_n)$ for all n-ary function symbols and all ground terms t_i.*

H-interpretations are uniquely determined by sets of ground atoms. We therefore denote H-interpretations $\mathcal{I}_H$ as subsets of the Herbrand base with the intended meaning that, for any ground atom $A = P(t_1, \ldots t_n)$: $A \in \mathcal{I}_H$ iff $(t_1, \ldots, t_n) \in \varphi_{\mathcal{I}_H}(P)$. It follows that $\mathcal{I}_H \subseteq HB(\mathcal{C})$ is an H-model for a clause set $\mathcal{C}$ iff for all ground instances $C\gamma$ of a clause $C \in \mathcal{C}$ there is either a positive literal A in $C\gamma$ s.t. A is in $\mathcal{I}_H$ or a negative literal $\neg A$ in $C\gamma$ s.t. $A \notin \mathcal{I}_H$.

3 Hyperresolution as Decision Procedure

All results in this paper are based on termination sets obtained by positive hyperresolution provers. That means that the model construction itself starts on finite sets of clauses $\mathcal{C}$ which are invariant under hyperresolution (and do not contain $\square$). In this section we present formal definitions of (variants of) hyperresolution in terms of resolution operators and present some classes of clause sets that are decidable by positive hyperresolution.

Definition 3.1 *A resolvent of two clauses C and D s.t. C is positive and factoring is only applied to C is called* defined under restrictive factoring.

Definition 3.2 *Let C be a non-positive clause and let the clauses D_i $(1 \leq i \leq n)$ be positive. Then $\Gamma = (C; D_1, \ldots, D_n)$ is called a* clash sequence.

Let $C_0 = C$ and C_{i+1} be a resolvent of C_i and D_{i+1} defined under restrictive factoring, for $i < n$. If C_n exists and is positive then it is called clash resolvent *defined by Γ.*

Hyperresolution only produces positive clauses (or $\square$).

It was shown in [Nol80] that positive hyperresolution under restrictive factoring is complete.

Throughout the paper we describe hyperresolution by an operator R_{HS} defined on sets of clauses.

Definition 3.3 *For any set of clauses $\mathcal{C}$: $R_{HS}(\mathcal{C})$ is the set of all clash resolvents defined by clash sequences consisting of clauses in $\mathcal{C}$, reduced under subsumption and condensing (i.e., $sub(cond(R_{HS}(\mathcal{C}))) = R_{HS}(\mathcal{C})$).*

To describe the set of all derivable clauses we define:

$$R^0_{HS}(\mathcal{C}) = sub(cond(\mathcal{C})),$$
$$R^{i+1}_{HS}(\mathcal{C}) = sub(R_{HS}(R^i_{HS}(\mathcal{C})) \cup R^i_{HS}(\mathcal{C})).$$
$$R^*_{HS}(\mathcal{C}) = \bigcap_{i=1}^{\infty} \bigcup_{j=i}^{\infty} R^j_{HS}(\mathcal{C}).$$

(Note that R_{HS} should not be identified with R^1_{HS}). We shall use $RP^*(\mathcal{C})$ as an abbreviation for $P(R_{HS}(\mathcal{C}))$ (see def. 2.1).

Remark. Observe that, since previously derived clauses that are subsumed by new resolvents may be deleted, R^i_{HS} is not monotonic; i.e., $R^i_{HS}(\mathcal{C}) - R^j_{HS}(\mathcal{C})$ may be non-empty for $i < j$. This is the reason for defining R^*_{HS} as the limes superior, instead of the union, of the R^i_{HS}.

R_{HS} enjoys two important properties:

(1) R_{HS} is complete, i.e. $\square \in R^*_{HS}(\mathcal{C})$ whenever $\mathcal{C}$ is unsatisfiable. Note that, by definition of R^*_{HS} (which rather has the character of a problem reduction operator), $\square \in R^*_{HS}(\mathcal{C})$ implies $R^*_{HS}(\mathcal{C}) = \{\square\}$ (since $\square \leq_s C$ for all clauses C).

(2) If there is an i s.t. $R^i_{HS}(\mathcal{C}) = R^{i+1}_{HS}(\mathcal{C})$ then $R^*_{HS}(\mathcal{C}) = R^i_{HS}(\mathcal{C})$ and $R^*_{HS}(\mathcal{C})$ is finite.

It is well known that positive hyperresolution with subsumption (as replacement rule) is complete (cf., e.g., [Lov78]). That condensing preserves completeness is shown in [FLTZ].

If for all $\mathcal{C}$ in a class of clause sets Γ there exists an i s.t. $R^*_{HS}(\mathcal{C}) = R^i_{HS}(\mathcal{C})$ then the computation of $R^*_{HS}(\mathcal{C})$ defines a decision procedure for Γ; we say that R_{HS} decides Γ. We now define two classes which can be decided by R_{HS}.

Definition 3.4 PVD_+ *is the set of all sets of clauses* $\mathcal{C}$ *s.t. for all* $C \in \mathcal{C}$*:*

(1) $V(C_+) \subseteq V(C_-)$, *and*

(2) for all $x \in V(C^+)$*:* $\tau_{\max}(x, C_+) \leq \tau_{\max}(x, C_-)$.

(Observe, that (1) implies that all positive clauses in $\mathcal{C}$ *must be ground).*

PVD_+ is a subclass of PVD, a class that has been demonstrated to be decidable in [Lei92] and [FLTZ]. But there is no loss of generality in investigating PVD_+ instead of PVD, since the decision procedure for PVD can be easily reduced to that for PVD_+: A clause set $\mathcal{C}$ is in PVD iff there is a renaming η of the signs of the literals s.t. $\eta(\mathcal{C}) \in \mathrm{PVD}_+$. To get a resolution decision procedure for PVD we first construct an adequate renaming η and then apply R_{HS} to $\eta(\mathcal{C})$. Instead of transforming $\mathcal{C}$ to $\eta(\mathcal{C})$ one may also apply a different semantical setting (see [Lov78]) to $\mathcal{C}$ itself.

PVD_+ can be considered as a generalization of DATALOG to clause forms containing function symbols. The decidability of PVD_+ restricted to Horn was shown in [Lei90].

Definition 3.5 $\mathrm{OCC1N}_+$ *is the set of all sets of clauses* $\mathcal{C}$ *s.t. for all* $C \in \mathcal{C}$*:*

(1) $OCC(x, C_+) = 1$ *for all* $x \in V(C_+)$, *and*

(2) $\tau_{\max}(x, C_+) \leq \tau_{\min}(x, C_-)$ *for all* $x \in V(C - +) \cap V(C_-)$*:*

Like in the case of PVD_+, $\mathcal{C} \in \mathrm{OCC1N}$ iff there exists a renaming η of the signs of the literals s.t. $\eta(\mathcal{C}) \in \mathrm{OCC1N}_+$. OCC1N is shown to be decidable in [FLTZ]; the decidability of $\mathrm{OCC1N}_+$ restricted to Horn clauses was proved in [Fer90].

PVD_+ is even decidable by positive hyperresolution without subsumption and condensing (see [Lei92]), however, in deciding $\mathrm{OCC1N}_+$ condensing and restricted factoring are essential (see [FLTZ]). As R_{HS} is based on condensing, restricted factoring and subsumption, R_{HS} decides $\mathrm{PVD}_+ \cup \mathrm{OCC1N}_+$.

In the following sections the sets $R^*_{HS}(\mathcal{C})$ will serve as "raw material" for model building. All results obtained in this paper hold for the decidable class $\mathrm{PVD}_+ \cup \mathrm{OCC1N}_+$, although some of them are more general. A wider range of decidable classes (generalizations of PVD) was obtained in [Lei92] with the abstract concept of atom complexity measures; all these classes are decidable by hyperresolution and admit the finite model building procedure introduced in section 5.

4 Construction of finite representations of Herbrand models

In this section we describe a method to specify H-models for sets of clauses $\mathcal{C}$ where $R^*_{HS}(\mathcal{C})$ is finite and its clauses are decomposed (i.e., their literals are variable disjoint). This is the case for $\mathcal{C} \in$ PVD $\cup$ OCC1N (and, of course, also for every class Δ where $R^*_{HS}(\mathcal{C})$, for all $\mathcal{C} \in \Delta$, is finite and ground).
We use sets of (general) atoms to describe H-interpretations:

Definition 4.1 *$\mathcal{A}$ is called a* standard H-interpretation representation (SHIR) *of an H-interpretation $\mathcal{I}$ for a clause set $\mathcal{C}$ iff $\mathcal{I} = \{P\gamma \mid P \in \mathcal{A}, \gamma \in \mathbf{GS}(\mathcal{C})\}$. If $\mathcal{I}$ is a model of $\mathcal{C}$ then $\mathcal{A}$ is a* standard H-model representation (SHMR).

Of specific interest are cases where an SHMR is finite, but the corresponding H-model $\mathcal{I}$ is not.
Example 4.1. Let $\mathcal{C} = \{P(f(x)),\ \neg P(x) \vee P(f(x)),\ \neg P(a)\}$. Then $\mathbf{HU}(\mathcal{C}) = \{a, f(a), f(f(a)), \ldots\}$ and $\mathcal{M} = \{P(f(t)) \mid t \in \mathbf{HU}(\mathcal{C})\}$ is an H-model of $\mathcal{C}$. Therefore $\mathcal{A} = \{P(f(x))\}$ is an SHMR of $\mathcal{M}$.
Observe, that if positive hyperresolution terminates on a set $\mathcal{C}$ of Horn clauses then it produces a finite SHMR of an H-model of $\mathcal{C}$. This is trivially so, since positive hyperresolution only produces positive unit clauses. Therefore the union over all hyperresolvents (i.e., the derivable positive clauses) constitutes an SHMR (in fact an SHMR of the minimal H-model of $\mathcal{C}$). We shall demonstrate that also for non-Horn clause sets hyperresolution allows to construct an SHMR algorithmically.

Definition 4.2 *A set of clauses $\mathcal{C}$ is (R_{HS}-)*stable *iff $R_{HS}(\mathcal{C}) = (\mathcal{C})$.*

Observe that, by definition, $R^*_{HS}(\mathcal{C})$ is always stable.

Lemma 4.1 *For a clause set $\mathcal{C}$ let $\mathcal{A}$ be a finite set of atoms (over the signature of $\mathcal{C}$) s.t. $NP(\mathcal{C}) \cup \mathcal{A}$ is stable and satisfiable. Then $\mathcal{A}$ is an SHMR for $NP(\mathcal{C}) \cup \mathcal{A}$.*

Proof. Suppose, on the contrary, that $\mathcal{A}$ is not an SHMR. Then the set $NP(\mathcal{C}) \cup \mathcal{A} \cup \{\neg P | P \in \mathbf{HB}(\mathcal{C}) - \mathcal{I}\}$, where $\mathcal{I}$ is the H-interpretation represented by the SHIR $\mathcal{A}$, is unsatisfiable.
By the compactness theorem there exists a finite $\mathcal{F}$, $\mathcal{F} \subseteq \{\neg P | P \in \mathbf{HB}(\mathcal{C}) - \mathcal{I}\}$ s.t. $NP(\mathcal{C}) \cup \mathcal{A} \cup \mathcal{F}$ is unsatisfiable. $\mathcal{F}$ must be non-empty as, by assumption, $NP(\mathcal{C}) \cup \mathcal{A}$ is satisfiable.
But because $\mathcal{F}$ consists of negative unit clauses only and since R_{HS} is complete $\square \in NP(\mathcal{C}) \cup \mathcal{A} \cup \mathcal{F}$ iff there exists a clause $Q \in \mathcal{A}$ s.t. $\neg Q\gamma \in \mathcal{F}$ for some $\gamma \in \mathbf{GS}(\mathcal{C})$. (Remember that $NP(\mathcal{C}) \cup \mathcal{A}$ is stable and thus $RP^*(NP(\mathcal{C}) \cup \mathcal{A}) = \mathcal{A}$). Since $Q\gamma \in \mathcal{I}$, $\neg Q\gamma \notin \mathcal{F}$. Thus such a clause Q cannot exist. It follows that $\mathcal{A}$ is an SHMR. ∎

Lemma 4.1 suggests the following idea to identify a model of $\mathcal{C}$: Suppose $R^*_{HS}(\mathcal{C})$ is finite. Find a finite set of atoms $\mathcal{A}$ s.t. $NP(\mathcal{C}) \cup \mathcal{A}$ is stable, satisfiable and implies $\mathcal{C}$. Then $\mathcal{A}$ is an SHMR of an H-model of $\mathcal{C}$.

Definition 4.3 *We call $\mathcal{C}$ positively decomposed and write $\mathcal{C} \in PDC$ iff $RP^*(\mathcal{C})$ is finite and all clauses in $RP^*(\mathcal{C})$ are decomposed.*

Lemma 4.2 *Let $\mathcal{C} \in PDC$ and suppose that $RP^*(\mathcal{C})$ contains a non-unit clause D s.t. P is an atom in D. Then $NP(\mathcal{C}) \cup (RP^*(\mathcal{C}) - \{D\}) \cup \{P\}$ is satisfiable iff $\mathcal{C}$ is satisfiable.*

Proof. ($\Rightarrow$): This part of the proof is trivial as $P \rightarrow D$ is valid.
($\Leftarrow$): Suppose that $\mathcal{C}_1 = NP(\mathcal{C}) \cup (RP^*(\mathcal{C}) - \{D\}) \cup \{P\}$ is unsatisfiable. Then there exists an R_{HS}-refutation Γ_1 of $\mathcal{C}_1$. We transform Γ_1 into a deduction on $\mathcal{C}$:
In every clash of Γ_1 replace the electron $P\eta$ (η being a variable renaming substitution) by $D\eta$ and resolve $P\eta$ out of $D\eta$. Observe that Γ is an R_{HS}-deduction, too.
Let C_1 be a clash resolvent in Γ_1 and let C be the corresponding clash resolvent C in Γ. As D is decomposed, the application of the m.g.u. of this resolution step does not change $D - P$. However, $D - P$ may appear in form of different variants. Thus C is a variant of $C_1 \vee (D - P)\eta_1 \vee \ldots \vee (D - P)\eta_k$ for some variable renaming substitutions $\eta_1, \ldots, \eta_k$.
Because Γ_1 is a refutation of $\mathcal{C}_1$, Γ derives a clause of the form $(D - P)\eta_1 \vee \ldots \vee (D - P)\eta_k$ out of $\mathcal{C}$. Since R^*_{HS} is condensed, we conclude $(D - P)\eta \in RP^*(\mathcal{C})$ for some renaming substitution η. But $(D - P)\eta \leq_s D$ (and not $D \leq_s (D - P)\eta$ as D is already condensed). By definition, $R^*_{HS}(\mathcal{C})$ cannot contain different clauses D', D s.t. $D' \leq_s D$. Thus D cannot be in $RP^*(\mathcal{C})$ which contradicts the assumption: It follows that $\mathcal{C}_1$ is satisfiable. ■

Remark. Because $\mathcal{C}_1 \rightarrow \mathcal{C}$ is valid we may replace $\mathcal{C}$ by $\mathcal{C}_1$ and construct a model $\mathcal{M}$ for $\mathcal{C}_1$. $\mathcal{M}$ is then a model of $\mathcal{C}$, too.
We may apply lemma 4.2 iteratively: After construction of $\mathcal{C}_1$ compute $R^*_{HS}(\mathcal{C}_1)$ and repeat the procedure with $R^*_{HS}(\mathcal{C}_1)$. We are only left with the problem of the termination of this process.
The following example shows that $\mathcal{C}_1$ need not be stable. (If $\mathcal{C}_1$ would always be stable, we could replace $RP^*(\mathcal{C})$ by a set of unit clauses at once).
Example 4.2 . Let $\mathcal{C} = \{P(x) \vee Q(x) \vee \neg R(x) \vee \neg S(x), R(a) \vee Q(a), S(a) \vee E(a), \neg P(a) \vee \neg Q(a)\}$. Then $RP^*(\mathcal{C}) = \{R(a) \vee Q(a), S(a) \vee E(a), P(a) \vee Q(a) \vee E(a)\}$. We replace $S(a) \vee E(a)$ by $S(a)$ and get $\mathcal{C}_1 = (RP^*(\mathcal{C}) - \{S(a) \vee E(a)\} \cup \{S(a)\}) \cup NP(\mathcal{C})$. From lemma 4.2 we know that C_1 is satisfiable if $\mathcal{C}$ is so. But $\mathcal{C}_1$ is not stable, because $RP(\mathcal{C}_1) = \{R(a) \vee Q(a), S(a), P(a) \vee Q(a)\}$. Thus we replace $NP(\mathcal{C}) \cup RP^*(\mathcal{C})$ by $NP(\mathcal{C}) \cup RP^*(\mathcal{C}_1)$. We define $\mathcal{C}_2 = (\mathcal{C}_1 - \{P(a) \vee Q(a)\}) \cup \{Q(a)\}$ and get $RP^*(\mathcal{C}_2) = \{Q(a),\ S(a)\}$. $NP(\mathcal{C}) \cup \{Q(a),\ S(a)\}$ is stable. Thus $\{Q(a),\ S(a)\}$ is an H-model for $\mathcal{C}$.
The replacement of $\mathcal{C}$ by $\mathcal{C}_1$ can be defined algorithmically: Let $NP(\mathcal{C}) \cup \mathcal{D}$ be stable then $\alpha(NP(\mathcal{C}) \cup \mathcal{D}) = NP(\mathcal{C}) \cup \alpha(\mathcal{D})$, where α replaces a non-unit clause in $\mathcal{D}$ by one of its atoms. For the model construction method it is important that $R^*_{HS}(\alpha(\mathcal{C}))$ is smaller than $\mathcal{C}$ for a stable set of clause $\mathcal{C}$.

Definition 4.4 *For clause sets $\mathcal{C}$, $\mathcal{C}'$: $\mathcal{C} < \mathcal{C}'$ iff for all $C \in \mathcal{C}$ there exists a $D \in \mathcal{C}'$ s.t. $C \leq_s D$ and there exists a $C \in \mathcal{C}$ s.t. $D \not\leq_s C$ for all $D \in \mathcal{C}'$.*

It is easy to verify that there is no infinite sequence $\mathcal{C}_1 > \ldots > \mathcal{C}_n > \mathcal{C}_{n+1} > \ldots$ s.t. all $\mathcal{C}_i$ are stable. Thus $<$ is Noetherian on stable sets of clauses. Our goal is to show that $R^*_{HS}(\alpha(\mathcal{C})) < R^*_{HS}(\mathcal{C})$ for all sets of clauses $\mathcal{C} \in PDC$.

Lemma 4.3 *Let $\mathcal{C}$ be a stable, satisfiable set in PDC s.t. there is a non-unit clause among the positive clauses in $\mathcal{C}$. Then $R^*_{HS}(\alpha(\mathcal{C})) < \mathcal{C}$.*

Proof. First we show that there exists a $C \in R^*_{HS}(\alpha(\mathcal{C}))$ with $D \not\leq_s C$ for all $D \in \mathcal{C}$. Let E be a positive clause in $\mathcal{C}$ with $|E| \geq 2$ and P an atom of E s.t. $P \in \alpha(\mathcal{C})$. P cannot be in $\mathcal{C}$ as $\mathcal{C}$ is stable and thus reduced under subsumption (i.e. E cannot be in $\mathcal{C}$ since it would be properly subsumed by P). Moreover P cannot be subsumed by a clause in $\mathcal{C}$. By definition of R^*_{HS}, either $P \in R^*_{HS}(\alpha(\mathcal{C}))$ or some $P_0 \in R^*_{HS}(\alpha(\mathcal{C}))$ s.t. $P_0 \leq_s P$. In any case there is some $Q \in R^*_{HS}(\alpha(\mathcal{C}))$ s.t. Q is not subsumed by any clause in $\mathcal{C}$.
It remains to show that

(1) for all $C \in R^*_{HS}(\alpha(\mathcal{C}))$ there is a $D \in \mathcal{C}$ s.t. $C\theta \subseteq D$ for a variable renaming substitution θ. (It follows $C \leq_s D$).

We prove by induction that (1) holds for $R^n_{HS}(\alpha(\mathcal{C}))$:

For $n = 0$ the property is trivially fulfilled by definition of α and by the stability of $\mathcal{C}$.

(IH) Suppose that (1) holds for $R^n_{HS}(\alpha(\mathcal{C}))$.

Let Γ be a clash from clauses in $R^n_{HS}(\alpha(\mathcal{C}))$ s.t. $\Gamma = (C; D_1, \ldots, D_n)$ for $C \in NP(\mathcal{C})$, $D_i \in R^n_{HS}(\alpha(\mathcal{C}))$. By (IH), for every D_i there exists an $E_i \in \mathcal{C}$ s.t. $D_i\theta_i \subseteq E_i$ for a variable renaming substitution θ_i. Replace Γ by $\Gamma' = (C; E_1, \ldots, E_n)$. For every selection of atoms Q_i in D_i (for resolution) we have $Q_i\theta_i \subseteq E_i$ (for a renaming substitution θ_i) and therefore we can "simulate" the clash for C (i.e. we can resolve with $Q_i\theta_i$ instead of Q_i).

Suppose now that D is the clash resolvent of Γ and E the corresponding clash resolvent of Γ'. Then $D\eta \subseteq E$ for a renaming substitution η. Thus (1) also holds for $n+1$. It follows that for all $C \in R^*_{HS}(\alpha(\mathcal{C}))$ there is a $D \in \mathcal{C}$ s.t. $C \leq_s D$. ∎

We are now prepared to define a model building algorithm MB: (Let PU denote the class of all $\mathcal{C}$ s.t. $RP^*(\mathcal{C})$ consists of unit clauses only.)

> **program** *MB*
> {**Input:** clause set $\mathcal{C}$ }
> {**Output:** SHMR $\mathcal{A}$ (or **failure**) }
> **begin**
> $\mathcal{D} := R^*_{HS}(\mathcal{C})$;
>   **if** $\square \in \mathcal{D}$ **then failure**;
>   **while** $\mathcal{D} \notin PU$ **do** $\mathcal{D} := R^*_{HS}(\alpha(\mathcal{D}))$;
> $\mathcal{A} := P(\mathcal{D})$; (positive clauses of $\mathcal{D}$)
>   **return** $\mathcal{A}$
> **end.**

Theorem 4.1 *For every $\mathcal{C} \in PDC$ algorithm MB terminates and for all satisfiable $\mathcal{C} \in PDC$ MB yields a (finite) SHMR.*

Proof. By lemma 4.3 MB terminates. By lemma 4.2 we know that for satisfiable $\mathcal{C}$, $\alpha(\mathcal{C})$ is satisfiable, too, and $\alpha(\mathcal{C}) \to \mathcal{C}$.

If $\mathcal{C}$ is unsatisfiable then, by completeness of R_{HS}, MB terminates with failure.

If $\mathcal{C}$ is satisfiable then, by lemma 4.3, MB also terminates. Let $\mathcal{C}'$ be the last value of $\mathcal{D}$. Then, by iterated application of lemma 4.2, $\mathcal{C}'$ is stable, $\mathcal{C}' \to \mathcal{C}$ and $\mathcal{C}'$ is satisfiable. Since $\mathcal{C}'$ is in PU, $\mathcal{A}$ is a finite set of unit clauses. Since $\mathcal{C}' = NP(\mathcal{C}') \cup \mathcal{A}$ is stable, lemma 4.1 implies that $\mathcal{A}$ is a finite SHMR for $\mathcal{C}$. ∎

5 Extracting finite models from SHMR's

In the last section we have presented an algorithm for the construction of finite descriptions of H-models for certain classes of clause sets. Since H-models are infinite (except for cases where there are no function symbols in the clause set) it is natural to ask whether also finite models can be constructed along the approach of section 4. In particular, we do not want to introduce equational reasoning methods but only want to rely on the data generated by hyperresolution. We shall see that for clause sets in $\mathrm{PVD}_+ \cup \mathrm{OCC1N}_+$ it is always possible to extract a description of a finite model from an SHMR generated by MB. This is essentially achieved by truth value preserving projections of the Herbrand base into finite subsets of it.

For an elaborated example covering all concepts of this section and the proofs of lemmata 5.1, 5.2 and 5.3 we refer to the technical report [FL92].

We first characterize a class of SHMR's that allow for such a construction:

Definition 5.1 *We call a (finite) SHMR $\mathcal{A}$* appropriate for finite model building *iff for all $A \in \mathcal{A}$ and $x \in V(A)$: $OCC(x, A) = 1$. For such SHMR's let $T(\mathcal{A})$ denotes the set of all non-variable subterms of atoms in $\mathcal{A}$.*

Observe that, by the properties of hyperresolvents and the results of the last section, any SHMR constructed by algorithm MB for a clause set in class PVD_+ $\cup$ $\mathrm{OCC1N}_+$ is appropriate for finite model building.

Note that, although the classes PVD_+ and $\mathrm{OCC1N}_+$ are incomparable, the termination sets $RP^*(\mathcal{C})$ for $\mathcal{C} \in \mathrm{PVD}_+$ form a proper subset of the termination sets w.r.t. $\mathrm{OCC1N}_+$. Since we only consider termination sets in this section, PVD_+ appears as a special case of $\mathrm{OCC1N}_+$.

From now on we shall always refer to a fixed SHMR $\mathcal{A}$, that is appropriate for finite model building. Thus the set of all ground terms under consideration is uniquely defined as $\mathbf{HU}(\mathcal{A})$. We shall also write $t \in \mathrm{OCC1}$ as a shortcut for the fact that t is a non-variable term s.t. $\forall x \in V(t) : OCC(x,t) = 1$. (Clearly, all ground terms are in OCC1.) Moreover, just as in example 5.1, we identify expressions that are equal up to renaming of variables.

Clearly, $T(\mathcal{A})$ represents the set $G(T(\mathcal{A}))$ of all (ground) terms occurring in the atoms of the H-model represented by $\mathcal{A}$ (cf. definition 5.2). However, to get an appropriate domain for a finite model, we have to extend $T(\mathcal{A})$ to a representation of $G(T(\mathcal{A}))$ that is functional:

Definition 5.2 *A set of terms T* orthogonally represents *the set of its ground instances $G(T)$ iff for all $s \in G(T)$ there is exactly one $t \in T$ s.t. s is an instance of t.*

T is an orthogonal extension of *a set of terms T' if T orthogonally represents $G(T') = G(T)$ and if for each $t' \in T'$ there is an instance t of t' in T.*

Example 5.2. $T = \{a,\ f(x, f(u, v)),\ f(a, y),\ f(x, a)\}$ represents $G(T)$, which is set of all terms built up from a and f only. T is not an orthogonal representation since, e.g., $f(a, f(a, a)$ is an instance of both, $f(x, f(u, v))$ and $f(a, y)$. $T_1 = \{a, f(x, y)\}$ orthogonally represents the $G(T)$ but it is not an orthogonal extension of T, since e.g., no instance of $f(a, y)$ occurs in T_1. However, $T_2 = \{a,\ f(x, f(u, v)),\ f(a, a),\ f(f(x, y), a)\}$ is an orthogonal extension of T.

Remark. Observe that T is an orthogonal representation iff the terms in T are pairwise not unifiable.

Below we present an algorithm that constructs for any finite set T of terms in OCC1 another finite set of terms in OCC1 that orthogonally extends T.

We need some additional terminology:

Definition 5.3 *The set of* base terms BT *is defined as the set of all constants and all terms of the form $f(x_1, \ldots, x_n)$ in OCC1. (I.e., $x_1, \ldots, x_n$ are pairwise distinct variables.)*

Remark. Remember that we implictely refer to a fixed (finite) alphabet. Therefore, BT is uniquely defined and finite.

We need to characterize the set of all terms that are not instances of a certain term:

Definition 5.4 *Let t be in OCC1. Then the set of terms $co(t)$ is inductively defined as follows:*

(i) If t is a constant then $co(t) = BT - \{t\}$.

(ii) If $t = f(t_1, \ldots, t_n)$ then $co(t) = (BT - \{f(x_1, \ldots, x_n)\}) \cup \{f(s_1, \ldots, s_n) \mid s_i = t_i$ or $s_i \in co(t_i)$ but at least one $s_i \neq t_i\}$, where $V(s_i) \cap V(s_j) = \emptyset$ for all $i \neq j$. (Observe that $co(t)$ is undefined for variables; thus if t_i is a variable then $s_i = t_i$.)

The relativation of $co(t)$ to a term s—$co(t|s)$—is defined as the subset of $co(t)$ that contains all terms in $co(t)$ that are proper instances of s.

Example 5.3. Let $t = f(x, a)$, and assume that, besides a and f, there is only the unary function symbol g in the alphabet. Then $co(a) = \{g(x),\ f(x, y)\}$ and therefore $co(t) = \{a,\ g(x),\ f(x, g(y)),\ f(x, f(y, z))\}$. $co(t|f(x, y)) = f(x, g(y)),\ f(x, f(y, z))\}$.

The following lemma serves to guarantee the correctness of our orthogonalization algorithm. It can be proved by induction on the term depth (see [FL92]).

Lemma 5.1 *For any $t \in$ OCC1 $co(t)$ is an orthogonal representation of all ground terms that are not instances of t. Moreover, for all $s \in co(t)$: $s \in$ OCC1.*

We are now prepared to present the orthogonalization algorithm.

> **program** *Orthogonalize*
> {**Input:** finite set T of terms in OCC1}
> {**Output:** orthogonal extension of T}
> **begin**
> $T' := T$;
> **while** $\exists t_1, t_2 \in T'(t_1 \neq t_2)$ s.t. t_1 and t_2 are unifiable **do**
> **begin**
> $\theta :=$ mgu of t_1 and t_2;
> $T' := (T' - \{t_1, t_2\}) \cup \{t_1\theta(= t_2\theta)\} \cup co(t_1\theta|t_1) \cup co(t_1\theta|t_2)$
> **end** { **while** };
> **return** T'
> **end**.

Remarks. Remember that we identify terms that are identical up to a renaming of variables. Therefore in our case, e.g., $f(x, g(y))$ is unifiable with $f(g(x), y)(= f(g(u), v))$.

Lemma 5.2 *For each finite set T of terms in OCC1 the algorithm Orthogonalize computes a finite, orthogonal extension T_O of T. Moreover, T_O only contains terms in OCC1.*

Again, the proof of lemma 5.2 can be found in [FL92].
We specify the domain $D_{\mathcal{A}}$ of the finite model as follows:

Definition 5.5 *Let $\mathcal{A}$ be an SHMR that is appropriate for finite model building. We introduce an additional constant d not occurring in $\mathbf{HU}(\mathcal{A})$. (Informally spoken d represents all ground terms that are not represented by any other term of our domain). Let γ_d be the substitution that assigns d to all variables; i.e. $\forall x \in V(\mathcal{A}) : \gamma_d(x) = d$. Moreover, let $T_O(\mathcal{A})$ be an orthogonal representation of $G(T(\mathcal{A}))$. Then $D_{\mathcal{A}} = \{d\} \cup \{t\gamma_d \mid t \in T_O(\mathcal{A}))\}$.*

The fact that—for the SHMR's $\mathcal{A}$ under consideration—there there are always orthogonal extensions of $T(\mathcal{A})$ allows us to define the following function that assigns some element of $D_{\mathcal{A}}$ to each ground term.

Definition 5.6 *Let $\mathcal{A}$ be an SHMR for a clause set C that is appropriate for finite model building and let $T_O(\mathcal{A})$ be an orthogonal extension of $T(\mathcal{A})$. Then for each $t \in \mathbf{HU}(C) \cup \mathcal{D}_{\mathcal{A}}$ we define $\Phi_{\mathcal{A}}(t)$ as follows:*

(i) If t is an instance of some $s \in T_O(\mathcal{A})$ then $\Phi_{\mathcal{A}}(t) = s\gamma_d$ (where γ_d is defined as in definition 5.5).

(ii) Otherwise, let $\Phi_{\mathcal{A}}(t) = d$.

The definition of $\Phi_{\mathcal{A}}$ enables us to specify concisely the interpretation of the function symbols: We are now in the position to present the complete specification of a finite model corresponding to an SHMR.

Definition 5.7 *Let $\mathcal{A}$ be an SHMR that is appropriate for finite model building. Then the (finite) interpretation $\mathcal{FM}_{\mathcal{A}} = \langle D, \varphi \rangle$ is defined as follows:*

(i) $D = D_{\mathcal{A}}$.

(ii) $\varphi(c) = c$ *for all constants* $\in D_{\mathcal{A}}$. *For all constants c' that do not occur in $\mathcal{D}(A)$ (thus also not in $\mathcal{A}$) we define* $\varphi(c') = d$.

(iii) $\varphi(f)(t_1, \ldots, t_n) = \Phi_{\mathcal{A}}(f(t_1, \ldots, t_n))$ *for all n-ary function symbols f.*

(iv) $\langle t_1, \ldots, t_n \rangle \in \varphi(P)$ *iff* $P(t_1, \ldots, t_n)$ *is an instance of some* $A \in \mathcal{A}$. *for all n-ary predicate symbols P and all terms* $t_1, \ldots t_n \in D_{\mathcal{A}}$.

Of course, instead of using the set of terms $D_{\mathcal{A}}$ as domain of discourse, we could map $D_{\mathcal{A}}$ bijectively into any domain of the same cardinality and define φ accordingly.

We remark that if there is an H-model $\mathcal{M}$ for some clause set $\mathcal{C}$, s.t. $\mathcal{M} = \{A \mid v_{\mathcal{M}}(A) = \mathbf{true}\}$ is finite then we have a simple subcase of our approach: In that case $\mathcal{M}$ itself is an SHMR that is appropriate for finite model building. Moreover, the set of subterms occurring in $\mathcal{M}$—$T(\mathcal{M})$—is an orthogonal representation of itself. Therefore $\mathcal{FM}_{\mathcal{M}}$ is a model of cardinality $|T(\mathcal{M}) + 1|$ (the additional element d stands for all ground terms that are not in $T(\mathcal{M})$). It can be demonstrated that in general this model is not minimal w.r.t. the cardinality of the domain.

The main lemma to justify our construction is the following: (It essentially follows from the definition of $\Phi_{\mathcal{A}}$ and the evaluation function.)

Lemma 5.3 *For all ground terms t:* $v_{\mathcal{FM}_{\mathcal{A}}}(t) = \Phi_{\mathcal{A}}(t)$.

Given lemma 5.3 it is an easy task to prove:

Theorem 5.1 *If $\mathcal{A}$ is an SHMR for some clause set $\mathcal{C}$ that is appropriate for finite model building then $\mathcal{FM}_{\mathcal{A}}$ is a finite model for $\mathcal{C}$.*

Proof. Clearly, $\mathcal{FM}_{\mathcal{A}} = \langle D_{\mathcal{A}}, \varphi \rangle$ is a finite interpretation for $\mathcal{C}$. Let $\mathcal{M}_{\mathcal{A}}$ the H-model represented by $\mathcal{A}$. By Herbrand's theorem it suffices to show that for any ground atom A $v_{\mathcal{M}_{\mathcal{A}}}(A) = \mathbf{true} \Longleftrightarrow v_{\mathcal{FM}_{\mathcal{A}}}(A) = \mathbf{true}$. Let $A = P(t_1, \ldots t_n)$ be in $\mathcal{M}_{\mathcal{A}}$. (I.e., A is true in $\mathcal{M}_{\mathcal{A}}$.) By lemma 5.3 $v_{\mathcal{FM}_{\mathcal{A}}}(A) = \mathbf{true}$ iff $\langle \Phi_{\mathcal{A}}(t_1), \ldots \Phi_{\mathcal{A}}(t_n) \rangle \in \varphi(P)$. But since $\Phi_{\mathcal{A}}$ is based on an orthogonal extension of $T(\mathcal{A})$ we have $P(\Phi_{\mathcal{A}}(t_1), \ldots \Phi_{\mathcal{A}}(t_n)) = A'\gamma_d$ for some $A' \in \mathcal{A}$. It follows that A is true in $\mathcal{FM}_{\mathcal{A}}$, too.

On the other hand assume that $\langle \Phi_{\mathcal{A}}(t_1), \ldots \Phi_{\mathcal{A}}(t_n) \rangle \in \varphi(P)$. It follows by the definition of $\Phi_{\mathcal{A}}$ that $A = P(t_1, \ldots t_n)$ is an instance of some $A' \in \mathcal{A}$ s.t. $P(\Phi_{\mathcal{A}}(t_1), \ldots \Phi_{\mathcal{A}}(t_n)) = A'\gamma_d$. In other words, A is true in $\mathcal{M}_{\mathcal{A}}$. ∎

We have already remarked that theorem 5.1 is applicable to all clause sets in class OCC1N. Given the results of section 4 it is clear that $\mathcal{FM}_{\mathcal{A}}$ can be constructed effectively. Indeed, since $\mathcal{A}$ can be found without any kind of backtracking this also holds for the finite model $\mathcal{FM}_{\mathcal{A}}$. This is in sharp contrast with methods relying on traditional model theoretic results for the decision problem (as e.g.

described in [DG79]). There only upper bounds for the cardinality of finite models are given and thus representations of models would have to be determined through exhaustive search in the—generally far too large—space of all interpretations of a certain cardinality.

The following example demonstrates that, using our approach for finite model building (i.e., projecting the infinite domain of an H-model into a finite set), one cannot go essentially beyond the termination type OCC1.

Proposition 5.1 *Let $\mathcal{A} = \{P(x, f(x))\}$ be an SHMR of an H-model $\mathcal{M}_{\mathcal{A}}$. Then there is no finite interpretation $\mathcal{M}_D$ of (the Herbrand base of) $\mathcal{A}$ s.t. for all ground atoms P: $v_{\mathcal{M}_D}(P) = \mathbf{true} \Longleftrightarrow v_{\mathcal{M}_{\mathcal{A}}}(P) = \mathbf{true}$.*

Proof. Let $\mathcal{M}_D = \langle D, \varphi_D \rangle$ be any finite interpretation of $\mathbf{HB}(\mathcal{A})$. By definition of the evaluation function, we have $v_{\mathcal{M}_D}(f^n(a)) = (\varphi_D(f))^n(\varphi_D(a))$, where a is the only constant in $\mathbf{HU}(\mathcal{A})$. Since $\varphi_D(f)$ can only take finitely many different values there must be some $n, m \geq 1$, s.t. $n \neq m$ but $v_{\mathcal{M}_D}(f^n(a)) = v_{\mathcal{M}_D}(f^m(a))$. This implies that $v_{\mathcal{M}_D}(P(f^n(a), f^{n+1}(a)) = v_{\mathcal{M}_D}(P(f^m(a), f^{n+1}(a))$. On the other hand, by definition of $\mathcal{M}_{\mathcal{A}}$, $v_{\mathcal{M}_{\mathcal{A}}}(P(f^k(a), f^{l+1}(a)) = \mathbf{true}$ iff $k = l$. It follows $v_{\mathcal{M}_{\mathcal{A}}}(P(f^m(a), f^{n+1}(a)) \neq v_{\mathcal{M}_{\mathcal{A}}}(P(f^m(a), f^{n+1}(a))$. Therefore $v_{\mathcal{M}_{\mathcal{A}}}(P)$ and $v_{\mathcal{M}_D}(P)$ are different for some ground atoms P. ∎

6 Conclusion

In this work we aimed at the construction of models for certain sets of clauses using the set of all hyperresolvents, given that hyperresolution terminates without deriving □. In particular, we investigated two classes of clause sets that were demonstrated to be decidable by hyperresolution in [FLTZ]. The process of model construction proceeds in two steps: First hyperresolution is employed to arrive at a finite set of atoms that represents a description of an Herbrand model. In a second step we extract from this set of atoms a full representation of a model with finite domain. We emphasize that the second step can be performed on every stable set of this specific syntax type, no matter how this set was obtained. In this sense, step 2 is independent of step 1.

We consider the work of T. Tammet and the presented results just as a starting point in the more general project to extract (information about) models from termination sets of resolution provers. For example, we conjecture that any clause set s.t. hyperresolution terminates without yielding contradiction has a finite model. Clearly, more refined methods than those of section 5 above have to be used to settle this problem. Another range of questions arises if we do not only consider hyperresolution but also ordering strategies and other resolution refinements, possibly combined with certain methods for equality reasoning.

References

[CL73] C.-L. CHANG AND R.C.-T. LEE, *Symbolic Logic and Mechanical Theorem Proving*. Academic Press, New York and London, 1973.

[CZ91] R. CAFERRA AND N. ZABEL, *Extending Resolution for Model Construction*. In: Logics in AI (JELIA '90). Springer Verlag, LNCS 478 (1991), pp. 153–169.

[DG79] B. DREBEN AND W.D. GOLDFARB, *The Decision Problem*. Addison-Wesley, Massachusetts 1979.

[Fer90] C.G. FERMÜLLER, *Deciding some Horn Clause Sets by Resolution*. In: Yearbook of the Kurt-Gödel-Society 1989, Vienna 1990, pp. 60–73.

[Fer91] C.G. FERMÜLLER, *Deciding Classes of Clause Sets by Resolution*. Ph.D. Thesis, Technical University Vienna, 1991.

[FL92] C.G. FERMÜLLER AND A. LEITSCH, *Model Building by Resolution*. Technical Report TUW-E185.2-FL1, TU Wien, 1992.

[FLTZ] C.G. FERMÜLLER, A. LEITSCH, T. TAMMET, AND N. ZAMOV, *Resolution Methods for the Decision Problem*. To appear in LNCS, Springer Verlag.

[Hul88] J.-M. HULLOT, *Canonical Forms and Unification*. In: 5th Conference on Automated Deduction. Springer Verlag, LNCS 87 (1980), pp. 250–263.

[Joy76] W.H. JOYNER, *Resolution Strategies as Decision Procedures*. J. ACM 23,1 (July 1976), pp. 398-417.

[Lei90] A. LEITSCH, *Deciding Horn Classes by Hyperresolution*. In: CSL'89. Springer Verlag, LNCS 440 (1990), pp. 225–241.

[Lei92] A. LEITSCH, *Deciding Clause Classes by Semantic Clash Resolution*. To appear in: Fundamenta Informaticae.

[Lov78] D. LOVELAND, *Automated Theorem Proving — A Logical Basis*. North Holland Publ. Comp. 1978.

[MB88] R. MANTHEY AND F. BRY, *SATCHMO: A theorem prover implemented in Prolog*. In: 9th Conference on Automated Deduction. Springer Verlag, LNCS 310 (1988), pp. 415–434.

[Nol80] H. NOLL, *A Note on Resolution: How to Get Rid of Factoring Without Loosing Completeness*. In: 5th Conference on Automated Deduction. Springer Verlag, LNCS 87 (1980), pp. 250–263.

[Tam91] T. TAMMET, *Using Resolution for Deciding Solvable Classes and Building Finite Models*. In: Baltic Computer Science. Springer Verlag, LNCS 502 (1991), pp. 33–64.

[Tam92] T. TAMMET, *Resolution Methods for Decision Problems and Finite Model Building*. Dissertation, Department of Computer Science, Chalmers University of Technology. Chalmers/Göteburg, 1992.

[Win82] S. WINKER, *Generation and Verification of Finite Models and Counterexamples Using an Automated Theorem Prover Answering Two Open Questions*. J. of the ACM, Vol. 29/2, April 1982, pp. 273–284.

Comparative transition system semantics

Tim Fernando*
fernando@cwi.nl

CWI, P.O. Box 4079, 1009 AB Amsterdam, The Netherlands

Abstract. Employing the notion of a transition system, programs, conceived as binary (transition) relations on states, are related to processes, viewed as dynamic states. The comparative study is carried out syntactically over rules for transitions, and semantically in terms of bisimulation equivalence. A certain form of transitions is studied, and a "logical" approach to the notion of a bisimulation is taken that are somewhat non-standard (but, it is hoped, illuminating). Sequential composition, non-deterministic choice, iteration, and interleaving are analyzed alongside a notion of data. Atomization and synchronization are also considered.

A *(labelled) transition system* is a triple $\langle L, S, \rightarrow\rangle$ consisting of a set L of *labels*, a set S of *states*, and a *transition relation* $\rightarrow \subseteq S \times L \times S$. This general scheme brings together various notions of "transitions" $s \xrightarrow{l} s'$ (encoded as $(s, l, s') \in \rightarrow$), where labels and states can be analyzed in terms of each other. In some studies, such as process algebra (see, for instance, Baeten and Weijland [2] and the references cited therein), a label is taken to be primitive, the main point being to analyze what a state — i.e., process — is. Another point of view reduces a label to the pairs of states that it relates via $\rightarrow$. Indeed, the analysis in dynamic logic (see, for example, Harel [11]) of a program as a binary relation on data-states yields a transition system whose labels are programs, and whose states are data-states. Now, although any family $\mathbf{R}$ of binary relations on some set X can be regarded as a transition system $\langle \mathbf{R}, X, \leadsto\rangle$ with

$$\leadsto \;=\; \{(x, R, y) \in X \times \mathbf{R} \times X \mid (x, y) \in R\}\ ,$$

it is not the case that every transition system can be obtained this way. A ternary relation $\rightarrow$ can support at least two different kinds of transitions that a set of binary relations cannot (under the usual encoding of relations as sets). First, $\rightarrow$ allows two distinct labels l and l' to relate the same states

$$\{(s, s') \mid s \xrightarrow{l} s'\} \;=\; \{(s, s') \mid s \xrightarrow{l'} s'\}\ .$$

* My thanks to Jan Willem Klop for suggesting that a report entitled "Comparative transition system semantics" be written, and for help along the way; also to Jan Rutten, Daniele Turi, Alban Ponse, Frits Vaandrager, Johan van Benthem, Prakash Panangaden, Fer-Jan de Vries, Jan van Eijck, and Franck van Breugel for useful discussions. The work was funded by the Netherlands Organization for Scientific Research (NWO project NF 102/62-356, 'Structural and Semantic Parallels in Natural Languages and Programming Languages').

Second, by the usual axioms of set theory, circular predicates such as $(l,l) \in l$ can never hold, whereas $l \xrightarrow{l} l$ might. Now, if labels are taken to be programs, then the possibility of distinct labels relating the same states is, in fact, just what the view of programs as relations on states forbids. In contrast, as will become clear below, a restriction to non-circular transitions would represent a severe limitation for an account of programs (inasmuch as that account requires an analysis of intermediate states in a possibly non-terminating computation). The restriction can be relaxed, but not without introducing complications of its own, revolving around equality — complications that can be overcome either by reading "programs as relations" as "programs isomorphic to relations", or, alternatively, by passing to the non-well-founded set theory presented in Aczel [1]. These "alternatives" turn out to be equivalent, and, what's more, related to a well-known notion of equivalence on transition systems based on so-called bisimulations (Park [13]).

Conceptually, then, the present paper is directed towards reconciling two points of view,

(**P1**) a program as a (binary) relation on states,

and

(**P2**) a process as a state in a transition system,

enriching the former by incorporating a program into its notion of state, and the latter by passing from atomic transitions to a finite sequence of such transitions. The comparison given of (**P1**) and (**P2**) is carried out for the regular program constructs (viz., sequential composition, non-deterministic choice and iteration) and interleaving over a notion of data. Atomization and synchronization are also discussed briefly in the final section and in an appendix.

The paper begins with an informal account of five transition systems from dynamic logic, process algebra and the ground in between (more specifically, transition systems from Groote and Ponse [9], and De Boer, Kok, Palamidessi, and Rutten [6], which have been modified to highlight their essential characteristics — at least from the point of view of the present paper). This is followed in section 2 by a syntactic study based on a single unifying transition system. A semantics for (**P1**) is worked out in section 3, providing an "alternative" approach (based on (**P1**)) to the notion of a bisimulation designed for (**P2**). Section 4 relates the different transition systems semantically by analyzing models that are in a suitable sense final. Some directions for further work are discussed in section 5, and an appendix containing certain technical details behind section 3 attached.

1 Five transition systems

The present section concentrates on essential intuitions, leaving a precise formulation of the transition systems to the next section. In addition to writing $\langle L, S, \rightarrow\rangle$ for a transition system (to be instantiated below in different ways), the following notation is adopted, and kept throughout the paper. D is a set of "data-states" $d, d', d_1, \ldots$; P_1 is the set of "program(term)s" $p, \ldots$ given by

$$p \ ::= \ a \mid p_1; p_2 \mid p_1 + p_2 \mid p_1 \| p_2 \mid p_1^*$$

over a set A of "atomic programs" a, each of which comes with an "interpretation" $I_a \subseteq D \times D$.

While there is general agreement as to what ;,+ and * mean, the "parallel" construct $\|$ is a bit more troublesome. A view of $\|$ is adopted throughout this paper, allowing the program $x := 1 \| (x := 0; x := x + 1)$ to end in a state where $x = 2$ (resulting from interleaving $x := 1$ between $x := 0$ and $x := x + 1$). Otherwise, the program $x := 0; x := x + 1$ should be "atomized" (see De Bakker and De Vink [3], the references cited therein, and also section 5 below) before composing it in parallel with $x := 1$. Beyond interleaving, however, a notion of synchronization is often incorporated into $\|$. Such complications are taken up in Appendix B.

1.1 Dynamic logic (semantics)

The first example is an extreme, extensional formulation of (**P1**) where states are inputs and/or outputs.

Example 1. Let P_0 be the set of programs in P_1 with no occurrences of $\|$. The semantic framework underlying dynamic logic provides a transition system where $L = P_0$ and $S = D$, with transitions written

$$d \ [\![p]\!]_0 \ d' \tag{1}$$

indicating that program p can yield (as output) d', given input d (and where $[\![a]\!]_0 = I_a$). (Note: the present paper concerns only the semantic component of dynamic logic, and not the modal logic for analyzing these structures.)

A well-known limitation of (1) is that it cannot support a compositional account of parallelism in that $[\![x := 1]\!]_0 = [\![x := 0; x := x + 1]\!]_0$, whereas $[\![x := 1 \| x := 1]\!]_0$ better not equal $[\![x := 1 \| (x := 0; x := x + 1)]\!]_0$. (The concurrent extension of dynamic logic in Peleg [14] where programs are interpreted as subsets of $D \times 2^D$ is inconsistent with this view.) Beyond the matter of compositionality, a direct analysis of intermediate states is suggested also by interest in non-terminating computations.

1.2 Process algebra

A fine-grained, intensional analysis of programs is developed in

Example 2. Taking for granted a translation into P_1 of the corresponding processes, an example of a transition system in process algebra is provided by $L = A + \{\surd\}$ and $S = P_1 + \{\surd\}$, with transitions

$$p \xrightarrow{a} p' \tag{2}$$

indicating that process p can execute the atomic program a, and in so doing transform itself into p'. The symbol $\surd$ denotes successful termination.

1.3 Process algebra with data

Comparing Example 1 with Example 2, a question that arises is what happened to the data-states in D? D is (re-)introduced into the picture in

Example 3. In Groote and Ponse [9], $L = A + \{\surd\}$, and $S = (P_1 + \{\surd\}) \times D$ (so-called "configurations"), with transitions of the form

$$(p, d) \xrightarrow{a}_3 (p', d') \tag{3}$$

extending Example 2 with the condition that $dI_a d'$. Under this account, it is clear that implicit in the basic axiom $a; p \xrightarrow{a} p$ for (2) is the assumption that for every "normal" $a \in A$, it is the case that $\forall d\ \exists d'\ \ d\ I_a\ d'$. Atomic programs that need not have this property are central to Groote and Ponse [9], where they are called *guards*.

1.4 Data-state pairs as labels

Abstracting away the atomic program a from (3) and promoting the data-states to the center stage, above the arrow, we arrive at

Example 4. In De Boer, Kok, Palamidessi, and Rutten [6], $L = D \times D$, and $S = P_1 + \{\surd\}$, with transitions

$$p \xrightarrow{d,d'}_4 p' \tag{4}$$

indicating that program p can turn into p' starting from an initial data-state d which is updated to d'. A program p can then be assigned denotations (following De Boer, Kok, Palamidessi, and Rutten [6]) based on sequences $p \xrightarrow{d_1,d_1'}_4 p_1 \xrightarrow{d_2,d_2'}_4 p_2 \cdots$ of transitions that may or may not be connected in the sense that d_i' may or may not equal d_{i+1}. (This, at least, is the case for programs without * and of bounded non-determinism.)

The passage from d_i' to d_{i+i} is made explicit in the next example, where symbols in (4) are moved around a bit, and transitions caused by programs possibly more complicated than $a \in A$ are allowed.

1.5 Programs as relations on data and programs

The principle that programs are only "visible" through their effects on data is developed further in our fifth example, which trades atomicity for transitions that intuitively may take more than a single step (consistent with the absence of a global clock).

Example 5. Let $L = P_1$, $S = D \times (P_1 + \{\surd\})$, and consider transitions

$$(d_1, p_1)\ [p\rangle\ (d_2, p_2)\ , \tag{5}$$

reflecting the intuition that an operating system with a data-state d_1 and a suspended job p_1 can respond to an external request p by updating its data-state to d_2 and its "jobs-to-do" to p_2. (5) reduces to (1) when $p_1 = p_2 = \surd$, and $p \in P_0$.

External requests are assumed to have priority equal to internal jobs. Very roughly (i.e., up to one-step transitions), (5) is related to (4) as follows

$$(d_1, p_1)\ [p\rangle\ (d_2, p_2) \quad \text{iff} \quad p_1 \| p \overset{d_1, d_2}{\longrightarrow_4} p_2$$
$$p \overset{d, d'}{\longrightarrow_4} p' \quad \text{iff} \quad (d, \surd)\ [p\rangle\ (d', p') \ .$$

While the idea that transitions can compose to produce transitions is a natural one, work on transition systems has tended to focus on atomic labels. A notable exception is Boudol and Castellani [5], where "at each step, the performed action is a compound one, namely a labelled poset, not just an atom" (p. 25). The work below is most definitely related to Boudol and Castellani [5], and, going back further, to Plotkin [15], where *resumptions* from a domain Y satisfying

$$Y \quad = \quad D_\perp \rightarrow Pow(D_\perp + (D_\perp \times Y))$$

are considered. A simple connection between transition systems and P_1 can be established by interpreting P_1 in a domain X satisfying the equation

$$X \quad = \quad Pow((D \times (X + \{\surd\})) \times (D \times (X + \{\surd\}))) \ . \tag{6}$$

A *rough* way (ignoring $\perp$) of relating (6) to Y is by a map from $y \in Y$ to $x_y \in X$ such that

$$y(d) \quad \approx \quad \{d' \mid (d, \surd)\ x_y\ (d', \surd)\} \ + \ \{(d', y') \in D \times X \mid (d, \surd)\ x_y\ (d', x_{y'})\} \ ,$$

following the associations

$$\begin{aligned} D \rightarrow Pow(D + (D \times Y)) \quad &\approx \quad (D \times \{\surd\}) \rightarrow Pow((D \times \{\surd\}) + (D \times Y)) \\ &\approx \quad (D \times \{\surd\}) \rightarrow Pow(D \times (Y + \{\surd\})) \ . \end{aligned}$$

Section 3 describes a "final" model for (6), without appealing to domain theory or introducing $\perp$. The essential complication addressed is the indefinite character of equality brought on by circularity — a difficulty resolved below by a quotient construction essentially going back (at least) to Milner [12].

Remark. To confuse the symbol $\surd$ with some "no-op" *skip* $\in A$ (relating identical data states) is to invite trouble in grounding the circular notion of a program as a relation on $S = D \times (P_1 + \{\surd\})$. (See part 1 of Theorem 2 below.) The second component of such a state is, intuitively, a (finite) "bag" of programs to be executed — of which any number of instances of *skip* may be included. The symbol $\surd$ is *not* a program, but is an indicator that the bag is empty. The identification of bags with elements of $P_1 + \{\surd\}$ is possible because the external request p in (5) is assumed to have as much right to be executed as any of the suspended internal programs (whence a non-empty bag can, through $\|$, be reduced to a single program). For a different scheduling policy (involving, for example, priorities), the rules given in the next section must be modified accordingly.

2 One transition system for five

To relate transitions (1) through (5), the transition system of Example 5 is extended below by taking $L = P_1 \times A^*$, with transitions of the form

$$(d_1, p_1) \; [p, \bar{a}\rangle \; (d_2, p_2)$$

where $\bar{a} \in A^*$ records the sequence $a_1 \ldots a_n$ of atomic programs actually executed. The idea is that

$$\begin{array}{ll} (1) \text{ becomes} & \exists \bar{a} \;\; (d, \surd) \; [p, \bar{a}\rangle (d', \surd) \\ (2) \text{ becomes} & \forall d \; \exists d' \;\; (d, \surd) \; [p, a\rangle \; (d', p') \\ (3) \text{ becomes} & (d, \surd) \; [p, a\rangle \; (d', p') \\ (4) \text{ becomes} & \exists a \;\; (d, \surd) \; [p, a\rangle \; (d', p') \\ (5) \text{ becomes} & \exists \bar{a} \;\; (d_1, p_1) \; [p, \bar{a}\rangle \; (d_2, p_2) \, . \end{array}$$

To be more precise, assume that among the a's in A is *skip*, with $I_{skip} = \{(d, d) \mid d \in D\}$, and fix the following collection of rules, abusing notation so that $d, p, \theta, a, \bar{a}, \ldots$ are understood as schematic variables ranging over D, P_1, $\Theta = P_1 \cup \{\surd\}, A, A^*, \ldots$, respectively, and $a \in A$ is identified with the sequence of length one consisting of a. For every $a \in A$ and $(d, d') \in I_a$, throw in the axiom

$$(d, a, d') \quad \frac{}{(d, \surd) \; [a, a\rangle \; (d', \surd)}$$

and close these transitions under

$$(;) \quad \frac{(d, \surd) \; [p_1, \bar{a}\rangle \; (d_1, \surd) \qquad (d_1, \surd) \; [p_2, \bar{b}\rangle \; (d', \theta)}{(d, \surd) \; [p_1 ; p_2, \bar{a}\bar{b}\rangle \; (d', \theta)}$$

$$(+\text{r}) \quad \frac{(d, \surd) \; [p_1, \bar{a}\rangle \; (d', \theta)}{(d, \surd) \; [p_1 + p_2, \bar{a}\rangle \; (d', \theta)} \qquad (\text{l}+) \quad \frac{(d, \surd) \; [p_2, \bar{b}\rangle \; (d', \theta)}{(d, \surd) \; [p_1 + p_2, \bar{b}\rangle \; (d', \theta)}$$

$$(*) \quad \frac{(d, \surd) \; [skip + p; p^*, \bar{a}\rangle \; (d', \theta)}{(d, \surd) \; [p^*, \bar{a}\rangle \; (d', \theta)} \qquad (\|i) \quad \frac{(d, \surd) \; [p, \bar{a}\rangle \; (d', \surd)}{(d, p') \; [p, \bar{a}\rangle \; (d', p')}$$

$$(\|\text{-r}) \quad \frac{(d, \surd) \; [p, \bar{a}\rangle \; (d', p_1)}{(d, p_2) \; [p, \bar{a}\rangle \; (d', p_1 \| p_2)} \qquad (\text{l-}\|) \quad \frac{(d, \surd) \; [p, \bar{a}\rangle \; (d', p_1)}{(d, p_2) \; [p, \bar{a}\rangle \; (d', p_2 \| p_1)}$$

$$(\text{sym}) \quad \frac{(d, p_1) \; [p_2, \bar{a}\rangle \; (d', \theta)}{(d, p_2) \; [p_1, \bar{a}\rangle \; (d', \theta)} \qquad (\text{shift}) \quad \frac{(d, p_1) \; [p_2, \bar{a}\rangle \; (d', \theta)}{(d, \surd) \; [p_1 \| p_2, \bar{a}\rangle \; (d', \theta)}$$

$$(\text{trans}) \quad \frac{(d, \theta) \; [p, \bar{a}\rangle \; (d_1, p_1) \qquad (d_1, \surd) \; [p_1, \bar{b}\rangle \; (d', \theta')}{(d, \theta) \; [p, \bar{a}\bar{b}\rangle \; (d', \theta')}$$

$$(;i)\ \frac{(d,\surd)\ \lfloor p_1,\bar{a}\rangle\ (d',p)}{(d,\surd)\ \lfloor p_1;p_2,\bar{a}\rangle\ (d',p;p_2)} \qquad (;i')\ \frac{(d,\surd)\ \lfloor p_1,\bar{a}\rangle\ (d',\surd)}{(d,\surd)\ \lfloor p_1;p_2,\bar{a}\rangle\ (d',p_2)}\ .$$

The rules (d,a,d'), $(;)$, $(+\mathrm{r})$, $(\mathrm{l}+)$ and $(*)$ correspond exactly to dynamic logic's semantic clauses for regular constructs. The remaining rules have been introduced to insure that $\|$ is closed under interleaving. This can be made precise, although (suppressing the sequences of atomic programs from the labels for the sake of simplicity) the following derivation of $(d,p')\ \lfloor p\rangle\ (d_2,p_1\|p_2)$ from $(d,\surd)\ \lfloor p\rangle\ (d_1,p_1)$ and $(d_1,\surd)\ \lfloor p'\rangle\ (d_2,p_2)$ should be sufficient to make the point

$$\dfrac{\dfrac{(d,\surd)\ \lfloor p\rangle\ (d_1,p_1)}{(d,p')\ \lfloor p\rangle\ (d_1,p_1\|p')} \qquad \dfrac{\dfrac{(d_1,\surd)\ \lfloor p'\rangle\ (d_2,p_2)}{(d_1,p_1)\ \lfloor p'\rangle\ (d_2,p_1\|p_2)}}{(d_1,\surd)\ \lfloor p_1\|p'\rangle\ (d_2,p_1\|p_2)}}{(d,p')\ \lfloor p\rangle\ (d_2,p_1\|p_2)}$$

The reader interested in notions of parallelism connected with synchronization is referred to Appendix B.

For the record, the transition system defined above is $\langle P_1\times A^*, D\times\Theta, \Longrightarrow\rangle$ where $\Longrightarrow$ is

$$\{((d,\theta),(p,\bar{a}),(d',\theta'))\mid (d,\theta)\ \lfloor p;\bar{a}\rangle\ (d',\theta')\text{ is derivable from the rule set above}\}\ .$$

Observe that whenever $(d_1,p_1)\ \lfloor p,a_1\ldots a_n\rangle\ (d_2,p_2)$ is derivable, then $d_1\ I_{a_1}\circ\cdots\circ I_{a_n}\ d_2$. Other than the axioms (d,a,d'), the only rules that change the sequences of atomic programs in a transition (or the data-states related) are the "transitive" rules $(;)$ and (trans), both of which increase the length of sequences. Neither has a counterpart in Examples 2 through 4, which are confined to sequences in A^* of length one. Thus, the rules $(;)$ and (trans) can be introduced "conservatively" into Examples 2 through 4, since these are rendered irrelevant by the restriction to transitions with atomic program sequences of length one. (The addition of these rules takes on some significance, however, if rules reducing the length of atomic program sequences in transitions are subsequently introduced — see the discussion in the concluding section.) Thus, as pointed out informally at the begining of this section, the sets of transitions for Examples 1, 3, 4 and 5 can be described by the following definitions, where an additional example lying between Examples 4 and 5 (and so denoted 4.5) is included for later reference

$$\begin{aligned}
[\![\cdot]\!]_0 &= \{(d,p,d')\mid (d,\surd)\ \lfloor p,\bar{a}\rangle\ (d',\surd)\text{ is derivable from axioms }(d_0,a_0,d_0')\\
&\qquad\qquad\text{by }(;),\ (+\mathrm{r}),\ (\mathrm{l}+),\text{ and }(*),\text{ for some }\bar{a}\}\\
\rightarrow_3 &= \{((p,d),a,(\theta',d'))\mid (d,\surd)\overset{p,a}{\Longrightarrow}(d',\theta')\}\\
\rightarrow_4 &= \{(p,(d,d'),\theta')\mid (d,\surd)\overset{p,a}{\Longrightarrow}(d',\theta')\text{ for some }a\}\\
\rightarrow_{4.5} &= \{(p,(d,d'),\theta')\mid (d,\surd)\overset{p,\bar{a}}{\Longrightarrow}(d',\theta')\text{ for some }\bar{a}\}\\
\rightarrow_5 &= \{((d,\theta),p,(d',\theta'))\mid (d,\theta)\overset{p,\bar{a}}{\Longrightarrow}(d',\theta')\text{ for some }\bar{a}\}\ .
\end{aligned}$$

We show next that $\Longrightarrow$ extends $[\![\cdot]\!]_0$ "conservatively" in the sense that for every $p \in P_0$ and all $d, d' \in D$, if $(d, \surd) \xRightarrow{p,\bar{a}} (d', \surd)$ for some $\bar{a}$, then $d[\![p]\!]_0 d'$. For this purpose, it is convenient to work with rules where the sequences of atomic programs in labels are erased, so that $L = P_1$ and transitions have the form of (5). Call the resulting set of rules Γ_1, and let Γ_0 be the subset of Γ_1 given by (d, a, d')'s, (;), (+r), (l+), and (*) with the atomic sequences dropped from the labels. It is easy to see that

$$\begin{aligned} [\![\cdot]\!]_0 &= \{(d,p,d') \mid (d,\surd)\ [p\rangle\ (d',\surd) \text{ is derivable from } \Gamma_0\} \\ \rightarrow_5 &= \{((d,\theta),p,(d',\theta')) \mid (d,\theta)\ [p\rangle\ (d',\theta') \text{ is derivable from } \Gamma_1\}\ . \end{aligned}$$

Theorem 1. $\Longrightarrow$ *is a conservative extension of dynamic logic: that is, for every program* $p \in P_0$,

$$(d,\surd)\ [p\rangle\ (d',\surd) \textit{ follows from } \Gamma_0 \textit{ iff } (d,\surd)\ [p\rangle\ (d',\surd) \textit{ follows from } \Gamma_1\ .$$

Proof. Only ($\Leftarrow$) requires an argument. First, establish by induction on the length of Γ_1-derivations that

(†) For every $p \in P_0$, and every Γ_1-derivation of $(d,\surd)\ [p\rangle\ (d',\theta)$, it is the case that $\theta \in P_0 \cup \{\surd\}$ and the rules (sym), (shift), ($\|i$), ($\|$-r), and (l-$\|$) do *not* occur in the derivation.

Next, to push through an induction on the length of Γ_1-derivations, it is useful to strengthen the induction hypothesis as follows. Given a Γ_1-derivation D of $(d,\surd)\ [p\rangle\ (d',\theta)$, let $\varphi(D)$ be the assertion

if $\theta = \surd$, then Γ_0 proves that $(d,\surd)\ [p'\rangle\ (d',\surd)$,
else if $\theta = p'$ and Γ_0 proves that $(d',\surd)\ [p'\rangle\ (d'',\surd)$, then Γ_0 proves that $(d,\surd)\ [p\rangle\ (d'',\surd)$.

Now, induct on the length of D, noting from (†) that $\theta \in P_0 \cup \{\surd\}$, and that (sym), (shift), ($\|i$), ($\|$-r), and (l-$\|$) do not occur in such a derivation. The argument breaks up into different cases, according to the last rule applied in D. For lack of space, this is left to the reader. ⊣

Having established Theorem 1, note that the rule (;) is derivable from (trans) and (;i').

3 Semantic foundations for 'programs as relations'

This section works out a semantics for programs based on (**P1**), "a program as a relation on states", over a transition system $\langle P, D \times \Theta, [\cdot\rangle_\Gamma\rangle$ (with $\Theta = P + \{\surd\}$) whose transition relation is given by some rule set Γ (such as Γ_0 or Γ_1) as follows

$$(d,\theta)\ [p\rangle_\Gamma\ (d',\theta') \quad \text{iff} \quad (d,\theta)\ [p\rangle\ (d',\theta') \text{ follows from } \Gamma\ .$$

Notice that "$(d,\theta)\ [p\rangle\ (d',\theta')$" is viewed in the right hand side above as a syntactic expression — a practice that we will adapt for the remainder of the paper. Formally, the idea is to work in a logical system without equality, but with (atomic) predicate symbols

$$(d,\cdot)\ [\cdot\rangle\ (d',\cdot)\ ,\ (d,\surd)\ [\cdot\rangle\ (d',\cdot)\ ,\ (d,\cdot)\ [\cdot\rangle\ (d',\surd)\ ,\ (d,\surd)\ [\cdot\rangle\ (d',\surd)$$

for every d and $d' \in D$. The holes $\cdot$ are to be filled by programs in P. Now, a "semantics" for the set P of programs is at the very least a map f with domain P. One might expect this map to be defined by induction on terms (as in first-order logic), but for the time being, we will simply take f to be given, and will return to the matter of compositionality later. For f to be, in any sense, a model of Γ, all transitions $(d,\theta_0)\ [p\rangle_\Gamma\ (d',\theta_0')$ must hold in f. Without saying anything about the co-domain of f, let us focus on the "equality" relation $\{(p,p') \in P \times P \mid f(p) = f(p')\}$ on P that f induces. While our list of predicate symbols does not include equality, the principle of "substituting equals for equals" can be implemented by closing $[\cdot\rangle_\Gamma$ under equality

$$(d,\theta)\ [p\rangle\ (d',\theta') \text{ is } \Gamma\textit{-true in } f \quad \text{iff} \quad \exists\theta_0,\theta_0' \ \ f(\theta_0) = f(\theta),\ \ f(\theta_0') = f(\theta') \text{ and } \\ (d,\theta_0)\ [p\rangle_\Gamma\ (d',\theta_0')\ ,$$

where, for notational convenience, f has been extended to Θ by setting $f(\surd) = \surd$. Now, a minimal[2] condition of soundness on f with respect to Γ is that if f identifies p with p' then for every $d, d' \in D$ and $\theta, \theta' \in \Theta$,

$$(d,\theta)\ [p\rangle\ (d',\theta') \text{ is } \Gamma\text{-true in } f \quad \text{iff} \quad (d,\theta)\ [p'\rangle\ (d',\theta') \text{ is } \Gamma\text{-true in } f\ .$$

In other words, if $[f]_\Gamma$ is defined as the map from P given by

$$[f]_\Gamma(p) \quad = \quad \{((d,\theta),(d',\theta')) \mid (d,\theta)\ [p\rangle\ (d',\theta') \text{ is } \Gamma\text{-true in } f\}\ ,$$

then

$$(\star)\quad (\forall p,p' \in P)\ f(p) = f(p') \text{ implies } [f]_\Gamma(p) = [f]_\Gamma(p')\ ,$$

or, equivalently,

$(\star)'$ for all $p, p', d_1, d_2, \theta_1, \theta_2$ such that $f(p) = f(p')$ and $(d_1,\theta_1)\ [p\rangle_\Gamma\ (d_2,\theta_2)$, there exist θ_1' and θ_2' such that $f(\theta_1') = f(\theta_1)$, $f(\theta_2') = f(\theta_2)$ and $(d_1,\theta_1')\ [p'\rangle_\Gamma\ (d_2,\theta_2')$.

---

[2] As pointed out to the author by P. Panangaden, coarser notions of equivalence may for various purposes be desirable. But not every equivalence can be considered an "equality" from the point of view of a logical system given by the (atomic) predicate symbols $(d,\cdot)\ [\cdot\rangle\ (d',\cdot)$, $(d,\surd)\ [\cdot\rangle\ (d',\cdot)$, $(d,\cdot)\ [\cdot\rangle\ (d',\surd)$, and $(d,\surd)[\cdot\rangle\ (d',\surd)$ for every d and $d' \in D$. Of course, one can, as in logic, consider restrictions (i.e., "reducts") of this language, or translate from this language to another — for example, translating a program p to its atomization $[p]$ (discussed in section 5) would abstract out intermediate states, and identify programs with the same input-output behavior. While Theorem 2 below holds quite generally, the congruence and soundness results in the appendix do not.

A second condition to impose on f is

$$(\star\star)\ \ (\forall\theta\in\Theta)\ \ f(\theta)=\surd \text{ iff } \theta=\surd\,.$$

This condition was justified conceptually by the final remark in section 1.5 above, and can be motivated technically by part 1 of Theorem 2 below. In any case, call a function f with domain Θ satisfying $(\star)$ and $(\star\star)$ *Γ-consistent*. Assuming that $\surd\notin P$ (as we do throughout the paper), the identity id on Θ is Γ-consistent. A more interesting example arises from attempting to satisfy $(\star)$ trivially by asking for an f which, restricted to P, is $[f]_\Gamma$.

There are two possible complications with this request. One is that such f's might fail to exist — which is precisely the case for the usual universe of sets satisfying the axiom of foundation. So let's drop the axiom of foundation. The second complication is that there may be different f's for which $f=[f]_\Gamma$. While this "complication" may not seem so terrible, it is this very point that makes the domain equation (6) problematic (as the axiom of extensionality is no longer sufficient to settle questions of identity once circularity is admitted). It turns out to be convenient to appeal to the *Anti-Foundation Axiom* (AFA) in Aczel [1] asserting the uniqueness of an f for which $f=[f]_\Gamma$. (Readers familiar with Aczel [1] can see this by noting that for every $p\in P$,

$$[f]_\Gamma(p) \;=\; \{((d,f(\theta)),(d',f(\theta'))) \mid (d,\theta)\ [p\rangle_\Gamma\ (d',\theta')\}\,,$$

and that $\surd$ can certainly be assumed not to be a set of ordered pairs.) Working in a universe of sets satisfying AFA and the usual set-theoretic axioms minus foundation, let's call that unique solution $[\![\cdot]\!]_\Gamma$. One way to see the importance of $[\![\cdot]\!]_\Gamma$ is through a little category theory.

Form a category $\mathbf{C}_\Gamma$ with Γ-consistent functions as objects as follows. For a Γ-consistent function f, a function α^f_Γ with domain $f''P$ (i.e., the image $\{f(p)\mid p\in P\}$ of P under f) can be defined by requiring

$$\alpha^f_\Gamma(f(p)) \;=\; [f]_\Gamma(p)$$

since f satisfies $(\star)$. The $\mathbf{C}_\Gamma$-morphisms from f to g are (by definition) the functions φ from $f''\Theta$ to $g''\Theta$ such that $\varphi(\surd)=\surd$ and for every $p\in P$,

$$\alpha^g_\Gamma(\varphi f p) \;=\; \{((d,\varphi f\theta),(d',\varphi f\theta')) \mid (d,\theta)\ [p\rangle_\Gamma\ (d',\theta')\}\,.$$

This equation can be pictured as follows

$$\begin{array}{ccccc}
P & \xrightarrow{\ f\ } & f''P & \xrightarrow{\ \alpha^f_\Gamma\ } & F(f''P)\\
 & & \Big\downarrow{\scriptstyle\varphi} & & \Big\downarrow{\scriptstyle F(\varphi)}\\
P & \xrightarrow{\ g\ } & g''P & \xrightarrow{\ \alpha^g_\Gamma\ } & F(g''P)
\end{array}$$

where F is the functor on classes X (and class maps) given by

$$F(X) \quad = \quad Pow((D \times (X + \{\surd\})) \times (D \times (X + \{\surd\})))$$

together with the obvious map on morphisms (Aczel [1]).

Next, let $\theta \sim_\Gamma \theta'$ abbreviate $[\![\theta]\!]_\Gamma = [\![\theta']\!]_\Gamma$, and for every Γ-consistent f, define the function $[f]^0_\Gamma$ with domain P to record the input-output behavior of a program

$$\begin{aligned} [f]^0_\Gamma(p) &= \{(d,d') \mid (d,\surd)\ [p\rangle\ (d',\surd) \text{ is } \Gamma\text{-true in } f\} \\ &= \{(d,d') \mid (d,\surd)\ [f]_\Gamma(p)\ (d',\surd)\}\ . \end{aligned}$$

Also, let us understand "the theory" of a Γ-consistent object f to mean the set of all expressions $(d,\theta)[p\rangle(d',\theta')$ Γ-true in f. The following theorem records some pleasant logical properties enjoyed by $\mathbf{C}_\Gamma$, the first of which expresses that the input-output behavior prescribed by Γ is respected.

Theorem 2.

1. *For every* $\mathbf{C}_\Gamma$*-object* f, $[f]^0_\Gamma(p) = [id]^0_\Gamma(p)$.
2. $\mathbf{C}_\Gamma$*-morphisms preserve truth; that is, if there is a* $\mathbf{C}_\Gamma$*-morphism from* f *to* g *mapping* $f(p)$ *to* $g(p')$ *then* $[f]_\Gamma(p) \subseteq [g]_\Gamma(p')$.
3. id *is initial in* $\mathbf{C}_\Gamma$*, and has the least theory of all objects in* $\mathbf{C}_\Gamma$.
4 (AFA). *At the other extreme,* $[\![\cdot]\!]_\Gamma$ *is final in the category* $\mathbf{C}_\Gamma$*. It has the largest theory of all objects in* $\mathbf{C}_\Gamma$.
5 (AFA). *For all* $p,p' \in P$, $p \sim_\Gamma p'$ *iff there is a* Γ*-consistent function* f *with* $f(p) = f(p')$.

Proof. Unwrap the definitions, which, in the case of $[\![\cdot]\!]_\Gamma$, includes a uniqueness property. In particular, note that for every $\mathbf{C}_\Gamma$-object f, φ is a $\mathbf{C}_\Gamma$-morphism from f to $[\![\cdot]\!]_\Gamma$ iff $\varphi \circ f = [\![\cdot]\!]_\Gamma$. ⊣

So long as one is content with "a program *isomorphic* to a relation on states", it is only the equivalence $\{(p,p') \mid f(p) = f(p')\}$ induced by an interpretation f that matters. Thus, having characterized $\sim_\Gamma$ above with the help of AFA, it is possible to forget $[\![\cdot]\!]_\Gamma$ and AFA, and "simply" show that the function sending $\surd$ to $\surd$ and p to $\{p' \in P \mid p \sim_\Gamma p'\}$ is final in $\mathbf{C}_\Gamma$. It is the categorical property of finality that is interesting, not only because it captures a certain maximality formulated in part 4 of the theorem above, but also because it turns out to be useful in establishing that the equality induced is a congruence, provided Γ has a "nice" form. The technical details have been relegated to an appendix, since, as it happens, the case of Γ_1 is reducible to results from Groote and Vaandrager [10]. One other point that Appendix A attends to is that Γ is indeed sound for $[\![\cdot]\!]_\Gamma$ (under certain assumptions on Γ).

4 Relating the examples semantically

Underlying Examples 1 and 2, on which the other examples build, are the conceptions (**P1**) and (**P2**). Preserving (**P1**), Example 5 extends Example 1 (conservatively) by attaching programs to data-states. Similarly, Example 3 attaches data-states to Example 2's process-states, while Example 4 replaces atomic program labels with data-state pairs (from the atomic programs). The question arises as to whether (**P2**) and (**P1**) are fundamentally different. To attack this question semantically, we consider "natural" models for the examples above, focussing on the relations of "equality" induced on the programs in P_1.

The case of Example 5 has already been treated in the previous section; summarizing (and suppressing Γ_1 from the notation for simplicity), define

$$p \sim p' \quad \text{iff} \quad \text{there is a } \Gamma_1\text{-consistent } f \text{ such that } f(p) = f(p')$$

where f is Γ_1-consistent iff it is a function with domain Θ meeting conditions

$(\star)'$ for all $p, p', d_1, d_2, \theta_1, \theta_2$ such that $f(p) = f(p')$ and $(d_1, \theta_1) \xrightarrow{p}_5 (d_2, \theta_2)$, there exist θ_1' and θ_2' such that $f(\theta_1') = f(\theta_1)$, $f(\theta_2') = f(\theta_2)$ and $(d_1, \theta_1') \xrightarrow{p'}_5 (d_2, \theta_2')$,

and

$(\star\star)$ for all $\theta \in \Theta$, $f(\theta) = \surd$ iff $\theta = \surd$.

As for (**P2**), a standard notion of equivalence due to Park [13] is given as follows (modified slightly to take $\surd$ into account). Define a *bisimulation* on a transition system $\langle L, S, \rightarrow\rangle$ to be a relation $R \subseteq S \times S$ such that whenever sRs', then for every $l \in L$,

$$(\forall t \xleftarrow{l} s)\ (\exists t' \xleftarrow{l} s')\ tRt' \quad \text{and} \quad (\forall t' \xleftarrow{l} s')\ (\exists t \xleftarrow{l} s)\ tRt' \ ,$$

and $s = \surd$ iff $s' = \surd$. The $\forall\exists$-conjuncts above should be compared with condition $(\star)'$ for Γ-consistent functions, noting that the symmetry of the predicate $f(p) = f(p')$ renders the other half of the back-and-forth clause for $(\star)'$ unnecessary. Next, define the *bisimilarity* relation $\underline{\leftrightarrow}$ on $\langle L, S, \rightarrow\rangle$ by

$$s \,\underline{\leftrightarrow}\, s' \quad \text{iff} \quad \text{there is a bisimulation on } \langle L, S, \rightarrow\rangle \text{ relating } s \text{ to } s' \ .$$

The semantics studied in BKPR [6] does not build on bisimulations, but, as pointed out to the author by J. Rutten, it follows from Groote and Vaandrager [10] (see also Rutten [16]) that the bisimilarity predicate $\underline{\leftrightarrow}_4$ for Example 4 is a congruence. This property is preserved by the addition of the rule (trans); that is, the predicate $\underline{\leftrightarrow}_{4.5}$ for $\langle D \times D, \Theta, \rightarrow_{4.5}\rangle$ is a congruence.

Theorem 3. *$\underline{\leftrightarrow}_{4.5}$ is equal to $\sim$.*

Proof. That $p \sim p'$ implies $p \,\underline{\leftrightarrow}_{4.5}\, p'$ follows from the fact that $\sim$ is a bisimulation for $\langle D \times D, \Theta, \rightarrow_{4.5}\rangle$. Conversely, define a function f with domain Θ by setting $f(\surd) = \surd$ and for $p \in P_1$, $f(p) = \{\theta \mid p \,\underline{\leftrightarrow}_{4.5}\, \theta\}$. It is easy to see that if $p \,\underline{\leftrightarrow}_{4.5}\, p'$ then f witnesses $p \sim p'$ because, as noted above, $\underline{\leftrightarrow}_{4.5}$ is a congruence with respect to $\|$. ⊣

Adding the rule (trans) to Example 4 represents a real change in that, for instance, $a + skip; a \ \sim\ skip; a$ (although $skip; skip \not\sim skip$) while Example 4 differentiates between these when $I_a \neq I_{skip}$. Whereas the step from Example 1 to Example 4 shows that bisimilarity can become finer with an increase in rules, the move from $\rightarrow_4$ to $\rightarrow_{4.5}$ shows that it can also become coarser when the rule set is extended.

(Note that the transition predicate occurs both positively and negatively in the definition of a bisimulation.)

Theorem 4. *$\leftrightarrow_4$ is finer than $\sim$, and, in general, strictly so.*

Proof. Following the previous proof, define a function f with domain Θ by setting $f(\surd) = \surd$ and for $p \in P_1$, $f(p) = \{\theta \mid p \leftrightarrow_4 \theta\}$. Next, to see that f is Γ_1-consistent, it suffices to show that

(†) if a transition $(d,\theta)\ [p\rangle\ (d',\theta')$ can be derived in Γ_1, then there are $d_1, p_1, d_2, p_2, \ldots, d_k, p_k$ such that $(d,\theta) \xrightarrow{p}_4 (d_1, p_1)$, $(d_1, \surd) \xrightarrow{p_1}_4 (d_2, p_2), \ldots, (d_k, \surd) \xrightarrow{p_k}_4 (d', \theta')$, whence repeated applications of (trans) yield $(d,\theta)\ [p\rangle_{\Gamma_1}\ (d',\theta')$.

The assertion (†) holds because (i) the rule (;) can be replaced by (trans) and (;i'), and (ii) the rule (trans) can always be moved below a rule applied after it — for example, by locally converting the result of (trans) and ($*$) applied in that order

$$\cfrac{\cfrac{(d,\theta)\ [skip + p; p^*\rangle\ (d_1, p') \qquad (d_1,\surd)\ [p'\rangle\ (d',\theta')}{(d,\theta)\ [skip + p; p^*\rangle\ (d',\theta')}}{(d,\theta)\ [p^*\rangle\ (d',\theta')}$$

to the result of ($*$) and then (trans)

$$\cfrac{\cfrac{(d,\theta)\ [skip + p; p^*\rangle\ (d_1, p')}{(d,\theta)\ [p^*\rangle\ (d_1, p')} \qquad (d_1,\surd)\ [p'\rangle\ (d',\theta')}{(d,\theta)\ [p^*\rangle\ (d',\theta')}$$

Note: a rule called (at) is introduced in the next section with which (trans) cannot commute as above. ⊣

Similarly, Example 3 is even less abstract than Example 4. A *global D-bisimulation* on $\langle A, P \times D, \rightarrow_3\rangle$ is defined in Groote and Ponse [9] to be a relation $R \subseteq (P \times D) \times (P \times D)$ such that whenever $(p,d)R(p',d)$, then for every $a \in A$,

$$\forall (p_1, d_1) \xleftarrow{a}_3 (p,d)\ \exists (p_1', d_1) \xleftarrow{a}_3 (p',d)\ \forall d_2\ (p_1,d_2)R(p_1',d_2) \quad \text{and}$$
$$\forall (p_1', d_1) \xleftarrow{a}_3 (p',d)\ \exists (p_1, d_1) \xleftarrow{a}_3 (p,d)\ \forall d_2\ (p_1,d_2)R(p_1',d_2)\ .$$

The difference in the treatment of data from that of a process might be defended by appealing to the intuition that data is "irreducible" in a sense that a process is not. As it turns out, however, even for (1), the notion of a bisimulation is interesting, a point to which we will return in the final section. In any case, let

$p \leftrightarrow_3 p'$ iff for all $d \in D$ there is a global D-bisimulation R such that $(p,d)R(p',d)$.

Theorem 5. *$\leftrightarrow_3$ is finer than $\leftrightarrow_4$, and, in general, strictly so.*

Proof. If $p \xrightarrow{d,d'}_4 p'$, then $(p,d) \xrightarrow{a}_3 (p',d')$ for some a with $dI_a d'$. So if R is a global D-bisimulation relating (p,d) to (p_0,d) then for some $(p_0', d') \xleftarrow{a}_3 (p_0, d)$, R relates (p', d'') to (p_0', d'') for every d''. But in that case, $p_0 \xrightarrow{d,d'}_4 p_0'$. Hence, $p \leftrightarrow_3 p'$ implies $p \leftrightarrow_4 p'$.

As for the reverse direction, consider $a_1 + a_2$ and a when $I_{a_1} \cup I_{a_2} = I_a$ (or, to see that strictness can result even when every I_a is a distinct total function, $a_3 + a_4$ where $I_{a_1} \cup I_{a_2} = I_{a_3} \cup I_{a_4}$). ⊣

5 Discussion

As is made precise by Theorems 3 and 4, the key element setting Example 5 (the synthesis of (**P1**) and (**P2**)) apart from Examples 2 through 4 is the explicit incorporation into transitions of the composition rule (trans). If the rule (trans) increases the length of an atomic sequence labelling a transition, then what about a dual rule decreasing the length of the sequence? An example is provided by adding a closure clause for the "atomization" $[p]$ of p to P_1, together with the rule

$$(\mathrm{at}) \quad \frac{(d, \surd)\ [p, \bar{a}\rangle\ (d', \surd)}{(d, \surd)\ [[p], [p]\rangle\ (d', \surd)}\ ,$$

resetting L to $P \times (A \cup \{[p] \mid p \in P\})^*$. (De Bakker and De Vink [3] provide a different treatment; the rule above is similar to one proposed independently by Franck van Breugel. It would be interesting to "eliminate" so-called τ- or silent steps by suitable applications of the atomization construct.) The congruence and soundness results described in Appendix A cover this extension.

Intuitively, the rule (trans) functions as the computational "glue" between transitions, possessing something of the transitive character of the *cut* rule in logic. Accordingly, a proof that a particular extension of the rule set of section 2 is, in some sense, conservative (e.g., Theorem 1) will likely involve an induction principle in which the rule (trans) plays a crucial role. The (i) inevitability of such extensions, and the (ii) measure of computational content (either of the initial system or of the extension) the induction principles provide would seem to be interesting topics to investigate. The line of thinking here is motivated largely by the prevailing view behind applications of proof theory to programming language semantics that computation is related in a deep sense to cut-elimination (or more broadly, to deduction). This view would be a bit more convincing if it can be shown to shed light on computations that need not terminate.

A final point (suggested by the reference in section 3 to the set theory in Aczel [1]) is that the step from syntactic presentations of transition systems to semantic (extensional) notions of identity might be analyzed against a generalization of ordinary (numerical) recursion theory to a computational theory for sets. In particular, the set-theoretic recursion theory described in Barwise [4] has proved fruitful for infinitary extensions of first-order logic, which might be employed for getting a logical grip on $\sim$ (Fernando [7] [8]).

References

1. Peter Aczel. *Non-Well-Founded Sets.* CSLI Lecture Notes Number 14, Stanford, 1988.
2. J.C.M. Baeten and W.P. Weijland. *Process Algebra.* Cambridge Tracts in Theoretical Computer Science 18. Cambridge University Press, 1990.
3. J.W. de Bakker and E.P. de Vink. Bisimulation semantics for concurrency with atomicity and action refinement. Technical Report CS-R9210, Centre for Mathematics and Computer Science, 1992.
4. Jon Barwise. *Admissible Sets and Structures.* Springer-Verlag, Berlin, 1975.

5. G. Boudol and I. Castellani. Concurrency and atomicity. *Theoretical Computer Science*, 59, 1988.
6. F.S. de Boer, J.N. Kok, C. Palamidessi, and J.J.M.M. Rutten. The failure of failures in a paradigm for asynchronous communication. In *Proc. Concur '91*, LNCS 527. Springer-Verlag, Berlin, 1991.
7. Tim Fernando. A primitive recursive set theory and AFA: on the logical complexity of the largest bisimulation. To appear in the proceedings of Computer Science Logic '91 (Berne).
8. Tim Fernando. Between programs and processes: absoluteness and open ended-ness. Technical Report IAM 92-011, Institut für Informatik, Universität Bern, 1992.
9. J.F. Groote and A. Ponse. Process algebra with guards: combining Hoare logic with process algebra. In *Proc. Concur '91*, LNCS 527. Springer-Verlag, Berlin, 1991.
10. J.F. Groote and F.W. Vaandrager. Structured operational semantics and bisimulation as a congruence. In *Proc. 16th ICALP*, LNCS 372. Springer-Verlag, Berlin, 1989.
11. David Harel. Dynamic logic. In Gabbay et al, editor, *Handbook of Philosophical Logic, Volume 2*. D. Reidel, 1984.
12. Robin Milner. Calculi for synchrony and asynchrony. *Theoretical Computer Science*, 25, 1983.
13. David Park. Concurrency and automata on infinite sequences. In P. Deussen, editor, *Proc. 5th GI Conference*, LNCS 104. Springer-Verlag, Berlin, 1981.
14. David Peleg. Concurrent dynamic logic. *J. Assoc. Computing Machinery*, 34(2), 1987.
15. Gordon D. Plotkin. A powerdomain construction. *SIAM J. Comput.*, 5(3), 1976.
16. J.J.M.M. Rutten. Processes as terms: Non-well-founded models for bisimulation, 1992. To appear in Mathematical Structures in Computer Science.

Appendix A: a final congruence sound for 'programs as relations'

This appendix addresses the matter of compositionality and soundness of Γ-consistent functions. Compositionality is trivial for the initial $\mathbf{C}_\Gamma$-object *id*, but is most interesting for the final $\mathbf{C}_\Gamma$-objects. The problem of interpreting the program constructs $;$, $+$, * and $\|$ over a Γ-consistent function f reduces to showing that the equality $\{(p,p') \in P_1 \times P_1 \mid f(p) = f(p')\}$ it induces is a congruence with respect to the constructs. The point is, for example, that ; can be interpreted under f in at most one way $\bullet$ — viz., $f(p) \bullet f(p') = f(p;p')$. This equation can be taken as a sound definition of $\bullet$ iff it can be shown to be independent of the choice of representatives p and p' (i.e., iff $\{(p,p') \mid f(p) = f(p')\}$ is a congruence with respect to ;). As is made clear in Groote and Vaandrager [10], this property becomes problematic for a semantic analysis based on syntactic rules, such as $(*)$ in section 2, without a "subformula property."

The constraints imposed by Γ apply more directly to $[f]_\Gamma(p)$ than to $f(p)$, suggesting that it may be useful to restrict attention to $\mathbf{C}_\Gamma$-objects f for which α^f is 1-1 (i.e., for all p and $p' \in P_1$, $[f]_\Gamma(p) = [f]_\Gamma(p')$ implies $f(p) = f(p')$). But even this property is not strong enough to overcome obstacles posed by rules such as $(;i)$ to an argument (by induction on rules) that

$$f(p_1) = f(p_1') \text{ and } f(p_2) = f(p_2') \ \text{ imply } \ [f]_\Gamma(p_1;p_2) = [f]_\Gamma(p_1';p_2') \,.$$

The "strong extensionality" of $[\cdot]_\Gamma$ is, however, sufficient for a wide collection of rules, as will be shown shortly.

Given a set X of program variables and a family $\mathcal{F} \supseteq A$ of function symbols of various arities including 0 (e.g., ;, *, $a \in A$), let $\mathcal{F}(X)$ be the resulting collection of program terms with program variables in X. (Thus, $P_1 = (A \cup \{+, ;, ^*, \|\})(\emptyset)$.) Define R to be the following relation on $\mathcal{F}(\emptyset)$

$$\{(t[p_1, \ldots, p_n], t[p'_1, \ldots, p'_n]) \mid n < \omega,\ t(x_1, \ldots, x_n) \in \mathcal{F}(\{x_1, \ldots, x_n\}) \text{ and } p_1 \sim_\Gamma p'_1, \ldots, p_n \sim_\Gamma p'_n\}.$$

Call a rule set Γ *nice* if for all $(p, p') \in R$, $\theta_1, \theta_2 \in \Theta$, and $d_1, d_2 \in D$ such that $(d_1, \theta_1)\ [p\rangle_\Gamma\ (d_2, \theta_2)$,

$$\exists \theta'_1, \theta'_2 \text{ s.t. } (\theta_1, \theta'_1) \in R \cup \{(\surd, \surd)\},\ (\theta_2, \theta'_2) \in R \cup \{(\surd, \surd)\} \text{ and } (d_1, \theta'_1)\ [p'\rangle_\Gamma\ (d_2, \theta'_2).$$

Lemma A (AFA). *For every nice rule set Γ, $\sim_\Gamma$ is a congruence with respect to every function symbol in $\mathcal{F}$.*

Proof. Since Γ is nice, it follows that the map f sending $\surd$ to $\surd$ and every $p \in \mathcal{F}(\emptyset)$ to $\{p' \mid pRp'\}$ is a C_Γ-object. But by Theorem 2, $[\cdot]_\Gamma$ is final, whence R is $\sim_\Gamma$, as desired. ⊣

Proving that the particular rule set Γ_1 is nice by induction on Γ_1-derivations is complicated by the rules (sym) and (shift). (Try it.) For this reason, it is convenient to consider the stronger property

(N) for all $(p, p') \in R$, $\theta_1, \theta_2 \in \Theta$, $d_1, d_2 \in D$, and θ'_1 s.t. $(\theta_1, \theta'_1) \in R \cup \{(\surd, \surd)\}$, if $(d_1, \theta_1)\ [p\rangle_\Gamma\ (d_2, \theta_2)$ then $\exists \theta'_2$ s.t. $(\theta_2, \theta'_2) \in R \cup \{(\surd, \surd)\}$ and $(d_1, \theta'_1)\ [p'\rangle_\Gamma\ (d_2, \theta'_2)$.

Property (N) can be proved by induction on the length of Γ-derivations assuming the rules in Γ have a certain form. To describe such a form, fix a countable set X of program variables, and call a program term in $\mathcal{F}(X)$ *primitive* if it is either (a constant denoting) an element of A or a program variable ($\in X$). For $\hat{\theta} \in \mathcal{F}(X) \cup \{\surd\}$, let $Var(\hat{\theta})$ be the set of variables in X occurring in $\hat{\theta}$. Now, consider a rule of the form

$$(r) \quad \frac{(d_i, \theta_i)\ [p_i\rangle\ (d'_i, \theta'_i) \quad (i < n)}{(d, \theta)\ [p\rangle\ (d', \theta')}$$

where

(c0) d_i, d'_i $(i < n)$, d, and d' are (constants denoting) elements of D or variables ranging over D,
θ_i $(i < n)$, and $\theta' \in \mathcal{F}(X) \cup \{\surd\}$,
for $i < n$, $p_i \in \mathcal{F}(X)$,
(c1) θ and every θ'_i $(i < n)$ are primitive program terms or $\surd$,
(c2) $Var(\theta) \cap Var(p) = \emptyset$,
(c3) $Var(\theta'_i) \cap Var(\theta'_j) = \emptyset$ for $i < j < n$,
(c4) $Var(p_i) \cap Var(\theta'_j) = Var(\theta_i) \cap Var(\theta'_j) = \emptyset$ for $i \leq j < n$,
(c5) $Var(p) \cap Var(\theta'_i) = Var(\theta) \cap Var(\theta'_i) = \emptyset$ for $i < n$, and
(c6) p is either a primitive program term or $f(x_1, \ldots, x_n)$ for some n-ary $f \in \mathcal{F}$.

Condition (c0) is not much of a restriction; the other conditions are best appreciated in the course of establishing congruence and soundness. Note that Γ_1 can be presented as a set of rules with the form of (r) — a rule with variables ranging over Θ can be broken up (by cases) into multiple rules of the form (r), using the fact that $\Theta = P_1 \cup \{\surd\}$.

Lemma B. *If all rules in Γ have the form of (r) above, then property* **(N)** *holds and Γ is therefore nice.*

Proof. Assume all rules in Γ have the form of (r) above. To prove that **(N)** holds, first establish

($\dagger$) whenever $(d_1, \theta_1)\ [p\rangle_\Gamma\ (d_2, \theta_2)$ and $(\theta_1, \theta'_1) \in R \cup \{(\surd, \surd)\}$, then there is a θ'_2 such that $(\theta_2, \theta'_2) \in R \cup \{(\surd, \surd)\}$ and $(d_1, \theta'_1)\ [p\rangle_\Gamma\ (d_2, \theta'_2)$.

A proof of ($\dagger$) proceeds by induction on the length of Γ-derivations. Consider the last rule (r) of a shortest Γ-derivation of $(d_1, \theta_1)\ [p\rangle\ (d_2, \theta_2)$. The point is that (c1) and (c2) allow the the premise to be re-instantiated appropriately with θ_1 replaced by θ'_1, using the induction hypothesis and conditions (c3) to (c5). Then a θ'_2 can be chosen such that $(\theta_2, \theta'_2) \in R \cup \{(\surd, \surd)\}$ and the conclusion of (r) can be instantiated by $(d_1, \theta'_1)\ [p\rangle\ (d_2, \theta'_2)$.

Now, suppose $(p, p') \in R$, $\theta_1, \theta_2 \in \Theta$, $d_1, d_2 \in D$, $(\theta_1, \theta'_1) \in R \cup \{(\surd, \surd)\}$, and $(d_1, \theta_1)\ [p\rangle_\Gamma\ (d_2, \theta_2)$. If $p \sim_\Gamma p'$ then appeal to ($\dagger$) above. Otherwise, adapt the induction argument for ($\dagger$) using (c6) to replace p by p'. $\dashv$

Lemmas A and B yield

Theorem C (AFA). *If Γ is a set of rules with the form of (r), then $\sim_\Gamma$ is a congruence with respect to every function symbol in $\mathcal{F}$.*

Note that the theorem applies to Γ_1, and the connection with the general congruence results of Groote and Vaandrager [10] is made above for the particular case of $;$, $+$, $\|$ and *, with interleaving playing a central role in reducing Example 5 to Example 4 (via 4.5).

Finally, observe that another consequence of Lemma B (appealing to **(N)**, (c3) and (c4)) is

Theorem D (AFA). *If Γ is a set of rules with the form of (r), then Γ is sound for $[\cdot]_\Gamma$.*

Appendix B: living with 'programs as relations'

The "parallel" construct $\|$ formulated above captures interleaving. What about (as F. Vaandrager has asked) notions of synchronization say, on a set $H \subseteq A$ of atomic programs? Consider the binary program construct $\|_H$, described in the usual transition system format of Example 2 (without data) by the rules

$$(\alpha)\ \frac{p_1 \xrightarrow{a} p'_1 \quad p_2 \xrightarrow{a} p'_2}{p_1 \|_H p_2 \xrightarrow{a} p'_1 \|_H p'_2}\ a \in H \qquad\qquad (\beta)\ \frac{p_1 \xrightarrow{a} p'_1}{p_1 \|_H p_2 \xrightarrow{a} p'_1 \|_H p_2}\ a \notin H\ .$$

Rules (α) and (β) do not specify how data-states are transformed. Presumably, (β) translates, in the format of section 2, to

$$\frac{(d, \surd)\ [p_1, a\rangle\ (d', p'_1)}{(d, \surd)\ [p_1 \|_H p_2, a\rangle\ (d', p'_1 \|_H p_2)}\ a \notin H$$

but how about (α)? One possibility is that there are functions i and j from D^4 to D, allowing (α) to be reformulated as

$$\frac{(d_1, \surd)\ [p_1, a\rangle\ (d'_1, p'_1) \qquad (d_2, \surd)\ [p_2, a\rangle\ (d'_2, p'_2)}{(i(d_1, d'_1, d_2, d'_2), \surd)\ [p_1 \|_H p_2, a\rangle\ (j(d_1, d'_1, d_2, d'_2), p'_1 \|_H p'_2)}\ a \in H\ .$$

These rules present a problem for **(P1)** in that they are dependent on the specific atomic programs executed.

One way around this problem is to assume a sufficiently rich notion of data-state so as to be able to synchronize on data. It is true enough that data-states in dynamic logic only give values of program variables. But in the abstract set-up above, one can expand the notion to include "synchronization information" which the regular programs and interleaving $\|$ will ignore. Pushing the operating system intuition mentioned in section 1.5 further, the idea is that some processes will operate on only a section of the computer's memory. That is because other parts of the computer's memory are devoted to matters of control.

More concretely, suppose D is the set of finite functions from some set *Store* to some set of values, and that for every $a \in A$, we have an $I_a \subseteq D \times D$. Now, form a new set $\hat{D}$ of data-states by adjoining a slot for A marked by some $\hat{s} \notin Store$ as follows

$$\hat{D} \quad = \quad \{d \cup \{(\hat{s}, a)\} \mid d \in D,\ a \in A\}\ .$$

Then for every $a \in A$, pass from I_a to

$$\hat{I}_a \quad = \quad \{(d \cup \{(\hat{s}, a')\}, d' \cup \{(\hat{s}, a)\}) \ \mid\ a' \in A,\ dI_a d'\}\ ,$$

so that the slot $\hat{s}$ records the atomic program executed. Now, given an $H \subseteq A$, the rule (α) can be formulated as

$$(\alpha)' \quad \frac{(d_1, \surd)\ [p_1\rangle\ (d_1', p_1') \qquad (d_2, \surd)\ [p_2\rangle\ (d_2', p_2')}{(d, \surd)\ [p_1 \|_H p_2\rangle\ (d', p_1' \|_H p_2')} \quad S_H(d_1, d_1', d_2, d_2', d, d')$$

where

$$S_H(d_1, d_1', d_2, d_2', d, d') \quad \text{iff} \quad d_1'(\hat{s}) = d_2'(\hat{s}) \in H,\ d = d_1 \cup d_2,\ d' = d_1' \cup d_2'\ .$$

Similarly, for (β), take

$$(\beta)' \quad \frac{(d, \surd)\ [p_1\rangle\ (d', p_1')}{(d, \surd)\ [p_1 \|_H p_2\rangle\ (d', p_1' \|_H p_2)} \quad d'(\hat{s}) \notin H\ .$$

The rule $(\alpha)'$ can be adapted for a synchronization construct & that is "continued" by a possibly different construct $[\cdot,\cdot]_{\&}$, but which is assumed to come with some predicate $S_{\&}$

$$(\alpha)'' \quad \frac{(d_1, \surd)\ [p_1\rangle\ (d_1', p_1') \qquad (d_2, \surd)\ [p_2\rangle\ (d_2', p_2')}{(d, \surd)\ [p_1 \& p_2\rangle\ (d', [p_1', p_2']_{\&})} \quad S_{\&}(d_1, d_1', d_2, d_2', d, d')\ .$$

(In Groote and Ponse [9], for example, note that the construct $|$ is continued by $\|$.) The possibilites are legion, but the essential point for the present paper is that rules of the form $(\alpha)'$, $(\beta)'$ and $(\alpha)''$ are within the scope of the theory described in section 3 and Appendix A. (Side conditions such as $S_{\&}(d_1, d_1', d_2, d_2', d, d')$ can be eliminated by splitting the rule into the-cardinality-of-$S_{\&}$-many rules.)

While the "trick" above is quite general, certainly not every notion of control we dream up can be accomodated *directly* in the formulation of (**P1**) by the domain equation (6). (And notice that *skip* is no longer interpreted as the identity on data-states by $\hat{I}_{skip}$, although the extension from D to $\hat{D}$ and I_a to $\hat{I}_a$ is, in a suitable sense, conservative.) The question is whether (**P1**) is conceptually natural enough that we should welcome the discipline it imposes on our thinking. A trade-off between (**P1**) and (**P2**) (especially as realized by Example 2) is mediated by data-states and atomic programs, with the passage $D, I_a \mapsto \hat{D}, \hat{I}_a$, suggesting a broad interpretation of "data" that includes atomic programs. But uniformity for uniformity's sake can obscure simple ideas, and to insist (beyond observing that it *can* be done in principle) that a transition rule be put in a form with $L = P$ and $S = D \times (P + \{\surd\})$ would be silly.

Reasoning with Higher Order Partial Functions*

A. Gavilanes-Franco[1] and F. Lucio-Carrasco[2] and M. Rodríguez-Artalejo[1]

[1] Dep. de Informática y Automática. Fac. de Matemáticas. Univ. Complutense. Madrid.

[2] Dep. de Lenguajes y Sistemas Informáticos. Fac. de Informática. Univ. del País Vasco. San Sebastián.

Abstract. In this paper we introduce the logic $PHOL$, which embodies higher-order functions through a simply-typed λ-calculus and deals with partial objects by using partially ordered domains and three truth values. We define a refutationally complete tableaux method for $PHOL$ and we show how to derive a sound and complete cut free sequent calculus through a systematic analysis of the rules for tableaux construction.

1 Introduction

The formal methodologies used for the specification and verification of software (and also hardware) systems have motivated during the last years the development of so called *partial logics*, where partial functions can be used to argue about errors, diverging computations, and similar phenomena. Following some earlier precedents, such as [16], [21] and [5], quite many recent papers (as e.g. [2], [24], [17], [14], [19], [13]) have proposed partial versions of first order predicate logic aiming at this field of application. Partiality has been also investigated within the field of *data type specification* [3], [20]. These works show quite different views of partiality, semantics and proof calculi. Some of them adopt a third truth value (*undefined*), which has been argued to be useful in certain cases for reasoning about refinements and implementations of specifications (see [24], [13])[3]. In a previous paper [10], we developped a partial first-order logic which works with three truth values and strict functions, whose main contributions were a clarification of the possible views of logical consequence in a partial logic, as well as systematic methods for building sound and complete proof systems.

Our aim in the present paper is to extend this work in the direction of *higher order logic*. More specifically, our goal is to define a simple extension of first order logic (Partial Higher Order Logic, PHOL for short) which allows to reason about higher order, partial, possibly nonstrict functions, and partially defined and/or infinite data structures, as needed for formal reasoning about the behaviour of programs in lazy functional languages. Another motivation is that PHOL should provide a framework for *logic programming* languages with higher order features, similar to λ-Prolog [22]. Finally, we wish to achieve a clear semantics and a methodic construction of sound and complete proof calculi. According to these aims, we shall use an extension of first order logic which will include the simply typed λ-calculus and an approximation ordering over objects of any type. Following a classical idea of Henkin [12], we shall allow a broad class of models in order to enable completeness of finitary proof calculi and recursive enumerability of validities.

* Research partially supported by the PRONTIC Project TIC 89/0104 and the ESPRIT BR Working Group nb. 6028 CCL.

[3] The undefined thruth value has proved to be useful also in a different context within Computer Science, namely, the semantics of logic programs with negation [8], [18].

Several logics and systems with higher order features have been proposed already for different Computer Science purposes. The logic underlying the interactive proof system LCF [25] (going back to earlier work by D. Scott) is perhaps best known and most influential. Many features of PHOL (e.g. λ-terms, approximation ordering) exist in LCF. On the other hand, LCF is two-valued, and its semantics is rather different (models are standard, and as a consequence, the validity problem is not r.e.). Similar comments apply to the comparison between PHOL and HOL [11] or other approaches based on more sophisticated type systems [23], [26]. A recent paper by Farmer [7] presents a partial version of Church's type theory [4] which is more similar to PHOL in the use of first-order-like "general models". However, this logic is two valued, and its approach to proof calculi is completely different.

The organization of the paper is as follows. Section 2 defines the syntax and semantics of PHOL. In section 3 we develop a tableaux method for the logic and prove its soundness and completeness as a refutation procedure. Section 4 presents the relation between the tableaux method and proof systems for PHOL, showing how to derive a sound and complete, cut free sequent calculus (with a clearly isolated intuitionistic fragment) from a systematic analysis of tableaux rules. Finally, in section 5 we summarize our conclusions and we point to possible lines of future work.

Due to severe space limitations, some proofs had to be omitted. A more detailed version of the paper can be requested to authors.

2 Syntax and Semantics of PHOL

Given a set of primitive types (or *sorts*) $S = \{s_1, \ldots, s_n\}$, the set $Types_S$ of allowed types τ for $PHOL$ terms is defined by the rules $\tau ::= s_i | \tau \rightarrow \tau'$. We assume that the type constructor $\rightarrow$ is right-associative and that S includes the sort *bool*, intended for boolean values. In the sequel we assume some fixed S and abbreviate $Types_S$ as $Type$. A signature Σ over S is a collection of at most countable sets of typed constants C^τ, $\tau \in Type$. We assume that any signature contains boolean constants **T** (for true), **F** (for false) $\in C^{bool}$ and constants denoting *undefined* $\perp^\tau \in C^\tau$, for every type $\tau \in Type$. We shall write **U** for $\perp^{bool}$.

Given a family of countable pairwise disjoint sets of typed variables $X = \{X^\tau\}_{\tau \in Type}$, we define the set of (simply) typed Σ-terms as follows.

Definition 1 (Typed λ-terms) *The set $Term_\Sigma$ of Σ-terms is the typed family of sets $Term^\tau_\Sigma$, $\tau \in Type$, defined by the rules:*

$$t^\tau ::= x^\tau (\in X^\tau) | c^\tau (\in C^\tau) | (t_1^{\tau' \rightarrow \tau} t_2^{\tau'}) | (\lambda x^{\tau'}.t^{\tau''}) \text{ (for } \tau = \tau' \rightarrow \tau'') \quad \blacksquare$$

Some higher-order logics (see e. g. [7]) identify formulas with certain terms of type *bool*. $PHOL$ allows to use boolean terms as atomic formulas, but builds compound formulas by means of logical connectives and quantifiers. Moreover, inequalities $t_1^\tau \sqsubseteq_\tau t_2^\tau$ between terms (where $\sqsubseteq_\tau$ is intended to denote an approximation ordering) are also atomic formulas, as in LCF ([25]).

Definition 2 (Formulas) *The set For_Σ of Σ-formulas is defined by the rules:*

$$\varphi ::= t_1^\tau \sqsubseteq_\tau t_2^\tau (t_1^\tau, t_2^\tau \in Term^\tau_\Sigma) | t^{bool} (\in Term^{bool}_\Sigma) | \varphi \vee \psi | \varphi \wedge \psi | \varphi \Rightarrow \psi | \exists x^\tau \varphi | \forall x^\tau \varphi$$

In the following, we use "b" as a metavariable for boolean terms and omit types of terms and formulas in most cases. Application will be assumed to be left associative. The abbreviations $\neg\varphi$, $\varphi \Leftrightarrow \psi$ will stand for $\varphi \Rightarrow$ **F** and $(\varphi \Rightarrow \psi) \wedge (\psi \Rightarrow \varphi)$, respectively. Moreover, $t_1^\tau =_\tau t_2^\tau$ will abbreviate $(t_1^\tau \sqsubseteq_\tau t_2^\tau) \wedge (t_2^\tau \sqsubseteq_\tau t_1^\tau)$. We also assume the usual notions of *free* and *bound* occurrences of variables, as well as a *substitution* operation that avoids clashes (see e. g. [15], [6]).

For the semantics of $PHOL$ we use typed λ-models (see e. g. [15]) whose carriers are preordered sets such that the application operators behave monotonically. We call *precongruences* to this kind of preorders. They induce equivalence relations (which behave as *congruences* w.r.t. application) and a partial ordering over the equivalence classes in the obvious way. Our motivation for using partial orders is to express partiality and "degrees of definedness", as explained in the introduction. Moreover, we have chosen preorders instead of orders for the sake of more simple model constructions (quotients can be avoided). The formal definition follows. Note that carriers of functional types $\tau \to \tau'$ are sets of *function intensions*.

Definition 3 (Σ-structures) *An ordered λ-structure of signature Σ (briefly, Σ-oλs) is any system* $\mathcal{D} =< \{\mathcal{D}_\tau\}_{\tau \in Type}, \{@_{\tau\tau'}\}_{\tau,\tau' \in Type}, [\![\,]\!]^\mathcal{D} >$ *such that:*

(1) $\mathcal{D}_\tau =< D_\tau, \sqsubseteq_\tau^\mathcal{D}, \perp_\tau^\mathcal{D}, \{c^\mathcal{D}\}_{c \in C^\tau \setminus \{\perp_\tau\}} >$ *is a non-empty set, preordered by* $\sqsubseteq_\tau^\mathcal{D}$ *with least element* $\perp_\tau^\mathcal{D}$ *and interpretations* $c^\mathcal{D} \in \mathcal{D}_\tau$ *for the other constants of type* τ. *We use* $\approx_\tau^\mathcal{D}$ *for the equivalence relation induced by* $\sqsubseteq_\tau^\mathcal{D}$:

$$d_1 \approx_\tau^\mathcal{D} d_2 \text{ iff } d_1 \sqsubseteq_\tau^\mathcal{D} d_2 \text{ and } d_2 \sqsubseteq_\tau^\mathcal{D} d_1$$

(2) $\mathcal{D}_{bool}$ *is partitioned by* $\approx_{bool}^\mathcal{D}$ *into three equivalence classes with representants* $\mathbf{T}^\mathcal{D}, \mathbf{F}^\mathcal{D}, \mathbf{U}^\mathcal{D}$. *The order induced by* $\sqsubseteq_{bool}^\mathcal{D}$ *on these classes is isomorphic to the partial order* T *of Kleene's truth values* $\underline{t}, \underline{f}, \underline{u}$, *with* $\underline{u} \sqsubseteq_{bool} \underline{t}$, $\underline{u} \sqsubseteq_{bool} \underline{f}$ *and incomparable* $\underline{t}$, $\underline{f}$.

(3) *Each* $@_{\tau\tau'}$ *is a mapping* $@_{\tau\tau'} : D_{\tau\to\tau'} \times D_\tau \to D_{\tau'}$, *called* application operator, *which must be monotonic and strict w.r.t. its first argument:*

- $d \sqsubseteq_{\tau\to\tau'}^\mathcal{D} d', d_0 \sqsubseteq_\tau^\mathcal{D} d_0'$ *implies* $@_{\tau\tau'}(d, d_0) \sqsubseteq_{\tau'}^\mathcal{D} @_{\tau\tau'}(d', d_0')$
- $@_{\tau\tau'}(\perp_{\tau\to\tau'}^\mathcal{D}, d_0) \sqsubseteq_{\tau'}^\mathcal{D} \perp_{\tau'}^\mathcal{D}$

The phrase " $\sqsubseteq_\tau^\mathcal{D}$ *must be* precongruences*" will abbreviate these requirements in the sequel.*

(4) $[\![\,]\!]^\mathcal{D}$ *is a mapping, called* evaluation operator, *which assigns to each term* t^τ *and each (well typed) valuation of variables* $\rho = \{\rho_\tau / \rho_\tau : X^\tau \to D_\tau\}_{\tau \in Type}$, *a value* $[\![t^\tau]\!]_\rho^\mathcal{D} \in D_\tau$ *in such a way that the following conditions hold:*

(a) $[\![x^\tau]\!]_\rho^\mathcal{D} = \rho_\tau(x^\tau)$ **(b)** $[\![c^\tau]\!]_\rho^\mathcal{D} = (c^\tau)^\mathcal{D}$

(c) $[\![t_1^{\tau\to\tau'} t_2^\tau]\!]_\rho^\mathcal{D} = @_{\tau\tau'}([\![t_1^{\tau\to\tau'}]\!]_\rho^\mathcal{D}, [\![t_2^\tau]\!]_\rho^\mathcal{D})$

(d) $@_{\tau\tau'}([\![\lambda x^\tau.t^{\tau'}]\!]_\rho^\mathcal{D}, d) \approx_{\tau'}^\mathcal{D} [\![t^{\tau'}]\!]_{\rho[d/x^\tau]}^\mathcal{D}$, *for every* $d \in D_\tau$[4]

(e) *If* $[\![t_1^{\tau'}]\!]_{\rho[d/x^\tau]}^\mathcal{D} \sqsubseteq_{\tau'}^\mathcal{D} [\![t_2^{\tau'}]\!]_{\sigma[d/y^\tau]}^\mathcal{D}$, *for every* $d \in D_\tau$, *then* $[\![\lambda x^\tau.t_1^{\tau'}]\!]_\rho^\mathcal{D}$ $\sqsubseteq_{\tau\to\tau'}^\mathcal{D} [\![\lambda y^\tau.t_2^{\tau'}]\!]_\sigma^\mathcal{D}$ ■

Note that (d) is intended to reflect β-conversion, while (e) implies *Berry's extensionality property* (also called *weak extensionality* cfr. [15]), which says that equality between λ-abstractions is extensional:

$$[\![t_1^{\tau'}]\!]_{\rho[d/x^\tau]}^\mathcal{D} \approx_{\tau'}^\mathcal{D} [\![t_2^{\tau'}]\!]_{\sigma[d/y^\tau]}^\mathcal{D}, \text{ for every } d \in D_\tau \Rightarrow [\![\lambda x^\tau.t_1^{\tau'}]\!]_\rho^\mathcal{D} \approx_{\tau\to\tau'}^\mathcal{D} [\![\lambda y^\tau.t_2^{\tau'}]\!]_\sigma^\mathcal{D}.$$

The next definition isolates a particularly interesting kind of Σ-oλs's.

[4] $\rho[d/x^\tau]$ is the valuation ρ' such that $\rho'(x^\tau) = d$ and $\rho'(y) = \rho(y)$ for all $y \not\equiv x$.

Definition 4 *A Σ-oλs $\mathcal{D}$ is called* standard *iff it satisfies the following conditions:*

(1) *For every $\tau, \tau' \in Type$, $D_{\tau \to \tau'}$ is the set of all mappings $f : D_\tau \to D_{\tau'}$ which are monotonic w.r.t. $\sqsubseteq^{\mathcal{D}}_{\tau}, \sqsubseteq^{\mathcal{D}}_{\tau'}$, and $\sqsubseteq^{\mathcal{D}}_{\tau \to \tau'}$ is the pointwise ordering induced by $\sqsubseteq^{\mathcal{D}}_{\tau'}$.*

(2) *For every $\tau, \tau' \in Type$, $@_{\tau\tau'}$ is the natural application operator $@_{\tau\tau'}(f, d) = f(d)$*

(3) *For every abstraction $\lambda x^\tau.t^{\tau'}$ and valuation ρ, $[\![\lambda x^\tau.t^{\tau'}]\!]^{\mathcal{D}}_\rho$ is defined as the monotonic function $f : D_\tau \to D_{\tau'}$ such that $f(d) = [\![t^{\tau'}]\!]^{\mathcal{D}}_{\rho[d/x^\tau]}$.*
It is easy to check that standard Σ-oλs's are indeed Σ-oλs's in the sense of Definition 3. ■

Σ-oλs's satisfy all the provable equations of λ-calculus without extensionality as stated by the following lemma.

Lemma 5 *Let $\mathcal{D}$ be a Σ-oλs then:*

(i) *(Coincidence lemma for terms)*
If $\rho(x) = \sigma(x)$, for all $x \in free(t)$, then $[\![t]\!]^{\mathcal{D}}_\rho \approx^{\mathcal{D}}_\tau [\![t]\!]^{\mathcal{D}}_\sigma$

(ii) *(Substitution lemma for terms)*
$[\![t[s/x]]\!]^{\mathcal{D}}_\rho \approx^{\mathcal{D}}_\tau [\![t]\!]^{\mathcal{D}}_{\rho[[\![s]\!]^{\mathcal{D}}_\rho/x]}$

(iii) *(α-conversion)*
If $y \notin free(t)$ and $y \not\equiv x$ then $[\![\lambda x.t]\!]^{\mathcal{D}}_\rho \approx^{\mathcal{D}}_\tau [\![\lambda y.t[y/x]]\!]^{\mathcal{D}}_\rho$

(iv) *(β-conversion)*
$[\![(\lambda x.t_1)t_2]\!]^{\mathcal{D}}_\rho \approx^{\mathcal{D}}_\tau [\![t_1[t_2/x]]\!]^{\mathcal{D}}_\rho$ ■

Now we know how to evaluate λ-terms in Σ-oλs's. Next we define the interpretation of formulas in such structures. Since $PHOL$ is a three-valued logic, truth values will be taken from T (see Definition 3.2).

Definition 6 (Σ-interpretations) *A Σ-interpretation $\mathcal{I}$ is a pair $(\mathcal{D}, \rho)$, where $\mathcal{D}$ is a Σ-oλs and ρ is a well typed valuation of variables. Given $\mathcal{I}$, the denotation of a formula φ in $\mathcal{I} = (\mathcal{D}, \rho)$, written $[\![\varphi]\!]^{\mathcal{D}}_\rho$ or $\mathcal{I}(\varphi)$, and belonging to* T *is defined by structural induction on φ as follows:*

(1) $[\![t_1 \sqsubseteq_\tau t_2]\!]^{\mathcal{D}}_\rho = \begin{cases} \underline{t} \text{ if } [\![t_1]\!]^{\mathcal{D}}_\rho \sqsubseteq^{\mathcal{D}}_\tau [\![t_2]\!]^{\mathcal{D}}_\rho \\ \underline{f} \text{ otherwise} \end{cases}$

(2) *The interpretation of formulas b is given by taking the formula as a boolean term and defining:*

$$[\![b]\!]^{\mathcal{D}}_\rho = \begin{cases} \underline{t} \text{ if } [\![b]\!]^{\mathcal{D}}_\rho \approx^{\mathcal{D}}_{bool} \mathbf{T}^{\mathcal{D}} \\ \underline{f} \text{ if } [\![b]\!]^{\mathcal{D}}_\rho \approx^{\mathcal{D}}_{bool} \mathbf{F}^{\mathcal{D}} \\ \underline{u} \text{ if } [\![b]\!]^{\mathcal{D}}_\rho \approx^{\mathcal{D}}_{bool} \mathbf{U}^{\mathcal{D}} \end{cases}$$

(3) *The interpretation of propositional connectives is that of [16], given by the following tables:*

$\varphi \vee \psi$	$\underline{t}$	$\underline{f}$	$\underline{u}$
$\underline{t}$	$\underline{t}$	$\underline{t}$	$\underline{t}$
$\underline{f}$	$\underline{t}$	$\underline{f}$	$\underline{u}$
$\underline{u}$	$\underline{t}$	$\underline{u}$	$\underline{u}$

$\varphi \wedge \psi$	$\underline{t}$	$\underline{f}$	$\underline{u}$
$\underline{t}$	$\underline{t}$	$\underline{f}$	$\underline{u}$
$\underline{f}$	$\underline{f}$	$\underline{f}$	$\underline{f}$
$\underline{u}$	$\underline{u}$	$\underline{f}$	$\underline{u}$

$\varphi \Rightarrow \psi$	$\underline{t}$	$\underline{f}$	$\underline{u}$
$\underline{t}$	$\underline{t}$	$\underline{f}$	$\underline{u}$
$\underline{f}$	$\underline{t}$	$\underline{t}$	$\underline{t}$
$\underline{u}$	$\underline{t}$	$\underline{u}$	$\underline{u}$

(4) $[\![\exists x^\tau \varphi]\!]^{\mathcal{D}}_\rho = \begin{cases} \underline{t} \text{ if there is } d \in \mathcal{D}_\tau \text{ such that } [\![\varphi]\!]^{\mathcal{D}}_{\rho[d/x^\tau]} = \underline{t} \\ \underline{f} \text{ if } [\![\varphi]\!]^{\mathcal{D}}_{\rho[d/x^\tau]} = \underline{f}, \text{ for all } d \in \mathcal{D}_\tau \\ \underline{u} \text{ otherwise} \end{cases}$

(5) $[\![\forall x^\tau \varphi]\!]^{\mathcal{D}}_\rho = \begin{cases} \underline{t} \text{ if } [\![\varphi]\!]^{\mathcal{D}}_{\rho[d/x^\tau]} = \underline{t}, \text{ for all } d \in D_\tau \\ \underline{f} \text{ if there is } d \in D_\tau \text{ such that } [\![\varphi]\!]^{\mathcal{D}}_{\rho[d/x^\tau]} = \underline{f} \\ \underline{u} \text{ otherwise} \end{cases}$ ■

Note that equalities and inequalities between terms have a bivalued non strict interpretation. More generally, if all the atomic subformulas of φ are inequalities, then φ takes a defined truth value in every interpretation. Formulas with this property are called *defined*. Existential and universal quantification are interpreted as infinite disjunction and conjunction, respectively, as in [16].

As for terms, it is not difficult to prove the following basic results.

Lemma 7 *Let $\mathcal{I} = (\mathcal{D}, \rho)$ be a Σ-interpretation then:*

(i) *(Coincidence lemma for formulas)*
If $\rho(x) = \sigma(x)$, for all $x \in free(\varphi)$, then $[\![\varphi]\!]^{\mathcal{D}}_\rho = [\![\varphi]\!]^{\mathcal{D}}_\sigma$

(ii) *(Substitution lemma for formulas)*
$[\![\varphi[t/x]]\!]^{\mathcal{D}}_\rho = [\![\varphi]\!]^{\mathcal{D}}_{\rho[[\![t]\!]^{\mathcal{D}}_\rho/x]}$

Proof. By structural induction on φ. Cfr. classical proofs in [6]. ■

The language of *PHOL* is rich enough to formalize many interesting statements. It is possible, for instance, to write axioms about fixpoint and recursion operators, natural numbers, lists and other data types, having the freedom to specify if data constructors (such as *nil* and *cons* for lists) are required to be strict or not. See e. g. [28] for formalizations of this kind. As a concrete example, we show here an *extensionality axiom* for the type $\tau \to \tau'$:

$$\forall f^{\tau\to\tau'} \forall g^{\tau\to\tau'} (\lambda x^\tau.(fx) \sqsubseteq_{\tau\to\tau'} \lambda x^\tau.(gx) \Rightarrow f \sqsubseteq_{\tau\to\tau'} g)$$

Σ-oλs which satisfy this axiom for all types $\tau \to \tau'$ are called *extensional*. Standard Σ- oλs's are extensional, but this is not true of all Σ-oλs.

When reasoning with 3-valued formulas, some semantical notions must be made precise. Here we have some of them (cfr. [10] for more details).

Definition 8 *Let $\mathcal{I}$ be a Σ-interpretation. We say that:*

(1) *$\mathcal{I}$ strongly satisfies a formula φ (or $\mathcal{I}$ is a strong model of φ) iff $\mathcal{I}(\varphi) = \underline{t}$ (written $\mathcal{I} \models_s \varphi$)*

(2) *$\mathcal{I}$ weakly satisfies a formula φ (or $\mathcal{I}$ is a weak model of φ) iff $\mathcal{I}(\varphi) \neq \underline{f}$ (written $\mathcal{I} \models_w \varphi$)*

(3) *A set of formulas Φ is strongly (resp. weakly) satisfiable iff it has an strong (resp. weak) model $\mathcal{I}$ such that $\mathcal{I} \models_s \varphi$ (resp. $\mathcal{I} \models_w \varphi$) , for all $\varphi \in \Phi$. In symbols: $\mathcal{I} \models_s \Phi$ (resp. $\mathcal{I} \models_w \Phi$).* ■

For each choice of $a, b \in \{s, w\}$ we can define a logical consequence $\models_{ab}$ by stipulating that:

$$\Phi \models_{ab} \varphi \text{ iff for every interpretation } \mathcal{I}\text{: } \mathcal{I} \models_a \Phi \text{ implies } \mathcal{I} \models_b \varphi$$

[10] contains a detailed discusion of these logical consequences and gives some reasons for choosing $\models_{sw}$, which will be adopted in this paper and noted simply as $\models$ in the sequel. Two simple but important facts about $\models$ are given in the next proposition.

Lemma 9
(i) *$\Phi \models \psi$ iff not $Sat_s \Phi \cup \{\neg\psi\}$.*
(ii) *$\Phi \models \varphi$ and $\Phi \cup \{\varphi\} \models \psi$ do not always imply that $\Phi \models \psi$.*
Proof. (i) is obvious from the definitions. For (ii), note that $\models \mathbf{U}$ and $\mathbf{U} \models \mathbf{F}$, but not $\models \mathbf{F}$. ■

According to this lemma, reasoning by contradiction is a sound inference rule for $PHOL$, but the cut rule is not sound. The proof calculus given in the present paper will be cut-free and will include a *reductio ad absurdum* rule. If desired, a sound restricted cut rule could be obtained by requiring the cut formula φ to be *defined* in the sense explained after Definition 6 above.

3 Tableaux for PHOL

This section presents a tableaux method for $PHOL$, following our previous work [10] where tableaux for 3-valued logics were defined in a formulation inspired by [27]. In order to define the method we extend a given signature Σ with countably infinite sets AC^τ, $\tau \in Type$, of new auxiliary constants. The new signature so obtained will be denoted by $\overline{\Sigma}$.

A tableau is a labelled tree with signed formulas at its nodes. Signing formulas consists in prefixing them with one of two possible signs T or F. Strong satisfiability of signed formulas is defined in the natural way: An interpretation $\mathcal{I}$ is a strong model of $T\varphi$ (resp. $F\varphi$) iff $\mathcal{I}(\varphi) = \underline{t}$ (resp. $\underline{f}$). A set Φ of signed formulas is called strongly satisfiable iff it has some strong model. By Lemma 9.(i), strong satisfiability corresponds to our notion of logical consequence. In the construction of a tableau, the tree grows and branches out by choosing a signed formula labelling some of the nodes and following a specific rule. The tableaux rules depend on a semantic classification of formulas, which is given by the following proposition:

Lemma 10 *The set of signed $\overline{\Sigma}$-formulas is partitioned in the following five pairwise disjoint classes defined by semantic conditions:*

(0) *BASIC: All basic formulas not belonging to the other classes, that is the set $\{T(t_1 \sqsubseteq t_2), F(t_1 \sqsubseteq t_2)/t_1, t_2 \in Term_{\overline{\Sigma}}\}$. They have no constituents.*

(A) *ALPHA: Conjunctive formulas α with one or two constituents α_1, α_2 satisfying the semantic condition: $Sat_s \Phi \cup \{\alpha\} \iff Sat_s \Phi \cup \{\alpha_1, \alpha_2\}$*

α	α_1	α_2
Tb	$T(\mathbf{T} \sqsubseteq b)$	
Fb	$T(\mathbf{F} \sqsubseteq b)$	
$T(\varphi \wedge \psi)$	$T\varphi$	$T\psi$
$F(\varphi \vee \psi)$	$F\varphi$	$F\psi$
$F(\varphi \Rightarrow \psi)$	$T\varphi$	$F\psi$

(B) *BETA: Disjunctive formulas β with two constituents β_1, β_2 satisfying the semantic condition: $Sat_s \Phi \cup \{\beta\} \Leftrightarrow Sat_s \Phi \cup \{\beta_1\}$ or $Sat_s \Phi \cup \{\beta_2\}$*

β	β_1	β_2
$F(\varphi \wedge \psi)$	$F\varphi$	$F\psi$
$T(\varphi \vee \psi)$	$T\varphi$	$T\psi$
$T(\varphi \Rightarrow \psi)$	$F\varphi$	$T\psi$

(C) *GAMMA: Universal formulas γ with a constituent $\gamma(t)$ for each well typed $t \in Term_{\overline{\Sigma}}$, satisfying the semantic condition: $Sat_s \Phi \cup \{\gamma\} \Leftrightarrow Sat_s \Phi \cup \{\gamma, \gamma(t)\}$*

γ	$\gamma(t)$
$T\forall x\varphi$	$T\varphi[t/x]$
$F\exists x\varphi$	$F\varphi[t/x]$

(D) *DELTA: Existential formulas δ with a constituent $\delta(c)$ satisfying the semantic condition: $Sat_s\Phi \cup \{\delta\} \Leftrightarrow Sat_s\Phi \cup \{\delta(c)\}$, for each well typed $c \in AC$, if c is chosen "new" (i.e., occurring neither in Φ nor in δ)*

δ	$\delta(c)$
$F\forall x\varphi$	$F\varphi[c/x]$
$T\exists x\varphi$	$T\varphi[c/x]$

■

We say that a branch in a tableau is *closed*, and then it is neither enlarged nor splitted anymore, when an "obvious contradiction" is detected at its labels.

Definition 11 *Given a set Φ of signed $\overline{\Sigma}$-formulas, we say that Φ is* incoherent *iff it satisfies someone of the following conditions:*

(i) *there is a formula φ such that $T\varphi, F\varphi \in \Phi$*
(ii) *$T\eta \in \Phi$ for some η belonging to the set of* boolean contradictions
$\mathbf{BCD} = \{\mathbf{T} \sqsubseteq \mathbf{F}, \mathbf{F} \sqsubseteq \mathbf{T}, \mathbf{T} \sqsubseteq \mathbf{U}, \mathbf{F} \sqsubseteq \mathbf{U}\}$

In case that condition (i) is required only for φ of the form $t_1 \sqsubseteq t_2$, we say that Φ is atomically incoherent.

A branch B of a tableau $\mathcal{T}$ with Φ being the set of signed formulas labelling its nodes, is closed iff Φ is an incoherent set of signed formulas. Otherwise we say that the branch is open. ■

It is clear that condition 11.(i) expresses a contradiction. The reason why condition 11.(ii) expresses contradiction is that we require the values of the three boolean constants to be order- isomorphic to T (cfr. Definition 3).

The construction of tableaux needs also some *logical axioms*. The logical axioms we have chosen for PHOL are given by the following result:

Lemma 12 *All the formulas of the eight kinds shown below are strongly valid (i.e., true in every interpretation):*
$\bot \sqsubseteq t$, $t \sqsubseteq t$, $t_1 \sqsubseteq t_2 \wedge t_2 \sqsubseteq t_3 \Rightarrow t_1 \sqsubseteq t_3$,
$t_1 \sqsubseteq t_1' \wedge t_2 \sqsubseteq t_2' \Rightarrow (t_1 t_2) \sqsubseteq (t_1' t_2')$, $\bot\, t \sqsubseteq \bot$,
$\forall z(t_1[z/x] \sqsubseteq t_2[z/y]) \Rightarrow \lambda x.t_1 \sqsubseteq \lambda y.t_2$
(where $z \not\equiv x, y$ and $z \notin free(t_1) \cup free(t_2)$; weak extensionality),
$(\lambda x.t_1)t_2 = t_1[t_2/x]$ *(β-conversion),*
$b = \mathbf{T} \vee b = \mathbf{F} \vee b \sqsubseteq \mathbf{U}$
These formulas make up the set **LAX** *of* logical axioms *for PHOL.*

Proof. Straightforward consequence of Definition 3 and Lemma 5. ■

Now we are in a position to define the tableaux method formally, following classical presentations (cfr. [27], [9]).

Definition 13 *Finite tableaux for a non-empty set Φ of signed Σ-formulas are inductively constructed using $\overline{\Sigma}$-formulas according to the following rules:*

(o) *If $\{\varphi_1, \ldots, \varphi_n\} \subseteq \Phi$ then the linear tree with a branch labelled by the signed formulas φ_i is a finite tableau for Φ.*
(a) *(resp.* **(c)**, **(d)***) If $\mathcal{T}$ is a finite tableau for Φ, B is an open branch of $\mathcal{T}$ and α (resp γ, δ) is a signed formula labelling some of B's nodes then the tree resulting from enlarging B with nodes labelled by the constituent(s) of α (resp. γ, δ), is a finite tableau for Φ.*

In the case (c), the constituent $\gamma(t)$ must satisfy that t is adequate *to the branch B. This means that all free variables of t must occur free in B, and all the constants occurring in t must occur in B or belong to the set* $\{\mathbf{T}, \mathbf{F}, \mathbf{U}\} \cup \{\perp_\tau / \tau \in Type\}$[5].

In the case (d), the constituent $\delta(c)$ must satisfy that c is a new *auxiliary constant not occurring in the set of signed formulas labelling B's nodes.*

(b) *If $\mathcal{T}$ is a finite tableau for Φ, B is a branch of $\mathcal{T}$ and β is a signed formula labelling some of its nodes then the tree resulting from splitting B with two nodes labelled by β_1 and β_2, is a finite tableau for Φ.*

(LAX) *If $\mathcal{T}$ is a finite tableau for Φ then the result of enlarging some open branch with a new node labelled by $T\theta$, where $\theta \in$ **LAX** is adequate to this branch*[6] *is a finite tableau for Φ.*

(IN) *If $\mathcal{T}$ is a finite tableau for Φ then the result of enlarging some open branch B with a signed formula $\varphi \in \Phi$, which is not yet in B, is again a finite tableau for Φ.* ■

When building a finite tableau, signed formulas of the kinds α, β and δ only need to be analysed one time for each branch where they occur, and can be regarded as *used* (for that branch) in the sequel. This notion of used formula will be applied in section 4.

The notion of finite tableau induces in a natural way that of infinite tableau.

Definition 14 *A tableaux sequence for a non-empty set of signed formulas Φ is any sequence $< \mathcal{T}_k / k \in N >$ where $\mathcal{T}_0$ is some linear tableau as defined in 13.(o) above, and every $\mathcal{T}_{k+1}$ arises from $\mathcal{T}_k$ through one of the tableaux rules 13.(a)-(IN).*

An infinite tableau for a non-empty set of formulas Φ is defined as the limit of some tableaux sequence. This notion can be made precise in terms of a metric space of trees [1]. ■

Next, we proceed to prove that the tableaux method just defined is (refutationally) sound and complete. Soundness is easy to prove.

Theorem 15 (Soundness of the tableaux method) *For every set Φ of signed Σ-formulas, if Φ has a closed tableau (i. e., a finite tableau whose branches are all closed) then Φ has some finite subset Φ_0 which is not strongly satisfiable.*

Proof. Let $\mathcal{T}$ be a closed tableau for Φ and take Φ_0 as the (finite) set of all $\varphi \in \Phi$ used by some application of the tableau rules (o) and (IN) during the construction of $\mathcal{T}$. If Φ_0 were strongly satisfiable, a straightforward induction on $\mathcal{T}$'s construction (using lemmata 10 and 12) would yield also the strong satisfiability of the set of signed formulas labelling one (closed) branch of $\mathcal{T}$. This is impossible, since incoherent sets (cfr. Definition 11) are never strongly satisfiable. ■

The proof of the reciprocal result to Theorem 15 (completeness of the tableaux method) will rely on the following idea: if the construction of a tableau for Φ is carried out in a systematic manner and the tableau does not close, then there will be some branch with special saturation properties (w.r.t. the tableaux rules), which will allow the construction of a strong model of Φ. Next, we define sets of signed formulas with these saturation properties and show how to use them for building strong models. After this, we shall come back to tableaux.

[5] Note that a universal formula γ can have infinitely many constituents, as in first-order logic.

[6] The concept of signed formula adequate to a branch is defined likewise that of adequate term.

Definition 16 *A set H of signed $\overline{\Sigma}$-formulas is a Hintikka set iff the following conditions hold:*

(A) *for every $\alpha \in ALPHA$: $\alpha \in H \Rightarrow \alpha_i \in H$, $i = 1, 2$*
(B) *for every $\beta \in BETA$: $\beta \in H \Rightarrow \beta_i \in H$, for some $i = 1, 2$*
(C) *for every $\gamma \in GAMMA$: $\gamma \in H \Rightarrow \gamma(t) \in H$, for all $\overline{\Sigma}$-term t adequate to H*
(D) *for every $\delta \in DELTA$: $\delta \in H \Rightarrow \delta(c) \in H$, for some $c \in AC$*
(LAX) *for every logical axiom $\theta \in$ **LAX** adequate to H, $T\theta$ belongs to H*
(COH) *H is atomically coherent.* ■

In the next theorem we restrict ourselves to Hintikka sets without free variables, in order to avoid technical problems related to variable renaming. The theorem holds for arbitrary Hintikka sets, but the proof is technically more complicated. As we shall see later, this restriction is harmless.

Theorem 17 (Existence of strong models for Hintikka sets) *Every Hintikka set without occurrences of free variables is strongly satisfiable.*

Proof. Assume a Hintikka set H. Let Σ_0 be the signature which contains exactly those constants that have some occurrence in H. We use information from H to define a Σ_0-oλs $\mathcal{D}$ as follows:

- $D_\tau = \{t \in Term^\tau_{\Sigma_0} / t$ is closed, i.e. free(t) is empty$\}$
- $t \sqsubseteq^{\mathcal{D}}_\tau t' \Longleftrightarrow T(t \sqsubseteq t') \in H$
- $c^{\mathcal{D}} = c$ for all $c \in \Sigma_0$
- $@_{\tau\tau'}(t^{\tau\to\tau'}, t_0^\tau) = (t^{\tau\to\tau'} t_0^\tau)$
- $[\![t]\!]^{\mathcal{D}}_\rho = t\rho$, where $t \in Term_{\Sigma_0}$ and $t\rho$ stands for the result of applying ρ, viewed as a substitution, to t. Note that this substitution will require no renaming of bound variables, since $\rho(x)$ is some closed $s \in Term_{\Sigma_0}$, for every variable x.

We have to check the conditions (1)-(4) from Definition 3. Conditions (1)-(3) follow easily from the saturation properties and atomic coherence of Hintikka sets.

Consider now condition (4) of Definition 3. Parts (a)-(c) are trivial to check. For part (d) we have to prove that:

$$((\lambda x^\tau.t^{\tau'})\rho)s^\tau \approx^{\mathcal{D}}_{\tau'} t^{\tau'}\rho[s^\tau/x^\tau]$$

where s^τ is an arbitrary closed Σ_0-term of type τ. Let $\rho|x$ stand for the substitution ρ' such that $\rho'(x) = x$ and $\rho'(y) = \rho(y)$ for all $y \not\equiv x$. Then

$((\lambda x^\tau.t^{\tau'})\rho)s^\tau \equiv (\lambda x^\tau.t^{\tau'}(\rho|x^\tau))s^\tau \approx^{\mathcal{D}}_{\tau'} t^{\tau'}(\rho|x^\tau)[s^\tau/x^\tau]$
(since $T((\lambda x^\tau.t^{\tau'}(\rho|x^\tau))s^\tau =_{\tau'} t^{\tau'}(\rho|x^\tau)[s^\tau/x^\tau]) \in H$ by **(LAX)**)
$\equiv t^{\tau'}\rho[s^\tau/x^\tau]$

Finally, to check part (e) we have to prove that:

$$(\lambda x^\tau.t_1^{\tau'})\rho \equiv \lambda x^\tau.t_1^{\tau'}(\rho|x^\tau) \sqsubseteq^{\mathcal{D}}_{\tau\to\tau'} \lambda y^\tau.t_2^{\tau'}(\sigma|y^\tau) \equiv (\lambda y^\tau.t_2^{\tau'})\sigma$$

under the assumption that $t_1^{\tau'}\rho[s^\tau/x^\tau] \sqsubseteq^{\mathcal{D}}_{\tau'} t_2^{\tau'}\sigma[s^\tau/y^\tau]$ holds for every closed $s^\tau \in Term_{\Sigma_0}$. By atomic coherence, this assumption implies that H has no member

$$F(t_1^{\tau'}\rho[c^\tau/x^\tau] \sqsubseteq_{\tau'} t_2^{\tau'}\sigma[c^\tau/y^\tau])$$

with $c^\tau \in AC$. By condition **(D)** of Hintikka sets, we infer that:

$$F\forall z^{\tau}(t_1^{\tau'}(\rho|x^{\tau})[z^{\tau}/x^{\tau}] \sqsubseteq_{\tau'} t_2^{\tau'}(\sigma|y^{\tau})[z^{\tau}/x^{\tau}]) \notin H$$

where we choose z^{τ} as any variable which occurs neither in $\lambda x^{\tau}.t_1^{\tau'}$ nor in $\lambda y^{\tau}.t_2^{\tau'}$. By **(LAX)**, $T(\forall z^{\tau}(t_1^{\tau'}(\rho|x^{\tau})[z^{\tau}/x^{\tau}] \sqsubseteq_{\tau'} t_2^{\tau'}(\sigma|y^{\tau})[z^{\tau}/x^{\tau}]) \Rightarrow \lambda x^{\tau}.t_1^{\tau'}(\rho|x^{\tau}) \sqsubseteq_{\tau\to\tau'} \lambda y^{\tau}.t_2^{\tau'}(\sigma|y^{\tau})) \in H$ and using condition **(B)** of Hintikka sets we obtain:

$$T(\lambda x^{\tau}.t_1^{\tau'}(\rho|x^{\tau}) \sqsubseteq_{\tau\to\tau'} \lambda y^{\tau}.t_2^{\tau'}(\sigma|y^{\tau})) \in H$$

thus proving what we wanted.

This finishes the construction of $\mathcal{D}$. To complete our theorem, we use induction on φ to prove that:

(i) $T\varphi \in H \Longrightarrow [\![\varphi]\!]^{\mathcal{D}}_{\rho} = \underline{t}$ (ii) $F\varphi \in H \Longrightarrow [\![\varphi]\!]^{\mathcal{D}}_{\rho} = \underline{f}$

(where ρ may be any valuation, since H has no free variables). The base cases for atomic φ follow easily from the properties of Hintikka sets and the construction of $\mathcal{D}$. The induction steps for compound φ can be performed exactly as in the construction of models from first-order Hintikka sets (cfr. [27], [9]). ■

Now we come back to the completeness of the tableaux method. To adapt ourselves to the "no free variables assumption" from Theorem 17, we need an auxiliary result.

Lemma 18 *Let Φ be a set of signed Σ-formulas. Let $\overline{\Phi}$ be the result of substituting new constants (from some auxiliary set NC, disjoint from the set AC of auxiliary constants used in the tableaux method) for all the variables with free occurrences in Φ. Then:*

(i) *If $\overline{\Phi}$ has a closed tableau then Φ has also a closed tableau.*

(ii) *If $Sat_s\overline{\Phi}$ then also $Sat_s\Phi$.*

Sketch of Proof. (i) Given a closed tableau $\overline{\mathcal{T}}$ for $\overline{\Phi}$, we can build a closed tableau $\mathcal{T}$ for Φ by induction on $\overline{\mathcal{T}}$'s construction. $\mathcal{T}$ will be the same as $\overline{\mathcal{T}}$ up to some renamings of bound variables, which will have to be perfomed in order to avoid clashes with variables occurring free in Φ.

(ii) Given a strong model $\overline{\mathcal{I}} = (\overline{\mathcal{D}}, \overline{\rho})$ of $\overline{\Phi}$, we consider the Σ-interpretation $\mathcal{I} = (\mathcal{D}, \rho)$ where $\mathcal{D}$ is obtained from $\overline{\mathcal{D}}$ by forgetting about $c^{\overline{\mathcal{D}}}$, for all $c \in NC$, and ρ is such that $\rho(x) = c^{\overline{\mathcal{D}}}$, for every x variable which had been replaced by $c \in NC$ when passing from Φ to $\overline{\Phi}$. Then, lemma 7 allows to prove that $\mathcal{I}$ is a strong model of Φ. ■

Finally, we come to the main result of this section.

Theorem 19 (Completeness of the tableaux method) *For every set Φ of signed Σ-formulas, if Φ is not strongly satisfiable then Φ has a closed tableau.*

Proof. Let us assume that Φ had no closed tableau and show how to build a strong model. By lemma 18.(i), $\overline{\Phi}$ will have no closed tableau, either. In particular, the so called *canonical* (or *systematic*) tableau $\overline{\mathcal{T}}$ for $\overline{\Phi}$ will be not closed. $\overline{\mathcal{T}}$ may be infinite, but it is constructed in such a way that the signed formulas labelling any of its open branches forms a Hintikka set which includes Φ as a subset. Systematic procedures for achieving this effect are well known (see [27], [9], [10]). Using König's lemma, if $\overline{\mathcal{T}}$ is infinite (which will usually be the case) we get an infinite (open) branch B of $\overline{\mathcal{T}}$. This gives us a Hintikka set $H \supseteq \overline{\Phi}$, which in turn supplies a strong model of $\overline{\Phi}$ according to theorem 17. By lemma 18.(ii), Φ has also a strong model. ■

As a consequence of theorems 19 and 15 we immediately get

Corollary 20 (Compactness of PHOL) *For every set Φ of signed formulas, if Φ is not strongly satisfiable then some finite subset $\Phi_0 \subseteq \Phi$ is not strongly satisfiable.* ■

The attentive reader may have noted that the model obtained in the proof of theorem 19 has signature Σ_0, corresponding to the smaller signature used in the proof of Theorem 17. The next technical lemma shows that this can be extended to a Σ-model.

Lemma 21 *Let Σ_0, Σ be two signatures over the same sorts, such that $\Sigma_0 \subseteq \Sigma$. Then every Σ_0-oλs $\mathcal{D}_0$ can be extended to a Σ-oλs $\mathcal{D}$ by providing interpretations of the constants $c \in \Sigma \backslash \Sigma_0$, in such a way that $\mathcal{D}_0$ and $\mathcal{D}$ will behave the same for Σ_0-terms and formulas.*

Proof. Let us define $\mathcal{D}$ as follows: $D_\tau, c^{\mathcal{D}}$, for $c \in \Sigma_0$, and $@_{\tau\tau'}$ are the same as in $\mathcal{D}_0$. For each $c^\tau \in \Sigma \backslash \Sigma_0, (c^\tau)^{\mathcal{D}}$ is defined as $\perp_\tau^{\mathcal{D}_0}$. Finally, $[\![\,]\!]^{\mathcal{D}}$ is defined by taking $[\![t]\!]_\rho^{\mathcal{D}} = [\![t_0]\!]_\rho^{\mathcal{D}_0}$ where t is any Σ-term and t_0 is the Σ_0-term obtained by replacing all the occurrences of each $c^\tau \in \Sigma \backslash \Sigma_0$ in t by $\perp^\tau$. Straightforward inductions on the structure of terms and formulas prove that $\mathcal{D}$ is a Σ-oλs with the desired properties. ■

4 From Tableaux to Proof Systems

In [GL 90] we exposed a methodology for building sound and complete proof calculi through an analysis of the rules of a sound and refutationally complete tableaux method. Here we are going to use a similar approach: for each rule of the tableaux method given in the previous section, we shall define two associated inference rules which will be *refutation-* and *deduction-oriented* respectively. This will give rise to a *refutation calculus* $\mathcal{R}$ and a *deduction calculus* $\mathcal{D}$. $\mathcal{R}$ will be sound and complete due to its obvious connection to the tableaux method. $\mathcal{D}$ will be clearly sound and its completeness will follow from the fact that all rules of $\mathcal{R}$ will be derivable in $\mathcal{D}$.

Both $\mathcal{R}$ and $\mathcal{D}$ will be presented as sequent calculi[7], using *sequents* of the form $\Gamma \rhd \psi$, where the *antecedent* Γ is any finite *set* of formulas and the *consequent* ψ is a formula. Let $\mathcal{C}$ stand for either $\mathcal{R}$ or $\mathcal{D}$. As usual, the provability relation $\Phi \vdash_{\mathcal{C}} \psi$ is defined to mean that some sequent $\Gamma \rhd \psi$ with $\Gamma \subseteq \Phi$ can be derived in $\mathcal{C}$. Auxiliary constants will be used in $\mathcal{C}$-derivations similarly as in tableaux.

In order to obtain $\mathcal{R}$, we mentally associate to any tableaux branch B with still unused signed formulas $\mathrm{T}\varphi_i$ $(1 \leq i \leq m)$, $\mathrm{F}\psi_j$ $(1 \leq j \leq n)$ the sequent

$$S_B \equiv \varphi_1, \ldots \varphi_m, \neg\psi_1, \ldots, \neg\psi_n \rhd \mathbf{F}^8$$

We want to design $\mathcal{R}$ in such a way that the process of closing B will correspond to deriving S_B. In particular, closing an initial tableau with a single branch $\mathrm{T}\varphi_1, \ldots \mathrm{T}\varphi_m, \mathrm{F}\psi$ will correspond to deriving the sequent $\varphi_1, \ldots \varphi_m, \neg\psi \rhd \mathbf{F}$. To enable the derivation of $\varphi_1, \ldots, \varphi_m \rhd \psi$, we include in $\mathcal{R}$ a "reductio ad absurdum" rule

[7] In spite of the term "sequent", our proof systems are not sequent calculi in the traditional sense. This remark applies specially to system $\mathcal{R}$.

[8] By considering instead the *symmetric sequent* $\varphi_1, \ldots \varphi_m \rhd \psi_1, \ldots, \psi_n$ we could straightforwardly obtain a sound and complete *symmetric* sequent calculus. For our purposes, this has the disadvantage that *intuitionistic* aspects of reasoning would be not clearly isolated within the proof calculus.

$$(RA)\ \frac{\Gamma, \neg\psi \rhd \mathbf{F}}{\Gamma \rhd \psi}$$

The $\mathcal{R}$-rules for connectives, boolean terms and quantifiers are listed below, together with the tableaux rules which give rise to them by following the $B - S_B$ analogy ($\mathcal{R}$-rules are best read backwards to understand their relation to tableaux rules):

Tableau rule	Corresponding $\mathcal{R}$-rule	Tableau rule	Corresponding $\mathcal{R}$-rule
$\dfrac{T(\varphi \wedge \psi)}{\begin{array}{c}T\varphi\\ T\psi\end{array}}$	$(\wedge \rhd)\ \dfrac{\Gamma, \varphi, \psi \rhd \chi}{\Gamma, \varphi \wedge \psi \rhd \chi}$	$\dfrac{F(\varphi \wedge \psi)}{F\varphi \quad F\psi}$	$(R_1)\ \dfrac{\Gamma, \neg\varphi \rhd \chi \quad \Gamma, \neg\psi \rhd \chi}{\Gamma, \neg(\varphi \wedge \psi) \rhd \chi}$
$\dfrac{T(\varphi \vee \psi)}{T\varphi \quad T\psi}$	$(\vee \rhd)\ \dfrac{\Gamma, \varphi \rhd \chi \quad \Gamma, \psi \rhd \chi}{\Gamma, \varphi \vee \psi \rhd \chi}$	$\dfrac{F(\varphi \vee \psi)}{\begin{array}{c}F\varphi\\ F\psi\end{array}}$	$(R_2)\ \dfrac{\Gamma, \neg\varphi, \neg\psi \rhd \chi}{\Gamma, \neg(\varphi \vee \psi) \rhd \chi}$
$\dfrac{T(\varphi \Rightarrow \psi)}{F\varphi \quad T\psi}$	$(R_3)\ \dfrac{\Gamma, \neg\varphi \rhd \chi \quad \Gamma, \psi \rhd \chi}{\Gamma, \varphi \Rightarrow \psi \rhd \chi}$	$\dfrac{F(\varphi \Rightarrow \psi)}{\begin{array}{c}T\varphi\\ F\psi\end{array}}$	$(R_4)\ \dfrac{\Gamma, \varphi, \neg\psi \rhd \chi}{\Gamma, \neg(\varphi \Rightarrow \psi) \rhd \chi}$
$\dfrac{Tb}{T(\mathbf{T} \sqsubseteq b)}$	$(b \rhd)\ \dfrac{\Gamma, \mathbf{T} \sqsubseteq b \rhd \chi}{\Gamma, b \rhd \chi}$	$\dfrac{Fb}{T(\mathbf{F} \sqsubseteq b)}$	$(R_5)\ \dfrac{\Gamma, \mathbf{F} \sqsubseteq b \rhd \chi}{\Gamma, \neg b \rhd \chi}$
$\dfrac{T\forall x \varphi}{T\varphi[t/x]}$	$(\forall \rhd)\ \dfrac{\Gamma, \forall x\varphi, \varphi[t/x] \rhd \chi}{\Gamma, \forall x\varphi \rhd \chi}$	$\dfrac{F\forall x\varphi}{F\varphi[c/x]}$	$(R_6)\ \dfrac{\Gamma, \neg\varphi[c/x] \rhd \chi}{\Gamma, \neg\forall x\varphi \rhd \chi}$
$\dfrac{T\exists x\varphi}{T\varphi[c/x]}$	$(\exists \rhd)\ \dfrac{\Gamma, \varphi[c/x] \rhd \chi}{\Gamma, \exists x\varphi \rhd \chi}$	$\dfrac{F\exists x\varphi}{F\varphi[t/x]}$	$(R_7)\ \dfrac{\Gamma, \neg\exists x\varphi, \neg\varphi[t/x] \rhd \chi}{\Gamma, \neg\exists x\varphi \rhd \chi}$

where t must be adequate to the sequent in the rules $(\forall \rhd)$ and (R_7), and c must be a new auxiliary constant in (R_6) and $(\exists \rhd)$. In order to facilitate the transition to $\mathcal{D}$, we have allowed for arbitrary consequents instead of $\mathbf{F}$, which would suffice for the $B - S_B$ analogy.

We still have to define the contributions of the tableaux rules **(o)**, **(IN)** and **(LAX)** (Definition 13) and the notion of closed branch (Definition 11) to the calculus $\mathcal{R}$. In the proof of Theorem 22 below, we shall see that there is no need to use **(IN)** and that **(o)** merely contributes to build the antecedent of a sequent to be derived. Consequently, we introduce no $\mathcal{R}$-rules corresponding to them. **(LAX)** gives rise to the $\mathcal{R}$-rule:

$(LAX)\ \dfrac{\Gamma, \theta \rhd \varphi}{\Gamma \rhd \varphi}$ for every $\theta \in \mathbf{LAX}$ (cfr. Lemma 12)

Closed branches correspond to the "obvious contradictions" explained in Definition 11; they give rise to the following $\mathcal{R}$-axioms (rules without premises):

$(CTD)\ \Gamma, \varphi, \neg\varphi \rhd \chi$ (contradiction)
$(BCD)\ \Gamma, \eta \rhd \chi$ for $\eta \in \mathbf{BCD}$ (Boolean contradiction, cfr. Def. 11)

This completes the definition of $\mathcal{R}$. Everything has been prepared to make the proof of the following result go through easily.

Theorem 22 (Adequacy of $\mathcal{R}$) *For every set $\Phi \cup \{\psi\}$ of PHOL-formulas, we have:*

(i) $\Phi \vdash_{\mathcal{R}} \psi \Longrightarrow \Phi \models \psi$ *(soundness).*
(ii) $\Phi \models \psi \Longrightarrow \Phi \vdash_{\mathcal{R}} \psi$ *(completeness).*

Proof. (i) is proved straightforwardly by checking the soundness of the single rules. To prove (ii), let us assume $\Phi \models \psi$. By lemma 9.(i), the set of signed formulas $T\Phi \cup \{F\psi\}$ (where $T\Phi = \{T\varphi / \varphi \in \Phi\}$) is not strongly satisfiable. By completeness of the tableaux method (theorem 19), this set has some closed tableau, whose construction can be assumed to use only finitely many signed formulas $T\varphi_1, .., T\varphi_m$ from $T\Phi$. By soundness of the tableaux method (theorem 15), the finite set $\{T\varphi_1, .., T\varphi_m, F\psi\}$ is not strongly satisfiable. By another application of theorem 19 we can conclude that there is a closed tableau $\mathcal{T}$ for this set whose construction starts with a linear branch B labelled by signed formulas $T\varphi_1, .., T\varphi_m, F\psi$, and makes no use of the tableau rule (IN). Due to the close relation between $\mathcal{R}$-rules and tableaux rules, we can build a $\mathcal{R}$-derivation of a sequent $\varphi_1, \ldots \varphi_m, \neg\psi \rhd \mathbf{F}$ (with $\varphi_i \in \Phi$) by induction on the construction of $\mathcal{T}$. By (RA) we can derive $\varphi_1, \ldots \varphi_m \rhd \psi$, and thus $\Phi \vdash_{\mathcal{R}} \psi$. ■

Several rules of $\mathcal{R}$ (in particular $(\wedge \rhd), (\vee \rhd), (b \rhd), (\forall \rhd), (\exists \rhd)$) are intuitionistically acceptable and well suited to search for proofs backwards, being guided by the form of the current sequent. Now we want to define a calculus $\mathcal{D}$ such that all its rules (except (RA), which is obviously non intuitionistic) will have this character. The rules of $\mathcal{D}$ will arise from a different view of tableau branches. To a given branch B whose unused signed formulas are Tφ_i $(1 \leq i \leq m)$, Fψ_j $(1 \leq j \leq n)$ we associate now any of the n possible sequents

$$\varphi_1, \ldots \varphi_m, \neg\psi_1, \ldots, \neg\psi_{j-1}, \neg\psi_{j+1}, \ldots, \neg\psi_n \rhd \psi_j$$

In particular, the single branch $T\varphi_1, \ldots, T\varphi_m, F\psi$ of an initial tableau for a refutation proof of the thesis ψ with hypothesis $\varphi_1, \ldots, \varphi_m$ will correspond to the sequent $\varphi_1, \ldots \varphi_m \rhd \psi$. According to this new analogy, those tableaux rules which involve only positively signed formulas will give rise to the same rules as before (namely $(\wedge \rhd)$, $(\vee \rhd)$, $(b \rhd)$, $(\forall \rhd)$, $(\exists \rhd)$), but $(R_1) - (R_7)$ and (CTD) will be replaced by new rules as indicated below (the reader is invited to investigate how they are related to the corresponding tableaux rules):

Old $\mathcal{R}$-rule	New $\mathcal{D}$-rule	Old $\mathcal{R}$-rule	New $\mathcal{D}$-rule
(R_1)	$(\rhd\wedge)\ \dfrac{\Gamma \rhd \varphi \quad \Gamma \rhd \psi}{\Gamma \rhd \varphi \wedge \psi}$	(R_2)	$(\rhd\vee)\ \dfrac{\Gamma \rhd \varphi}{\Gamma \rhd \varphi \vee \psi} \quad \dfrac{\Gamma \rhd \psi}{\Gamma \rhd \varphi \vee \psi}$
(R_3)	$(\Rightarrow \rhd)\ \dfrac{\Gamma \rhd \varphi \quad \Gamma, \psi \rhd \chi}{\Gamma, \varphi \Rightarrow \psi \rhd \chi}$	(R_4)	$(\rhd \Rightarrow)\ \dfrac{\Gamma, \varphi \rhd \psi}{\Gamma \rhd \varphi \Rightarrow \psi}$
(R_5)	$(\rhd b)\ \dfrac{\Gamma, \mathbf{F} \sqsubseteq b \rhd \mathbf{F}}{\Gamma \rhd b}$	(R_6)	$(\rhd\forall)\ \dfrac{\Gamma \rhd \varphi[c/x]}{\Gamma \rhd \forall x\varphi}$
(R_7)	$(\rhd\exists)\ \dfrac{\Gamma \rhd \varphi[t/x]}{\Gamma \rhd \exists x\varphi}$	(CTD)	$(HIP)\ \Gamma, \varphi \rhd \varphi$

where c must be new in $(\rhd\forall)$ and t must be adequate in $(\rhd\exists)$. To understand rule $(\rhd b)$, remember that a consequence b behaves as $\neg(\mathbf{F} \sqsubseteq b)$ in our semantics.

In summary, $\mathcal{D}$ consists of two rules $(\rhd\circ)$ and $(\circ \rhd)$ for each propositional connective $\circ$, two rules $(\rhd Q)$ and $(Q \rhd)$ for each quantifier Q, the rules (HIP), (LAX) and (BCD), and the rule (RA). Note that there is no cut rule. The reasons for this were discussed at the end of Section 2.

The analogy between sequents and tableau branches which has given birth to $\mathcal{D}$ is not enough to obtain a direct analogue of Theorem 22. However, we can prove the following:

Lemma 23 *The rules of $\mathcal{R}$ are derivable in $\mathcal{D}$. That is: for each instance of a rule of $\mathcal{R}$, there is a derivation in $\mathcal{D}$ of its conclusion from its premises* [9].

Proof. First we derive the *contraposition rule* (CP) $\dfrac{\Gamma, \neg\varphi \rhd \psi}{\Gamma, \neg\psi \rhd \varphi}$ as follows (we display derivations as trees growing downwards, with the sequent to be derived at the root):

[9] This notion of *derivable rule* conforms to the definition in [15], pg. 70.

$$\dfrac{\dfrac{\Gamma,\neg\psi \rhd \varphi}{\Gamma,\neg\psi,\neg\varphi \rhd \mathbf{F}}(RA)}{\dfrac{\Gamma,\neg\varphi \rhd \psi}{Premise} \qquad \dfrac{\Gamma\neg\varphi,\mathbf{F} \rhd \mathbf{F}}{(HIP)}}(\Rightarrow \rhd)$$

With the help of (CP), (R_1) – (R_7) and (CTD) are easy to derive in $\mathcal{D}$. All other rules of $\mathcal{R}$ do already belong to $\mathcal{D}$. ■

It is straigthforward to check that all the rules in $\mathcal{D}$ are sound for *PHOL*'s semantic consequence. Hence, the following is an immediate corollary of Theorem 22 and Lemma 23.

Theorem 24 (Adequacy of $\mathcal{D}$) *$\mathcal{D}$ is a sound and complete calculus for PHOL.*

5 Conclusions and Future Work

We have designed the three-valued, higher order logic PHOL, which supports formal reasoning involving simply typed λ-terms and an approximation ordering for objects of any type. We have obtained a clean semantics, based on first-order-like structures, inspired by λ-models. Finally, we have defined a tableaux method and a sound and complete proof calculus, which was produced through a methodic analysis of the tableaux rules.

We view PHOL as a very simple and elementary, but also very general framework for higher order reasoning, much in the same sense as classical first order logic, which is regarded as expressive and useful because it has already supported the formalization of many interesting theories. Needless to say, this is not the case for PHOL. Within the intentions and limits of this paper, we could not demonstrate PHOL's expressiveness. We plan, however, to investigate this by applying PHOL to the formalization of reasoning about functional programs and testing its ability to mimic a special purpose logic, which was designed for the functional language Miranda^{TM} [28]. As a different field of application, we think of higher order functional logic programming.

Along a different line, we plan to investigate the intuitionistic fragment of PHOL, focusing on a Kripke semantics and on the relations to more traditional intuitionistic systems.

Acknowledgements.- We thank the anonymous referees for their constructive criticisms, which helped (as we hope) to improve the quality of the presentation.

References

1. A. Arnold, M. Nivat. *The metric space of infinite trees. Algebraic and topological properties.* Ann. Soc. Math. Polon. Ser. IV Fund. Inform. 4, 445-476, 1980.
2. H. Barringer, J. H. Cheng, C. B. Jones. *A logic covering undefinedness in program proofs.* Acta Informatica 22, 251-269, 1984.
3. M. Broy, M. Wirsing. *Partial Abstract Types.* Acta Informatica 18, 47-64, 1982.
4. A. Church. *A formulation of the simple theory of types.* J. Symbolic Logic 5, 56-68, 1940.
5. H.-D- Ebbinghaus. *Über eine Prädikatenlogik mit partiell definierten Prädikaten und Funktionen.* Arch. Math. Logik 12, 39-53, 1969.
6. H.-D. Ebbinghaus, J. Flum, W. Thomas. *Mathematical Logic.* Springer-Verlag, 1984.
7. W. M. Farmer. *A partial function version of Church's simple theory of types.* J. Symbolic Logic 55 (3), 1269-1291, 1990.
8. M. Fitting. *A Kripke-Kleene semantics for logic programs.* J. Logic Programming 2 (4), 295-312, 1985.
9. M. Fitting. *First-Order Logic and Automated Theorem Proving.* Springer-Verlag, 1990.

10. A. Gavilanes-Franco, F. Lucio-Carrasco. *A first order logic for partial functions.* Theoretical Computer Science 74, 37-69, 1990.
11. M. J. Gordon. *HOL: a proof generating system for higher-order logic.* In *VLSI specification, verification, and synthesis.* (G. Birtwistle, P. A. Subrahmanyam, eds.). Kluwer, Dordrecht, 73-128, 1987.
12. L. Henkin. *Completeness in the theory of types.* J. Symbolic Logic 15, 81-91, 1950.
13. M. Holden. *Weak logic theory.* Theoretical Computer Science 79, 295-321, 1991.
14. A. Hoogewijs. *Partial-predicate logic in computer science.* Acta Informatica 24, 281-293, 1987.
15. J. R. Hindley, J. P. Seldin. *Introduction to Combinators and λ-calculus.* Cambridge University Press, 1986.
16. S. C. Kleene. *Introduction to Metamathematics.* North-Holland, 1952.
17. B. Krieg-Brückner et al. *PROSPECTRA Project Summary.* University of Bremen, 1985.
18. K. Kunen. *Negation in logic programming.* J. Logic Programming 4, 289-308, 1987.
19. B. Konikowska, A. Tarlecki and A. Blikle. *A three-valued logic for software specification and validation.* In Proc. *VDM'88, VDM-The Way Ahead.* LNCS 328, 218-242, 1988.
20. J. Loeckx. *Algorithmic specifications: A constructive specification method for abstract data types.* ACM Trans. Prog. Lang. 9 (4), 646-685, 1987.
21. J. McCarthy. *A basis for a mathematical theory of computation.* In Computer Programming and Formal Systems. North-Holland, 33-70, 1963.
22. G. Nadathur, D. Miller. *An overview of λ-prolog.* Procs. *ICLP'88.* The MIT Press, 810-827, 1988.
23. B. Nordström, K. Peterson, J. M. Smith. *Programming in Martin-Löf's Type Theory.* Oxford Science Publications, 1990.
24. O. Owe. *An approach to program reasoning based on a first-order logic for partial functions.* Comp. Sc. Techn. Report No CS-081. Dep. Elect. Engien. and Comp. Sc., Univ. of California, 1984.
25. L. C. Paulson. *Logic and computation. Interactive proof with Cambridge LCF.* Cambridge University Press, 1987.
26. The PRL group (Constable et al.). *Implementing mathematics with the Nuprl proof development system.* Prentice Hall, 1986.
27. R. M. Smullyan. *First-Order Logic.* Springer, Berlin, 1968.
28. S. Thompson. *A Logic for Miranda.* Formal Aspects of Computing 1, 339-365, 1989.

Communicating Evolving Algebras

PAOLA GLAVAN   DEAN ROSENZWEIG

University of Zagreb
FSB, Salajeva 5
41000 Zagreb, Croatia
dean.rosenzweig@uni-zg.ac.mail.yu
dean@cromath.math.hr

Abstract. We develop the first steps of a theory of concurrency within the framework of evolving algebras of Gurevich, with the aim of investigating its suitability for the role of a general framework for modeling concurrent computation. As a basic tool we introduce a 'modal' logic of transition rules and runs, which is, in the context of evolving algebras, just a definitional extension of ordinary first order logic. A notion of independence of rules and runs enables us to introduce a notion of (and notation for) concurrent runs, on which a logical theory of ('true') concurrency may be based. The notion of concurrent run also has (but does not depend on) an interleaving interpretation. Some basic constructs (concurrent composition, addition of guards and updates) and some derived constructs (internal and external choice, sequential composition) on evolving algebras are introduced and investigated. The power of the framework is demonstrated by developing simple and transparent evolving algebra models for the Chemical Abstract Machine of Berry and Boudol and for the π–calculus of Milner. Their respective notions of parallelism map directly and faithfully to native concurrency of evolving algebras.

Introduction

The notion of *evolving algebra* of [Gurevich 88, Gurevich 91] captures two basic ideas which should help logic meet 'the challenge of computer science'.

The *basic objects* (of the system to be modelled) had better be expressed directly, without encoding, as abstract entities, such as they appear in the system. Abstract data types should thus be freely available—abstract domains of objects with operations. The signature of abstract data types should contain no less, but also no more, than what is present or implicit in the system.

The *dynamics*, as reflected by the intuition of 'actions–in–time', need not be reduced to something else, but may well be left 'modelled by itself'. Most (all?) actions, which in computing appear as basic at the given level of abstraction, consist in *local modifications*, with clear local preconditions and effects. It is through interactions in a complex system that complexity appears. The actions could then be represented as local modifications, guarded by simple conditions.

These two ideas are captured by putting together a first–order *signature* and a finite set of *transition rules*, i.e. sets of local modifications guarded by a simple formula (cf. [Gurevich 88, Gurevich 91] and also Section 1). This determines a *transition system*, whose states are first-order structures (of the signature), and the transition relation is given by the rules, understood as *structure transformers*. The metaphor of a 'daemon', attempting to fire a rule selected nondeterministically from those allowed by their guards, may be useful in forming an intuitive picture.

This simple framework has shown remarkable success in modeling complex computing situations, such as Modula 2 [GurMor 88], full Prolog [BörRos 92b], the Warren Abstract Machine [BörRos 92a], C [GurHug 93] ..., avoiding the combinatorial explosion of formalism and/or mathematics, too characteristical of attempts to give a formal semantics of real systems. In [GurMos 90] the framework has for the first time accomodated the notion of concurrency, in special form of Occam, to be continued with studies of Parlog and Concurrent Prolog [BörRic 93, BörRic 92].

Generalizing this work, we undertake to develop a *theory* of concurrency within the framework of evolving algebras, with the aim of investigating its suitability for the role of a general framework for modeling concurrent computation.

To pursue the daemon metaphor, we want to see what happens if we let several independent daemons loose in an evolving algebra. Some of them may be fast, some may be very slow—we do not rely on any notion of 'global time', explicit or implied. We view the daemons as being truly independent, i.e. as having no way of communicating whatsoever, except for feeling the effect of others on the underlying structure, as reflected in their own ability to fire rules.

We investigate the notion of *independence* of rules and runs. What is independent, may happen concurrently. Since firing of a rule is, in this framework, an indivisible atomic action, we may view a deamon as *locking* the part of the world his action depends on, for the duration of that action, preventing anyone else from modifying it during that time. For instance, while updates $a := 0$, $a := 1$ are *inconsistent* and cannot conceivably be executed simultaneously, the updates $a := b$, $b := a$ are consistent but *not independent*. They can be executed simultaneously by one daemon, as actions within the same rule, but cannot be conceived as executed concurrently by independent daemons without *some* sequencing in time—the sequencing can be *derived* from their joint effect. 'Locking daemons' are thus prevented from executed them 'simultaneously'. This kind of enforced sequentialization is our synchronization mechanism.

Given the notion of independence, we can describe possible *concurrent runs* of an evolving algebra. Any notion of 'time', i.e. of 'sequencing of events', derivable from a concurrent run, is only a partial order. In order to say all this precisely, without relying on daemon, locking, or any other metaphor, we have to develop a *logic* of evolving algebras, up to a point, not unlike the modal logics of processes developed in the context of CCS and π-calculus of Milner [Milner 89, MiPaWa 91, Milner 92]. Happily, the needed fragment of the logic of evolving algebras turns out to be (a definitional extension of) ordinary first order logic.

Having fixed our model of concurrency, we define some basic *operations* on evolving algebras, starting from (concurrent) *composition*. Their properties are nicely expressed by *inference rules* for the underlying transition system, not unlike the SOS rules through which algebraic calculi of processes are usually formulated. It might be emphasized that we *do not*, at this point, organize these rules into a calculus, nor do we introduce any relation of 'operational ordering' or 'equivalence' on evolving algebras. Our attitude here is rather *prealgebraic*—study of such relations is bound to be very fruitful, in particular when they relate algebras constructed at different levels of abstraction. We are however not compelled to introduce 'semantic' equivalence relations early, in order to *create* our objects of study by passing to quotient: evolving algebras are not just a syntax. As 'very abstract machines', they are interesting objects in their own right.

The power of the framework is demonstrated through two applications, providing simple and transparent models of the the Chemical Abstract Machine of [BerBou 92] and the π–calculus of [Milner 92]. The Chemical Abstract Machine simply *is* a special evolving algebra, and our model of the π calculus is built very much in the 'chemical' spirit. Additional flexibility of evolving algebras allows us to give 'heating' and 'cooling' some direction, however. The notions of parallelism, expressed by the cham and the π–calculus respectively, correspond precisely, in the models, to concurrency of evolving algebra runs.

Our study is confined to 'local structure' of the transition systems given by evolving algebra descriptions. This structure seems to be much more complex than the local structure of graphs given by SOS rules —it is, in general, intended (and has succeeded to a degree, as witnessed by work done so far) to model, directly and faithfully, all kinds of computational phenomena, and not one or another carefully isolated aspect. As a consequence, however, the passage to the study of global structure—homomorphisms, invariants, quotients and their algebras—cannot be swift, and requires a lot of insight into the local structure. Our aim here is to provide some additional insight, in rather general terms. For instance, the (local) notion of independence, introduced below, seems to be an important invariant, and some class of independence preserving homomorphisms is likely to play a role. Since the local structure is determined by elementary (meta)theory of first–order logic, we shall investigate it in such terms.

The paper is organized as follows.

In Section 1 we say rather pedantically what we mean by evolving algebras and how exactly they evolve[1]. Here we develop the modal logic of simple rules and sequential runs.

In Section 2 we develop the notions of independence of rules and sequential runs, and introduce concurrent runs, laying down the basics of a *logical theory of 'true' concurrency*. We show that every concurrent run can be collapsed to an equivalent sequential one, which provides their interleaving semantics as well.

In Section 3 we display the basic and derived constructs on evolving alge-

[1] Since the notion of evolving algebra is understood by its creator as being 'open ended' [Gurevich 91], as a methodology rather than formalism, these definitions may be taken with a grain of salt—yet we have to start somewhere.

bras and investigate their properties. Some useful extensions of the language are discussed here, in order to show their conservativity wrt to the above results.

Section 4 contains the applications. The model of the Chemical Abstract Machine is little more than a rewriting of the *reaction law* of [BerBou 92]—other laws are valid automatically. The model of the π–calculus is also very simple and natural, but in order to *prove* that it captures the computational content of the calculus, some 'chemical' analysis (of that content) is required.

A Notation and Prerequisites section makes some underlying technical assumptions explicit. On a first reading it might as well be skimmed with a quick glance, and returned to as needed.

Main results of sections 1,2,3 are part of the first author's Master Thesis at the University of Zagreb. We are indebted to Egon Börger and Simone Martini, who have been helpful in many ways in, for us, a difficult time, to Yuri Gurevich for inspiring conversations and very useful comments on an early version, and to Janos Makowsky for a good hint. The comments of two anonymous referees have helped us to improve the exposition of several crucial points. Both authors have been supported by the Ministry of Science of the Republic of Croatia.

Notation and prerequisites

Our framework is first–order logic of equality and partial functions. The language is the usual one, built from variables, function symbols with fixed arities, true, false, $=, \neg, \wedge, \vee, \Rightarrow, \forall, \exists$.

There are several ways to formalize first–order reasoning with function symbols understood as representing *partial* functions.

One is adopted by [Gurevich 91]: assume an 'undefined' element of the model, and a corresponding constant *undef* in the signature. An interpretation I in a universe U then determines, for any n–ary function symbol f, a *total* function $I(f) : (U \setminus \{I(undef)\})^n \rightarrow U$, from which a *partial* function $I'(f)$ of the same type can be extracted, so that

$$I'(f)(\mathbf{y}) = \begin{cases} I(f)(\mathbf{y}) & \text{if } I(f)(\mathbf{y}) \neq I(undef) \\ \text{undefined} & \text{otherwise,} \end{cases}$$

where $\mathbf{y}$ is any n–tuple of values in $U \setminus \{I(undef)\}$.

Since we are going to talk explicitly about models and interpretations, we prefer to have our partial functions directly at hand, and to minimize the assumptions to be made about (the universe of) the model. Thus we adopt an alternative formulation, the *logic of partial terms* of [Beeson 85] (cf. also 'E^+–logic' in [TroDal 88] and 'partial algebras' in [Wirsing 90]). There are no special 'undefined' objects in the model, but not all terms need denote something there. Those which do are distinguished by an additional unary predicate $\downarrow$, such that $\downarrow t$ intuitively means 't is defined', or 't denotes'. Then only defined terms may instantiate a universal quantification (get substituted for a variable) or witness an existential one. Equality implies that both sides are defined, and $\downarrow t \Leftrightarrow t = t$. *Kleene equality* is a defined notion:

$$s \simeq t \stackrel{\mathrm{def}}{=} (\downarrow s \vee \downarrow t) \Rightarrow t = s.$$

A *partial interpretation* in a universe U determines a *partial function*

$$I(f) : U^n \rightharpoonup U$$

for any n–ary function symbol f, and the interpretation of terms as elements of U is partial indeed—it is defined only on terms distinguished by $\downarrow$. If we abuse slightly the notation, and apply $\simeq, \downarrow$ also to 'pseudoterms' built from partial functions over U, we can say that $I(f(t_1, \ldots, t_n)) \simeq I(f)(I(t_1), \ldots, I(t_n))$, and $I \models \downarrow s \stackrel{\mathrm{def}}{=} \downarrow I(s)$. This notion of interpretation makes the *strictness* condition $\downarrow f(t_1, \ldots, t_n) \Rightarrow \downarrow t_1 \wedge \ldots \wedge \downarrow t_n$ valid. In particular, for a 0–ary symbol ('distinguished element') c we need to have an $I(c) \in U$ only when we want $I \models \downarrow c$. The usual 'substitutivity of equals' schema may be strengthened to $a \simeq b \wedge \phi(a) \Rightarrow \phi(b)$, as can be easily verified by induction over ϕ.

Syntactical identity of function symbols and terms will be denoted by $\doteq$.

For the purpose of following the present paper, the above remarks on the logic of partial terms should suffice. A full account of the logic (and its reducibility to ordinary first–order logic) may however be found in [Beeson 85, TroDal 88]. The reader familiar with the *undef* approach of [Gurevich 91] may easily translate our notation to that framework according to the following table

$$\begin{aligned}
\downarrow t &\longrightarrow t \neq \mathit{undef} \\
s \simeq t &\longrightarrow s = t \\
s = t &\longrightarrow s \neq \mathit{undef} \ \wedge\ s = t \\
I(f)\ :\ U^n \rightharpoonup U &\longrightarrow I(f)\ :\ (U \setminus \{I(\mathit{undef})\})^n \rightharpoonup U \\
\exists x \phi &\longrightarrow \exists x (x \neq \mathit{undef} \ \wedge\ \phi) \\
\forall x \phi &\longrightarrow \forall x (x \neq \mathit{undef} \ \Rightarrow\ \phi),
\end{aligned}$$

assuming, further, $x \neq \mathit{undef}$ (on the right) for any free variable x used (on the left).

We shall also freely use 'terms' of form $\mathsf{if}\ \phi\ \mathsf{then}\ a\ \mathsf{else}\ b$ for any quantifier–free formula ϕ and terms a, b, understanding that

$$I(\mathsf{if}\ \phi\ \mathsf{then}\ a\ \mathsf{else}\ b) \simeq \begin{cases} I(a) & \text{if } I \models \phi \\ I(b) & \text{otherwise.} \end{cases}$$

This usage may be formalized either by adopting, in the syntax, a simultaneous inductive definition of both terms and formulae, or by assuming to have *booleans*, in every signature and in every model, together with function symbols which suffice to *encode* quantifier–free formulae by boolean terms[2]. We shall suppress such encoding.

[2] In that case, however, if_then_else should not be understood as a function symbol, but as a 'term constructor'—it is namely not strict in all arguments.

We shall define several syntactical operators on terms (*term transformers*), and *extend* them to operators on formulae 'by homomorphism wrt formula constructors'. If Δ is a term transformer, its homomorphic extension will be Δ', as given by clauses

$$\Delta'(s = t) \stackrel{\text{def}}{=} \Delta s = \Delta t \qquad \Delta' \downarrow s \stackrel{\text{def}}{=} \downarrow \Delta s$$
$$\Delta' f(\ldots\phi\ldots) \stackrel{\text{def}}{=} f(\ldots\Delta'\phi\ldots) \qquad \Delta' Qx\phi \stackrel{\text{def}}{=} Qx\Delta'\phi$$

where f is any 0–ary, unary or binary logical connective, Q a quantifier. We understand that Δ (if ϕ then a else b) $\simeq$ if $\Delta'\phi$ then Δa else Δb.

We shall also use *partial* sentential operators, well defined only under some conditions. If $\nabla\phi$ is well defined when ψ is true, we shall use 'formulae' of form $\psi \Rightarrow \nabla\phi$ and $\psi \wedge\!\Rightarrow \nabla\phi$ in the following sense:

$$\psi \Rightarrow \nabla\phi \stackrel{\text{def}}{=} \begin{cases} \text{true} & \text{if } \neg\psi \\ \nabla\phi & \text{otherwise} \end{cases} \qquad \psi \wedge\!\Rightarrow \nabla\phi \stackrel{\text{def}}{=} \begin{cases} \text{false} & \text{if } \neg\psi \\ \nabla\phi & \text{otherwise} \end{cases}$$

(see also [Aczel 80]).

1 (Sequential) evolving algebras

1.1 Basic definitions—the logic of simple rules

Definition 1.1.1. If Σ is a first order signature,

– a *function update* over Σ is an expression of the form $f(t_1, \ldots, t_n) := t$, where $f(t_1, \ldots, t_n), t$ are terms of S;

– a *transition rule* over Σ is an expression R of the form **if** $R?$ **then** $R!$, where $R?$ is a quantifier–free formula (*guard* of R), and $R!$ is a finite set of its function *updates*.

Definition 1.1.2. An *evolving algebra* is given by a finite set of transition rules.

The signature of a rule, or that of an evolving algebra, can always be reconstructed, as the set of function symbols occurring there. In applications of evolving algebras one usually encounters a *heterogenous* signature with several universes, which may in general grow and shrink in time—update forms are provided to extend or restrict a universe. [Gurevich 91] has however shown how to reduce such setups to the above basic model of a homogenous signature (with one universe) and function updates only (see also Section 3).

Assumption. For now (until the end of Section 3) we additionally assume that all terms occurring in transition rules are *closed*, i.e. contain no variables.

Definition 1.1.3. Let A be an evolving algebra. A *static algebra* of A is any algebra of $\Sigma(A)$, i.e. a pair (U, I) where U is a set and I is a partial interpretation of $\Sigma(A)$ in U.

In applications an evolving algebra usually comes together with a set of *integrity constraints*, i.e. extralogical axioms and/or rules of inference, specifying the intended domains. We tacitly understand the notion of interpretation as validating any integrity constraints imposed. Although in applications only *some* static algebras will be of interest, such as those reachable from designated *initial states*, in first 2 sections we shall consider arbitrary static algebras.

In the rest of this section we shall denote by R a transition rule of form **if** $R?$ **then** $R!$, where $R!$ is the set of updates $f_i(\mathbf{s}_i) := t_i$, and $\mathbf{s}_i$ a tuple of terms of f_i's arity, $i = 1, \ldots, n$.

Since a set of updates is viewed as being executed simultaneously, it has to be *consistent* in order to be executable at all, i.e. it may not simultaneously update the same function at equal arguments to different values[3]. It also has to be able to evaluate everything that needs to be evaluated, i.e. we shall regard updating a function at 'undefined' arguments as inconsistent. We allow however updating to an 'undefined value'—should the term in rhs be such.

Definition 1.1.4. (*Consistency* of updates)

$$\text{cons}(R!) \stackrel{\text{def}}{=} (\bigwedge_i \downarrow \mathbf{s}_i) \wedge (\bigwedge_{\substack{i \neq j \\ f_i \doteq f_j}} (\mathbf{s}_i = \mathbf{s}_j \Rightarrow t_i = t_j))$$

We shall always assume all transition rules to be *consistent*, i.e. the formula $R? \Rightarrow \text{cons}(R!)$ to be valid for every rule R. In case of doubt, this assumption can be syntactically enforced by appending $\text{cons}(R!)$ to the guards.

Under the assumption of consistency, we can now say in model theoretical terms what is meant by 'simultaneous execution' of a set of updates.

Definition 1.1.5. The *effect* of updates $R! = \{f_i(\mathbf{s}_i) := t_i\}$, consistent in an algebra (U, I), is to transform it to $(U, I_{R!})$, where

$$I_{R!}(f)(\mathbf{y}) \stackrel{\text{def}}{=} \begin{cases} I(t_i) & \text{if } f \doteq f_i,\ \mathbf{y} = I(\mathbf{s}_i),\ i = 1, \ldots, n \\ I(f)(\mathbf{y}) & \text{otherwise} \end{cases}$$

where $\mathbf{y}$ is any tuple of values in U of f's arity.

The assumption of consistency ensures that $I_{R!}$ is well defined. In view of how interpretation of compound terms is built, we have

Proposition 1.1.1.

$$I_{R!}(f(\mathbf{u})) \simeq \begin{cases} I(t_i) & \text{if } f \doteq f_i,\ I_{R!}(\mathbf{u}) = I(\mathbf{s}_i),\ i = 1, \ldots, n \\ I(f)(I_{R!}(\mathbf{u})) & \text{otherwise} \end{cases}$$

[3] [Gurevich 91] allows such contradictory updates, imagining the daemon as selecting nondeterministically what to execute. We confine nondeterminism to selection of a *rule*, all of whose updates must be executed simultaneously once the guard has been checked, going back to the concept of [GurMos 90].

The proposition motivates the following *term transform* $[R!]$, associated to a set of updates $R! = \{f_i(\mathbf{s}_i) := t_i\}$, simulating its effect by syntactical means.

$$\begin{aligned}[R!]f(\mathbf{u}) \stackrel{\text{def}}{=} & \ \mathsf{if}\ [R!]\mathbf{u} = \mathbf{s}_{i_1}\ \mathsf{then}\ t_{i_1} \\ & \ \ldots \\ & \ \mathsf{elsif}\ [R!]\mathbf{u} = \mathbf{s}_{i_m}\ \mathsf{then}\ t_{i_m} \\ & \ \mathsf{else}\ f([R!]\mathbf{u})\end{aligned}$$

where $i_1, \ldots, i_m$ are all i's such that $f_i \doteq f$. It might be useful for the reader to spell out the case of 0–ary f. For instance, given $\downarrow a$, a simple calculation shows that

$$\begin{bmatrix} a := 1 \\ f(a) := 2 \end{bmatrix} f(f(a)) \simeq \begin{cases} \mathsf{if}\ a = 1\ \mathsf{then}\ f(2) \\ \mathsf{elsif}\ a = f(1)\ \mathsf{then}\ 2 \\ \mathsf{else}\ f(f(1)). \end{cases}$$

By construction, we have a generalization of *substitution lemma* of first–order logic:

Proposition 1.1.2. (Update Lemma) If cons$(R!)$ holds under I, then, for any term t, $I_{R!}(t) \simeq I([R!]t)$.

Since the definition of $[R!]$ assumes consistency, (only) *possible* rules may be seen as term transformers.

Definition 1.1.6. (*Possibility* of rules and the associated *term transform*)

$$\text{pos}(R) \stackrel{\text{def}}{=} R? \qquad [R]t \stackrel{\text{def}}{=} [R!]t$$

where $[R]$ is well defined as soon as pos(R) holds.

By homomorphic extension, $[R]'$ is a well defined sentential operator as soon as R is possible. We want $[R]\phi$ to be a formula for any first order formula ϕ, with the meaning 'if R can be executed, then after execution ϕ will hold', and dually, a formula $\langle R\rangle\phi$, meaning 'it is possible to execute R and have ϕ hold after execution'. They can also be understood as 'weakest precondition operators'.

Definition 1.1.7. (*Modal operators* associated to rules)

$$[R]\phi \stackrel{\text{def}}{=} \text{pos}(R) \Rightarrow [R]'\phi \qquad \langle R\rangle\phi \stackrel{\text{def}}{=} \text{pos}(R) \wedge\!\Rightarrow [R]'\phi \ .$$

Unsurprisingly, it is easy to see that $\langle R\rangle\phi \Leftrightarrow \neg[R]\neg\phi$, and that, given pos$(R)$, both modal operators are homomorphisms wrt to logical connectives. Simple special cases are the well known *Hoare's rule*, $[a := b]\,\phi \Leftrightarrow \phi(b/a)$, and the *array assignment rule* of dynamic logic [Harel 84], whose simplest basis–of–induction case takes the form

$$[f(a) := b]\,(f(x) = y) \Leftrightarrow (x = a \Rightarrow y = b) \wedge (x \neq a \Rightarrow y = f(x))$$

(if $\downarrow a$, f doesn't occur in x, y). In fact, it is a remark of Janos Makowsky, that the array assignment rule might be relevant for evolving algebras, which prompted us to look into $[R]$ in the first place.

Since pos$(R) \Leftrightarrow \langle R\rangle$true, we shall, abusing slightly the notation, abbreviate both formulae by $\langle R\rangle$.

To a rule R possible under I we can now associate the interpretation $I_R \stackrel{\text{def}}{=} I_{R^!}$. If $\mathcal{A} = (U, I)$, we shall denote (U, I_R) by $\mathcal{A}_R$. The necessity operator enables us to simulate the logic of $\mathcal{A}_R$ within $\mathcal{A}$. Note that the modal operators do not introduce any quantifiers, i.e. they preserve the class of formulae which may be used in conditional terms and in guards of rules.

Theorem 1.1.3. If $\mathcal{A} \models \langle R\rangle$, then $\mathcal{A}_R \models \phi$ iff $\mathcal{A} \models [R]\phi$.

We have now laid down precisely the way in which transition rules transform first order structures. Evolving algebras can then be understood as *transition systems*[4], whose states are first order structures, and the transition relation is given by possible rules.

Definition 1.1.8. $\mathcal{A} \xrightarrow{R} \mathcal{A}_R$ whenever $\mathcal{A} \models \langle R\rangle$.

When the form of $\mathcal{A}_R$ doesn't tell us much, we shall drop it (since it is determined by $\mathcal{A}$ and R anyway), writing just $\mathcal{A} \xrightarrow{R}$.

This is then 'what the daemon does'—it selects one of maybe several possible rules, jumping thereby to one of maybe several successor states (static algebras). In applications, we shall be interested in *connected components* of this huge graph, determined by an evolving algebra specification, or in the subsystems of states *reachable* from given *initial states*. Note that the universe will be preserved in the interesting cases—it is the interpretation which evolves. It may be also noted that the interpretations, in any two connected states, will differ just at finitely many points. Hence, if the initial states are chosen so that interpretation is, say, (feasibly) computable, it will remain such ever after—regardless of strange mathematical creatures residing in other unconnected nodes of the graph. This explains why computability of interpretations, or cardinality of universes, is not an issue for evolving algebras.

1.2 Sequential runs

Here we extend the notion of possibility, as well as the association of term transformers and modal operators, from simple rules to *sequential runs*, i.e. sequences of rules to be consecutively executed.

Definition 1.2.1. Every rule (name) is a sequential *run expression*. If ρ, σ are sequential run expressions, so is $\rho\sigma$.

[4] By transition system we mean here a directed graph, whose nodes are understood as *states* and (labelled) arrows as *transitions*—the labels are usually viewed as representing basic actions.

Possibility of sequential run expressions is defined inductively: $\rho\sigma$ is possible if ρ is, and after it σ is possible—since the effect is defined simultaneously, by its homomorphic extension to formulae and inductive assumption we already know what the last formula means.

Definition 1.2.2. (*Possibility* and *effect*)

$$\mathrm{pos}(\rho\sigma) \stackrel{\mathrm{def}}{=} \mathrm{pos}(\rho) \wedge \Rightarrow [\rho]'\mathrm{pos}(\sigma)$$
$$[\rho\sigma]s \stackrel{\mathrm{def}}{=} [\rho][\sigma]s \qquad \text{given } \mathrm{pos}(\rho\sigma)$$

where $[\rho]'$ is homomorphic extension of $[\rho]$. It is easy to see that concatenation is associative (as far as possibility and effect are concerned).

Definition 1.2.3. (Modal operators associated to run expressions)

$$[\rho]\phi \stackrel{\mathrm{def}}{=} \mathrm{pos}(\rho) \Rightarrow [\rho]'\phi \qquad \langle\rho\rangle\phi \stackrel{\mathrm{def}}{=} \mathrm{pos}(\rho) \wedge \Rightarrow [\rho]'\phi$$

We extend our notational abuse of 1.1 to abbreviating $\mathrm{pos}(\rho)$ and $\langle\rho\rangle\mathrm{true}$ by $\langle\rho\rangle$. The modal operators could as well have been defined by induction over run expressions.

Proposition 1.2.1. $[\rho\sigma]\phi \Leftrightarrow [\rho][\sigma]\phi$
$\langle\rho\sigma\rangle\phi \Leftrightarrow \langle\rho\rangle\langle\sigma\rangle\phi$.

Definition 1.2.4. A (sequential) *run* of a static algebra $\mathcal{A}$ is given by a run expression ρ such that $\mathcal{A} \models \langle\rho\rangle$.

As with a simple rule, to any run ρ of $\mathcal{A} = (U, I)$ we can associate $\mathcal{A}_\rho = (U, I_\rho)$ with $I_\rho(s) \simeq I([\rho]s)$. The integrity is assured by

Theorem 1.2.2. If ρ is a run of $\mathcal{A}$, then $\mathcal{A}_\rho \models \phi$ iff $\mathcal{A} \models [\rho]\phi$.

Proof by induction over ρ, simultaneously for all static algebras (i.e. passing to $\mathcal{A}_{\rho\sigma}$, we apply the inductive hypothesis both to $\mathcal{A}$ and to $\mathcal{A}_\rho$).

Definition 1.2.5. $\mathcal{A} \xrightarrow{\rho} \mathcal{A}_\rho$ whenever $\mathcal{A} \models \langle\rho\rangle$.

Like before, we shall skip $\mathcal{A}_\rho$ when its form is not particularly interesting, writing just $\mathcal{A} \xrightarrow{\rho}$. Adding an empty run, we obtain the *free category* generated by the graph (transition system) of simple rules. This is then what the daemon does in the long run—it executes long runs.

2 Independence and concurrency

In this section we introduce the basic building blocks of a *logical theory of concurrency*—of what can be conceived as happening concurrently, without any notion implying a concept of a 'global time', such as 'simultaneuity' or 'interleaving'. We investigate the notion of *independence* of simple rules and sequential runs. Intuitively, two runs are independent if they could be thought of as being executed by two independent daemons without any direct communication or synchronization whatsoever, each working at its own pace, affecting each other only by affecting the underlying structure. Since execution of a rule in an evolving algebra is considered to be a primitive, atomic 'timeless' action, this is conceivable only if they do not interfere—if one doesn't modify those elements of the structure the other needs to see. We don't have to prohibit them from modifying the same points, as long as they do that consistently. A useful intuitive criterion is the requirement that no conclusion related to sequencing ('global time') might be drawn from the joint effect of their action. A consequence is that all *interleavings* of independent runs should have equivalent effect. In fact, two daemons may be viewed as being synchronized by *not* being able to do anything independently, i.e. by one having to wait for an action of the other to be completed.

Our notion of *extreme* ('true') *concurrency* then allows concurrent execution of any set of pairwise independent runs[5]. We develop the notion of possibility and the term transformers and modal operators associated to *concurrent runs*, and show that they can be reduced (in sense of effect, but not, say, complexity) to equivalent sequential runs, i.e. we provide an interleaving semantics as well. The notion of concurrency remains however independent of it.

2.1 Independence of rules

Two simple rules are independent if they could be consistently executed together, so that one doesn't touch what the other has to use. We might think of two independent daemons, each *locking* (for *write*) all the points he needs to see before executing a rule.

The points 'a daemon needs to see' are given, syntactically, by

Definition 2.1.1. (*Syntactic domains* of terms, formulae, updates, rules)

$$Df(t_1,\ldots,t_n) \stackrel{\text{def}}{=} (\bigcup_{i=1}^{n} Dt_i) \cup \{f(t_1,\ldots,t_n)\} \quad (n \geq 0)$$

$$D\phi \stackrel{\text{def}}{=} \bigcup\{Dt \mid t \text{ occurs in } \phi\}$$

$$D(f(t_1,\ldots,t_n) := t) \stackrel{\text{def}}{=} (\bigcup_{i=1}^{n} Dt_i) \cup Dt$$

[5] We are more stringent here than [Gurevich 91], who allows simultaneous execution of any set of rules allowed by their guards, 'leaving it to the daemon' to select between eventual contradictory updates.

$$DR! \stackrel{\text{def}}{=} \bigcup_{u \in R!} Du$$

$$DR \stackrel{\text{def}}{=} DR? \cup DR!$$

Then the following condition will allow two rules to act concurrently without conflict.

Definition 2.1.2. (*Independence* of rules)

$$\begin{aligned} I(R,S) \stackrel{\text{def}}{=} & \langle R\rangle \wedge \langle S\rangle \wedge \text{cons}(R! \cup S!) \\ & \wedge \bigwedge_{r \in DR} ([S]r \simeq r) \\ & \wedge \bigwedge_{s \in DS} ([R]s \simeq s) \end{aligned}$$

The first clause ensures that R and S can be executed simultaneously, i.e. that a rule of form **if** $R? \wedge S?$ **then** $R! \cup S!$ would be possible. The other clauses ensure that R and S could be, with the same effect, executed in any ordering.

Note that $I(R,S)$ is always a quantifier–free formula. Of course, rules independent in one static algebra may be dependent in another—it is a 'run–time property'. Independence entails that both are possible, and remain so after execution of the other rule; also that their effect is commutative (as far as it can be detected by a formula).

Proposition 2.1.1. $I(R,S)$ entails, for any formula ϕ,

$$\begin{array}{ll} (i) & \langle R\rangle \wedge \langle S\rangle \\ (ii) & [R]\langle S\rangle \wedge [S]\langle R\rangle \\ (iii) & [RS]\phi \Leftrightarrow [SR]\phi \\ (iv) & \langle RS\rangle\phi \Leftrightarrow \langle SR\rangle\phi \end{array}$$

Proof. (ii) follows from $D\langle S\rangle = DS? \subset DS$, and likewise for R. Let the function symbol f be updated in R by updates $f(\mathbf{u}_i) := r_i,\ i = 1,\ldots,n$, and in S by updates $f(\mathbf{v}_j) := s_j,\ j = 1,\ldots,m$ $(n, m \geq 0)$.

By independence, $[S!]\mathbf{u}_i \simeq \mathbf{u}_i$, $[S!]r_i \simeq r_i$, $[R!]\mathbf{v}_j \simeq \mathbf{v}_j$, $[R!]s_j \simeq s_j$. Then

$$\begin{aligned} [R!][S!]f(\mathbf{t}) \simeq [R!](\ & \mathsf{if}\ [S!]\mathbf{t} = \mathbf{v}_1\ \mathsf{then}\ s_1 \\ & \ldots \\ & \mathsf{elsif}\ [S!]\mathbf{t} = \mathbf{v}_m\ \mathsf{then}\ s_m \\ & \mathsf{else}\ f([S!]\mathbf{t})\ ) \end{aligned}$$

$$
\begin{aligned}
\simeq\ & \mathsf{if}\ [R!][S!]\mathsf{t} = \mathsf{v}_1\ \mathsf{then}\ s_1 \\
& \dots \\
& \mathsf{elsif}\ [R!][S!]\mathsf{t} = \mathsf{v}_m\ \mathsf{then}\ s_m \\
& \mathsf{elsif}\ [R!][S!]\mathsf{t} = \mathsf{u}_1\ \mathsf{then}\ r_1 \\
& \dots \\
& \mathsf{elsif}\ [R!][S!]\mathsf{t} = \mathsf{u}_n\ \mathsf{then}\ r_n \\
& \mathsf{else}\ f([R!][S!]\mathsf{t})
\end{aligned}
$$

By inductive hypothesis, $[R!][S!]\mathsf{t} \simeq [S!][R!]\mathsf{t}$, and by symmetry$[R!][S!]f(\mathsf{t}) \simeq [S!][R!]f(\mathsf{t})$. Statements (*iii*) and (*iv*) now follow immediately[6].

2.2 Independence of sequential runs

It is rather awkward to extend the notions of 'points seen' and 'points touched' to sequential runs. Luckily, an inductive definition turns out to work. The idea is that two runs are independent if they are both possible, their arbitrary initial segments are independent, and both properties are preserved by execution of any initial segment of either run. It is concisely expressed by the following inductive definition, based on Def. 2.1.2, of independence of simple rules.

Definition 2.2.1. (*Independence* of runs)

$$
\begin{aligned}
I(R,\tau) &\stackrel{\text{def}}{=} I(\tau, R) \qquad \text{for composite } \tau \\
I(\rho\sigma,\tau) &\stackrel{\text{def}}{=} I(\rho,\tau) \wedge [\rho] I(\sigma,\tau)
\end{aligned}
$$

Note how, when the second clause exhausts the first argument, first clause reapplies it to the second argument. This definition suffices to derive the properties expected of independence.

Proposition 2.2.1. $I(\rho,\sigma)$ entails, for any formula ϕ,

$$
\begin{array}{ll}
(i) & \langle\rho\rangle \wedge \langle\sigma\rangle \\
(ii) & [\rho]\langle\sigma\rangle \wedge [\sigma]\langle\rho\rangle \\
(iii) & [\rho\sigma]\phi \Leftrightarrow [\sigma\rho]\phi \\
(iv) & \langle\rho\sigma\rangle\phi \Leftrightarrow \langle\sigma\rho\rangle\phi .
\end{array}
$$

Proof by induction over $I(\rho,\sigma)$, simultaneously for for all five statements and all static algebras[7].

---

[6] We have in fact proved a stronger statement than (*iii*), i.e. effect commutes literally: $[R][S]t \simeq [S][R]t$ for any term t. However, (*iii*) is all we need, and it will carry over smoothly to extensions of the language in 3.3, whereas the stronger statement would not.

[7] The previous footnote applies here as well—we could have a corresponding stronger statement than (*iii*).

By routine induction over $I(\rho,\sigma)$ we obtain its symmetry, lacking so conspiciously in the definition.

Corollary 2.2.2. $I(\rho,\sigma) \Leftrightarrow I(\sigma,\rho)$.

Independence of two runs does not depend on execution of a third one, independent from both:

Proposition 2.2.3. $I(\rho,\tau) \wedge I(\sigma,\tau)$ entails

$$I(\rho,\sigma) \Leftrightarrow [\tau]I(\rho,\sigma).$$

Proof is by induction over the sum of lengths of ρ,σ,τ, simultaneously for all static algebras. The basic case, of three simple rules, follows immediately noting that the formula $I(R,S)$ mentions only terms in domains of R,S. Induction step over τ is straightforward. Induction step over ρ has a somewhat weird propositional structure. Note that, in order to prove $A \wedge B \wedge C \Rightarrow (D \wedge E \Leftrightarrow F \wedge G)$, it suffices to prove the following three statements

$$A \wedge C \Rightarrow (D \Leftrightarrow F) \qquad A \wedge B \wedge H \Rightarrow (E \Leftrightarrow G) \qquad A \wedge D \Rightarrow (C \Leftrightarrow H)\ .$$

What we need to prove is that, in any algebra $\mathcal{A}$,

$$I(\mu\nu,\tau) \wedge I(\sigma,\tau) \Rightarrow (I(\mu\nu,\sigma) \Leftrightarrow [\tau]I(\mu\nu,\sigma)),$$

$$\text{i.e.}\quad \begin{array}{lll} A \stackrel{\text{def}}{=} I(\mu,\tau) & B \stackrel{\text{def}}{=} [\mu]I(\nu,\tau) & C \stackrel{\text{def}}{=} I(\sigma,\tau) \\ D \stackrel{\text{def}}{=} I(\mu,\sigma) & E \stackrel{\text{def}}{=} [\mu]I(\nu,\sigma) & F \stackrel{\text{def}}{=} [\tau]I(\mu,\sigma) \\ G \stackrel{\text{def}}{=} [\tau][\mu]I(\nu,\sigma) & & H \stackrel{\text{def}}{=} [\mu]I(\sigma,\tau) \end{array}$$

Under these definitions, all three statements follow by induction hypothesis applied to $\mathcal{A},\mathcal{A}_\mu,\mathcal{A}$ respectively.

2.3 Concurrency

Independent runs can be executed concurrently. We introduce then run expression which explicitly indicate concurrency— $\rho(\sigma|\tau)(\mu|\nu)$ means executing ρ first, then concurrently σ and τ, and, when that is completed, concurrently μ and ν. Such concurrent runs provide a proper generalization of the notion of run used in [GurMos 90], by providing a partial ordering of states involved.

We say when is a concurrent run possible, and characterize its effect by term transformer and modal operators. We show how to collapse a concurrent run to an equivalent sequential one, providing an interleaving semantics as well. All results proved for sequential runs carry thus over to concurrent runs. A concurrent run might be seen as a 'cannonical representative' of a class of equivalent sequential runs.

Definition 2.3.1. (*Run expressions*) Definition 2.2.1, of sequential run expressions, gets expanded by clause 'and so is $\rho|\sigma$'.

Concatenation is understood here as binding more strongly than concurrency, so that $\rho\sigma|\tau$ is parsed as $(\rho\sigma)|\tau$

Definition 2.3.2. (*Possibility, independence, effect*) Definitions 1.2.2 (of possibility and effect of sequential runs) and 2.2.1 (of their independence) get expanded by clauses

$$\begin{aligned} \mathrm{pos}(\rho|\sigma) &\stackrel{\mathrm{def}}{=} I(\rho,\sigma) \\ I(\rho|\sigma,\tau) &\stackrel{\mathrm{def}}{=} I(\rho,\tau) \wedge I(\sigma,\tau) \wedge I(\rho,\sigma) \\ [\rho|\sigma]s &\stackrel{\mathrm{def}}{=} [\rho][\sigma]s \end{aligned}$$

We shall likewise abbreviate both equivalent formulae $\mathrm{pos}(\rho)$ and $\langle\rho\rangle$true by $\langle\rho\rangle$.

Definition 2.3.3. (*Sequentialization* of concurrent runs)

$$R^* \stackrel{\mathrm{def}}{=} R \quad (\rho\sigma)^* \stackrel{\mathrm{def}}{=} \rho^*\sigma^* \quad (\rho|\sigma)^* \stackrel{\mathrm{def}}{=} \rho^*\sigma^*$$

Assuming Definition 1.2.4, of the modal operators, extended to concurrent runs, we have

Theorem 2.3.1. $\langle\rho\rangle$ entails, for any formula ϕ,

$$\begin{gathered} \langle\rho^*\rangle \\ I(\rho,\sigma) \Leftrightarrow I(\rho^*,\sigma^*) \\ [\rho]\phi \Leftrightarrow [\rho^*]\phi \end{gathered}$$

Proof by routine induction over ρ, simultaneously for all static algebras.

As far as effect and modal operators are concerned, a concurrent run can thus be replaced with an arbitrary interleaving of its components—all results proved previously for sequential runs carry over to their concurrent equivalents. Note however that the notion of concurrent run *does not presuppose* interleaving, or any other sequential 'semantics'—no statement related to a 'global time' can be derived from it, just some information on synchronization of 'local times'.

Corollary 2.3.2. If $\langle\rho|\sigma\rangle$, and τ is an arbitrary interleaving of ρ^*,σ^*, then, for any formula ϕ,

$$\begin{gathered} [\rho|\sigma]\phi \Leftrightarrow [\tau]\phi \\ \langle\rho|\sigma\rangle\phi \Leftrightarrow \langle\tau\rangle\phi \end{gathered}$$

3 Constructs on evolving algebras

Here we introduce some cannonical constructs, creating new evolving algebras from preexisting ones. The basic construct is (concurrent) *composition*, which, together with addition of guards and updates, obviously suffices to construct any evolving algebra.

With some management of function symbols, the basic constructs allow us to *define* sequential composition and internal and external choice.

By looking at corresponding static algebras, we derive compositional *inference rules* for the transition system, not unlike the SOS rules used in process calculi. We do *not*, however, at this point, formulate a calculus of evolving algebras—the inference rules are just useful derived properties.

We finally show how our rudimentary language can be enriched in several useful ways, preserving all properties proved so far.

3.1 Basic constructs

Remember that, by Def. 1.1.2, we identify an evolving algebra with its program—a finite set of transition rules[8]. The basic construct is (concurrent) composition, allowing two different dynamics to interact.

Definition 3.1.1. If A, B are evolving algebras, their *composition* $A|B$ is given as disjoint sum (of rule sets)[9].

Static algebras are slightly trickier to compose (remember that a static algebra is, by Def. 1.1.3, a universe–interpretation pair). Since the integrity of a static algebra of A is to be preserved by executing rules of B and vice versa, we have to assume the universes to be the same—were they different, updating terms denoting something in the intersection could spoil preservation of universe. Since universes can always be blown up, nothing is lost by that restriction. However, if there is some shared signature, it has to be interpreted in the same way.

Definition 3.1.2. (*consistency* of interpretations and static algebras). If J, K are interpretations of signatures Σ_A, Σ_B respectively, in the same universe U, they are *consistent* if they agree on $\Sigma_A \cap \Sigma_B$. Static algebras $\mathcal{A} = (U, J)$, $\mathcal{B} = (U, K)$ are *consistent* if J, K are.

Consistent static algebras can be reasonably composed, by taking the union of (graphs of) interpretations.

[8] When writing expressions for composite evolving algebras, we shall often confuse a singleton algebra with its single rule, suppressing the curly brackets.

[9] Two rules coming from different component algebras are thus 'different' by *fiat*, in order to avoid explicit naming, or even considering the question when are two rules 'the same'.

Definition 3.1.3. (*composition* of static algebras). If $\mathcal{A} = (U, J)$, $\mathcal{B} = (U, K)$ are consistent static algebras of, respectively, A, B,

$$\mathcal{A}|\mathcal{B} \equiv (U, J \cup K) \ .$$

Consistency ensures that $J \cup K$ is well defined. Composing consistent static algebras of A, B obviously gives a static algebra of $A|B$; by restricting the interpretation to $\Sigma(A), \Sigma(B)$ we see that all such algebras are of this form.

As an immediate consequence of definitions, we have 'preservation of local truth' in the composed algebra:

Proposition 3.1.1 If $\mathcal{A}, \mathcal{B}$ are consistent static algebras of A, B, and ϕ is a formula,

$$\begin{array}{ll} \mathcal{A} \models \phi \Leftrightarrow \mathcal{A}|\mathcal{B} \models \phi & \text{if } \Sigma(\phi) \subset \Sigma(A) \\ \mathcal{B} \models \phi \Leftrightarrow \mathcal{A}|\mathcal{B} \models \phi & \text{if } \Sigma(\phi) \subset \Sigma(B) \end{array}$$

In particular it applies to consistency, possibility and independence of rules. Executing a rule R of A in $\mathcal{A}|\mathcal{B}$ will modify $\mathcal{A}$ to $\mathcal{A}_R$. It will however, by modifying interpretation of shared signature, in general also modify $\mathcal{B}$. This modification is the *basic communication mechanism* of evolving algebras:

Theorem 3.1.2. If $\mathcal{A} = (U, J)$, $\mathcal{B} = (U, K)$ are consistent static algebras of, respectively, A, B, and R (resp. S) is a possible rule of A (resp. B), then

$$(\mathcal{A}|\mathcal{B})_R = \mathcal{A}_R|\mathcal{B}^R \qquad (\mathcal{A}|\mathcal{B})_S = \mathcal{A}^S|\mathcal{B}_S$$

where

$$\mathcal{B}^R = (U, (J \sqcup K)_R\lceil\Sigma(B)) \qquad \mathcal{A}^S = (U, (J \sqcup K)_S\lceil\Sigma(A))$$

($\lceil$ denotes restriction). Extending the above notation to runs, inference rules about the transition system of static algebras and runs (of Section 1.2) follow immediately.

Corrollary 3.1.3. If $\mathcal{A}, \mathcal{B}$ are consistent static algebras, the following inference rule is valid.

$$\frac{\mathcal{A} \xrightarrow{\rho} \mathcal{A}_\rho \quad \mathcal{B} \xrightarrow{\sigma} \mathcal{B}_\sigma \quad \mathcal{A}|\mathcal{B} \models I(\rho, \sigma)}{\mathcal{A}|\mathcal{B} \xrightarrow{\rho|\sigma} \mathcal{A}_\rho^\sigma \,|\, \mathcal{B}_\sigma^\rho}$$

Note that properties of independence make it irrelevant whether $\mathcal{A}_\rho^\sigma$ is read as $(\mathcal{A}_\rho)^\sigma$ or the other way around. If we allow an empty run (independent of any other), the following special cases arise, reminiscent of SOS treatment of 'parallelism–as–interleaving' (of inference).

$$\frac{\mathcal{A} \xrightarrow{\rho} \mathcal{A}_\rho}{\mathcal{A}|\mathcal{B} \xrightarrow{\rho} \mathcal{A}_\rho|\mathcal{B}^\rho} \qquad \frac{\mathcal{B} \xrightarrow{\sigma} \mathcal{B}_\sigma}{\mathcal{A}|\mathcal{B} \xrightarrow{\sigma} \mathcal{A}^\sigma|\mathcal{B}_\sigma}$$

Definition 3.1.4 (*adding* guards)[10]. If A is an evolving algebra, b a boolean term, $b?A$ denotes the algebra A with every rule R modified to

$$b?R \equiv \textbf{if } b \wedge R? \textbf{ then } R! \ .$$

Static algebras of $b?A$ will be denoted by $\beta?\mathcal{A}$.

Here we make full use of the liberty we have allowed ourselves, to confuse boolean terms and quantifier–free formulae—b in the guard stands for $b = tt$. $\mathcal{A}$ is obtained form $\beta?\mathcal{A}$ by restricting the interpretation to $\Sigma(A)$, and β reminds us that also b has to be interpreted. The following rule of inference is obviously valid (upwards as well as downwards)

$$\frac{\mathcal{A} \xrightarrow{R} \quad \beta?\mathcal{A} \models b}{\beta?\mathcal{A} \quad \xrightarrow{b?R}}$$

We shall often use the pseudoconstruct $\{f\}A$, where f is a function symbol, indicating the *intention* to keep f *internal* [Gurevich 91], private to A: whenever $\{f\}A$ is composed with some B, we assume $f \notin \Sigma(B)$. Then f in $\{f\}A$ assumes the syntactical properties of a *bound variable* in logic or λ–calculus, or of a *restricted name* in π–calculus: it can be renamed to a fresh symbol, and it must be kept distinct from all function symbols appearing elsewhere. Since we are not quite at the point of formulating a *calculus* of evolving algebras, we shall not formalize the properties of $\{f\}A$—keep in mind, however, that updates of f do not affect the environment, and that anything that happens in the environment cannot affect f (by *environment* of A we mean here anything A is composed with).

3.2 Derived constructs

Some constructs, close to those of process calculi, can be easily derived in our framework: *internal* and *external choice, sequential composition*. Since ? is a `while`–loop, taken together with (concurrent) composition this constitutes already a reasonable concurrent programming language.

It is convenient to assume a distinction between *dynamic* and *static* function symbols—those which are modified, either within the algebra or by the environment, and those which are not (cf. also [Gurevich 91]). We shall assume here, without further mention, static booleans *tt*, *ff*.

It is also convenient to assume that initially all dynamic functions are empty ('everywhere undefined')—composing (sequentially, cf. below) with appropriate initialization updates we may still obtain any (finite) initial states we want. Note that this assumption just means restricting our attention to static algebras reachable from this kind of initial state.

[10] We could symmetrically introduce an operation of adding updates, but we have no use for it here.

Definition 3.2.1. (*internal choice*). Let A, B be evolving algebras, and $c \notin \Sigma(A) \cup \Sigma(B)$ a fresh distinguished element. Then

$$A + B \equiv \{c\}(T(f) \mid F(c) \mid c?A \mid (\neg c)?B)$$

$$\text{where} \quad T(c) \equiv \textbf{if } \uparrow c \textbf{ then } c := tt$$
$$F(c) \equiv \textbf{if } \uparrow c \textbf{ then } c := ff$$

where $\uparrow$ is, of course, the negation of $\downarrow$, and $\neg c$ represents $c = ff$. Given that c is initially undefined, only T or F can fire. Since they are very much dependent, only one of them can fire—the choice between them is the nondeterministic choice between A and B. Since c is private, the environment cannot influence the choice. Since c doesn't appear either in A or in B, after the choice its value is frozen, and the algebra behaves like an algebra of A (resp. B). The reader may have some fun drawing various derived rules about $\mathcal{A} + \mathcal{B}$.

Definition 3.2.2. (*explicit external choice*). Let A, B be evolving algebras, $c \notin \Sigma(A) \cup \Sigma(B)$ a fresh distinguished element, b a boolean term. Then

$$A \oplus_b B \equiv \{c\}(S(b,c) \mid c?A \mid (\neg c)?B)$$

$$S(b,c) \equiv \textbf{if } \uparrow c \wedge \downarrow b \textbf{ then } c := b$$

Here the environment may, by setting b, choose between A and B. The role of c is to make the choice definitive—once the choice is made ($S(b,c)$ is executed and c gets defined), the environment cannot take it back by resetting b. The reader may amuse himself with derived rules for $\oplus_b$.

This kind of choice is 'explicitly expressed' by b. There are many other ways in which an environment can select between several algebras—in Section 4 we shall see some of them.

Sequential composition can be defined by a trick known in process calculi. Let us say that *termination* of an evolving algebra A is *witnessed* by boolean s_A if s_A 'expresses that A has stopped', i.e. the following is valid.

$$\bigwedge_{R \in A} \neg\langle R \rangle \Leftrightarrow s_A$$

Definition 3.2.3.(*sequential composition*). Let A, B be evolving algebras, where termination of A is witnessed by some $s_A \notin \Sigma(B)$. Then

$$A; B \equiv \{s_A\}(A \mid s_A?B).$$

Its static algebras are denoted by $\mathcal{A};\mathcal{B}$, where $\mathcal{A}, \mathcal{B}$ are obtained by restricting the interpretation.

The usual kind of rules for sequential composition can now be easily derived.

$$\frac{\mathcal{A} \xrightarrow{\rho} \mathcal{A}_\rho}{\mathcal{A};\mathcal{B} \xrightarrow{\rho} \mathcal{A}_\rho;\mathcal{B}^\rho} \qquad \frac{\mathcal{A} \models s_A \quad \mathcal{B} \xrightarrow{\sigma}}{\mathcal{A};\mathcal{B} \xrightarrow{s_A?\sigma}}$$

It is through modifying $\mathcal{B}$ to $\mathcal{B}^\rho$ that the effect of $\mathcal{A}$ is passed over. With these constructs arbitrary finite initialization of dynamic functions can be realized by setting up a corresponding set of updates (a rule without guards) *Init*—an initialized static algebra is then realized by $Init\,;\mathcal{A}$, which will behave like $\mathcal{A}^{Init}$.

Taken together with known techniques for modelling *assignment*, such as in [GurMos 90, GurHug 93, BörRos 92a] and *channel communication* [GurMos 90], (cf. also [Boudol 87] and Section 4), we obtain a trivial way to construct an evolving algebra model for any given Occam program.

3.3 Extending the language—counters, universes, variables

We have so far relied on a very rudimentary, spartan notion of evolving algebra. We show here how that framework can be extended in several useful ways, taking care however to preserve the results proved so far.

It is often desirable to *count* some actions performed. In sequential evolving algebras nothing is easier than that: take a distinguished integer for a counter, and add a counter incrementing update $count := count + w$ to any action you want to count. Since *count* would, according to Def. 2.1.1, be in the domain of such an update, two actions updating the same counter would never be independent, according to Def. 2.1.2, hence could not be executed concurrently, which would pose a silly obstacle to parallelism. Under the metaphor of several concurrent daemons, a counting daemon should lock the counter first, forcing others to wait. If they increment the counter by different weights, their simultaneous action would even be inconsistent, according to Def 1.1.5 (even if they were by chance consistent, the effect would not be cumulative). Such a counter is a splendid example of the standard trap awaiting uncareful designers of concurrent systems—it is easy to build in an inocuous little *bottleneck*, which can cause degeneration of elaborately construed parallelism.

But it is perfectly consistent for several daemons to throw balls simultaneously into a jar, and to do that independently, at various speeds, with equal overall effect—only when someone wants to *count* the balls in the jar, i.e. to use the *value* of the counter, the ball throwers should be asked to wait. Definitions of consistency (1.1.4) and independence (2.1.2) don't know about the special jar–like nature of counters.

We shall assist them by introducing a special form of update $count \mathrel{+}= w$, for integer terms *count* and w, under

Counter convention (*consistency, effect and domain of counter increments*). Let

$$C = \{\, count \mathrel{+}= w_1, \ldots, count \mathrel{+}= w_k \,\}$$

and let $R!$ be a set of updates free of *count* increments. Then

$$\mathrm{cons}(C \cup R!) = \downarrow count \;\wedge\; \mathrm{cons}(R!) \;\wedge\; ([R!]count = count)$$

$$[C \cup R!]\,t = \begin{cases} count + w_1 + \ldots + w_k & \text{if } t = count \\ \text{by recursion of sec.1} & \text{otherwise} \end{cases}$$

$$D(C \cup R!) = Dw_1 \cup \ldots \cup Dw_k \cup DR!$$

Several increments of the same counter are then independent, according to Def. 2.1.2, and the theory of Section 2 carries over to concurrent increments.

A 'higher order' kind of jar is provided by *dynamic universes* of [Gurevich 88, Gurevich 91]. Several universes of a heterogenous signature are seen as embedded in a *superuniverse* U. Whenever we mention $S \subset U$, we are talking about its characteristic function ϕ_S, and $a \in S$ is just shorthand for $\phi_S(a) = tt$. When we talk about 'extending' or 'contracting' a universe $S \subset U$, we assume a special 'reserve' universe $R \subset U \setminus S$, and move an element from R to S or out of S. R, as well as the universe S getting modified, are dynamic universes, i.e. ϕ_R, ϕ_S are dynamic functions. Domains and codomains of all 'ordinary' functions are understood as being disjoint from R. In the jargon of algebraic specification, this means that reserve objects are *junk*—they are not denoted by terms. ϕ_R is also an exception to the convention that dynamic functions should be initially empty—it should be infinite.

Discarding an element is simple—we use an update

$$\text{discard } t \text{ from } S \stackrel{\text{def}}{=} \phi_S(t) := \mathit{ff}\,.$$

For extension we shall use an update form

$$\begin{aligned}\text{extend } S \text{ by } x \text{ with } u \stackrel{\text{def}}{=} \{\, & \phi_R(x) := \mathit{ff}, \\ & \phi_S(x) := tt\} \cup u\end{aligned}$$

where x is an *extension variable* and u a set of *extension updates* depending, in general, on x. Extension variables are similar to bound variables in that they have to be seen as distinct from each other. Such updates will be used under

Extension convention. If an evolving algebra A contains universe extensions, we assume that, to every static algebra of A, a fixed valuation of (sufficiently many) extension variables with distinct values in (current) R is associated. Extension updates u are interpreted under this valuation.

Further, updates distinct from extend or discard do not mention ϕ_R at all, and do not modify ϕ_S for any dynamic universe S.

We don't really care what valuation of extension variables is used, since the effect will be the same—up to a trivial isomorphism of structures. Under the convention, concurrent extensions will be independent (if respective extension updates are), and the theory of Section 2 is preserved, since we have, in the language, no means to tell one reserve object from another—junk has its uses after all.

In the extension update a variable has appeared in our language for the first time. We shall in the sequel often use *free variables* in rules, in two different ways.

Usage of variables starting with capital letters is just an abbreviational device, which enhances readability, but is otherwise eliminable. Say,

$$\begin{array}{l}\textbf{if } \ldots a = < X, Y > \ldots \\ \textbf{then } \ldots X \ldots Y \ldots\end{array}$$

abbreviates

$$\begin{array}{l}\textbf{if } \ldots \text{ispair}(a) \ldots \\ \textbf{then } \ldots \text{fst}(a) \ldots \text{snd}(a) \ldots,\end{array}$$

sparing us the need to write explicitly the recognizers and the selectors[11], such as 'ispair', 'fst', 'snd' above.

Using variables starting with small letters *does* properly extend the language. They will always occur guarded by a condition of the form $x \in S$, where S is a dynamic universe. Since dynamic universes, apart from R, are always finite by our conventions, a rule containing such a variable will represent the finite set of its possible instances. Replacing the finite set of possible rules with the finite set of possible instances of rules preserves our theory of runs, independence and concurrency, and its modal logic[12], although the set of instances depends on the static algebra ('computation state'). The number of rules namely, or quantification over rules, just doesn't occur anywhere in the theory—even if it occurred, we could always use counters to keep track of the cardinality of relevant dynamic universes. Note also that it is the set of possible rules that counts, and it has been 'state dependent' all the time.

4 Applications

4.1 The Chemical Abstract Machine

A Chemical Abstract Machine [BerBou 92] ('cham' in the sequel) operates on *solutions*. Solutions are finite multisets of *molecules*, which are specified, for each cham, by giving a set of function symbols (*molecule constructors*) with arities—molecules are just terms built using the constructors. Since any solution may itself be considered as a 0-ary molecule constructor, the definition of solutions and molecules is mutually recursive—solutions may occur as components of complex molecules. More precisely,

1. every 0-ary constructor is a single molecule;
2. $f(\mathbf{t})$ is a single molecule, if $\mathbf{t}$ is an n-tuple of molecules, and f is an n-ary constructor;
3. a single molecule is a molecule;
4. a finite multiset of molecules is a solution;
5. a solution is a molecule.

---

[11] One might say that it allows us to use a 'logic programming' or 'pattern matching' rather than 'functional' style.

[12] Distinct instances of the same extension update have, of course, distinct extension variables.

Solutions are transformed according to a finite set of *transformation rules* of form

$$e_1, \ldots, e_k \mapsto e'_1, \ldots, e'_l$$

where e_i, e'_j are *molecule expressions*, built by same clauses as molecules, but using also *solution variables* $S, S', \ldots$ and *single molecule variables* $m, m', \ldots$. Transformation rules are used as axiom schemata, to be applied anywhere inside a solution. The transformation relation between solutions is made precise by the following SOS rules (where $\uplus$ denotes multiset union, and multisets given by enumeration are delimited by $\{\!|\ ,\ |\!\}$, denoting a *membrane*):

Reaction law

$$\{\!|\, M_1, \ldots, M_k \,|\!\} \mapsto \{\!|\, M'_1, \ldots, M'_l \,|\!\}$$

where $M_1, \ldots, M_k \mapsto M'_1, \ldots, M'_l$ is an instance of a transformation rule;

Chemical Law

$$\frac{S \mapsto S'}{S \uplus S'' \mapsto S' \uplus S''}$$

Membrane Law

$$\frac{S \mapsto S'}{\{\!|\, C[S] \,|\!\} \mapsto \{\!|\, C[S'] \,|\!\}}$$

where $C[\,]$ is any context.

The chemical and membrane law express just the *principle of locality*: transformation rules may be applied anywhere in a solution, even when it occurs deeply inside a term—transforming a part will always transform the whole.

Background, motivation and explanation can be found in [BerBou 92]—let us just remark that cham models of TCCS and CCS are given there, as well as an embedding of a higher-order concurrent λ-calculus, and that these models have had considerable influence on subsequent development of the π-calculus [MiPaWa 92, Milner 92]. [Abramsky 92] has provided a cham-based computational interpretation of classical linear logic.

We argue that chams are just special evolving algebras. In view of inductive definition of molecules and solutions, they are trees, and this has just to be explicated with appropriate evolving algebra signatures. Multisets can be readily modelled, by sets of occurrences, or decorated objects—repetition will be reflected by presence of objects with identical decorations. Once the representation of solutions is given, the cham laws will be translated into evolving algebra rules—but only the reaction law: as the chemical and the membrane law say just that modifying a part modifies the whole, they will simply be true of our model.

We postulate a dynamic universe M of (occurrences of) molecules–in–the–tree. The tree structure will be given by a (dynamic) function

$$father : M \rightarrow M$$

and the type of molecules will be given by

$$type : M \longrightarrow \{s, \ldots\}$$

where ... stand for the designated molecule constructors whose arities we can access by *arity*, and s is the type of solutions. The 'top–level' solution will be rooted in a distinguished object *world* (with undefined *father*). By chemical view of the world, $type(world) = s$. Elements of solutions are not ordered, but components of other molecules are; hence the component-selecting function

$$\arg : M \times \mathbb{N} \rightharpoonup M$$

under the constraint

$$\downarrow arg(m,i) \Leftrightarrow type(m) \neq s \ \wedge \ i \leq arity(type(m)) \ \wedge \ father(arg(m,i)) = m$$

Since transformation rules have to check whether a solution is singleton, we have to keep count of the cardinality of solutions; the signature is hence concluded with a function

$$card : M \rightharpoonup \mathbb{N}$$

under the constraint

$$\downarrow card(m) \Leftrightarrow type(m) = s.$$

The values of *card* will be, after initialization, updated as counters (cf. 3.3), in order to enable independent evolution within a solution.

[BerBou 92] constrain the solution–subexpressions occurring inside molecule–expressions of transformation rules to be either a solution variable, or of form $\{\!|\, m \,|\!\}$, where m is a single molecule. The variables occurring in a transformation rule have to be 'single', i.e. have a single occurrence in its lhs and at most one in its rhs. The purpose of this constraint is 'to avoid multiset matching'. We might add: also to avoid duplicating complex structures (which may be seen as falling under biology rather than chemistry).

Transformation rules can now be easily *compiled* to evolving algebra rules—the guards should check the lhs, and the updates should destroy the lhs and construct the rhs. Molecules corresponding to variables on the left which also occur on the right should not be destroyed, but preserved and reatttached at appropriate places on the right, updating the cardinalities as needed. We shall not spell out the straightforward (but tedious) compilation algorithm[13]—the following example should suffice.

---

[13] For the reader familiar with compilation of logic languages, let us just remark that checking of lhs corresponds to (a simplification of) *getting* instructions of the WAM, construction of rhs to (a simplification of) *putting* instructions, and preserving with reattachment to some aspects of *variable classification*—cf. also [BörRos 92a].

In the model of TCCS of [BerBou 92] the following rule appears:

$$< \{\!|\, \alpha\,.\,m \,|\!\},\, S > \mapsto \alpha\,.\,l :< m,\, S >$$

where $\alpha\,.\,\_$ is unary, and $< \_\,,\,\_ >$, $l{:}< \_\,,\,\_ >$ are binary molecule constructors–more pictorially,

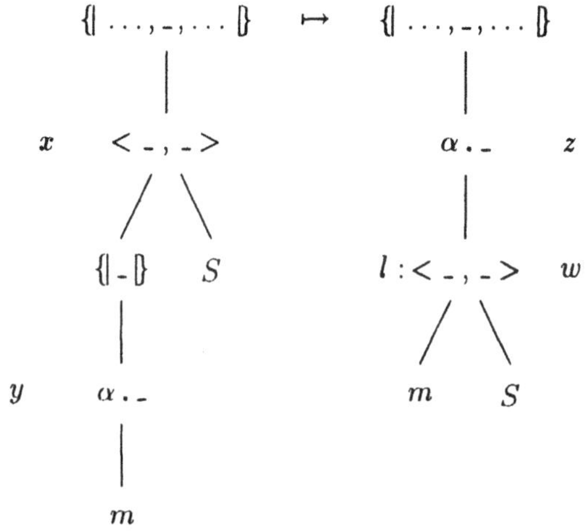

(where annotations x, y, z, w serve as a legend to the rule below).

The corresponding evolving algebra rule, 'describing the picture', would be

$$
\begin{array}{l}
\textbf{if } x, y \in M \\
\quad \wedge\ type(father(x)) = s \ \wedge\ type(x) =< \_\,,\,\_ > \\
\quad \wedge\ type(fst(x)) = s \ \wedge\ card(fst(x)) = 1 \\
\quad \wedge\ type(snd(x)) = s \\
\quad \wedge\ father(y) = fst(x) \ \wedge\ type(y) = \alpha.\,\_ \\
\quad \wedge\ type(fst(y)) \neq s \\
\textbf{then } \text{discard } x, y, fst(x) \text{ from } M \\
\qquad \text{extend } M \text{ by } z, w \text{ with} \\
\qquad\quad father(z) := father(x) \\
\qquad\quad type(z) := \alpha\,.\,\_ \\
\qquad\quad \text{attach } w \text{ as } fst(z) \\
\qquad\quad type(w) := l{:}< \_\,,\,\_ > \\
\qquad\quad \text{attach } fst(y) \text{ as } fst(w) \\
\qquad\quad \text{attach } snd(x) \text{ as } snd(w)
\end{array}
$$

where *fst*, *snd*, attach_as are obvious abbreviations. Note that, since x gets discarded and z gets created, update of the cardinality of $father(x)$ is 'optimized away': it should be $card(father(x)) += 1 - 1$.

In spite of its size, there is nothing in this rule which is not present or implicit in the cham transformation rule it represents. Wherever (in a solution in the tree) this kind of rule finds x, y which satisfy the guards, it will transform the corresponding part of the tree—the chemical law and the membrane law are valid by construction.

Of course, given a syntactically distinguished class of molecules which cannot contain solutions as subterms (such as 'agents' in the above mentioned cham model of TCCS), their structure need not be analyzed into the tree—a molecule of that kind, however complex, might be associated to a node as a single decoration.

[BerBou 92] say:

> 'A cham is an intrinsically parallel machine: one can simultaneously apply several rules to a solution provided that no molecule is involved in more than one rule; one can also transform subsolutions in parallel ...(the expressive power of chams) does not depend on parallel evaluation, since a non-conflicting parallel application of rules is equivalent to any sequence of the individual rules.'

There is little to add here; we might only notice that, when 'no molecule is involved in more than one rule', these rules will be *independent* in our model. Our claim that the cham just *is* a special evolving algebra rests not so much on the simple and direct representation–by–explication (in spite of apparent tension between 'geometry' of cham pictures and our 'arithmetization' of it), as in the circumstance that the intended cham notion of parallelism maps so precisely to evolving algebra notion of concurrency.

4.2 The π–calculus

The π–calculus of [Milner 92] relates *processes*, i.e. expressions built from a set of *names* (to be understood as names of *communication ports or links*) according to the following syntax:

$$P \longrightarrow \sum_{i \in I} \pi_i.P_i \mid (P|P) \mid !P \mid \nu x P$$

where I a finite index set, each π_i a *prefix* of form $x(y)$ or $\overline{x}y$, x, y are names. The *restriction operator* ν binds x in $\nu x P$ and the *input prefix* $x(y)$ binds y in $x(y).P$—other names are free. The empty sum is denoted by 0, and the singleton sum is abbreviated by its only summand. The operator | is called *composition*, ! is *replication*, and $\overline{x}y$ is *output prefix*. Expressions which differ just in names of their bound names are not distinguished. Note that, with this syntax, $\sum_i \pi_i$ is an atomic constructor, represented, say, by a vector of prefixes, and acting, in the fashion of scalar product, on a vector of processes.

For (rich) background, motivation and explanation the reader is referred to [Milner 92, MiPaWa 92] and their references. It will have to suffice here to say that, intuitively,

0 is a process which cannot do a thing;

$\overline{x}z.P$ is a process which is ready to output name z on port named x, and to behave subsequently like P;

$x(y).P$ is a process which is ready to input name y on port named x, and to behave subsequently like P;

$\nu x P$ is a process which behaves like P but keeps name x private;
$!P$ is a process which can spawn new copies of P;
$P \mid Q$ is parallel composition of P and Q;
$\sum_i \pi_i$ is an operator of guarded choice.

Precise meaning of these remarks is given by relations of *reduction* and *structural congruence*. The latter, denoted by $\equiv$, is the smallest congruence relation (wrt all process constructors) over the set $\mathcal{P}$ of processes satisfying

1. $(\mathcal{P}/\equiv, \mid, 0)$ is a symmetric monoid
2. $!P \equiv P \mid !P$
3. $\nu x 0 \equiv 0, \quad \nu x \nu y P \equiv \nu y \nu x P$
4. $P \mid \nu x Q \equiv \nu x (P \mid Q)$ if x is not free in P

The corresponding *reduction relation*,$\rightarrow$, is given by the following SOS rules:

$$\text{COMM:}\ (\ldots + x(y).P + \ldots) \mid (\ldots + \overline{x}z.Q + \ldots) \rightarrow P\{z/y\} \mid Q$$

$$\text{PAR:}\ \frac{P \rightarrow P'}{P \mid Q \rightarrow P' \mid Q} \qquad \text{RES:}\ \frac{P \rightarrow P'}{\nu x P \rightarrow \nu x P'}$$

$$\text{STRUCT:}\ \frac{Q \equiv P \quad P \rightarrow P' \quad P' \equiv Q'}{Q \rightarrow Q'}$$

The calculus specifies a *notion of computation*, whose basic unit consists in an application of COMM rule—P inputs the value offered on port x, the choice is made thereby, and component processes, which succeeded to communicate, proceed. Note that, through name z, P has now obtained 'access' to other processes it may had not had before. Rules PAR and RES express a (partial) principle of locality, while STRUCT allows $\equiv$ to enable computation (communication) by 'bringing agents into juxtaposition'.

Attempting to model this notion of computation with an abstract machine, there are essentially two ways to proceed:

(a) reflect every step of inference by some machine action(s), or
(b) obtain first a more explicit rendering of the intended notion of computation, and model it in a more direct fashion.

[BerBou 92] have, developing cham models of TCCS and CCS, pursued course (a), obtaining an analysis of the 'fine structure' of the calculi. The price paid is that their simulators may cycle forever between inverse local operations even when a 'meaningful' action is possible, and can thus be viewed as just partially correct. In presence of a symmetric relation such as $\equiv$ this is inevitable—under requirement (a).

We attempt, through a 'coarse-grained' analysis of the calculus, to achieve (b). Observe first that reduction is *lazy* wrt to sum and replication—it is not applied inside them—while congruence may be inferred anywhere. This indicates that $\equiv$ plays a double role: to enable computation, and also to lay down the minimal requirements (of compatibility with the syntax) on any semantical notion to be derived from the calculus. We are interested here in the first role of $\equiv$, which may be isolated as follows.

Define a subrelation $\equiv_l$ by the very same clauses 1.–4. that define $\equiv$, requiring of it however to be a congruence only wrt to composition and restriction. Replacing $\equiv$ by $\equiv_l$ in rule STRUCT, we obtain a lazy version of reduction, $\rightarrow_l$, and a lazy version of the calculus, which treats sums and replications as true black boxes, icecubes—nothing whatsoever is inferred about their contents.

Note a technical property of reduction rules, common to $\rightarrow, \rightarrow_l$.

Lemma 4.2.1. If $Q \rightarrow Q'$ or $Q \rightarrow_l Q'$, it has a derivation with at most one inference by STRUCT, which immediately derives the conclusion.

Proof. Note that STRUCT commutes with PAR and RES, while COMM has no premisses; all inferences by STRUCT can be then pushed down, and merged into one final STRUCT.

An inference of $Q \rightarrow Q'$ can thus be brought to the form

$$\dfrac{\begin{array}{ccc} & \text{COMM} & \\ \Theta & \Pi & \Sigma \\ Q \equiv P & P \rightarrow P' & P' \equiv Q' \end{array}}{Q \rightarrow Q'}$$

where Θ, Σ are derivations by $\equiv$, and Π is a chain of inferences by RES and PAR.

Lemma 4.2.2. If $Q \rightarrow Q'$, then $Q \rightarrow_l Q''$ for some $Q'' \equiv Q'$.

Proof sketch. The proof has some flavor of cut–elimination. Inferences by $\equiv$ within an icecube can be postponed till after the icecube gets melted (by COMM or replication). This can be seen by tracing the replaced icecube down θ and up Π (on the left of $\rightarrow$). If the icecube melts or vanishes on the way, the replacement can be postponed till there or dropped, respectively. If the icecube never melts, and jumps over to the right side of $\rightarrow$, it can be traced further, down Π (on the right of $\rightarrow$) till it, appearing in P', jumps over to Σ. The replacement can be, in such a case, postponed till there. Repeating this process as many times as needed modifies P, P' and the derivations, but, if executed with some care, leaves Q, Q' untouched—modified Θ will be a derivation of $\equiv_l$, while modified P' will be the Q'' of the lemma. Unwinding these remarks to a detailed proof is quite straightforward (though somewhat tedious).

Of course, $\overset{*}{\rightarrow}_l \subset \overset{*}{\rightarrow}$ (where $*$ denotes reflexive and transitive closure). On the other hand, Lemma 1 permits us to iterate Lemma 2, obtaining

Proposition 4.2.1. If $Q \overset{*}{\rightarrow} Q'$, then $Q \overset{*}{\rightarrow}_l Q''$ for some $Q'' \equiv Q'$.

Note that no communications are gained or lost thereby. Computation–as–communication is all in the $Q \overset{*}{\rightarrow}_l Q''$ part—the $Q'' \equiv Q'$ part may be viewed as just 'beautifying' the result. Such decomposition is made possible by Milner's decision to stick to *guarded sums* in the latest version [Milner 92] of the π–calculus, feeling that '...forms such as $(P|Q)+R$ have very little significance'. While some might strongly object to it, this decision enables us to construct a particularly simple and transparent evolving algebra model—a different notion of sum (preferably orthogonal) may be added or refined to, once the essentials of π–calculus are well represented.

But the process of distilling 'computational content' from the calculus is not yet complete—a (pseudo)cham model will help us make the next step.

In the final version of [BerBou 92] the following cham model of the π–calculus is presented (without establishing a formal relationship to the calculus). Molecules of the cham are processes. Assuming that the initial process was closed, (lazy) congruence is rendered by the following transformation rules for composition, replication and 0

$$P \mid Q \rightleftharpoons P,\ Q \qquad !P \rightleftharpoons P,\ !P \qquad 0 \rightleftharpoons$$

and communication ('reaction') by the rule

$$(\ldots + x(y).P + \ldots),\ (\ldots + \overline{x}z.Q + \ldots) \mapsto P\{z/y\},\ Q\ .$$

The rules denoted by $\rightharpoonup$ are *heating rules*, since they make reaction possible; their inverses are *cooling rules*, denoted by $\rightharpoondown$. Process constructors can be seen here as 'encapsulating computations', which might trigger when heated. Note how replication heats into spawning a replicant, and how 0 evaporates under heat, and gets created *ex nihilo* by cooling. For restriction, which [Milner 92] explains saying that 'νxP – "new x in P" ...declares a new unique name x ...for use in P', [BerBou 92] introduce an integer 'name–server' n and a rule

$$\nu xP,\ n \rightharpoonup P\{n/x\},\ n+1\ .$$

In a solution containing $\nu xP,\ \nu yQ,\ n$ the following results of restriction–heating are conceivable

$$P\{n/x\},\ Q\{n+1/x\},\ n+2 \qquad P\{n+1/x\},\ Q\{n/x\},\ n+2\ .$$

But they are both *sequential*—the name–server is a splendid example of an inocuous little bottleneck. While formally not forbidden by SOS 'parallelism–as–interleaving' of cham definition, this certainly violates the intended notion of cham parallelism, as can be clearly seen if the cham is viewed as an evolving algebra—the two runs are formally not independent. Treating n as a counter doesn't help here, since both agents have to use its *value*. It is now easy to concoct examples of 'highly parallel' processes which would degenerate to completely serial execution because of the name–server.

Taking Milner's explanation of restriction literally would result in 'heating' and 'cooling' rules of form

$$\nu xP \rightleftharpoons P\{y/x\} \qquad (y \text{ fresh})$$

where condition on y means that it may not appear free elsewhere. Such a pair of rules may appear as problematic for several reasons:

1. cham definition doesn't seem to allow creating new objects, which is the idea underlying the heating rule;
2. the condition on y is, in case of cooling, *not local*—it refers to all molecules in the *world*;
3. the rules don't preserve equivalence of processes, at least locally.

Objection 1. is of no import if we view the cham as an evolving algebra, there we may create new objects at will, and do that concurrently. Objection 3. doesn't pertain to *global* equivalence—assuming that the initial process was closed, we may view a solution $\{\!|\, P_1, \ldots, P_n \,|\!\}$ as representing the closed process

$$\nu\mathbf{x}(P_1 \mid \ldots \mid P_n)$$

where $\mathbf{x}$ lists all names occurring free in P_i's, without loss of generality. Restriction–heating should then be seen as 'telescoped' extrusion of ν[14] to the top level.

Objection 2. is however quite valid (both from cham and evolving algebra viewpoints), and this is why we are talking about a **pseudo**cham model. Since that model is not our goal, but just a useful analytic tool, we shall still employ it for an analysis of $\rightarrow_l$ without worrying too much about definitions—for a while.

To summarize, together with the above communication rule, we assume for the pseudocham the following communication, heating and cooling rules:

$$(\ldots + x(y).P + \ldots),\ (\ldots + \overline{x}z.Q + \ldots) \mapsto P\{z/y\},\ Q$$

$$P \mid Q \rightleftharpoons P,\ Q \qquad !P \rightleftharpoons P,\ !P \qquad 0 \rightleftharpoons \qquad \nu x P \rightleftharpoons P\{y/x\}$$

If $(\rightharpoonup\ \cup\ \rightharpoondown)^*$ is denoted as $\stackrel{*}{\rightleftharpoons}$, and $(\rightharpoonup\ \cup\ \rightharpoondown\ \cup\ \mapsto)^*$, by abuse, as $\stackrel{*}{\mapsto}$, the discussion may be summarized by

Lemma 4.2.3. If P, Q are closed, $P \equiv_l Q$ iff $\{\!|\, P \,|\!\} \stackrel{*}{\rightleftharpoons} \{\!|\, Q \,|\!\}$.

Proposition 4.2.2. If P, Q are closed, $P \stackrel{*}{\rightarrow}_l Q$ iff $\{\!|\, P \,|\!\} \stackrel{*}{\mapsto} \{\!|\, Q \,|\!\}$.

Now cooling commutes both with heating and with communication. This observation is essential for development of the evolving algebra model.

Lemma 4.2.4. If $S \rightharpoondown S' \rightharpoonup S''$, then $S'' = S$ or, for some solution S''', $S \rightharpoonup S''' \rightharpoondown S''$.

Proof consists in the observation that a molecule which can be heated cannot be an immediate result of cooling, unless this is by exactly the inverse rule. Otherwise the rules are applied independently.

Lemma 4.2.5. If $S \rightharpoondown S' \mapsto S''$, then, for some solution S''', $S \mapsto S''' \rightharpoondown S''$.

[14] 'Extrusion' (of scope) is line 4. of $\equiv$ definition, cf. [Milner 92] for discussion.

Proof. A sum is never a result of cooling—the rules are applied independently.

Possibility to postpone cooling is succinctly expressed by a further decomposition theorem (if $\stackrel{*}{\rightsquigarrow} \stackrel{\text{def}}{=} (\rightharpoonup \cup \mapsto)^*$):

Proposition 4.2.3. If P, Q are closed, $P \stackrel{*}{\rightarrow}_l Q$ iff $\{\!|\, P \,|\!\} \stackrel{*}{\rightsquigarrow} S \stackrel{*}{\rightharpoonup} \{\!|\, Q \,|\!\}$ for some solution S.

The $\{\!|\, P \,|\!\} \stackrel{*}{\rightsquigarrow} S$ part may be seen as containing computation—the cooling part just imposes 'form' on the hot solution, 'formats' it into a process expression.

It is perfectly simple (in view of Section 4.1) to provide an evolving algebra account of $\stackrel{*}{\rightsquigarrow}$. We shall assume the signature needed to represent the syntax of processess—we skip here the straightforward details of syntax modelling. Note only that $\ldots + \_ + \ldots$ will represent, strictly speaking, a list-membership predicate, and that a dynamic universe *Name* is needed in order to create fresh names as required by restriction-heating. Since molecules contain no solutions as subterms, the treatment of section 4.1 can be simplified in the following ways.

- since *father* of everything is *world*, they can both be dropped;
- since molecules need not be further analyzed, *type*, *arg* can be dropped;
- since only *world* has a cardinality, *card* becomes a distinguished integer (which will be needed later, but not for $\stackrel{*}{\rightsquigarrow}$);
- every molecule m comes decorated with an $agent(m)$ representing a process.

The communication and cooling rules will be called, in turn, $cm, |h, !h, 0h, \nu h$. Noting that $!h$ may retain the $!P$ molecule, and create just a replicant, the reader might by now be able to write down the evolving algebra rules himself. Nevertheless, we list them for the record.

if $x \in M \ \wedge \ agent(x) = (\ldots + X(Y).P + \ldots)$
  $y \in M \ \wedge \ agent(y) = (\ldots + \overline{X}Z.Q + \ldots)$
then $agent(x) := P\{Z/X\}$
  $agent(y) := Q$

if $x \in M \ \wedge \ agent(x) = P \mid Q$
then discard x from M
  extend M by y, z with
    $agent(y) := P$
    $agent(z) := Q$
    $card += 2 - 1$

if $x \in M \ \wedge \ agent(x) = !P$
then extend M by y with
    $agent(y) := P$
    $card += 1$

if $x \in M \ \wedge \ agent(x) = 0$
then discard x from M
    $card += -1$

if $x \in M \ \wedge \ agent(x) = \nu X P$
then extend *Name* by y with
    $agent(x) := P\{y/X\}$

Let us name the evolving algebras of heating rules and heating with communication, respectively, as

$$heat \stackrel{\text{def}}{=} ((|h) \mid !h \mid 0h \mid \nu h) \qquad \pi \stackrel{\text{def}}{=} (cm \mid heat) \ .$$

Since π simulates $\stackrel{*}{\rightsquigarrow}$ step by step, denoting by alg(P) the static algebra corresponding to the solution $\{\!|\, P \,|\!\}$, we have

Corrolary 4.2.4. If P, Q are closed, $P \stackrel{*}{\rightarrow}_l Q$ iff $\text{alg}(P) \stackrel{\rho}{\longrightarrow} \mathcal{A}$ where ρ is a run of π and $\mathcal{A}$ a heating of alg(Q).

We can also cool a solution by evolving algebra rules. Since we are talking about form here, in view of Proposition 1 not much is lost if we drop the $0c$ rule, while the (problematic) νc rule may take the (unproblematic) form

$$\begin{array}{l}\textbf{if } card = 1 \wedge agent(x) = P \wedge Y \text{ free in } P \\ \textbf{then } agent(x) := \nu Y P\ ,\end{array}$$

applicable only to a (final) singleton solution. With it we are firmly back in the realm of the official evolving algebra definition (we leave it to the reader to complete the *cool* algebra, by obvious rules $|c$ and $!c$).

Simply putting π and *cool* together would yield a dumb simulator, which could cycle forever even in presence of possibility for communication. With evolving algebras it is however perfectly straightforward to *steer* the reaction by turning the heat on and off, either through the environment or through a daemon; adding a (fresh) boolean h, we can set up for instance

$$\Pi(h) \stackrel{\text{def}}{=} (h?\pi \mid (\neg h)?cool)$$
$$\Pi \stackrel{\text{def}}{=} \{h\}(\textbf{if } \uparrow h \textbf{ then } h := tt \mid \textbf{if } h \textbf{ then } h := f\!f \mid \Pi(h)).$$

(where $\neg h$ stands for $h = f\!f$). Π can at any time nondeterministically choose to *cool*[15]. It is related to Milner's calculus by

Proposition 4.2.5. If P, Q are closed, $P \stackrel{*}{\rightarrow} Q$ iff $\text{alg}(P) \stackrel{\rho}{\longrightarrow} \text{alg}(Q')$ for some run ρ of Π and some $Q' \equiv Q$.

$\Pi(h), \Pi$ still have a *Schönheitsfehler*: a solution may never be really hot, in the sense that no heating rule would be applicable—a replication can keep creating (spurious) replicants, even when some meaningful action is available. But these models can, once at hand, be forced to behave sensibly in many ways. A simple minded way would be to bound (and count) the number of unconsumed replicants (of each replication).

---

[15] This capability would be lost were $\Pi(h)$ defined just as $\pi \oplus_h$ *cool*. A construct like $\Pi(h)$, 'perpetual external choice', might be the right building block to define several other kinds of choice, such as $+$, $\oplus_h$ and constructs like Π—'π until choosing to *cool*'.

A reader familiar with [Milner 92] would probably by now find it a straightforward excercise to adapt our model to the *polyadic* calculus, refining the syntax of processes with *abstractions* and *concretions*, and decomposing our communication rule to *handshake* and *value-passing*—other evolving algebra rules might remain. The carefully designed rules of [Milner 92], regulating when and with what a restriction may commute in this context, would be automatically validated by the present restriction-heating rule, when applied to arbitrary abstractions and conretions.

References

[Abramsky 92] S.Abramsky, *Computational Interpretations of Linear Logic*, to appear in Theoretical Computer Science.

[Aczel 80] P.Aczel, *Frege Structures and the Notions of Proposition, Truth and Set*, in The Kleene Symposium, North-Holland 1980, pp. 31–59

[Beeson 85] M.J.Beeson, *Foundations of Constructive Mathematics*, Springer 1985.

[BerBou 92] G.Berry, G.Boudol, *The Chemical Abstract Machine*, Theoretical Computer Science 96 (1992), pp. 217–248 (an earlier version also in POPL'90).

[Boudol 87] G.Boudol, *Communication is an Abstraction*, INRIA Rapport de Recherche 636 (1987).

[BörRic 92] E.Börger, E.Riccobene, *A Mathematical Model of Concurrent Prolog*, CSTR-92-15, Dept. of Computer Science, University of Bristol, 1992 .

[BörRic 93] E.Börger, E.Riccobene, *A formal specification of Parlog*. In: M.Droste, Y.Gurevich (Eds.): Semantics of programming languages and model theory, 1993 (to appear).

[BörRos 92a] E.Börger, D.Rosenzweig, *The WAM—Definition and Compiler Correctness*, TR-14/92, Dipartimento di Informatica, Università di Pisa, June 1992.

[BörRos 92b] E.Börger, D.Rosenzweig, *A Simple Mathematical Model for Full Prolog*, TR-33/92, Dipartimento di Informatica, Università di Pisa, October 1992.

[Gurevich 88] Y.Gurevich, *Logic and the challenge of computer science.* in: E.Börger (Ed.), Trends in Theoretical Computer Science, Computer Science Press, Rockville MA 1988, pp.1-57.

[Gurevich 91] Y.Gurevich, *Evolving Algebras. A Tutorial Introduction.* in: Bulletin of the European Association for Theoretical Computer Science, no.43, February 1991, pp. 264-284

[GurHug 93] Y.Gurevich, J.Huggins, *The Semantics of the C Programming Language*, these Proceedings.

[GurMor 88] Y.Gurevich, J.Morris, *Algebraic operational semantics and Modula-2*, in CSL'87, 1st Workshop on Computer Science Logic, Springer LNCS 329 (1988), 81–101.

[GurMos 90] Y.Gurevich, L.S.Moss, *Algebraic Operational Semantics and Occam*, in CSL'89, Proceedings of 3rd Workshop on Computer Science Logic, Springer LNCS 440 (1990), pp. 176–196.

[Harel 84] D.Harel, *Dynamic Logic*, in D.M.Gabbay, F.Guenthner (Eds.): Handbook of Philosophical Logic 2, Reidel 1984, pp. 497–604.

[Milner 89] R.Milner, *Communication and Concurrency*, Prentice Hall 1989.

[Milner 92] R.Milner, *The Polyadixc π-Calculus: A Tutorial*, in Proceedings of the International Summer School on Logic and Algebra of Specification, Marktoberdorf 1991 (to appear).

[MiPaWa 91] R.Milner, J. Parrow, D.Walker, *Modal Logics for Mobile Processes*, in Proceedings of CONCUR'91, Springer LNCS 527 (1991), pp. 45-60.

[MiPaWa 92] R.Milner, J.Parrow, D.Walker, *A Calculus of Mobile Processes 1,2*, Information and Computation 100, 1(1992), pp. 1-77.

[TroDal 88] A.S.Troelstra, D. van Dalen, *Constructivism in Mathematics*, North-Holland 1988.

[Wirsing 90] M. Wirsing, *Algebraic Specification*, in J. van Leeuwen (Ed.): Handbook of Theoretical Computer Science B, Elsevier 1990, pp. 675-788.

On the Completeness of Narrowing as the Operational Semantics of Functional Logic Programming*

Juan Carlos González-Moreno, María Teresa Hortalá-González
Mario Rodríguez-Artalejo

Facultad de Matemáticas (UCM), Dpto. de Informática y Automática,
Ciudad Universitaria, Av. Complutense s/n, E-28040 Madrid, SPAIN

Abstract. This paper is a continuation of [10]. It presents soundness and completeness results for a higher-order (HO) functional logic language which has a domain-based declarative semantics and uses conditional narrowing (for applicative, constructor-based rewriting systems) as operational semantics. HO-unification is avoided by forbidding λ-abstractions in the language. However, narrowing must include a mechanism for binding HO logic variables to simple functional patterns built by partial application. A deeper investigation of lazy strategiees and infinite narrowing derivations is foreseen.

1 Introduction and related work

In spite of the amount of research devoted to the integration of functional and logic programming (see e.g. [6, 3, 1] for surveys), there are still some combination problems worth of investigation. More precisely, we are interested in the integration of partial and higher-order (HO) functions and predicates within a programming language which supports lazy evaluation, logic variables and goal solving.

Partial functions and lazy evaluation are provided by some first-order (FO) functional logic languages such as K-LEAF [8] and BABEL [12], both of which use Scott domains for their declarative semantics and narrowing for their operational semantics. K-LEAF has been extended to the HO language IDEAL [5, 4], whose semantics was specified in terms of a translation of IDEAL programs into K-LEAF programs. Another relevant HO logic language is λ-PROLOG [13], which works with HO Horn clauses and allows a very powerful use of typed λ-terms as data structures, supported by HO-unification. However, defining functions by rewrite rules is not possible in λ-PROLOG. More progress is still needed in defining HO functional logic languages which combine the different features found in these examples. The present paper is intended as a first step in this direction.

Taking [10] as a basis, we introduce a Simple Functional Logic Language (SFL for short) where programs are given as conditional, applicative rewriting systems (with a constructor discipline). A declarative semantics, based on applicative algebras over Scott domains, can be borrowed from [10]. As operational semantics, we specify conditional narrowing with two peculiarities: strict behaviour of equality when solving conditions, and a special mechanism for binding HO-variables. We can avoid HO-unification due to the absence of λ-abstractions in our syntax. Our main results are soundness and completeness of narrowing for computing finite solutions, w.r.t. the declarative semantics. Investigation of infinite narrowing derivations (in analogy to our previous results on infinite rewriting) and lazy strategies for narrowing are left for future research.

The organization of the paper is as follows. In section 2 we recall some algebraic notions from [10], which we need as a basis for our declarative semantics.

* Research supported by the national project TIC 89/0104 and the ESPRIT BR Working Group N. 6028 CCL

Sections 3, 4, and 5 develop the formal presentation of SFL together with its *declarative* and *operational semantics*. Section 6, the core of the paper, presents our soundness and completeness results. Finally, section 7 summarizes our conclusions. The reader is assumed to be familiar with the basic theory of logic programming [2] and term rewriting systems [7].

2 Algebraic preliminaries

Our approach to functional logic programming relies on some algebraic notions which were exposed in [10]. In this section we recall them and we also include some extensions needed for supporting *tuples*.

2.1 Signatures and expresions

A *signature with constructors* is any pair $\Sigma = (DC_\Sigma, FS_\Sigma)$ where $DC_\Sigma = \cup DC^n_\Sigma$ and $FS_\Sigma = \cup FS^n_\Sigma$ are mutually disjoint sets of *ranked constructors* and *function symbols*, respectively. We write *rank*(ϕ) for the rank of any symbol $\phi \in DC_\Sigma \cup FS_\Sigma$. Assuming a countably infinite set of *variables* X, Y, Z $\in$ Var, we define the set of *applicative expressions* e $\in$ Exp_Σ of signature Σ by the following syntax:

$$
\begin{array}{llll}
e ::= X & \% \text{ variable} & \mid c & \% \text{ constructor} \\
\quad\mid f & \% \text{ function symbol} & \mid (e_0\ e_1) & \% \text{ application} \\
\quad\mid \lt e_1, .., e_n \gt & \% \text{ n-tuple, } n \geq 0 & &
\end{array}
$$

As usual, we assume that application associates to the left and omit parenthesis accordingly. Note that this explains the notation (e_0 e_1 .. e_m), which is often used with $e_0 \in Var \cup DC_\Sigma \cup FS_\Sigma$. We distinguish three relevant kinds of expressions, namely:

Patterns, s,t $\in Ptr_\Sigma$:

$$
\begin{array}{ll}
t ::= X & \% X \in Var \\
\quad\mid (c\ t_1 .. t_m) & \% c \in DC^n_\Sigma,\ 0 \leq m \leq n \\
\quad\mid (f\ t_1 .. t_m) & \% f \in FS^n_\Sigma,\ 0 \leq m < n \\
\quad\mid \lt t_1, .., t_n \gt & \% n \geq 0
\end{array}
$$

FO expressions[2], e $\in FOExp_\Sigma$:

$$
\begin{array}{ll}
e ::= X & \% X \in Var \\
\quad\mid (c\ e_1 .. e_n) & \% c \in DC^n_\Sigma \\
\quad\mid (f\ e_1 .. e_n) & \% f \in FS^n_\Sigma \\
\quad\mid \lt e_1, .., e_n \gt & \% n \geq 0
\end{array}
$$

First-order patterns, s,t $\in FOPtr_\Sigma$ where $FOPtr_\Sigma =_{def} Ptr_\Sigma \cap FOExp_\Sigma$.

In the sequel, *var(e)* stands for the set of all variables occurring in expression e.

2.2 Continuous applicative algebras and expression evaluation

Definition 1 (Cartesian Scott domains). We say that a Scott domain [14] D with ordering $\sqsubseteq$ and bottom $\perp$ is *cartesian* iff D is closed under the construction of arbitrary finite tuples and the ordering $\sqsubseteq$ satisfies the following two conditions:

- If $x \sqsubseteq y$ then $x = \lt x_1, .., x_n \gt$ for some $x_1, .., x_n \in D$ iff $y = \lt y_1, .., y_n \gt$ for some $y_1, .., y_n \in D$.
- $\lt x_1, .., x_n \gt \sqsubseteq \lt y_1, .., y_n \gt$ iff $x_i \sqsubseteq y_i$, $1 \leq i \leq n$

Definition 2 (C.A.A.s). A *cartesian continuous applicative algebra* (shortly, C.A.A.) of signature Σ is any algebraic structure $\mathcal{A}$ consisting of the following items:

- a cartesian Scott domain $D_\mathcal{A}$ with ordering $\sqsubseteq_\mathcal{A}$ and bottom $\perp_\mathcal{A}$.

[3] This terminology turns out to be convenient for exposition. However, the reader is warned that our first order expressions may sometimes denote functions.

- a continuous *appplication operation* $\circ_{\mathcal{A}} \in [D_{\mathcal{A}} \times D_{\mathcal{A}} \rightarrow D_{\mathcal{A}}]$ which must be strict w.r.t. its first argument.
- *interpretations* $\phi_{\mathcal{A}} \in D_{\mathcal{A}}$ for all symbols $\phi \in DC_{\Sigma} \cup FS_{\Sigma}$.

For a given C.A.A. $\mathcal{A}$ we can define the Scott domains $Fun^n(D_{\mathcal{A}})$ of *continuous functions* over $D_{\mathcal{A}}$ recursively:

- $Fun^0(D_{\mathcal{A}}) = D_{\mathcal{A}}$
- $Fun^{n+1}(D_{\mathcal{A}}) = [D_{\mathcal{A}} \rightarrow Fun^n(D_{\mathcal{A}})]$

Any $x \in D_{\mathcal{A}}$ induces functions $\{ x \}^n_{\mathcal{A}} \in Fun^n(D_{\mathcal{A}})$ which are defined as follows:

- $\{ x \}^0_{\mathcal{A}} = x$,
- $\{ x \}^{n+1}_{\mathcal{A}}(y) = \{ x \circ_{\mathcal{A}} y \}^n_{\mathcal{A}}$.

Here and in the sequel we use infix notation for the application operation. For $\phi \in DC^n_{\Sigma} \cup FS^n_{\Sigma}$, we abbreviate $\{\phi_{\mathcal{A}}\}^n_{\mathcal{A}}$ as $\phi^n_{\mathcal{A}}$. A *valuation* over a $\mathcal{A}$ is any mapping η: Var $\rightarrow D_{\mathcal{A}}$. The *evaluation* of $e \in Exp_{\Sigma}$ in $\mathcal{A}$ under η yields a value $[\![e]\!]^{\mathcal{A}}\eta \in D_{\mathcal{A}}$ which is defined recursively:

$$[\![X]\!]^{\mathcal{A}}\eta = \eta\,(X), \text{ for } X \in Var, \quad [\![<e_1, .., e_n>]\!]^{\mathcal{A}}\eta = < [\![e_1]\!]^{\mathcal{A}}\eta, .., [\![e_n]\!]^{\mathcal{A}}\eta >$$
$$[\![\phi]\!]^{\mathcal{A}}\eta = \phi_{\mathcal{A}}, \text{ for } \phi \in DC_{\Sigma} \cup FS_{\Sigma}, \quad [\![e_0\ e_1]\!]^{\mathcal{A}}\eta = [\![e_0]\!]^{\mathcal{A}}\eta \circ_{\mathcal{A}} [\![e_1]\!]^{\mathcal{A}}\eta$$

For *ground expressions* e without variables, $[\![e]\!]^{\mathcal{A}}\eta$ does not depend on η and can be written simply as $[\![e]\!]^{\mathcal{A}}$.

Expression evaluation allows to define the meaning of *equations*. We need to make a distinction between *nonstrict equations* "$e_1 = e_2$" and *strict equations* "$e_1 == e_2$", whose satisfaction in a given $\mathcal{A}$ is defined as follows:

- $\mathcal{A} \models (e_1 = e_2)\eta$ iff $[\![e_1]\!]^{\mathcal{A}}\eta = [\![e_2]\!]^{\mathcal{A}}\eta$
- $\mathcal{A} \models (e_1 == e_2)\eta$ iff $[\![e_1]\!]^{\mathcal{A}}\eta = [\![e_2]\!]^{\mathcal{A}}\eta = x$ *finite and maximal* w.r.t. $\sqsubseteq_{\mathcal{A}}$.

In the sequel we say that a finite and maximal element is *totally defined*[4]. We also use the following notations:

- **fin** x to mean "x is finite".
- **def** x to mean "x is totally defined".

2.3 Special C.A.A.s

Some particular kinds of C.A.A.s play an important role in our framework.

Definition 3 (Liberal C.A.A.s). We say that $\mathcal{A}$ is *liberal* iff it satisfies the following axioms:

- $\forall X_1 .. X_m$ (**def** $X_1 \wedge .. \wedge$ **def** $X_m \Rightarrow$ **def** $(c\ X_1 .. X_m)$) for all $c \in DC^n_{\Sigma}$, $0 \le m \le n$.
- $\forall X_1 .. X_m$ (**def** $X_1 \wedge .. \wedge$ **def** $X_m \Rightarrow$ **def** $(f\ X_1 .. X_m)$) for all $f \in FS^n_{\Sigma}$, $0 \le m < n$.
- $\forall X_1 .. X_n$ (**def** $X_1 \wedge .. \wedge$ **def** $X_n \Rightarrow$ **def** $< X_1, .., X_n >$) for all $n \ge 0$.

Definition 4 (Minimally Free C.A.A.s). $\mathcal{A}$ is called *minimally free* iff it verifies the following:

- $\forall X_1 .. X_m\ Y_1 .. Y_m\ ((c\ X_1 .. X_m) = (c\ Y_1 .. Y_m) \Rightarrow X_1 = Y_1 \wedge .. \wedge X_m = Y_m)$ for all $c \in DC^n_{\Sigma}$, $1 \le m \le n$.
- $\forall X_1 .. X_k\ Y_1 .. Y_l\ ((c\ X_1 .. X_k) \ne (d\ Y_1 .. Y_l))$ for all $c \in DC^n_{\Sigma}$, $d \in DC^m_{\Sigma}$, $0 \le k \le n$, $0 \le l \le m$.
- $\forall X_1 .. X_k\ Y_1 .. Y_l\ ((c\ X_1 .. X_k) \ne < Y_1, .., Y_l >)$ for all $c \in DC^n_{\Sigma}$, $0 \le k \le n$, $0 \le l$.
- $\forall X_1 .. X_n$ (**fin** $(c\ X_1 .. X_n) \Rightarrow$ **fin** $X_1 \wedge .. \wedge$ **fin** X_n) for all $c \in DC^n_{\Sigma}$, $n \ge 1$.
- $\forall X_1 .. X_n$ (**fin** $X_i \Rightarrow X_i \ne t$) for every $t \in FOPtr_{\Sigma}$ such that $var(t) = \{ X_1, .., X_n \}$ and $X_i \not\equiv t$, $1 \le i \le n$.

The axioms of minimally free algebras reflect the essential properties of free constructors, while those of liberal algebras have been chosen so that the following result holds trivially:

[4] *totally defined* elements were called *finished* in our previous paper [10]

Proposition 5. *Let $t \in Ptr_\Sigma$. Then $[\![t]\!]^{\mathcal{A}}\eta$ is totally defined, provided that $\mathcal{A}$ is liberal and $\eta(X)$ is totally defined for every variable X which occurs in t.* □

Next, we define Herbrand domains and Herbrand algebras. Let $V \subseteq Var$ be any set of variables. Let $\Sigma_\perp = \Sigma \dot{\cup} \{ \perp \}$, where the new nullary constructor $\perp$ is intended to denote an *undefined* value. The *Herbrand domain* $H_\Sigma(V)$ is defined as the CPO-completion of the set of all patterns $t \in Ptr_{\Sigma_\perp}$ with $var(t) \subseteq V$, partially ordered by the least ordering $\sqsubseteq$ that satisfies:

- $\perp \sqsubseteq t$ for all t.
- $s_i \sqsubseteq t_i \ (1 \le i \le m) \Longrightarrow \begin{cases} (c\ s_1\ ..\ s_m) \sqsubseteq (c\ t_1\ ..\ t_m) \text{ for all } c \in DC^n_\Sigma,\ 0 \le m \le n \\ (f\ s_1\ ..\ s_m) \sqsubseteq (f\ t_1\ ..\ t_m) \text{ for all } f \in FS^n_\Sigma,\ 0 \le m < n \\ < s_1, .., s_m > \sqsubseteq < t_1, .., t_m > \end{cases}$

Elements of $H_\Sigma(V)$ can be represented as trees, similarly as in the construction of free continuous algebras [9]. It is easy to check that $H_\Sigma(V)$ is a Scott domain. By abuse of notation and terminology, we assume that $Ptr_\Sigma(V) = \{ t \in Ptr_\Sigma \mid var(t) \subseteq V \} \subseteq H_\Sigma(V)$ and we use the name *partial patterns* for elements of $H_\Sigma(V)$.

Definition 6 (Herbrand C.A.A.s). [5] $\mathcal{A}$ is called a *Herbrand algebra* iff $D_\mathcal{A}$ is a Herbrand domain and the following conditions are satisfied:

- $c_\mathcal{A} \equiv c$ for all $c \in DC_\Sigma$; $f_\mathcal{A} \equiv f$ for all $f \in FS^n_\Sigma$, $n > 0$,
- $t_0 \circ_\mathcal{A} t_1 = (t_0\ t_1)$, whenever $(t_0\ t_1)$ is a partial pattern; $t_0 \circ_\mathcal{A} t_1 = \perp$, whenever $(t_0\ t_1)$ is neither a partial pattern nor of the form $(f\ s_1\ ..\ s_n)$, $f \in FS^n_\Sigma$, $n > 0$.

Note that the previous definition leaves free only the following options for the definition of a Herbrand algebra: $f_\mathcal{A}$, for $f \in FS^0_\Sigma$, $f \circ_\mathcal{A} t$, for $f \in FS^1_\Sigma$, and $(f\ t_1\ ..\ t_{n-1}) \circ_\mathcal{A} t$, for $f \in FS^n_\Sigma$ and $n > 1$. In terms of the notation given in subsection 2.2, these values can be written as $f^0_\mathcal{A}$, $(f^1_\mathcal{A}\ t)$ and $(f^n_\mathcal{A}\ t_1\ ..\ t_{n-1}\ t)$, respectively. The next result follows easily from this observation and our definitions.

Proposition 7. *Let $H = H_\Sigma(V)$ for some fixed Σ and $V \subseteq Var$. Then:*

1. *Every Herbrand algebra $\mathcal{A}$ over H is univocally determined by the family of continuous functions $\{ f^n_\mathcal{A} \mid f \in FS^n_\Sigma, n \in \mathbb{N}\}$*
2. *The collection HALG of all the Herbrand algebras over H is a Scott domain under the ordering $\sqsubseteq_{HALG}$ defined by $\mathcal{A} \sqsubseteq_{HALG} \mathcal{B}$ iff $f^n_\mathcal{A} \sqsubseteq f^n_\mathcal{B}$ for all $f \in FS^n_\Sigma$.*
3. *Herbrand algebras are liberal and minimally free.* □

3 SFL-programs

In this Section we introduce the language of *S*imple *F*unctional *L*ogic (shortly, SFL) programs as an extension of SFL as presented in [10]. Now, extra variables in conditions will be allowed. A *defining rule* for $f \in FS^n_\Sigma$ is any conditional equation:

$$\underbrace{f\ t_1\ ..\ t_n}_{lhs} = \underbrace{r}_{rhs} \Leftarrow \underbrace{C}_{condition}$$

where the lhs is *linear* (without multiple ocurrences of variables), t_i are FO *patterns*, the rhs r is any expression such that $var(r) \subseteq var(f\ t_1\ ..\ t_n)$, and the *condition* C is any system of strict equations between expressions: $C \equiv l_1 == r_1, .., l_m == r_m$, where C may include some *extra variables* which do not occur in $(f\ t_1\ ..\ t_n)$. These have an existential reading, as in bodies of Horn clauses.

[5] Note that, according to this definition, each variable $X \in V$ behaves as a constant constructor in Herbrand algebras with domain $H_\Sigma(V)$.

Definition 8 (SFL programs). A SFL-program is any set $\mathcal{R}$ of defining rules which satisfies the following *nonambiguity condition*: given any two rules for the same f: R_1: f t_1 .. t_n = $r_1 \Leftarrow C_1$, R_2: f s_1 .. s_n = $r_2 \Leftarrow C_2$ (renamed apart to share no variables) there are no minimally free algebra $\mathcal{A}$ and valuation η which satisfy: $\mathcal{A} \models (t_i = s_i)\eta$ for all $1 \le i \le n$, $\mathcal{A} \models (C_1, C_2)\eta$, and $[\![r_1]\!]^{\mathcal{A}}\eta \neq [\![r_2]\!]^{\mathcal{A}}\eta$

The nonambiguity condition plays in our framework a role similar to the *confluence hypothesis* which are usually needed for proving completeness of narrowing (cfr. e. g. [11]). It can be proved that nonambiguity of a given set of SFL rules is undecidable. However, in [10] we presented a decidable condition which implies nonambiguity and is pragmatically satisfactory. It can be straightforwardly extended to deal with tuples, thus covering the present formulation of our language.

To close this section we present a small example of SFL-program. The symbols for which no defining rule is given are intended as constructors. List constructors are written in PROLOG notation.

Example 1.

```
plus 0 Y = Y.                          map F [ ] = [ ].
plus (suc X) Y = suc (plus X Y).       map F [X | Xs] = [F X | map F Xs].
double X = plus X X.                   composition G F X = G (F X).
solution F = true ⇐ (map F [(suc 0), 0, (suc 0)]) == [(suc³ 0), (suc 0), (suc³ 0)].
```

4 Declarative semantics

The *declarative semantics* of SFL is in the spirit of logic programming. A given program $\mathcal{R}$ is viewed as a set of logical assertions, to be interpreted in CAAs. This gives rise to *models*, in particular *least Herbrand models*, of programs. In our present development we can closely follow [10], with some minor modifications for taking extra variables into account. The reader should note that our approach does not correspond to the *denotational semantics* usually adopted for HO functional languages. A comparison of both approaches, as well as results that may justify the choice of the declarative semantics, can be also found in [10].

Definition 9 (Models). Let $\mathcal{R}$ be a SFL-program. The notion of *model* for $\mathcal{R}$ is exactly the same as in [10]:

- $\mathcal{A}$ is a *model* (resp. *exact model*) of $\mathcal{R}$ iff $\mathcal{A}$ satisfies (resp. exactly satisfies) all the rules in $\mathcal{R}$.
- $\mathcal{A}$ *satisfies* (resp. *exactly satisfies*) a rule $l = r \Leftarrow C$ iff every valuation η such that $\mathcal{A} \models C\eta$ verifies $[\![l]\!]^{\mathcal{A}}\eta \sqsupseteq_{\mathcal{A}} [\![r]\!]^{\mathcal{A}}\eta$ (resp. $[\![l]\!]^{\mathcal{A}}\eta = [\![r]\!]^{\mathcal{A}}\eta$).

We use the notations $\mathcal{A} \models \mathcal{R}$, $\mathcal{A} \models_= \mathcal{R}$ for models and exact models, respectively.

4.1 Least Herbrand models

Let H be some Herbrand domain (i.e. $H = H_\Sigma(V)$ for some $V \subseteq Var$). From Proposition 7, we have a Scott domain HALG consisting of all Herbrand algebras, and we know that any particular Herbrand algebra $\mathcal{A}$ is determined by the functions $f^n_{\mathcal{A}}$. Assuming a given SFL-program $\mathcal{R}$, we are going to define an operator $\mathcal{T}_{\mathcal{R}}$: HALG $\rightarrow$ HALG analogous to the "immediate consequences" operator known from Horn clause logic programming.

Definition 10 (Operator $\mathcal{T_R}$). For a given $\mathcal{A} \in$ HALG, we define $\mathcal{T_R}(\mathcal{A}) = \mathcal{B} \in$ HALG such that for all f $\in FS^n_\Sigma$ and all $s_1, .., s_n \in$ H;

$f^n_{\mathcal{B}}\ s_1\ ..\ s_n = \bigsqcup R_{\mathcal{A}}(f\ s_1\ ..\ s_n)$ where:

$$R_{\mathcal{A}}(f\ s_1\ ..\ s_n) =_{def} \{\ s \in H \mid \text{there are a rule } f\ t_1\ ..\ t_n = r \Leftarrow C \text{ in } \mathcal{R}, \text{ and a valuation } \eta\text{: Var} \rightarrow H \text{ such that } [\![\ t_i\ ]\!]^{\mathcal{A}}\eta = s_i\ (1 \leq i \leq n),\ [\![\ r\ ]\!]^{\mathcal{A}}\eta = s \text{ and } \mathcal{A} \models C\eta\ \}$$

Note that the set $R_{\mathcal{A}}(f\ s_1\ ..\ s_n)$ used in the previous definition is either empty or a singleton, since $\mathcal{R}$ is nonambiguous and Herbrand algebras are minimally free by Proposition 7. Moreover, it can be checked that the functions $f^n_{\mathcal{B}}$ obtained according to Definition 10 are continuous. Hence, $\mathcal{T_R}$ is well defined. Moreover:

Theorem 11 (Least Herbrand models). *The operator $\mathcal{T_R}$ is continuous and satisfies the following properties:*

1. *For any $\mathcal{A} \in HALG_{\mathcal{A}} : \mathcal{A} \models \mathcal{R}$ iff $\mathcal{T_R}(\mathcal{A}) \sqsubseteq_{HALG} \mathcal{A}$.*
2. *$\mathcal{T_R}$ has a least fixpoint $\mathcal{H_R} = \bigsqcup_k \mathcal{H}^k_{\mathcal{R}}$, where $\mathcal{H}^0_{\mathcal{R}} = \bot_{HALG}$ and $\mathcal{H}^{k+1}_{\mathcal{R}} = \mathcal{T_R}(\mathcal{H}^k_{\mathcal{R}})$.*
3. *$\mathcal{H_R} \models_{=} \mathcal{R}$ and $\mathcal{H_R} \sqsubseteq_{HALG} \mathcal{A}$ for every $\mathcal{A} \in HALG$ such that $\mathcal{A} \models \mathcal{R}$ (i.e. $\mathcal{H_R}$ is the least model of $\mathcal{R}$ within HALG).*

Proof Idea. We can argue as in the proof of Theorem 4.1 in [10], since extra variables in the conditions of rules introduce no additional complications. □

Our notion of Herbrand model in the present paper is more general than in [10], because now we allow for Herbrand domains with variables. In the particular case V = $\emptyset \subseteq$ Var, we obtain the *ground Herbrand domain* H = $H_\Sigma(\emptyset)$, which gives rise to *ground Herbrand models*. Again, extra variables in conditions place no additional difficulty, and we are able to obtain the following result by borrowing the proof of Theorem 4.2 from [10]:

Theorem 12. *The least ground Herbrand model $\mathcal{H_R}$ of any SFL-program $\mathcal{R}$ is an initial object in the category $LIBMOD_{\mathcal{R}}$[6] of liberal models of $\mathcal{R}$.* □

5 Narrowing semantics

5.1 Motivation

We are going to use *conditional narrowing* as the operational semantics of SFL. This means that each defining rule l =r ⇐ C will be used operationally as a conditional rewrite rule l → r ⇐ C ↓. Here, the notation "C ↓" indicates that the rule will be applied for rewriting l to r only if the condition C has been solved. We first discuss the main ideas about how to solve a simple condition e == e', before giving a formal definition of narrowing. To solve e == e', we shall require that the pair < e, e' > can be narrowed to a new pair < s, t > consisting of two unifiable *patterns*. Because of our semantics for strict equality, to require that s, t were unifiable *expressions* (instead of patterns), as it is done in most approaches to narrowing, would be not sufficient to guarantee correctness. For the same reason, we are forced to restrict the variables involved in the unification of s and t to take totally defined values. A simple example may clarify this. Given the defining rule:
f (suc X) = true ⇐ X == Y, we can narrow the expression (f Z) by binding Z to (suc X) and obtain true, but since we have solved the condition X == Y, the variable X is restricted to take totally defined values.

[6] The definition of $LIBMOD_{\mathcal{R}}$ can be found in [10]

5.2 Formal specification of narrowing

For our formal definition we need some concepts and notations. Given a finite substitution $\sigma = \{ X_1/e_1, .., X_r/e_r \}$ of expressions for variables (where we assume $X_i \not\equiv e_i$), the *domain* and the *range* of σ are defined as:

• $dom(\sigma) = \{X_1, .., X_r\}$ • $ran(\sigma) = var(e_1) \cup .. \cup var(e_r)$

The unification of a rule lhs $(f\ t_1\ ..\ t_n)$ with some expression of the form $(f\ e_1\ ..\ e_n)$ can be treated as a first-order unification problem. Following [2], we assume a unification algorithm which computes ***idempotent most general unifiers*** (m.g.u.'s) σ with the following properties:

- $\theta = \sigma\theta$ for every other unifier θ (where $\sigma\theta$ indicates the composition of σ followed by θ)
- $dom(\sigma) \cup ran(\sigma)$ is included in the set of variables of the expressions that have been unified.

The most general unifiers with these two special properties will be called *canonical* in the sequel. Due to left linearity of lhs of rules, we can easily prove the following:

Lemma 13. *If σ is a canonical m.g.u of a rule lhs $l \equiv (f\ t_1\ ..\ t_n)$ and an expression $e \equiv (f\ e_1\ ..\ e_n)$ sharing no variables with l, then $X\sigma$ is a subpattern of some t_i for every variable $X \in var(e) \cap dom(\sigma)$.* □

Some other notations that we need are as follows:

$\mathcal{C}[e]$: expression e in *context* $\mathcal{C}$; i.e., some expression with a distinguished occurrence of e as a subexpression.

Δ: finite set of variables, constrained to take totally defined values.

<C>: the expression $<< l_1, .., l_m>, <r_1, .., r_m>>$, built from a given condition $C \equiv l_1 == r_1, .., l_m == r_m$.

Our formal definition of conditional narrowing is given by a calculus whose rules specify three mutually dependent relations:

$e \leadsto_{\sigma,\Delta} e'$ "e can be narrowed to e' in one top-level step[7], via σ and Δ".

$e \leadsto^*_{\sigma,\Delta} e'$ "e can be narrowed to e' in finitely many steps, via σ and Δ".

$C \downarrow_{\sigma,\Delta}$ "C can be solved by narrowing in finitely many steps, via σ and Δ".

When using these notations, σ and/or Δ can be omitted if they are empty. Assuming a given SFL-program $\mathcal{R}$, the rules of our narrowing calculus are as follows:

(N1) $$\frac{(C\sigma) \downarrow_{\sigma',\Delta'}}{\mathcal{C}[f\ e_1\ ..\ e_n] \leadsto_{\sigma\sigma',\Delta'} \mathcal{C}[r]\ \sigma\sigma'}$$ if $f\ t_1\ ..\ t_n = r \Leftarrow C$ is a variant of a rule in $\mathcal{R}$, with *fresh variables* and σ is a canonical m.g.u. of $(f\ t_1\ ..\ t_n)$ and $(f\ e_1\ ..\ e_n)$.

(N2) $\mathcal{C}[X\ e] \leadsto_{\{ X / t \}} \mathcal{C}[X\ e]\ \{ X / t \}$ if t is some pattern of the form $(c\ X_1\ ..\ X_m)$ or $(f\ X_1\ ..\ X_m)$, with *fresh variables* X_i.

(N*1) $e \leadsto^* e$

(N*2) $$\frac{e \leadsto_{\sigma,\Delta} e' \quad e' \leadsto^*_{\sigma',\Delta'} e''}{e \leadsto^*_{\sigma\sigma',\Delta''} e''}$$ where $\Delta'' = (\Delta \backslash dom(\sigma')) \cup var(\Delta\sigma') \cup \Delta'$.

(S) $$\frac{< C > \leadsto^*_{\sigma,\Delta} <s, t>}{C \downarrow_{\sigma\sigma',\Delta'}}$$ if s, t are patterns, σ' is a canonical m.g.u. such that $s\sigma' \equiv t\sigma' \equiv u$, and $\Delta' = (\Delta \backslash dom(\sigma')) \cup var(u)$.

[7] We say "top-level step" because it may involve a conditional rule whose condition has to be solved by a narrowing subcomputation "at a lower level".

In rules (N1) and (N2), "fresh variable" means a variable which has not been used so far in the narrowing derivation. In particular, it will not occur in the expression to be narrowed.

Example 2. According to our narrowing calculus, one of the possible narrowing derivations for the program in the example of section 3 is as follows:

$$\text{solution F} \leadsto^*_{\{\text{F / (composition suc double)}\}} \text{true}$$

5.3 Computations and solutions

The derivability of assertions $e \leadsto^*_{\sigma,\Delta} e'$ resp. $C \downarrow_{\sigma,\Delta}$ in our narrowing calculus means that we can build some *narrowing computation* which *narrows* e to e' resp. *solves* C. In both cases, we say that σ and Δ are the *computed answer substitution* and the *computed restriction*, respectively. In the first case, we also speak of the *computed result* $|\ e'\ |$, which must be understood according to the following definition:

Definition 14 (Shell). The *shell* of an expression e is a partial pattern $|\ e\ |$ which represents the outer part of e already evaluated, and is defined recursively: $|\ X\ | = X$ for $X \in Var$; $|\ \varphi\ | = \varphi$ for $\varphi \in DC_\Sigma \cup FS^n_\Sigma$ with $n > 0$; $|\ (\varphi\ e_1\ ..\ e_k)\ | = (\varphi\ |\ e_1\ |\ ..\ |\ e_k\ |)$ for $\varphi \in DC^n_\Sigma \cup FS^m_\Sigma$, $1 \le k \le n$, $1 \le k < m$; $|\ <e_1, .., e_n>\ | = <|\ e_1\ |, .., |\ e_n\ |>$ for $n \ge 0$; $|\ e\ | = \bot$, otherwise

A very useful property of finite narrowing computations is that it is legitimate to reason by induction on their structure (or, in other words, by induction on the application of the rules of the narrowing calculus). This technique will be exploited in the next section.

6 Soundness and completeness results

In this section we shall prove that narrowing is sound and complete w.r.t. declarative semantics in the following sense: computed solutions are valid in all liberal models of the program (*soundness*); and every solution which is valid in all the liberal models of the program is subsumed by some computed solution (*completeness*).

6.1 Preliminaries

For the rest of the paper, we assume some given SFL-program $\mathcal{R}$ of signature Σ. H and $\mathcal{H}$ will denote the Herbrand universe $H_\Sigma(Var)$ and the least Herbrand model of $\mathcal{R}$ over H, respectively. The ordering of H will be noted as $\sqsubseteq$. Because of theorem 11, we know that $\mathcal{H} = \bigsqcup_k \mathcal{H}^k$. Moreover, we shall also write $\circ^k$ for the application operation of $\mathcal{H}^k$, and we shall abbreviate $[\![\ ]\!]^{\mathcal{H}^k}$ as $[\![\ ]\!]^k$.

Lemma 15. *If* $e \leadsto^*_{\sigma,\Delta} e'$ *then* $var(e') \subseteq var(e\sigma)$ *and* $|\ e\sigma\ | \sqsubseteq |\ e'\ |$.

Proof. Straightforward induction on the structure of computations. □

Lemma 16 (Substitution Lemma). *Let $\mathcal{A}$ be a C.A.A., η: $Var \to D_\mathcal{A}$ a valuation, and σ a substitution. Let $(\sigma\eta)^\mathcal{A}$ be the valuation η': $Var \to D_\mathcal{A}$ defined by $\eta'(X) = [\![X\sigma]\!]^\mathcal{A}\eta$. Then:*

1. $[\![e\sigma]\!]^\mathcal{A}\eta = [\![e]\!]^\mathcal{A}(\sigma\eta)^\mathcal{A}$ *for every expression e.*
2. $\mathcal{A} \models (C\sigma)\eta \Leftrightarrow \mathcal{A} \models C(\sigma\eta)^\mathcal{A}$ *for every condition C.*

Proof. 1) is easily checked by induction on e, and 2) follows immediately. □

Lemma 17 (Continuity Lemma). *Given a valuation η: $Var \rightarrow H$, we say that η_0 is a* finite approximation *of η (in symbols, $\eta_0 \sqsubseteq_{\mathrm{fin}} \eta$) iff for all $X \in Var$ $\eta_0(X)$ is finite and such that $\eta_0(X) \sqsubseteq \eta(X)$.*

1. $[\![e]\!]^{\mathcal{H}}\eta \sqsupseteq s$ *and* **fin** $s \Longrightarrow$ *for some* $k \in \mathbb{N}$ *and some* $\eta_0 \sqsubseteq_{\mathrm{fin}} \eta$: $[\![e]\!]^k\eta_0 \sqsupseteq s$.
2. $\mathcal{H} \models C\eta \Longrightarrow$ *for some* $k \in \mathbb{N}$ *and some* $\eta_0 \sqsubseteq_{\mathrm{fin}} \eta$: $\mathcal{H}^k \models C\eta_0$.
3. *For every fixed* $k \in \mathbb{N}$: $[\![e]\!]^k\eta \sqsupseteq s$ *and* **fin** $s \Longrightarrow$ *for some* $\eta_0 \sqsubseteq_{\mathrm{fin}} \eta$: $[\![e]\!]^k\eta_0 \sqsupseteq s$.
4. *For every fixed* $k \in \mathbb{N}$: $\mathcal{H}^k \models C\eta \Longrightarrow$ *for some* $\eta_0 \sqsubseteq_{\mathrm{fin}} \eta$: $\mathcal{H}^k \models C\eta_0$

Proof. 1) can be proved by a continuity argument, using the fact that $\mathcal{H} = \bigsqcup_k \mathcal{H}^k$. 2) follows easily from 1) Parts 3) and 4) are proved by straightforward continuity arguments within $\mathcal{H}^k$. □

Definition 18. Consider a set $\Delta \subseteq$ Var of restricted variables and a valuation η: Var $\rightarrow \mathcal{A}$. We say that η *satisfies* Δ (in symbols: η **sat** Δ) iff **def** η(X) for all X $\in \Delta$.

6.2 Soundness

Theorem 19 (Soundness of conditional narrowing). *Computed solutions are correct in the following sense: for any liberal model $\mathcal{A}$ of $\mathcal{R}$ we have:*

(N) $e \leadsto^*_{\sigma,\Delta} e' \Rightarrow$ *for all η such that η* **sat** Δ: $[\![e\sigma]\!]^{\mathcal{A}}\eta \sqsupseteq_{\mathcal{A}} [\![e']\!]^{\mathcal{A}}\eta$
Moreover: exact equality holds if $\mathcal{A}$ is an exact model.

(S) $C \downarrow_{\sigma,\Delta} \Rightarrow$ *for all η such that η* **sat** Δ: $\mathcal{A} \models (C\sigma)\eta$

Proof. (N) and (S) are proved simultaneously by induction on the structure of computations. We must consider one case for each rule of the narrowing calculus.

(N1): By induction hypothesis for (S), applied to the premise of (N1), we can assume that $\mathcal{A} \models (C\sigma\sigma')\eta$, that is (by lemma 16 $\mathcal{A} \models$ C $(\sigma\sigma'\eta)^{\mathcal{A}}$. Since $\mathcal{A} \models \mathcal{R}$, we get that $[\![\mathrm{f}\ \mathrm{t}_1 \ldots \mathrm{t}_n]\!]^{\mathcal{A}}\ (\sigma\sigma'\eta)^{\mathcal{A}} \sqsupseteq_{\mathcal{A}} [\![\mathrm{r}]\!]^{\mathcal{A}}\ (\sigma\sigma'\eta)^{\mathcal{A}}$ (or even exact equality if $\mathcal{A} \models_= \mathcal{R}$) and we can reason as follows:

$$
\begin{aligned}
[\![(\mathrm{f}\ \mathrm{e}_1 \ldots \mathrm{e}_n)\ \sigma\sigma']\!]^{\mathcal{A}}\eta &= [\![(\mathrm{f}\ \mathrm{e}_1 \ldots \mathrm{e}_n)\ \sigma]\!]^{\mathcal{A}}(\sigma'\eta)^{\mathcal{A}} && \text{\% by lemma 16}\\
&= [\![(\mathrm{f}\ \mathrm{t}_1 \ldots \mathrm{t}_n)\ \sigma]\!]^{\mathcal{A}}(\sigma'\eta)^{\mathcal{A}} && \text{\% because } \sigma \text{ is m.g.u.}\\
&= [\![\mathrm{f}\ \mathrm{t}_1 \ldots \mathrm{t}_n]\!]^{\mathcal{A}}(\sigma\sigma'\eta)^{\mathcal{A}} && \text{\% by lemma 16}\\
&\sqsupseteq_{\mathcal{A}} [\![\mathrm{r}]\!]^{\mathcal{A}}(\sigma\sigma'\eta)^{\mathcal{A}} && \text{\% as explained above}\\
&= [\![\mathrm{r}\sigma\sigma']\!]^{\mathcal{A}}\eta && \text{\% by lemma 16}
\end{aligned}
$$

Then we also have: $[\![\mathcal{C}[\mathrm{f}\ \mathrm{e}_1 \ldots \mathrm{e}_n]\ \sigma\sigma']\!]^{\mathcal{A}}\eta \sqsupseteq [\![\mathcal{C}[\mathrm{r}]\sigma\sigma']\!]^{\mathcal{A}}\eta$, and we obviously obtain exact equality if $\mathcal{A} \models_= \mathcal{R}$.

(N2): Trivial, since in this case e$\sigma \equiv$ e'.

(N*1): Trivial, since e $\equiv$ e', $\sigma = \{\ \}$, $\Delta = \{\ \}$.

(N*2): Let us look at the computation which occurs as conclusion of rule (N*2). We have to prove (N) for it, and we may assume as induction hypothesis that (N) already holds for the computations of premises.

Assume that η **sat** $\Delta'' = (\Delta \setminus \mathrm{dom}(\sigma')) \cup \mathrm{var}(\Delta\sigma') \cup \Delta'$. We claim that $(\sigma'\eta)^{\mathcal{A}}$ **sat** Δ; this can be proved by using lemma 13 and proposition 7. Also, η **sat** Δ' because $\Delta' \subseteq \Delta''$. Then we get:

$$
\begin{aligned}
[\![\mathrm{e}\ \sigma\sigma']\!]^{\mathcal{A}}\eta &= [\![\mathrm{e}\ \sigma]\!]^{\mathcal{A}}(\sigma'\eta)^{\mathcal{A}} && \text{\% by lemma 16}\\
&\sqsupseteq_{\mathcal{A}} [\![\mathrm{e'}]\!]^{\mathcal{A}}(\sigma'\eta)^{\mathcal{A}} && \text{\% by induction hypothesis since } (\sigma'\eta)^{\mathcal{A}} \textbf{ sat } \Delta\\
&= [\![\mathrm{e'}\sigma']\!]^{\mathcal{A}}\eta && \text{\% by lemma 16}\\
&\sqsupseteq_{\mathcal{A}} [\![\mathrm{e''}]\!]^{\mathcal{A}}\eta && \text{\% by induction hypothesis since } \eta \textbf{ sat } \Delta'
\end{aligned}
$$

(S): Now we must prove (S) for the conclusion of rule (S) from the narrowing calculus, and we may assume as induction hypothesis that (N) holds for the premise. Assume that η **sat** $\Delta' = (\Delta \setminus \mathrm{dom}(\sigma')) \cup \mathrm{var}(u)$. For each $X \in \mathrm{dom}(\sigma')$, $X\sigma'$ is a subpattern of u, so that $\mathrm{var}(X\sigma') \subseteq \mathrm{var}(u) \subseteq \Delta'$. From this observation and proposition 7 we conclude that $(\sigma'\eta)^{\mathcal{A}}$ **sat** Δ. Now we can apply our induction hyphotesis for the premise of rule (S) and we infer that:

$$[\![< C > \sigma]\!]^{\mathcal{A}}(\sigma'\eta)^{\mathcal{A}} \sqsupseteq_{\mathcal{A}} [\![< s, t >]\!]^{\mathcal{A}}(\sigma'\eta)^{\mathcal{A}}$$

Then, by lemma 16 and $s\sigma' \equiv t\sigma' \equiv u$, we get $[\![< C > \sigma\sigma']\!]^{\mathcal{A}}\eta \sqsupseteq_{\mathcal{A}} [\![< u, u >]\!]^{\mathcal{A}}\eta$. Since η **sat** $\Delta' \supseteq \mathrm{var}(u)$ and u is a pattern, proposition 7 tells us that $[\![u]\!]^{\mathcal{A}}\eta$ is totally defined, and we obtain $[\![< C > \sigma\sigma']\!]^{\mathcal{A}}\eta = [\![< u, u >]\!]^{\mathcal{A}}\eta$ which, in view of the relationship between < C > and C, means that $\mathcal{A} \models (C\sigma\sigma')\eta$. □

From the soundness theorem we can derive

Corollary 20. *Let* id *be the identity valuation over $\mathcal{H}$, defined by $id(X) = X$ for all $X \in Var$. Then:*

(N) $e \leadsto^*_{\sigma,\Delta} e' \Rightarrow [\![e\sigma]\!]^{\mathcal{H}} id \sqsupseteq |\, e' \,|$.

(S) $C \downarrow_{\sigma,\Delta} \Rightarrow \mathcal{H} \models (C\sigma)id$

Proof. It suffices to apply theorem 19 and to observe that: id **sat** Δ, since **def** X holds in $\mathcal{H}$ for every variable X. And $[\![e']\!]^{\mathcal{H}} id \sqsupseteq [\![\, |\, e' \,|\,]\!]^{\mathcal{H}} id = |\, e' \,|$. □

6.3 Completeness

For our discussion of completeness we need a notion of *correct solution.*

Definition 21 (Correct Solution).

1. A *finite d-substitution* is any finite substitution of *finite partial patterns* for variables.
2. A *solution correct in* $\mathcal{A}$ for a given expression e is any triple $< \theta, \Delta, s >$ such that: θ is a finite d-substitution with $\mathrm{dom}(\theta) \subseteq \mathrm{var}(e)$; $\Delta \subseteq \mathrm{var}(e\theta)$ is a finite set of restricted variables; $s \in H$ is a finite partial pattern, with $\mathrm{var}(s) \subseteq \mathrm{var}(e\theta)$; and $[\![e\theta]\!]^{\mathcal{A}}\eta \sqsupseteq [\![s]\!]^{\mathcal{A}}\eta$ for every η such that η **sat** Δ.
3. A *solution correct in* $\mathcal{A}$ for a given condition C is any pair $< \theta, \Delta >$ such that: θ is a finite d-substitution with $\mathrm{dom}(\theta) \subseteq \mathrm{var}(C)$; $\Delta \subseteq \mathrm{var}(C\theta)$ is a finite set of restricted variables; and $\mathcal{A} \models (C\theta)\eta$ for every η such that η **sat** Δ

Example 3.
If we assume that $\mathcal{R}$ includes: foo [X | Xs] = < X, 0 > ⇐ X == Y
then a correct solution for (foo Zs) in $\mathcal{H}$ is given by the triple:

$\theta = \{$ Zs / [(suc M), N | ⊥] $\}$, $\Delta = \{$ M, N $\}$, s = < (suc M), ⊥ >

Intuitively, the solution just shown in the example is too restrictive. We would like conditional narrowing to compute for us a more general solution; which in this case can be achieved, e.g. by the computation:

$$\text{foo Zs} \leadsto^*_{\{Zs\, /\, [X \mid Xs],\ Y/X\},\{X\}} < X, 0 >.$$

More generally, we would like computed solutions to be "better" than given correct solutions in several respects: they should compute a more general answer substitution, produce a more defined result and restrict less variables to be totally defined. We are looking for a completeness theorem which will make these ideas precise. It turns out that it is difficult to control the generality of computed restrictions. In general, it may happen that different computations are mutually incomparable in this respect. Let us illustrate this point.

Example 4. Assume that $\mathcal{R}$ includes the two rules
(R1) fun X Y = < X, Y > $\Leftarrow$ X == Z **(R2)** fun X Y = < X, Y > $\Leftarrow$ Y == Z
Then, for the expression (fun A B) there are two possible computations:

fun A B $\rightsquigarrow^{*}_{\{Z/A\},\{A\}}$ < A, B > (using **(R1)**)
fun A B $\rightsquigarrow^{*}_{\{Z/B\},\{B\}}$ < A, B > (using **(R2)**)

For the moment we can only present a completeness theorem which does not give any information about the generality of computed restrictions. It will be a consequence of the following key lemma:

Lemma 22 (Completeness Lemma). *Assume an expression e, a finite d-substitution θ and a finite set of variables $V \supseteq var(e) \cup dom(\theta)$. Let $k \in \mathbb{N}$ be given. Then we can find some narrowing computation $e \rightsquigarrow^{*}_{\sigma,\Delta}$ e'and some finite d-substitution θ' such that: $dom(\theta') \subseteq V' =_{def} var(V\sigma)$[8]; $\theta =_V \sigma\theta'$; $|\ e'\theta'\ | \sqsupseteq [\![\ e\theta\ ]\!]^k$.* □

Explanation "$\theta =_V \sigma\theta'$" means that $X\theta \equiv X(\sigma\theta')$ for all X $\in$ V, and "$[\![\ e\theta\ ]\!]^k$" is meant as $[\![\ e\theta\ ]\!]^{\mathcal{H}^k}$id.

Note that the interplay between θ and θ' in the completeness lemma plays essentially the same role as the *lifting lemma* which is used in most completeness proofs for narrowing and SLD-resolution. The proof of the lemma is very technical and has been left as an appendix. Here we show how to derive completeness results from it.

Theorem 23 (Completeness of Conditional Narrowing).
Solutions which are correct in $\mathcal{H}$ are subsumed by computed solutions in the following sense:

(N) *< θ, Δ_0, s > is correct for e in $\mathcal{H} \Longrightarrow e \rightsquigarrow^{*}_{\sigma,\Delta}$ e' and there is some partial d-substitution θ' such that: $dom(\theta') \subseteq var(e\sigma)$, $\theta =_{var(e)} \sigma\theta'$, $|\ e'\theta'\ | \sqsupseteq s$*

(S) *< θ, Δ_0 > is correct for C in $\mathcal{H} \Longrightarrow C \downarrow_{\sigma,\Delta}$ and there is some partial d-substitution θ' such that: $dom(\theta') \subseteq var(C\sigma)$, $\theta =_{var(C)} \sigma\theta'$* [9]

Proof.
<u>(N)</u>: Correctness of < θ, Δ_0, s > for e in $\mathcal{H}$ means in particular that $[\![\ e\theta\ ]\!]^{\mathcal{H}}$id $\sqsupseteq$ s, because id **sat** Δ_0. By lemma 17, we can find some k $\in \mathbb{N}$ such that $[\![\ e\theta\ ]\!]^k \sqsupseteq$ s. Then we can apply lemma 22 with V = var(e) and we are done.
<u>(S)</u>: Correctness of < θ, Δ_0 > for C in $\mathcal{H}$ means in particular that $\mathcal{H} \models (C\theta)$id, again because id **sat** Δ_0. Due to the relationship between C and < C >, we can conclude that $[\![\ < C > \theta\ ]\!]^{\mathcal{H}}$id = < u, u > for some *finite* and *total* pattern u $\in \mathcal{H}$. By lemma 17, we can find k $\in \mathbb{N}$ such that $[\![\ < C > \theta\ ]\!]^k$ = < u, u > and we can apply lemma 22 with V = var(< C >) = var(C). We obtain a computation < C > $\rightsquigarrow^{*}_{\sigma\Delta}$ < s, t > and a finite d-substitution θ' such that:

•$dom(\theta') \subseteq var(C\sigma)$ • $\theta =_{var(C)} \sigma\theta'$ • $|$ < s, t > θ' $| \sqsupseteq$ < u, u >

Since u is totally defined, the last condition implies that s, t must be patterns and that < s, t > θ' = < u, u >. Then s, t must have a canonical m.g.u. σ', and because θ' is another unifier we know that $\theta' = \sigma'\theta'$. According to the rules in our narrowing calculus, it follows that $C\downarrow_{\sigma\sigma',\Delta'}$ (for a suitable Δ'). Moreover, $\theta =_{var(C)} \sigma\theta' = \sigma\sigma'\theta'$ and we are done. □

[8] By lemma 15 we also know that var(e') $\subseteq$ V'.
[9] Note that we claim nothing about the relationship between Δ_0 and Δ.

From theorem 23 we can immediately derive the following result, which is almost the reciprocal of theorem 19 (except for the lack of information about the generality of computed restrictions).

Corollary 24. *Any solution which is correct in all the liberal models of $\mathcal{R}$ is subsumed by some computed solution (in the technical sense explained in the statement of theorem 23).*

Proof. Obvious from theorem 23, because $\mathcal{H}$ is liberal. □

7 Conclusions

We have presented soundness and completeness results for *conditional narrowing* with respect to a *declarative semantics* which provides a well founded framework for HO functional logic programming, on the basis of conditional applicative rewriting systems with extra variables in the condition and a constructor discipline. We believe that this framework contributes to a better understanding of some existing functional logic languages, as e.g. K-LEAF [8], IDEAL [5, 4] and BABEL [12].

This work is a natural continuation of [10] and it leaves open several lines for future research. In the first place, the completeness theorem should be improved so that it will provide information on the *generality of computed restrictions*. The behaviour of *infinite computations*, that was investigated in [10] for the rewriting case, could be also considered in the present framework. The investigation of *complete lazy narrowing strategies* is another important issue, also in relation to eventual implementations, since the computations provided by the current completeness theorem may perform unneeded narrowing steps. Finally, it seems very interesting to look for some extension of the present framework which incorporates *types* and *HO unification*.

References

1. H.Aït-Kaci, R.Nasr: Integrating Logic and Functional Programming. In Lisp and Symbolic Computation, 2, 1989, pp. 51-89.
2. K.R.Apt: Logic Programming. In J.van Leeuwen (ed.), Handbook of Theoretical Computer Science, vol. B, Elsevier Science Publishers, 1990, pp. 493-574.
3. M.Bellia, G.Levi: The Relation between Logic and Functional Languages: a Survey. In J. Logic Programming 3, 1986, pp. 217-236.
4. M.Bellia, P.G.Bosco, E.Giovannetti, G.Levi, C.Moiso, C.Palamidessi: A two-level approach to logic and functional programming. In Procs. PARLE '87, Springer Verlag, 1987, pp. 374-393.
5. P.G. Bosco and E. Giovannetti: IDEAL: An Ideal Deductive Applicative Language. In Procs. 1986 Symp. on Logic Programming, IEEE Comp. Soc. Press, 1986, pp. 89-94.
6. D. de Groot, G. Lindstrom (eds): Logic Programming: Functions, Relations and Equations, Prentice Hall, 1986.
7. N.Dershowitz, J.P.Jouannaud: Rewrite Systems. In J. van Leeuwen (ed.), Handbook of Theor. Comp. Sci., vol B, Elsevier Publishers, 1990, pp. 243-320.
8. E. Giovannetti, G. Levi, C. Moiso and C. Palamidessi: Kernel-LEAF: a Logic plus Functional Language. In JCSS, 42, 2, 1992, pp. 139-185.
9. J.A.Goguen, J.W.Thatcher, E.G.Wagner, J.B. Wright: On Initial Algebra Semantics and Continuous Algebras. In J. ACM 24, 1, 1977, pp. 68-95.
10. J.C.González-Moreno, M.T.Hortalá-González, M.Rodríguez-Artalejo: Denotational Versus Declarative Semantics for Functional Programming. In Procs. CSL'91, Springer LNCS, 1992, pp. 134-148.
11. A.Middeldorp, E.Hamoen: Counterexamples to Completeness Results for Basic Narrowing. In Procs. ALP'92, Springer LNCS 632, 1992, pp. 244-258

12. J.J.Moreno-Navarro & M.Rodríguez-Artalejo: Logic Programming with Functions and Predicates: The Language BABEL. In J.Logic Programming 12, 1992, pp. 191-223.
13. G.Nadathur, D.Miller: An overview of λ-PROLOG. In Procs. ICLP'88, the MIT Press, 1988, pp. 810-827.
14. D.S.Scott: Domains for denotational semantics. In Procs. ICALP'82, Springer LNCS 140, 1982, pp. 577-613.

A Appendix

Proof Lemma 22 (Completeness Lemma).

We reason by induction on the pair $< e\theta, k >$ with respect to the following well founded ordering over $\mathrm{Exp}_\Sigma \times \mathbb{N}$:

$$< b, l > \prec < a, k > \Leftrightarrow_{def} l < k \text{ or } (l = k \text{ and b is a proper subexpression of a}).$$

Since we are not going to prove anything about computed restrictions, we shall ignore them in order to simplify the notation. Given e, θ, V and k as in the statement of the lemma, we distinguish five cases according to the form of e (remember the syntax of expressions from subsection 2.1). We start with the most difficult case.

<u>Case 1</u>: $e \equiv (e_0\ e_1)$. We assume that $[\![e\theta]\!]^k \neq \bot$ (otherwise, the proof is trivial). By applying the induction hypothesis to e_0, θ, V, k we obtain $e_0 \rightsquigarrow^*_{\sigma_1} e'_0$ and θ_1 such that: $\mathrm{dom}(\theta_1) \subseteq V_1 = \mathrm{var}(V\sigma_1)$, $\theta =_V \sigma_1\theta_1$, and $| e'_0\theta_1 | \sqsupseteq [\![e_0\theta]\!]^k$

We have that $\mathrm{var}(e_1\sigma_1) \subseteq V_1$ and also that $(e_1\sigma_1)\theta_1 \equiv e_1\theta$ (because of $\theta =_V \sigma_1\theta_1$). Thus, we are in a position to apply the induction hypothesis to $e_1\sigma_1$, θ_1, V_1, k and we get $e_1\sigma_1 \rightsquigarrow^*_{\sigma_2} e'_1$, θ_2 such that: $\mathrm{dom}(\theta_2) \subseteq V_2 = \mathrm{var}(V_1\sigma_2)$, $\theta_1 =_{V_1} \sigma_2\theta_2$ and $| e'_1\theta_2 | \sqsupseteq [\![(e_1\sigma_1)\theta_1]\!]^k = [\![e_1\theta]\!]^k$.

By combining the two computations: $e_0 e_1 \rightsquigarrow^*_{\sigma_1} e'_0(e_1\sigma_1) \rightsquigarrow^*_{\sigma_2} (e'_0\sigma_2)e'_1$ and θ_2 such that: $\mathrm{dom}(\theta_2) \subseteq V_2 = \mathrm{var}(V\sigma_1\sigma_2)$, $\theta =_V \sigma_1\sigma_2\theta_2$, $| (e'_0\sigma_2)\theta_2 | \sqsupseteq [\![e_0\theta]\!]^k$ (since $(e'_0\sigma_2)\theta_2 \equiv e'_0\theta_1$), and $| e'_1\theta_2 | \sqsupseteq [\![e_1\theta]\!]^k$.

That is, we have for $(e'_0\sigma_2)e'_1$, θ_2, V_2, k a situation of the same kind as the initial one, and in addition we know that:

(1) $\qquad \bot \neq [\![(e_0\ e_1)\theta]\!]^k = [\![e_0\theta]\!]^k \circ^k [\![e_1\theta]\!]^k \sqsubseteq | (e'_0\sigma_2)\theta_2 | \circ^k | e'_1\theta_2 |$

which also implies (because of the strictness of $\circ^k$ w.r.t. its first argument and the definition of shell):

(2) $\qquad | e'_0\sigma_2\theta_2 | \neq \bot \qquad | e'_0\sigma_2 | \neq \bot \qquad | e'_0 | \neq \bot$

We must distinguish subcases according to the form of $e'_0\sigma_2$. In view of (1), (2) there are only three possibilities: <u>1.1</u>: $e'_0\sigma_2 \equiv c\ a_1 .. a_m$, $0 \leq m < \mathrm{rank}(c)$ <u>1.2</u>: $e'_0\sigma_2 \equiv f\ a_1 .. a_m$, $0 \leq m < \mathrm{rank}(f)$ <u>1.3</u>: $e'_0\sigma_2 \equiv X\ a_1 .. a_m$, $0 \leq m$

Let us go through these subcases.

<u>Subcase 1.1</u>: In this case we have: $|(e'_0\sigma_2)\theta_2| \circ^k |e'_1\theta_2| = c\ |a_1\theta_2| .. |a_m\theta_2|\ |e'_1\theta_2| = |((e'_0\sigma_2)\ e'_1)\theta_2|$ and we are done because of (1).

<u>Subcase 1.2</u>: If m+1 < rank(f), we can reason as in subcase 1.1. Otherwise, it must be the case that m+1 = rank(f), because of (2) and the definition of shell. Let us take n = m+1 and rename e'_1 as a_n for uniformity of notation. Because of (1) and the behaviour of application in Herbrand algebras, we have:

(3) $\qquad \bot \neq [\![(e_0\ e_1)\ \theta]\!]^k \sqsubseteq f^n_k\ | a_1\theta_2 | .. | a_n\theta_2 |$

where f^n_k abbreviates $f^n_{\mathcal{H}^k}$, and we must continue some computation from the expression that we have reached already, that is $(e'_0\sigma_2)e'_1 \equiv f\ a_1 .. a_n$.

Because of (3) it must be k > 0, and since $\mathcal{H}^k = \mathcal{T}_\mathcal{R}(\mathcal{H}^{k-1})$ we can choose a variant of some rule of $\mathcal{R}$ using fresh variables (in particular *variable disjoint* from V_2) R: $f\ t_1 .. t_n = r \Leftarrow C$ such that for a suitable valuation η: Var → H the following will hold (remember definition 9): $[\![t_i]\!]^{k-1}\eta = |a_i\theta_2|$ $(1 \leq i \leq n)$, $\mathcal{H}^{k-1} \models C\eta$ and $f^n_k\ |a_1\theta_2| .. |a_n\theta_2| = [\![r]\!]^{k-1}\eta$.

Let LV = var(f t_1 .. t_n), CV = var(C) and EV = CV \ LV. By the *continuity lemma 17* there is no loss of generality in assuming **fin** η(Y) for all Y $\in$ EV. On the other hand, since t_i and $|\ a_i\theta_2\ |$ are finite patterns, $[\![\ t_i\ ]\!]^{k-1}\eta = |\ a_i\theta_2\ |$ means that η(X) must be a finite partial pattern (and even a subpattern of some $|\ a_i\theta_2\ |$) for every X $\in$ LV. These considerations allow us to define a finite d-substitution μ as follows:

$\mu =_{def} \{\ Z\ /\ \eta(Z)\ |\ Z \in var(R),\ \eta(Z) \neq Z\ \}$

We observe that μ inherits the following properties from η:

(4) $t_i\mu \equiv |a_i\theta_2|$,, $(1 \leq i \leq n)$, $\mathcal{H}^{k-1} \models (C\mu)$ id, $f_k^n\ |a_1\theta_2|\ ..\ |a_n\theta_2| = [\![\ r\mu\ ]\!]^{k-1}$.

The first of these conditions holds because $t_i\mu \equiv [\![\ t_i\ ]\!]^{k-1}\eta$. The second and third are justified by the substitution lemma 16, noting that $(\mu id)^{\mathcal{H}^{k-1}} = \eta$. Remember that $[\![\ r\mu\ ]\!]^{k-1}$ is meant as $[\![\ r\mu\ ]\!]^{\mathcal{H}^{k-1}}$id.

For each occurrence of a variable X $\in$ dom(μ) in some t_i, the first condition in (4) ensures that we can find a subexpression a_X of $a_i\theta_2$ such that $X\mu = |\ a_X\ |$. Since rule R is left linear, each X $\in$ LV has one single occurrence in some t_i and we can define a finite substitution $\hat{\mu}$ with dom($\hat{\mu}$) = dom(μ) and such that:

(5) $\hat{\mu}(X) \equiv a_X$ for all X $\in$ dom(μ) $\cap$ LV, $\hat{\mu}(X) \equiv \mu(X)$ for all X $\in$ dom(μ) \ LV.

In general, $\hat{\mu}$ will not behave as a d-substitution over variables belonging to LV. However, the construction of $\hat{\mu}$ guarantees that $(f\ t_1\ ..\ t_n)\hat{\mu} \equiv (f\ a_1\ ..\ a_n)\theta_2$ This means that the finite substitution $\hat{\mu}\ \dot{\cup}\ \theta_2$[10] unifies (f t_1 .. t_n) and (f a_1 .. a_n), and we can consider some canonical m.g.u. σ that will satisfy the following:

(6) $$(\hat{\mu}\ \dot{\cup}\ \theta_2) = \sigma(\hat{\mu}\ \dot{\cup}\ \theta_2)$$

We are going to see that condition $C\sigma$ can be solved by some narrowing computation, in order to apply rule R, which will allow us to complete our argument for subcase 1.2. We need the following:

<u>**Claim**</u>.- For every expression b such that var(b) $\subseteq$ var(R) and for every l $\in$ IN (in particular, l = k-1): $[\![\ b\hat{\mu}\ ]\!]^l \sqsupseteq [\![\ b\mu\ ]\!]^l$

<u>**Justification**</u>.- By lemma 16, the assertion of the claim is equivalent to $[\![\ b\ ]\!]^{\mathcal{H}^l}\ (\hat{\mu}\ id)^{\mathcal{H}^l} \sqsupseteq [\![\ b\ ]\!]^{\mathcal{H}^l}\ (\mu\ id)^{\mathcal{H}^l}$. Hence, it is enough to prove that $[\![\ X\hat{\mu}\ ]\!]^l \sqsupseteq [\![\ X\mu\ ]\!]^l$ for all X $\in$ var(R). This is obvious for X $\notin$ dom(μ) as well as for X $\in$ dom(μ)\LV, due to (5). For X $\in$ dom(μ) $\cap$ LV we know by construction of $\hat{\mu}$ that $X\hat{\mu} \equiv a_X$ while $X\mu = |\ a_X\ |$; then $[\![\ X\hat{\mu}\ ]\!]^l = [\![\ a_X\ ]\!]^l \sqsupseteq [\![\ |\ a_X\ |\ ]\!]^l = [\![\ X\mu\ ]\!]^l$. Now, if we define:

(7) $$\hat{V} =_{def} var(CV\sigma) \cup var(V_2\sigma) \qquad \hat{\theta} =_{def} (\hat{\mu}\ \dot{\cup}\ \theta_2)\ |\ \hat{V}$$

we obviously have that $\hat{V} \supseteq var(C\sigma) \cup dom(\hat{\theta})$. Moreover, using that $\mathcal{H}^{k-1} \models (C\mu)$id (cfr. (4)) we know that $[\![< C >\mu\ ]\!]^{k-1} = <u, u>$ for some u $\in$ H, **def** u. It follows that also $[\![< C\sigma > \hat{\theta}\ ]\!]^{k-1} = <u, u>$, since u is maximal w.r.t. $\sqsubseteq$ and:

$$\begin{aligned}
[\![< C\sigma > \hat{\theta}\ ]\!]^{k-1} &= [\![< C > \sigma\ (\hat{\mu}\dot{\cup}\theta_2)\ ]\!]^{k-1} && \%\ \text{by (7)}\\
&= [\![< C > (\hat{\mu}\dot{\cup}\theta_2)\ ]\!]^{k-1} && \%\ \text{by (6)}\\
&= [\![< C > \hat{\mu}\ ]\!]^{k-1} && \%\ \text{because CV} \cap \text{dom}(\theta_2) = \emptyset\\
&\sqsupseteq [\![< C > \mu\ ]\!]^{k-1} && \%\ \text{by the claim}\\
&= <u, u>
\end{aligned}$$

We are now in a position to apply the induction hypothesis to $< C\ \sigma >$, $\hat{\theta}$, $\hat{V}$, k-1 and we obtain $< C\sigma > \leadsto^*_{\sigma'} < s, t >$ and θ' such that:

(8) $$dom(\theta') \subseteq var(\hat{V}\sigma') \qquad \hat{\theta} =_{\hat{V}} \sigma'\theta' \qquad |< s, t >\theta'\ | \sqsupseteq <u, u>$$

since u is a totally defined pattern, it follows that s, t must be patterns which are unified by θ'. By taking a canonical m.g.u. σ'', we get:

[10] Note that $\hat{\mu}$ and θ_2 have disjoint domains, because dom(θ_2) $\subseteq$ V_2 and dom($\hat{\mu}$) $\subseteq$ var(R).

(9) $(C\sigma)\downarrow_{\sigma'\sigma''}$ $\theta' = \sigma''\theta'$

According to our narrowing calculus, this in turn allows to apply R to obtain the following computation: $f\ a_1 \ldots a_n \leadsto_{\sigma\sigma'\sigma''} r\sigma\sigma'\sigma''$

In order to complete our proof for subcase 1.2, we still have to prove that the induction hypothesis can be applied to $r\sigma\sigma'\sigma''$. For this purpose we define:

(10) $\sigma_3 =_{def} \sigma\sigma'\sigma''$ $V_3 =_{def} var(V_2\sigma_3)$ $\theta_3 =_{def} \theta' \mid V_3$

Using the different properties that we have stablished above, it can be checked that

(11) $V_3 \supseteq var(r\sigma_3) \cup dom(\theta_3)$ $\theta_2 =_{V_2} \sigma_3\theta_3$

the first of these conditions legitimates us to apply the induction hypothesis to $r\sigma_3$, θ_3, V_3 and k-1, thus obtaining: $r\sigma_3 \leadsto^*_{\sigma_4} e'$ and θ_4 such that:

(12) $dom(\theta_4) \subseteq V_4 =_{def} var(V_3\sigma_3)$, $\theta_3 =_{V_3} \sigma_4\theta_4$, and $\mid e'\theta_4 \mid \sqsupseteq [\![r\ \sigma_3\theta_3]\!]^{k-1}$

From previous conditions it follows very easily that $\theta =_V \sigma_1\sigma_2\sigma_3\sigma_4\theta_4$ and, in view of (3), we can close subcase 1.2 if we prove that $[\![r\ \sigma_3\theta_3]\!]^{k-1} \sqsupseteq f^n_k \mid a_1\theta_2 \mid \ldots \mid a_n\theta_2 \mid$. Let us do this:

$$
\begin{aligned}
[\![r\sigma_3\theta_3]\!]^{k-1} &= [\![r\ \sigma_3\theta']\!]^{k-1} && \text{\% by (10),(11)}\\
&= [\![r\ \sigma\sigma'\sigma''\theta']\!]^{k-1} && \text{\% by (10)}\\
&= [\![r\ \sigma\sigma'\theta']\!]^{k-1} && \text{\% by (9)}\\
&= [\![r\ \sigma(\hat{\mu}\dot{\cup}\theta_2)]\!]^{k-1} && \text{\% using (7), (8) and } var(r\sigma) \subseteq \hat{V}\ ^{11}\\
&= [\![r\ \hat{\mu}]\!]^{k-1} && \text{\% since } dom(\theta_2) \subseteq V_2 \text{ and } V_2 \cap var(r) = \emptyset\\
&\sqsupseteq [\![r\mu]\!]^{k-1} && \text{\% by the claim}\\
&= f^n_k \mid a_1\theta_2 \mid \ldots \mid a_n\theta_2 \mid && \text{\%by (4)}
\end{aligned}
$$

Now we go back to the remaining subcase of case 1, which is much easier to handle.

<u>**Subcase 1.3**</u>: Because of (2) we can claim that $X\theta_2$ must have one of the three following forms: <u>**1.3.1**</u>: $X\theta_2 \equiv f\ b_1 \ldots b_k$, $0 \leq k+m < rank(f)$
<u>**1.3.2**</u>: $X\theta_2 \equiv c\ b_1 \ldots b_k$, $0 \leq k+m < rank(c)$ <u>**1.3.3**</u>: $X\theta_2 \equiv F\ b_1 \ldots b_k$, $F \in Var$, $k \geq 0$

In the case 1.3.3 we are already done, similarly as in subcase 1.1.

In the case 1.3.2 we can use the rule **(N2)** of the narrowing calculus to perform one narrowing step: $X\ a_1 \ldots a_m\ \sigma_1 \leadsto_{\sigma_3} (c\ X_1 \ldots X_k\ a_1 \ldots a_m\ e'_1)\sigma_3$, where $\sigma_3 = \{ X / (c\ X_1 \ldots X_k) \}$ with fresh variables X_i (in particular, $X_i \notin V_2$). Now, if we take $\theta_3 =_{def} (\theta_2 \setminus X) \dot{\cup} \{ X_1 / b_1, \ldots, X_k / b_k \}$ [12], and $V_3 =_{def} var(V_2\sigma_3)$ it turns out that $V_3 \supseteq var(c\ X_1 \ldots X_k\ a_1 \ldots X_m\ e'_1)\sigma_3 \cup dom(\theta_3)$ and also $\theta_2 =_{V_2} \sigma_3\theta_3$. We are then in a position to continue reasoning as in subscase 1.1.

The case 1.3.1 can be reduced to subcase 1.2 by similar considerations. This completes case 1. Fortunately, the remaining four cases can be disposed very quickly.

<u>**Cases 2, 3**</u>: $e \equiv X$ or $e \equiv c$. We are done trivially with $e \leadsto^* e$ and $\theta' = \theta$.

<u>**Case 4**</u>: $e \equiv f$. $[\![e\theta]\!]^k$ is simply $[\![f]\!]^k$. If $rank(f) > 0$ or $[\![f]\!]^k = \perp$ we are done with $f \leadsto^* f$ and $\theta' = \theta$. Otherwise, we can reason in a similar way as in subcase 1.2 with $m+1 = rank(f)$, by using a suitable rule $f = r \Leftarrow C$ of $\mathcal{R}$.

<u>**Case 5**</u>: $e \equiv \langle e_1, \ldots, e_n \rangle$. This case can be handled by reiteration of the argument which was used in case 1 for getting the computation $e_0\ e_1 \leadsto^*_{\sigma_1} e'_0\ (e_1\sigma_1) \leadsto_{\sigma_2} (e'_0\sigma_2)e'_1$. Now, it will suffice to repeat the same kind of reasoning n times, for successively narrowing $e_1, \ldots, e_n$. □

---

[11] $var(r\sigma) \subseteq var((f\ t_1 \ldots t_n)\sigma) = var((f\ a_1 \ldots a_n)\sigma) \subseteq var(V_2\sigma) \subseteq \hat{V}$

[12] $(\theta_2 \setminus X)$ stands for the substitution that does not affect X and acts as θ_2 over any other variable

Inductive Definability with Counting on Finite Structures

Erich Grädel*   Martin Otto*

1 Introduction

We study the properties and the expressive power of logical languages that include both a mechanism for inductive definitions and the ability to count. The most important of these languages is fixpoint logic with counting terms, denoted (FP + C).

To motivate the study of these logics, we recall that the expressive power of first-order logic (FO) is limited by two main reasons: It lacks the power to express anything that requires recursion (the most notable example is the transitive closure query) and it cannot count (the best-known example here is that no first-order formula is true precisely on the structures with even cardinality). There are several well-studied logics and database query languages that add recursion in one way or another to FO (or part of it), notably the various forms of fixpoint logics, the query language Datalog and its extensions.

On ordered finite structures, some of these languages express precisely the queries that are computable in PTIME or PSPACE. However, on arbitrary finite structures they do not, and almost all known examples showing this involve counting. While in the presence of an ordering, the ability to count is inherent e.g. in fixpoint logic, hardly any of it is retained in its absence. Thus, Immerman [15] proposed to add counting quantifiers to fixpoint logic and asked whether this would suffice to capture PTIME. Cai, Fürer and Immerman [5] answered this question negatively; in fact (FP + C) does not even express all LOGSPACE-computable queries. Nevertheless we argue that fixpoint logic with counting is an important language that deserves more attention.

We will show that in the presence of counting terms (or counting quantifiers) inductive definability on arbitrary finite structures has nice properties that it retains without counting only in the case of ordered structures. The organization of the paper is summed up in the following:

Equivalent characterizations of inductive definability with counting. In section 2, we define *(inflationary) fixpoint logic with counting* (FP + C) and *partial fixpoint logic with counting* (PFP + C) by closing a two-sorted version of first-order logic under counting terms and the usual rules for building inflationary or partial fixpoints. We also investigate several other logical and algorithmical versions of inductive definability with counting and show that they all have the same expressive power. This supports our belief that (FP + C) is a natural and robust class. In particular we consider:

*Lehrgebiet Mathematische Grundlagen der Informatik, RWTH Aachen, Ahornstr. 55, D-5100 Aachen, email: {graedel,otto}@mjoli.informatik.rwth-aachen.de

Datalog with counting. Counting terms can be added also to Datalog. We show that (Datalog + C) = (FP + C). In particular, (Datalog + C) is closed under negation and therefore, in the presence of counting, all the many extensions of Datalog, notably Stratified Datalog are equivalent to Datalog. This is not at all the case without counting.

A functional logic. Gurevich has proved that the usual schemes, that define — over N — the recursive functions, characterize, when interpreted over ordered finite structures, precisely the functions computable in polynomial time. We define a class of global functions, defined by a similar scheme, which has precisely the power of (FP + C).

A machine theoretic characterization. An algorithmic definition of (FP + C) is provided by a "relational machine" inspired by [3]. This machine operates like a Turing machine but interacts with a relational store in a "generic" way so that the machine treats algebraically indistinguishable structures and tuples in the same way. The queries computed in polynomial-time by these machines are precisely the queries in (FP + C).

Infinitary logic with counting. Let $C^k_{\infty\omega}$ be infinitary logic with k variables, $L^k_{\infty\omega}$, extended by the quantifiers $\exists^{\geq m}$ ("there exist at least m") for all $m \in \mathbb{N}$. It is easy to see that (FP + C) $\subseteq \bigcup_k C^k_{\infty\omega}$. We investigate the relationship of fixpoint logic and infinitary logic in the presence of counting. In particular, we show that the $C^k_{\infty\omega}$-types can be uniformly ordered by a formula in (FP + C). These results parallel those in [3, 8] on FP and $L^\omega_{\infty\omega}$. As a consequence, a functor can be introduced which associates with every structure an arithmetical invariant (i.e. essentially a collection of numerical predicates), which characterizes the original structure up to $C^k_{\infty\omega}$-equivalence. Furthermore the (FC + C) definable properties of the original structures exactly correspond to the FP definable (or PTIME) properties of the associated invariants.

We also obtain a characterization of (FP + C) as a sublogic of $\bigcup_k C^k_{\infty\omega}$ in terms of families $(\varphi_n)_{n\in\omega}$ of finitary formulae in $C^k_{\infty\omega}$ (in which φ_n applies to structures of size n). The uniformity condition which exactly restricts the expressive power of these to (FP + C) is just PTIME constructibility of the sequence of the φ_n. This is quite unlike the situation for FP and the $L^k_{\infty\omega}$.

Inflationary vs. partial fixpoints. Abiteboul and Vianu [3] recently proved that partial fixpoint logic collapses to fixpoint logic if and only if PTIME = PSPACE. The analogous result in the presence of counting is also true: PTIME = PSPACE $\Longleftrightarrow$ (FP + C) = (PFP + C). On the other side, Abiteboul and Vianu also proved that already a collapse of PFP$|_P$ to fixpoint logic is equivalent to PTIME = PSPACE. (The logic PFP$|_P$ is partial fixpoint logic restricted to PFP operators that close after at most polynomially many iterations on every structure.) In the presence of counting we can prove instead that (FP + C) = (PFP$|_P$ + C). In fact it turns out that the PSPACE restriction of the above machine model corresponds to (PFP + C), or While with counting.

The algorithmic characterization of (FP + C), which is fully expanded in [19], inspired some of the results concerning the relation with infinitary logic and that with partial fixpoints. We here present these results in a self-contained way, which does not rely on the machine model in an essential way. While the present approach might be methodically more concise, we also would like to refer the interested reader to the treatment fully exploiting the algorithmic characterization in [19]. For some arguments it remains the intuitively more appealing.

2 Characterizations of inductive definability with counting

2.1 Fixpoint logics with counting

Logics with counting are two-sorted. With any one-sorted finite structure $\mathfrak{A}$ with universe A, we associate the two-sorted structure $\mathfrak{A}^* := \mathfrak{A} \dot{\cup} \langle n; \leq, 0, e\rangle$ with $n = |A| + 1$, where $\leq$ is the canonical ordering on $n = \{0, \ldots, n-1\}$ and 0 and e stand for the first and the last element of n. Thus, we have taken the disjoint union of $\mathfrak{A}$ with a canonical ordered structure of size n. We take $n = |A| + 1$ rather than $n = |A|$ to be able to represent the cardinalities of all subsets of $|A|$ within n.

We start with first-order logic over two-sorted vocabularies $\sigma \cup \{\leq, 0, e\}$, with semantics over structures $\mathfrak{A}^*$ defined in the obvious way. We will use latin letters $x, y, z, \ldots$ for the variables over the first sort, and greek letters $\lambda, \mu, \nu, \ldots$ for variables over the second sort.

The two sorts are related by *counting terms*, defined by the following rule: Let $\varphi(x)$ be a formula with variable x (over the first sort) among its free variables. Then $\#_x[\varphi]$ is a term in the second sort, with the set of free variables $\text{free}(\#_x[\varphi]) = \text{free}(\varphi) - \{x\}$. The value of $\#_x[\varphi]$ is the number of elements a that satisfy $\varphi(a)$.

First-order logic with counting, denoted (FO + C), is the closure of two-sorted first-order logic under counting terms.

Example. To illustrates the use of counting terms we present a formula $\psi(E_1, E_2) \in$ (FO + C) which expresses that two equivalence relations E_1 and E_2 are isomorphic; of course a necessary and sufficient condition for this is that for every i, they have the same number of equivalence classes of size i:

$$\psi(E_1, E_2) \equiv (\forall \mu)(\#_x[\#_y[E_1 xy] = \mu] = \#_x[\#_y[E_2 xy] = \mu]).$$

We obtain *(inflationary) fixpoint logic with counting* (FP + C) and *partial fixpoint logic with counting* (PFP + C) by closing first-order logic under counting terms and the usual rule for building inflationary or partial fixpoints:

Let $R \notin \sigma$ be a relational variable of mixed arity (k, ℓ), i.e. with arity k over the first and arity ℓ over the second sort. A formula $\psi(R, \bar{x}, \bar{\mu})$ of vocabulary $\sigma \cup \{R\}$, defines for every σ-structure $\mathfrak{A}$ an operator $\psi^{\mathfrak{A}}$ on the class of (k, ℓ)-ary predicates over $\mathfrak{A}^*$, namely

$$\psi^{\mathfrak{A}} : R \longmapsto R \cup \{(\bar{u}, \bar{\nu}) \mid \mathfrak{A}^* \models \psi(R, \bar{u}, \bar{\nu})\}.$$

Since this operator is *inflationary* — i.e. $\psi^{\mathfrak{A}}(R) \supseteq R$ — it has a fixed point that is constructed inductively, starting with the empty relation: Set $\Psi^0 = \varnothing$ and $\Psi^{i+1} = \psi^{\mathfrak{A}}(\Psi^i)$. At some stage i, this process reaches a stable predicate $\Psi^i = \Psi^{i+1}$, which is called the *inflationary fixed point* of ψ on $\mathfrak{A}$, and denoted by Ψ^∞. The *stage* of a tuple $(\bar{u}, \bar{\nu})$ in Ψ^∞ is the minimal i such that $(\bar{u}, \bar{\nu}) \in \Psi^i$ (if $(\bar{u}, \bar{\nu}) \notin \Psi^\infty$ then $\text{stage}_R(\bar{u}, \bar{\nu}) = \infty$).

Definition 2.1 *Fixpoint logic with counting*, denoted (FP + C), is the closure of (two-sorted) first-order logic under

(i) the rule for building counting terms;

(ii) the usual rules of first-order logic for building terms and formulae;

(iii) the fixpoint formation rule: Suppose that $\psi(R, \bar{x}, \bar{\mu})$ is a formula of vocabulary $\sigma \cup \{R\}$ where $\bar{x} = x_1, \ldots, x_k$, $\bar{\mu} = \mu_1, \ldots, \mu_\ell$, and R has mixed arity (k, ℓ), that $\bar{u}$ is a k-tuple of terms over the first sort and $\bar{\nu}$ an ℓ-tuple of terms over the second sort. Then

$$[\mathrm{IFP}_{R,\bar{x},\bar{\mu}}\ \psi(R, \bar{x}, \bar{\mu})](\bar{u}, \bar{\nu})$$

is a formula of vocabulary σ.

The interpretation of the fixpoint formula in *(iii)* in a structure $\mathfrak{A}$ is that $(\bar{u}, \bar{\nu})$ is included in the fixpoint Ψ^∞.

An interesting example for a (FP + C)-query is the method of *stable colourings* for graph-canonization. Given a graph G with a colouring $f : V \longrightarrow 0, \ldots, r$ of its vertices, we define a refinement f' of f, where vertex x has the new colour $f'x = (fx, n_1, \ldots, n_r)$ where $n_i = \#y[Exy \wedge (fy = i)]$. The new colours can be sorted lexicographically so that they form again an initial subset of $\mathbb{N}$. Then the process can be iterated until a fixpoint, the *stable colouring* of G is reached. It is known that almost all graphs have the property that no two vertices have the same stable colour. Thus stable colourings provide a polynomial-time graph-canonization algorithm for a dense class of graphs. It should be clear that the stable colouring of a graph is definable in (FP + C) (see [16] for more details).

Remarks. A slightly different definition of fixpoint logic with counting has been proposed by Immerman [15], who related the two sorts by *counting quantifiers* rather than counting terms. Counting quantifiers have the form $(\exists i\ x)$ for "there exist at least i x", where i is a second-sort variable. It is obvious, that the two definitions are equivalent. In fact, (FP + C) is a very robust logic. For instance, it would not change its expressive power if we would allow counting over tuples, even of mixed type, i.e. terms of the form $\#_{\bar{x},\bar{\mu}}\varphi$. We will illustrate this for the case of (Datalog + C), a language that is equivalent to (FP + C) (see section 2.2). However, this is not true for (FO + C) which is very sensitive to the precise definition that is chosen.

Different variants of counting logics have been investigated by Grumbach and Tollu [10]; their languages are weaker since they do not allow nested counting (i.e. counting over formulae, that do themselves contain counting terms). For further results on fixed point logic with counting we refer to [9].

Partial fixpoints. Besides the inflationary operator defined above, a formula $\psi(R, \bar{x}, \bar{\mu})$ gives us also the operator

$$\tilde{\psi}^{\mathfrak{A}} : R \longmapsto \{(\bar{u}, \bar{\nu}) \mid \mathfrak{A}^* \models \psi(R, \bar{u}, \bar{\nu})\}.$$

Starting with $\tilde{\Psi}^0 = \varnothing$, we can also define the iterative stages $\tilde{\Psi}^m$ of this operator, but since it is in general neither inflationary nor monotone, this process is not guaranteed to reach a fixed point. (As an example consider the formula $\psi(R, x) \equiv \neg Rx$, whose stages oscillate between the empty and the full relation.) Define the *partial fixpoint* of ψ on $\mathfrak{A}$ to be $\tilde{\Psi}^m$ if there is an m with $\tilde{\Psi}^{m+1} = \tilde{\Psi}^m$, and to be empty otherwise.

Partial fixpoint logic with counting, denoted (PFP + C), is defined in the same way as (FP + C), but with the inflationary fixpoint operation replaced by a partial fixpoint operation.

Fixpoint logics without counting. There is an extensive literature on various forms of fixpoint logics without counting terms (see e.g. [2, 4, 6, 7, 13, 14, 17, 18, 21]). We just recall a few important facts about fixpoint logic:

First, on every class of finite structures, the expressive power of fixpoint logic is between first-order definability and PTIME-computability. There are two extreme cases: On sets, fixpoint logic collapses to first-order logic, and on ordered structures, fixpoint logic coincides with PTIME [14, 21]. On most other classes, and in particular on the class of all finite structures, fixpoint logic is more powerful than first-order logic, but does not capture all PTIME-queries.

Abiteboul and Vianu have shown, that in the absence of counting, partial fixpoint logic has the same expressive power as the query language *while*, which is first-order logic extended by while-loops as an iteration construct. Together with a result of Vardi [21], this proves that on ordered structures, partial fixpoint logic captures PSPACE.

2.2 Datalog with counting

Let us first recall the basic definitions of Datalog:

Definition 2.2 A *Datalog program* Π consists of a finite set of rules of the form

$$H \leftarrow B_1 \wedge \cdots \wedge B_m$$

where H is an atomic formula $S(x_1, \ldots, x_r)$, called the *head* of the rule, and $B_1 \wedge \cdots \wedge B_m$ is a conjunction of atomic formulae $R(x_1, \ldots, x_m)$ and equalities $x_i = x_j$. Every x_i is either a variable or a constant. A predicate that occurs in the head of some rule is called an *intensional database predicate*, abbreviated IDB predicate; a predicate occuring only in the bodies of the rules is called an *extensional database predicate*, or EDB predicate. One of the IDB predicates is the *goal predicate* of the program. The extensional vocabulary of Π is formed by all EDB predicates and by all constants occurring in Π; a finite structure of this vocabulary is called an *extensional database* EDB for Π. Given any extensional database $\mathfrak{B}$, the program computes intensional relations, by the usual fixpoint semantics (or, equivalently, minimum model semantics). The result of Π on $\mathfrak{B}$ is the value of the goal predicate after the computation has terminated. If Π is a Datalog program with goal predicate Q we denote by (Π, Q) the query computed by this program.

On unordered databases, Datalog, in fact even Stratified Datalog, is a proper subset of fixpoint logic [7, 17]. On ordered structures, Datalog is also weaker than fixpoint logic. However, Datalog($\neg$), i.e. Datalog with negations over the EDB predicates is equivalent to fixpoint logic over ordered structures and therefore captures polynomial time [4]. Here we show that in the presence of counting, the situation is different: Datalog with counting terms has the same expressive power as fixpoint logic with counting.

We can increase the power of Datalog by a counting mechanism in a very similar way as we did with fixpoint logic:

Definition 2.3 *Datalog with counting*, denoted (Datalog + C), extends Datalog by allowing two-sorted IDB predicates and counting terms. The two-sorted IDB predicates have the form $R(\bar{x}, \bar{\mu})$, where $\bar{x}$ range over the first sort, i.e. over elements of the extensional database $\mathfrak{A}$, and $\bar{\mu}$ range over the second sort, i.e. over n. For any atom $R(x, \bar{y}, \bar{\mu})$ we have a counting term $\#_x[R(x, \bar{y}, \bar{\mu})]$. A term over the second sort is called an *arithmetical term*. The arithmetical terms are either 0, e, counting terms or $t + 1$, where t is also an arithmetical term. Thus, a program in (Datalog + C) is a finite set of clauses of the form

$$H \leftarrow B_1 \wedge \cdots \wedge B_m$$

where the head H is an atomic formula $R(\bar{x}, \bar{\mu})$, and $B_1, \ldots, B_m$ are atomic formulae $R(\bar{x}, \bar{\mu})$ or equalities of terms (over the first or the second sort).

On every extensional database, the program computes intensional relations by inflationary fixpoint semantics. Note that for classical Datalog programs, it makes no difference whether the fixpoint semantics is defined to be inflationary or not, since the underlying operator is monotone anyway. However, for programs in (Datalog + C), the semantic has to be inflationary, since otherwise, the equalities of arithmetical terms give rise to non-monotone operators. For the same reason, minimum model semantics will no longer be defined. Since inflationary fixpoint semantics is one of the various equivalent ways to define the semantics of Datalog, we feel that both syntax and semantics of (Datalog + C) generalize Datalog in a natural way.

We could also introduce counting in a (at first sight) more general form, namely by allowing counting terms of the form $\#_{\bar{x},\bar{\mu}}[R(\bar{x}, \bar{\mu}, \bar{y}, \bar{\nu})]$. While this may be convenient to write a program in shorter and more understandable form, it does not affect the power of (Datalog + C):

Lemma 2.4 *Counting over tuples, even of mixed type, does not increase the expressive power of* (Datalog + C).

PROOF. To illustrate the argument we show that counting over pairs can be reduced to counting over single elements. Note that

$$\#_{x,y}[Rxy] = \sum_x \#_y[Rxy] = \sum_\mu \mu \cdot \#_x[\#_y[Rxy] = \mu].$$

Using the two clauses

$$\begin{aligned} Ax\mu &\leftarrow (\#_y[Rxy] = \mu) \\ B\mu\nu &\leftarrow (\#_x[Ax\mu] = \nu) \end{aligned}$$

we can build up a purely arithmetical predicate B, which is the graph of the function $f(\mu) = \#_x[\#_y[Rxy] = \mu]$. We then have to build a program for the expression $\sum_\mu \mu f(\mu)$, which is no problem since, over the second, ordered sort, arithmetical expressions of this kind can be handled in Datalog. Counting over elements and tuples of the second sort can also be easily simulated using similar arithmetical expressions. ∎

Hence cardinalities of arbitrary predicates can be equated in a Datalog program: we take the liberty to write equalities like $|Q| = |R|$ in the body of a rule, for simplicity.

Lemma 2.5 *Let* Π *be a* (Datalog + C)*-program with IDB predicates* $Q_1, \ldots, Q_r$. *There exists another* (Datalog + C)*-program* Π', *whose IDB predicates include* $Q_1, \ldots, Q_r$ *and a Boolean control predicate* C^* *such that*

- $(\Pi', Q_i) = (\Pi, Q_i)$ *for all* i;
- (Π', C^*) *is true on all databases and* C^* *becomes true only at the last stage of the evaluation of* Π'.

PROOF. Besides C^*, we add a unary IDB predicate C^0 and, for every IDB predicate Q_i of Π a new IDB predicate Q_i' of the same arity. Then, Π' is obtained by adding the following clauses to Π:

$$\begin{array}{rcl} C^0(x) & & \\ Q'_i(\bar{x},\bar{\mu}) & \leftarrow & Q_i(\bar{x},\bar{\mu}) \quad \text{for } 1 \le i \le r \\ C^* & \leftarrow & C^0(x) \wedge (|Q_1| = |Q'_1|) \wedge \cdots \wedge (|Q_r| = |Q'_r|) \end{array}$$

Observe that Q'_i just lags one step behind Q_i up to the point of saturation. The atom $C^0(x)$ is necessary to avoid that C^* is set to true right at the first stage. ∎

Lemma 2.5 essentially says, that we can attach to any program a Boolean control predicate which becomes true when the evaluation of the program is terminated. We can then compose two Datalog programs while making sure that the evaluation of the second program starts only after the first has been terminated. As a first application, we show that (Datalog + C) is closed under negation.

Lemma 2.6 *The complement of a* (Datalog + C)*-query is also a* (Datalog + C)*-query.*

PROOF. Let (Π, Q) be a (Datalog + C)-query, and let Π' be the program as specified in Lemma 2.5. Take a new variable z and new IDB predicates $\tilde{Q}$ and R with $\text{arity}(R) = \text{arity}(Q)$ and $\text{arity}(\tilde{Q}) = \text{arity}(Q) + 1$. Construct Π'' by adding to Π' the rules:

$$\begin{array}{rcl} \tilde{Q}(\bar{x},\bar{\mu},z) & \leftarrow & Q(\bar{x},\bar{\mu}) \\ R(\bar{x},\bar{\mu}) & \leftarrow & C^* \wedge (\#_z[\tilde{Q}(\bar{x},\bar{\mu},z)] = 0) \end{array}$$

The query (Π'', R) is the complement of (Π, Q). ∎

Difficulties to express negation are the reason why, in the absence of counting (or of an ordering), Datalog is weaker than fixpoint logic. Also the limited form of negation that is available in Stratified Datalog (which does not allow for "recursion through negation") does not suffice to express all fixpoint queries [7, 17]. We now prove that (Datalog + C) does not have these limitations and is equally expressive as (FP + C):

Theorem 2.7 (Datalog + C) = (FP + C).

PROOF. It is obvious that (Datalog + C) $\subseteq$ (FP + C). Conversely, we show that for every formula $\psi \in$ (FP + C), there exists a (Datalog + C)-program Π_ψ with goal predicate Q_ψ such that (Π_ψ, Q_ψ) is equivalent to ψ.

If ψ is atomic, this is trivial. Also, both Datalog and (Datalog + C) are trivially closed under disjunction and existential quantification. Closure under negation has already been proved.

We next consider counting terms: Suppose that we have a program Π_φ describing the formula $\varphi(x, \bar{y}, \bar{\mu})$ in the way just specified. By Lemma 2.5, we can assume that Π_φ contains a control predicate C^*_φ becoming true at the last stage of the evaluation of Π_φ. We need a program Π_ψ for the formula $\psi(\bar{y},\bar{\mu},\nu) \equiv \#_x[\varphi(\bar{x},\bar{y},\bar{\mu})] = \nu$. This can be obtained by adding to Π_φ the clause

$$Q_\psi(\bar{y},\bar{\mu},\nu) \leftarrow C^*_\varphi \wedge \#_x[Q_\varphi(\bar{x},\bar{y},\bar{\mu})] = \nu.$$

Finally, we explain how to simulate fixpoint formulae $\psi \equiv [\text{IFP}_{R,\bar{x},\bar{\mu}}\, \varphi(R,\bar{x},\bar{\mu})](\bar{y},\bar{\nu})$. By induction hypothesis, there exists a program Π_φ with goal predicate Q_φ and control predicate C^*_φ which defines φ. Of course R is an EDB predicate of Π_φ.

Construct Π_ψ as follows: For every (EDB or IDB) predicate Q in Π_φ (including the control predicate), take a predicate Q' whose arity over the second sort is increased by a sufficiently large number ℓ; intuitively $Q'(\bar{x},\bar{\mu},\bar{\lambda})$ shall mean that $Q(\bar{x},\bar{\mu})$ is true at stage $\bar{\lambda}$ of the evaluation of the fixpoint operator.

Start with the clauses

$$C^*_\varphi(\bar{0})$$
$$Q'(\bar{x},\bar{\mu},0) \leftarrow Q(\bar{x},\bar{\mu}) \qquad \text{for all predicates } Q$$

Add to this the program obtained from Π_φ as follows

1. Replace all EDB atoms $Q(\bar{x},\bar{\mu})$ by $Q'(\bar{x},\bar{\mu},\bar{\alpha})$ (this includes the casé that $Q = R$).
2. Replace all IDB atoms $Q_\varphi(\bar{x},\bar{\mu})$ by $R'(\bar{x},\bar{\mu},\bar{\alpha}+1)$ and all other IDB atoms $P(\bar{x},\bar{\mu})$ by $P'(\bar{x},\bar{\mu},\bar{\alpha}+1)$.
3. Add to the body of every clause the predicate $C^*_\varphi(\bar{\alpha})$.
4. Add the clauses
$$R'(\bar{x},\bar{\mu},\bar{\alpha}+1) \leftarrow R'(\bar{x},\bar{\mu},\bar{\alpha})$$
$$Q_\psi(\bar{x},\bar{\mu}) \leftarrow R'(\bar{x},\bar{\mu},\bar{\alpha})$$

This programs simulates the evalution of the fixpoint operator stage per stage. ∎

To illustrate the expressive power of (Datalog + C) we exhibit a simple program for the GAME query. The GAME query is complete for fixpoint logic with respect to quantifier-free interpretations and is the canonical example that separates fixpoint logic from Stratified Datalog [7, 17]. Given a structure $\mathfrak{A} = (A,S,M)$ with universe A to be interpreted as a set of positions in a two-player game, with a monadic predicate S defining the positions where Player I has won, and binary predicate M describing the possible moves. The GAME query asks for the set Z of positions from which Player I has a winning strategy when he is to make the next move. GAME is definable in fixpoint logic, since

$$Z = [\mathrm{IFP}_{W,x}\, Sx \vee \exists y \forall z (Mxy \wedge (Myz \rightarrow Wz))].$$

Here is a (Datalog + C)–program with goal predicate Z defining GAME:

$$Wx0 \leftarrow Sx$$
$$Uyz\mu \leftarrow Myz \wedge Wz\mu$$
$$Vy\mu \leftarrow \#_z[Myz] = \#_z[Uyz\mu]$$
$$Wx\lambda \leftarrow Mxy \wedge Vy\mu \wedge \lambda = \mu + 1$$
$$Zx \leftarrow Wx\mu$$

The evaluation of this program on a game structure $\mathfrak{A}$ assigns to W (respectively V) the set of pairs $(x,\mu) \in A \times |A|$, such that Player I has a winning strategy from position x in μ moves when he (respectively Player II) begins the game.

2.3 A functional logic

Gurevich showed that the usual schemes for defining recursive functions on the natural numbers define, when interpreted over finite ordered structures, precisely the polynomial-time computable global functions [11, 12]. In particular, both fixpoint logic and recursive functions coincide over ordered finite structures with polynomial-time computability.

In this section we define a similar functional scheme that works over the two-sorted structures $\mathfrak{A}^*$. We obtain a class of global functions which corresponds to (FP + C).

Definition 2.8 A global function f of arity (j,k), coarity (m,ℓ) and vocabulary σ, is a functor that defines for every σ-structure $\mathfrak{A}$ a function

$$f^{\mathfrak{A}} : A^j \times n^k \longrightarrow A^m \times n^\ell$$

where, as above, $n = |A| + 1$.

It should be kept in mind that there is a natural bijection of numbers up to $n^m - 1$ with m-tuples of numbers up to $n-1$. So actually we consider the values in the second sort of these functions to be in an initial subset of $\mathbb{N}$, and we will take the liberty to perform the usual arithmetical operations on these values.

Definition 2.9 The *basic σ-functions* are

(i) the functions in σ and the characteristic functions χ_R of the predicates $R \in \sigma$ (the coarity of χ_R is of course $(0,1)$);

(ii) the characteristic functions of equality (over either sort)

$$\begin{aligned} \mathbf{eq}(x,y) &= (\textbf{if } x = y \textbf{ then } 1 \textbf{ else } 0) \\ \mathbf{eq}(\mu,\nu) &= (\textbf{if } \mu = \nu \textbf{ then } 1 \textbf{ else } 0) \end{aligned}$$

(iii) the constants 0, e (considered as functions of arity $(0,0)$ and coarity $(0,1)$);

(iv) the successor function S, with the convention that $Se = e$;

(v) the function $1 \dot{-} \mu = (\textbf{if } \mu = 0 \textbf{ then } 1 \textbf{ else } 0)$ (of arity and coarity $(0,1)$);

(vi) the projections.

Composition is defined in the obvious way: if $H_1, \dots H_m$ are global functions of the same arity, with coarities summing up to the arity of the global function F, then we can build the global function $F(\bar{x},\bar{\mu}) = G(H_1(\bar{x},\bar{\mu}), \dots, H_m(\bar{x},\bar{\mu}))$.

The summation operator allows to define from a given global function $G(x,\bar{y},\bar{\mu})$ of arity $(1+j,k)$ and coarity $(0,\ell)$ a new global function

$$F(\bar{y},\bar{\mu}) = \sum_x G(x,\bar{y},\bar{\mu})$$

of arity (j,k) and coarity $(0,\ell+1)$, with the obvious meaning. (Note that coarity $(0,\ell+1)$ suffices: the values of $G(x,\bar{y},\bar{\mu})$ are smaller than n^ℓ, so the value of $F(\bar{x},\bar{\mu})$ does not exceed $(n-1)(n^\ell - 1) < n^{\ell+1}$.)

Recursion. Let G be a global function of arity (j,k), coarity (m,ℓ) and vocabulary σ; let H be a global function of the same arity and coarity, but over the vocabulary $\sigma \cup \{h\}$ where h is a function symbol that has the same arity and coarity as G. Then a new global function of vocabulary σ, arity $(j,k+1)$ and coarity (m,ℓ) is defined by the scheme

$$\begin{aligned} F(\bar{x},\bar{\mu},0) &= G(\bar{x},\bar{\mu}) \\ F(\bar{x},\bar{\mu},\nu+1) &= [h(\bar{y},\bar{\lambda}) \equiv F(\bar{y},\bar{\lambda},\nu)]H(\bar{x},\bar{\mu}). \end{aligned}$$

The notation $[h(\bar{y},\bar{\lambda}) \equiv F(\bar{y},\bar{\lambda},\nu)]$ is to be understood as an explicit definition of h: to compute $F(\bar{x},\bar{\mu},\nu+1)$, one has to evaluate H, replacing every occurrence of a term $h(\bar{y},\bar{\lambda})$ by $F(\bar{y},\bar{\lambda},\nu)$.

Definition 2.10 Let $\mathcal{S}$ be the closure of the basic functions under composition, the summation operator and the recursion scheme given above.

Theorem 2.11 *$\mathcal{S}$ has the same expressive power as* (FP + C).

PROOF. It is easy to see that the graph of every function in $\mathcal{S}$ can be expressed by a formula in (FP + C). For the converse, we show that for every formula φ of arity (k, ℓ) in (FP + C), there exists a function $F_\varphi : A^k \times n^\ell \rightarrow \{0,1\}$ in $\mathcal{S}$, equivalent to the characteristic function of φ, and that every arithmetical term ν of arity (k, ℓ) is equivalent to a function $G_\nu : A^k \times n^\ell \rightarrow n$ in $\mathcal{S}$. These simulations are presented in the following table:

$x = y$	$\mathbf{eq}(x, y)$	$(\exists x)\psi$	$1 \dot{-} (1 \dot{-} \sum_x F_\psi)$
$R\bar{x}$	$\chi_R(\bar{x})$	$\#_x[\psi]$	$\sum_x F_\psi$
$\psi \wedge \varphi$	$F_\psi \cdot F_\varphi$	$\mu \leq \nu$	$1 \dot{-} (G_\mu \dot{-} G_\nu)$
$\neg\psi$	$1 \dot{-} F_\psi$		

Finally, a fixpoint formula $\psi \equiv [\text{IFP}_{R,\bar{y},\bar{\nu}}\ \varphi](\bar{x}, \bar{\mu})$ is described by $F_\psi = F(\bar{x}, \bar{\mu}, \bar{e})$ where F is defined from F_φ by the scheme

$$\begin{aligned} F(\bar{x}, \bar{\mu}, 0) &= [\chi_R \equiv 0] F_\varphi(\bar{x}, \bar{\mu}) \\ F(\bar{x}, \bar{\mu}, \lambda + 1) &= [\chi_R(\bar{y}, \bar{\nu}) \equiv F(\bar{y}, \bar{\nu}, \lambda)] F_\varphi(\bar{x}, \bar{\mu}). \end{aligned}$$

■

2.4 An algorithmic characterization

We introduce a machine model which leads to yet another characterization of (FP + C) and (PFP + C), respectively. The model itself is inspired by the corresponding approach to FP developed in [3, 1]. It is a "generic" or, as we prefer to say, isomorphism-preserving model of computation in the sense that the performance of the machine only depends on the isomorphism type of the input structure (and not on a particular choice of representation which adds extra structure, and hence reduces symmetry).

The intended algorithms can be performed by the following type of machines which consist of two components:

- A relational store: a sequence of relational registers, $(R_i)_{i \in \omega}$, all of some fixed arity r. Every register R_i can hold a collection of r-tuples of elements from the universe of the input structure. The register contents should be thought of as current interpretations of auxiliary r-ary relations over the input domain.

- A Turing component, with a work tape with read/write head as usual, and also extra tapes (with write heads only), which are used for communication from the Turing component to the relational store.

The interaction of the components is twofold:

(i) the Turing component can trigger the execution of certain algebraic operations on the contents of the relational registers, and

(ii) the performance of the Turing control (apart from the usual dependency) also depends on the information whether or not one particular relational register, R_0 say, is empty.

The finite set of operations which can be applied to the relational store in a transition, $\mathcal{O}$ say, depends on the relational type of the input structures, and on the arity r of the relational registers. $\mathcal{O}$ contains the boolean operations, load operations for the input relations and for equality, and the crucial "counting projections". The latter correspond to updates $R_{i_1} := \{(x_1, \ldots, x_r) \mid \exists^{\geq i_0} x_j\ R_{i_2} x_1 \ldots x_r\}$ in which the relational indices i_1 and i_2 as well as the counting threshold i_0 are all to be regarded as parameters provided from the comunication tapes whenever the operation is called.

Definition 2.12 Let $\mathcal{M}^r[\tau]$ be the class of machines as characterized, with arity r for the registers in the relational store, and formatted for loading finite τ-structures. Let $\mathcal{M}[\tau] := \bigcup_r \mathcal{M}^r[\tau]$. The time or space complexity of an $\mathcal{M}$-computation is defined in terms of the corresponding resource of the Turing component.

The proof of the following theorem is to be found in [19].

Theorem 2.13 *The* PTIME *and* PSPACE *restrictions of the computational model $\mathcal{M}$ correspond to* (FP + C) *and* (PFP + C) *resp., in the sense that e.g. exactly those classes of finite τ-structures are definable by a sentence of* (FP + C)$[\tau]$, *which are accepted by a machine in $\mathcal{M}[\tau]$ in polynomial time.*

3 Fixpoint logic and infinitary logic with counting

Definition 3.1 Let $L^r_{\omega\omega}$ be first-order logic with only r variable symbols. Let $L^r_{\infty\omega}$ be the corresponding infinitary logic, i.e. with infinite disjunctions and conjunctions. Extending these by the set of all the counting quantifiers $\exists^{\geq m}$, for $m \in \omega$, we obtain $C^r_{\omega\omega} := L^r_{\omega\omega}(\exists^{\geq m})_{m\in\omega}$ and $C^r_{\infty\omega} := L^r_{\infty\omega}(\exists^{\geq m})_{m\in\omega}$, respectively. Also put $L^\omega_{\omega\omega} := \bigcup_r L^r_{\omega\omega}$, and similarly for the others: $C^\omega_{\omega\omega}$, $L^\omega_{\infty\omega}$, and $C^\omega_{\infty\omega}$.

Lemma 3.2 (Immerman) $\quad$ (FP + C) $\subsetneqq C^\omega_{\infty\omega}$.

PROOF. We want to show that every formula of (FP + C) without free variables of the second sort can be translated into a formula in $C^\omega_{\infty\omega}$. Both the fixpoint generation and the occurences of variables of the second sort have to be expanded in a uniform way which takes into account the size of the structure under consideration. So the formulae we get in $C^\omega_{\infty\omega}$ will be of the form

$$\bigvee_{n\in\omega} (\exists^{\geq n} x\,(x=x) \wedge \neg\exists^{\geq n+1} x\,(x=x) \wedge \varphi_n),$$

so that the φ_n just have to capture the meaning of the given formula over structures of size n.

Remark. It is not difficult to see that this is a normal form for $C^\omega_{\infty\omega}$, with the $\varphi_n \in C_{\omega\omega}$ even, in the sense that any formula in $C^\omega_{\infty\omega}$ is equivalent (in the finite) to one of these. (Use the fact that on every fixed finite structure, every formula of $C^\omega_{\infty\omega}$ is equivalent to one in $C^\omega_{\omega\omega}$.)

We only briefly sketch how to deal with second-sort variables, since the expansion of fixpoint constructions is standard. Consider a formula $\psi(x,\nu)$. W.r.t. ν we expand to a sequence of formulae $(\psi_\nu(x))_{\nu\in\omega}$ such that the mixed-type global relation given by $\psi(x,\nu)$ on structures of size n is just $\bigcup_{\nu\leq n}\{x|\psi_\nu(x)\} \times \{\nu\}$. In this way the translation is straightforward. To indicate one typical step in the inductive treatment, let $\psi(x,\nu) \equiv \nu \leq \#_y\chi(x,y)$. Suppose $\chi'_n(x,y)$ captures the meaning of $\chi(x,y)$ on size n

structures. Then the desired family for ψ consists of $\psi_{n,\nu} :\equiv \exists^{\geq\nu} y\, \chi'_n(x,y)$, for $\nu \leq n$, n the parameter for the size of the structure.

$C^{\omega}_{\infty\omega}$ is much stronger than (FP + C), since it expresses non-recursive properties: For any predicate $R \subset \omega$ consider the sentence $\bigvee_{n\in R}(\exists^{\geq n}x\,(x=x) \wedge \neg\exists^{\geq n+1}x\,(x=x))$.
■

It is possible to isolate the fragment of $C^{\omega}_{\infty\omega}$ which corresponds to (FP + C) precisely. The required restriction is given in terms of a very natural uniformity condition:

Definition 3.3 Let $P\text{-}C^{\omega}_{\infty\omega}$ be the sublogic of $C^{\omega}_{\infty\omega}$ which consists of all formulae

$$\bigvee_{n\in\omega} (\exists^{\geq n}x\,(x=x) \wedge \neg\exists^{\geq n+1}x\,(x=x) \wedge \varphi_n),$$

for sequences $(\varphi_n)_{n\in\omega}$ in some $C^{r}_{\omega\omega}$ such that the map $n \mapsto \varphi_n$ is PTIME constructible.

Proposition 3.4 (FP + C) $= P\text{-}C^{\omega}_{\infty\omega}$.

PROOF. One inclusion is settled by the proof of the previous lemma. The proof of the remaining inclusion, from right to left, is easiest as an application of the algorithmic characterization of (FP + C). The machines in $\mathcal{M}^r$ can evaluate formulae corresponding to PTIME sequences in $C^{r}_{\infty\omega}$ quite naturally, see [19]. Alternatively, it can be shown that the PTIME construction of the relevant formula can – by means of a suitable encoding – be simulated over the second, ordered sort in (FP + C); similarly the evaluation of the encoded formula over the first sort is rendered definable in (FP + C). ■

Remark. This is quite unlike the situation for FP itself. Let $P\text{-}L^{\omega}_{\infty\omega}$ be the logic consisting of all formulae $\bigvee_{n\in\omega}(\exists^{\geq n}x\, x=x \wedge \neg\exists^{\geq n+1}x\, x=x \wedge \varphi_n)$, for PTIME constructible families in some $L^{r}_{\omega\omega}$, $r \in \omega$. This logic can be shown to be strictly more expressive than FP. Consider the following example. Let $R \subset \omega$ be a numerical predicate. Associate with R the class K_R of structures represented by $\{\mathfrak{A}_{mn} := (n, m, <\!|_m) \mid m < \log n, m \in R\}$. There are at most $(m+r)^r$ distinct $L^{r}_{\infty\omega}$-types realized over $\mathfrak{A}_{mn}$. It follows from the considerations in [3] that any FP formula over $\mathfrak{A}_{mn}$ closes in time polynomial in $(m+r)^r$ and hence in m. For R such that $m \in R$ is decidable in time 2^m but not in time $p(m)$ for any polynomial p, K_R can thus be shown to be definable in $P\text{-}L^{\omega}_{\infty\omega}$, but not in FP.

One of the nice features of the logics $C^{r}_{\infty\omega}$ is the simple Ehrenfeucht-Fraïssé type of characterization of $\equiv_{C^{r}_{\infty\omega}}$, the relation of indistinguishability of structures w.r.t. the logic $C^{r}_{\infty\omega}$. This characterization is best understood in its game formulation, in terms of suitably adapted pebble games. These were introduced in [16].

We fix a relational vocabulary τ. The game is played by two players, **I** and **II**, on a pair of finite τ-structures $(\mathfrak{A}, \mathfrak{A}')$. There are two sets of r pebbles each, one for each structure. Denote them by $p_1, \ldots, p_r$ and $p'_1, \ldots, p'_r$ resp. A move in the game comprises the following steps:

- **I** chooses one of the structures, $\mathfrak{A}$ say, a pair of corresponding pebbles (possibly a pair already placed on the sructures), say (p_j, p'_j), and a non-empty subset of the universe of the chosen structure, say $A_0 \subset A$ — **II** has to answer with a subset of the same size in the other structure, $A'_0 \subset A'$ say, with $|A'_0| = |A_0|$.

- **I** places the pebble p'_j within A'_0 — **II** must place the pebble p_j in A_0.

The game continues as long as the second player can maintain the following condition:

Let $\overline{a}$ and $\overline{a}'$ be the positions of the pebbles in $\mathfrak{A}$ and $\mathfrak{A}'$ resp., then the map taking a_j to a'_j for $j < r$, is a partial isomorphism, i.e. these tuples have the same atomic types in their respective structures.

I wins the game as soon as **II** violates this condition.

It can be shown (cf. [16, 5]) that **II** has a winning strategy in the game on $(\mathfrak{A}, \mathfrak{A}')$ (i.e. a way to maintain the above condition indefinitely) iff $\mathfrak{A}$ and $\mathfrak{A}'$ satisfy exactly the same sentences in $C^r_{\infty\omega}$. Or, that $\overline{a}$ satisfies exactly the same formulae in $\mathfrak{A}$ as does $\overline{a}'$ in $\mathfrak{A}'$ iff **II** has a strategy to continue the game on $(\mathfrak{A}, \mathfrak{A}')$ indefinitely from the situation in which the pebbles are placed on $\overline{a}$ and $\overline{a}'$, respectively. We shall denote the game played in this latter fashion, from starting positions with pebbles initially placed on $\overline{a}$ and $\overline{a}'$, as the game on $(\mathfrak{A}, \overline{a}; \mathfrak{A}', \overline{a}')$.

In the form presented here, the number of possible moves in a given situation is not polynomial since the choice of a set introduces exponential variation. In this sense the games are not suited for an exhaustive search for a strategy (as can be done directly for $L^r_{\infty\omega}$ and the associated ordinary pebble game). For a closer analysis, define the following equivalence relations:

Definition 3.5 Let $\sim^r_i$ be the equivalence relation defined (on the class of finite relational structures with a distinguished r-tuple of elements) through:

$$(\mathfrak{A}, \overline{a}) \sim^r_i (\mathfrak{B}, \overline{b})$$

iff in the game on $(\mathfrak{A}, \overline{a}; \mathfrak{B}, \overline{b})$, player **II** has a strategy to maintain the winning condition for at least i moves. Let $\sim^r$ be the corresponding relation for a strategy in the infinite game, or the common refinement of all the $\sim^r_i$.

It has to be checked of course that this relation is actually transitive. We point out that $\sim^r_0$ is just equality of atomic types.

Lemma 3.6 (Immerman/Lander) $(\mathfrak{A}, \overline{a}) \sim^r (\mathfrak{B}, \overline{b})$ *iff* $(\mathfrak{A}, \overline{a}) \equiv_{C^r_{\infty\omega}} (\mathfrak{B}, \overline{b})$. [1]

The proof is an application of the usual techniques related to Ehrenfeucht-Fraïssé games, see [16].

Another consequence of the transitivity of this notion of equivalence is that $\sim^r_i$ itself factorizes w.r.t. its own restriction to one and the same structure: For a single structure $\mathfrak{A}$, also denote by $\sim^r_i$ the equivalence relation on A^r induced by

$$\overline{a} \sim^r_i \overline{a}' \quad \text{iff} \quad (\mathfrak{A}, \overline{a}) \sim^r_i (\mathfrak{A}, \overline{a}').$$

Writing α, β for arbitrary elements of the quotients $A^r / \sim^r_i$ and $B^r / \sim^r_i$ resp., we use the notation $(\mathfrak{A}, \alpha) \sim^r_i (\mathfrak{B}, \beta)$ to say that for some (and hence all) $\overline{a} \in \alpha$ and $\overline{b} \in \beta$ we have $(\mathfrak{A}, \overline{a}) \sim^r_i (\mathfrak{B}, \overline{b})$.

Say that for $\alpha \in A^r / \sim^r_i$ and $\overline{a} = (a_1, \ldots, a_r) \in A$, α is j-close to $\overline{a}$ if there is an a in A such that $\overline{a}\frac{a}{a_j} \in \alpha$. The crucial point in our analysis of the game is the following claim:

$(\mathfrak{A}, \overline{a}) \sim^r_{i+1} (\mathfrak{B}, \overline{b})$ if and only if the following condition $(*)$ as well as its counterpart, with the roles of the structures exchanged, hold

[1]The notation on the right is to say that the tuples satisfy the same formulae over their respective structures (i.e. it should not be read in the sense of first expanding the structures with constants).

(*) For all $1 \le j \le r$, for all $\alpha \in A^r/\sim_i^r$, which are j-close to $\bar{a}$, there is some $\beta \in B^r/\sim_i^r$, which is j-close to $\bar{b}$, such that:
$(\mathfrak{A},\alpha) \sim_i^r (\mathfrak{B},\beta)$ and $\left| \{a \in A \mid \bar{a}\frac{a}{a_j} \in \alpha\} \right| = \left| \{b \in B \mid \bar{b}\frac{b}{b_j} \in \beta\} \right|$.

Note that the first of the conditions in (*) already determines the appropriate β uniquely.

The claim follows from an analysis of a single exchange of moves in the game. E.g., in the situation $(\mathfrak{A},\bar{a};\mathfrak{B},\bar{b})$ with pebble pair j and a subset $B_0 \subset B$ chosen by **I**, decompose B_0 into disjoint parts according to: $B_0 = \bigcup_{\beta \in B^r/\sim_i^r} \{b \in B_0 \mid \bar{b}\frac{b}{b_j} \in \beta\}$. The correct response of **II** then is to take portions of matching sizes from the respective $\sim_i^r$-equivalent classes over A to form A_0.

For future reference, we also exhibit the special simplified form which (*) takes for the case that we are working in just one structure $\mathfrak{A}$:

(**) $(\mathfrak{A},\bar{a}) \sim_{i+1}^r (\mathfrak{A},\bar{a}')$ iff for all $1 \le j \le r$, and for all $\alpha \in A^r/\sim_i^r$:
$\left| \{a \in A \mid \bar{a}\frac{a}{a_j} \in \alpha\} \right| = \left| \{a \in A \mid \bar{a}'\frac{a}{a'_j} \in \alpha\} \right|$.

This criterion suggests an inductive process for the generation of the limit $\sim^r$ of the $\sim_i^r$ on a single given structure $\mathfrak{A}$, and even of a linear ordering between the classes. Inductively we define pre-orderings $\preceq_i^r$ on the r-th power of the universe, which induce the equivalence relation $\sim^r$ on that domain (i.e.: $\bar{a} \sim_i^r \bar{a}'$ iff $\bar{a} \preceq_i^r \bar{a}' \wedge \bar{a}' \preceq_i^r \bar{a}$); the pre-ordering is best thought of as a strict linear ordering of the equivalence classes. Let $\prec_i^r$ denote the accompanying strict ordering: $\bar{a} \prec_i^r \bar{a}'$ iff $\bar{a} \preceq_i^r \bar{a}' \wedge \neg \bar{a}' \preceq_i^r \bar{a}$.

The following procedure inductively defines the stages $\prec_i^r$ uniformly on finite $\mathfrak{A}$:

- On the zero level, just fix any linear ordering of the atomic types.
- To pass from $\prec_i^r$ to $\prec_{i+1}^r$, we can inductively suppose that $\prec_i^r$ already induces a strict linear ordering on $A^r/\sim_i^r$, and put for $\bar{a},\bar{a}' \in A^r$: $\bar{a} \prec_{i+1}^r \bar{a}'$ iff not $\bar{a} \sim_{i+1}^r \bar{a}'$, and for the least j such that condition (**) is violated, and for the $\prec_i^r$-least class $\alpha \in A^r/\sim_i^r$ violating (**) for that j we have: $\left| \{a \in A \mid \bar{a}\frac{a}{a_j} \in \alpha\} \right| < \left| \{a \in A \mid \bar{a}'\frac{a}{a'_j} \in \alpha\} \right|$.

This process will, in polynomially many steps, lead to saturation. The limit $\prec^r$ of the $\prec_i^r$ thus obtained will be a linear ordering of the equivalence classes w.r.t. $\sim^r$. In other words, this procedure uniformly gives a linear ordering of the $C_{\infty\omega}^r$-types.

A closer analysis of this procedure shows that its outcome $\prec^r$, regarded as a global relation of arity $2r$ on the universe, is almost definable by a formula of FP; the only ingredient FP lacks for this is simple cardinality comparison. The Rescher quantifier exactly introduces this extra expressive power, [20]:

Definition 3.7 The Rescher quantifier Q_R is the Lindström quantifier, which binds two formulae, via one single variable each: From $\varphi_1(x,\bar{z})$ and $\varphi_2(y,\bar{z})$ the new formula $\psi(\bar{z}) \equiv Q_R xy\ (\varphi_1(x,\bar{z}),\varphi_2(y,\bar{z}))$ is formed. Its semantics is defined such that $\psi(\bar{z})$ expresses that $|\{x|\varphi_1(x,\bar{z})\}| < |\{y|\varphi_2(y,\bar{z})\}|$. Let FP($Q_R$) be the extension of FP obtained by closing w.r.t. quantification with Q_R.

Lemma 3.8 *The linear ordering $\prec^r$ can be defined as a global relation of arity $2r$, by a formula in the extension of FP by just the Rescher quantifier, FP(Q_R).*

In particular, $\prec^r$ is definable in (FP + C). Coming back to the analysis of (FP +C), we want to combine the structural information contained in $(\mathfrak{A},\prec^r)$, in one derived structure associated with $\mathfrak{A}$. Since this abstracted structure will be of a purely numerical nature, we apply the term "arithmetical invariant" to the functor which takes $\mathfrak{A}$ to this derived structure. It is an invariant in the sense that the value depends only on the isomorphism type of $\mathfrak{A}$ (as it should, of course). To prepare the definition of this functor introduce the following notation: For $1 \leq k \leq k' \leq r$, the global relations $\Delta_{kk'} = \{\bar{x} | x_k = x_{k'}\}$ and for $P \in \tau, 1 \leq \bar{k} \leq r$ the global relations $P_{\bar{k}} = \{\bar{x} | P x_{k_1} \dots x_{k_r}\}$ are defined in terms of atomic types. It follows that they factorize w.r.t. $\sim^r$. As these classes have been definably ordered, the above relations can be presented as numerical relations. For $\mathfrak{A}$, let $\alpha_0, \alpha_1, \dots, \alpha_s$ be an enumeration of $A^r / \sim^r$, ordered w.r.t. $\prec^r$. Then put

$$\underline{P_{\bar{k}}}(\mathfrak{A}) := \{i \leq s \,|\, \alpha_i \subset P_{\bar{k}}\},$$

similarly for the Δ. In an analogous way, abstract the relevant counting information from $\mathfrak{A}$, putting, for $1 \leq j \leq r$:

$$\underline{C_j}(\mathfrak{A}) := \Big\{(i, i', k) \;\Big|\; \text{for } \bar{a} \in \alpha_i : |\{a | \bar{a}\frac{a}{a_j} \in \alpha_{i'}\}| = k\Big\}.$$

Definition 3.9 Let I^r be the following functor on the class of all finite τ-structures:

$$I^r\colon \quad \mathfrak{A} \longmapsto \Big(n^r, <, (\underline{\Delta_{kk'}})_{1 \leq k \leq k' \leq r}, (\underline{P_{\bar{k}}})_{P \in \tau, 1 \leq \bar{k} \leq r}, (\underline{C_j})_{1 \leq j \leq r}\Big)(\mathfrak{A}),$$

where n is the cardinality of the domain A.

Theorem 3.10 *For the arithmetical invariant I^r:*

(i) I^r characterizes all finite τ-structures exactly up to $\equiv_{C^r_{\infty\omega}}$:
$\mathfrak{A} \equiv_{C^r_{\infty\omega}} \mathfrak{A}'$ iff $I^r(\mathfrak{A}) = I^r(\mathfrak{A}')$.

(ii) I^r can be computed in PTIME.

(iii) There is a formula of (FP + C) *which globally interprets (an encoding of) $I^r(\mathfrak{A})$ in the second sort of $\mathfrak{A}^*$.*

(iv) (FP + C) = *FP(I^r). Here FP(I^r) is shorthand for the class of those properties of finite τ-structures $\mathfrak{A}$, which are FP-definable over the corresponding $I^r(\mathfrak{A})$.*

PROOF. *(i)* follows from the game characterization of $\equiv_{C^r_{\infty\omega}}$: if $I^r(\mathfrak{A}) = I^r(\mathfrak{B})$ then a strategy for player **I** in the game on $(\mathfrak{A}; \mathfrak{B})$ can be extracted; recall how the equivalence classes of $\sim^r$ relate to the game and how they are represented in I^r.

(ii) is obvious from the FP(Q_R)-definability of $\prec^r$, and the way in which $I^r(\mathfrak{A})$ can be obtained from $(\mathfrak{A}, \prec^r)$ by simple counting.

(iii) again follows directly from observations already made, or from the proof of *(ii)*, together with the fact that $I^r(\mathfrak{A})$ is encoded as a numerical predicate within n^r. In fact, also the projection $\pi : A^r \longrightarrow A^r / \sim^r$ is definable in this sense.

(iv): the inclusion from right to left is a consequence of *(iii)*. One way to obtain the converse is presented in [19], where it is shown that the transition to the arithmetical invariant I^r provides a normal form for computations in $\mathcal{M}^r$. Then the result follows from the fact that PTIME(I^r) = FP(I^r) and Theorem 2.13 above.

We here want to indicate a direct proof very briefly: Since (FP + C) $\subset C^{\omega}_{\infty\omega}$ by Lemma 3.2, any relation definable in (FP + C) over $\mathfrak{A}$ will be a union of equivalence classes in $A^r / \sim^r$, for some r. (Here we treat second-sort variables as external parameters as we did in Lemma 3.2.) It is thus possible to replace any such relation R (w.l.o.g. of arity r) by its representation over $I^r(\mathfrak{A})$: $\underline{R} := \{i < n^r \,|\, \alpha_i \subset R\}$, where, as in the definition of I^r, α_i refers to the ordered enumerastion of $A^r / \sim^r$. In this way, any (FP + C) formula can be "simulated" over I^r in a uniform way. The arithmetic governing the variables of the second sort is all PTIME and hence FP-definable over the linearly ordered structures I^r (note that $|I^r(\mathfrak{A})| = n^r$). The evaluation of counting terms, finally, uses just the information encoded in the $\underline{C_j}$ in I^r. ■

4 Inflationary and partial fixpoints

As pointed out above, in Theorem 2.13, (FP + C) and (PFP + C), or the extensions of *fixpoint* and *while* by counting, correspond to the PTIME and PSPACE restrictions, resp., of the machine model $\mathcal{M}$. In particular, we have that the PTIME restriction of $\mathcal{M}$ captures exactly those properties which are expressible in (PFP + C) under the additional restriction that the PFP-generations must all close in polynomial time.

Definition 4.1 Let PFP $|_P$ stand for this extension of first-order logic by partial fixpoint operators which must close within polynomially many steps. Similarly define (PFP + C) $|_P$.

As already mentioned, from the algorithmic characterization, we directly get:

Proposition 4.2 (PFP + C) $|_P$= (FP + C) .

No generic model is known with equally nice restrictions to *fixpoint* and *while* in the absence of counting. In fact the following result of Abiteboul and Vianu, [3], makes it rather unlikely that a natural model with these restriction properties exists:

Theorem 4.3 (Abiteboul/Vianu) PFP $|_P$= FP *iff* PTIME = PSPACE

Another important result of [3] survives the extension to the case with counting:

Lemma 4.4 *If* PSPACE = PTIME, *then* (PFP + C) = (FP + C).

PROOF. As already indicated above, and excplicitly proved in [19], the properties definable in (PFP + C) or (FP + C) are just those which can be computed in PSPACE or PTIME from the invariants I^r (by an ordinary Turing machine, say). This immediately proves the claim. ■

Together with the obvious converse to this lemma (consider linearly ordered structures; here (PFP + C) coincides with PFP or PSPACE, similarly (FP + C) = FP = PTIME), we get an extension of the above-mentioned theorem of Abiteboul/Vianu:

Theorem 4.5 (PFP + C) = (FP + C) *iff* PSPACE = PTIME .

References

[1] S. Abiteboul, M. Vardi and V. Vianu, *Fixpoint Logics, Relational Machines and Computational Complexity*, Proceedings of 7th IEEE Symposium on Structure in Complexity Theory (1992).

[2] S. Abiteboul and V. Vianu, *Datalog Extensions for Database Queries and Updates*, J. Comp. Syst. Sciences **43** (1991), 62–124.

[3] S. Abiteboul and V. Vianu, *Generic Computation and Its Complexity*, Proceedings of 23rd ACM Symposium on Theory of Computing (1991).

[4] A. Blass and Y. Gurevich, *Existential fixed-point logic*, in: "Computation Theory and Logic" (E. Börger, Ed.), Lecture Notes in Computer Science Nr. 270, Springer 1987, 20–36.

[5] J. Cai, M. Fürer and N. Immerman, *An Optimal Lower Bound on the Number of Variables for Graph Identification*, Proceedings of 30th IEEE Symposium on Foundations of Computer Science (1989), 612–617.

[6] A. Chandra and D. Harel, *Structure and Complexity of Relational Queries*, J. Comp. Syst. Sciences **25** (1982), 99–128.

[7] E. Dahlhaus, *Skolem Normal Forms Concerning the Least Fixed Point*, in: "Computation Theory and Logic" (E. Börger, Ed.), Lecture Notes in Computer Science Nr. 270, Springer 1987, 101–106.

[8] A. Dawar, S. Lindell and S. Weinstein, *Infinitary Logic and Inductive Definability over Finite Structures*, Technical Report, University of Pennsylvania, (1991).

[9] P. Dublish and S. Maheshwari, *Query Languages which Express all PTIME Queries for Trees*, Hildesheimer Informatik-Berichte 2 (1989).

[10] S. Grumbach and C. Tollu, *The Generic Complexity of Query Languages with Counters*, Proceedings of Third Workshop on Foundations of Models and Languages for Data and Objects, Aigen (Austria) 1991.

[11] Y. Gurevich, *Algebras of feasible functions*, Proceedings of 24th IEEE Symposium on Foundations of Computer Science 1983, 210–214.

[12] Y. Gurevich, *Logic and the Challenge of Computer Science*, in: E. Börger (Ed), Trends in Theoretical Computer Science, Computer Science Press (1988), 1–57.

[13] Y. Gurevich and S. Shelah, *Fixed Point Extensions of First Order Logic*, Annals of Pure and Applied Logic **32** (1986), 265–280.

[14] N. Immerman, *Relational Queries Computable in Polynomial Time*, Information and Control **68** (1986), 86–104.

[15] N. Immerman, *Expressibility as a Complexity Measure: Results and Directions*, Proc. of 2nd Conf. on Structure in Complexity Theory (1987), 194–202.

[16] N. Immerman and E. Lander, *Describing Graphs: A First Order Approach to Graph Canonization*, in: A. Selman (Ed), Complexity Theory Retrospective. (In Honor of Juris Hartmanis), Springer, New York 1990, 59–81.

[17] Ph. Kolaitis, *The Expressive Power of Stratified Logic Programs*, Information and Computation **90** (1991), 50–66.

[18] D. Leivant, *Inductive Definitions over Finite Structures*, Information and Computation **89** (1990), 95–108.

[19] M. Otto, *The Expressive Power of Fixed-Point Logic with Counting*, submitted for publication.

[20] N. Rescher, *Plurality Quantification*, JSL **27** (1962), 373–374.

[21] M. Vardi, *Complexity of Relational Query Languages*, Proc. of 14th ACM Symposium on Theory of Computing (1982), 137–146.

Linear time algorithms and NP-complete problems

by Etienne Grandjean
Université de Caen, LIUC
14032 Caen Cedex, FRANCE

Abstract. We define and study a machine model (random access machine with powerful input/output instructions) and show that for this model the classes, DLINEAR and NLINEAR, of problems computable in deterministic (resp. nondeterministic) linear time are robust and powerful. In particular DLINEAR includes most of the concrete problems commonly regarded as computable in linear time (such as graph problems : topological sorting, strong connectivity...) and most combinatorial NP-complete problems belong to NLINEAR. The interest of NLINEAR class is enhanced by the following fact : some natural NP-complete problems, for example "reduction of incompletely specified automata" (in short : RISA), are NLINEAR-complete (consequently, NLINEAR $\neq$ DLINEAR iff RISA $\notin$ DLINEAR). That notion probably strengthens NP-completeness since we argue that propositional satisfiability is not NLINEAR-complete.

I - Introduction and discussion

One commonly admits that the most efficient algorithms run in linear time. However, to our knowledge, there is no robust and canonical definition of linear time in literature. (On the contrary, polynomial time is a very robust notion : it is not sensitive neither to machine model nor to presentation of the input, provided that they are "reasonable".) Papers and books (for example [AHU1, AHU2, FGS, Kn, Me, Se]) concerning algorithms explicitly or - more often- implicitly use a variant of random access machine (RAM of Cook and Reckhow [CoRe]) but do not precisely explain or do not justify what "the good model" should be : what should be elementary instructions (addition, substraction, multiplication ?...) ? the input/output process ? (A notable exception is paper [AnVa] of Angluin and Valiant which describes and justifies the RAC model ; however they have studied it only partially.)

- One generally adopts the uniform time criterion : each instruction requires one time unit ; but it is not clear what an elementary instruction should be.

- An algorithm runs in linear time if it executes O(n) instructions on each input of "length" (or "size") n. As the notion of instruction depends on the chosen machine, the notion of input length depends on the nature of the input. The "length" of a word $w \in \Sigma^n$ in a finite alphabet Σ is n. The "size" of a graph (represented by its list of edges or by the successor lists of its vertices) is m + p where m (resp. p) is the number of vertices (resp. edges) : for example [Ta] and [AHU1] compute the strongly connected components of a directed graph in "linear time" O(m + p). The "length" of a propositional formula is often (but not always) the number n of its symbols : occurrences of connectives, parentheses (example : the "length" of formula $(p_0 \vee \neg p_1) \wedge p_{10}$ is 8) ; for example, [DoGa, ItMa1, ItMa2, Mi] describe some decision algorithms for satisfiability of Horn propositional formulas which run in "linear time" O(n).

Gurevich and Shelah [GuSh] ("nearly linear time") and Grädel [Gl] gave two "robust closures" of linear time. Previously, Schnorr [Sr] had similarly defined "quasilinear time" (that is time $O(n (\log n)^{O(1)})$ on some Turing machine) and proved that many NP-complete problems belong to nondeterministic quasilinear time class and are complete for this class. Those authors define extensions of linear time because, as [GuSh] explains, "it is possible that there is no universal notion of linear time and different versions of linear time are appropriate to different applications".

The present paper adopts the opposite point of view : we define and justify a unified, robust and powerful notion of linear time both in deterministic and nondeterministic cases. In a precedent paper [Gr 3], we define a complexity class, denoted LINEAR, by using a variant of classical RAM of [CoRe] under logarithmic cost criterion (time of each instruction is the total number of bits it manipulates, i.e. the sum of lengths of the integers it involves). Our RAM extends the classical model in some points and restricts it in some other ones. (Our purpose is to get a more realistic and more robust machine). Let us describe informally the differences.

- The input process of classical RAM is very restrictive : the input can only be read one bit at a time (the output instruction of [CoRe] is more powerful : the RAM writes the whole contents of a register onto the output tape) ; Schönhage has proved that this RAM cannot compute the very simple function $w \mapsto ww$ (repetition of the input word w !) in linear time under logarithmic cost ; we allow our RAM to read its input in blocks by an instruction, denoted $READ_v$ (u), which stores in register u a subword of the input as long as contents of register v. Note that our input and output instructions are symmetrical each other (in some manner) .

- In the classical model, the RAM memory can be utilized in a very scattered way since addresses and register contents are unbounded ; on this point, our RAM is similar to the RAC ("random access computer") of Angluin and Valiant [AnVa, GuSh] : a computation is allowed to use only integers polynomially bounded into the input length n (i.e. their length is O(logn)).

- The only allowed operations of the RAM of [CoRe] are addition + and substraction - ; we allow our RAM to use any operation computable in linear time on a Turing machine (i.e. intuitively : "easy" operation) : +, -, concatenation but neither multiplication nor division. (Invariance properties of RAM_S with compact memory and linear time Turing computable operations have been stated in [GrRo].)

Let LINEAR denote the class of functions computable in linear time on the RAM described above ; more precisely, a function $f : \Sigma^* \to \Sigma^*$ belongs to LINEAR iff such a RAM computes for each input $w \in \Sigma^*$, the value f(w) in time O(length(w)) under logarithmic cost. In [Gr3] we have shown that LINEAR is a robust complexity class. However we are unable to prove that it is closed under composition (i.e. if $f,g \in$ LINEAR then $f \circ g \in$ LINEAR) unless RAM_S are allowed to use files (to contain intermediate results).

In the present paper we still study the RAM above defined but use the uniform cost criterion. This does not change the invariance properties of the machine model but allows to prove that the class, denoted DLINEAR, of functions computable by RAMs without files within time O(n/logn) (under uniform cost : n is the length of the input) is closed under composition. Note that DLINEAR $\subseteq$ LINEAR. The O(n/logn) time bound of DLINEAR seems to be very restrictive but is justified by the following facts :

(i) Katajainen and others [KvLP] have proved that any function f computable in linear time O(n) on a Turing machine can be computed in time O(n log logn) on the classical RAM of [CoRe] (under logarithmic cost) and they have mentioned that "with better input/output pattern the simulation could perhaps be sped up further". We can prove (cf. Section III below) that function f belongs to DLINEAR (i.e. is computable by our RAM in time O(n/logn) under uniform cost).

(ii) Many (most of ?) concrete problems commonly regarded as computable in linear time belong to DLINEAR : that contrasts with the (above-mentioned) lower bound result of Schönhage and shows the power of our input pattern.

The rest of the present paper is dedicated to the nondeterministic version of linear time. As mentioned above, Schnorr [Sr] has proved that many NP-complete problems are complete in nondeterministic quasilinear time class (via deterministic quasilinear time reduction). His machine model was Turing machine. We improve his results as follows. Let NLINEAR denote the nondeterministic version of the DLINEAR class : the RAM has a nondeterministic instruction : it can "guess" an integer. Note the inclusions NTIME(n) $\subseteq$ NLINEAR $\subseteq$ NTIME(n logn).

We show that many combinatorial NP-complete problems (20 among the 21 problems listed by Karp [Ka]) belong to NLINEAR. Moreover we prove that some NP-complete problems (for example R.I.S.A. : "reduction of incompletely specified automata") are complete in NLINEAR via DLINEAR reductions (more strongly, reductions are computable in linear time on Turing machines). That improves results of [Gr1, Gr2, Ra1, Ra2] which state that each problem in NTIME(n) is reducible to RISA (and to some other NP-complete problems) in deterministic linear time. The proof of the strengthened result is similar to the original proof. That entails the following equivalence

$$\text{RISA} \notin \text{DLINEAR iff DLINEAR} \neq \text{NLINEAR}$$

which can be compared with the well-known equivalence: SAT $\notin$ P iff P $\neq$ NP.

Let us mention some related results. To our knowledge, Dewdney [De] was the first author who defined and exhibited linear time reductions : in particular he proved that several NP-complete problems, including 3-SAT (i.e. satisfiability for clauses with 3 literals) and 3-COLORABILITY are mutually reducible via linear time reductions (on Turing machines); he assumes that inputs are previously normalized, for example the set of variables which occur in a propositional formula must be exactly $\{p_1, p_2, \ldots, p_m\}$. More recently, Hunt and Stearns [HuSt] prove that a number of problems, including SAT, are mutually reducible via some comparable but weaker reduction (linear length bounded and computable in quasilinear time). Note that if two problems are reducible each other in linear time (for example, two NLINEAR-complete problems) then they have the same time complexity.

Our paper is organized as follows. After the definition of our machine model in Section II (Preliminaries), we show that the deterministic and nondeterministic linear time classes it defines (DLINEAR and NLINEAR) are very robust (Section III). In Section IV we exhibit many problems in DLINEAR which are useful in the proof that most concrete problems usually regarded as computable in linear time belong to DLINEAR. Section V states properties of linear time reductions. In Section VI we show that most combinatorial NP problems belong to NLINEAR. In Section VII we prove that several NP-Complete problems (for example, RISA) are NLINEAR-complete but we argue that propositional satisfiability must not be NLINEAR-complete. We also state other characterizations of the NLINEAR class. Section VIII gives conclusions and open problems.

II - Preliminaries : definitions of our machine model and of linear time complexity.

- Let logn denote the logarithm of n in base 2.
- Let $\Gamma = \{1,2,\ldots,d\}$ be a fixed alphabet, with $d \geq 2$. For convenience we identify a word $w = w_0 w_1 \ldots w_{n-1} \in \Gamma^n$ with the integer it represents in d-adic notation, that is

$$w = \sum_{i<n} w_i d^i$$

(In particular, the empty word is identified with zero). Let length(w) = n ; in particular length(0) = 0.

- Let c,d be fixed integers and let Γ, Σ, be alphabets such that $\Gamma = \{1,2,\ldots,d\}$ and $\Sigma \subseteq \Gamma$ (Σ will be called the input/output alphabet ; sometimes we will use more usual alphabets including, for example, parentheses). Our machine model, denoted DRAM, is the deterministic random access machine of [CoRe] (its sequence of registers are denoted R(0), R(1), R(2)...) with the following changes (i-iv).

(i) An input (resp. output) is a word $w \in \Sigma^n$ (resp. $v \in \Sigma^m$) which is contained (resp. which will be written) in a special one-way read-only (resp. write-only) tape called the input (resp. output) tape.

(ii) The contents of each register R(i) is a nonnegative integer (which is initially zero) whose length, in d-adic notation, is at most c logn (that is similar to the memory condition of the RAC in [AnVa]) ; consequently the total number of registers utilized by a DRAM is polynomially bounded in n.

(iii) A DRAM uses a fixed set of operations ; all of them have to be computable in linear time on a deterministic Turing machine (for example, +, -, concatenation, shift, length), denoted "linear time Turing computable" (in short, LTTC).

(iv) The input/output instructions of a DRAM are the following ones : WRITE(R(i)) writes the contents of register R(i) onto the output tape ; $READ_{R(i)}$ (R(j)) stores in register R(j) a portion of the input of length L where L is the length of the integer that register R(i) contains at this moment (a special case occurs when the nonread part of the input has length less than L : then store in R(j) the whole remainder).

• An NRAM is the nondeterministic version of the previous DRAM : it works similarly but has the additional ability of guessing an integer by the nondeterministic instruction GUESS(R(i)) which stores any integer of length at most c logn in register R(i).

• We are now ready to define our time complexity classes.

Definition. Let $T(n) \geq n/\log n$ be any time function. A function $f : \Sigma^* \to \Sigma^*$ (resp. a language $A \subseteq \Sigma^*$) belongs to class DRAM (T(n)) if f is computable (resp. A is recognizable) on a DRAM which executes O(T(n)) instructions (uniform time criterion) where n denotes the length of the input w ($w \in \Sigma^n$). The class of languages NRAM (T(n)) is defined similarly.

Convention : In this paper, when we write that a word w of length n is accepted by a nondeterministic machine within some ressource (time O(T(n)), space O(S(n))...), we only mean that there is an accepting computation of the machine (on that input w) which respects that ressource (time O(T(n)),...).

Notations: DLINEAR = DRAM(n/logn); NLINEAR = NRAM(n/logn).

Remark : These notations are justified by the fact that the time required by a DRAM(n/logn) (resp. NRAM(n/logn)) computation is linear (i.e. O(n)) under logarithmic cost.

Conventions : In the sequel, in each case when the time cost criterion is not mentioned, we shall implicitly assume that the RAM uses uniform cost criterion.
• Letter n will always denote the length of the input.

III - Classes DLINEAR and NLINEAR are robust.

On one hand Katajainen and others [KvLP] have shown that any function computable in linear time O(n) on a Turing machine can be computed in time O(n log logn) on an ordinary RAM (of [CoRe]) under logarithmic cost. (Recall that in one instruction, such a RAM can read at most one bit of the input.) On the other hand Schönhage [Sc2] has proved that time bound O(n log logn) cannot be replaced by O(n). In this section we will show that that impossibility originates from the weakness of the input/output process of usual RAM and we will state several robustness properties of the DRAM (resp. NRAM).

• For technical reasons, it is useful to define a variant of DRAM without input/output instructions. Instead, the input is divided and stored in a standard way at the start of a computation. (That can be compared with the start condition of RACs in [AnVa] ; however that paper does not state precisely how the input is stored in memory.).

Notation. Let $k \geq 1$ be a fixed integer. Let $b_k(n) = \lceil 1/k \log (n+1) \rceil$

Definition. Let $k \geq 1$ be a fixed integer. A $DRAM_k$ is like a DRAM except for input/output :

(i) each input word $w \in \Sigma^n$ of length n is divided into nonempty words w_1, $w_2, \ldots, w_q$ initially stored in respective registers R(1), R(2),...,R(q) :

$w = w_1 \hat{} \, w_2 \ldots \hat{} \, w_q$ (concatenation)

where $\text{length}(w_i) = b_k(n)$ for $i<q$ and $0 < \text{length}(w_q) \leq b_k(n)$ (the other registers R(q+1),... initially contain zero) ; $w_1, w_2, \ldots w_q$ are called the $b_k(n)$-blocks of w ;

(ii) an output word v (if any) is similarly divided into nonempty $b_k(n)$-blocks $v_1, v_2, \ldots, v_s$ (the last block v_s may be shorter) respectively stored in registers R(1), R(2),...,R(s) at the end of a computation (with R(s+1) = 0) ; note that n is not the length of the output but is still the length of the input.

• We similarly define a NRAM$_k$.

Definition. A function (resp. a language) belongs to DRAM$_k$(n/logn) (resp. NRAM$_k$(n/logn)) if it is computed (resp. recognized) by a DRAM$_k$ (resp. NRAM$_k$) in time O(n/logn) under uniform cost.

Remarks. The important things about the block length $b_k(n)$ is the fact that it is easy to compute and that $b_k(n) = \Theta(\log n)$ for $n \geq 1$ and then the number q (resp. s) of blocks of the input (resp. output) is O(n/logn). We shall also use the fact that $\lim_{k \to +\infty} b_k(n) = 0$.

The following proposition shows how robust the classes DLINEAR = DRAM(n/logn) and NLINEAR = NRAM(n/logn) are.

Proposition 3.1. Let $f : \Sigma^* \to \Sigma^*$ be a function.

(i) For all integers k,k', $f \in$ DRAM$_k$ (n/logn) iff $f \in$ DRAM$_{k'}$ (n/logn)

(ii) $f \in$ DRAM(n/logn) iff $f \in$ DRAM$_k$(n/logn) for some k.

(iii) If f is computable in linear time O(n) on a deterministic Turing machine then $f \in$ DRAM(n/logn).

(iv) Assertions (i-iii) hold for languages (instead of functions) and the nondeterministic classes NRAM(n/logn) and NRAM$_k$(n/logn).

(v) Let $f, g : \Sigma^* \to \Sigma^*$ be two functions such that $f, g \in$ DRAM(n/logn) ; then $g \circ f \in$ DRAM (n/logn).

(vi) Let $f \in$ DRAM (n/logn) and $\varepsilon > 0$ any real number ; then there exists an integer K such that for each $k \geq K$, there is a DRAM$_k$ which computes f in time O(n/logn) and uses exclusively the operation + (resp. concatenation) and integers $O(n^{1+\varepsilon})$.

(vii) Let $A \in$ NRAM (n/logn) ; then there exists an integer K such that for each $k \geq K$, there is a NRAM$_k$ which recognizes A in time O(n/logn) and uses exclusively the successor operation (i.e. function $x \mapsto x+1$) and integers O(n/logn).

Sketch of proof. (i) and (ii) are roughly proved as follows : the simulating machine reorganizes blocks of the input (resp. output) by using new LTTC operations (for example : the length function).

To prove (i), it is sufficient to demonstrate

(i') $f \in$ DRAM$_k$(n/logn) iff $f \in$ DRAM$_{k+1}$(n/logn).

To prove (i') note that $b_{k+1}(n) \leq b_k(n) \leq 2b_{k+1}(n)$

$$\text{(since } \frac{1}{k+1} \leq \frac{1}{k} \leq \frac{2}{k+1}\text{)}$$

and thus an input organized in $b_{k+1}(n)$-blocks is reorganized into $b_k(n)$-blocks by concatenating fragments of two consecutive $b_{k+1}(n)$-blocks to obtain each $b_k(n)$-blocks. Note that values n, $b_k(n)$ and $b_{k+1}(n)$ have to be previously computed.

To prove (ii), exploit the fact that (by definition) a DRAM executes only O(n/logn) read (resp. write) instructions on integers of length O(logn) : for example, a read instruction is simulated on a DRAM$_k$ by the concatenation of O(1) (fragments of) $b_k(n)$-blocks. This proves the implication: $f \in$ DRAM(n/logn) implies $f \in$ DRAM$_k$(n/logn). Let us prove the converse. The simulating DRAM first reads the input by blocks by the following subroutine (it stores into registers R(1), R(2),...,R(m) the respective subwords $w_1, w_2 \ldots w_m$ of the input word w in such a way that $w = w_1 \hat{}\ w_2 \hat{} \ldots w_m$ where length(w_i)= length(i) for i<m and 0 < length(w_m) ≤ length(m)) :

begin

R(0) ← 1

while the input is not entirely stored do

begin

$READ_{R(0)}(R(R(0)))$

R(0) ← R(0) + 1

end

end ;

We easily see that $m = \Theta(n/\log n)$. Then the simulating machine reorganizes the stored input into $b_k(n)$-blocks as above. This proves (ii).

(iii) easily follows from (ii) and the classical technics of blocks (see for example [HPV] and [KVLP] : represent each block of $b_k(n)$ consecutive cells of a tape (of the Turing machine) by one register of a $DRAM_k$.

(iv) is proved exactly as (i-iii).

• To prove (v), use (ii) and distinguish two cases according to the length, denoted L, of the intermediate value f(w) :

- $n^{1/2} \leq L = O(n)$ (where n = length(w)) : it is easy to convert the $b_k(n)$-blocks of f(w) into $b_k(L)$-blocks and to convert the $b_k(L)$-blocks of g o f(w) into $b_k(n)$-blocks (since $\log L = \Theta(\log n)$) ;

- $L < n^{1/2}$: block conversion of words f(w) and g o f(w) is easy since they have only $O(n^{1/2})$ blocks of length O(logn).

• If we use (ii), assertion (vi) is essentially the same as Corollary 3.3 in [Gr3] with uniform time cost instead of logarithmic time cost ; moreover the proof can be simplified (because we do no longer need to distinguish "little addresses"). It essentially combines two technics :

- division of registers into "small registers" of length $b_k(n)$;

- hashing technics (and pairing functions) to encode "big addresses" with "small addresses" $O(n^{1+\varepsilon})$. More precisely, a "big address" u is first divided into b-blocks (i.e. blocks of length $b = b_k(n)$): $u = u_m \ldots u_1 u_0$ (concatenation). Then we successively construct the numbers $p_1 = PAIR(u_0, u_1)$, $p_2 = PAIR(p_1, u_2)$,... $p_m = PAIR(p_{m-1}, u_m)$. The "big address" u is roughly "simulated" by p_m which is O(n). (PAIR is a dynamical pairing function whose values are O(n): it can be stored in a two-dimensional array whose first index is O(n) and second one is a b-block and then is $O(2^b) = O(n^{\varepsilon/2})$; it can be simulated by a one-dimensional array of indices $O(n^{1+\varepsilon})$; for more details, see [Gr3] or see below the proof of Proposition 4.1 where these technics are reused.)

• (vii) is proved by variants of the above technics (simplified):

- registers are divided into "little registers" of length $b_k(n)$;

- addressing is simulated by an essential use of nondeterminism as described below.

Let $a_1 < a_2 \ldots < a_m$ denote the list of distinct addresses (in increasing order) used by the simulated $NRAM_k$ denoted M. First, the simulating $NRAM_k$ M' guesses that list. Note that

$\sum_{i \leq m}$ length $(a_i) \leq n$ and then $m = O(n/\log n)$.

The main idea of the simulation consists in representing the register (of M) of address a_i by a register (of M') of address i. M' has two one-dimensional arrays, denoted ADDRESS(1...m) and CONTENTS(1...m), such that at each moment of the simulation ADDRESS(i) = a_i ($1 \leq i \leq m$) and CONTENTS(i) = contents of register with address a_i (in M). (M' is a "multimemory" $NRAM_k$). An access (by M) to some register of address α is simulated by M' as follows :

- guess an integer $i \in \{1,2,\ldots,m\}$;
- check if ADDRESS(i) (that is a_i) is equal to α (otherwise reject) ;
- use CONTENTS(i).

Note that our "multimemory" $NRAM_k$ M' can be simulated in linear time by an ordinary $NRAM_k$ (interleave the arrays : see [CoRe] for details) and that we exclusively use addresses $O(m) = O(n/\log n)$. □

Remarks : the exact function $b_k(n)$ we choose (for block length) is not essential. Proposition 3.1 and its proof still hold for other "easily computable" functions $\Theta(\log n)$, for example : $\lfloor 1/k \log n \rfloor$ (with $n \geq 2^k$) or $\lceil 1/k \text{ length}(n) \rceil$.

Identity DLINEAR = $DRAM_k(n/\log n)$ is essential in our proof that DLINEAR class is closed under composition. That characterization (resp. the similar one for NLINEAR) will be useful in proofs that concrete problems belong to DLINEAR (resp. NLINEAR).

IV Many problems computable in linear time

In this section we will prove that most problems usually regarded as computable in linear time belong to DLINEAR. In the proofs we shall need several standard DLINEAR subroutines, for example a normalization algorithm.

Remark. For convenience, our DRAMs (or $DRAM_k$) will sometimes use multidimensional arrays. It is not an essential extension of our machine model since a DRAM using a multidimensional array, for example A(i,j), can be simulated (in linear time) by an usual DRAM: register A(i,j) is represented by register R(p) where p = Pair(i,j) and Pair is an LTTC pairing function, that is (for example) : p = Pair(i,j) = i ^ 12 ^ rep(j) (concatenation), where rep(j) is obtained by repeating twice each digit of j (see [Gr3] or [GrRo,Ro1] for more details).

IV.1. A normalization algorithm

In most of "linear time bounded algorithms" on concrete data structures, such as graphs, formulas,... (cf. books [AHU1, AHU2, FGS, Kn, Me, Se]), one implicitly assumes that input structures are prealably normalized, for example :
- the set of vertices of a graph is the interval of integers $\{1,2,\ldots,m\}$;
- Dowling and Gallier [DoGa] and Minoux [Mi] assume that their satisfiability algorithm works on propositional Horn formulas whose variables are exactly $p_1, p_2,\ldots,p_m$.

However it is not clear how to transform an input, for example a propositional formula into such a normalized one : of course we can list all the occurrences of variables p_i and then renumber them by sorting but no linear time algorithm is known for the sorting problem. (Note that Itai and Makowsky [ItMa1] also mentioned the necessity of replacing large indices by small ones.)

We formalize the normalization problem as follows. (Assume we have fixed a finite alphabet $\Sigma = \{1,2,\ldots,d\}$ which does not include symbols □ and :)

Problem NORMALIZE :

Input : A string S of the form $\square A_1 \square A_2 \ldots \square A_m$ where each $A_i \in \Sigma^*$. Let n denote the length of S ($n = m + \sum_{i \leq m} \text{length}(A_i)$)

Output : A string S' of the form $\square A_1 : a_1 \square A_2 : a_2 \ldots \square A_m : a_m$ where
(i) for all i,j, $a_i = a_j$ iff $A_i = A_j$;
(ii) the a_i are integers such that $\{a_1, a_2,\ldots,a_m\} = \{1,2,\ldots,M\}$
(a consequence of (i-ii) is $a_i = O(n/\log n)$ for each i) ;
(iii) length (S') = O(n).

Remark. For each fixed alphabet Σ , the problem NORMALIZE is not unique. Our purpose is only to construct a DLINEAR algorithm which computes a string S' which satisfies (i-iii).

Proposition 4.1. Problem NORMALIZE belongs to DLINEAR.

Proof. It is sufficient to prove that NORMALIZE belongs to $DRAM_k$ (n/logn) for sufficiently large k (we shall take $k \geq 2 \log(d+1)$). Input S is prealably divided into $b_k(n)$-blocks (in short, b-blocks with $b = b_k(n)$). Our DLINEAR algorithm will consist of two parts :

- first : lexical analysis of S;
- secondly : computation of the a_i (by hashing technics).

Lexical analysis of S :

In the sequel, a subword $\Box A_i$ of S which includes at least the first symbol of a b-blocks in S will be called a "mile-stone". Lexical analysis consists in determining the "mile-stones" in S and dividing S as follows :

$$S = \Box A_{h(1)} B_1 \Box A_{h(2)} B_2 \dots \Box A_{h(q)} B_q$$

where $\Box A_{h(1)}, \Box A_{h(2)}, \dots, \Box A_{h(q)}$ are the successive "mile-stones" (notice that $h(1) = 1$) and subwords $B_i \in (\Sigma \cup \{\Box\})^*$ are the "remainders" : for example B_1 is the string of symbols (maybe empty) which lie between the first and second "miles-stones" A_1 and $A_{h(2)}$. We easily see that length(B_i) < b (otherwise, B_i would contain the first symbol of a b-block and then it would include a "mile-stone") and $q \leq$ (numbers of b-blocks) $= \lceil n/b \rceil = O(n/\log n)$.

It is easy to compute and store the list of "mile-stones" $\Box A_{h(1)}, \dots, \Box A_{h(q)}$ (resp. the list $B_1, B_2, \dots B_q$) in a bidimensional (resp. one-dimensional) array where each element has length $\leq b = \Theta(\log n)$. Notice that the total number of b-block in all the $A_{h(i)}$ is $O(n/b) = O(n/\log n)$

Clearly a $DRAM_k$ can execute the above computation in time $O(n/\log n)$. The lexical analysis of S is not complete since it does not separate the A_i contained in the B_j.

Computation of the a_i :

We now use some technics of [Gr3, paragraph III.3] (cf. [Wi2] for some comparable technics). That requires some definitions : a subword A_i in S is "big" if length $(A_i) \geq b$; otherwise it is "small". Note that each subword A_i of some B_j is "small" and that a word $A_{h(j)}$ (in a "mile-stone" $\Box A_{h(j)}$) may be "big" or "small". We do not separately encode each occurrence of a "small" A_i (it would require too many steps in case too many "small" A_i occur in S) but precompute them as follows.

• In a first phase, recapitulate the small A_i that occur in S (use a boolean array, denoted PRESENT, indexed by words of Σ^* of length less than b) and then encode them with an array denoted CODE (we will take a_i = CODE(A_i)). The subroutine of this first phase is the following :

```
begin
        for each w ∈ Σ* such that length (w) < b do
              PRESENT(w) : = false {initialization}
        for i : = 1 to q do
              if length (A_h(i)) < b then PRESENT (A_h(i)) : = true
        for i : = 1 to q do
              if B_i is "new" (i.e. B_i ≠ B_j for each j < i)
              then for each A_j which occurs in B_i
                    do PRESENT(A_j) : = true
        {note : we check if B_i is "new" by means of another boolean array indexed by words
        of length less than b}
```

{we now encode "small" A_i (that occur in S) with consecutive integers}
a ← 1
for each w ∈ Σ^* such that length (w) < b do
if PRESENT(w) then
begin
CODE(w) : = a {meaning : if w = A_i then a = CODE(A_i) = a_i}
a : = a + 1
end
end ;

Note that at each moment the global variable a represents the least integer that is not an encoding (it will be utilized in the third phase below).

• In a second phase, encode all $B_i \in (\Sigma \cup \{\Box\})^*$ in S as follows. If $B_i = \Box\, \alpha_1 \Box\, \alpha_2 \ldots \Box \alpha_s$ where each $\alpha_j \in \Sigma^*$ then take CODE(B_i) := $\Box\alpha_1$: CODE (α_1) ... $\Box$ α_s: CODE(α_s) (these encodings can be stored in the above CODE arrays, indexed by words of $(\Sigma \cup \{\Box\})^*$ of length < b).

• The reader should be convinced that first and second phases above require only O(n/logn) steps for sufficiently small b = $b_k(n)$, i.e. for sufficiently large k : in particular, the analysis of the A_j included in the "new" B_i requires time $O(b.(d+1)^b) = O((\log n)\, n^u)$ where $u = (1/k) \log(d+1)$ that is $O(n^{1/2}. \log n)$ if $k \geq 2 \log(d+1)$. We can now compute the parts of the output S' that concern the subwords B_i and the "mile-stones" $\Box A_{h(i)}$ for "small" $A_{h(i)}$. The third phase consists in encoding "big mile-stones" $\Box A_j$ (i.e. for which length(A_j) ≥b) by technics of [Gr3, paragraph III.3]. Let us describe it to have a self-contained proof.

Examine and encode successively the "big" words A_j by using two global variables denoted Nextp and Nexta and two "dynamic" "pairing functions" denoted PAIR and ENCODE (here we define a "pairing function" as an injective partial function $\mathbb{N} \times \mathbb{N} \rightarrow \mathbb{N}-\{0\}$) : these functions are stored in bidimensional arrays, also denoted PAIR and ENCODE, and each one is completed if necessary, i.e. when it is undefined on some useful arguments. At each moment, values of function PAIR (resp. ENCODE) have to form an initial segment [1, Nextp - 1] (resp. a segment [a, Nexta - 1] where a is the least positive integer that does not encode a "little" A_i : see first phase) and value Nextp (resp. Nexta) is kept in a special register.

Let us explain how functions PAIR and ENCODE are computed and utilized. Let $u_1,u_2,\ldots,u_r$ denote the successive b-blocks of a "big" word A_j (length (A_j) ≥ b) we want to encode: more precisely, $u = u_1\wedge u_2 \ldots \wedge u_r$ with r ≥ 1, length (u_j) = b for each i < r and 0 < length(u_r) ≤ b. Compute successively $p_2 = \text{PAIR}(p_1,u_2)$ where $p_1 = u_1$, $p_3 = \text{PAIR}(p_2,u_3),\ldots$, $p_r = \text{PAIR}(p_{r-1},u_r)$ and e = ENCODE(p_r,r) in respecting the following rule: each time we need to use a pair of arguments (x, y) for which function PAIR is undefined (i.e. register PAIR(x,y) has never been visited and contains zero for that reason), execute the following assignments (and similarly for ENCODE) :

PAIR(x,y) : = Nextp ; Nextp : = Nextp + 1 ;

At the start of the third phase, initialize variables by:

Nextp : = 1 ; Nexta : = a ;

Obviously, a nonzero integer is never contained in two distinct locations of the array PAIR (resp. ENCODE) : hence PAIR and ENCODE are pairing functions. Let us define function $\text{PAIR}^i : \mathbb{N}^i \rightarrow \mathbb{N}$ by recurrence. $\text{PAIR}^i(x_1, x_2,\ldots,x_i)$ is equal (if it is defined) to

• x_1 if i = 1
• $\text{PAIR}(\text{PAIR}^{i-1}(x_1,\ldots,x_{i-1}), x_i)$ if i ≥ 2.

$\text{PAIR}^i : \mathbb{N}^i \rightarrow \mathbb{N}$ is clearly an injective partial function. In particular we have considered (above) the following values associated with some "big" word $A_j = u_1 \wedge u_2 \ldots \wedge u_r$:
$p_2 = \text{PAIR}^2(u_1,u_2)$, $p_3 = \text{PAIR}^3(u_1,u_2,u_3)$, ... $p_r = \text{PAIR}^r(u_1,u_2,\ldots,u_r)$ and
$e = \text{ENCODE}(\text{PAIR}^r(u_1,\ldots,u_r),r)$.

We will take $a_i := e$. Condition (i) of problem NORMALIZE is clearly respected. Notice that the last value of variable Nexta is M + 1 where M is the integer involved in Condition (ii).

We have now achieved the construction of output S'. One can easily convince himself that the third phase runs within O(n/logn) steps (each b-blocks is manipulated O(1) times) and only involves words of length O(logn). □

IV.2. Other problems computable in linear time

In proofs that concrete problems belong to DLINEAR or NLINEAR we will often need the above NORMALIZE algorithm but also algorithms for the following problems.

Problem LIST-COMPRESS :

Input : A string $S = \square A_1 \square A_2 \ldots \square A_m$ where each $A_i \in \Sigma^*$ and $\square \notin \Sigma$.

Output : The string $S' = \square A'_1 \square A'_2 \ldots \square A'_p$ such that :

(i) $\{A_1, A_2, \ldots, A_m\} = \{A'_1, A'_2, \ldots, A'_p\}$;

(ii) there is no repetition in S', i.e. $A'_i \neq A'_j$ if $i \neq j$;

(iii) A'_i occurs before A'_j in S' (that is : $i < j$) iff the first occurrence of A'_i in S precedes the first occurrence of A'_j.

Problem MULTISET-COMPRESS :

Input : A string $S = \square A_1 \square A_2 \ldots \square A_m$ where each $A_i \in \Sigma^*$ and symbols □ and : do not belong to Σ.

Output : A string S' of the form $\square A'_1 : n_1 \square A'_2 : n_2 \ldots \square A'_p : n_p$ such that

(i) $\{A_1, A_2, \ldots, A_m\} = \{A'_1, A'_2, \ldots, A'_p\}$;

(ii) $A'_i \neq A'_j$ if $i \neq j$;

(iii) for each $i \leq p$, n_i is the number of indices j such that $A_j = A'_i$

Problem ELEMENT-DISTINCTNESS :

Input : A string $\square A_1 \square A_2 \ldots \square A_m$ where each $A_i \in \Sigma^*$ and $\square \notin \Sigma$.

Question : Are the elements A_i all distinct ?

Remark The computational complexity of ELEMENT-DISTINCTNESS has been studied in some recent papers : for example, Yao [Ya] proves a near optimal time-space tradeoff for it.

Problem PERMUTATION :

Input : A string $\square A_1 \square A_2 \ldots \square A_m \, \Delta \, A'_1 \square A'_2 \ldots \square A'_m$ where A_i, A'_i are words of Σ^* ($\square, \Delta \notin \Sigma$).

Question : Is there a permutation π of $\{1,2,\ldots,m\}$ such that $A'_i = A_{\pi(i)}$ for each $i \leq m$?

Problem SET-INCLUSION (resp. SET-EQUALITY) :

Input : Two lists of words $A_1, A_2, \ldots A_m$ and $A'_1, A'_2, \ldots A'_p$.

Question : Does the inclusion $\{A_1, \ldots A_m\} \subseteq \{A'_1, \ldots A'_p\}$ (resp. equality $\{A_1, \ldots A_m\} = \{A'_1, \ldots A'_p\}$) hold ?

Remark : The input is presented above in a natural structured manner but formally, the input of a DRAM is a string ; the exact input of SET-INCLUSION have to be, for example, of the form $A_1 \square A_2 \ldots \square A_m \, \Delta \, A'_1 \square A'_2 \ldots \square A'_p$ (with new symbols □ and Δ). For convenience, in the sequel our problems will be presented in a structured manner.

Problem CHECK-SORT :
Input : Two lists of integers (or words) $A_1, A_2, \ldots, A_m$ and $A'_1, A'_2, \ldots, A'_m$ such that $A'_1 \leq A'_2, \ldots \leq A'_m$.
Question : Is the second list the sorted version of the first list, i.e. is there a permutation π of $\{1,2,\ldots,m\}$ such that $A'_i = A_{\pi(i)}$ for each $i \leq m$?

Problem PARTITION :
Input : A list P of "sets" $S_1, S_2, \ldots, S_m$, where each S_i is of the form: $S_i = \{e_1, e_2, \ldots, e_p\}$.
Question : Is P a "set partition", i.e. are the sets S_i mutually disjoint ?

Problem SET-INTERSECT (resp. SET-DIFFERENCE) :
Input : A set S and a list L of sets $S_1, S_2, \ldots, S_m$.
Output : The list L' of intersection sets $S_1 \cap S, \ldots, S_m \cap S$
(resp. difference sets $S_1 - S, \ldots, S_m - S$).

Problem ADDITION :
Input : A list of integers $A_1, A_2, \ldots, A_m$.
Output : The integer $A_1 + A_2 + \ldots + A_m$.

Problem FUNCTION-DEF :
Input : A list of "arrows" $x_1 \to y_1, x_2 \to y_2, \ldots x_m \to y_m$ where the x_i, y_i are words.
Question : Does that list define a function, i.e. does the implication $x_i = x_j \Rightarrow y_i = y_j$ hold, for all i,j ?

Problem FUNCTION-APPLY : (Assume that c is a fixed integer.)
Input : A "finite" partial function F on integers, i.e. a list of "arrows" $x_1 \to y_1, x_2 \to y_2, \ldots x_m \to y_m$ where words x_i are all distinct and $\text{length}(y_i) \leq c\ \text{length}(x_i)$ and a string of the form $\Box z_1 \Box z_2 \ldots \Box z_p$.
Output : The string $\Box z'_1 \Box z'_2 \ldots \Box z'_p$ where $z'_i = F(z_i)$ if $F(z_i)$ is defined (i.e. z_i is some x_j and $F(z_i) = y_j$) and $z'_i = z_i$ otherwise.

Proposition 4.2. Problems LIST-COMPRESS, MULTISET-COMPRESS, ELEMENT-DISTINCTNESS, PERMUTATION, SET-INCLUSION and SET-EQUALITY all belong to DLINEAR.

Sketch of proof : We only mention the differences with the proof of proposition 4.1. After the lexical analysis of string S which produces (for instances of LIST-COMPRESS, MULTISET-COMPRESS and ELEMENT-DISTINCTNESS) the decomposition
$S = \Box A_{h(1)} B_1 \ldots \Box A_{h(q)} B_q$
(as above), the output of LIST-COMPRESS is computed by the following loop:

For i : = 1 to q do
begin
    if $A_{h(i)}$ is "new" {i.e., that is the first occurrence of $A_{h(i)}$ in S}
        then write $\Box A_{h(i)}$

    if B_i is "new" {i.e. $B_i \neq B_j$ for each $j < i$}
        then for each A_j that occurs in B_i do
            if A_j is "new" then write $\Box A_j$
end ;

Note that we check if a "small" A_j is "new" with the boolean array PRESENT (of the proof of proposition 4.1) indexed by words w such that length(w) < b, and that a "big" A_j is "new" iff register ENCODE(PAIR^r($u_1,\ldots,u_r$),r) contains zero at that moment.

For MULTISET-COMPRESS, we compute the part of output S' concerning the "big" A_i by acceding to its encoding a_i (computed as usual) and to a counter COUNT(a_i) indexed by a_i. To produce the part of S' concerning the "small" A_i, we first compute for each word $w \in (\Sigma \cup \{\Box\})^*$ of length less than b, the number, denoted NUMBER(w), of indexes i such that B_i = w. Then for each word w of length less than b and of the form $\Box\alpha_1 \Box\alpha_2\ldots\Box\alpha_s$ with $\alpha_j \in \Sigma^*$, we count its contribution to the numbers of "small" A_i by the loop

for j : = 1 to s do COUNTER(α_j) := COUNTER(α_j) + NUMBER(w).

Lastly count the "small" $A_{h(i)}$ by the loop

for i : = 1 to q do

if length ($A_{h(i)}$) < b then COUNTER($A_{h(i)}$) := COUNTER($A_{h(i)}$) + 1

This proves the proposition for MULTISET-COMPRESS.

Problem ELEMENT-DISTINCTNESS is obviously decided by MULTISET-COMPRESS algorithm. The PERMUTATION algorithm is similar to the previous one. The main difference is that we count separately the A_i and the A'_j. At the end we check if the corresponding counters are equal. Problems SET-INCLUSION and SET-EQUALITY have similar algorithms. □

Proposition 4.3. Problem ADDITION belongs to DLINEAR.

Proof. Without loss of generality, assume that the input is $\Box A_1 \Box A_2\ldots\Box A_m$ .First transform that input (as follows) in such a way that the number m of summands A_i becomes O(n/logn). After the lexical analysis which gives $\Box A_{h(1)}\ B_1\ \Box A_{h(2)}\ B_2 \ldots \Box A_{h(q)}\ B_q$ (cf. Proof of proposition 4.1), each string B_i is replaced by the sum of the integers that it contains, i.e. more precisely, if $B_i = \Box\alpha_1\Box\alpha_2\ldots\Box\alpha_s$, then B_i is replaced by string A'_i where A'_i is the result of the sum $\alpha_1+\alpha_2 \ldots+\alpha_s$.That can be done efficiently by computing previously for each w of the form $w = \Box\alpha_1 \Box\alpha_2 \ldots\Box\alpha_s$, (where the α_i are integers) such that length(w) < b, the integer $\alpha_1+\alpha_2 \ldots+\alpha_s$ and storing it in a table SUM(w).

Secondly, add the m = O(n/logn) integers A_i by the usual classroom algorithm on b-blocks, i.e. with integers in base d^b (recall that integers are originally written in d-adic notation). The reader should easily convince himself that it requires only O(n/logn) steps and involves integers of length O(logn) (because b = Θ(logn)). □

Proposition 4.4 Problems PARTITION, SET-INTERSECT and SET-DIFFERENCE belong to DLINEAR.

Proof. Let us consider PARTITION. Without loss of generality, assume that m = O(n/logn) (For that purpose, apply ELEMENT-DISTINCTNESS algorithm to the list P where each S_i is regarded as a string, i.e. an ordered list ; reject if two lists S_i, S_j are equal). Choose a sufficiently large k to get a sufficiently small block length $b = b_k(n)$ (that will be precisely stated below).

Then use a deterministic Turing machine which separates the list P into two lists P' and P" such that $P' = (S'_1, S'_2,\ldots,S'_m)$, $P'' = (S''_1, S''_2,\ldots,S''_m)$ and each $S'_i = \{e \in S_i : \text{length}(e) \geq b\}$ and each $S''_i = \{e \in S_i : \text{length}(e) < b\}$.

Clearly P is a partition iff both P' and P" are partitions. The Turing machine runs in linear time O(n) and can be simulated by a DRAM within time O(n/logn) (by proposition 3.1 (iii)).

It is easy to check if P' is a partition by the above hashing technics. Therefore, without loss of generality, assume that in $P = (S_1, S_2, \ldots, S_m)$, each S_i is a list of elements e such that length(e) < b. Then divide each string S_i into b-block, i.e. into substrings $S_i = B^1{}_i B^2{}_i \ldots B^{p_i}{}_i$ (concatenation) where each block $B^j{}_i$ is a sublist of S_i of the form $e_1, e_2, \ldots, e_s$ (each e_u is an "elementary element") such that length($B^j{}_i$) $\leq$ b. This can be done in such a way that the total number of blocks $B^j{}_i$ (for all $i \leq m$ and $j \leq p_i$) is O(n/logn) (since m = O(n/logn)).

Then for each block $B^j{}_i$, store (in an array denoted SET) the index i = SET($B^j{}_i$) of the set S_i where $B^j{}_i$ is included ; reject in case there is a collision, i.e. $B^j{}_i = B^{j'}{}_{i'}$, for $i \neq i'$ (that means that S_i and $S_{i'}$ have common elements). Lastly for each possible pair of blocks B, B' which contain at least one common element e and such that SET(B) $\neq$ 0 and SET(B') $\neq$ 0 (i.e. both B and B' occur in P), check if SET(B) = SET(B') : that can be done in time o(n/logn) for sufficiently small block length b.

This proves that PARTITION belongs to DLINEAR. The algorithms and proofs for SET-INTERSECT and SET-DIFFERENCE are similar and then are left to the reader. □

Proposition 4.5 Problems CHECK-SORT, FUNCTION-DEF and FUNCTION-APPLY belong to DLINEAR.

Proof Problem CHECK-SORT is a restriction of the DLINEAR problem PERMUTATION.

Let us consider FUNCTION-DEF. Apply successively to its input the LIST-COMPRESS algorithm (to avoid repetitions of "arrows") and the NORMALIZE algorithm. This gives a string of the form $a_1 \rightarrow b_1, a_2 \rightarrow b_2, \ldots, a_p \rightarrow b_p$ where p = O(n/logn) and each a_i, b_i is O(n/logn). Now by storing the "arrows" in an array F by assignments $F(a_i) := b_i$ check if there is no collision.

The construction of an algorithm for FUNCTION-APPLY is left to the reader. □

From the NORMALIZE algorithm, we deduce the following "Thesis"

"Assertion 4.6 ("Linear Time Thesis"). "Concrete algorithms" usually regarded as "computable in linear time" and concerning

- (weighted, directed, nondirected) graphs given by their lists of edges or
- propositional formulas in clausal form

belong to DLINEAR.

Remark : This statement is called "Assertion" or "Thesis" for it appears to be vague. It can be *proved* only for specific problems, for example:

Problem HORN-SAT [DoGa, ItMa1, ItMa2, Mi] :
Input : a set S of propositional Horn clauses $\{C_1, C_2, \ldots C_m\}$.
Question : Is S satisfiable ?

Problem SCC ("strongly connected components" [Ta, AHU1])
Input : a directed graph G = (V,E) given by its list of vertices $v_1, \ldots v_m$ and its list of arcs $e_1, \ldots e_p$ ($e_i \in V^2$).
Output : The set of strongly connected components of graph G.

Proof of the "Assertion" 4.6 for SCC :
First normalize the input: apply LIST-COMPRESS algorithm to remove repetitions in the list of vertices and the list of arcs: so we can assume that m+p = O(n/logn); apply NORMALIZE algorithm to these lists so that the set of vertices becomes V = {1,2,...m}. Then implement the usual algorithm of [Ta] (see also [AHU1, AHU2]) on the DRAM model : it runs in time O(m+p) = O(n/logn). □

Proof of the "Assertion" 4.6 for HORN-SAT:
The algorithms of [DoGa, ItMa1, ItMa2, Mi] can be trivially implemented on a DRAM in time O(n/log n) if in the list of clauses S

(i) each propositional variable has length O(log n) and

(ii) the total number of occurrences of variables is O(n/logn).

For that reason it is sufficient to exhibit a DRAM(n/logn) algorithm which computes from each input S a new set of Horn clauses S' for which (i-ii) hold and which is satisfiable iff S is satisfiable.

We can assume that each clause of S has one of the two following forms:

- $(v_1,...,v_s \rightarrow v_0)$: pure Horn clause;
- $(v_1,..., v_s \rightarrow \text{FALSE})$: negative clause

where each v_i is any propositional variable. The list of variables $v_1,...,v_s$ is called the *hypothesis list* of the clause. Let $C_1,C_2,....C_m$ denote the list of clauses of S : by the LIST-COMPRESS algorithm, we can assume that there is no repeated clause so that $m = O(n/\log n)$.

Let us fix a sufficiently large integer k. We now separate in each hypothesis list the "big hypotheses", i.e. variables v_i such that $\text{length}(v_i) \geq b = b_k(n)$ and the "small hypotheses", i.e. the v_i such that $\text{length}(v_i) < b$. A linear time bounded Turing machine can transform each clause into the form $(v_1,...,v_r, v_{r+1},...,v_s \rightarrow u)$ where u is either a variable or the FALSE value and for $i \leq r$ (resp. $i > r$), v_i is a big (resp. small) hypothesis.

Then for each clause C_i ($i \leq m$), divide its list of small hypotheses $L_i = v_{r+1},...,v_s$ into "blocks": $L_i = B^1{}_i B^2{}_i .... B^{q_i}{}_i$ (concatenation) where each "block" $B^j{}_i$ is a sublist of L_i such that $\text{length}(B^j{}_i) \leq b$. This can be done in such a way that the total number of sublists $B^j{}_i$ (for $i \leq m$ and $j \leq q_i$) is $O(n / \log n)$ (since $m = O(n / \log n)$).

The next idea is to replace each $B^j{}_i$ (in each clause C_i) by a new propositional variable, denoted $v^j{}_i$ which will be interpreted as the conjunction of the variables of $B^j{}_i$. We store the new variable $v^j{}_i$ in an array $\text{VARIABLE}(B^j{}_i)$ so that if $B^j{}_i = B^{j'}{}_{i'}$ then $v^j{}_i$ and $v^{j'}{}_{i'}$ denote the same variable. let C'_i denote the clause C_i so transformed, i.e.

$$C'_i = (v_1,....,v_r, v^1{}_i,v^2{}_i,....,v^{q_i}{}_i \rightarrow u).$$

Let S' be the set of clauses

$$S' = \{C'_1,...,C'_m\} \cup \{ B^j{}_i \rightarrow v^j{}_i \}_{i \leq m, j \leq q_i}$$

where repetitions among the clauses are removed. Note that there are as many distinct clauses $B^j{}_i \rightarrow v^j{}_i$ as distinct "blocks" $B^j{}_i$, i.e. less than d^b (where d is the cardinality of the alphabet of clauses) which is $O(n^{1/2})$ for sufficiently small $b = b_k(n)$.

It remains to prove the two following claims.

Claim 1: the number of occurrences of propositional variables in S' is O(n/logn);

Claim 2: S is satisfiable iff S' is satisfiable.

Claim 1 follows from the following three facts:

- $m = O(n/\log n)$;
- there are $O(n/b)$ occurrences of hypotheses in $C'_1,...,C'_m$;
- there are $O(d^b).O(b) = O(n^{1/2} . \log n)$ occurrences of variables in all the clauses of the form $B^j{}_i \rightarrow v^j{}_i$.

Claim 2 follows from the following easy facts:

- S' implies S by transitivity of implication ;
- conversely, if an interpretation satisfies S then S' is trivially satisfied by the extended interpretation where each $v^j{}_i$ is interpreted as the conjunction of the variables of $B^j{}_i$ (i.e. $v^j{}_i$ is true iff each variable in $B^j{}_i$ is true). □

V - Linear time bounded reductions

In the next sections VI and VII, we will state that many NP-complete problems belong to NLINEAR or are complete in this class. In the proofs we will use the following notions.

Definitions and notations :

• Let $T(n) \geq n$ be a time function. A function on words (or a language) belongs to DTIME(T(n)) (resp. NTIME(T(n))) if it is computable (resp. recognizable) on a deterministic (resp. nondeterministic) Turing machine in time O(T(n)) where n is the length of the input word.

• Let $A, B \subseteq \Sigma^*$ be two languages. We write $B \in \mathrm{NTIME}^A(n)$ (resp. $B \in \mathrm{NRAM}^A(n/\log n)$) or $B \leq_{NTM} A$ (resp. $B \leq_{NRAM} A$) if there is a nondeterministic Turing machine (resp. a NRAM) M such that for each word $w \in \Sigma^n$:

- each computation of M on input w runs within time O(n) (resp. time O(n/logn)) ;
- w belongs to B iff some computation of M on w computes a word of A.

In the special case when machine M is deterministic, we write $B \in \mathrm{DTIME}^A(n)$ (resp. $B \in \mathrm{DRAM}^A(n/\log n)$) or $B \leq_{DTM} A$ (resp. $B \leq_{DRAM} A$) and the function computed by M is called a reduction of B to A

Proposition 5.1.

(i) $\mathrm{DTIME}(n) \subseteq \mathrm{DRAM}(n/\log n)$; in consequence, if $A \leq_{DTM} B$ then $A \leq_{DRAM} B$.

(ii) For each language A, $\mathrm{NTIME}^A(n) \subseteq \mathrm{NRAM}^A(n/\log n)$.

(iii) If A, B are languages such that $A \in \mathrm{NTIME}(n)$ (resp. $A \in \mathrm{NRAM}(n/\log n)$) and $B \in \mathrm{NTIME}^A(n)$ (resp. $B \in \mathrm{NRAM}^A(n/\log n)$) then $B \in \mathrm{NTIME}(n)$ (resp. $B \in \mathrm{NRAM}(n/\log n)$).

(iv) Relations $\leq_{DTM}$, $\leq_{DRAM}$, $\leq_{NTM}$ and $\leq_{NRAM}$ are transitive.

(v) Let $X \in \{D,N\}$ be a letter for "determinism" or "nondeterminism" and $T(n) \geq n$ (resp. $T(n) \geq n/\log n$) be a nondecreasing function. If $A \leq_{XTM} B$ and $B \in \mathrm{XTIME}(T(n))$ (resp. $A \leq_{XRAM} B$ and $B \in \mathrm{XRAM}(T(n))$) then there is a constant c such that : $A \in \mathrm{XTIME}(T(cn))$ (resp. $A \in \mathrm{XRAM}(T(cn))$).

Proof. Assertions (i-v) either are classical results concerning Turing machines (see for example [HoUl,Mo]), or are restatements or easy generalizations of assertions of proposition 3.1 : the proofs are similar. □

VI Most combinatorial NP problems belong to NLINEAR.

We feel that the twenty-one NP-complete problems 1-21 listed by R. Karp [Ka] are "combinatorial problems" (a possible exception is problem 19, called "job sequencing" in [Ka] and "sequencing to minimize tardy task weight" in [GaJo]: it appears to be more algebraic). The notion of "combinatorial problem" is as intuitive and at least as difficult to formulate as the notions of "computability" (captured by Turing computability) or "feasible computability" (captured by class P of polynomial time Turing computable problems). In the present section, we prove that many NP problems, and, in particular, the twenty NP-complete problems 1-18 and 20-21 of [Ka] belong to NLINEAR : this justifies our conviction that combinatorial NP problems are NLINEAR problems. We choose to give explicit proofs for the following six significant problems.

SUBSET-SUM (problem 18 in [Ka])

Input : Integer B and list of integers $A_1, A_2, \ldots, A_m$

Question : Is there a subset I of $\{1,2,\ldots,m\}$ such that $\sum_{i \in I} A_i = B$?

EXACT-COVER (problem 14 in [Ka])
Input : List of sets $S_1, S_2, \ldots, S_m$ and set S.
Question : Is there a subset I of $\{1,2,\ldots,m\}$ such that the family $(S_i)_{i \in I}$ is a partition of S ?

HITTING-SET (problem 15 in [Ka] ; a variant numbered [SP8] appears in [GaJo]).
Input : List of sets $S_1, S_2, \ldots, S_m$
Question : Is there a set S such that for each $i \leq m$, $S \cap S_i$ has exactly one element ?

FEEDBACK-ARC-SET (problem 8 in [Ka])
Input : Integer K and directed graph G = (V,A).
Question : Is there a subset $A' \subseteq A$ of cardinality at most K such that A' contains at least one arc from every directed cycle in G ?

TRAVELING-SALESMAN (problem [ND22] in [GaJo])
Input : Integer K and weighted nondirected graph G = (V,E) given by its list of weighted edges $e_1, e_1, \ldots, e_p$ where $e_i = (a_i, b_i, w_i)$, $a_i, b_i \in V$ and w_i is a positive integer.
Question : Has G a Hamilton circuit of weight not greater than K ?

Pfleeger [Pf] has proved the NP-completeness of the following problem.
Problem RISA ("Reduction of Incompletely Specified Automata" : [AL7] in [GaJo])
Input : An integer K and an incompletely specified deterministic finite state automaton $\mathcal{A} = (Q, \Sigma, \delta, q_0, F)$ where Q is the set of states, Σ is the input alphabet, $q_0 \in Q$ is the initial state, $F \subseteq Q$ is the set of final states and δ is a partial function : $Q \times \Sigma \rightarrow Q$.
Question : Can δ be extended to a total function ; $Q \times \Sigma \rightarrow Q$ in such a way that the minimal automaton of the resulting completely specified automaton has no more than K states ?

Remark (RISA characterization). It is easy to see that the question above is equivalent to the following one (where Q' denotes the set of states reachable from q_0). Is there a map $h : Q' \rightarrow \{1,2,\ldots,K\}$ which respects the following conditions (i-ii) ?

(i) for all $q, q' \in Q'$ such that $h(q) = h(q')$ we have $q \in F$ iff $q' \in F$. (i.e. h respects the distinction "final/nonfinal states") ;

(ii) for all $q, q' \in Q'$, $\sigma \in \Sigma$ such that $h(q)$ and $h(q')$ are equal and $\delta(q,\sigma)$ and $\delta(q',\sigma)$ are both defined, we have $h(\delta(q,\sigma)) = h(\delta(q',\sigma))$ (i.e. h respects the deterministic nature of the automaton).

Theorem 6.1. Twenty among the twenty-one NP-complete problems given by [Ka] that are problems 1-18 and 20-21, and problems TRAVELING-SALESMAN and R.I.S.A. belong to NLINEAR.

Remark. When an input involves a (weighted, directed, nondirected) graph we assume that it is represented by its list of arcs.

Proof of Theorem 6.1. We prove the theorem for the six problems explicitly given above. The reader is invited to adapt the proofs to the other ones.

•TRAVELING-SALESMAN : Apply the FUNCTION-DEF algorithm (to check if no edge has two distinct weights) and the LIST-COMPRESS one (to suppress edge repetitions). By the NORMALIZE algorithm, we can assume that the set of vertices is $V = \{1,2,\ldots,m\}$ with $m = O(n/\log n)$. Then guess a cycle $e'_1, e'_2, \ldots, e'_m$ where e'_i is a weighted edge :

- check if the cycle is included in the list of edges of G and includes each vertex (by two applications of SET-INCLUSION algorithm) ;
- check that no vertex is repeated in the cycle (by ELEMENT-DISTINCTNESS algorithm) ;
- check if the weight of the cycle does not exceed K (by ADDITION algorithm).

• RISA : By successive applications of FUNCTION-DEF, LIST-COMPRESS and NORMALIZE algorithms, normalize the input as above : in particular we can assume that $\Sigma = \{1,2,\ldots,r\}$ and $Q = \{1,2,\ldots,s\}$ with $r + s = O(n/\log n)$ and that the number t of transitions of δ is $O(n/\log n)$. By a classical algorithm, a DRAM can compute the set Q' of reachable states within time $O(t) = O(n/\log n)$ and suppress the nonreachable states and the transitions which involve them. We now use the RISA characterization given in the remark above.

Guess for each reachable state $q \leq s$ a value $h(q) \leq K < s$ with which q is identified (function h can be stored in an array). Transform each transition $\delta(q,\sigma) = q'$ into transition $\delta'(h(q),\sigma) = h(q')$ and check if the resulting automaton (with set of states included in $\{1,2,\ldots,K\}$ and transition function δ') is deterministic (with FUNCTION-DEF algorithm) and has a well-defined set F of final states : $h(q) \in F$ iff $q \in F$.

• SUBSET-SUM : A linear time bounded nondeterministic Turing machine (simulated by a NRAM in time $O(n/\log n)$) can guess a sublist $(A_i)_{i \in I}$. Then check if $\Sigma_{i \in I} A_i = B$ (by ADDITION algorithm).

• EXACT-COVER : Guess the sublist $(S_i)_{i \in I}$ as above and then check if it is a partition (by PARTITION algorithm) and if S is equal to the union set $\bigcup_{i \in I} S_i$ (by SET-EQUALITY algorithm).

• HITTING-SET : Guess a set S (of course, $\text{length}(S) \leq n$) and then check if for each $i \leq m$, set $S_i \cap S$ has exactly one element (by SET-INTERSECT algorithm).

• FEEDBACK-ARC-SET : The question is equivalent to the following one. Is there a subset $A' \subseteq A$ of cardinality at most K such that the directed graph $G' = (V, A-A')$ is acyclic ? The recognizability of acyclicity of directed graphs belongs to DLINEAR (use some classical algorithm). □

The reader is invited to prove similarly that many other combinatorial NP problems (subgraph isomorphism, graph isomorphism, partition into cliques,...) belong to NLINEAR.

VII - Some NP-complete problems are NLINEAR-complete.

As seen in Section VI, the class NLINEAR = NRAM(n/logn) that refines the NP class includes many natural NP-complete problems. The following notion is comparable to NP-completeness.

Definition. A problem A is NLINEAR-complete if
(i) $A \in$ NLINEAR and
(ii) A is NLINEAR-hard, i.e. for each problem B in NLINEAR, $B \leq_{DTM} A$.

Remarks. It is clear that a NLINEAR-complete problem is also NP-complete.
• Recall that $B \leq_{DTM} A$ implies $B \leq_{DRAM} A$.

The following proposition is an immediate consequence of the definition above.

Proposition 7.1. Let A be a NLINEAR-complete problem. The following assertions are equivalent :
(i) NLINEAR = DLINEAR ;
(ii) $A \in$ DLINEAR.

Remark. We do not know whether NLINEAR ≠ DLINEAR. However Paul and others [PPST] have proved a similar assertion for Turing machines, that is NTIME(n) ≠ DTIME(n). It implies that no DTIME(n) algorithm exists for any NLINEAR-hard problem.

The following proposition means that any efficient algorithm for a NLINEAR-complete problem can be used to solve any other NLINEAR problem without loss of efficiency.

Proposition 7.2. Let $T(n) \geq n$ (resp. $T(n) \geq n/\log n$) be any nondecreasing function. If any NLINEAR-complete (or NLINEAR-hard) problem belongs to DTIME(T(n)) (resp. DRAM(T(n)), then $\mathrm{NLINEAR} \subseteq \bigcup_c \mathrm{DTIME}(T(cn))$ (resp. $\mathrm{NLINEAR} \subseteq \bigcup_c \mathrm{DRAM}(T(cn))$)

Proof. Immediate. □

Corollary 7.3. Let A, B be two NLINEAR-complete problems and $T(n) \geq n$ and $T'(n) \geq n/\log n$ be two nondecreasing functions. Then

(i) $A \in \bigcup_c \mathrm{DTIME}(T(cn))$ iff $B \in \bigcup_c \mathrm{DTIME}(T(cn))$;

(ii) $A \in \bigcup_c \mathrm{DRAM}(T'(cn))$ iff $B \in \bigcup_c \mathrm{DRAM}(T'(cn))$.

It is well-known that the problem SAT of Cook ("satisfiability of propositional formulas in clausal form") is a "generic" NP-complete problem [Co1, Co2, HoUl]. Similarly we are going to use the following problem, denoted CONTRACT ("contraction of partial functional structures"), as a tool to prove NLINEAR-completeness of other problems.

Problem CONTRACT [Ra1, Ra2].
Input. Set C of "constants", set X of "variables", set $\mathfrak{F}$ of unary function symbols (elements of sets C, X, $\mathfrak{F}$ are assumed to be presented in lexicographical order) and conjunction Γ of equalities of the form $f(u) = v$ where $f \in \mathfrak{F}$, $u, v \in C \cup X$ and such that terms $f(u)$ are all distinct in Γ and occur in lexicographical order.
Question: Is conjunction Γ satisfiable on set C, i.e. are there a function $VAL : C \cup X \to C$ which is the identity on C and, for each $f \in \mathfrak{F}$, an interpretation f', i.e. a partial function $f'' : C \to C$ such that if $f(u) = v$ is an equality in Γ, then $f'(VAL(u)) = VAL(v)$ holds ?.

The following logical notion (a variant of a concept introduced by Scholz [Sz] and generalized by Fagin [Fa]) will be an essential tool in our proof that CONTRACT is a NLINEAR-complete problem.

Definition. Let φ be a first order sentence (with equality) with type $\mathfrak{T} = \{f_1, f_2, \ldots f_p, g_1, g_2, \ldots, g_q\}$ where the f_i, g_j are unary function symbols, respectively called specified and unspecified function symbols. The generalized spectrum of φ, denoted GenSPECTRUM(φ), is the set of structures $(m, f_1, f_2, \ldots f_p)$ (where m is a positive integer identified with set $\{0,1,\ldots,m-1\}$) that have an expanded structure $(m, f_1, f_2, \ldots f_p, g_1, g_2, \ldots, g_q)$ which satisfies φ.

Remark. For convenience, an interpretation of a function symbol f_i, g_j will be often denoted f_i, g_j (i.e. without change).

Fagin in his well-known seminal paper [Fa] proved that the class of generalized first-order spectra (in other words, the existential second-order logic) exactly captures the class NP. The following proposition refines Fagin's result in some manner: it shows that the subclass NLINEAR can be similarly captured by the existential second-order logic with only one (first-order) variable.

Proposition 7.4. If $A \in$ NLINEAR then there is a first-order sentence φ of unary type $\mathfrak{T}$ as above, such that

(i) $A \leq_{DTM}$ GenSPECTRUM(φ);

(ii) φ is of the form $\forall x\ \psi(x)$ where $\psi(x)$ has only one variable and is a conjunction $\bigwedge_i \sigma_i(x) = \tau_i(x)$ of equalities of the form

$$G_r \ldots G_2\, G_1(x) = H_s \ldots H_2\, H_1(x)$$

where $r \geq 1$, $s \geq 0$ and each $G_i, H_j \in \mathfrak{T}$, and such that

(ii.1) no term or subterm of the form $f_i(x)$ (where f_i is a specified function symbol) occurs in ψ ;

(ii.2) no first member $\sigma_i(x)$ is a subterm of another first member $\sigma_j(x)$ (for $i \neq j$) or of a second member $\tau_j(x)$ (here "subterm" means "proper subterm or equal term").

Sketch of proof. The proof is rather long and technical. It is essentially given in [Gr2, Ra1, Ra2] where the result is stated in a slightly weaker form : language A is assumed to be in NTIME(n) (because the authors focus on Turing machines) but as a matter of fact, they prove the stronger result $A \leq_{DTM}$ GenSPECTRUM(φ) for all $A \in NRAM_k(n/\log n)$ (for sufficiently large k) and use the inclusion $NTIME(n) \subseteq NRAM_k(n/\log n)$. Let us roughly recall the stages of the proof (for details, the reader is invited to read the complete proof in [Gr2, Ra1, Ra2]).

- First, construct a prenex first-order sentence φ which describes the computation of the $NRAM_k$ (which recognizes A in time $O(n/\log n)$) : φ has to be interpreted on domain $m = \{0,1,\ldots,m-1\}$ where $0,1,\ldots, m-1$ intuitively represent the m successive instants of the computation (in particular $m = O(n/\log n)$) ; moreover the prenex sentence φ has only one universally quantified variable and its only nonlogical symbols are =, < (equality and natural order on m) and unary function symbols.

- Secondly, transform formula φ in such a way that it satisfies conditions (ii) above : in particular, natural order <, negations and disjunctions are suppressed by some technics comparable to Skolemization ; new specified or unspecified function symbols are introduced for that purpose.

Remark. We can assume that an output of the reduction $A \leq_{DTM}$ GenSPECTRUM (in proposition 7.4) which is a structure $\mathcal{S} = (m, f_1, f_2,\ldots, f_p)$ with length($\mathcal{S}$) = O(n), is presented in the following manner : each function f_j is given in increasing order of its arguments, i.e. by a list of the form $0 \rightarrow f_j(0)$, $1 \rightarrow f_j(1)$,,..., $m-1 \rightarrow f_j(m-1)$ (recall that $m = O(n/\log n)$ and the number p of functions does not depend on input w).

Lemma 7.5. Let φ be the sentence $\forall x\ \psi(x)$ of proposition 7.4. We have

GenSPECTRUM(φ) $\leq_{DTM}$ CONTRACT

where each input structure $\mathcal{S} = (m, f_1,\ldots,f_p)$ is assumed to be presented in increasing order (cf. remark above).

Proof. The main idea of the DTIME(n) reduction is the following one : unroll the sentence $\varphi = \forall x\ \psi(x)$ on its domain m; i.e. replace φ by conjunction $\bigwedge_{e<m} \psi(e)$ where $\psi(e)$ is obtained by substituting constant e for each occurrence of x in ψ . From the definition of generalized spectrum, it is clear that a structure $\mathcal{S}$ (on domain m) belongs to GenSPECTRUM(φ) if and only if the following conjunction of equalities, denoted γ, is satisfiable on domain m :

$$\gamma = \bigwedge_{e<m} \psi(e) \wedge \bigwedge_{j \leq p} \mathrm{Diagram}(f_j)$$

where Diagram(f_j) denotes the conjunction of equalities $\bigwedge_{e<m} f_j(e) = e_j$ (sorted in increasing order of argument e) and e_j denotes the value of the term $f_j(e)$ in $\mathcal{A}$.

The conjunction γ is almost an instance of problem CONTRACT (with constants 0,1,..., m-1). We now have to "break" the compositions of functions in each conjunct $\psi(e)$ ($\psi(e)$ is of the form $\bigwedge_i \sigma_i(e) = \tau_i(e)$). For that purpose introduce for each (sub)term $\Theta(e)$ in γ (except for constants $e < m$) a "variable" denoted $\underline{\Theta(e)}$ (underlined). The instance of problem CONTRACT corresponding to $\mathcal{A}$ will be the following one :

- $C = m = \{0,1,\ldots,m-1\}$;
- $X = \{\underline{\Theta(e)} : \Theta(e)$ is a (sub)term in γ, distinct from e$\}$;
- $\mathcal{F} = \{f_1,\ldots,f_p, g_1,\ldots,g_q\}$;
- conjunction Γ is obtained by unrolling each (sub)term in γ ; for example, if γ contains the equality $f_1\, g_3 = g_1(2)$ we replace that equality by the following conjunction $g_1(2) = \underline{g_1(2)} \wedge g_3(2) = \underline{g_3(2)} \wedge f_1(\underline{g_3(2)}) = \underline{g_1(2)}$

We easily see that conjunction γ is satisfiable on m iff Γ is satisfiable on m. Hence the correspondence $\mathcal{A} \mapsto (C, X, \mathcal{F}, \Gamma)$ is a reduction of GenSPECTRUM(φ) to problem CONTRACT. That reduction is computable in linear time on a deterministic Turing machine. □

Corollary 7.6. Problem CONTRACT is NLINEAR-hard.

Proposition 7.7. CONTRACT $\leq_{DTM}$RISA.

Proof. An instance $\mathfrak{C} = (C, X, \mathcal{F}, \Gamma)$ of problem CONTRACT (without loss of generality, we assume $C = \{0,1,\ldots,m-1\}$) is reduced to the following instance ($\mathcal{Q}$,K) of problem RISA :

$\mathcal{Q} = (Q, \Sigma, \delta, q_0, F)$ where

- $Q = C \cup \{m\} \cup X$;
- $\Sigma = \mathcal{F} \cup \{Succ\}$ (Succ is intuitively the successor function on integers);
- partial transition function $\delta : Q \times \Sigma \rightarrow Q$ is defined by :
  - $\delta(u,f) = v$ for each conjunct $f(u) = v$ in Γ (note that $\delta(u,f)$ is not defined twice since term f(u) occurs at most once in Γ) ;
  - $\delta(u,Succ) = u+1$ for each $u \in C = \{0,1,\ldots,m-1\}$ (δ(m,Succ) is undefined) ;
- $q_0 = 0$; • $F = \{m\}$; • $K = m+1$ (K is the cardinality of set $C \cup \{m\}$).

We easily see that each state $q \in Q$ is reachable from $q_0 = 0$ and that a deterministic Turing machine can compute the instance ($\mathcal{Q}$,K) in linear time O(n). There remains to prove the following equivalence : $\mathfrak{C} \in$ CONTRACT iff ($\mathcal{Q}$,K) $\in$ RISA. From a solution h for instance $\mathfrak{C}$ (i.e. a "contraction" of $\mathfrak{C}$), we easily deduce a K-reduction of the incomplete automaton $\mathcal{Q}$: that is the contraction h of $\mathfrak{C}$ on C that is extended by taking h(m) = m. The converse, i.e. ($\mathcal{Q}$,K) $\in$ RISA implies $\mathfrak{C} \in$ CONTRACT is an easy consequence of the following lemma (which concludes the proof of Proposition 7.7).

Lemma 7.8. Let $\mathcal{Q}' = (Q, \Sigma, \delta', q_0, F)$ be a completely specified finite automaton which completes the above $\mathcal{Q}$, i.e. whose transition function δ' extends δ. In $\mathcal{Q}'$

(i) all the states are reachable (from q_0) ;
(ii) the equivalence class of state m is $\{m\}$;

(iii) if $\mathcal{A}'$ is K-reducible (recall $K = m+1$) then its K equivalence classes are exactly the class of 0, the class of 1,... the class of m-1 and {m}.

Proof. (i) holds for $\mathcal{A}$ and then for $\mathcal{A}'$.(ii) follows from the fact that m is the only final state. (iii) : we have $\delta(j, Succ^{m-j}) = m \in F$ and $\delta(i, Succ^{m-j}) = m - j + i \notin F$ for all $i<j\leq m$; hence states i,j are not equivalent. □

Corollary 7.9. Problems CONTRACT and RISA are NLINEAR-complete.
Proof. It follows from the fact that for each A in NLINEAR,
$A \leq_{DTM}$ CONTRACT $\leq_{DTM}$ RISA and RISA $\in$ NLINEAR. □

Remark. Ranaivoson [Ra1, Ra2] has strengthened Corollary 7.8. : problem RISA (resp. CONTRACT) remains NLINEAR-complete when it is restricted to a binary alphabet $\Sigma = \{0,1\}$ (resp. to type $\mathcal{T} = \{f_0, f_1\}$ with only two function symbols) ; he also shows that RISA is in P when $\Sigma = \{1\}$. Moreover he exhibits several NLINEAR-complete problems concerning acyclic directed graphs. Note that [Ra1,Ra2] shows the NLINEAR-completeness of each above problem A by proving CONTRACT $\leq_{DTM}$A but the reduction itself and the proof of its correctness are much longer and more technical than ours for A = RISA. Our student N. Creignou [Cr] has recently proved the NLINEAR completeness of problems of directed or nondirected graphs by a simpler method.

Proposition 7.4 will be useful in the proof of the following characterization of class NLINEAR.
Proposition 7.10. NLINEAR = $NTIME^A(n)$ for all A $\in$ {FUNCTION-DEF, CHECK-SORT, PERMUTATION, SET-EQUALITY, SET-INCLUSION}.

Proof. We have seen that each problem A above belongs to DLINEAR. Hence $NTIME^A(n) \subseteq$ NLINEAR. To get the converse inclusion for A = FUNCTION-DEF, it suffices to prove that for each first-order sentence $\varphi = \forall x\ \psi(x)$ described in proposition 7.4, we have GenSPECTRUM(φ) $\leq_{NTM}$ FUNCTION-DEF. We show it by a variant of the proof of lemma 7.5 (that states GenSPECTRUM(φ) $\leq_{DTM}$ CONTRACT). Let us describe only the differences with that proof. Instead of unrolling the conjunction γ (associated with structure $\mathcal{S}$) by the introduction of "variables" for terms and subterms in γ, we now guess the intermediate values (less than m), i.e. values of concerned (sub)terms ; for example if m=5 and if γ contains the equality $f_1g_3(2) = g_1(2)$, this equality may be replaced by the following conjunction $g_1(2) = \underline{4} \wedge g_3(2) = \underline{0} \wedge f_1(0) = 4$ (convention : underlined $\underline{4}$ and $\underline{0}$ are "guessed"; the other occurrences 2, 0 and 4 are only "copied"). Let Γ denote the conjunction γ so transformed (there are as many possible Γ as possible guessed assignments). γ is satisfiable on m iff some associated Γ is satisfiable. Γ is almost an instance of FUNCTION-DEF but it concerns not one but several functions $f_1,...,f_p, g_1,...g_q$. For that reason, each equality $f_i(u) = v$ (resp. $g_j(u) = v$) in Γ is replaced by "arrow" $u' \rightarrow v$ where u' is the integer $(p + q)\ u + i$ (resp. $(p + q)\ u + p + j$) : u' encodes the term $f_i(u)$ (resp. $g_j(u)$). Let Γ' denote the resulting instance of FUNCTION-DEF. Clearly Γ is satisfiable iff $\Gamma' \in$ FUNCTION-DEF.

We can construct a nondeterministic Turing machine with input $\mathcal{S}$, that computes successively γ, Γ and Γ' within linear time. That proves GenSPECTRUM(φ) $\leq_{NTM}$ FUNCTION-DEF.

To obtain the same results for any of the other four problems A, it suffices to show FUNCTION-DEF $\leq_{NTM}$ A. For any instance $L = (x_1 \rightarrow y_1, x_2 \rightarrow y_2,..., x_s \rightarrow y_s)$ of FUNCTION-DEF, guess some list $L = (x'_1 \rightarrow y'_1,..., x'_s \rightarrow y'_s)$ where for each $i < s$: either $x'_i < x'_{i+1}$ or $x'_i = x'_{i+1}$ and $y'_i = y'_{i+1}$ (intuitively : L' defines a function and is sorted).

Clearly L $\in$ FUNCTION-DEF if there is some list L' such that (L,L') $\in$ CHECK-SORT (L' is the sorted version of list L). Moreover a nondeterministic Turing machine can guess L' in linear time. Hence FUNCTION-DEF $\leq_{NTM}$ CHECK-SORT

The same result is easily proved for problems PERMUTATION, SET-EQUALITY and SET-INCLUSION with the same reduction. □

Remark. Proposition 7.10 holds for many other problems A. The above-mentioned problems are the simplest ones we have found : in our opinion FUNCTION-DEF is the most significant. It indicates what computational power a nondeterministic linear time bounded Turing machine lacks to capture NLINEAR.

Corollary 7.11. NTIME(n) $\subseteq$ NLINEAR $\subseteq$ NTIME(nlogn).
Proof. The first inclusion has been proved above. The second one results from the following facts : NLINEAR $\leq_{NTM}$ CHECK-SORT and CHECK-SORT $\in$ DTIME(nlogn) (by the usual merge-sort algorithm). □

We have seen that several NP-complete problems (RISA, CONTRACT,...) are NLINEAR-complete. The reader may ask whether most of natural NP-complete problems in NLINEAR are NLINEAR-complete. We suggest a partial answer by studying problem SAT (i.e. satisfiability of propositional formulas in clausal form). We will need the following mixed complexity class.

Definition : Let A be a language and T(n), T'(n) be time functions. We write

$$A \in \text{NDRAM}(T(n), T'(n))$$

if A is recognized by a NRAM in such a way that for any input $w \in A$ of length n there is an accepting computation of the NRAM with only O(T(n)) GUESS instructions (i.e. nondeterministic steps) and O(T'(n)) other instructions (i.e. deterministic steps).

Proposition 7.12. Problem SAT belongs to NDRAM$(n/(\log n)^2, n/\log n)$.

Proof. By running the NORMALIZE algorithm, we can assume that the set of propositional variables is $\{p_1, p_2, \ldots, p_r\}$ where $r = O(n/\log n)$. Guess a truth assignment $\alpha = (\alpha_1, \alpha_2, \ldots, \alpha_r) \in \{1,2\}^r$ for the r variables (convention : 1 = TRUE, 2 = FALSE) by blocks of length $b = \ulcorner c \log(n+1) \urcorner$ (for a constant integer c) ; more precisely, a string $\alpha = B_1 B_2 \ldots B_s$ where each $B_i \in \{1,2\}^*$, length$(B_i) = b$ for $i < s$ and $0 < \text{length}(B_s) \leq b$ ($s = \ulcorner r/b \urcorner$) is guessed and stored by s instructions GUESS(R(i)) (B_i is stored in register R(i)).

The stored assignment α is analysed by O(t) = O(n/logn) deterministic instructions which compute the list of pairs $(p_1, \alpha_1), (p_2, \alpha_2), \ldots, (p_r, \alpha_r)$.

Then run FUNCTION-APPLY algorithm to substitute for each variable p_i in each clause its truth value α_i. Lastly a finite automaton can evaluate all the clauses. Note that the number of nondeterministic instructions is $s = \ulcorner r/b \urcorner = O(n/(\log n)^2)$. □

Corollary 7.13. The NLINEAR-completeness of problem SAT would imply the following inclusion : NRAM$(n/\log n) \subseteq$ NDRAM$(n/(\log n)^2, n/\log n)$.

That inclusion is very inlikely : it would mean that any nondeterministic algorithm could be simulated by executing fewer nondeterministic steps and the same number (up to a multiplicative constant) of deterministic steps. We conjecture that that inclusion implies P = NP. (Indeed it seems stronger than P = NP).

Remarks. Proposition 7.12 holds not only for SAT but also for many other NP-complete problems, in particular it can be proved for five among the six problems studied in Theorem 6.1: TRAVELING-SALESMAN, SUBSET-SUM, EXACT-COVER, HITTING-SET and FEEDBACK-ARC-SET (the proof is not trivial for TRAVELING-SALESMAN and SUBSET-SUM). The reader is invited to prove that and to discover other such problems.

Notice that Schnorr [Sr] and Cook [Co2] (resp. Robson [Ro2]) have proved that any problem in NTIME(n) (resp. computable by some nondeteterministic RAM in linear time under logarithmic cost) is reducible to problem SAT in time $O(n(\log n)^2)$ on a deterministic Turing machine.

VIII - Conclusions and open problems.

We have defined and studied two robust complexity classes : DLINEAR and NLINEAR. We think that these classes satisfyingly capture the intuitive notions of deterministic and nondeterministic linear time.

- Each DLINEAR algorithm runs within linear time under logarithmic cost (hence DLINEAR is included in the complexity class LINEAR defined in [Gr3]) ; and similarly, for NLINEAR.
- DLINEAR contains most (all ?) classical "linear time bounded" algorithms (problems of graphs represented by their list of arcs, propositional Horn satisfiability,...) and several others (CHECK-SORT, LIST-COMPRESS,...).
- NLINEAR includes most (all ?) combinatorial NP-complete problems.
- The concepts we have introduced are natural and fruitful since there are natural NLINEAR-complete problems (the notion of NLINEAR-completeness strengthens that of NP-completeness) and then most combinatorial NP-complete problems are reducible to these problems in deterministic linear time.

The fact that many concrete algorithms run in linear time can be formalized in a uniform manner (i.e. independently of the nature of the inputs : graphs, lists, words, formulas...) by using the DLINEAR concept. Similarly we are convinced that our study of the class NLINEAR contributes to a better understanding of the theory of NP-completeness of natural problems. In particular, it is interesting to notice that most of combinatorial NP-complete problems belong to NLINEAR, i.e. belong to a low (nondeterministic) complexity class.

We conclude this paper by describing some open problems and conjectures.

• Problem CHECK-SORT belongs to DLINEAR but we have not been able to prove the same result for the sorting problem (our hashing technics does not respect the natural order of integers).

• We strongly conjecture that DLINEAR includes the classical "linear time bounded" algorithms when each graph is represented by its list of successors (resp. predecessors) i.e. by $(v_1, L_1), ..., (v_m, L_m)$ where L_i denotes the list of successors of vertex v_i. The difficult point is the fact that if such a representation has length n, then the number of distinct arcs of the graph is not necessarily $O(n / \log n)$ but may be $\Omega(n / \log \log n)$. We think to get over that difficulty by using the following remarks:

- it seems that each natural "linear time" algorithm on graphs is controlled either by depth-first search, or by breadth-first search, or by topological sorting;
- it is very likely that the technics of grouping "little vertices" into blocks can be naturally applied to these three runnings in such a way that they work in time $O(n/\log n)$ on a DRAM.

• JOB-SEQUENCING is the only problem among the 21 NP-complete problems presented by Karp [Ka] for which we have not succeeded in proving that it belongs to NLINEAR. However we strongly conjecture that it does (it seems to be more complicate and more algebraic than the twenty other ones : for example, partial sums occur in its statement).

• Class DLINEAR (resp. NLINEAR) is contained in the class of problems computable in deterministic (resp. nondeterministic) linear time under logarithmic cost. We conjecture that these classes are equal. Wiedermann [Wi2] has proved the following weaker result : a RAM which runs in time T(n) under logarithmic cost can be simulated by a RAM which runs within time $O(T(n)/\log \log T(n))$ under uniform cost.

• We have proved that it can be required that NLINEAR (resp. DLINEAR) computations involve only integers O(n/logn) (resp. $O(n^{1+\epsilon})$) : therefore in the deterministic case, the DRAM may use somewhat "scattered" addresses $O(n^{1+\epsilon})$. It would be of great interest to show that the memory of a DRAM can be utilized in the most "compact" manner, i.e. to require a DLINEAR computation to only involve addresses O(n/logn) as in the nondeterministic case.

• We have exhibited natural NLINEAR-complete problems. Similarly, the discovery of a natural DLINEAR-complete problem (i.e. a problem A such that DLINEAR = $DTIME^A(n)$) would be a nice breakthrough ; it would mean that one algorithm (the DLINEAR algorithm of A) would be the "pattern" of all the other DLINEAR algorithms (and they are numerous !). We think that such a result would follow from a logical characterization of class DLINEAR exactly as the NLINEAR-completeness of problems CONTRACT and RISA has been deduced from our characterization of class NLINEAR by generalized spectra of first-order sentences.

• All the NLINEAR-complete problems we know concern contractions of structures (can a given structure be transformed by identification of some elements in such a way that the resulting structure respects some given constraint ?) or satisfiability of formulas (cf. [Gr1]): in fact, contraction problems can be regarded as satisfiability problems, and conversely. Are there some NLINEAR-complete problems of some very different nature (for example, subgraph isomorphism) ?

• We conjecture that a careful analysis of the $\leq_{DTM}$ reductions presented in Section VII (to show that CONTRACT, RISA... are NLINEAR-hard) would prove that they are reset-loglin : this very strict reduction has been introduced by Compton and Henson [CoHe] ; it strengthens the notion of logarithmic space linear time bounded reduction ; moreover the class of reset-loglin reductions is closed under composition.

• Does the NLINEAR-completeness of problem SAT imply P = NP ?

Aknowledgments : Thanks to Yuri Gurevich for helpful discussions. Many thanks to Sylvie Hunout-Déroff for her nice typing of the manuscript.

References

[AHU1] A.V. AHO, J.E. HOPCROFT and J.D. ULLMAN, The design and analysis of computer algorithms, Addison-Wesley, Reading 1974.

[AHU2] A.V. AHO, J.E. HOPCROFT and J.D. ULLMAN, Data structures and algorithms, Addison-Wesley, Reading 1983.

[AnVa] D. ANGLUIN and L. VALIANT, Fast probabilistic algorithms for hamiltonian circuits and matchings, J. Comput. System Sci. 18 (1979), pp. 155-193.

[BMS] A. BERTONI, G. MAURI and N. SABADINI, Simulations among classes of random access machines and equivalence among numbers succinctly represented, Annals Discrete Math. 25 (1985) pp. 65-90.

[BuGo] J.F. BUSS and J. GOLDSMITH, Nondeterminism within P, STACS '91, Lect. Notes Comput. Sci. 480 (1991), pp. 348-359.

[CoHe] K.J. COMPTON and C.W. HENSON, A uniform method for proving lower bounds on the computational complexity of logical theories, Annals Pure and Applied Logic 48 (1990), pp. 1-79.

[Co1] S. A. COOK, The complexity of theorem proving procedures, 3rd STOC (1971), pp. 151-158.

[Co2]S.A. COOK, Short propositional formulas represent nondeterministic computations, Inform. Process. Lett. (1987/88), pp. 269-270.

[CoRe] S.A. COOK and R.A. RECKHOW, Time-bounded random access machines, J. Comput. System Sci. 7 (1973), pp. 354-375.

[Cr] N. CREIGNOU, Ph. D. Thesis, in preparation.

[De] A.K. DEWDNEY, Linear time transformations between combinatorial problems, Internat. J. Comput. Math. 11 (1982), pp. 91-110.

[DoGa] W. F. DOWLING and J.H. GALLIER, Linear-time algorithms for testing the satisfiability of propositional Horn formulas, J. Logic Progr. 3 (1984), pp. 267-284.

[Fa] R. FAGIN, Generalized first-order spectra and polynomial-time recognizable sets. In Compl. of Comp., Proc. Symp., New York (1973), SIAM-AMS Proc. Vol. VII (1974), pp. 43-73.
[FGS] C. FROIDEVAUX, M.C. GAUDEL and M. SORIA, Types de données et algorithmes, Mc Graw Hill.
[GaJo] M.R. GAREY and D.S. JOHNSON, Computers and intractability: a guide to the theory of NP-completeness, Freeman, San Francisco (1979).
[Gl] E. GRAEDEL, On the notion of linear time computability, Proc. 3rd Italian Conf. Theoret. Comput. Sci. 1989, World Scientific Publ. Co. pp. 323-334, also appears in Internat. J. Foundations Comput. Sci. 1, 1990, pp. 295-307.
[Gr1] E. GRANDJEAN, A natural NP-complete problem with a nontrivial lower bound, SIAM J. Comput. 17 (1988), pp. 786-809.
[Gr2] E. GRANDJEAN, A nontrivial lower bound for an NP problem on automata, SIAM J. Comput. 19 (1990), pp. 438-451.
[Gr3] E. GRANDJEAN, Invariance properties of RAMs and linear time, T.R. Univ. de Caen, France, Cahiers du LIUC 90-11 (1990), to appear in revised form in journal: Computational Complexity.
[GrRo] E. GRANDJEAN and J.M. ROBSON, RAM with compact memory: a realistic and robust model of computation, T.R. Univ. de Caen, France, Cahiers du LIUC 90-8 (1990), appears in revised form in CSL 90, Lect. Notes Comput. Sci. 533 (1991) pp. 195-233.
[GuSh]Y. GUREVICH and S. SHELAH, Nearly linear time, Lect. Notes Comput. Sci. 363 (1989), Springer-Verlag, pp. 108-118.
[HaSi] J. HARTMANIS and J. SIMON, On the power of multiplication in random access machines, 15th IEE Symp. on Switching and Automata Theory (1974), pp. 13-23.
[HPV] J.E. HOPCROFT, W. PAUL and L. VALIANT, On time versus space and related problems, JACM 24 (1977), pp. 332-337.
[HoUl] J.E. HOPCROFT and J.D. ULLMAN, Introduction to automata theory, languages and computation (1969), Addison-Wesley, Reading.
[HuSt] H.B. HUNT and R.E. STEARNS, The complexity of very simple boolean formulas with applications, SIAM J. Comput. 19 (1990), pp. 44-70.
[ItMa1] A. ITAI and J.A. MAKOWSKY, On the complexity of Herbrand's Theorem, Computer Science Dpt, Technion, Haifa, Israel, T.R. 243, May 1982.
[ItMa2] A. ITAI and J.A. MAKOWSKY, Unification as a complexity measure for logic programming, Computer Science Dpt, Technion, Haifa, Israel, T.R. 301, November 1983, appears (in a revised form) in J. Logic Programming 4 (1987), pp. 105-117.
[Ka] R.M. KARP, Reducibility among combinatorial problems, IBM Symp. 1972, Complexity of Computer Computations, Plenum Press, New York 1972.
[KvLP] J. KATAJAINEN, J. van LEUWEN and M. PENTTONEN, Fast simulation of Turing machines by random access machines, SIAM J. Comput. 17 (1988), pp. 77-88.
[Kn] D.E. KNUTH, The art of programming, Vol 1,2,3 (1968, 1969, 1973), Addison-Wesley, Reading.
[Me] K. MELHORN, Data structures and algorithms (3 vol.), Springer-Verlag (1984).
[Mi] M. MINOUX, LTUR: a simplified linear-time resolution algorithm for Horn formulae and computer implementation, Inf. Proc. Letters 29 (1988), pp. 1-12.
[Mo] B. MONIEN, About the derivation languages of grammars and machines, 4th ICALP 1977, pp. 337-351, Lecture Notes Comput. Sci. 52, Springer-Verlag.
[PPST] W. PAUL, N. PIPPENGER, E. SZEMEREDI and W.T. TROTTER, On determinism versus non-determinism and related problems, 24th Symp. Found. of Comput. Sci. (1983), pp. 429-438.
[Pf] C.P. PFLEEGER, State reduction in incompletely specified finite-state machines, IEEE Trans. Comput. 22 (1973), pp. 1099-1102.
[Ra1] S. RANAIVOSON, Ph. D. Thesis (1990), Université de Caen, France.
[Ra2] S. RANAIVOSON, Nontrivial lower bounds for some NP-complete problems on directed graphs, CSL 90, Lecture Notes Comput. Sci. 533, pp. 318-339.
[Ro1] J.M. ROBSON, Random access machines and multi-dimensional memories, Inf. Proc. Letters 34 (1990), pp. 265-266.
[Ro2] J.M. ROBSON, An O(T log T) reduction from RAM computations to satisfiability, Theoret. Comput. Sci. 82 (1991), pp. 141-149.

[Sr] C.P. SCHNORR, Satisfiability is quasilinear complete in NQL, JACM 25 (1978), pp. 136-145.
[Sc1] A. SCHOENHAGE, Storage modifications machines, SIAM J. Comput. 9 (1980), pp. 490-508.
[Sc2] A. SCHOENHAGE, A nonlinear lower bound for random access machines undes logarithmic cost, J. ACM 35 (1988), pp. 748-754.
[Sz] H. SCHOLZ, J. Symbolic Logic 17 (1952), p. 160.
[Se] R. SEDGEWICK, Algorithms, Addison-Wesley.
[SlvEB] C. SLOT and P. van EMDE BOAS, The problem of space invariance for sequential machines, Inform. and Comput. 77 (1988), pp. 93-122.
[Ta] R.E. TARJAN, Depth-first search and linear time algorithms, SIAM J. Comput. 1 (1972), pp. 146-160.
[WaWe] K. WAGNER and G. WECHSUNG, Computational complexity, Reidel ed. Berlin (1986).
[Wi1] J. WIEDERMANN, Deterministic and nondeterministic simulation of the RAM by the Turing machine, in Proc. IFIP Congress 83, R.E.A. Mason ed., North Holland, Amsterdam 1983, pp. 163-168.
[Wi2] J. WIEDERMANN, Normalizing and accelerating RAM computations and the problem of reasonable space measures, TR OPS-3 / 1990 (June 1990), Dpt of Programming Systems, Bratislava, Czechoslovakia, and Proc. 17th ICALP, Lecture Notes Comput. Sci. 443, Springer-Verlag (1990).
[Ya] A.C. YAO, Near optimal time-space tradeoff for element distinctness, Proc. Symp. Found. of Comput. Sci. 1988, pp. 91-97

The Semantics of the C Programming Language

Yuri Gurevich* and James K. Huggins*

EECS Department, University of Michigan, Ann Arbor, MI 48109-2122, USA

0 Introduction

We present formal operational semantics for the C programming language. Our starting point is the ANSI standard for C as described in [KR]. Knowledge of C is not necessary (though it may be helpful) for comprehension, since we explain all relevant aspects of C as we proceed.

Our operational semantics is based on evolving algebras. An exposition on evolving algebras can be found in the tutorial [Gu]. In order to make this paper self-contained, we recall the notion of a (sequential) evolving algebra in Sect. 0.1.

Our primary concern here is with semantics, not syntax. Consequently, we assume that all syntactic information regarding a given program is available to us at the beginning of the computation (via static functions). We intended to cover all constructs of the C programming language, but not the C standard library functions (*e.g.* `fprintf()`, `fscanf()`). It is not difficult to extend our description of C to include any desired library function or functions.

Evolving algebra semantic specifications may be provided on several abstraction levels for the same language. Having several such algebras is useful, for one can examine the semantics of a particular feature of a programming language at the desired level of abstraction, with unnecessary details omitted. It also makes comprehension easier. We present a series of four evolving algebras, each a refinement of the previous one. The final algebra describes the C programming language in full detail.

Our four algebras focus on the following topics respectively:

1. Statements (*e.g.* `if`, `for`)
2. Expressions
3. Memory allocation and initialization
4. Function invocation and return

What about possible errors, *i.e.*, division by zero or de-referencing a pointer to an invalid address? These issues are very implementation-dependent. Even what constitutes an error is implementation-dependent. If an external function does not produce any value in a state where a value is expected, the evolving algebra will be stalled in that state forever. It is natural to suppose that if an external function does produce a value, it is of the appropriate type. One may want to augment the guards of transition rules to check for errors; in this way,

* Partially supported by ONR grant N00014-91-J-1861 and NSF grants CCR-89-04728 and CCR-92-04742.

the evolving algebra will halt on error conditions (and may even output an error message if desired). There are more subtle ways to handle errors. We ignore the issue here.

To reflect the possibility of different implementations, our evolving algebras contain implementation-dependent parameters. For example, the set of values "storable" in a pointer variable is implementation-dependent. Thus, each of our four evolving algebras gives rise to a family of different evolving algebras.

0.1 Evolving Algebras

An evolving algebra $\mathcal{A}$ is an abstract machine. Here we restrict attention to sequential evolving algebras. The ***signature*** of $\mathcal{A}$ is a (finite) collection of function names, each name having a fixed arity. A state of $\mathcal{A}$ is a set, the ***superuniverse***, together with interpretations of the function names in the signature. These interpretations are called ***basic functions*** of the state. The superuniverse does not change as $\mathcal{A}$ evolves; the basic functions may.

Formally, a basic function of arity r (*i.e.* the interpretation of a function name of arity r) is an r-ary operation on the superuniverse. (We often use basic functions with $r = 0$; such basic functions will be called ***distinguished elements***.) But functions naturally arising in applications may be defined only on a part of the superuniverse. Such partial functions are represented by total functions in the following manner.

The superuniverse contains distinct elements ***true***, ***false***, ***undef*** which allow us to deal with relations (viewed as binary functions with values ***true*** or ***false***) and partial functions (where $f(\overline{a}) = \textit{undef}$ means f is undefined at the tuple $\overline{a}$). These three elements are ***logical constants***. Their names do not appear in the signature; this is similar to the situation in first-order logic with equality where equality is a logical constant and the sign of equality does not appear in the signature. In fact, we use equality as a logical constant as well.

Further, a ***universe*** U is a special type of basic function: a unary relation usually identified with the set $\{x : U(x)\}$. The universe $Bool = \{true, false\}$ is another logical constant. When we speak about, say, a function f from a universe U to a universe V, we mean that formally f is a unary operation on the superuniverse such that $f(a) \in V$ for all $a \in U$ and $f(a) = \textit{undef}$ otherwise. We use self-explanatory notations like $f : U \to V$, $f : U_1 \times U_2 \to V$, and $f : V$. The last means that the distinguished element f belongs to V.

In principle, a program of $\mathcal{A}$ is a finite collection of transition rules of the form

$$\textbf{if } t_0 \textbf{ then } f(t_1, \ldots, t_r) := t_{r+1} \textbf{ endif} \tag{1}$$

where t_0, $f(t_1, \ldots, t_r)$, and t_{r+1} are closed terms (*i.e.* terms containing no free variables) in the signature of $\mathcal{A}$. An example of such a term is $g(h_1, h_2)$ where g is binary and h_1 and h_2 are zero-ary. The meaning of the rule shown above is this: Evaluate all the terms t_i in the given state; if t_0 evaluates to ***true*** then change the value of the basic function f at the value of the tuple $(t_1, .., t_r)$ to the value of t_{r+1}, otherwise do nothing.

In fact, rules are defined in a slightly more liberal way; if k is a natural number, $b_0, \ldots, b_k$ are terms and $C_0, \ldots, C_{k+1}$ are sets of rules then both of the following are rules:

if b_0 **then** C_0
elseif b_1 **then** C_1
⋮
elseif b_k **then** C_k
else C_{k+1}
endif

if b_0 **then** C_0
elseif b_1 **then** C_1
⋮
elseif b_k **then** C_k
endif

Since the C_i are sets of rules, nested transition rules are allowed (and occur frequently). To save space, we abbreviate the series of **endif**'s at the tail of a transition rule by **ENDIF**.

A program is a set of rules. It is easy to transform a program to an equivalent program comprising only rules of the stricter form (1). We use rules of the more liberal form, as well as macros (textual abbreviations), for brevity.

How does $\mathcal{A}$ evolve from one state to another? In a given state, the demon (or interpreter) evaluates all the relevant terms and then makes all the necessary updates. If several updates contradict each other (trying to assign different values to the same basic function at the same place), then the demon chooses nondeterministically one of those updates to execute.

We call a function (name) f *dynamic* if the demon (interpreter) may change f as the algebra evolves; *i.e.* if an assignment of the form $f(t_1, \ldots, t_r) := t_0$ appears anywhere in the transition rules. Functions which are not dynamic are called *static*. To allow our algebras to interact conveniently with the outside world, we also make use of *external* functions within our algebra. External functions are syntactically static (that is, never changed by rules), but have their values determined by an oracle. Thus, an external function may have different values for the same arguments as the algebra evolves.

0.2 Acknowledgements

An earlier version of this paper appeared as a technical report [GH]. We gratefully acknowledge comments made by Egon Börger, Andre Burago, Martin J. Dürst, Stefano Guerrini, Raghu Mani, Arnd Poetzsch-Heffter, Dean Rosenzweig, and Marcus Vale.

1 Algebra One: Handling C Statements

Our first evolving algebra models the control structures of C.

1.1 Some Basic Functions

A universe *tasks* consists of elements representing tasks to be accomplished by the program interpreter. The notion of task is a general one: *e.g.*, a task may be the execution of a statement, initialization of a variable, or the evaluation of an expression. The elements of this universe are dependent on the particular C program being executed by the abstract machine. It is often useful to mark a given task with tags indicating its nature. This gives rise to a universe of ***tags***.

A distinguished element *CurTask: **tasks*** indicates the current task. In order to execute tasks in the proper order, a static function *NextTask: **tasks*** $\rightarrow$ ***tasks*** indicates the next task to be performed once the current task has been completed. A static function *TaskType: **tasks*** $\rightarrow$ ***tags*** indicates the action to be performed by the task.

A universe ***results*** contains values which may appear as the result of a computation.

1.2 Macro: Moveto

Often, we transfer control to a particular task, modifying *CurTask* to indicate the transfer of control. In Algebra Two, the rules for modifying *CurTask* will change somewhat; in order to facilitate this change, we will use the *Moveto(Task)* macro each time that we wish to transfer control. For now, the definition of *Moveto(Task)* is:

Moveto(Task)

CurTask := *Task*

1.3 Statement Classification in C

According to [KR], there are six categories of statements in C:

1. Expression statements, which evaluate the associated expression.
2. Selection statements (`if` and `switch`).
3. Iteration statements (`for`, `while`, and `do-while`).
4. Jump statements (`goto`, `continue`, `break`, and `return`).
5. Labeled statements (`case` and `default` statements used within the scope of a `switch` statement, and targets of `goto` statements).
6. Compound statements, consisting of a (possibly empty) list of local variable declarations and a (possibly empty) list of statements.

1.4 Expression Statements

An expression statement has one of the following forms:

expression-statement $\rightarrow$;
expression-statement $\rightarrow$ *expression* ;

To execute an expression statement, evaluate the attached expression (if any), even though the resulting value will not be used. While this may seem unnecessary, note that the evaluation of an expression in C may generate side-effects (such as assigning a value to a variable). Note also that the evaluation of an expression may not halt. In this algebra, the evaluation of expressions is handled by an external function *TestValue: tasks → results*.

Since expression statements perform no additional work, the algebra simply proceeds to the next task. The transition rule for expression tasks is:

---

if *TaskType(CurTask) = expression* **then**
  Moveto(NextTask(CurTask))
endif

1.5 `if` Statements

There are two types of selection statements in C: `if` statements and `switch` statements. An `if` statement has one of the following forms:

if-statement → `if` (*expression*) *statement1*
if-statement → `if` (*expression*) *statement1* `else` *statement2*

where *statement1* and *statement2* are statements.

To execute an `if` statement, begin by evaluating the guard expression. If the resulting value is non-zero, execute *statement1*. If the resulting value is zero and an `else` clause is present, execute *statement2*. otherwise, execute the statement following the `if` statement. Static partial functions *TrueTask: tasks → tasks* and *FalseTask: tasks → tasks* indicate the task to be performed if the guard of the `if` statement evaluates to a non-zero value or zero, respectively.

The branching decision made in the `if` statement is represented by an element of the *tasks* universe for which the *TaskType* function returns *branch*. We illustrate a typical `if` statement with the graph in Fig. 1, where ovals represent tasks, labeled arcs represent the corresponding unary functions, and boxes represent subgraphs. If an `else` clause is not present in an `if` statement, the corresponding task graph omits the lower portion of Fig. 1, with the *FalseTask* function connecting the *branch* task to the task following the `if` statement. The transition rule for *branch* tasks is:

---

if *TaskType(CurTask) = branch* **then**
  if *TestValue(CurTask)* $\neq$ *0* **then**
    Moveto(TrueTask(CurTask))
  elseif *TestValue(CurTask) = 0* **then**
    Moveto(FalseTask(CurTask))
ENDIF

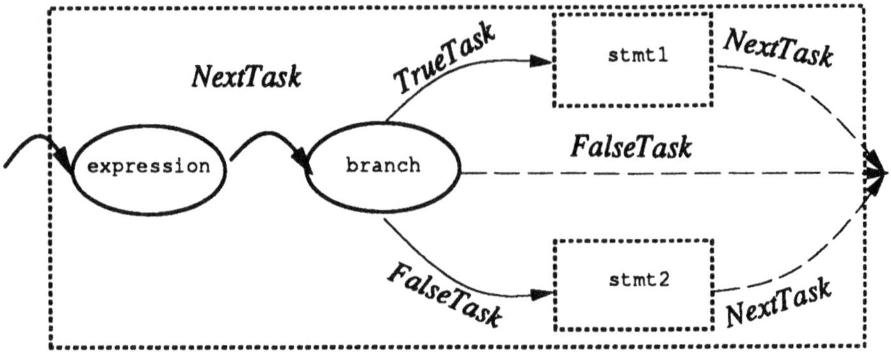

Fig. 1. A typical **if** statement.

Remark. Figure 1 shows a typical `if` statement, though it does not represent all `if` statements. The presence of a jump statement in *statement1* or *statement2* may cause *NextTask* to point to a different task than the one which immediately follows the `if` statement.

1.6 `switch` Statements

A `switch` statement has the following form:

switch-statement → **`switch`** (*expression*) *body*

where *body* is a statement, usually compound.

Within the body of a `switch` statement there are (usually) labeled `case` and `default` statements. Each `case` or `default` is associated with the smallest enclosing `switch` statement.

To execute a `switch` statement, evaluate the guard expression, and within the body of the switch, transfer control to the `case` statement for the `switch` whose labeled value matches the value of the expression, or to the `default` statement for the `switch`, whichever comes first. If no such statement is found, transfer control to the statement following the `switch` statement.

Labels on `case` statements are required to be unique within a `switch`, and a `switch` may not have more than one `default` statement. Thus, for a given expression value, there is exactly one statement to which control should be passed. A static partial function *SwitchTask: tasks* × *results* → *tasks* indicates the next task to be executed for a given expression value. We illustrate a typical `switch` statement with the graph in Fig. 2; with regard to the possible effects of embedded jump statements, see the remark in Sect. 1.5. The rule for `switch` tasks is:

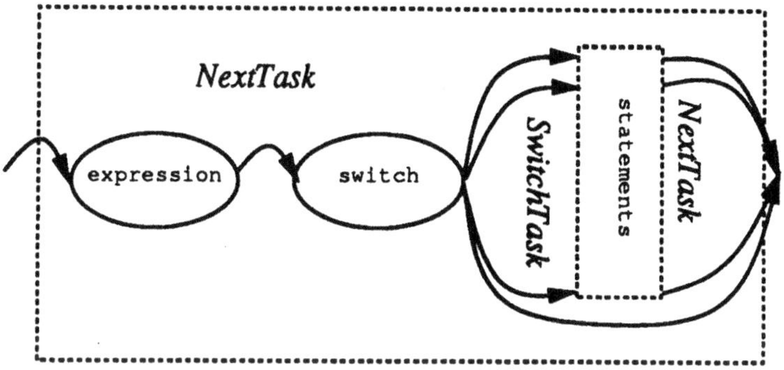

Fig. 2. A typical `switch` statement.

---

if *TaskType(CurTask) = switch* **then**
  Moveto(SwitchTask(CurTask, TestValue(CurTask)))
endif

1.7 `while` Statements

A `while` statement has the following form:

while-statement → `while` (*expression*) *body*

where *body* is a statement.

To execute a `while` statement, keep evaluating the guard expression until the value of the expression becomes zero. Each time that the value of the guard expression is not zero, execute *body*.

We illustrate a typical `while` statement with the graph in Fig. 3; with regard to the possible effects of embedded jump statements, see the remark in Sect. 1.5. Since the only types of tasks used to represent `while` statements are the *expression* and *branch* tasks, our previously-presented transition rules are sufficient to model the behavior of `while` statements.

Note that it is possible to enter a `while` loop by means of a `goto` statement, thus circumventing the initial test of the expression at the beginning of the loop. The ANSI standard [KR] does not give specific semantics for such behavior. In such a situation, our abstract machine would continue as if the loop had been entered normally (*i.e.*, after completion of the statement body, control returns to the guard expression to be evaluated). We believe this is a reasonable interpretation of such an event.

1.8 `do-while` Statements

A `do-while` statement has the following form:

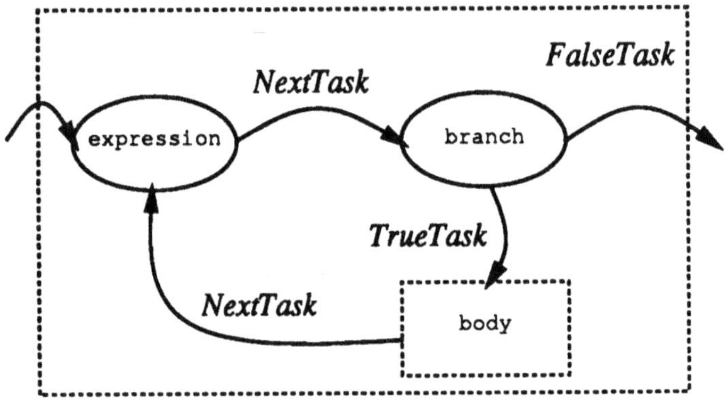

Fig. 3. A typical **while** statement.

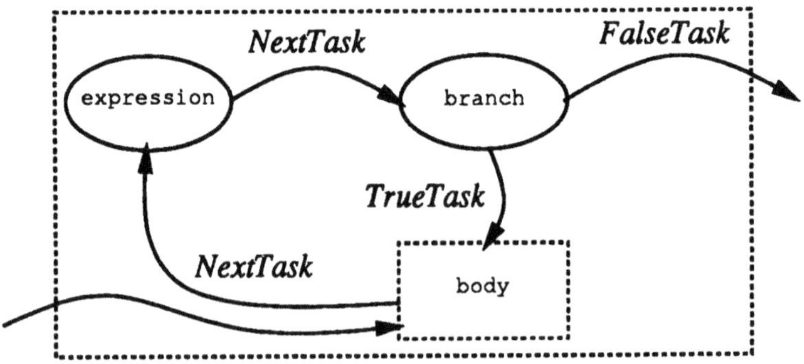

Fig. 4. A typical **do-while** statement.

do-while-statement → **do** *body* **while** (*expression*) ;

where *body* is a statement.

do-while statements are identical to **while** statements except that the guard expression and statement body are visited in the opposite order. We illustrate a typical **do-while** with the graph shown in Fig. 4; with regard to the possible effects of embedded jump statements, see the remark in Sect. 1.5. (Note the similarity between this graph and that of the **while** loop.) As with **while** loops, no new transition rules are required to model the behavior of **do-while** statements.

1.9 for Statements

The most complete form of the **for** statement is:

for-statement → **for** (*initializer* ; *test* ; *update*) *body*

where *initializer*, *test*, and *update* are expressions, any of which may be omitted, and *body* is a statement, usually compound. We begin by describing the behavior and representation of a **for** statement when all expressions are present.

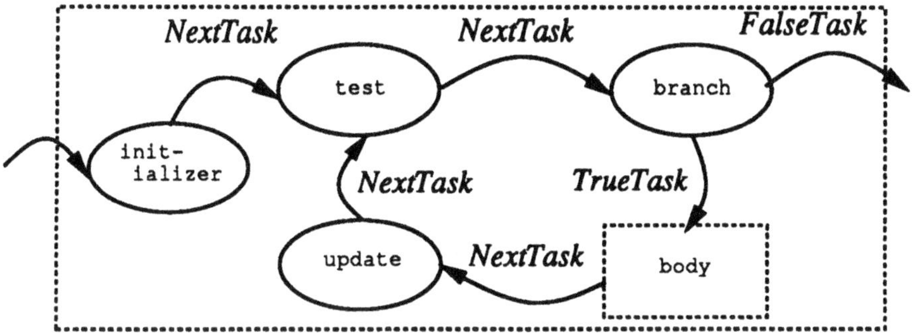

Fig. 5. A typical **for** statement.

In executing a **for** statement, begin by evaluating the initializer. Evaluate the test next; if the result is non-zero, execute the body and evaluate the update (in that order) and re-evaluate the test. If the value of the test is zero, transfer control to the statement following the **for** loop.

We illustrate a typical **for** statement with the graph in Fig. 5; with regard to the possible effects of embedded jump statements, see the remark in Sect. 1.5. Again, no new transition rules are required to model the behavior of **for** statements.

The graphs for **for** loops missing one or more of the three expressions (initializer, test, and update) omit the corresponding tasks, with *NextTask* pointing to the next task in the graph sequence. If the test is omitted, both the test and the *branch* task are omitted, which creates an infinite loop (which may still be broken through the use of jump statements).

1.10 Jump Statements

A jump statement has one of the following forms:

jump-statement → **goto** *identifier* ;
jump-statement → **continue** ;
jump-statement → **break** ;
jump-statement → **return** ;
jump-statement → **return** *expression* ;

Each of these jump statements is a command indicating that control should be unconditionally transferred to another task in the task graph:

- **goto** statements indicate directly the task to which control passes.
- **continue** statements may only occur within the body of an iteration statement. For a given **continue** statement C, let S be the smallest iteration statement which includes C. Executing C transfers control to the task within S following the statement body of S: *e.g.*, for **for** statements, control passes to the update expression, while for **while** statements, control passes to the guard expression.

- **break** statements may only occur within the body of an iteration statement. For a given **break** statement B, let S be the smallest iteration statement which includes B. Executing B transfers control to the first task following S.
- **return** statements occur within the body of function abstractions, indicating that the current function execution should be terminated. A more complete discussion of **return** statements will be presented in Algebra Four, where function abstractions are presented. For now, we assert that executing a **return** statement should set *CurTask* to *undef*, which will bring a halt to the algebra, since we only have one function (**main**) being executed.

The *NextTask* function contains the above (static) information for jump statement tasks. Thus, the transition rule for jump statements is trivial:

---

if *TaskType(CurTask) = jump* **then**
  Moveto(NextTask(CurTask))
endif

1.11 Labeled Statements

A labeled statement has one of the following forms:

labeled-statement → *identifier* : *statement*
labeled-statement → **case** *constant-expression* : *statement*
labeled-statement → **default** : *statement*

Statement labels identify the targets for control transfer in **goto** and **switch** statements. *NextTask* and *SwitchTask* return the appropriate tasks in each case; no further transition rules are needed.

1.12 Compound Statements

A compound statement has the following form:

compound-statement → { *declaration-list statement-list* }

where the declaration and/or statement lists may be empty.)

Since *NextTask* indicates the order in which tasks are processed, we have no need for rules concerning compound statements. Each statement or declaration in a compound statement is linked to its successor via *NextTask*. (Declarations are not treated until Algebra Three; nonetheless, the same principle holds for declaration tasks.)

1.13 Initial and Final States

We assert that initially, *CurTask* indicates the first task of the first statement of the program.

A final state in our algebra is any state in which *CurTask = undef*. In this state, no rules will be executed, since *TaskType(undef) = undef.*

2 Algebra Two: Evaluating Expressions

Our second evolving algebra refines the first and focuses on the evaluation of expressions.

We replace each occurrence of a task of type *expression* from the first algebra with numerous tasks reflecting the structure of the expression. Also, *TestValue* is now an internal, dynamic function.

In Algebra Two, we treat the evaluation of expressions at a relatively high level of abstraction. We map variable identifiers to memory locations through external functions. We also treat function invocations as expressions whose values are provided by external functions. In Algebras Three and Four we will eliminate these abstractions.

2.1 New Basic Functions Related To Memory Management

In C, one may (re)cast types. For example, one may cast a pointer to a structure into a pointer to an array of characters. Thus, one can access the individual bytes of most values which might exist during the execution of the program.[2]

A static function *Size: typename* $\rightarrow$ *integer* indicates how many bytes are used by a particular value type in memory. A dynamic function *Memory: addresses* $\rightarrow$ *bytes* indicates the values stored in memory at a given byte. Since most values of interest are larger than a byte, we need a means for storing members of *results* as individual bytes. For example, assume that the `int` value 258 is represented in the memory of a particular system by the four (eight-bit) bytes 0, 0, 1, and 2, stored consecutively. We need a way to go from a value in *results* (*e.g.*, 258) to its component bytes (0, 0, 1, 2) and vice versa.

A static partial (n+1)-ary function *ByteToResult: typename* $\times$ *byte*^{n} $\rightarrow$ *results* converts the memory representation of a value of the specified basic type into its corresponding value in the *results* universe. Here n is the maximum number of bytes used by the memory representation of any particular basic type (and is implementation-dependent). For types whose memory representations are less than n bytes in length, we ignore any unused parameters. In our example above, *ByteToResult(int,0,0,1,2)* = *258*.

A static partial function *ResultToByte: results* $\times$ *integer* $\times$ *typename* $\rightarrow$ *byte* yields the specified byte of the memory representation of the specified value from the specified universe. This function can be thought of as the inverse of *ByteToResult*. In our example above, *ResultToByte(258,3,int)* = *2*. (We assume tacitly that the arguments of *ResultToByte* uniquely define the value of the function, which is the case in all the implementations that we know.)

We define an abbreviation *MemoryValue: address* $\times$ *typename* $\rightarrow$ *results*, which indicates the value of the specified type being stored in memory beginning at the indicated address. *MemoryValue (addr,type)* abbreviates *ByteToResult (type, Memory(addr), Memory(addr+1), ..., Memory(addr + Size(type) - 1))*.

---

[2] The distinguished value `void` is an example of a value which cannot be accessed in this manner.

2.2 Other New Basic Functions

Two sub-universes of *results*, the universe of computational results, are of particular interest in Algebra Two. A universe *bytes* contains those values which may be "stored" in a **char** variable. (This universe is usually identical to $\{0,1,\ldots,255\}$, but we prefer the more general definition.) A universe *addresses* contains positive integers corresponding to valid memory locations. This is also the universe of values which may be stored in a pointer-type variable. (Of course, these two universes are implementation-dependent.)

A universe *typename* contains elements representing the different types of storable values. A static partial function *ValueType: tasks* $\rightarrow$ *typename* indicates the type of the resulting value when an expression has been evaluated.

Static partial functions *LeftTask, RightTask: tasks* $\rightarrow$ *tasks* indicate the left and right operands of binary operators whose order of evaluation is not defined within C (*e.g.*, +). A static partial function *Parent: tasks* $\rightarrow$ *tasks* indicates the parent (*i.e.*, closest enclosing) expression for a given expression. For expressions which are not contained in any other expressions, *Parent* returns the corresponding *branch* task which uses the expression (if one exists) or *undef* (if none exists). A static partial function *WhichChild: tasks* $\rightarrow$ {*left, right, only, test, ...*} (where *left*, *etc.* are members of the *tags* universe) indicates the relationship between a task and its parent.

Dynamic partial functions *LeftValue, RightValue: tasks* $\rightarrow$ *results* indicate the results of evaluating the left and right operands of binary operators with ambiguous evaluation order. Similarly, a dynamic partial function *OnlyValue: tasks* $\rightarrow$ *results* indicates the result of evaluating the single operand of a unary operator. A static partial function *ConstVal: tasks* $\rightarrow$ *results* indicates the values of program constants.

2.3 Macros: EvaluateOperands and Moveto

In C, as described by [KR], many binary operators (such as the assignment operator "=") do not have a defined order of evaluation: either operand may be evaluated first. When one or more operands of such an operator generate side effects (as in "`a[i] = i++`"), the value or the side-effects generated by the expression may depend upon the order of evaluation. Writing such code is usually unwise, since such code may not be portable; however, an optimizing compiler may take advantage of this ambiguity to generate code which minimizes the resources required to perform a particular computation [ASU].

For expressions involving such operators, our algebra must be flexible enough to reflect any possible evaluation order of an expression's operands, even if this decision is made at run-time. While we believe most compilers make this decision at compile-time, we must still provide a mechanism for making this decision dynamically. (If this decision is always made statically in a particular system, the algebra may be explicitly structured to incorporate those static decisions into the task graph). We will use an external function *ChooseTask* to represent this decision.

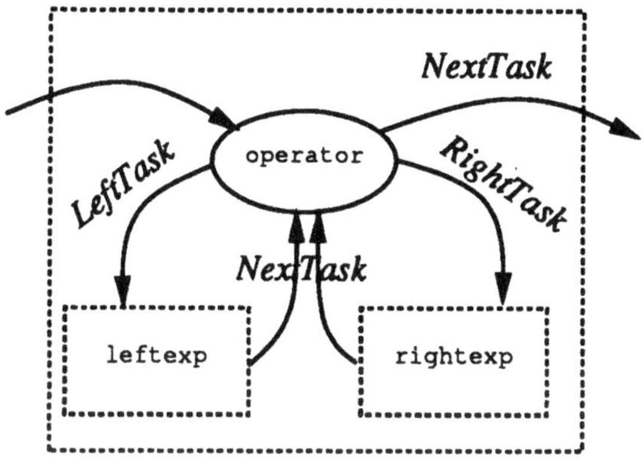

Fig. 6. Binary operators.

We illustrate how expressions with such operators are represented in our algebra by the graph in Fig. 6. A dynamic function *Visited: tasks* $\rightarrow$ {*left, right, both, neither*} indicates which subexpressions have been evaluated at a given moment. Initially, *Visited* has the value *neither* for all tasks.

To evaluate expressions of this type, begin by evaluating the sub-expression indicated by *ChooseTask*. When that sub-expression has been evaluated, evaluate the other sub-expression. Finally, when both expressions have been evaluated, perform the desired operation.

To handle the portion of this behavior dealing with the operator task, we use the *EVALUATE OPERANDS* macro. Informally, the macro means "Evaluate both operands in the order given by *ChooseTask*; when both operands are evaluated, then do ...". To handle the portion of this behavior dealing with movement between subtasks, we redefine the *Moveto(Task)* macro to jump directly between the subtasks of an operator of this type. The definitions for *EVALUATE OPERANDS* and *Moveto* are shown below.

2.4 Macro: DoAssign

Our rules for assignment to memory are a little complicated, since a given assignment may require an arbitrarily large number of updates to the *Memory* function. We need rules which perform a loop to make those arbitrarily large number of updates in a systematic fashion.[3]

[3] Most computer systems provide a means for memory assignments in units larger than a byte, but the particular sizes available are implementation-dependent. We thus present rules using the lowest common denominator, the byte.

```
EVALUATE OPERANDS WITH
    statements
END EVALUATE

if Visited(CurTask) = neither then
    if ChooseTask(CurTask) = LeftTask(CurTask) then
        Visited(CurTask) := left
    elseif ChooseTask(CurTask) = RightTask(CurTask) then
        Visited(CurTask) := right
    endif
    Moveto(ChooseTask(CurTask))
elseif Visited(CurTask) = both then
    Visited(CurTask) := neither
        statements
endif
```

```
Moveto(Task)

if Visited(Task) = neither then
    CurTask := Task
elseif Visited(Task) = both then
    CurTask := Task
elseif Visited(Task) = left then
    CurTask := RightTask(Task)
    Visited(Task) := both
elseif Visited(Task) = right then
    CurTask := LeftTask(Task)
    Visited(Task) := both
endif
```

To facilitate this loop, we use several distinguished elements. *CopyValue: results* denotes the value to be copied. *CopyType: typename* denotes the type of value to be copied. *CopyLocation: address* denotes the location to which the value is to be copied. *CopyByte: integer* denotes which byte of the representation of *CopyValue* is being copied into memory. *OldTask: tasks* denotes the task which invoked the memory copying procedure. *CopyTask: tasks* is a static distinguished element used to indicate that the copying procedure should begin.

We will invoke the copying procedure using the *DoAssign(address, value, type)* macro, defined below. The copying process itself is relatively straightforward. We utilize the distinguished element *CopyByte* to denote which byte of the memory representation of *CopyValue* we are copying into memory at a given moment in time. We copy bytes singly, incrementing the value of *CopyByte* after each assignment to memory, halting when all bytes have been copied. The transition rule for copying to memory is shown below.

DoAssign(address, value, type)

CopyValue := value
CopyType := type
CopyLocation := address
CopyByte := 0
OldTask := CurTask
CurTask := CopyTask

```
if CurTask = CopyTask then
    if CopyByte < Size(CopyType) then
        Memory(CopyLocation + CopyByte) :=
            ResultToByte(CopyValue, CopyByte, CopyType)
        CopyByte := CopyByte + 1
    elseif CopyByte = Size(CopyType) then
        CurTask := NextTask(OldTask)
    endif
endif
```

2.5 Macro: ReportValue

When we process tasks corresponding to expression evaluation, we assign the value of an evaluated expression to the appropriate storage function in the parent expression (*e.g. LeftValue(Parent(CurTask))*). We use the *ReportValue* macro to accomplish this, shown below.

ReportValue(value)

```
if WhichChild(CurTask) = left then
    LeftValue(Parent(CurTask)) := value
elseif WhichChild(CurTask) = right then
    RightValue(Parent(CurTask)) := value
elseif WhichChild(CurTask) = only then
    OnlyValue(Parent(CurTask)) := value
elseif WhichChild(CurTask) = test then
    TestValue(Parent(CurTask)) := value
endif
```

2.6 Assignment Expressions

A simple assignment has the following form:

assignment-expression → *expr1* = *expr2*

where *expr1* and *expr2* are expressions.

To evaluate a simple assignment expression, copy the value of *expr2* into the memory location given by *expr1*, returning that value as the value of the parent expression.

This is the first occurrence of an operator in Algebra Two with an ambiguous evaluation order, as discussed in Sect. 2.3. Thus, the expression is represented in our algebra as in Fig. 6. The transition rule for assignment operators is shown below.

Within C, there are other assignment operators ("+=", "*=", *etc.*) which perform a mathematic operation on the value of *expr2* and the value stored in the memory location given by *expr1*. The result is copied into the memory location given by *expr1*. For example, "`i *= 2`" has the same value and effect as "`i = i * 2`". Since expressions involving these operators may be replaced with equivalent expressions involving only simple assignments, we assert that all expressions involving these operators are replaced by the appropriate equivalent expressions. Thus, we do not need to give rules for handling such operators.

---

if *TaskType(CurTask)* = *simple-assignment* **then**
  EVALUATE OPERANDS WITH
    DoAssign(LeftValue(CurTask),RightValue(CurTask),
        ValueType(CurTask))
    ReportValue(RightValue(CurTask))
  END EVALUATE
endif

2.7 Comma Operators

A comma expression has the following form:

comma-expression → *expr1* , *expr2*

where *expr1* and *expr2* are expressions.

To evaluate a comma expression, evaluate *expr1* and *expr2*, left to right, returning the value of *expr2* as the value of the parent expression. (Though it may seem unnecessary to evaluate the first expression since we ignore its value, recall that expressions in C may generate side-effects.) We represent comma expressions as a sequence of two expressions linked by the *NextTask* function. Thus, no additional transition rules are needed to process comma operators.

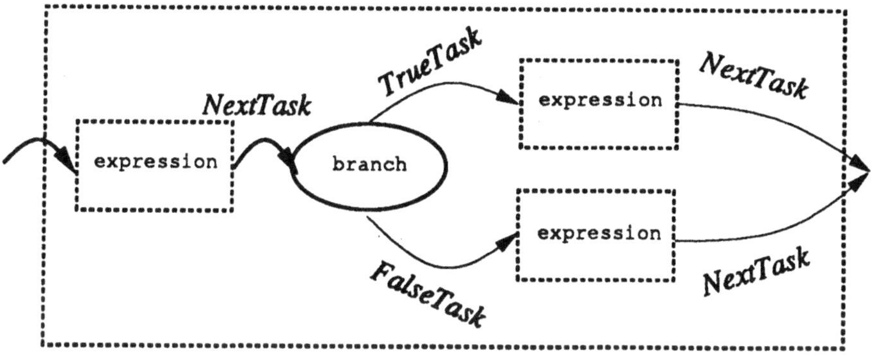

Fig. 7. Conditional expression.

2.8 Conditional Expressions

A conditional expression has the following form:

conditional-expression → *expr1* ? *expr2* : *expr3*

where *expr1*, *expr2*, and *expr3* are expressions.

To evaluate a conditional expression, evaluate *expr1*. If the resulting value is non-zero, evaluate *expr2* and return its value as the value of the parent expression; otherwise, evaluate *expr3* and return its value as the value of the parent expression.

We will represent conditional expressions in our algebra in a manner similar to that in which we represent conditional statements, as illustrated in Fig. 7. The tasks corresponding to the center and right sub-expressions will update the appropriate *Value* function for the parent expression upon completion of the evaluation of the subexpression. No new transition rules are needed to handle conditional expressions.

2.9 Logical OR Expressions

A logical OR expression has the following form:

logical-OR-expression → *expr1* || *expr2*

where *expr1* and *expr2* are expressions.

To evaluate a logical OR expression, start by evaluating *expr1*. If the result is non-zero, the value of the parent expression is 1 and *expr2* is not evaluated. Otherwise, the value of the parent expression is the value of *expr2*, with non-zero values coerced to 1.

To represent a logical OR expression, we will introduce two new task types, *OR* and *makeBool*. We illustrate how logical OR expressions are represented in our algebra by the graph in Fig. 8.

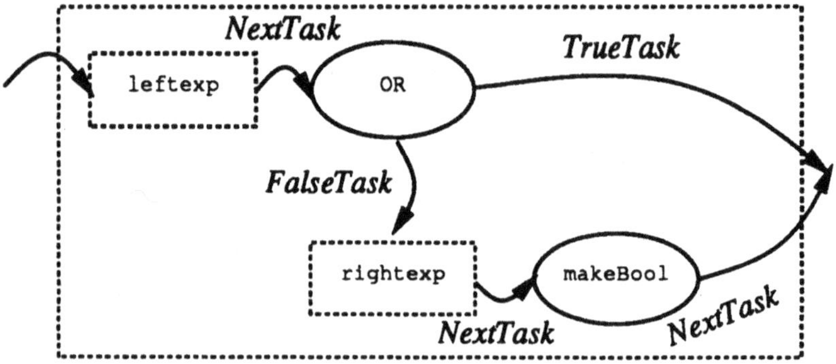

Fig. 8. Logical OR expression.

In processing the *OR* task, the value of *expr1* will be examined. If the value is not zero, the rules for *OR* tasks will set the value of the expression to 1 and end processing of the parent expression. Otherwise, the rules will pass control to the tasks which evaluate *expr2*. The rules for *makeBool* tasks will examine the value of *expr2* and coerce it to 0 or 1. The transition rules for *OR* and *makeBool* tasks are:

---

if *TaskType(CurTask) = OR* **then**
    if *OnlyValue(CurTask) ≠ 0* **then**
        ReportValue(1)
        Moveto(TrueTask(CurTask))
    elseif *OnlyValue(CurTask) = 0* **then**
        Moveto(FalseTask(CurTask))
ENDIF

---

if *TaskType(CurTask) = makeBool* **then**
    if *OnlyValue(CurTask) ≠ 0* **then**
        ReportValue(1)
    elseif *OnlyValue(CurTask) = 0* **then**
        ReportValue(0)
    endif
    Moveto(NextTask(CurTask))
endif

2.10 Logical AND expressions

A logical AND expression has the following form:

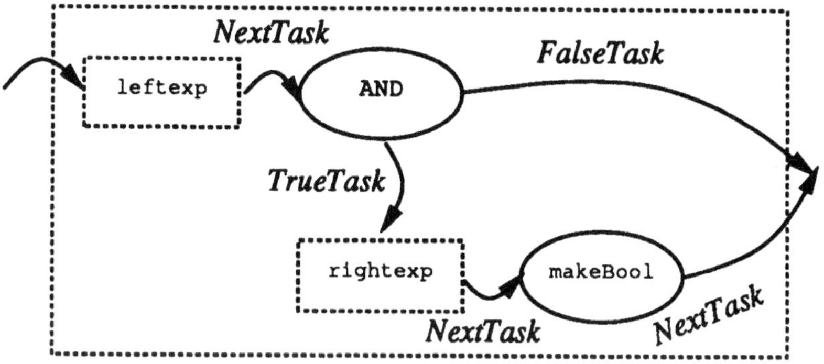

Fig. 9. Logical AND expression.

logical-AND-expression → *expr1* **&&** *expr2*

where *expr1* and *expr2* are expressions.

To evaluate an AND expression, begin by evaluating *expr1*. If the resulting value is 0, the value of the parent expression is 0. Otherwise, the value of *expr2* (coerced to 0 or 1) is the value of the parent expression. The representation of logical AND expressions is similar to that of logical OR expressions, as illustrated in Fig. 9. The transition rule for *AND* tasks (used in such representations) is:

---

if *TaskType(CurTask) = AND* **then**
    if *OnlyValue(CurTask) = 0* **then**
        ReportValue(0)
        Moveto(FalseTask(CurTask))
    elseif *OnlyValue(CurTask) ≠ 0* **then**
        Moveto(TrueTask(CurTask))
ENDIF

2.11 The sizeof Operator

Expressions involving **sizeof** have one of the following forms:

sizeof-expression → **sizeof** *expression*
sizeof-expression → **sizeof** (*type-name*)

The *ValueType* function converts either operand into a value in the *typename* universe. With that information, we may use the *Size* function to determine the size, in bytes, of an element of that particular type and return that number as the value of the unary expression. The transition rule for size operators is:

```
if TaskType(CurTask) = sizeof then
    ReportValue(Size(ValueType(CurTask)))
    Moveto(NextTask(CurTask))
endif
```

2.12 General Mathematic Expressions

There are a large number of mathematic expressions in C involving binary operators ("*", "+", "-", *etc.*) whose behaviors are similar. (We treat the bit-wise operators (*e.g.*, |, &) as ordinary mathematic operators.) We assume that infix functions corresponding to these C functions are present within our algebra. Thus, to evaluate one of these expressions, evaluate both operand expressions and apply the appropriate function. We present the transition rule for multiplication below as a representative of this category of expressions, and omit the rules for other binary operators of this form for brevity.

```
if TaskType(CurTask) = multiplication then
    EVALUATE OPERANDS WITH
        ReportValue(LeftValue(CurTask) * RightValue(CurTask))
        Moveto(NextTask(CurTask))
    END EVALUATE
endif
```

Rules for the addition and subtraction operators are slightly more complicated, since one may add or subtract an integer to a pointer. One may add an integer i to a pointer variable p with the result being a pointer which is i units forward in memory from p. Consider, for example, a pointer p to an `int` in a system where `int`s require 4 bytes in memory. The expression "$p+1$" refers to the location in memory 4 bytes after p.

Similarly, one may subtract an integer i from a pointer variable p, with the result being a pointer i units in memory preceding p. (In our example, "$p-1$" refers to the location in memory 4 bytes before p.) Further, one may subtract two pointers of the same type, resulting in the number of units of memory lying between the two pointers. (Thus, $((p+i)-p)=i$.) In each case, the size of a "unit" of memory is determined by the size of the object type to which the pointer points.

This requires specialized rules for the addition and subtraction operators. As we process each of these operators, it now becomes necessary to know whether or not a given variable is a pointer; a static partial function *PointerType: tasks* $\rightarrow$ {*true, false*} contains this information. For tasks for which *PointerType* returns *true*, a static partial function *PointsToType : tasks* $\rightarrow$ *typename* indicates the object type to which the pointer points.

The transition rules for the addition and subtraction operators are:

```
if TaskType(CurTask) = addition then
    if PointerType(LeftTask(CurTask)) = true then
        ReportValue(LeftValue(CurTask)
            + (Size(PointsToType(CurTask)) * RightValue(CurTask)))
    elseif PointerType(RightTask(CurTask)) = true then
        ReportValue(RightValue(CurTask)
            + (Size(PointsToType(CurTask)) * LeftValue(CurTask)))
    else
        ReportValue(LeftValue(CurTask) + RightValue(CurTask))
    endif
    Moveto(NextTask(CurTask))
endif
```

```
if TaskType(CurTask) = subtraction then
    if PointerType(LeftTask(CurTask)) = true and
            PointerType(RightTask(CurTask)) = true then
        ReportValue((LeftValue(CurTask) - RightValue(CurTask))
                / Size(PointsToType(CurTask)))
    elseif PointerType(LeftTask(CurTask)) = true and
            PointerType(RightTask(CurTask)) = false then
        ReportValue(LeftValue(CurTask)
            - (Size(PointsToType(CurTask)) * RightValue(CurTask)))
    else
        ReportValue(LeftValue(CurTask) - RightValue(CurTask))
    endif
    Moveto(NextTask(CurTask))
endif
```

2.13 Mathematical Unary Operators

Unary operator expressions have one of the following forms:

unary-expression → + *expression*
unary-expression → - *expression*
unary-expression → ~ *expression*
unary-expression → ! *expression*

Evaluating these expressions takes a form similar to that for binary mathematical operators. We present the transition rule for the negation operator below as a representative example.

if *TaskType(CurTask) = negation* **then**
    ReportValue(- OnlyValue(CurTask))
    Moveto(NextTask(CurTask))
endif

2.14 Casting Expressions

A casting expression has the following form:

cast-expression → (*type-name*) *expression*

A static function *CastType: tasks* → *typename* indicates the old type from which the value of the expression is to be cast; *ValueType* indicates the new type into which the value will be cast. A static function *Convert: typename* × *typename* × *values* → *values* converts elements from one universe into the corresponding elements of another universe. For example, *Convert(float,int,X)* is the closest integer to X (assuming X is a floating-point value). Note that the meaning of "closest" is implementation-defined.

To perform a cast, evaluates the argument expression and use the *Convert* function to generate the proper return value. Our task sequence places the expression to be cast before the task which performs the casting; thus, the argument of the cast has been evaluated already and its value is available. The transition rule for casting expressions is:

if *TaskType(CurTask) = cast* **then**
    ReportValue(Convert(CastType(CurTask),
        ValueType(CurTask), OnlyValue(CurTask)))
    Moveto(NextTask(CurTask))
endif

2.15 Pre-Increment and Pre-Decrement

A pre-increment or pre-decrement expression has the following form:

pre-incr-expression → ++ *expression*
pre-decr-expression → -- *expression*

To evaluate a pre-increment (resp. pre-decrement) expression, increment (decrement) the value stored at the indicated memory location by one and store the new value into that memory location; the incremented (decremented) value is the value of the parent expression. Note that the expression to be modified may be a pointer; in this case, the value in memory is incremented (decremented) by the size of the object to which the pointer points (as with normal pointer addition and subtraction). The transition rule for pre-increment expressions is shown below. The transition rules for pre-decrement expressions are similar to those presented here and thus omitted.

---

if *TaskType(CurTask) = pre-increment* **then**
  if *PointerType(CurTask) = true* **then**
    DoAssign(OnlyValue(CurTask),
      MemoryValue(OnlyValue(CurTask), ValueType(CurTask))
        + Size(PointsToType(CurTask)),
      ValueType(CurTask))
    ReportValue(MemoryValue(OnlyValue(CurTask),
                ValueType(CurTask))
          + Size(PointsToType(CurTask)))
  else
    DoAssign(OnlyValue(CurTask),
      MemoryValue(OnlyValue(CurTask), ValueType(CurTask)) + 1,
      ValueType(CurTask))
    ReportValue(MemoryValue(OnlyValue(CurTask),
                ValueType(CurTask))+1)
ENDIF

2.16 Post-Increment and Post-Decrement

A post-increment or post-decrement expression has the following form:

postincr-expression → *expression* ++
postdecr-expression → *expression* --

Post-increment (resp. post-decrement) operators are handled in the same manner as pre-increment (pre-decrement) operators except that the sequence of operations is reversed: *i.e.*, the value of the parent expression is established before the incrementing (decrementing) takes place. The transition rule for the post-increment operator is shown below. (As before, the transition rules for the post-decrement operator are similar and thus omitted.)

2.17 Addresses

An addressing expression has the following form:

addressing-expression → & *expr1*

where *expr1* is an expression.

The & operator in C passes back as its result the address of the memory location indicated by the argument expression.

if *TaskType(CurTask) = post-increment* **then**
  if *PointerType(CurTask) = true* **then**
    DoAssign(OnlyValue(CurTask),
      MemoryValue(OnlyValue(CurTask), ValueType(CurTask))
        + Size(PointsToType(CurTask)),
      ValueType(CurTask))
  else
    DoAssign(OnlyValue(CurTask),
      MemoryValue(OnlyValue(CurTask), ValueType(CurTask))+1,
      ValueType(CurTask))
  endif
  ReportValue(MemoryValue(OnlyValue(CurTask),
                ValueType(CurTask)))
endif

As we evaluate expressions which refer to objects in memory, we need to know whether we need to use the address of an object or the object itself in our calculations. (For example, in the assignment statement "`a = b;`", the address or *lvalue* of variable `a` is needed, but the object being referenced or *rvalue* of variable `b` is needed.) A static partial function *ValueMode: tasks* → {*lvalue, rvalue*} indicates which of the two pieces of information should be computed for a given task.

We assert that *ValueMode(e) = lvalue* for the sub-expressions of *expr1*; thus, the value returned through evaluation of the argument expression is the address (and not the value) of the argument expression in memory. We simply pass this address up the task graph. The resulting simple transition rule for the addressing operator is:

if *TaskType(CurTask) = address* **then**
  ReportValue(OnlyValue(CurTask))
  Moveto(NextTask(CurTask))
endif

2.18 De-Referencing

A de-referencing expression has the following form:

de-reference-expression → * *expression*

If the parent expression is an rvalue, evaluate the argument and use the *Memory* function to return the value stored in memory at the indicated location. Otherwise, return the address indicated by the argument (since the expression is an lvalue and requires that a pointer be returned to the parent expression). The transition rule for de-referencing is:

---

if *TaskType(CurTask) = de-referencing* **then**
  if *ValueMode(CurTask) = rvalue* **then**
    ReportValue(MemoryValue(OnlyValue(CurTask),
                ValueType(CurTask)))
  elseif *ValueMode(CurTask) = lvalue* **then**
    ReportValue(OnlyValue(CurTask))
  endif
  Moveto(NextTask(CurTask))
endif

2.19 Array References

An array reference has the following form:

array-ref-expression → *expr1* `[` *expr2* `]`

where *expr1* and *expr2* are expressions.

According to [KR], an array reference of the form `a[b]` is identical, by definition, to the expression `*((a)+(b))`.[4] This definition is valid because the name of an array in C may be used as a pointer to the first element of the array.

We assert that any array references present in the program being modeled in our algebra are represented as an expression of equivalent form involving addition and de-referencing. Thus, we do not need to present any additional rules to handle array references.

One may prefer to think of arrays as objects in their own right, and present an algebra intermediate to Algebras One and Two where expressions like `a[b]` could be treated at a higher level of abstraction. While such a presentation is possible, we choose not to do so here.

2.20 Function Invocations

A function invocation has the following form:

func-invocation-expression → *expression* (*expression-list*)

Since we have disallowed function invocations for the moment, we will obtain the value of a function invocation expression from an external function *FunctionValue: tasks* → *results*. Thus, the transition rules for function invocations are:

[4] Note that this means that `a[b]` and `b[a]` evaluate to the same value.

if *TaskType(CurTask) = function-invocation* **then**
    ReportValue(FunctionValue(CurTask))
    Moveto(NextTask(CurTask))
endif

2.21 Identifiers

An external function *FindID: tasks → addresses* maps identifier expression tasks to the corresponding memory location used by the associated variable. (In Algebra Three we shall eliminate the use of this function.) Thus, to handle an identifier expression, one returns the appropriate address or value from memory, as specified by the *ValueMode* function. The transition rule for identifiers is:

if *TaskType(CurTask) = identifier* **then**
    if *ValueMode(CurTask) = lvalue* **then**
        ReportValue(FindID(CurTask))
    elseif *ValueMode(CurTask) = rvalue* **then**
        ReportValue(MemoryValue(FindID(CurTask),
                                            ValueType(CurTask)))
    endif
    Moveto(NextTask(CurTask))
endif

2.22 `struct` or `union` References

A `struct` or `union` reference has the following form:

struct-expression → expr1 . identifier

where *expr1* is an expression evaluating to a `struct` or `union`. There is also another form:

struct-expression → expr2 -> identifier

where *expr2* is an expression evaluating to a pointer to a `struct` or `union`. Expressions of the form "`a->b`" are equivalent to those of the form "`(*a).b`". Thus, we will only consider references of the form "`a.b`", asserting that references of the other form are represented using their equivalent expansions.

The *ConstVal* function applied to the `struct` reference task returns the offset in memory to be used in obtaining the address or value of the specified field of the structure. The transition rules for struct references are:

if *TaskType(CurTask) = struct-reference* **then**
    if *ValueMode(CurTask) = lvalue* **then**
        ReportValue(OnlyValue(CurTask))
    elseif *ValueMode(CurTask) = rvalue* **then**
        ReportValue(MemoryValue(OnlyValue(CurTask)
                                        + ConstVal(CurTask),
                                    ValueType(CurTask)))
    endif
    Moveto(NextTask(CurTask))
endif

2.23 Bit Fields

Bit fields are members of **structs** which use a user-specified number of bits for their representations. Bit fields are used to minimize the space used by a **struct** or to represent accurately input or output values with bit-level significance. Much about bit fields behave is implementation-dependent: *e.g.* how bit fields are packed into adjacent bytes, whether or not unnamed "holes" will appear in **structs** between bit fields, or whether bit fields are read left-to-right or right-to-left. As a rule, operaptions are done with bytes (even bit operations); we show how to handle bit-fields in a byte-based model of memory.

An example of a **struct** using bit fields is shown below. Here, the field `b` holds a 4-bit unsigned integer and `c` holds a 15-bit unsigned integer. Since bytes usually comprise 8 bits, `c` will probably lie in two or three consecutive bytes in memory, possibly sharing a byte with `b`.

```
struct {
   unsigned int a;
   unsigned int b:4;
   unsigned int c:15;
   } bitty;
```

Suppose that we want to execute `bitty.c = 12`. We need to obtain the bytes which hold `c`'s bits and modify those bits accordingly while leaving all other bits unchanged. This gives rise to a static function *BitAssign: results* × *typename* × *results* → *results* which indicates the change occurring in the value of the appropriate collection of contiguous bytes when a bit field is modified. Given the content *oldval* of an appropriate piece of memory, *BitAssign(oldval, bittype, 12)* returns the new value; here *bittype* is the type of `c` in `bitty`.

Assigning to bit-fields is thus slightly different than usual. We present the transition rule for simple assignments to bit fields below:

```
if TaskType(CurTask) = bit-assignment then
   EVALUATE OPERANDS WITH
      DoAssign(LeftValue(CurTask),
         BitAssign(MemoryValue(    LeftValue(CurTask),
                                   ValueType(CurTask)),
               ValueType(CurTask),
               RightValue(CurTask)),
         ValueType(CurTask))
      ReportValue(RightValue(CurTask))
   END EVALUATE
endif
```

Evaluating bit-field references is also slightly different, since we need to extract the value of the bit field from the (usually) larger enclosing value. A static function *BitExtract: results* × *typename* → *results* performs this extraction. A static partial function *BitType: tasks* → *typename* indicates the type of bit field being references in such situations.

The transition rules for structure bit-field references are:

```
if TaskType(CurTask) = struct-bit-reference then
   if ValueMode(CurTask) = lvalue then
      ReportValue(OnlyValue(CurTask) + ConstVal(CurTask))
   elseif ValueMode(CurTask) = rvalue then
      ReportValue(BitExtract(
            MemoryValue(OnlyValue(CurTask) + ConstVal(CurTask),
            BitType(CurTask)),
         ValueType(CurTask)))
   endif
   Moveto(NextTask(CurTask))
endif
```

3 Algebra Three: Allocating and Initializing Memory

Algebra Three refines Algebra Two and focuses on memory allocation and initialization of variables.

3.1 Declarations

We represent declarations in C as elements of the *tasks* universe, linked in the proper order with statement tasks by *NextTask*.

C distinguishes between so-called *static variables* and other variables. The difference between static and non-static variables arises when control is passed to the declaration task for a variable. If the variable is not static, new memory is always allocated to the variable and its initializing expression (if it exists) is evaluated with the value of the expression being assigned to the new memory location. If the variable is static, the above allocation and initialization is performed only the first time that the declaration is executed; should the declaration become the focus of control once again, the same memory segment is allotted to the variable.

A static partial function *DecType: tasks* → {*static, non-static*} indicates what type of variable is being declared. (Note that there are also **extern** variable declarations in C which do not reserve memory but serve as syntactic linkage between variables. We omit consideration of such declarations since their function is wholly syntactic in nature.) A static partial function *Initializer: tasks* → *tasks* indicates the appropriate initializing expression (if any). We will store the value of the initializing expression using *RightValue*.

A partial function *StaticAddr: tasks* → *addresses* stores the current address (if any) that has been assigned to a static variable. In Algebra Three, *StaticAddr* is not really needed, since the *OnlyValue* function would provide the proper storage for the address of the static variable. However, it will simplify other rules to be presented. An external function *NewMemory: tasks* → *addresses* returns an address in memory to be used for the given declaration task. The transition rules for declarations are shown below.

---

if *TaskType(CurTask) = declaration* **and** *DecType(CurTask) = static* **then**
    if *StaticAddr(CurTask)* $\neq$ *undef* **then**
        OnlyValue(CurTask) := StaticAddr(CurTask)
        Moveto(NextTask(CurTask))
    elseif *StaticAddr(CurTask) = undef* **then**
        if *Initializer(CurTask)* $\neq$ *undef* **and**
            RightValue(CurTask) = undef **then**
            Moveto(Initializer(CurTask))
        else
            OnlyValue(CurTask) := NewMemory(CurTask)
            StaticAddr(CurTask) := NewMemory(CurTask)
            if *Initializer(CurTask)* $\neq$ *undef* **then**
                DoAssign(NewMemory(CurTask),
                    RightValue(CurTask), ValueType(CurTask))
            else
                Moveto(NextTask(CurTask))
ENDIF

3.2 Initializers

Initializers in C come in two forms: expressions (for variables of the basic types) and lists of expressions (for variables representing arrays and structures).

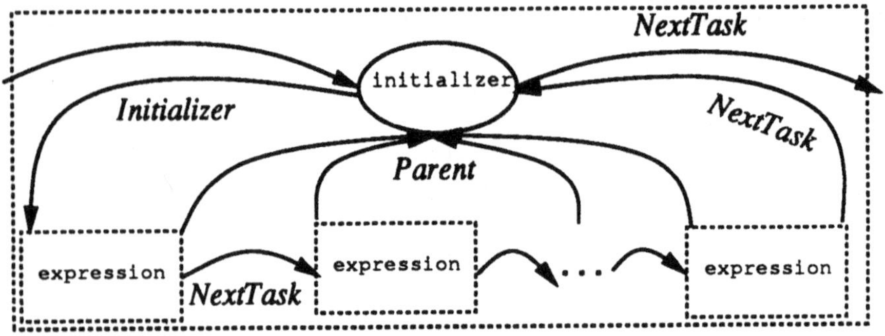

Fig. 10. Aggregate initializer.

---

if *TaskType(CurTask) = declaration* **and** *DecType(CurTask) ≠ static* **then**
  if *Initializer(CurTask) ≠ undef* **and**
    RightValue(CurTask) = undef **then**
    Moveto(Initializer(CurTask))
  else
    OnlyValue(CurTask) := NewMemory(CurTask)
    if *Initializer(CurTask) ≠ undef* **then**
      DoAssign(NewMemory(CurTask),
        RightValue(CurTask), ValueType(CurTask))
    else
      Moveto(NextTask(CurTask))
ENDIF

Our previous rules for evaluating expressions will handle initializers for simple expressions. To assist in handling aggregate expressions, a static function *AddTo: typename* × *results* × *results* → *results* appends a value onto the end of an aggregate structure of the specified type. (For example, if [42, 8] is an integer array, then *AddTo*(*array*, [42, 8], 4) = [42, 8, 4].)

We illustrate how expressions with aggregate initializers (*i.e.* expressions whose initializer is an expression list) are represented in our algebra by the graph in Fig. 10. Our previous transition rules will insure that each expression in the initializer list will be evaluated; we need to provide rules that combine the results of these evaluations into the proper aggregate value.

The *WhichChild* function returns the value *aggregate* when the expression being evaluated is a component of an aggregate initializer. We extend our *ReportValue* macro as shown below to correctly combine aggregate expressions.

if *WhichChild(CurTask) = aggregate* **then**
    OnlyValue(Parent(CurTask)) :=
        AddTo(ValueType(Parent(CurTask)),
                Value(Parent(CurTask)),value)
endif

3.3 Revision: Identifiers

A static partial function *Decl: tasks* → *tasks* maps tasks corresponding to occurrences of an identifier to the task corresponding to the declaration task for that variable. The revised rule for identifiers (generated by replacing each previous occurrence of *FindID()* with *OnlyValue(Decl())*) is shown below.

if *TaskType(CurTask) = identifier* **then**
    if *ValueMode(CurTask) = lvalue* **then**
        ReportValue(OnlyValue(Decl(CurTask)))
    elseif *ValueMode(CurTask) = rvalue* **then**
        ReportValue(MemoryValue(OnlyValue(Decl(CurTask)),
                                    ValueType(CurTask)))
    endif
    Moveto(NextTask(CurTask))
endif

4 Algebra Four: Handling Function Definitions

Algebra Four revises Algebra Three and focuses on C function definitions. In this way, we also (implicitly) present rules for starting a C program, since the starting function `main` is an ordinary C function (with externally provided parameters).

(There are also function declarations in C, which are used to specify syntactic information. Since their purpose is wholly syntactic in nature, we ignore them.)

4.1 Modeling The Stack

C functions may have several active incarnations at a given moment. Thus, we must have some means for storing multiple values of a function for a given task.

The universe *stack* comprises the positive integers, with a distinguished element *StackRoot = 1*. Static functions *StackPrev: stack* → *stack* and *StackNext: stack* → *stack* are the predecessor and successor functions on the positive integers. A dynamic distinguished element *StackTop* indicates the current top of the stack.

To store state-associated information on the stack, we modify the various *Value* functions *LeftValue, RightValue, OnlyValue,* and *TestValue* to be binary functions from ***tasks*** × ***stack*** to ***results***. This requires us to rewrite almost every rule that has appeared previously; we simplify matters by stating that every previous reference to *V(X)* should be replaced by *V(X, StackTop)*, where *V* is one of the *Value* functions listed above.

4.2 Function Invocations: Caller's Story

A function invocation has the following form:

func-invocation → ***func-name*** (***expression-list***)

In Algebra Two we used an external function *FunctionValue* to obtain the value of a function invocation. Here, we eliminate the use of this function.

The name of a function is often an identifier, but in general it is an expression referring to the address of the function. (What resides in memory at that address is implementation-dependent.) A static partial function *AddrToFunc: **addresses*** → ***tasks*** maps function addresses to the first task of the function definition.

While processing a function invocation, we wish to copy the value[5] of each parameter to an appropriate place for the callee to process. The *Parent* function (utilized by our *ReportValue* macro) maps each argument-expression task to its corresponding function parameter task. A partial function *ParamValue: **tasks*** × ***stack*** → ***results*** indicates the values of parameters being passed. We append the following rule to the *ReportValue* macro:

---

if *WhichChild(CurTask)* = *param* **then**
    ParamValue(Parent(CurTask),StackNext(StackTop)) := *value*
endif

We need to store the current task in order to resume execution at this point after the callee has finished. A dynamic function *ReturnTask: **stack*** → ***tasks*** indicates the new value of *CurTask* when the execution of the current function terminates. We assert that when *CurTask* = *1*, *ReturnTask(CurTask)* = *undef*, which will cause the algebra to terminate when the top-level function terminates.

To process a function invocation, evaluate the name of the function along with all of the arguments in the expression list, and then transfer control to the specified function. At the same time, "push another frame onto the stack"; *i.e.*, increment *StackTop* by 1. When control returns from the function, the stack value will be "popped" (*i.e.*, decremented by 1) and the function's return value will be passed to the parent expression.

As with the operands to most arithmetic operators, the ANSI standard [KR] does not specify the order in which arguments to a function are evaluated. We

[5] In C, all function parameters are call-by-value.

thus must present specialized rules for evaluating the expressions associated with a function invocation. The external function *ChooseTask* will indicate at each moment which expression associated with a function invocation should be evaluated next. Thus, our transition rules will simply make repeated calls to *ChooseTask* until all expressions have been evaluated, which will occur when *ChooseTask* returns *undef*. The transition rule for function invocation is shown below.

```
if TaskType(CurTask) = function-invocation then
    if ChooseTask(CurTask) ≠ undef then
        Moveto(ChooseTask(CurTask))
    elseif ChooseTask(CurTask) = undef then
        if OnlyValue(CurTask,StackTop) = undef then
            StackTop := StackNext(StackTop)
            ReturnTask(StackNext(StackTop)) := CurTask
            Moveto(AddrToFunc(LeftValue(CurTask,StackTop)))
        elseif OnlyValue(CurTask,StackTop) ≠ undef then
            ReportValue(OnlyValue(CurTask,StackTop))
            Moveto(NextTask(CurTask))
ENDIF
```

4.3 Function Invocations: Callee's Story

A function definition in C consists of a list of parameter declarations and a compound statement. To process a parameter declaration, we allocate new memory for each parameter and assign the appropriate value (stored here by the function invocation transition rules) to that new memory location. The transition rule for parameter declarations is shown below.

```
if TaskType(CurTask) = parameter-declaration then
    DoAssign(NewMemory(CurTask),
        ParamValue(CurTask,StackTop), ValueType(CurTask))
    OnlyValue(CurTask,StackTop) := NewMemory(CurTask)
endif
```

A `return` statement has one of the following forms:

return-statement → `return` ;
return-statement → `return` *expression* ;

If an expression is present, copy the value of the expression to the task which invoked the current function (as indicated by *ReturnTask* and *StackPrev*). Whether or not an expression is present, return control to the invoking task. The transition rule for `return` statements is:

```
if TaskType(CurTask) = return then
    OnlyValue(ReturnTask(StackTop),StackPrev(StackTop)) :=
        OnlyValue(CurTask,StackTop)
    StackTop := StackPrev(StackTop)
    CurTask := ReturnTask(StackTop)
endif
```

If a **return** statement is not explicitly present at the end of a function, our algebra will still contain a *return* task as the last task of the function, as if the statement "`return ;`" was present as the last statement of the original C function.

4.4 Global Variables

Since function definitions may not contain other function definitions, a given variable identifier refers either to a variable local to the current function or to a variable declared outside any function. A static partial function *GlobalVar: tasks* → *Bool* indicates whether or not a given identifier refers to a global variable. We present the modified transition rule for identifiers:

```
if TaskType(CurTask) = identifier then
    if ValueMode(CurTask) = lvalue then
        if GlobalVar(CurTask) = true then
            ReportValue(OnlyValue(Decl(CurTask),StackRoot))
        elseif GlobalVar(CurTask) = false then
            ReportValue(OnlyValue(Decl(CurTask),StackTop))
        endif
    elseif ValueMode(CurTask) = rvalue then
        if GlobalVar(CurTask) = true then
            ReportValue(MemoryValue(OnlyValue
                (Decl(CurTask),StackRoot), ValueType(CurTask)))
        elseif GlobalVar(CurTask) = false then
            ReportValue(MemoryValue(OnlyValue
                (Decl(CurTask),StackTop), ValueType(CurTask)))
        endif
    endif
    Moveto(NextTask(CurTask))
endif
```

References

[ASU] Alfred V. Aho, Ravi Sethi, and Jeffrey D. Ullman, "Compilers: Principles, Techniques, and Tools", Addison-Wesley, 1988.

[Gu] Yuri Gurevich, "Evolving Algebras: An Introductory Tutorial", Bulletin of European Assocation for Theoretical Computer Science, February 1991. (A slightly updated version will appear in the EATCS Book of Columns, World Scientific Publishers.)

[GH] Yuri Gurevich and James K. Huggins, "The Evolving Algebra Semantics of C: Preliminary Version", CSE-TR-141-92, EECS Department, University of Michigan, 1992.

[KR] Brian W. Kernighan and Dennis M. Ritchie, "The C Programming Language", 2nd edition, Prentice Hall, 1988.

A Theory of Classes for a Functional Language with Effects

Furio Honsell
Udine University
honsell@uduniv.cineca.it

Ian A. Mason
Stanford University
iam@cs.stanford.edu

Scott Smith
Johns Hopkins University
scott@cs.jhu.edu

Carolyn Talcott
Stanford University
clt@sail.stanford.edu

1. Introduction

It is well known that the addition of references or other mutable data to a functional programming language complicates matters. Adding operations for manipulating references to the simply typed lambda calculus causes the failure of most of the nice mathematical properties. For example strong normalization fails because it is possible to construct a fixed-point combinator for any functional type:

$$\mathtt{Y}_m = \lambda p.\mathtt{let}\{z := \mathtt{mk}(g)\}\mathtt{seq}(\mathtt{set}(z, \lambda x.\mathtt{app}(\mathtt{app}(p, \mathtt{get}(z)), x)), \mathtt{get}(z))$$

where $\mathtt{mk}(v)$ allocates a cell with contents v, $\mathtt{get}(z)$ gets the current contents of the cell z, and $\mathtt{set}(z, v)$ sets the contents of the cell z to be v, and g is any variable. In addition, references are problematic for polymorphic type systems [28, 29]. References are also troublesome from a denotational point of view as illustrated by the absence of fully abstract models. For example, in [17] Meyer and Sieber give a series of examples of programs that are operationally equivalent (according to the intended semantics of block-structured Algol-like programs) but which are not given equivalent denotations in traditional denotational semantics. They propose various modifications to the denotational semantics which solve some of these discrepancies, but not all. In [21, 20] a denotational semantics that overcomes some of these problems is presented. However variations on the seventh example remain problematic. Since numerous proof systems for Algol are sound for the denotational models in question, [8, 7, 25, 22, 12, 21, 20], these equivalences, if expressible, must be independent of these systems.

In this paper we introduce a variable typed logic of effects (i.e. a logic of effects where classes can be defined and quantified over) inspired by the variable type systems of Feferman [3, 4] for purely functional languages. A similar extension incorporating non-local control operations was introduced in [27]. The logic we present provides an expressive language for defining specifications and constraints and for studying properties and program equivalences, in a uniform framework. Thus it has an advantage over a plethora of systems in the literature that aim to capture solitary aspects of computation. The theory also allows for the construction of inductively defined sets and derivation of the corresponding induction principles. Classes can be used to express, inter alia, the non-expansiveness of terms [29]. Other effects can also be represented within the system. These include read/write effects and various forms of interference [24]. The first order fragment is described in [16] where it is used to resolve the denotationally problematic examples of [17].

In our language atoms, references and lambda abstractions are all first class values and as such are storable. This has several consequences. Firstly, mutation and variable binding are separate and so we avoid the problems that typically arise (e.g. in Hoare's and dynamic logic) from the conflation of program variables and logical variables. Secondly, the equality and sharing of references (aliasing) is easily expressed and reasoned about. Thirdly, the combination of mutable references and lambda abstractions allows us to study object based programming within our framework. Our atomic formulas express the (operational or observational) equivalence of programs à la Plotkin [23]. Neither Hoare's logic nor Dynamic logic incorporate this ability, or make use of such equivalences (e.g. by replacing one piece of program text by another without altering the overall meaning).

The terms of our language are simply the terms of the call-by-value lambda calculus extended by the reference primitives `mk`, `set`, `get`. We also include a collection of operations[1] and basic constants or atoms $\mathbb{A}$, (such as the Lisp booleans `t` and `nil` as well as the integers $\mathbb{Z}$). We can think of this language as an untyped dialect of ML. The atomic formulas of our language assert class membership and the operational equivalence of expressions. In addition to the usual first-order formula constructions and quantification over class variables, we add a mechanism for annotating points in programs with formulas. Namely, *contextual assertions:* if Φ is a formula and U is a *univalent context*, then $U[\![\Phi]\!]$ is a formula. The formula, $U[\![\Phi]\!]$ expresses the fact that the assertion Φ holds at the point in the program text, U, when and if the hole requires evaluation. Univalent contexts are the largest natural class of contexts (expressions with a unique hole) whose symbolic evaluation is unproblematic. Contextual assertions generalize Hoare's triples in that they can be nested, used as assumptions, and their free variables may be quantified. They are similar in spirit to program modalities in dynamic logic. Using contextual assertions we can express the axioms concerning the effects of `mk` and `set` simply and elegantly. This improves the complete system (for quantifier/recursion free expressions) presented in [15] where the corresponding rules had complicated side-conditions.

The semantics of expressions is a call-by-value evaluation relation given by a reduction relation on syntactic entities. In [14] we used this approach to establish a *useful* characterization of operational equivalence. This characterization reduces the number of contexts that need to be considered. The class of contexts that need to be considered correspond naturally to states of an abstract machine. The logic is a partial term logic with variables ranging over values. The characterization of operational equivalence allows for a natural notion of satisfaction of first order formulas relative to a memory state and assignment of values to variables. Our style of operational semantics naturally provides for the symbolic evaluation of contexts, which is the key to defining the semantics of contextual assertions. Classes range over sets of values closed under operational equivalence.

---

[1] In our work operations come in three flavors: algebraic operations which act on atomic data, and whose properties are given by algebraic equations; structural operations which act uniformly on specific kinds of data (other than atomic) such as pairs, records, finite sets; and computational operations which provide access to computation state, these include memory operations, and control operations. Algebraic and structural operations are *context free* – their action/meaning is independent of computation state, whereas the meaning of computation primitives is effected by and can effect computation state.

In the presence of effects several notions split into spectrums of variations. We give two examples:

Firstly, there are many possible notions of "function space" according to how the effects of a computation are accounted for. One example is the class of memory functions, $X_1, \ldots, X_n \xrightarrow{\mu} Y$ with arguments in $X_1, \ldots, X_n$ and result in Y allowing for the possible modification of memory in the process. This can be refined by making the possible effects explicit in the spirit of [10, 11]. At the other end of the spectrum there is the function space that corresponds to those operations that return appropriate values without even enlarging memory, let alone altering existing memory.

Secondly, in the presence of effects there are several degrees of "definedness". They are all easily expressible in our system. The weakest notion is that of computational definedness. An expression is computationally defined, if (for any assignment of free variables) it returns a value. A stronger notion is that of an an expression evaluating to a value, without altering (but possibly enlarging) memory. An even stronger notion of definedness is that of an expression evaluating to a value, without altering or enlarging memory. The strongest notion of definedness is that of evaluating to a value independently of the memory. The following terms exemplify these degrees: $\mathtt{mk}(x)$, $\mathtt{seq}(\mathtt{mk}(x), 1)$, $\mathtt{get}(x)$, 1.

Feferman [4] proposes an explanation of ML types in the variable type framework. This gives a natural semantics to ML type expressions, but there are problems with polymorphism, even in the purely functional case. For example the fixed point operator can be typed in ML as $(\forall X, Y)([X \to Y] \to [X \to Y] \to [X \to Y])$ but this is false in the variable type framework as there are types (classes) not closed under "limits of chains". The situation becomes more problematic when references are added. Naive attempts to represent ML types as classes fails in sense that ML inference rules are not valid. It seems that the essential feature of ML type system, in addition to the inference rules, is the preservation of types during the execution of well-typed programs. In this sense they are more syntactic than semantic.

The remainder of this paper is organized as follows. Section 2. reviews the syntax and semantics of the underlying computational language and summarizes the main results of previous work. The unfamiliar reader may be advised to consult [14, 16] for a more detailed treatment. VTLoE (Variable Typed Logic of Effects) is introduced in two stages. The first stage is the first-order theory of individuals built on assertions of equality (operational equivalence), and contextual assertions. This is presented in Section 3. The second stage extends the logic to include classes and class membership. This is presented in Section 4. In Section 5. we present our conclusions and suggest further directions of research.

Notation Let X, Y, Y_0, Y_1 be sets. We specify meta-variable conventions in the form: let x range over X, which should be read as: the meta-variable x and decorated variants such as x', x_0, ..., range over the set X. We use the usual notation for set membership and function application. Y^n is the set of sequences of elements of Y of length n. Y^* is the set of finite sequences of elements of Y. $\bar{y} = [y_1, \ldots, y_n]$ is the sequence of length n with ith element y_i. $\mathbf{P}_\omega(Y)$ is the set of finite subsets of

Y. $[Y_0 \to Y_1]$ is the set of total functions f with domain Y_0 and range contained in Y_1. We write $\mathrm{Dom}(f)$ for the domain of a function and $\mathrm{Rng}(f)$ for its range. For any function f, $f\{y := y'\}$ is the function f' such that $\mathrm{Dom}(f') = \mathrm{Dom}(f) \cup \{y\}$, $f'(y) = y'$, and $f'(z) = f(z)$ for $z \neq y, z \in \mathrm{Dom}(f)$. $\mathbb{N} = \{0, 1, 2, \ldots\}$ is the natural numbers and $i, j, n, n_0, \ldots$ range over $\mathbb{N}$.

2. The Syntax and Semantics of Terms

The syntax of the terms of our language is a simple extension of that of the lambda calculus to include basic constants or atoms $\mathbb{A}$, (such as the Lisp booleans `t` and `nil` as well as the integers $\mathbb{Z}$). Together with a collection of primitive operations, $\mathbb{F}$, which include the the memory operations $\{\mathtt{get}, \mathtt{set}, \mathtt{mk}\}$, and the *context free* operations $\{\mathtt{cell}, \mathtt{eq}, \mathtt{br}\}$. The branching primitive `br` is a strict version of the Lisp conditional, `if`. We assume an infinite set of variables, $\mathbb{X}$ and use these to define, by mutual induction, the set of λ-abstractions, $\mathbb{L}$, the set of value expressions, $\mathbb{V}$, and the set of expressions, $\mathbb{E}$ as the least sets satisfying the following equations:

$$\mathbb{L} = \lambda\mathbb{X}.\mathbb{E} \qquad \mathbb{V} = \mathbb{X} + \mathbb{A} + \mathbb{L} \qquad \mathbb{E} = \mathbb{V} + \mathsf{app}(\mathbb{E}, \mathbb{E}) + \mathbb{F}_n(\mathbb{E}^n)$$

λ is a binding operator and free and bound variables of expressions are defined as usual. $\mathrm{FV}(e)$ is the set of free variables of e. For any syntactic domain Y and set of variables X we let Y_X be the elements of Y with free variables in X. A *value substitution* is a finite map σ from variables to value expressions, we let σ range over value substitutions. e^σ is the result of simultaneous substitution of free occurrences of $x \in \mathrm{Dom}(\sigma)$ in e by $\sigma(x)$. We represent the function which maps x to v by $\{x := v\}$. Thus $e^{\{x:=v\}}$ is the result of replacing free occurences of x in e by v (avoiding the capture of free variables in v).

We use the syntactic sugar `let`, `if`, `seq` (a sequencing construct akin to `progn`, `begin` or `;`) to make programs more readable. For example

$\mathtt{let}\{x := e_0\}e_1$ abbreviates $\mathsf{app}(\lambda x.e_1, e_0)$

$\mathtt{seq}(e_0, e_1)$ abbreviates $\mathsf{app}(\lambda z.e_1, e_0)$ $\quad$ z fresh

$\mathtt{if}(e_0, e_1, e_2)$ abbreviates $\mathsf{app}(\mathtt{br}(\lambda z.e_1, \lambda z.e_2, e_0), \mathtt{nil})$ $\quad$ z fresh

$\mathsf{app}(f, x_1, \ldots, x_n)$ abbreviates $\mathsf{app}(\ldots(\mathsf{app}(f, x_1), x_2), \ldots, x_n)$

Contexts are expressions with holes. We use $\bullet$ to denote a hole. The set of contexts, $\mathbb{C}$, is defined by

$$\mathbb{C} = \{\bullet\} + \mathbb{X} + \mathbb{A} + \lambda\mathbb{X}.\mathbb{C} + \mathsf{app}(\mathbb{C}, \mathbb{C}) + \mathbb{F}_n(\mathbb{C}^n)$$

We let C range over $\mathbb{C}$. $C[e]$ denotes the result of replacing any holes in C by e. Free variables of e may become bound in this process. The finite set of variable which may be trapped by filling the context C are called its traps and denoted by $\mathrm{Traps}(C)$.

The operational semantics of expressions is given by a reduction relation $\overset{*}{\mapsto}$ on a syntactic representation of the state of an abstract machine, called *descriptions*.

A state has three components: the current state of memory, the current continuation, and the current instruction. Their syntactic counterparts are *memory contexts*, *reduction contexts* and *redexes* respectively. Redexes describe the primitive computation steps (β-reduction or the application of a primitive operation to a sequence of value expressions). Reduction contexts identify the subexpression of an expression that is to be evaluated next. They correspond to the left-first, call-by-value reduction strategy of Plotkin [23] and were first introduced in [5].

$$\mathbb{R} = \{\bullet\} + \mathsf{app}(\mathbb{R}, \mathbb{E}) + \mathsf{app}(\mathbb{V}, \mathbb{R}) + \mathbb{F}_{m+n+1}(\mathbb{V}^m, \mathbb{R}, \mathbb{E}^n)$$

R ranges over $\mathbb{R}$. The crucial fact to note is that an arbitrary expression is either a value expression, or *decomposes uniquely* into a redex placed in a reduction context. We represent the state of memory using memory contexts. A memory context Γ is a context of the form

$$\mathtt{let}\{z_1 := \mathtt{mk}(\mathtt{nil})\} \ldots \mathtt{let}\{z_n := \mathtt{mk}(\mathtt{nil})\}\mathtt{seq}(\mathtt{set}(z_1, v_1), \ldots, \mathtt{set}(z_n, v_n), \bullet)$$

where $z_i \neq z_j$ when $i \neq j$. We have divided the context into allocation, followed by assignment to allow for the construction of cycles. Thus, any state of memory is constructible by such an expression. We let $\Gamma, \Gamma_0, \ldots$ range over memory contexts. We can view memory contexts as *finite maps from variables to value expressions*. Thus we refer to their domain, $\mathrm{Dom}(\Gamma)$; modify them, $\Gamma\{z := \mathtt{mk}(v)\}$, when $z \in \mathrm{Dom}(\Gamma)$; extend them, $\Gamma\{z := \mathtt{mk}(v)\}$, when $z \notin \mathrm{Dom}(\Gamma)$; and form the disjoint union of two of them, $(\Gamma_0 \cup \Gamma_1)$.

A *description* is a *pair*, $\Gamma; e$, with first component a memory context and second component an arbitrary expression (As mentioned above, this arbitrary expression is either a value expression, or *decomposes uniquely* into a redex placed in a reduction context.) *Value descriptions* are descriptions whose expression is a value expression, $\Gamma; v$. We use the convention that an expression used as a description means that the memory context is empty, thus e abbreviates $\emptyset; e$. Note that descriptions may have free variables (i.e. it need not be the case that $\mathrm{FV}(e) \subseteq \mathrm{Dom}(\Gamma)$).

The reduction relation $\overset{*}{\mapsto}$ is the reflexive transitive closure of $\mapsto$. The interesting clauses are:

(beta) $\Gamma; R[\mathsf{app}(\lambda x.e, v)] \mapsto \Gamma; R[e^{\{x:=v\}}]$

(mk) $\Gamma; R[\mathtt{mk}(v)] \mapsto \Gamma\{z := \mathtt{mk}(v)\}; R[z]$ $\quad z \notin \mathrm{Dom}(\Gamma) \cup \mathrm{FV}(R[v])$

(get) $\Gamma; R[\mathtt{get}(z)] \mapsto \Gamma; R[v]$ $\quad$ assuming $z \in \mathrm{Dom}(\Gamma)$ and $\Gamma(z) = v$

(set) $\Gamma; R[\mathtt{set}(z, v)] \mapsto \Gamma\{z := \mathtt{mk}(v)\}; R[\mathtt{nil}]$ $\quad$ assuming $z \in \mathrm{Dom}(\Gamma)$

(cell) $\Gamma; R[\mathtt{cell}(v)] \mapsto \begin{cases} \Gamma; R[\mathtt{t}] & \text{if } v \in \mathrm{Dom}(\Gamma) \\ \Gamma; R[\mathtt{nil}] & \text{if } v \in \mathbb{L} \cup \mathbb{A} \end{cases}$

A description, $\Gamma; e$ is *defined* (written $\downarrow\Gamma; e$) if it evaluates to a value description. For closed expressions e, we write $\downarrow e$ to mean $\downarrow\emptyset; e$. We say two expressions are *eqidefined*, $e_0 \updownarrow e_1$, to mean that $(\downarrow e_0)$ iff $(\downarrow e_1)$ Note that in the `cell` rule if one of the arguments is a variable not in the domain of the memory context, then no reduction step is possible.

Operational (or observational) equivalence formalizes the notion of programs as black-boxes. Treating programs as black boxes requires only observing what effects and values they produce, and not how they produce them. In our framework the allowed observations are those made by closing program contexts. Two expressions are *operationally equivalent*, written $e_0 \cong e_1$, if for any closing context C, $C[e_0]$ is defined iff $C[e_1]$ is defined. This definition extends the extensional equivalence relations defined by Morris [19] and Plotkin [23] to computation over memory structures.

In general it is very difficult to establish the operational equivalence of expressions. Thus it is desirable to have a simpler characterization of $\cong$, one that limits the class of contexts (or observations) that must be considered. We define a *closed instantiation* of an expression e to be a memory context, Γ, together with a value substitution, σ, such that $\Gamma[e^\sigma]$ is closed. A *use* of an expression e is the placement of e into a reduction context. The desired result is then that two expressions are operationally equivalent just if all closed instances of all uses are equidefined.

Theorem (ciu):

$$e_0 \cong e_1 \;\Leftrightarrow\; (\forall \Gamma, \sigma, R)(\mathrm{FV}(\Gamma[R[e_i^\sigma]]) = \emptyset \;\Rightarrow\; (\Gamma[R[e_0^\sigma]] \updownarrow \Gamma[R[e_1^\sigma]]))$$

The proof of (**ciu**) appears in [14]. Using this theorem we can easily establish, for example, the validity of the let-rules of the lambda-c calculus [18].

3. The Syntax and Semantics of Formulas I. — Individuals

In addition to being a useful tool for establishing laws of operational equivalence, (**ciu**) can be used to define a satisfaction relation between memory contexts and equivalence assertions. In an obvious analogy with the usual first-order Tarskian definition of satisfaction this can be extended to define a satisfaction relation $\Gamma \models \Phi[\sigma]$.

The atomic formulas of our language assert the operational equivalence of two expressions. In addition to the usual first-order formula constructions we add *contextual assertions:* if Φ is a formula and U is a certain type of context, then $U[\![\Phi]\!]$ is a formula. This form of formula expresses the fact that the assertion Φ holds at the point in the program text marked by the hole in U, if execution of the program reached that point. The contexts allowed in contextual assertions are called *univalent contexts*, (U-contexts). They are the largest natural class of contexts whose symbolic evaluation is unproblematic. The key restriction is that we forbid the hole to appear in the scope of a (non-`let`) lambda, thus preventing the proliferation of holes. The class of U-contexts, $\mathbb{U}$, is defined as follows.

Definition ($\mathbb{U}$):

$$\mathbb{U} = \{\bullet\} + \mathtt{let}\{\mathbb{X} := \mathbb{E}\}\mathbb{U} + \mathtt{if}(\mathbb{E}, \mathbb{U}, \mathbb{U}) + \mathsf{app}(\mathbb{U}, \mathbb{E}) + \mathsf{app}(\mathbb{E}, \mathbb{U}) + \mathbb{F}_{m+n+1}(\mathbb{E}^m, \mathbb{U}, \mathbb{E}^n)$$

The well-formed formulas, $\mathbb{W}$, of (the first order part of) our logic are defined as follows:

Definition ($\mathbb{W}$):

$$\mathbb{W} = (\mathbb{E} \cong \mathbb{E}) + (\mathbb{W} \Rightarrow \mathbb{W}) + (\mathbb{U}[\![\mathbb{W}]\!]) + (\forall \mathbb{X})(\mathbb{W})$$

Note that the context U will in general bind free variables in Φ. A simple example is the axiom which expresses the effects of mk:

$$(\forall y)(\mathtt{let}\{x := \mathtt{mk}(v)\}[\![\neg(x \cong y) \wedge \mathtt{cell}(x) \cong \mathtt{t} \wedge \mathtt{get}(x) \cong v]\!])$$

In order to define the semantics of contextual assertions, we need to extend computation to univalent contexts. The idea here is quite simple, to compute with contexts we need to keep track of the β-conversions that have taken place with the hole in the scope of the λ. To indicate that the substitution σ has taken place at the hole in U we write $U[\![\sigma]\!]$. Computation is then written as $\Gamma; U[\![\sigma]\!] \overset{*}{\mapsto} \Gamma'; U'[\![\sigma']\!]$ and is defined as follows:

Definition ($\Gamma; U[\![\sigma]\!] \overset{*}{\mapsto} \Gamma'; U'[\![\sigma']\!]$): Let $U \in \mathbb{U}$ be such that $\mathrm{Traps}(U) = \{x_1, \ldots x_n\}$, assume $\mathrm{Dom}(\sigma) \cap \mathrm{Traps}(U) = \emptyset$ and let z be a fresh variable. We write

$$\Gamma; U[\![\sigma]\!] \overset{*}{\mapsto} \Gamma'; U'[\![\sigma']\!]$$

to mean,

$$\Gamma; (U[\mathsf{app}(\ldots \mathsf{app}(z, x_1), \ldots, x_n)])^{\sigma} \overset{*}{\mapsto} \Gamma'; (U'[\mathsf{app}(\ldots \mathsf{app}(z, x_1), \ldots, x_n)])^{\sigma'}$$

and $\mathrm{Dom}(\sigma') = \mathrm{Dom}(\sigma) \cup (\mathrm{Traps}(U) - \mathrm{Traps}(U'))$. Note that σ and σ' will agree on the domain of σ.

The Tarskian definition of satisfaction $\Gamma \models \Phi[\sigma]$ is given by a simple induction on the structure of Φ.

Definition ($\Gamma \models \Phi[\sigma]$): $(\forall \Gamma, \sigma, \Phi, e_j)$ such that $\mathrm{FV}(\Phi^{\sigma}) \cup \mathrm{FV}(e_j^{\sigma}) \subseteq \mathrm{Dom}(\Gamma)$ for $j < 2$ we define satisfaction:

$$\Gamma \models (e_0 \cong e_1)[\sigma] \quad \text{iff} \quad (\forall R \in \mathbb{R}_{\mathrm{Dom}(\Gamma)})(\downarrow\Gamma[R[e_0^{\sigma}]] \quad \text{iff} \quad \downarrow\Gamma[R[e_1^{\sigma}]])$$

$$\Gamma \models (\Phi_0 \Rightarrow \Phi_1)[\sigma] \quad \text{iff} \quad (\Gamma \models \Phi_0[\sigma]) \quad \text{implies} \quad (\Gamma \models \Phi_1[\sigma])$$

$$\Gamma \models U[\![\Phi]\!][\sigma] \quad \text{iff} \quad (\forall \Gamma', R, \sigma')((\Gamma; U[\![\sigma]\!] \overset{*}{\mapsto} \Gamma'; R[\![\sigma']\!]) \quad \text{implies} \quad \Gamma' \models \Phi[\sigma'])$$

$$\Gamma \models (\forall x)\Phi[\sigma] \quad \text{iff} \quad (\forall v \in \mathbb{V}_{\mathrm{Dom}(\Gamma)})(\Gamma \models \Phi[\sigma\{x := v\}])$$

The atomic clause in the definition of satisfaction is justified by the (**ciu**) theorem. Negation is definable, $\neg\Phi$ is just $\Phi \Rightarrow \mathtt{False}$, where $\mathtt{False}$ is any unsatisfiable assertion, such as $\mathtt{t} \cong \mathtt{nil}$. We can also express the computational definedness of expressions by the following assertion$\neg\mathbf{seq}(e, [\![\mathtt{False}]\!])$. We let $\Downarrow e$ abbreviate this expression and $\Uparrow e$ abbreviate its negation. We say that a formula is valid, written $\models \Phi$, if $\Gamma \models \Phi[\sigma]$ for Γ, σ such that $\mathrm{FV}(\Phi^{\sigma}) \subseteq \mathrm{Dom}(\Gamma)$. We define $\Phi_{\neg\mathsf{write}}(e)$ to abbreviate:

$$\Downarrow e \wedge (\forall z)(\forall y)((\mathtt{cell}(z) \cong \mathtt{t} \wedge \mathtt{get}(z) \cong y) \Rightarrow \mathtt{let}\{x := e\}[\![\mathtt{get}(z) \cong y]\!]).$$

This asserts that the evaluation of the expression e does not visibly alter the contents of any pre-existing cell. In VTLoE we cannot express that an expression e does not modify the contents of any cells. This can be seen by considering the following expression:

$$\texttt{if}(\texttt{or}(\texttt{atom}(\texttt{get}(x)), \texttt{cell}(\texttt{get}(x))), x, \texttt{let}\{y := \texttt{get}(x)\}\texttt{set}(x, \lambda z.\texttt{app}(y, z)))$$

which makes no detectable changes to any cells. Consequently we can only express that the expression does not modify the contents of cells modulo operational equivalence.

The theorem (**ca**) provides three principles for reasoning about contextual assertions: a general principle for introducing contextual assertions (akin to the rule of necessitation in modal logic); a principle for propagating contextual assertions through equations; and a principle for composing contexts (or collapsing nested contextual assertions).

Theorem (ca):

(i) $$\frac{\models \Phi}{\models U[\![\Phi]\!]}$$

(ii) $U[\![e_0 \cong e_1]\!] \Rightarrow U[e_0] \cong U[e_1]$

(iii) $U_0[\![U_1[\![\Phi]\!]]\!] \Leftrightarrow (U_0[U_1])[\![\Phi]\!]$

It is in general false that $\Phi \Rightarrow U[\![\Phi]\!]$ holds, a simple counter-example being

$$\texttt{get}(x) \cong 2 \Rightarrow \texttt{let}\{x := \texttt{mk}(3)\}[\![\texttt{get}(x) \cong 2]\!]$$

The converse of (**ca.ii**) is false. as can be seen by the following:

$$\texttt{let}\{x := \texttt{mk}(0)\}\texttt{let}\{y := \texttt{mk}(0)\}[x] \cong \texttt{let}\{x := \texttt{mk}(0)\}\texttt{let}\{y := \texttt{mk}(0)\}[y]$$

but $\texttt{let}\{x := \texttt{mk}(0)\}\texttt{let}\{y := \texttt{mk}(0)\}[\![\neg(x \cong y)]\!]$. Note that (**ca.iii**) is false for general contexts; a simple counterexample is when $C_0 = \texttt{app}(\bullet, v)$, $C_1 = \lambda x.\bullet$ and $\Phi = \texttt{t} \cong \texttt{nil}$. Contextual assertions also interact nicely with the propositional connectives, if we take proper account of assertions that are true for the trivial reason that during execution, the point in the program text marked by the context hole is never reached.

Lemma (con.prop):

(triv) $U[\![\texttt{False}]\!] \Rightarrow U[\![\Phi]\!]$

(not) $U[\![\neg\Phi]\!] \Leftrightarrow (U[\![\texttt{False}]\!] \vee \neg U[\![\Phi]\!])$

(imp) $U[\![\Phi_0 \Rightarrow \Phi_1]\!] \Leftrightarrow (U[\![\Phi_0]\!] \Rightarrow U[\![\Phi_1]\!])$

The case of the quantifier is a little less simple.

Lemma (con.∀):

(∀) $U[\![\forall x \Phi]\!] \Rightarrow \forall x\, U[\![\Phi]\!]$

The converse to (∀) is easily shown to be false by considering U to be $\texttt{let}\{y := \texttt{mk}(\texttt{t})\}\bullet$ and Φ to be $\neg(x \cong y)$. Contextual assertions do interact nicely with evaluation.

Lemma (eval):

If $\Gamma_0; U_0[\![\sigma_0]\!] \overset{*}{\mapsto} \Gamma_1; U_1[\![\sigma_1]\!]$, then $\Gamma_0 \models U_0[\![\Phi]\!][\sigma_0]$ iff $\Gamma_1 \models U_1[\![\Phi]\!][\sigma_1]$

This logic extends and improves the complete first order system presented in [13, 15]. There certain reasoning principles were established as basic, and from these all others, suitably restricted, could be derived using simple equational reasoning. The system presented there had several defects. In particular the rules concerning the effects of mk and set had complicated side-conditions. Using contextual assertions we can express them simply and elegantly. Their justification is also unproblematic. The contextual assertions and axioms involving mk, and set are:

Definition (mk axioms):

(**mk**.i) $\mathtt{let}\{x := \mathtt{mk}(v)\}[\neg(x \cong y) \wedge \mathtt{cell}(x) \cong \mathtt{t} \wedge \mathtt{get}(x) \cong v]$ x fresh

(**mk**.ii) $y \cong \mathtt{get}(z) \Rightarrow \mathtt{let}\{x := \mathtt{mk}(v)\}[y \cong \mathtt{get}(z)]$ x fresh

(**mk**.iii) $\Downarrow \mathtt{mk}(z)$

(**mk**.iv) $\mathtt{let}\{y := e_0\}\mathtt{let}\{x := \mathtt{mk}(v)\}e_1 \cong \mathtt{let}\{x := \mathtt{mk}(v)\}\mathtt{let}\{y := e_0\}e_1$

$x \notin \mathrm{FV}(e_0),\ y \notin \mathrm{FV}(v)$

The assertion, (**mk.i**), describes the allocation effect of a call to mk. While (**mk.ii**) expresses what is unaffected by a call to mk. The assertion, (**mk.iii**), expresses the totality of mk. The mk delay axiom, (**mk.iv**), asserts that the time of allocation has no discernable effect on the resulting cell.

Definition (set axioms):

(set.i) $\mathtt{cell}(z) \cong \mathtt{t} \Rightarrow \mathtt{let}\{x := \mathtt{set}(z, y)\}[\mathtt{get}(z) \cong y \wedge x \cong \mathtt{nil}]$

(set.ii) $(y \cong \mathtt{get}(z) \wedge \neg(w \cong z)) \Rightarrow \mathtt{let}\{x := \mathtt{set}(w, v)\}[y \cong \mathtt{get}(z)]$

(set.iii) $\mathtt{cell}(z) \cong \mathtt{t} \Rightarrow \Downarrow \mathtt{set}(z, x)$

(set.iv) $\neg(x_0 \cong x_2) \Rightarrow \mathtt{seq}(\mathtt{set}(x_0, x_1), \mathtt{set}(x_2, x_3)) \cong \mathtt{seq}(\mathtt{set}(x_2, x_3), \mathtt{set}(x_0, x_1))$

(set.v) $\mathtt{seq}(\mathtt{set}(x, y_0), \mathtt{set}(x, y_1)) \cong \mathtt{set}(x, y_1)$

(set.vi) $\mathtt{let}\{z := \mathtt{mk}(x)\}\mathtt{seq}(\mathtt{set}(z, w), e) \cong \mathtt{let}\{z := \mathtt{mk}(w)\}e$ z not free in w

The first three contextual assertions regarding set are analogous to those of mk. They describe what is returned and what is altered, what is not altered as well as when the operation is defined. The remaining three principles involve the commuting, cancellation, absorption of calls to set. For example the set absorption principle, (**set.vi**), expresses that under certain simple conditions allocation followed by assignment may be replaced by a suitably altered allocation.

4. The Syntax and Semantics of Formulas II. — Classes

Using methods of Feferman [2, 4] and Talcott [27], we extend our theory to include a general theory of classifications (classes for short). With the introduction of classes, principles such as structural induction, as well as principles accounting for the effects of an expression can easily be expressed. Classes serve as a starting point for studying semantic notions of type. As will be seen direct representation of type inference systems can be problematic, and additional notions may be required to provide a formal semantics. Even here classes are likely to play an important role.

We extend the syntax to include class terms. Class terms are either class variables, $\mathbf{X}^c$, class constants, $\mathbb{A}^c$, or comprehension terms, $\{x \mid \Phi\}$.

Definition ($\mathbb{K}$): The set $\mathbb{K}$ of class terms is defined by

$$\mathbb{K} = \mathbf{X}^c + \mathbb{A}^c + \{\mathbf{X} \mid \mathbb{W}\}$$

We extend the set $\mathbb{W}$ of formulas to include class membership and quantification over class variables. We should point out that $\mathbb{K}$ and $\mathbb{W}$ form a mutual recursive definition. The definition of expressions remains unchanged.

Definition ($\mathbb{W}$):

$$\mathbb{W} = (\mathbb{E} \cong \mathbb{E}) + (\mathbb{E} \in \mathbb{K}) + (\mathbb{W} \Rightarrow \mathbb{W}) + (\forall \mathbf{X})\mathbb{W} + (\forall \mathbf{X}^c)\mathbb{W} + \mathbb{U}[\mathbb{W}]$$

We let $A, B, C, \ldots X, Y, Z$ range over $\mathbf{X}^c$ and K range over $\mathbb{K}$. We will use identifiers beginning with an upper case letter in **This** font (for example **Val**) for class constants.

To give semantics to the extended language, we extend the satisfaction relation as follows. Firstly we let $\mathbb{K}_{\mathsf{Dom}(\Gamma)}$, the set of class values over Γ, be the set of subsets of $\mathbb{V}_{\mathsf{Dom}(\Gamma)}$ closed under $\cong$. We extend value substitutions to map class variables to class values. This is used to define $[K]_\Gamma^\sigma$, the value of a class term, K, relative to the given memory context, Γ, and the closing value substitution σ. In principle, the class term evaluation is relative to a valuation for class constants, but since all of our class constants are introduced by definitional extension, this can be ignored.[2]

Definition ($[\mathbb{K}]_\Gamma^\sigma$):

$$[X]_\Gamma^\sigma = \sigma(X)$$

$$[\{x \mid \Phi\}]_\Gamma^\sigma = \{v \in \mathbb{V}_{\mathsf{Dom}(\Gamma)} \mid \Gamma \models \Phi[\sigma\{x := v\}]\}$$

We then extend the satisfaction relation to formulas involving class terms and quantifiers.

Definition ($\Gamma \models \Phi[\sigma]$): The new clauses in the inductive definition of satisfaction are:

$$\Gamma \models e \in K[\sigma] \Leftrightarrow (\exists v \in \mathbb{V}_{\mathsf{Dom}(\Gamma)})(\Gamma; e^\sigma \overset{*}{\mapsto} \Gamma; v \wedge v \in [K]_\Gamma^\sigma)$$

$$\Gamma \models (\forall X)\Phi[\sigma] \Leftrightarrow (\forall C \in \mathbb{K}_{\mathsf{Dom}(\Gamma)})(\Gamma \models \Phi[\sigma\{X := C\}])$$

It is important to note that if $\Gamma \models e \in K[\sigma]$, then e evaluates (in the appropriate state) to a value without altering memory, the so-called non-expansive expressions [28, 29]. We define (extensional) equality and subset relations on classes in the usual manner.

$$K_0 \subseteq K_1 \Leftrightarrow (\forall x)(x \in K_0 \Rightarrow x \in K_1)$$

$$K_0 \equiv K_1 \Leftrightarrow K_0 \subseteq K_1 \wedge K_1 \subseteq K_0$$

---

[2] Some class *constants* are absolute (have meaning independent of memory), but most have meaning that varies with memory (even when they are closed). Thus the semantics of classes should be parameterized by a constant interpretation mapping constants to functions that map a memory to a set of values existing in that memory. As class constants only occur in our language as definitional extensions, this issue has been swept under the rug in the definition of $[K]_\Gamma^\sigma$.

As a consequence of the semantics of classifications the following are valid.

Lemma (class):

(def) $e \in K \Rightarrow \Downarrow e$

(allE) $(\forall X)\Phi[X] \Rightarrow \Phi[K]$ where Φ contains no contextual assertions

(ca) $(\forall x)(x \in \{x \mid \Phi\} \Leftrightarrow \Phi)$

The usual form of (**allE**) is false, $\neg((\forall X)\Phi[X] \Rightarrow \Phi[K])$. A counter example will be given below after some additional notation has been introduced.

We introduce the following class constants. $\emptyset$ is the empty class, $\emptyset = \{x \mid x \not\cong x\}$. **Val** is the class of all values, $\mathbf{Val} = \{x \mid x \cong x\}$. **Nil** is the class containing the single element nil, $\mathbf{Nil} = \{x \mid x \cong \mathtt{nil}\}$. **Cell** is the class of memory cells, $\mathbf{Cell} = \{x \mid \mathtt{cell}(x) \cong \mathtt{t}\}$. Note that the interpretation of $\emptyset$ and **Nil** is independent of memory contexts, while $[\mathbf{Val}]_\Gamma = \mathbb{V}_{\mathrm{Dom}(\Gamma)}$ and $[\mathbf{Cell}]_\Gamma = \mathrm{Dom}(\Gamma)$. A class operator is a class term with a distinguished class variable. We write $T[X]$ making the variable explicit and $T[K]$ for the result of replacing the distinguished variable X by K (with suitable renaming of bound variables to avoid capture). We can refine the class of cells to reflect the class of their contents.

$\mathbf{Cell}[Z] = \{x \mid \mathtt{cell}(x) \cong \mathtt{t} \wedge \mathtt{get}(x) \in Z\}$

Thus $\mathbf{Cell} \equiv \mathbf{Cell}[\mathbf{Val}]$.

Definition ($\rightarrow \overset{p}{\rightarrow} \overset{\mu}{\rightarrow}$): There are a number of function spaces in our world. The three simplest are total, partial and memory.

$X \rightarrow Y = \{f \mid (\forall x \in X)(\exists y \in Y)\mathtt{app}(f,x) \cong y\}$

$X \overset{p}{\rightarrow} Y = \{f \mid (\forall x \in X)(\forall y)(\mathtt{app}(f,x) \cong y \Rightarrow y \in Y)\}$

$X \overset{\mu}{\rightarrow} Y = \{f \mid (\forall x \in X)(\mathtt{let}\{y := \mathtt{app}(f,x)\}[\![y \in Y]\!])\}$

The first set denotes the set of total functions which do not alter existing memory (they may enlarge the domain of the memory by producing garbage (i.e. unreachable cells), but they may not alter existing cells); the second set denotes partial functions which also do not alter existing memory (again they may produce garbage); while the third set denotes partial functions which can possibly alter memory.

Now we can characterize the functionality of operations such as mk,and get, as follows.

(mk) $(\forall X)(\forall x \in X)(\mathtt{let}\{z := \mathtt{mk}(x)\}[\![z \in \mathbf{Cell}[X]]\!])$

(get) $(\forall X)(\forall x \in \mathbf{Cell}[X])(\mathtt{get}(x) \in X)$

However (as we shall see in the last example of §4.) the analagous fact concerning set is false.

(set) $\neg(\forall X)(\forall x,y)(x \in X \wedge y \in \mathbf{Cell} \Rightarrow \mathtt{let}\{z := \mathtt{set}(y,x)\}[\![z \in \mathbf{Nil} \wedge y \in \mathbf{Cell}[X]]\!])$

The (**mk**), (**get**) and (**set**) can be restated as

$\lambda x.\mathtt{mk}(x) \in X \overset{\mu}{\rightarrow} \mathbf{Cell}[X]$

$\lambda x.\mathtt{get}(x) \in \mathbf{Cell}[X] \rightarrow X$

$\lambda xy.\mathtt{set}(x,y) \in \mathbf{Cell}[X] \rightarrow Y \overset{\mu}{\rightarrow} \mathbf{Nil}$

Now we give the promised counter example for classical (allE). Let

$$K = \{x \mid \Phi_{\text{1-cell}}\}$$

$$\Phi_{\text{1-cell}} \Leftrightarrow (\exists x)(\mathtt{cell}(x) \cong \mathtt{t}) \wedge (\forall x, y)(\mathtt{cell}(x) \cong \mathtt{t} \wedge \mathtt{cell}(y) \cong \mathtt{t} \Rightarrow x \cong y)$$

$$\Phi = (\forall x \in X)(\mathtt{let}\{z := \mathtt{mk}(x)\}[\![z \in \mathbf{Cell}[X]]\!])$$

Note that Φ is the body of the mk functionality formula. Now $(\forall X)\Phi$ holds, but $\Phi[K]$ fails for any memory with singleton domain. In fact $(\forall X)\Phi$ holds since the meaning of the class variable is determined at the outset, and remains unchanged throughout. The meaning of the class K is instead computed twice, with respect to two different memories. The second time it denotes the empty class.

Class membership expresses a very restricted form of non-expansiveness, allowing neither expansion of memory domain nor change in contents of existing cells. Let $\Phi_{\neg\mathbf{expand}}(e)$ stand for the formula

$$(\forall X)(X \equiv \mathbf{Cell} \Rightarrow \mathtt{seq}(e, [\![X \equiv \mathbf{Cell}]\!]).$$

Then $\Phi_{\neg\mathbf{expand}}(e)$ says that execution of e does non expand the memory, although it might modify contents of existing cells.

To illustrate some of the subtleties regarding class membership, and notions of expansiveness, consider the following expressions:

$$e_0 = \lambda x.\mathtt{mk}(\mathtt{nil})$$

$$e_1 = \mathtt{let}\{z := \mathtt{mk}(\mathtt{nil})\}\lambda x.z$$

$$e_2 = \mathtt{seq}(\mathtt{if}(\mathtt{cell}(y), \mathtt{set}(y, \mathtt{nil}), \mathtt{nil}), \lambda x.\mathtt{mk}(\mathtt{nil}))$$

$$e_3 = \mathtt{seq}(\mathtt{if}(\mathtt{cell}(y), \mathtt{set}(y, \mathtt{nil}), \mathtt{nil}), \mathtt{let}\{z := \mathtt{mk}(\mathtt{nil})\}\lambda x.z)$$

Then each of these expressions evaluates to a memory function mapping arbitrary values to cells containing nil. But they differ in the effects they have. e_0 is a value (and as such neither expands nor modifies memory). e_1 is not a value and is expansive (its evaluation enlarges the domain of memory) but does not modify existing memory. e_2 may modify existing memory, but does not expand it. e_3 is expansive, and it may modify existing memory. These observations can be expressed in the theory as follows. Let T be $\mathbf{Val} \overset{\mu}{\rightharpoonup} \mathbf{Cell}[\mathbf{Nil}]$, then

$$e_0 \in T \subseteq \mathbf{Val}$$

$$e_j \notin \mathbf{Val} \quad \text{for} \quad 1 \le j \le 3$$

$$\mathtt{let}\{x := e_j\}[\![x \in T]\!] \quad \text{for} \quad 0 \le j \le 3$$

$$\Phi_{\neg\mathbf{write}}(e_j) \quad \text{for} \quad 0 \le j \le 1$$

$$\Phi_{\neg\mathbf{expand}}(e_j) \quad \text{for} \quad j \in \{0, 2\}$$

Feferman [4] proposes an explanation of ML types in the variable type framework. This gives a natural semantics to ML type expressions, but there are problems with polymorphism, even in the purely functional case. The collection of classes is much to rich to be considered a type system. One problem that arises is that fixed-point combinators can not be uniformly typed over all classes. This problem arises even in the absence of memory [26, 27].

Theorem (FixTypeFails): Let $\mathtt{Y}_v$ by any fixed-point combinator (such that $f(\mathtt{Y}_v(f)) \cong \mathtt{Y}_v(f))$. Then it is not the case that

$$f \in ((A \overset{\mathrm{P}}{\to} B) \to (A \overset{\mathrm{P}}{\to} B)) \Rightarrow \mathtt{Y}_v(f) \in (A \overset{\mathrm{P}}{\to} B)$$

for all classes A, B.

Proof (FixTypeFails): Define the P to be the class of *strictly* partial maps from N to N, $P = \{g \in \mathrm{N} \overset{\mathrm{P}}{\to} \mathrm{N} \mid (\exists n \in \mathrm{N})(\neg \Downarrow g(n))\}$. Let $f \cong \lambda p.\lambda n.\mathtt{if}(\mathtt{eq}(n, 0), n, p(n - 1))$. Then we can prove

$$\textbf{(1)} \quad f \in P \to P \qquad \textbf{(2)} \quad \mathtt{Y}_v(f) \in \mathrm{N} \to \mathrm{N}$$

(**1**) follows by simple properties of `if`, eq and arithmetic (**2**) follows by induction on N using the fixed point property of Y. Consequently, $\neg(\mathtt{Y}_v(f) \in P)$ □

The situation becomes more problematic when references are added, even in the simply typed (or monomorphic) case. Naive attempts to represent ML types as classes fail in the sense that the ML inference rules are not valid. The essential feature of the ML type system, in addition to the inference rules, is the preservation of types during the execution of well-typed programs. In this sense they are more syntactic than semantic.

In the following we illustrate the problems that arise in trying to encode the monomorphic type system with higher-order functions and references (cf. [28, 29, 9, 6]). In this system types are built from base types, **N** and **Nil**, using the reference construction and a suitable function space constructor (provisionally denoted by $\overset{\lambda}{\to}$).

$$\mathbb{T} = \mathbf{N} + \mathbf{Nil} + (\mathbb{T} \overset{\lambda}{\to} \mathbb{T}) + \mathbf{ref}(\mathbb{T})$$

The typing judgement in this system is of the form $\{x_i : \tau_i \mid i < n\} \vdash e : \tau$ and the constants, for each $\tau \in \mathbb{T}$, have the following type

$$\mathtt{mk}_\tau : \tau \overset{\lambda}{\to} \mathbf{ref}(\tau) \quad \mathbf{get}_\tau : \mathbf{ref}(\tau) \overset{\lambda}{\to} \tau \quad \mathbf{set}_\tau : \mathbf{ref}(\tau) \overset{\lambda}{\to} \tau \overset{\lambda}{\to} \mathbf{Nil}$$

which will be encoded by the corresponding η-ized operations

$$\lambda x.\mathtt{mk}(x) \quad \lambda x.\mathtt{get}(x) \quad \lambda xy.\mathtt{set}(x, y)$$

To encode this system requires a class term $\underline{\tau}$ corresponding to the type expression τ, and a formula $\Phi_:(e, \underline{\tau})$ encoding the typing judgement $e : \tau$. Using these we can represent the judgement $\{x_i : \tau_i \mid i < n\} \vdash e : \tau$ by the formula $\bigwedge_{i<n} \Phi_:(x_i, \underline{\tau_i}) \Rightarrow \Phi_:(e, \underline{\tau})$ To simplify matters we shall assume that base types are represented by their corresponding class constants, in other words $\underline{\mathbf{N}} = \mathbf{N}$ and $\underline{\mathbf{Nil}} = \mathbf{Nil}$. We also require that belonging to a class via $\Phi_:$ implies membership in that class:

$$\Phi_:(e, \underline{\tau}) \Rightarrow \mathtt{let}\{z := e\}[\![z \in \underline{\tau}]\!].$$

That this encoding is faithful amounts to requiring several conditions. These include:

(1a) $\Phi_:(x, \underline{\tau}) \Rightarrow \Phi_:(\mathtt{mk}(x), \underline{\mathbf{ref}(\tau)})$

(1b) $\Phi_:(x, \underline{\mathbf{ref}(\tau)}) \Rightarrow \Phi_:(\mathtt{get}(x), \underline{\tau})$

(1c) $(\Phi_:(x, \underline{\mathbf{ref}(\tau)}) \wedge \Phi_:(y, \underline{\tau})) \Rightarrow \Phi_:(\mathtt{set}(x, y), \mathbf{Nil})$

(2) $\bigwedge_{i<n} \Phi_:(x_i, \underline{\tau_i}) \Rightarrow ((\Phi_:(x, \underline{\tau_a}) \Rightarrow \Phi_:(e, \underline{\tau_b})) \Rightarrow \Phi_:(\lambda x.e, \underline{\tau_a \overset{\lambda}{\to} \tau_b}))$

(3) $\bigwedge_{i<n} \Phi_:(x_i, \underline{\tau_i}) \Rightarrow ((\Phi_:(e, \underline{\tau_a \overset{\lambda}{\to} \tau_b}) \wedge \Phi_:(e_a, \underline{\tau_a})) \Rightarrow (\Phi_:(\mathrm{app}(e, e_a), \underline{\tau_b})$

The principle of *type faithfulness* [1]

$$e_a \cong e_b \Rightarrow ((\bigwedge_{i<n} \Phi_:(x_i, \underline{\tau_i}) \Rightarrow \Phi_:(e_a, \underline{\tau})) \Rightarrow (\bigwedge_{i<n} \Phi_:(x_i, \underline{\tau_i}) \Rightarrow \Phi_:(e_b, \underline{\tau})))$$

is also a desirable property. Note that in our framework this implies that a subterm of a typable term need not be typable. This can be regarded as an advantage of a semantic approach to types over the syntactic approach.

The simplest encoding would be to take

$$\underline{ref(\tau)} = \mathbf{Cell}[\underline{\tau}] \quad \underline{\tau_a \overset{\lambda}{\to} \tau_b} = \underline{\tau_a} \overset{\mu}{\to} \underline{\tau_b} \quad \Phi_:(e, \underline{\tau}) = \mathtt{let}\{z := e\}[\![z \in \underline{\tau}]\!]$$

However the usual inference rules are not sound under this encoding. One source of trouble is that in (3) the evaluation of e_a may invalidate the assumptions that $\bigwedge_{i<n} \Phi_:(x_i, \underline{\tau_i})$. A counterexample to this is the following:

$$y \in \mathbf{Cell}[\mathbf{N}] \Rightarrow \mathtt{let}\{x := \mathtt{set}(y, \mathtt{t})\}[\![x \in \mathbf{Nil}]\!]$$

$$y \in \mathbf{Cell}[\mathbf{N}] \Rightarrow \mathtt{let}\{z := \lambda w.\mathtt{get}(y)\}[\![z \in \mathbf{Nil} \to \mathbf{N}]\!]$$

But the resulting conclusion

$$y \in \mathbf{Cell}[\mathbf{N}] \Rightarrow \mathtt{let}\{z := \mathrm{app}(\lambda w.\mathtt{get}(y), \mathtt{set}(y, \mathtt{t}))\}[\![z \in \mathbf{N}]\!]$$

is clearly false.

To ensure that this phenomenon is ruled out, one would like $\Phi_:$ to have the following property: if e is typed, and its evaluation alters the contents of a cell, then any type that cell had before evaluation it has after evaluation. Similarly for values of functional type. We can express this as the following schema:

$$(\forall z)(\bigwedge_{i<n} \Phi_:(x_i, \underline{\tau_i}) \Rightarrow ((\Phi_:(e, \underline{\tau}) \wedge \Phi_:(z, \underline{\tau'})) \Rightarrow \mathtt{let}\{w := e\}[\![\Phi_:(w, \underline{\tau}) \wedge \Phi_:(z, \underline{\tau'})]\!]))$$

Since this property is not true for general classes the existence of such a $\Phi_:$ seems doubtful.

To see that classes are too rich to be preserved by evaluation, consider the following class:

$$A = \{x \in \mathbf{Cell} \mid (\exists n \in \mathbf{N})(\mathtt{get}^n(x) \cong \mathtt{nil})\}$$

where $\mathtt{get}^n(x)$ is the nth iterate of $\mathtt{get}$, definable by simple recursion. Now observe that

$$x \in \mathbf{Cell}[A] \Rightarrow x \in A$$

Consequently

$$x \in \mathbf{Cell}[A] \Rightarrow \mathtt{let}\{z := \mathtt{set}(x, x)\}[\![z \in \mathbf{Nil}]\!]$$

$$x \in \mathbf{Cell}[A] \Rightarrow \lambda y.x \in \mathbf{Nil} \to \mathbf{Cell}[A]$$

But also note that

$$(\forall x \in \mathbf{Cell}[A])(\mathtt{let}\{z := \mathtt{app}(\lambda y.x, \mathtt{set}(x, x))\}[\![z \notin \mathbf{Cell}[A]]\!])$$

5. Issues and Conclusions

In this paper we have presented a logic, VTLoE, for specifying and reasoning about programs with effects. The semantics of this logic is based on a notion of program equivalence relative to a given memory. VTLoE goes well beyond traditional programming logics, such as Hoare's and Dynamic logic:

(1) The underlying programming language is a rich language based on the call-by-value lambda calculus extended by the reference primitives `mk`, `set`, `get`, as well as constants representing traditional forms of atomic and structured data.

(2) In our language atoms, references and lambda abstractions are all first class values and as such are storable.

(3) The separation of mutation and variable binding allows us to avoid the problems that typically arise (e.g. in Hoare's and dynamic logic) from the conflation of program variables and logical variables.

(4) The equality and sharing of references (aliasing) may be directly expressed and reasoned about.

(5) The combination of mutable references and lambda abstractions allows us to study object-based programming within VTLoE.

(6) Central to VTLoE is the ability to express the operational equivalence of programs, a very general notion of program equivalence.

(7) In addition to the usual first-order formula constructions and quantification over class variables, the logic includes contextual assertions. This allows for direct reasoning about changes in state, and subsumes the state-based reasoning methods possible in both Hoare's and Dynamic logic.

(8) Class membership and quantification allow us to express a wide variety of useful notions including presence of effects, functionality, and structural induction principles (the latter issue is not dealt here).

The logic presented here is perhaps best viewed as a starting point for further research rather than a final product. In particular there are at least four directions for further research.

(A) There seems to be good evidence to support a *more localized* semantics for contextual assertions. One indication is the failure of the following principle:

$$e_0 \cong e_1 \Rightarrow \mathtt{let}\{x := e_0\}[\![\Phi]\!] \Leftrightarrow \mathtt{let}\{x := e_1\}[\![\Phi]\!]$$

This principle is false even for quantifier free Φ. In particular operational equivalence does not preserve membership in classes. One counterexample is:

$$e_0 = \lambda y.y \quad e_1 = \mathtt{let}\{z := \mathtt{mk}(\lambda y.y)\}\lambda w.\mathtt{app}(\mathtt{get}(z), w) \quad \Phi = x \in \{x \mid x \cong \lambda y.y\}$$

The problem is that the reduction contexts allowed in the atomic clause are allowed to alter the contents of any cell in memory, regardless of whether or not that cell is local. Similarly it may be that by restricting the quantifiers to range over *visible* or *non-local* values, the resulting logic will have nicer metatheoretic properties.

(B) At present there are very simple *valid* principles that VTLoE as axiomatized herein does not establish. One simple example is:

$$\mathtt{let}\{z := \mathtt{get}(y)\}\mathtt{let}\{x := \mathtt{mk}(v)\}[\![\mathtt{get}(y) \cong z]\!]$$

Even though it appears to be related to (**get.i**) and (**mk.iv**), it does not follow from them. What is lacking is any way of reasoning about the equivalence of contexts, not just expressions. It will probably be fruitful to extend the formal system to include assertions concerning the equivalence of contexts as well as expressions. For example we could introduce a new judgement of the form

$$U_0 \cong_X U_1$$

to mean that the contexts are equivalent with respect to a set X of variables allowed trapped.

Two important concepts that appear to lie outside the realm of VTLoE are the ability to express type information, and the ability to perform some sort of computation induction.

(C) Since it appears that *types* cannot be encoded as *classes* it may be that some type structure, via a new form of judgment, should be built in from the beginning.

(D) A powerful semantic method for establishing laws of program equivalence is computation induction, induction on the length of computation. Unfortunately, by its very nature, computation induction does not yield readily to axiomatization in a formal theory that admits non-trivial equivalences. This is due to the difficulty of maintaining a dual view of programs as descriptions of computations and programs as black boxes within a single formal theory. One approach to solving this problem is to extend the logic to include an ordering $e_0 \sqsubseteq e_1$ that expresses the operational approximation of computations. Finite projection operations modeled on the finite projections of domain theory can hopefully be used to prove inductive properties of $\sqsubseteq$.

Acknowledgements

This research was partially supported by DARPA contract NAG2-703, NSF grants CCR-8917606, CCR-8915663, and CCR-9109007, and Italian grant CNR-Stanford n.89.00002.26.

6. References

[1] M. Abadi, B. Pierce, and G. Plotkin. Faithful ideal models for recursive polymorphic types. *International Journal of Foundations of Computer Science*, 2(1):1–21, 1991.

[2] S. Feferman. A language and axioms for explicit mathematics. In *Algebra and Logic*, volume 450 of *Springer Lecture Notes in Mathematics*, pages 87–139. Springer Verlag, 1975.

[3] S. Feferman. A theory of variable types. *Revista Colombiana de Matématicas*, 19:95–105, 1985.

[4] S. Feferman. Polymorphic typed lambda-calculi in a type-free axiomatic framework. In *Logic and Computation*, volume 106 of *Contemporary Mathematics*, pages 101–136. A.M.S., Providence R. I., 1990.

[5] M. Felleisen. *The Calculi of Lambda-v-cs Conversion: A Syntactic Theory of Control and State in Imperative Higher-Order Programming Languages.* PhD thesis, Indiana University, 1987.

[6] M. Felleisen and A. K. Wright. A syntactic approach to type soundness. Technical Report Rice COMP TR91-160, Rice University Computer Science Department, 1991.

[7] J.Y. Halpern, A. R. Meyer, and B. A. Trakhtenbrot. The semantics of local storage, or what makes the free-list free? In *11th ACM Symposium on Principles of Programming Languages*, pages 245–257, 1983.

[8] J.Y. Halpern, A. R. Meyer, and B. A. Trakhtenbrot. From denotational to operational and axiomatic semantics for ALGOL-like languages: An overview. In E. Clarke and D. Kozen, editors, *Logics of Programs, Proceedings 1983*, volume 164 of *Lecture Notes in Computer Science*. Springer, Berlin, 1984.

[9] X. Leroy and P. Wies. Polymorphic type inference and assignment. In *Proceedings of the 18th Annual Symposium on Principles of Programming Languages*, pages 291–302. ACM, 1990.

[10] J. M. Lucassen. *Types and Effects, towards the integration of functional and imperative programming.* PhD thesis, MIT, 1987. Also available as LCS TR-408.

[11] J. M. Lucassen and D. K. Gifford. Polymorphic effect systems. In *16th annual ACM Symposium on Principles of Programming Languages*, pages 47–57, 1988.

[12] Z. Manna and R Waldinger. Problematic features of programming languages. *Acta Informatica*, 16:371–426, 1981.

[13] I. A. Mason and C. L. Talcott. Axiomatizing operational equivalence in the presence of side effects. In *Fourth Annual Symposium on Logic in Computer Science*. IEEE, 1989.

[14] I. A. Mason and C. L. Talcott. Equivalence in functional languages with effects. *Journal of Functional Programming*, 1:287–327, 1991.

[15] I. A. Mason and C. L. Talcott. Inferring the equivalence of functional programs that mutate data. *Theoretical Computer Science*, 105(2):167–215, 1992.

[16] I. A. Mason and C. L. Talcott. References, local variables and operational reasoning. In *Seventh Annual Symposium on Logic in Computer Science*, pages 186–197. IEEE, 1992.

[17] A. R. Meyer and K. Sieber. Towards fully abstract semantics for local variables: Preliminary report. In *15th ACM Symposium on Principles of Programming Languages*, pages 191–208, 1988.

[18] E. Moggi. Computational lambda-calculus and monads. In *Fourth Annual Symposium on Logic in Computer Science*. IEEE, 1989.

[19] J. H. Morris. *Lambda calculus models of programming languages.* PhD thesis, Massachusetts Institute of Technology, 1968.

[20] P.W. O'Hearn and R.D. Tennent. Semantic Analysis of Specification Logic, Part 2. *Information and Computation*, ?:?–?, 199?

[21] P.W. O'Hearn and R.D. Tennent. Semantics of Local Variables. Technical Report ECS-LFCS-92-192, Laboratory for foundations of computer science, University of Edinburgh, 1992.

[22] E. Olderog. Hoare's logic for programs with procedures–what has been accomplished. In E. Clarke and D. Kozen, editors, *Logics of Programs, Proceedings 1983*, volume 164 of *Lecture Notes in Computer Science*. Springer, Berlin, 1984.

[23] G. Plotkin. Call-by-name, call-by-value and the lambda-v-calculus. *Theoretical Computer Science*, 1:125–159, 1975.

[24] J. C. Reynolds. Syntactic control of interference. In *Conference record of the 5th annual ACM Symposium on Principles of Programming Languages*, pages 39–46, 1978.

[25] K. Sieber. A partial correctness logic for programs (in an Algol-like language). In R. Parikh, editor, *Logics of Programs*, volume 193 of *Lecture Notes in Computer Science*. Springer, Berlin, 1985.

[26] Scott Fraser Smith. Partial objects in type theory. Technical Report TR 88-938, Department of Computer Science, Cornell University, 1988. Ph. D. thesis.

[27] C. L. Talcott. A theory for program and data specification. In *Design and Implementation of Symbolic Computation Systems, DISCO'90*, volume 429 of *Lecture Notes in Computer Science*, pages 91–100. Springer-Verlag, 1990. Full version to appear in TCS special issue.

[28] M. Tofte. *Operational Semantics and Polymorphic Type Inference*. PhD thesis, Edinburgh University, 1988.

[29] M. Tofte. Type inference for polymorphic references. *Information and Computation*, 89:1–34, 1990.

Logical Definability of NP–Optimisation Problems with Monadic Auxiliary Predicates

Clemens Lautemann

Institut für Informatik, Johannes Gutenberg–Universität Mainz

Abstract. Given a first–order formula φ with predicate symbols $e_1 \dots e_l$, $s_0, \dots, s_r$, an NP–optimisation problem on $<e_1, \dots, e_l>$–structures can be defined as follows:
for every $<e_1, \dots, e_l>$–structure G, a sequence $<S_0, \dots, S_r>$ of relations on G is a feasible solution iff $<G, S_0, \dots .S_r>$ satisfies φ, and the value of such a solution is defined to be $|S_0|$. In a strong sense, every polynomially bounded NP–optimisation problem has such a representation, however, it is shown here that this is no longer true if the predicates $s_1, \dots, s_r$ are restricted to be monadic. The result is proved by an Ehrenfeucht–Fraïssé game and remains true in several more general situations.

1 Introduction

With his seminal paper [Fa74], Ronald Fagin introduced finite model theory into computational complexity theory. He provided a non–computational characterisation of NP by proving that, for any finite signature σ, a set L of finite σ–structures is in NP iff there is an extension σ' of σ and a first–order sentence φ over σ' such that a σ–structure G belongs to L if and only if it can be extended to a σ'–structure which satisfies φ.
This result has since been developed further, mainly in two directions. On the one hand similar characterisations have been found for many other complexity classes (for a survey see [Im89]), on the other hand the possibility of restricting the syntactical form of the formula φ has been used to differentiate between problems which cannot be distinguished by their computational complexity (cf.[dR87]).
One motivation for this line of research is the fact that in finite model theory there are means of showing *separation results*, something we do not seem to be able to do in computational complexity. However, so far, none of the model theoretic separations have carried over to a separation of complexity classes.
Recently, Papadimitriou and Yannakakis took this approach of looking at restricted formulae one step further, and applied it to NP–optimisation problems [PY91]. Their aim was to find an explanation for the different behaviour of such problems with respect to polynomial–time approximation algorithms. The main result of their paper is the approximability to within a constant factor of all those optimisation problems which are defined in a certain way by a formula of the form $\forall \bar{x} \exists \bar{y} \varphi$.[1] Their way of defining optimisation problems by first–order formulae can be rephrased as follows.

[1] Here $\forall \bar{x}$ is short for $\forall x_1, \dots, \forall x_m$, for some $m \geq 0$, similarly, $\exists \bar{y}$.

Let σ be a signature and σ' an extension of σ, such that $\sigma' - \sigma = <s_0, s_1, \ldots, s_r>$. Then every first order sentence φ over σ' gives rise to the following optimisation problem:

Given A σ–structure G.
Wanted A σ'–extension $<G, S_0, \ldots, S_r>$ of G which maximises (or minimises) $|S_0|$, subject to the condition that $<G, S_0, \ldots, S_r> \models \varphi$.

As an example consider the problem MAXCLIQUE:

Given A graph $G = (U, E)$.
Wanted A maximum–size subset $V \subseteq U$ such that G induces a complete graph on V.

If we take σ to consist of one binary predicate symbol e, $\sigma' - \sigma$ of one unary symbol v, we can express the fact that the elements of V form a clique by

$$\varphi \equiv \forall x \forall y\, (x \neq y \wedge v(x) \wedge v(y)) \rightarrow (e(x, y) \vee e(y, x))\,.$$

Consequently the clique number $\omega(G)$ can be expressed as:

$$\omega(G) = max\{|V| \,/ <G, V> \models \varphi\}.$$

Given this definition of optimisation problems by formulae, Papadimitriou and Yannakakis' result can be stated in the following way.

1.1 Theorem ([PY91])
Let $\sigma' - \sigma = <s_0, \ldots, s_r>$, and let ψ be a quantifier–free first–order formula over $\sigma' - \{s_0\}$ with free variables $\bar{x}, \bar{y}$. If $\varphi = \forall \bar{x} \exists \bar{y} s_0(\bar{x}) \rightarrow \psi$, then the maximisation problem defined by φ can be solved in polynomial time approximatively to within a constant factor. □

How strong is this result? Although the restriction for φ seems rather severe compared to full first–order logic, the example above shows that we cannot hope for much better results along these lines. In fact the formula for the clique problem violates the restrictions of the theorem "only just"; however, there seems to be little hope of finding a polynomial–time approximation algorithm with guaranteed performance for MAXCLIQUE[2]. On the other hand, at first sight, it is not clear which class of optimisation problems can be defined at all by first–order formulae in the way described above. A first answer to this question was given by Kolaitis and Thakur in [KT90], where they showed that the optimal value of *every* NP–optimisation problem (with polynomially bounded values) equals the optimal value of an optimisation problem which is defined, in the way sketched above, by some first–order formula φ, in fact by a formula of the form $\forall \bar{x} \exists \bar{y} \psi$. Their result is, however, unsatisfactory in that the logically defined version of a given NP–optimisation problem might only have the same *optimal*

[2] In fact, in a recent paper [ALMSS], Arora et al show that approximation to within n^ϵ would imply P = NP.

values, not necessarily the same range of *approximate* values. In section 3, I will strengthen their result and show that the structure of any (polynomially bounded) NP–optimisation problem can be completely modeled by a logically defined one, in fact by one which is defined by a sentence of the form $\forall\bar{x}\exists\bar{y}\psi$. Although this is not a deep result (its proof consists mainly of a detailed analysis of Fagin's construction in [Fa74], I consider it important since it shows that we can move from *any* (polynomially bounded) NP–optimisation problem to a logically defined one, without losing any information. In particular, approximate solutions of the logically defined version represent approximate solutions of the original problem.

Given that with $\varphi = \forall\bar{x}\exists\bar{y}\psi$ we can define any NP–optimisation problem, but with $\varphi = \forall\bar{x}\exists\bar{y}s_0(\bar{x}) \rightarrow \psi$ only those with a constant factor approximation, the question arises as to the impact of other syntactical restrictions. A number of authors have looked at restrictions of the quantifier–free part of $\forall\bar{x}\exists\bar{y}\psi$, e.g., concerning positive or negative occurrences of predicates, cf.[KT90], [KT91], [PR90], [Ih90].

I want to complement their work by considering restrictions on the signature extension $\sigma' - \sigma$, rather than the syntactic structure of φ. The main result here is that auxiliary non–monadic predicates are necessary, i.e., that there are NP–optimisation problems (e.g. the vertex coloring problem) which cannot be defined in the way described above with the predicate symbols $s_1, \ldots, s_r$ all being unary, even if φ can make use of full first–order. The proof is by an Ehrenfeucht–Fraïssé game, and it can be extended to show an even stronger non–expressibility result: the chromatic number of a graph (and other optimisation problems) cannot be expressed as

$$min\{h(|S_0|, \ldots, |S_r|) \,/ <G, S_0, \ldots, S_r> \models \varphi\}$$

for a large class of functions h, with a monadic second–order sentence φ, no matter what the arities of $S_0, \ldots, S_r$.

2 Preliminaries

2.1 NP–Optimisation Problems

The notion of an NP–optimisation problem is not always made precise in the literature. I will use the following definition, which is equivalent to the one found in [PR90].

Let Σ be a finite alphabet (w.l.o.g. Σ can be assumed to be $\{0, 1\}$). An *NP–optimisation problem* is given by a triple $<R, f, opt>$, where

- For all $x, y \in \Sigma^* : <x, y> \in R \Rightarrow |y| \leq p(|x|)$, for some polynomial p;
- $R \in P$;
- $f : R \rightarrow I\!N$ is computable in polynomial time;
- $opt \in \{min, max\}$.

For every NP–optimisation problem $<R, f, opt>$ there is a mapping $opt_{R,f}$, defined by $opt_{R,f}(x) = opt\{f(x, y) / <x, y> \in R\}$.

If there is a polynomial q such that $f(x,y) \leq q(|x|)$ for every $<x,y> \in R$, the optimisation problem $<R,f,opt>$ is called *polynomially bounded.*
For every $<R,f>$, the sets
$L^{\leq} = \{(x,k)/\exists y : <x,y> \in R \land f(x,y) \leq k\}$, and
$L^{\geq} = \{(x,k)/\exists y : <x,y> \in R \land f(x,y) \geq k\}$
are in NP, and it is mainly in this form that optimisation problems have been treated in complexity theory (cf. [GJ79]), other approaches are surveyed in [BJY90].

2.2 Logic

A ***signature*** is a finite sequence of pairwise distinct *predicate symbols.* Each predicate symbol s has an *arity,* $a(s) \in I\!N \setminus \{0\}$; if $a(s) = k$ s is called *k-ary.* 1–ary predicate symbols are also called *monadic,* 2–ary ones *binary.* For the rest of this section, fix a signature $\sigma = <e_1, \ldots, e_m>$.
Let U be a set, $k \in I\!N \setminus \{0\}$. A k–ary *relation* on U is a subset of U^k. A *σ–sequence on* U is a sequence $<E_1, \ldots, E_m>$ of relations on U such that E_i is $a(e_i)$–ary, for $i = 1, \ldots, m$. A *σ–structure* $G = <U_G, E_1, \ldots, E_m>$ consists of a nonempty finite[3] set U_G together with a σ–sequence on U_G. For $U \subseteq U_G$, the σ–structure *induced* by G on U, $G[U]$, is defined as $<U, E_1(U), \ldots, E_m(U)>$, where $E_i(U) = E_i \cap U^{a(e_i)}$, for $i = 1, \ldots, m$.
For any other signature $\tau = <s_0, \ldots, s_r>$, a *τ–extension* of G is a στ–structure[4] $<U_G, E_1, \ldots, E_m, S_0, \ldots, S_r>$, also denoted by $<G, S_0, \ldots, S_r>$.
A *first-order σ–formula* (also: *f.o. formula over* σ) is a first–order formula φ all of whose atoms are of one of the forms

- $x = y$, or
- $e_i(x_1, \ldots, x_{a(e_i)})$, for some i $\in \{1, \ldots, m\}$.

Here $x, y, x_1, \ldots, x_{a(e_i)} \in X$, where X is the set of (f.o.) variables. φ is called a *sentence* if all its variables are bound by quantifiers. Let φ be a f.o. σ–sentence, G a σ–structure. G is a *model* of φ, written $G \models \varphi$, if φ holds true in G when every predicate symbol is interpreted by the corresponding relation.

3 Logically Defined NP–Optimisation Problems

Let $\sigma = <e_1, \ldots, e_m>$ be a signature. A σ–structure G can be represented by a string over $\{0,1\}$ in some canonical way, for instance as $1^n0w_1 \ldots w_m$, where U_G is identified with $\{0, \ldots, n-1\}$, and $|w_i| = n^{a(e_i)}$, for $i = 1, \ldots, m$. For every $\alpha_1, \ldots, \alpha_{a(e_i)} \in U_G$, if $<\alpha_1, \ldots, \alpha_{a(e_i)}> \in E_i$ then w_i has 1 in position $j := \Sigma_{l=1}^{a(e_i)} \alpha_l n^{l-1} + 1$, otherwise 0.
On the other hand, every string $w \in \{0,1\}^+$ can be viewed as the representation of a finite structure. Either w is of the form described above, or it represents the

[3] All structures in this paper are finite.
[4] We can assume the symbols of τ to be distinct from those of σ.

structure $<U, E>$, where $U = \{0, \ldots, |w| - 1\}$ and $E \subseteq U$, such that $i \in E \iff w_{i+1} = 1$.
It is thus possible to view every set of strings as the representation of a set of σ–structures, for some σ.
The starting point for the following development is Fagin's celebrated theorem.

3.1 Theorem ([Fa74])
A set L of (representations of) σ–structures belongs to NP if and only if there is a signature τ and a f.o. sentence φ over $\sigma\tau$ such that for every σ–structure G: G $\in$ L $\iff$ there is a τ–extension G' of G such that G' $\models \varphi$. □

Let σ, τ be signatures, φ a f.o. sentence over $\sigma\tau$. By Fagin's theorem, the pair $<\tau, \varphi>$ defines a set in NP, but we can also derive an NP–optimisation problem from it. Let $\tau = <s_0, \ldots, s_r>$, and $opt \in \{min, max\}$. Then the tuple $<\tau, \varphi, opt>$ defines the NP–optimisation problem $<R_{\tau,\varphi}, f_{\tau,\varphi}, opt>$, where

- $R_{\tau,\varphi} := \{<G, <S_0, \ldots, S_r>> / <G, S_0, \ldots, S_r> \models \varphi\}$,
- $f_{\tau,\varphi}(G, <S_0, \ldots, S_r>) = |S_0|$.

An NP–optimisation problem of this form is called *defined by* φ, or *defined in first-order logic.*
Clearly, every NP–optimisation problem of the form $<R_{\tau,\varphi}, f_{\tau,\varphi}, opt>$ is polynomially bounded; but can every polynomially bounded NP–optimisation problem $<R, f, opt>$ be defined in first–order logic? Kolaitis and Thakur showed in [KT90] that for every such $<R, f, opt>$ there are τ,φ and opt' such that $opt_{R,f} = opt'_{R_{\tau,\varphi}, f_{\tau,\varphi}}$; here φ can be chosen to be of the form $\forall\bar{x}\exists\bar{y}\psi$, where ψ is quantifier–free. However, a careful analysis of (a modified version of) Fagin's proof shows that a stronger[5] result holds. A proof can be found in [Kn92].

3.2 Theorem
Let $<R, f, opt>$ be a polynomially bounded NP–optimisation problem. There are signatures σ, τ and a first–order sentence φ over $\sigma\tau$ such that the following hold.

1. There is a polynomial–time computable mapping which, for every pair $<G, y>$ constructs a τ–sequence $<S_0, \ldots, S_r>$ such that
   - $<G, y> \in R \iff <G, S_0, \ldots, S_r> \models \varphi$, and
   - $|S_0| = f(G, y)$, if $<G, y> \in R$.
2. There is a polynomial–time computable mapping which, for all $<G, S_0, \ldots, S_r>$ constructs y such that
   - $<G, S_0, \ldots, S_r> \models \varphi \iff <G, y> \in R$, and
   - $f(G, y) = |S_0|$, if $<G, S_0, \ldots, S_r> \models \varphi$.

Moreover, φ can be chosen to be of the form $\forall\bar{x}\exists\bar{y}\psi$, with quantifier–free ψ. □

Thus in a very detailed way, logically defined NP–optimisation problems represent the structure of all (polynomially bounded) NP–optimisation problems.

[5] In the context of [KT90], perhaps, no better result is possible, since there a weaker notion of NP–optimisation problem is used. In particular, R is not required to be decidable in polynomial time.

4 Ehrenfeucht–Fraïssé Games for First–Order Definability of NP–Optimisation Problems

Fagin's theorem gives rise to a hierarchy within NP, by restricting the arities of the auxiliary predicate symbols in τ. It is well-known that the first level of this hierarchy, although it does contain some NP-complete problems, does not even include all of P (or even NL): Fagin showed in [Fa75] that the class of connected graphs cannot be defined by $<\tau, \varphi>$, if τ contains only monadic predicates. Related results can also be found in [dR87], [AF90]. I will ask a similar question about optimisation problems: Given $<R, f, opt>$, can the function $opt_{R,f}$ be expressed[6] as

$$opt'\{|S_0|/<G, S_0, \ldots, S_r> \models \varphi\}^7$$

with *monadic* auxiliary relations $S_1, \ldots, S_r$? (I call these relations *auxiliary* because they do not contribute to the function value.)
The answer is, in general, no, irrespective of the arity of S_0 or the syntactical form of the first-order sentence φ.
The main tool in the proof will be a variant of the Ehrenfeucht–Fraïssé game. This game is played by two players, called Spoiler and Duplicator respectively, on two finite σ'-structures G, H, according to the following rules.

1. There is a fixed number n of *rounds*. Each round consists of one move of Spoiler, followed by one move of Duplicator.
2. In each round, Spoiler moves by choosing an element from U_G, or U_H. Duplicator then moves by choosing an element from the other structure.

Thus, in every round, one element from each structure is chosen. Let $a_i \in U_G$, $b_i \in U_H$ be the elements chosen in round i, for $i = 1, \ldots, n$. Duplicator has won, if the mapping $a_i \mapsto b_i$, $i = 1, \ldots, n$, is an isomorphism between the induced structures $G[\{a_1, \ldots, a_n\}]$ and $H[\{b_1, \ldots, b_n\}]$. Otherwise Spoiler has won.
The following fact is well-known.

4.1 Fact
Assume that there is a f.o. sentence φ with at most n quantifiers such that $G \models \varphi$, but $H \not\models \varphi$. Then Spoiler has a winning strategy in the n-round Ehrenfeucht–Fraïssé game on G, H. □

From this fact, one can easily deduce the following lemma, which is similar to Theorem 4.5 of [AF90].

4.2 Lemma
Let L be a class of σ-structures, let $n \in \mathbb{N}$, and let $\tau = <s_0, \ldots, s_r>$ be a signature. Assume that there is a σ-structure $G \in L$ such that the following holds:

---

[6] Note that here, a strong correspondence of $<R, f, opt>$ and $<R_{\tau,\varphi}, f_{\tau,\varphi}, opt'>$ as discussed in the last section is not required, only the coincidence of the optimal values.

[7] More precisely, this should be written as $\{|S_0| \,/ \exists S_1, \ldots, S_r : <G, S_0, \ldots, S_r> \models \varphi\}$, but I will use the abridged notation throughout.

For every τ–extension G' of G there is a σ–structure $H \notin L$, and a τ–extension H' of H such that Duplicator has a winning strategy in the n–round Ehrenfeucht–Fraïssé game on G', H'.
Then there is no first–order $\sigma\tau$–sentence φ with at most n quantifiers such that $L = \{G / \exists S_0, \ldots, S_r \langle G, S_0, \ldots, S_r\rangle \models \varphi\}$.
Proof: Assume to the contrary that $L = \{G / \exists S_0, \ldots, S_r \langle G, S_0, \ldots, S_r\rangle \models \varphi\}$, for some f.o. sentence φ over $\sigma\tau$ with at most n quantifiers. Let $G \in L$ be as in the premiss of the lemma, and let G' be a τ–extension of G such that $G' \models \varphi$. Then the premiss of the lemma asserts the existence of $H \notin L$, and a τ–extension H' of H, such that Duplicator has a winning strategy in the n–round Ehrenfeucht–Fraïssé game on G', H'. On the other hand $H' \not\models \varphi$, since $H \notin L$. But then, by Fact 4.1, Spoiler has a winning strategy on G', H', a contradiction. □

This lemma can easily be adapted so as to provide a tool for proving nondefinability of functions.

4.3 Lemma

Let f be a mapping on σ–structures with values in $\mathbb{N}$, let $n \in \mathbb{N}$, and let $\tau = \langle s_0, \ldots, s_r\rangle$ be a signature. Assume that there is a σ–structure G such that the following holds:
For every τ–sequence $\langle S_0, \ldots, S_r\rangle$ on U_G for which $|S_0| = f(G)$ there is a σ–structure H, and there is a τ–sequence $\langle T_0, \ldots, T_r\rangle$ on U_H such that

- $f(H) > f(G)$
- $|T_0| = |S_0|$
- Duplicator has a winning strategy in the n–round Ehrenfeucht–Fraïssé game on $\langle G, S_0, \ldots, S_r\rangle$, $\langle H, T_0, \ldots, T_r\rangle$.

Then there is no f.o. $\sigma\tau$–sentence φ with at most n quantifiers such that for every σ–structure $G, f(G) = min\{|S_0| \,/\, \langle G, S_0, \ldots, S_r\rangle \models \varphi\}$.[8]
Proof: Assume to the contrary that $\forall G\, f(G) = min\{|S_0| \,/ \langle G, S_0, \ldots, S_r\rangle \models \varphi\}$, for some $\sigma\tau$–sentence φ with at most n quantifiers. Let G be as in the premiss of the lemma, and let $S_0, \ldots, S_r$ be given such that $\langle G, S_0, \ldots, S_r\rangle \models \varphi$, and $|S_0| = f(G)$. By assumption, there are $H, T_0, \ldots, T_r$ such that $f(H) > f(G)$, $|T_0| = |S_0|$, and Duplicator has a winning strategy in the n–round Ehrenfeucht–Fraïssé game on $\langle G, S_0, \ldots, S_r\rangle, \langle H, T_0, \ldots, T_r\rangle$. On the other hand, since $|T_0| = |S_0| = f(G) < f(H)$, and $f(H) = min\{|T_0'| \;/ \langle H, T_0', \ldots, T_r'\rangle \models \varphi\}$, it follows that $\langle H, T_0, \ldots, T_r\rangle \not\models \varphi$. This implies that Spoiler has a winning strategy on $\langle G, S_0, \ldots, S_r\rangle, \langle H, T_0, \ldots, T_r\rangle$, a contradiction. □

This lemma will now be used to prove that certain functions cannot be expressed over restricted signatures. Let $\chi(G)$ and $\omega(G)$ denote the chromatic number, respectively the clique number of a graph G, thus χ and ω map $\langle e\rangle$–structures onto natural numbers.

---

[8] Of course, a similar lemma can be proved for *max* instead of *min*.

4.4 Theorem

Let $\tau = <s_0, \ldots, s_r>$ be a signature in which the predicate symbols $s_1, \ldots, s_r$ are monadic. There is no f.o. $<e, s_0, \ldots, s_r>$–sentence φ such that for every graph G

$$\chi(G) = min\{|S_0| \,/ <G, S_0, \ldots, S_r> \models \varphi\}.$$

Proof: Let $k = a(s_0)$, and let n be given. Let $m > n \cdot 2^r$, and $p > k \cdot m$. In order to apply Lemma 4.3 chose G to be the disjoint union of p copies of K_m, the complete graph on m vertices. Now, let $S_0 \subseteq V_G^k$, $\quad S_1, \ldots, S_r \subseteq V_G$ be given, such that $|S_0| = \chi(G) = m$. Consider the set

$$U(S_0) := \bigcup_{<a_1,\ldots,a_k> \in S_0} \{a_1, \ldots, a_k\}.$$

Since $|S_0| = m$, $\quad |U(S_0)| \leq k \cdot m < p$, so there must be some component, G_0, of G with $U_{G_0} \cap U(S_0) = \emptyset$. Then G can be written as the disjoint union of G_0 with some graph G'. Now, for every $\alpha \in \{0,1\}^r$, define U_α to be the set

$$U_\alpha := U_{G_0} \cap \bigcap_{\alpha_i = 1} S_i \cap \bigcap_{\alpha_i = 0} (U_G \setminus S_i).$$

Then the sets U_α form a partition of U_{G_0} into at most 2^r parts, so one of them, say U_β, must contain more than n elements. Let W be a set with $|U_\beta| + 1$ elements, and let H_0 be the complete graph on $W \uplus (U_{G_0} \setminus U_\beta)$, i.e., H_0 is a copy of K_{m+1}. Define the graph H to be the disjoint union of H_0 and G'. Then $\chi(H) = m{+}1 > \chi(G)$, as required. Finally, $T_0, \ldots, T_r$ are defined as follows:
$T_0 := S_0$, this is possible since $S_0 \subseteq U_{G'}^k \subseteq U_H^k$.
For $i = 1, \ldots, r$, set
$T_i \cap (U_H \setminus W) := S_i \cap (U_G \setminus U_\beta)$, and
$$T_i \cap (W) = \begin{cases} W, \; if \; \beta_i = 1 \\ \emptyset, \; \text{if} \; \beta_i = 0. \end{cases}$$
With this definition, $|T_0| = |S_0|$, and it remains to show that Duplicator has a winning strategy in the n–round Ehrenfeucht–Fraïssé game on $<G, S_0, \ldots, S_r>$, $<H, T_0, \ldots, T_r>$. The strategy is the following: in round j,

- if Spoiler plays a vertex which was already played in some round $l < j$, then Duplicator plays the other vertex played in round l;
- if Spoiler plays a vertex $v \in U_G \setminus U_\beta (= U_H \setminus W)$, then Duplicator plays the same vertex v;
- if Spoiler plays a new vertex $v \in U_\beta$, then Duplicator plays a new vertex $w \in W$;
- if Spoiler plays a new vertex $w \in W$, then Duplicator plays a new vertex $v \in U_\beta$.

It is not difficult to show that with this strategy Duplicator always wins the game, thus all assumptions of Lemma 4.3 are satisfied, and the theorem follows.

□

It is easy to see that the graph colouring problem can be logically defined, using a signature $\tau = <s_0, s_1>$ where s_0 is monadic and s_1 is binary. In fact, let φ be the $<e, s_0, s_1>$–sentence

$$(\forall x \exists y s_0(y) \wedge s_1(x,y)) \wedge (\forall x \forall y \forall z (s_1(x,z) \wedge s_1(y,z) \rightarrow \neg\, e(x,y))).$$

Then, whenever $<G, S_0, S_1> \models \varphi$, the binary relation S_1 includes an admissible colouring of G with colours from S_0, and $\chi(G) = min\{|S_0| \,/ <G, S_0, S_1> \models \varphi\}$. However, Theorem 4.4 says that $\chi(G)$ cannot be defined logically with only monadic auxiliary predicates even if full first–order logic is available for φ. Here, the predicates $s_1, \ldots, s_r$ are called *auxiliary* since the cardinalities of $S_1, \ldots, S_r$ do not contribute to the value of a solution $<G, S_0, \ldots, S_r>$. [9] In particular, there is no φ for which $\chi(G) = min\{|S_0| \,/ <G, S_0> \models \varphi\}$. This proves a conjecture of Kolaitis and Thakur in [KT91][10].
In the proof of Theorem 4.4 all graphs are unions of cliques. For such graphs chromatic number and clique number coincide, and it follows that the theorem remains true if χ is replaced by ω in its statement.

4.4.1 Corollary
Let $\tau = <s_0, \ldots, s_r>$ be a signature in which the predicate symbols $s_1, \ldots, s_r$ are monadic. There is no first–order sentence φ over $<e, s_0, \ldots, s_r>$ such that for every graph G, $\omega(G) = min\{|S_0| \,/ <G, S_0, \ldots, S_r> \models \varphi\}$. □

Why should it be interesting to have a logical definition of the clique number as a *minimum*? After all, the problem MAXCLIQUE seems inherently a maximisation problem, and, in fact, can easily be defined as such without auxiliary predicates. However, in [KT91], Kolaitis and Thakur showed that NP = CoNP if and only if the clique number has a f.o. definition as a minimum. They also proved for certain restrictions on the form of the defining f.o. sentence that this is indeed not possible.
Corollary 4.4.1 now complements their results (leaving, however, the NP = CoNP question wide open).

5 Extensions

The main ingredient in the proof of Theorem 4.4 is an Ehrenfeucht–Fraïssé game on two complete graphs of different sizes. It is well known that Duplicator can win such a game even in more complicated situations, and accordingly, Theorem 4.4 can be extended in several ways. In particular, I will present extensions to

- monadic second–order logic,
- ordered structures, and
- a more powerful graph logic.

[9] Note that the arity of s_0 is not restricted.
[10] In fact, the conjecture in [KT91] was weaker, in that it involved only sentences φ of rather restricted form.

Furthermore, the restriction that $f_{\tau,\varphi}(G, <S_0, \ldots, S_r>) = |S_0|$ can be weakened to $f_{\tau,\varphi}(G, <S_0, \ldots, S_r>) = h(|S_0|, \ldots, |S_r|)$, for certain functions h.
Let us first consider the extension to monadic second-order logic. φ is a *mso formula* over τ if

- φ is a f.o. formula over τ, or
- φ is of one of the forms $\forall s\psi$, or $\exists s\psi$, where s is a monadic predicate symbol, and ψ is a mso formula over $\tau<s>$, or
- φ is of one of the forms $\forall x\psi$, or $\exists x\psi$, where x is an individual variable, and ψ is a mso formula over τ.

Let $MSO^n(\sigma)$ denote the set of mso sentences over σ with at most n quantifiers (first or second order). The following fact will be basic for the rest of this paper.

5.1 Fact
Let $\sim$ be the equivalence relation on $MSO^n(\sigma)$ defined by
$\varphi \sim \psi$ iff $M \models \varphi \Leftrightarrow M \models \psi$, for every σ-structure M.
Then $\sim$ has finite index (i.e., there are only finitely many equivalence classes).
□

The n-round mso-Ehrenfeucht-Fraïssé game on two structures G, H is played in the same way as the game defined in Section 4, with the additional possibility to play sets instead of elements. More precisely, in every round, Spoiler first decides whether this is to be a set round or an element round. In an element round, both players play according to the rules defined in Section 4. In a set round, Spoiler chooses a set $U \subseteq U_G$, or a set $V \subseteq U_H$, and Duplicator replies by choosing a subset of the universe of the other structure.
Let A_i be the subset of U_G, B_i the subset of U_H played in set round i, for $i = 1, \ldots, l$, and let $a_j \in U_G$, $b_j \in U_H$ be the elements played in element round j, for $j = 1, \ldots, m$, where $l + m = n$. Duplicator wins, if the mapping $a_j \mapsto b_j$, $j = 1, \ldots, m$, defines an isomorphism between the two structures $<G, A_1, \ldots, A_l>[\{a_1, \ldots, a_m\}]$ and $<H, B_1, \ldots, B_l>[\{b_1, \ldots, b_m\}]$. In analogy to the first-order game we have the following fact.

5.2 Fact
Assume that there is a mso sentence φ with at most n quantifiers such that $G \models \varphi$, but $H \not\models \varphi$. Then Spoiler has a winning strategy in the n-round mso-Ehrenfeucht-Fraïssé game on G, H. □

Fact 5.1 will be used in the following form which can easily be derived from Fact 5.2.

5.3 Lemma
Let $\mathcal{M}$ be an infinite set of σ-structures. For every n there are M_1, $M_2 \in \mathcal{M}$ such that Duplicator has a winning strategy in the n-round mso-Ehrenfeucht-Fraïssé game on M_1, M_2. □

The following lemma corresponds to Lemma 4.3. It can be proved in much the same way as Lemma 4.3, using Fact 5.2 instead of Fact 4.1.

5.4 Lemma
Let $h : \mathbb{N}^{r+1} \to \mathbb{N}$, let f be a mapping on σ–structures with values in $\mathbb{N}$, let $n \in \mathbb{N}$, and let $\tau = <s_0, \ldots, s_r>$ be a signature. Assume that there is a σ–structure G, such that the following holds:
For every τ–sequence $<S_0, \ldots, S_r>$ on U_G for which $h(|S_0|, \ldots, |S_r|) = f(G)$ there is a σ–structure H, and there is a τ–sequence $<T_0, \ldots, T_r>$ on U_H such that

- $f(H) > f(G)$
- $h(|T_0|, \ldots, |T_r|) = h(|S_0|, \ldots, |S_r|)$
- Duplicator has a winning strategy in the n–round mso–Ehrenfeucht–Fraïssé game on $<G, S_0, \ldots, S_r>$, $<H, T_0, \ldots, T_r>$.

Then there is no mso $\sigma\tau$–sentence φ with at most n quantifiers such that for every σ–structure G, $f(G) = min\{h(|S_0|, \ldots, |S_r|) \,/ <G, S_0, \ldots, S_r> \models \varphi\}$. □

The existence of a linear order can make a crucial difference in expressive power. It is easy, e.g., to find a mso sentence which holds for an *ordered* structure if and only if its universe has an even number of elements, whereas without an order relation no such sentence exists. However, for the problems considered here, the addition of a linear order does not help, as will now be shown.
An *ordered graph* is a $<e, o>$–structure $<V, E, O>$, where $<V, E>$ is a graph and O is a linear order on V.

5.5 Theorem
Let $\sigma = <s_0, \ldots, s_r>$ be a signature and let $h : \mathbb{N}^{r+1} \to \mathbb{N}$ be such that for every $p \in \mathbb{N}$ the set $h^{-1}(p)$ is finite. Then there is no mso sentence φ over $<e, o>\tau$ such that for every ordered graph G
$\chi(G) = min\{h(|S_0|, \ldots, |S_r|) \,/ <G, S_0, \ldots, S_r> \models \varphi\}$.
Proof: Let n be given and let, for every $p \in \mathbb{N}$, K_p be an ordered complete graph on p vertices. By Lemma 5.3 there are $p, q \in \mathbb{N}$, $p < q$ such that Duplicator has a winning strategy in the n–round mso–Ehrenfeucht–Fraïssé game on K_p, K_q.
Let k_i be the arity of s_i, for $i = 0, \ldots, r$, and let m be such that $m > \sum_{i=0}^{r} m_i k_i$, whenever $h(m_0, \ldots, m_r) = p$. Let, for $i = 1, \ldots, m$, G_i be a copy of K_p and let G be the disjoint union of the G_i, ordered in such a way that $u_i < u_{i+1}$, for all $u_i \in U_{G_i}, u_{i+1} \in U_{G_{i+1}}, i = 1, \ldots, m-1$. Now, if $S_0, \ldots, S_r$ is a τ–sequence on G with $h(|S_0|, \ldots, |S_r|) = p (= \chi(G))$, then one of G's components, say G_{i_0}, has no vertex in common with the set

$$\bigcup_{i=0}^{r} \bigcup_{<u_1, \ldots, u_{k_i}> \in S_i} \{u_1, \ldots, u_{k_i}\}.$$

In order to apply Lemma 5.4, let H be G with G_{i_0} replaced by H_{i_0}, a copy of K_q, and let $T_i := S_i$, for $i = 0, \ldots, r$. Then $\chi(H) = q > p = \chi(G)$, and $|T_i| = |S_i|$, for $i = 0, \ldots, r$, so $h(|T_0|, \ldots, |T_r|) = h(|S_0|, \ldots, |S_r|)$.
All that remains to show is that Duplicator has a winning strategy in the n–round mso–Ehrenfeucht–Fraïssé game on G, H. But this follows from her strategy on K_p, K_q: all Duplicator has to do is mimic Spoilers behaviour on $G \setminus G_{i_0}, H \setminus H_{i_0}$

(which is no problem because these structures are identical), and play according to her strategy for K_p, K_q on G_{i_0}, H_{i_0}.
The theorem now follows from Lemma 5.4. □

As a final application I want to show that the problems CHROMATIC NUMBER and CHROMATIC INDEX[11] are not "linear emso extremum problems" as defined by Arnborg et al. in [ALS91]. Here emso stands for "extended monadic second-order" and with the notation used in this paper, linear emso extremum problems can be defined as those optimisation problems $<R_\varphi, f, opt>$, where

- φ is a mso sentence over $\sigma\tau$, and $\tau = <s_0, \ldots, s_r>$ consists of monadic predicate symbols only, and
- $f(<G, S_0, \ldots, S_r>) = a + \sum_{i=o}^{r} a_i |S_i|$, for some $a, a_i \in Q$.

In the given context, graphs are represented by their *incidences* rather than their *adjacencies*. More precisely, the signature σ consists of one binary predicate *inc*, and a graph with vertex set V and edge set E is represented as the $<inc>$–structure $<U, I>$, where

- $U = V \cup E$, and
- $<u_1, u_2> \in I \Leftrightarrow u_1 \in V, u_2 \in E$, and u_1 is an endpoint of u_2.

This representation of graphs allows us to quantify over *edge sets* in mso sentences, hence we can talk about *arbitrary*, not only *induced* subgraphs, which considerably increases the expressive power of mso sentences (for a more detailed discussion, see [Co90]).

5.6 Theorem
CHROMATIC NUMBER is not a linear emso minimum problem. □

The proof has to be omitted, due to space restrictions. Again, its main part is an Ehrenfeucht–Fraïssé game on cliques of different sizes. Since for cliques with an odd number of vertices chromatic number and chromatic index are the same, we get the following corollary.

5.6.1 Corollary
CHROMATIC INDEX is not a linear emso minimum problem. □

Of course, Theorem 5.6 can also be extended to ordered graphs.

6 Conclusion

The main result of this paper, the fact that chromatic number cannot be defined in first–order logic with only monadic auxiliary predicates, is in accordance with known results about decision problems. However, whereas in the presence of a linear order on inputs, the proofs there are quite involved ([dR87, AF90]), here the extension of the proof to ordered structures is rather straightforward. In the light of Lynch's results in [Ly82], it would be interesting to see if it can even be extended to addition structures.

[11] The chromatic index of a graph is the chromatic number of the corresponding edge graph.

Acknowledgements

I would like to thank Elias Dahlhaus for his insightful comments and suggestions.

References

[AF90] Miklos Ajtai, Ronald Fagin, *Reachability is harder for directed than for undirected finite graphs.* JSL 55, pp. 113-150.

[ALMSS] Sanjeev Arora, Carsten Lund, Rajeev Motwani, Madhu Sudan, Mario Szegedy, *Proof verification and intractability of approximation problems (preliminary version).* Preprint, Computer Science Division, University of California, Berkeley, April 1992.

[ALS91] Stefan Arnborg, Jens Lagergren, Detlev Seese, *Easy problems for tree decomposable graphs.* J. of Algorithms, 12, pp.308–340.

[BJY90] Danilo Bruschi, Deborah Joseph, Paul Young, *A structural overview of NP optimization problems.* Rapporto Interno 75/90, Dipartimento di Scienze dell'Informazione, Universitá degli Studi di Milano.

[Co90] Bruno Courcelle, *On the expression of monadic second-order graph properties without quantifications over sets of edges.* Proc. 5th Ann. IEEE Symposium on Logic in Computer Science, pp. 190–196.

[dR87] Michel de Rougemont, *Second-order and inductive definability on finite structures.* Zeitschrift f. math. Logik und Grundlagen d. Math. 33, pp. 47–63.

[Fa74] Ronald Fagin, *Generalized first-order spectra and polynomial-time recognizable sets.* In Richard Karp, (Ed.):"Complexity of Computation", SIAM-AMS Proc. 7, pp. 43–73.

[Fa75] Ronald Fagin, *Monadic generalized spectra.* Zeitschr. f. math. Logik und Grundlagen d. Math. 21, pp. 89–96.

[GJ79] Michael R. Garey, David S. Johnson, *Computers and Intractability.* Freeman, N.Y.

[Ih90] Edmund Ihler, *Approximation and existential second-order logic.* Bericht 26, Institut für Informatik, Universität Freiburg.

[Im89] Neil Immerman, *Descriptive and computational complexity.* In: Juris Hartmanis (Ed.):"Computational Complexity Theory", Proc. AMS Symp. Appl. Math. 38, pp. 75–91.

[Kn92] Dieter Knobloch, *Zur Komplexität kombinatorischer Optimierungsprobleme.* Diplomarbeit, FB Mathematik, Johannes Gutenberg–Universität Mainz.

[KT90] Phokion G. Kolaitis, Madhukar N. Thakur, *Logical definability of NP optimization problems.* Technical Report UCSC-CRL-90-48, Computer Research Laboratory, University of California, Santa Cruz.

[KT91] Phokion G. Kolaitis, Madhukar N. Thakur, *Approximation properties of NP minimization classes.* Proc. 7th Structure in Complexity Theory Conference, pp. 353–366.

[Ly82] James F. Lynch, *Complexity Classes and Theories of Finite Models.* Math. Syst. Theory 15, pp. 127–144.

[PR90] Alessandro Panconesi, Desh Ranjan, *Quantifiers and approximation (extended abstract).* Proc. 22nd ACM STOC, pp. 446–456.

[PY91] Christos H. Papadimitriou, Mihalis Yannakakis, *Optimization, approximation, and complexity classes.* JCSS 43, pp. 425–440.

Universes in the theories of types and names

Markus Marzetta

Institut für Informatik und angewandte Mathematik
University of Berne, Switzerland

Abstract. In this paper we recall the basic ideas of the theories of types and names, which were introduced by Jäger and which are closely related to Feferman's systems of explicit mathematics. We start off from the elementary theory of types and names with the datatype of the natural numbers, which is equivalent to EM_0 or to $EM_0{\upharpoonright}$ depending on the form of induction taken. The elementary theory is extended by ***universes***, i.e. types enjoying special closure properties, to a theory *UTN*. We show how new type constructions become possible in presence of the universes and prove a lower bound for its proof theoretic strength. We sketch the proof of the upper bound of the theory, which shows that *UTN* does not exceed the limits of predicativity.

1 Introduction

Theories of types and names were introduced by Jäger in [11] and can be situated between subsystems of analysis and Feferman's systems for explicit mathematics: in fact they have their origins in the work of Feferman on foundations of constructive (in particular Bishop-style) mathematics, which issued in theories like e.g. T_0 [2, 3].

Besides from the foundational point of view, nowadays they are relevant to computer science for at least two kinds of reasons. One reason is related to the wide use of type theories in computer science as a base for programming languages of every sort: imperative, relational and functional; in this context theories of types and names with their great flexibility provide a fairly uniform approach to type theories with flexible typing and of a controllable proof-theoretic strength. The second reason is that these theories seem to be quite adequate for representing and proving properties of functional programs of a Turing complete programming language. For further motivation for the use of theories of this kind we refer to Feferman [6, 5], Talcott [19] and Kahle [12].

The idea of *universes* is already present in Martin-Löf's type theories [13, 14] and has been introduced also in explicit mathematics for the proof of Hancock's conjecture [4]. It has also been studied extensively in the field of computer science for example by Palmgreen [17], by Mitchell and Harper [15] in order to handle polymorphism, by Nordström, Petersson and Smith [16], recently by Harper and Pollack [9] and by many many others.

The aim of this paper is to consider the rôle of universes in the framework of types and names.

Theories of types and names are formulated in a two-sorted language: objects and types, where the objects form a partial combinatory algebra (we might think of them as λ-terms, LISP-programs, bitstrings in a computer memory, ...) and the types are just collections of objects. There are two relations between objects s and types T: the obvious membership relation "$s \in T$" and a *naming* relation "$n(s, T)$". The idea is that types, i.e. abstract classes of objects, have concrete representatives among the objects. Every type has at least a name (i.e. we are only interested in types, which have a concrete representation), an object has not necessarily to be the name of a type, but, if this is the case, the denoted type is determined unequivocally. This coding mechanism makes it possible to mimic some constructions of set theory such that e.g. building types whose members are themselves (names of) types, etc.

The types are obtained basically by comprehension axioms, i.e. according to properties of the objects expressible in the language, but also include, as a sort of paradigmatic example of inductively generated datatype, the type of the natural numbers $\mathbf{N}$. Depending on the forms of comprehension and on the principles of induction the proof-theoretic strength can be varied in a wide range.

In this paper we will expand the language by a monadic predicate U on types for the universes and introduce a theory *UTN* about universes, types and names. Universes contain only names of types and are closed under elementary comprehension and join. Instead of taking some type constants for the universes $V_0, V_1, \ldots$, we will require that every type has a name in some universe, by the *limit axiom*.

We prove a lower bound for the proof-theoretic strength of *UTN* with the induction axiom, by embedding Friedman's theory ATR_0 [7]; to that extent we show how to construct the hyper-jump hierarchy along a well ordering in our theory. We give also a sketch of the proof of the upper bound; the two results together show that the ordinal of *UTN* with the induction axiom is Γ_0, the ordinal of predicative analysis.

2 Preliminaries

In this section we give the basic ingredients of a formal system, which reflects the ideas exposed above, specifying the syntax, the logic and some axioms.

2.1 Syntax, logic

The language $\mathcal{L}_{TN}$ is two-sorted: objects and types, each one with countable lists of free and bound variables. We are going to write $a, b, c, \ldots, x, y, z, \ldots$ for object variables and $A, B, C, \ldots, X, Y, Z, \ldots$ for type variables. Underscored letters will denote sequences of syntactical objects of the corresponding kind. Apart from object constants the only function symbol of $\mathcal{L}_{TN}$ is the binary application $\cdot$ (written infix, assumed left associative) between objects.

There are two unary relation symbols on objects: $\downarrow$, for the defined terms, and N, for the naturals; furthermore there is one unary relation on types, U, for universes; the binary relation symbols are $=$ for equality between objects and also for equality between types, $\in$ for the membership relation and n for the naming relation between objects and types.

Let us look a bit more precisely to the constants. There are constants for zero 0, successor $\mathbf{s}_N$, predecessor $\mathbf{p}_N$, pairing $\mathbf{p}$ and projections $\mathbf{p}_0, \mathbf{p}_1$, definition by cases $\mathbf{d}$, join $\mathbf{j}$ and the combinators $\mathbf{s}$ and $\mathbf{k}$; furthermore we use a constant $\mathbf{c}_{\varphi[a,\underset{\sim}{b},\underset{\sim}{C}]}$ in order to obtain names for the type defined by a formula $\varphi[a, \underset{\sim}{b}, \underset{\sim}{C}]$. Hence object terms and formulae have to be defined by simultaneous induction.

Definition 1. The *type terms* of $\mathcal{L}_{TN}$ are the free type variables.
The *object terms* and the *formulae* are defined inductively as follows:

1) the free object variables and the constants $0, \mathbf{s}_N, \mathbf{p}_N, \mathbf{p}, \mathbf{p}_0, \mathbf{p}_1, \mathbf{d}, \mathbf{j}, \mathbf{s}, \mathbf{k}$ are object terms,
2) if $\varphi[a, \underset{\sim}{b}, \underset{\sim}{C}]$ is a formula, then the constant $\mathbf{c}_{\varphi[a,\underset{\sim}{b},\underset{\sim}{C}]}$ is an object term,[1]
3) if s, t are object terms, then also $(s \cdot t)$ is.
4) if s, t are object terms, then $s = t$, $s\downarrow$ and $N(s)$ are (atomic) formulae,
5) if A is a type term, then $U(A)$ is an (atomic) formula,
6) if s is an object term and A is a type term, then $s \in A$ and $n(s, A)$ are (atomic) formulae,
7) if A and B are type terms, then $A = B$ is an (atomic) formula,
8) if φ, ψ are formulae, then also $\neg\varphi, \varphi \wedge \psi, \varphi \vee \psi$.
9) if $\varphi[a], \psi[B]$ are formulae, then also $\forall x \varphi[x], \exists x \varphi[x], \forall X \psi[X], \exists X \psi[X]$.

The logical connectives $\rightarrow, \leftrightarrow$ are defined from $\neg, \vee$ and $\wedge$ as usual. We put $A \subset B :\equiv \forall x(x \in A \rightarrow x \in B)$. Let $\underset{\sim}{s} \equiv s_1 \ldots s_m$ be a sequence of object terms and $\underset{\sim}{A} \equiv A_1 \ldots A_m$ be a sequence of type terms of the same length $m \geq 0$. We will use the following shorthand: $n(\underset{\sim}{s}, \underset{\sim}{A}) :\equiv n(s_1, A_1) \wedge \ldots \wedge n(s_m, A_m)$.

There are basically two possibilities to express the fact that a type A is member of another type (typically a universe) B: either we say that every name of A is an element of B, or that some name of A is in B: the latter turns out to be the right one. Hence we define: $A \dot{\in} B :\equiv \exists x(n(x, A) \wedge x \in B)$.

Since objects will carry the structure of a partial combinatory algebra, we must take into account, that some object terms do not denote any element of the carrier, and hence use some suitable extension of classical predicate calculus; on the contrary, since we are not interested in possibly undefined type terms, usual classical predicate logic is adequate for the sort of types. We have chosen Beeson's *logic of partial terms* [1] for the sort of the objects.

Here are the main differences with respect to classical predicate calculus: in addition to equality there is a second predicate $\downarrow$ for "denotes", "is defined", to whom is deserved a special treatment. Additional axioms state that variables and constants are always defined: $a\downarrow, k\downarrow$; the quantifier axioms are slightly different: $\varphi[t] \wedge t\downarrow \rightarrow \exists x \varphi[x], \forall x \varphi[x] \wedge t\downarrow \rightarrow \varphi[t]$; and *strictness* is postulated for function symbols f and for relation symbols R: $f(t)\downarrow \rightarrow t\downarrow$, $R(t) \rightarrow t\downarrow$. Also the equality axioms for objects are submitted to minor and rather obvious changes.

[1] In order to avoid overloading the exposition with details we have cheated a little bit: when building a constant $\mathbf{c}_{\varphi[a,\underset{\sim}{b},\underset{\sim}{C}]}$ we should have replaced the free variables $a, \underset{\sim}{b}, \underset{\sim}{C}$ by bound ones, in order to make clear that substitutions should leave these constants unchanged.

2.2 Axioms about the objects

Combinatorial axioms. These axioms state that the objects form a partial combinatory algebra.

$$\mathbf{k} \neq \mathbf{s}, \qquad \mathbf{k}ab = a, \qquad \mathbf{s}abc \simeq ac(bc), \qquad \mathbf{s}ab\downarrow.$$

Pairing and projections.

$$\mathbf{p}_0(\mathbf{p}ab) = a, \qquad \mathbf{p}_1(\mathbf{p}ab) = b.$$

Natural numbers.

$$N(0), \qquad (N(a) \rightarrow N(\mathbf{s}_N a) \wedge \mathbf{s}_N a \neq 0 \wedge \mathbf{p}_N(\mathbf{s}_N a) = a).$$

Definition by cases on the natural numbers. The special combinator $\mathbf{d}$ tests two natural numbers on equality.

$$N(a) \wedge N(b) \wedge a = b \rightarrow \mathbf{d}abc_1c_2 = c_1, \quad N(a) \wedge N(b) \wedge a \neq b \rightarrow \mathbf{d}abc_1c_2 = c_2.$$

The combinators introduced above can be used to define *abstraction terms* and to find a *fixed point combinator* (see Strahm [18] for a precise proof).

Lemma 2 (Fixed-point theorem). *There is an object term* $\mathbf{f}$ *such that the axioms given above prove:* $\forall x \forall g(\mathbf{f}g\downarrow \wedge (g(\mathbf{f}g)x \simeq (\mathbf{f}g)x))$.

2.3 Ontological axioms

We did not say anything about the predicate n till now. There are two axioms which reflect our basic understanding of types and names. the first is that, when an object "names" or "represents" a type, this type should be uniquely determined

(O_1) $\quad n(a, B_0) \wedge n(a, B_1) \rightarrow B_0 = B_1,$

the second is related to the constructive point of view and requires all types to have a representation as an object

(O_2) $\quad \forall X \exists y n(y, X).$

The idea of types and names is that the "names" should reflect the *intention* in building a class of objects, whereas the types represent just the *extension* of that class; this becomes clear by the axiom of *extensionality*

(E) $\quad \forall x(x \in A \leftrightarrow x \in B) \rightarrow A = B.$

In Feferman's theories extensionality (of classes) has an entirely different meaning, namely that intention and extension coincide! Feferman's $\in$ relation behaves very much like $a \tilde{\in} b :\equiv \exists Y(n(b, Y) \wedge a \in Y)$, but the extensionality of $\tilde{\in}$ means that n is one-to-one.

3 The elementary theory

Although the language $\mathcal{L}_{TN}$ looks like the language of analysis, and might hence give the impression to be a priori safe from any set-theoretic consistency problems, one must be careful in the choice of type existence axioms because of the naming predicate n, which enables to express the existence of types of higher level, e.g. the powerset of a given A by $x \in \text{Pow}(A) \leftrightarrow \exists Y(n(x, Y) \wedge Y \subset A)$. In fact, assuming unbounded comprehension, it is possible to mimic Russell's paradox as follows.

Example 1. Assume the existence of a type $\mathbf{R}$ containing all objects, which are not elements of the type they represent: $\forall x\,(x \in \mathbf{R} \leftrightarrow \exists X(n(x, X) \wedge x \notin X))$. By (O_2) $\mathbf{R}$ has a name, say r, which represents no other types but $\mathbf{R}$; it follows that $r \in \mathbf{R} \leftrightarrow r \notin \mathbf{R}$.

We could avoid these inconsistency problems limiting comprehension to so-called *stratified* formulae: those formulae of $\mathcal{L}_{TN}$ which do not contain the naming predicate n neither in subformulae, nor in indices of constants $\mathbf{c}_\varphi$, etc. But this restriction is not strong enough to avoid impredicative type definitions.

So let us define a sufficiently small class of formulae, the *elementary* formulae, with respect to which we are going to do comprehension both without running into such inconsistencies and without leaving the field of predicativity. Loosely speaking elementary formulae are those $\mathcal{L}_{TN}$-formulae φ, where bound type variables and formulae of the form $n(s, A)$ or $U(A)$ do not appear neither as subformulae of φ, nor in the indices of constants $\mathbf{c}_\chi$, nor ...

Definition 3. The *elementary object terms* and the *elementary formulae* are defined inductively as follows:

1) the free variables and the constants $0, \mathbf{s}_N, \mathbf{p}_N, \mathbf{p}, \mathbf{p}_0, \mathbf{p}_1, \mathbf{d}, \mathbf{j}, \mathbf{s}, \mathbf{k}$ are elementary object terms,
2) if φ is an elementary formula, then $\mathbf{c}_\varphi$ is an elementary object term,
3) elementary object terms are closed against application,
4) if s, t are elementary object terms, then $s = t$, $s\downarrow$ and $N(s)$ are elementary formulae,
5) if s is an elementary object term and A is a type term, then $s \in A$ is an elementary formula,
6) if A and B are type terms, then $A = B$ is an elementary formula,
7) elementary formulae are closed under propositional connectives and quantification over objects.

Definition 4. If $\varphi \equiv \varphi_0[a_0, \underline{b}_0, \underline{C}_0]$ is an elementary formula with all free variables among the indicated, then the *comprehension axiom* for φ has the form

$$(CA_\varphi) \qquad n(\underline{c}, \underline{C}) \rightarrow \exists Z\,(n(\mathbf{c}_\varphi \underline{b}\underline{c}, Z) \wedge \forall x(x \in Z \leftrightarrow \varphi_0[x, \underline{b}, \underline{C}])).$$

We refer to the whole schema for all elementary formulae by ECA. [2] The basic theory about types and names we will start with is

$$ETN :\equiv \text{Logical axioms and rules for } \mathcal{L}_{TN} + \text{Combinatorial axioms} +$$

[2] Here a slight oversimplification becomes visible: take for example the formula $\varphi \equiv a = 0 \wedge b = 1$ with its associated constant $\mathbf{c}_\varphi$; it is not clear from our definitions whether $\mathbf{c}_\varphi 0$ should be a name of $\{ x \mid x = 0 \wedge 0 = 1 \} = \emptyset$ or of $\{ x \mid 0 = 0 \wedge x = 1 \} = \{1\}$. In a precise development some information about the comprehension variable and about the order of the parameters must be supplyed to the $\mathbf{c}_\varphi$.

Pairing and projections axioms + Natural numbers axioms +
Definition by cases axioms + Ontological axioms +
ECA

Example 2. Let $\nu \equiv \nu_0[a]$ be the formula $N(a)$. ν is elementary and has no parameters. Hence *ETN* proves the existence of a type $\mathbf{N}$ containing exactly the natural numbers, which is represented by $\mathbf{c}_\nu$: $n(\mathbf{c}_\nu, \mathbf{N}) \wedge \forall x(x \in \mathbf{N} \leftrightarrow N(x))$.

Remark. The consistency of *ETN* can be checked building what we might call the *standard model* of *ETN*: the sort of the object is interpreted as the set of the natural numbers ω, application is defined as $a \cdot b :\simeq \{a\}(b)$; N is the whole carrier ω; the sort of the types is interpreted as the class of arithmetical subsets of ω, $\in$ as membership and $n(a, B)$ means that the number a codes an arithmetical definition of the set B.

3.1 Type constructions

The only type terms of $\mathcal{L}_{TN}$ are type variables; we are not allowed to build terms of the form $\{\, x \mid \varphi[x] \,\}$. This makes metamathematical analysis easier, but forces us to use long periphrases for certain simple statements: instead of $\chi[\{\, x \mid \varphi[x] \,\}]$ we have to write $\exists X(\forall x(x \in X \leftrightarrow \varphi[x]) \wedge \chi[X])$. In presence of the elementary comprehension scheme it is easy to show that the language $\mathcal{L}_{TN}$ can be expanded to a language $\mathcal{L}'_{ETN}$ with type terms of the form $\{\, x \mid \varphi[x] \,\}$ for elementary formulae $\varphi[a]$, such that *ETN'*, i.e. *ETN* formulated in $\mathcal{L}'_{ETN}$ is a conservative extension of *ETN*. In the following we will switch between $\mathcal{L}'_{ETN}$ and $\mathcal{L}_{TN}$ according to what seems more convenient in the specific circumstances.

We have already seen that we have the type of natural numbers, here are some types whose existence can be proven in *ETN* ($S, T, \ldots$ are syntactic variables for $\mathcal{L}'_{ETN}$ type terms): the all-type $\mathbf{V} :\equiv \{\, x \mid x = x \,\}$, the cartesian product $S \times T$, the partial function space $S \rightharpoonup T :\equiv \{\, z \mid \forall x(x \in S \wedge zx{\downarrow} \to zx \in T) \,\}$, the bounded product $\prod_{x:S} T[x]$ and the bounded sum $\sum_{x:S} T[x]$, the bounded intersection $\bigcap_{x:S} T[x]$ and the bounded union $\bigcup_{x:S} T[x]$, etc.

4 Induction

When proving properties of programs involving some inductively generated data type as the natural numbers, which we have chosen as an example, the main tool we can use is induction. We will consider two well-known ways to formulate induction: the induction axiom

$$(T\text{-}ind) \qquad 0 \in A \wedge \forall x(x \in A \to \mathbf{s}_N x \in A) \to \forall x(N(x) \to x \in A)$$

and the induction scheme *F-ind*, which consists of

$$(F\text{-}ind_\varphi) \qquad \varphi[0/a] \wedge \forall x(\varphi[x/a] \to \varphi[\mathbf{s}_N x/a]) \to \forall x(N(x) \to \varphi[x/a])$$

for all $\mathcal{L}_{TN}$ formulae φ.

Lemma 5. *For all $\mathcal{L}_{TN}$ formulae χ the following holds:*

1. $ETN + T\text{-}ind \vdash \chi \implies ETN + F\text{-}ind \vdash \chi$,
2. $ETN + T\text{-}ind \vdash \chi \iff ETN + (F\text{-}ind_\varphi)_{\varphi \text{ elementary}} \vdash \chi$.

Starting with 0, successor $\mathbf{s}_N$, predecessor $\mathbf{p}_N$ and definition by cases $\mathbf{d}$, and using the fixed-point theorem 2, in order to solve the recursion equations arising from the defining equations, we yield with *T-ind*:

Lemma 6. *Every primitive recursive function can be represented by an object term of* $\mathcal{L}_{TN}$.

4.1 Metamathematical results

The following results can easily be established (see [11]).

Lemma 7. *1.* $ETN + T\text{-}ind \equiv (\Pi^1_0\text{-}CA){\restriction}$,
2. $ETN + F\text{-}ind \equiv (\Pi^1_0\text{-}CA)$.

Due to the equivalence between Feferman's theory EM_0 [1, 3] and $ETN + F\text{-}ind$ lemma 7 has the following corollary.

Corollary 8. $EM_0 \equiv (\Pi^1_0\text{-}CA)$.

5 Universes

Feferman has introduced universes in his theories for explicit mathematics in order to prove Hancock's conjecture: in [4] he introduces theories $T'_{0,n}$ with n universes $V_0, \ldots, V_{n-1}$ where ML_n can be simulated.

The admissible sets required axiomatically in the subsystems of set theory studied by Jäger in [10] also can be viewed as universes. In KPi^0, for example, the admissibles are introduced not via constants of the language, but by a so-called *limit axiom*, which states that every set is element of an admissible.

In the following we present a theory of universes, types and names, *UTN*, which treats universes in an explicit framework, by means of a limit axiom.

5.1 The theory *UTN*

UTN is an extension of *ETN* by additional axioms (J), (O_3)–(O_6), (C_φ), (C_j) and (Lim). The first is *join*, which says that, if we are given an index type A and a construction f such that for all $x \in A$ fx is the name of a type, then we can build the disjoint union of this family of types. Let $\Sigma(A, f, B)$ be the formula

$$\forall x(x \in B \leftrightarrow x = \mathbf{p}(\mathbf{p}_0 x)(\mathbf{p}_1 x) \wedge \mathbf{p}_0 x \in A \wedge \exists X(n(f(\mathbf{p}_0 x), X) \wedge \mathbf{p}_1 x \in X)).$$

The join axiom has the form

$$(J) \qquad n(a, A) \wedge \forall x \in A \exists Y n(fx, Y) \rightarrow \exists Z(n(\mathrm{j}af, Z) \wedge \Sigma(A, f, Z)).$$

Ontological axioms. The following axiom expresses the fact that the elements of universes are (names of) types:

(O_3) $U(A) \wedge b \in A \to \exists X n(b, X)$,

without this we could interpret the predicate U trivially as $\{\mathbf{V}\}$. The universes are pairwise comparable with respect to the modified membership $\dot{\in}$:

(O_4) $U(A) \wedge U(B) \to A \dot{\in} B \vee A = B \vee B \dot{\in} A$.

Also a kind transitivity holds: if a universe A is element of a universe B in the sense of $\dot{\in}$, then A is also included B,

(O_5) $U(A) \wedge U(B) \wedge A \dot{\in} B \to A \subset B$.

Furthermore there is no empty universe, each one contains the canonical name for the natural numbers:

(O_6) $U(A) \to \mathrm{c}_\nu \in A$.

Closure properties. Universes are closed against elementary comprehension and join: we have for all elementary formulae $\varphi \equiv \varphi_0[a, \underline{b}_0, \underline{C}_0]$

(C_φ) $U(D) \wedge \underline{c} \in D \to \mathrm{c}_\varphi \underline{b}\underline{c} \in D$

(C_j) $U(D) \wedge a \in D \wedge n(a, A) \wedge (\forall x \in A) f x \in D \to \mathbf{j} a f \in D$

Remark. From the ontological axioms and the closure properties it follows that the universes are linearly ordered by $\dot{\in}$. The main point is the irreflexivity of $\dot{\in}$: assume we have $D \dot{\in} D$ for a universe D. Then we can join the identity **skk** over D to get the restriction of the relation $\tilde{\in}$ to D as a type in D itself. This leads to a contradiction in the same way as the assumption of the existence of the type of all types in explicit mathematics [2].

In fact, as far as the proof-theoretic strength is concerned, the linear ordering of the universes could be replaced by upper semilattice. This means that axiom (O_4) might be replaced by a weaker one, which would still allow us to prove the lower bound of the theory. We state this axiom as a corollary:

Corollary 9. $U(D_0) \wedge U(D_1) \wedge a_0 \in D_0 \wedge a_1 \in D_1 \to \exists Z(U(Z) \wedge a_0 \in Z \wedge a_1 \in Z)$

The limit axiom. None of the axioms of *UTN* given above really requires the existence of universes, for we could satisfy them all even setting U always false. Here is the axiom which forces the existence of universes: every type is $\dot{\in}$-element of some universe

(Lim) $\exists Z\,(U(Z) \wedge A \dot{\in} Z)$.

This completes the list of the axioms of *UTN*.

5.2 A lower bound for *UTN* + *T-ind*: *ATR*$_0$

The goal of this section is to give a lower bound for the proof-theoretic strength of *UTN* + *T-ind* by means of an interpretation of Friedman's theory ATR_0 of arithmetical transfinite recursion [7]; a great deal of classical mathematics can be formalised in ATR_0, e.g. the reduction theorem for Π_1^1 relations, the fact that every Σ_1^1 set of reals is Lebesgue measurable, etc. The proof-theoretic analysis ATR_0 is done in [8] and shows that its ordinal strength is Γ_0.

ATR_0 is formulated in the language $\mathcal{L}_{pr}^2$ of second order arithmetic, which contains function symbols for primitive recursive sequence numbers: the sequence number built out of numbers $x_1, \ldots, x_n$ will be denoted by $\langle x_1, \ldots, x_n \rangle$ and the i-th component of a sequence number y by $(y)_i$; accordingly a sequence of sets can be coded into a single one and $t \in (C)_s$ will be used as a shorthand for $\langle s, t \rangle \in C$.

Let $WO(C)$ be a Π_1^1 formula expressing that C codes a well ordering with least element 0. The concrete form of $WO(C)$ is not important for the following: we will just need that the corresponding formulae $s \in \text{supp}(C)$ and $s \prec_C t$, expressing respectively that the term s belongs to the support of the well ordering C and that s precedes t in the well ordering C, are arithmetical. Also we will write $(A)_{\prec_C m}$ for $\{ y \mid \exists x (x \prec_C m \wedge \langle x, y \rangle \in A \}$.

The mathematical axioms of ATR_0 are those of primitive recursive arithmetic PRA, the schema of arithmetical comprehension, the induction axiom and the schema of arithmetical transfinite recursion:

$$\forall Z \, (WO(Z) \rightarrow \forall X \exists Y \, H_{\varphi,a,B}(Z, X, Y)) \quad (\varphi \text{ arithmetical, } a, B \text{ free variables})$$

where $H_{\varphi,a,B}(Z, X, Y)$ means

1. $(Y)_0 = X$ and
2. $0 \prec_Z m \rightarrow (Y)_m = \{ x \mid \varphi[x/a, (Y)_{\prec_Z m}/B] \}$.

The embedding of ATR_0 into $UTN + T\text{-}ind$ We are going now to define the translation "*" of the language of ATR_0 into that of $UTN + T\text{-}ind$ starting from the standard translation of the object terms $s \mapsto s^*$ and assuming both languages to have the same variables. Let $SN(A)$ be an abbreviation of $\forall z (z \in A \rightarrow N(z))$, which will be useful for in order to relativise the type quantifiers to the subtypes of $\mathbf{N}$. The clauses for atomic formulae and propositional connectives are straightforward and the quantifiers are treated as follows:

$$(\forall x \varphi[x/a])^* :\equiv \forall x (N(x) \rightarrow \varphi^*[x/a]) \quad (\forall X \varphi[X/A])^* :\equiv \forall X (SN(X) \rightarrow \varphi^*[X/A])$$
$$(\exists x \varphi[x/a])^* :\equiv \exists x (N(x) \wedge \varphi^*[x/a]) \quad (\exists X \varphi[X/A])^* :\equiv \exists X (SN(X) \wedge \varphi^*[X/A]).$$

Lemma 10. *Let φ be an arithmetical formula of $\mathcal{L}_{pr}^2$. Then φ^* is an elementary formula of $\mathcal{L}_{TN}$.*

Lemma 11. *For all elementary $\mathcal{L}_{TN}$-formulae φ $UTN + T\text{-}ind$ proves:*

$$SN(C) \wedge WO^*(C) \wedge (\forall x \in \mathbf{N} (\forall y (y \prec_C x \rightarrow \varphi[y/a]) \rightarrow \varphi[x/a]) \rightarrow \forall x \in \mathbf{N} \varphi[x/a]).$$

Proof. Since φ is elementary there is a type $\{ x \in \mathbf{N} \mid \varphi[x/a] \} \subset \mathbf{N}$. Now, since $WO^*(Z)$ means that Z codes a well ordering with respect to the subtypes of the natural numbers, the claim is easily established.

Theorem 12. *For every formula φ of $\mathcal{L}^2_{pr}$ with all its free variables among $\underline{a}, \underline{B}$, if $ATR_0 \overset{n}{\vdash} \varphi$ then $UTN + T\text{-}ind \vdash N(\underline{a}) \wedge SN(\underline{B}) \to \varphi^*$.*

Proof. By induction on n. The translation of the axioms of PRA is obviously provable by the properties of our standard translation of the terms. Arithmetical comprehension in $\mathcal{L}^2_{pr}$ translates into elementary comprehension in $\mathcal{L}_{TN}$, the induction axiom of the first language into the induction axiom of the second. Let us now turn to the schema of arithmetical transfinite recursion: suppose for simplicity that φ is an arithmetical formula with parameters a, B; we have to show that the following is provable in $UTN + T\text{-}ind$

$$\forall Z\left(SN(Z) \wedge WO^*(Z) \to \forall X\left(SN(X) \to \exists Y(SN(Y) \wedge H^*_{\varphi^*,a,B}(Z,X,Y))\right)\right).$$

The idea of the proof is to construct a name for the hyperjump hierarchy Y by means of the fixed-point theorem 2. Assume that Z is a subtype of the natural numbers and a well ordering. By the ontological axiom (O_2) there is a name z of Z and by the limit axiom a universe D_0 such that $z \in D_0$. Let X be also a type with $SN(X)$, x such that $n(x, X)$ and D_1 with $U(D_1) \wedge x \in D_1$. Hence by corollary 9 there exists a universe D such that $x \in D \wedge z \in D$.

Since $a \prec_Z b$, $\mathrm{supp}(Z)$ and φ are arithmetical formulae, their translations are elementary, and the elementary comprehension scheme gives us constants $\mathbf{c}_\rho, \mathbf{c}_\sigma, \mathbf{c}_\varphi$ such that:

$$n(\mathbf{c}_\rho mz, \{\, u \in \mathbf{N} \mid u \prec_Z m \,\})$$
$$n(\mathbf{c}_\sigma z, \mathrm{supp}(Z))$$
$$n(c, C) \to n(\mathbf{c}_\varphi c, \{\, u \in \mathbf{N} \mid \varphi^*[u/a, C/B] \,\}).$$

What we are looking for is an object f, such that for all $m \in \mathrm{supp}(Z)$ fm is a name for $(Y)_m$; this means that f should satisfy

$$fm \simeq \mathbf{d}m0x(c_\varphi(\mathbf{j}(c_\rho mz)f)),$$

in order to take the join of it over $\mathrm{supp}(Z)$ and obtain a name of the hyperjump hierarchy

$$y := \mathbf{j}(\mathbf{c}_\sigma z)f.$$

Actually the point is not the existence of such an f, which follows immediately from the fixed-point theorem, but the fact that y must represent a type; in order to show it using (J), we have to prove

$$(\star) \qquad \forall m \in \mathrm{supp}(Z) \exists Y\, n(fm, Y).$$

But the necessary induction can not be performed since the induction formula is not elementary. Here the universes come into play: we can prove by elementary induction along Z that

$$\forall m \in \mathrm{supp}(Z) fm \in D,$$

which by (O_3) implies $(\star)$.

For $m = 0$ we have $fm \simeq x \in D$.

For $0 \prec_z m$ we know from the induction hypothesis that $(\forall n \prec_z m) fn \in D$. Because of the closure of the universes against elementary comprehension and join we get $c_\rho mz \in D$ and $\mathbf{j}(c_\rho mz)f \in D$. It follows from the closure axiom (C_φ) that

$$fm \simeq \mathbf{d}m0x(c_\sigma(\mathbf{j}(c_\rho mz)f)) \simeq c_\sigma(\mathbf{j}(c_\rho mz)f) \in D,$$

which completes the induction along Z.

Now we can apply $(C_\mathbf{j})$ to obtain $y \in D$ and, by (O_3), the existence of a Y such that $n(y, Y)$ and $\Sigma(\mathrm{supp}(Z), f, Y)$. It is easy to verify that

$$(Y)_0 = X \text{ and } 0 \prec_z m \rightarrow (Y)_m = \{\, n \mid \varphi[n/a, (Y)_{\prec_z m}/B] \,\},$$

hence the type Y, that we have constructed is in fact the hyperjump hierarchy of φ along Z.

5.3 An upper bound for *UTN* + *T-ind*

In the following we outline the proof of the upper bound for the proof-theoretic strength of *UTN* + *T-ind*, which does not fit here in full detail. The idea is similar to that of Jäger's analysis of KPi^0 [10].

In the first step we formulate a Tait calculus for *UTN* + *T-ind*. We say that a formula φ of the language of the Tait calculus is Σ^1 (Π^1), if it does not contain universal (resp. existential) quantifiers ranging over types; the axioms of *UTN* + *T-ind* are all translated into the language of the Tait calculus, except the join axiom, which is split up into three rules in such a way that cut elimination up to Σ^1 / Π^1 formulae becomes provable.

For the second step we have to introduce variants of *UTN* with finitely many universes. To formulate them expand the language by a countable list of type constants $\mathbf{D}_0, \mathbf{D}_1, \mathbf{D}_2, \ldots$ Now let $UTN(k)$ + *T-ind* be *UTN* + *T-ind* where the limit axiom is replaced by axioms:

$$U(\mathbf{D}_0), \ldots, U(\mathbf{D}_{k-1}) \text{ and } \mathbf{D}_0 \,\dot{\in}\, \mathbf{D}_1 \,\dot{\in}\, \ldots \,\dot{\in}\, \mathbf{D}_{k-1}.$$

For arbitrary natural numbers m, n we define an asymmetric interpretation $\varphi^{(m,n)}$ of a formula φ, bounding the universal type quantifiers by $\mathbf{D}_m$ and the existential type quantifiers by $\mathbf{D}_n$, as follows

$$\forall X \varphi[X] \mapsto \forall X \,\dot{\in}\, \mathbf{D}_m \varphi^{(m,n)}[X], \qquad \exists X \varphi[X] \mapsto \exists X \,\dot{\in}\, \mathbf{D}_n \varphi^{(m,n)}[X].$$

Then the following crucial result can be proved for $s > m + 2^n$:

$$UTN + T\text{-}ind \mid\!\frac{n}{1}\; \Gamma[\underline{A}] \implies UTN(s) + T\text{-}ind \vdash \underline{A} \,\dot{\notin}\, \mathbf{D}_m, \Gamma[\underline{A}]^{(m,m+2^n)},$$

i.e. if a set of formulae $\Gamma[\underline{A}]$ with all free type variables among $\underline{A}$ can be proved in *UTN* + *T-ind* with length n and at most Σ^1 / Π^1 cuts, then the asymmetric interpretation of $\Gamma[\underline{A}]$ can already be proved in the theory $UTN(s)$ + *T-ind* with s universes for a sufficiently large s.

In the third step we reduce the theory $UTN(s)$ + *T-ind* to $\widehat{ID}_{s+1}$, the theory of $s+1$ times iterated inductive definitions, in which only the fixed-point property is asserted (instead of closure + induction principle), in analogy to [4].

Hence $|UTN + T\text{-}ind| \leq \sup_{s \in \omega} |\widehat{ID}_{s+1}| = \Gamma_0.$

Together with the result of section 5.2, this shows that

Theorem 13. $|UTN + T\text{-}ind| = \Gamma_0.$

References

1. M. J. Beeson. *Foundations of Constructive Mathematics.* Springer, Berlin, 1984.
2. S. Feferman. A language and axioms for explicit mathematics. In *Algebra and Logic*, volume 450 of *LNM*, Berlin, 1975. Springer.
3. S. Feferman. Constructive theories of functions and classes. In Boffa, editor, *Logic Colloquium '78*, Amsterdam, 1978. North-Holland.
4. S. Feferman. Iterated inductive fixed-point theories: application to Hancock's conjecture. In Metakides, editor, *Patras Logic Symposion*, Amsterdam, 1982. North-Holland.
5. S. Feferman. Logics for termination and correctness of functional programs, II. Logics of Strength PRA. Lecture notes for the Summer School and Conference on Proof Theory in Leeds, 30 July - 2 August 1990.
6. S. Feferman. Logics for termination and correctness of functional programs. In Moschovakis, editor, *Logic from Computer Science*, New York, 1992. Springer.
7. H. Friedman. Systems of second order arithmetic with restricted induction (abstracts). *Journal for Symbolic Logic*, 41, 1976.
8. H. M. Friedman, K. McAloon, and S. G. Simpson. A finite combinatorial principle which is equivalent to the 1-consistency of predicative analysis. In Metakides, editor, *Patras Logic Symposion*, Amsterdam, 1982. North-Holland.
9. R. Harper and R. Pollack. Type checking with universes. *Theoretical Computer Science*, 89, 1991.
10. G. Jäger. The strength of admissibility without foundation. *Journal for Symbolic Logic*, 49(3), 1984.
11. G. Jäger. Induction in the elementary theory of types and names. In Börger, editor, *Proceedings CSL '87.* Springer, 1987. LNCS 329.
12. R. Kahle. Theorien von Operationen und Klassen zur Beschreibung von ML-Programmen. Master's thesis, Mathematisches Institut, Universität München, 1993.
13. P. Martin-Löf. An intuitionistic theory of types: predicative part. In Rose and Shepherdson, editors, *Logic Colloquium '73*, Amsterdam, 1975. North-Holland.
14. P. Martin-Löf. *Intutionistic Type Theory.* Bibliopolis, 1984.
15. J. C. Mitchell and R. Harper. The essence of ML. In *Proc. of the 15th Annual ACM SIGACT-SIGPLAN Symp. on Principles of Programming Languages.* ACM, 1988.
16. B. Nordström, K. Petersson, and J. M. Smith. *Programming in Martin-Löf's Type Theory: An Introduction.* Clarendon Press, Oxford, 1990.
17. E. Palmgreen. A domain interpretation of Martin-Löf's partial type theories with universes. Technical Report UUDM 1989:11, University of Uppsala, Department of Mathematics, 1989.
18. T. Strahm. Theorien mit Selbst-Applikation. Technical Report IAM-PR-90350, Institut für Informatik und angewandte Mathematik, Universität Bern, oct 1990.
19. C. Talcott. A theory for program and data type specification. *Theoretical Computer Science*, 104(1), Oct. 1992.

Notes on Sconing and Relators

John C. Mitchell* and Andre Scedrov**

[1] Department of Computer Science
Stanford University
Stanford, CA 94305
[2] Department of Mathematics
University of Pennsylvania
Philadelphia, PA 19104-6395

Abstract. This paper describes a semantics of typed lambda calculi based on relations. The main mathematical tool is a category-theoretic method of sconing, also called glueing or Freyd covers. Its correspondence to logical relations is also examined.

1 Introduction

Many modern programming languages feature rather sophisticated typing mechanisms. In particular, languages such as ML include *polymorphic* data types, which allow considerable programming flexibility. Several notions of polymorphism were introduced into computer science by Strachey [Str67], among them the important notion of *parametric* polymorphism. Strachey's intuitive definition is that a polymorphic function is parametric if it has a uniformly given algorithm in all types, that is, if the function's behavior is independent of the type at which the function is instantiated. Reynolds [Rey83] proposed a mathematical definition of parametric polymorphic functions by means of invariance with respect to certain relations induced by types. Unfortunately, the mathematical tools used in [Rey83] were limited to a standard set-theoretic framework, which Reynolds later showed to be vacuous in the case of so-called second order polymorphic lambda calculus (see Pitts [Pit87]).

We describe a general mathematical framework for semantics of type disciplines based on relations. This framework, which is also proposed independently by Ma and Reynolds [MaR92], may be used to give a mathematically correct notion of relational invariance proposed in [Rey83] and a rigorous framework for the considerations in Wadler [Wad89]. The relational semantic framework we describe is also useful in investigating other aspects of programming languages. Abramsky and Jensen [AbJ91] use it for strictness analysis and O'Hearn and Tennent [OHT93] use it in studying semantics of local variables.

The main mathematical tool used in this work is the category-theoretic method of *sconing* described in Freyd and Scedrov [FrS90] and also called glueing or Freyd covers, see Lambek and Scott [LS86]. To our knowledge, the first application of this method to

* jcm@cs.stanford.edu Supported in part by an NSF PYI Award, matching funds from Digital Equipment Corporation, the Powell Foundation, and Xerox Corporation; and the Wallace F. and Lucille M. Davis Faculty Scholarship.

** andre@cis.upenn.edu Scedrov is an American Mathematical Society Centennial Research Fellow. He is partially supported by NSF Grant CCR-91-02753 and by ONR Grants N00014-88-K-0635 and N00014-92-J-1916.

type disciplines is given in Appendix C of Lafont [Laf88]. In the case of simple types, this method corresponds closely to so-called *logical relations*, described for instance in Plotkin [Plo80], Statman [Sta85], and Mitchell [Mit90]. This correspondence is examined in detail. In the case of polymorphic types, a central role is played by *relators*, *i.e.*maps that take objects to objects and relations to relations. Sconing may be used to express relators as functors.

In order to make the methods and arguments accessible to readers with minimal background in category theory, we choose a much more relaxed style of presentation than would be usual in a research paper. In this vein, here we discuss only the basic framework for simple types and for implicit polymorphism. The extension to explicit polymorphism should be routine to readers thoroughly familiar with [Pit87, See87, Gir86, Wad89]. (Note to specialists: the extension to second order polymorphism differs significantly from second order logical relations proposed by Mitchell and Meyer [MM85]; we shall take this up elsewhere.)

It is a pleasure to acknowledge the collaboration with Samson Abramsky and Philip Wadler in the initial stages of this work. We would like to thank John Reynolds for many inspiring discussions of polymorphism.

2 Simply typed lambda calculus

2.1 Types and terms

We will consider typed lambda calculus with terminator (terminal object) $\mathbf{1}$, product and function types, and any set S of type constants $b_1, b_2, \ldots$. We will use the symbol "$\Rightarrow$" in function types, and reserve the single arrow "$\rightarrow$" for morphisms in a category (including functions between sets). The well-formed terms over any set of types are defined using the subsidiary notion of type assignment. A *type assignment* Γ is a finite set of formulas $x{:}\sigma$ associating types to variables, with no variable x occurring twice. We write $\Gamma, x{:}\sigma$ for the type assignment

$$\Gamma, x{:}\sigma = \Gamma \cup \{x{:}\sigma\},$$

where, in writing this, we assume that x does not appear in Γ. Terms will be written in the form $\Gamma \triangleright M{:}\sigma$, which may be read, "$M$ has type σ relative to Γ." The well-formed terms are given by the following rules, where we assume $*$ is a constant symbol of type $\mathbf{1}$. For simplicity, we will assume that this is the only constant symbol.

(*var*) $$x{:}\sigma \triangleright x{:}\sigma$$

(*cst*) $$\emptyset \triangleright *{:}\mathbf{1}$$

($\Rightarrow E$) $$\frac{\Gamma \triangleright M{:}\sigma \Rightarrow \tau,\ \Gamma \triangleright N{:}\sigma}{\Gamma \triangleright MN{:}\tau}$$

($\Rightarrow I$) $$\frac{\Gamma, x{:}\sigma\ \triangleright M{:}\tau}{\Gamma \triangleright \lambda x{:}\sigma.M{:}\sigma \Rightarrow \tau}$$

($\times\ E$) $$\frac{\Gamma \triangleright M{:}\sigma \times \tau}{\Gamma \triangleright \mathbf{Proj}_1^{\sigma,\tau} M{:}\sigma,\ \Gamma \triangleright \mathbf{Proj}_2^{\sigma,\tau} M{:}\tau}$$

($\times\ I$) $$\frac{\Gamma \triangleright M{:}\sigma,\ \Gamma \triangleright N{:}\tau}{\Gamma \triangleright \langle M, N\rangle{:}\sigma \times \tau}$$

(*add hyp*) $$\frac{\Gamma \triangleright M{:}\tau}{\Gamma, x{:}\sigma\ \triangleright M{:}\tau}$$

2.2 Equations

Given our formulation of terms, it is natural to write equations in the form

$$\Gamma \triangleright M = N : \tau$$

where we assume that $\Gamma \triangleright M{:}\,\tau$ and $\Gamma \triangleright N{:}\,\tau$. The equational axioms and inference rules are as follows, where $[N/x]M$ denotes substitution of N for x in M.

(*one*) $$\Gamma \triangleright x = * : \mathbf{1}$$

($\mathbf{Proj}_1$) $$\Gamma \triangleright \mathbf{Proj}_1\langle M, N\rangle = M : \sigma$$

($\mathbf{Proj}_2$) $$\Gamma \triangleright \mathbf{Proj}_2\langle M, N\rangle = N : \tau$$

(**Pair**) $$\Gamma \triangleright \langle \mathbf{Proj}_1\, M, \mathbf{Proj}_2\, M\rangle = M : \sigma \times \tau$$

(α) $$\Gamma \triangleright \lambda x{:}\,\sigma.M = \lambda y{:}\,\sigma.[y/x]M : \sigma \Rightarrow \tau, \text{ provided } y \notin FV(M)$$

(β) $$\Gamma \triangleright (\lambda x{:}\,\sigma.M)N = [N/x]M : \tau$$

(η) $$\Gamma \triangleright \lambda x{:}\,\sigma.(Mx) = M : \sigma \Rightarrow \tau, \text{ provided } x \notin FV(M)$$

(*ref*) $$\Gamma \triangleright M = M : \sigma$$

(*sym*) $$\frac{\Gamma \triangleright M = N : \sigma}{\Gamma \triangleright N = M : \sigma}$$

(*trans*) $$\frac{\Gamma \triangleright M = N : \sigma, \ \Gamma \triangleright N = P : \sigma}{\Gamma \triangleright M = P : \sigma}$$

(ξ) $$\frac{\Gamma, x{:}\,\sigma \triangleright M = N : \tau}{\Gamma \triangleright \lambda x{:}\,\sigma.M = \lambda x{:}\,\sigma.N : \sigma \Rightarrow \tau} \cdot$$

(μ) $$\frac{\Gamma \triangleright M_1 = M_2 : \sigma \Rightarrow \tau, \ \Gamma \triangleright N_1 = N_2 : \sigma}{\Gamma \triangleright M_1 N_1 = M_2 N_2 : \tau}$$

(*add hyp*) $$\frac{\Gamma \triangleright M = N : \sigma}{\Gamma, x{:}\,\tau \triangleright M = N : \sigma}$$

3 Cartesian closed categories

A *cartesian closed category* (or *ccc*) is a category with specified terminal object, products and exponentials. This means that a category **C** is cartesian closed only if the following additional data are provided and regarded as part of the structure:

- An object $\mathbf{1}$ with unique arrow $O^A{:}\,A \to \mathbf{1}$ for each object A,
- A binary object map $\times$ with, for any objects A, B and C, specified arrows $\mathbf{Proj}_1^{A,B}$, $\mathbf{Proj}_2^{A,B}$, and map $\langle \cdot, \cdot \rangle_{C,A,B}$ on arrows such that for every $f{:}\,C \to A$ and $g{:}\,C \to B$, the arrow $\langle f, g\rangle_{C,A,B}{:}\,C \to A \times B$ is the unique h satisfying

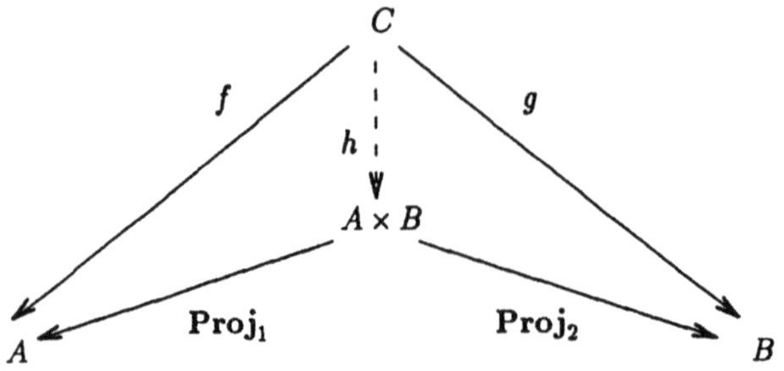

- A binary object map $\Rightarrow$ with, for any objects A, B and C, a specified arrow $\mathbf{App}^{A,B}$ and map $\mathbf{Curry}^{A,B,C}$ on arrows such that for every $f: C \times A \rightarrow B$, the arrow $\mathbf{Curry}^{A,B,C}(f)$ is the unique h satisfying

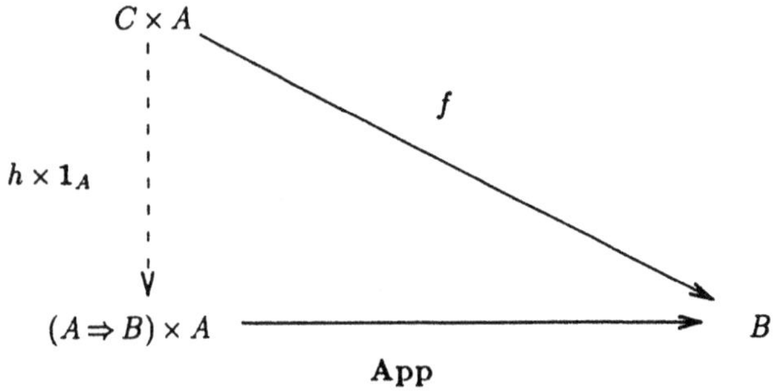

A category $\mathbf{C}$ is *well-pointed* if, for any $f, g: A \rightarrow B$, we have $f = g$ iff for every $x: \mathbf{1} \rightarrow A$, we have $f \circ x = g \circ x$. Another phrase for "well-pointed" is *generated by* $\mathbf{1}$.

Functors that preserve cartesian closed structure will play an important role in our discussion. A *representation of cartesian closed categories* (or simply *ccc-representation*) is a functor $\mathbf{F}: \mathbf{C} \longrightarrow \mathbf{D}$ from one cartesian closed category to another that preserves terminator, products and exponentials (function spaces). In the literature, ccc-representations are sometimes called cartesian closed functors, or cc-functors.

4 Sconing

Let $\mathbf{C}$ be a category with a terminator (terminal object) $\mathbf{1}$. We will define a functor $|\cdot|$ from $\mathbf{C}$ to *Set*, the category of sets and functions, which associates to each object its *set of global elements.* For any object A in $\mathbf{C}$, let $|A|$ be the set of all morphisms $\mathbf{1} \rightarrow A$ in $\mathbf{C}$. For any morphism $x: A \rightarrow A'$ in $\mathbf{C}$ let $|x|: |A| \rightarrow |A'|$ be the function defined by composition, according to the following diagram.

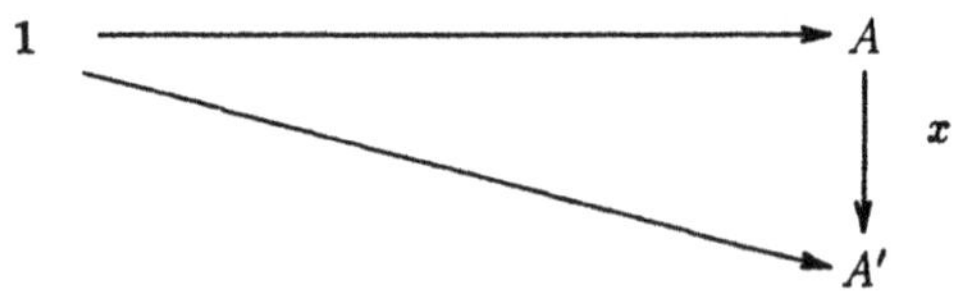

It is not hard to see that the functor $|\cdot|$ from $\mathbf{C}$ to *Set* preserves products up to isomorphism. (The functor also preserves equalizers up to isomorphism, but we shall not use this fact.)

The category $\hat{\mathbf{C}}$, called the *scone of* $\mathbf{C}$, is defined as follows. The objects of $\hat{\mathbf{C}}$ are triples $\langle S, f, A\rangle$, where S is a set, A is an object of $\mathbf{C}$, and $f\colon S \to |A|$ is a function (in *Set*). The morphisms $\langle S, f, A\rangle \to \langle S', f', A'\rangle$ in $\hat{\mathbf{C}}$ are pairs $\langle t, x\rangle$, where $t\colon S \to S'$ is a function and $x\colon A \to A'$ is a morphism in $\mathbf{C}$ such that:

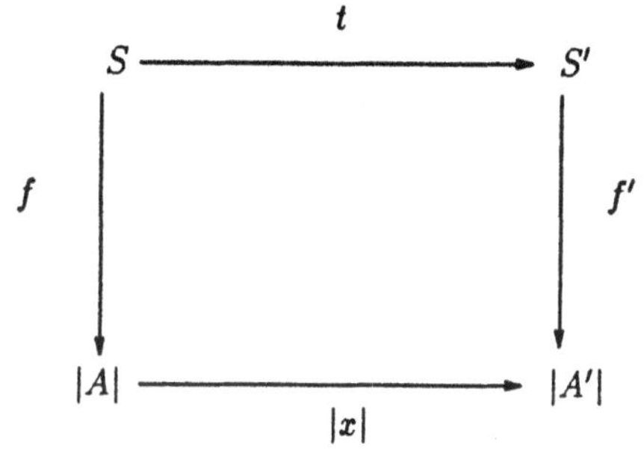

The reader may easily verify that $\langle |\mathbf{1}|, id, \mathbf{1}\rangle$ is a terminator in $\hat{\mathbf{C}}$. There is a canonical "forgetful" functor $\pi\colon \hat{\mathbf{C}} \longrightarrow \mathbf{C}$ which takes $\langle S, f, A\rangle$ to A and $\langle t, x\rangle$ to x. Category theorists will recognize the scone as a special case of a *comma category*, namely $\hat{\mathbf{C}} = (id \downarrow |\cdot|)$, where id is the identity functor on *Set*; see, *e.g.*, [McL71]. It is common to overload notation and use the symbol *Set* for the identity functor from sets to sets (and similarly for other categories). With this convention, we may write $\hat{\mathbf{C}} = (Set \downarrow |\cdot|)$.

In order to make a connection with logical predicates and logical relations described *e.g.*, in [Mit90], we will be interested in a subcategory $\tilde{\mathbf{C}}$ of the scone which we call the *subscone* of $\mathbf{C}$. The objects of the subscone $\tilde{\mathbf{C}}$ are the triples $\langle S, f, A\rangle$ with $f\colon S \hookrightarrow |A|$ an inclusion of sets. The morphisms $\langle S, f, A\rangle \to \langle S', f', A'\rangle$ of $\tilde{\mathbf{C}}$ are the same as for $\hat{\mathbf{C}}$, so $\tilde{\mathbf{C}}$ is a *full* subcategory of $\hat{\mathbf{C}}$. Notice, however, that if $\langle t, x\rangle$ is a morphism in $\tilde{\mathbf{C}}$, then the set function $t\colon S \to S'$ is uniquely determined by x and S. Specifically, t is the restriction of $|x|$ to S.

An illustrative first example is the subscone of a partially ordered set $\mathbf{C}$ with a top element *top*. In this case, $\tilde{\mathbf{C}}$ is again partially ordered, with a "new" element (indicated by $*$ below) just below *top*.

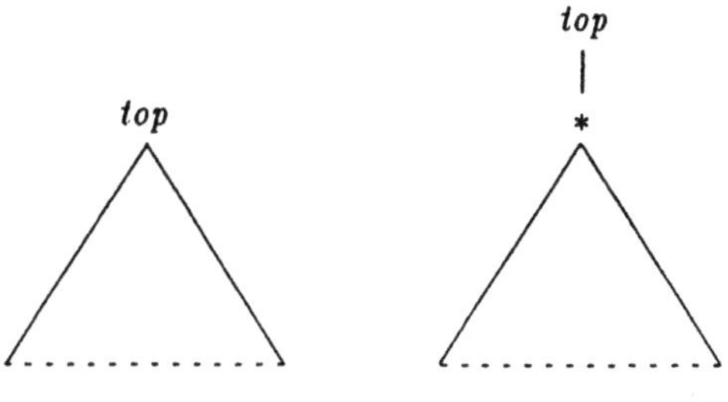

$\mathbf{C}$ $\qquad$ $\tilde{\mathbf{C}}$

A remarkable feature of *sconing* in general is that it preserves almost any additional categorical structure that $\mathbf{C}$ might have. For further discussion, see [FrS90], for example.

The sconing construction is also discussed in [LS86] and [ScS82], where $\hat{\mathbf{C}}$ is called the *Freyd cover of* $\mathbf{C}$. We will be concerned with sconing and cartesian closed structure.

Proposition 1. *If* $\mathbf{C}$ *is a cartesian closed category, then* $\hat{\mathbf{C}}$ *is cartesian closed and the canonical functor* $\hat{\mathbf{C}} \longrightarrow \mathbf{C}$ *is a representation of cartesian closed categories.*

Proof. Let $X = \langle S, f, A\rangle$ and $Y = \langle S', f', A'\rangle$ be objects of $\hat{\mathbf{C}}$. Then the object

$$X \times Y = \langle S \times S', f \times f', A \times A'\rangle,$$

is a product of X and Y, where $(f \times f')\langle s, s'\rangle = \langle f(s), f'(s')\rangle_{1,A,A'}$. An exponential of X and Y (an object of functions from X to Y) is given by

$$X \Rightarrow Y = \langle M, h, A \Rightarrow A'\rangle,$$

where M is the set of morphisms $X \rightarrow Y$ in $\hat{\mathbf{C}}$ and $h(\langle t, x\rangle)\colon \mathbf{1} \rightarrow A \Rightarrow A'$ is the morphism in $\mathbf{C}$ obtained from $x\colon A \rightarrow A'$ by currying.

The subcategory $\tilde{\mathbf{C}}$ inherits (an isomorphic copy of) the cartesian closed structure of $\hat{\mathbf{C}}$. More precisely, $\tilde{\mathbf{C}}$ is isomorphic to a full sub-cartesian-closed category of $\hat{\mathbf{C}}$. The reader may enjoy constructing exponentials in the subscone. In Section 4.1, we will work out the special case where $\mathbf{C}$ is a product of two ccc's.

It is category-theoretic folklore that Proposition 1 (and its proof) extend to a more general setting with *Set* replaced by any cartesian closed category $\mathbf{C}'$ with equalizers and the functor $|\cdot|\colon \mathbf{C} \longrightarrow Set$ replaced by any functor $\mathbf{F}\colon \mathbf{C} \longrightarrow \mathbf{C}'$ that preserves products up to isomorphism. This generalization takes us from the specific comma category $\hat{\mathbf{C}} = (Set \downarrow |\cdot|)$ to a more general case of the form $(\mathbf{C}' \downarrow \mathbf{F})$, see [Laf88, MaR92].

Proposition 2. *Let* $\mathbf{C}$ *and* $\mathbf{C}'$ *be cartesian closed categories and assume* $\mathbf{C}'$ *has equalizers. Let* $\mathbf{F}\colon \mathbf{C} \longrightarrow \mathbf{C}'$ *be a functor that preserves finite products up to isomorphism. Then the comma category* $(\mathbf{C}' \downarrow \mathbf{F})$ *is a cartesian closed category and the canonical functor* $(\mathbf{C}' \downarrow \mathbf{F}) \longrightarrow \mathbf{C}$ *is a representation of cartesian closed categories.*

4.1 A special case: sconing with relations

An illuminating case is the scone of a product category. Let us consider a product category $\mathbf{C} = \mathbf{A} \times \mathbf{B}$, where both categories $\mathbf{A}$ and $\mathbf{B}$ have a terminator. We remind the reader that the objects of $\mathbf{A} \times \mathbf{B}$ are ordered pairs of objects from $\mathbf{A}$ and $\mathbf{B}$ and the morphisms $\langle A, B\rangle \rightarrow \langle A', B'\rangle$ are pairs $\langle x, y\rangle$, where $x\colon A \rightarrow A'$ is a morphism in $\mathbf{A}$ and $y\colon B \rightarrow B'$ is a morphism in $\mathbf{B}$. When $\mathbf{A}$ and $\mathbf{B}$ are cartesian closed, so is $\mathbf{A} \times \mathbf{B}$, with coordinatewise cartesian closed structure. A terminator of $\mathbf{A} \times \mathbf{B}$ is $\langle \mathbf{1}, \mathbf{1}\rangle$. It is easy to see that $|\langle A, B\rangle|_{\mathbf{C}} = |A|_{\mathbf{A}} \times |B|_{\mathbf{B}}$, where the cartesian product of $|A|_{\mathbf{A}}$ and $|B|_{\mathbf{B}}$ is taken in sets. We will often omit the subscripts from various $|\cdot|$'s, since these are generally clear from context. We will focus on the scone of product categories for the rest of this section.

The objects of the scone $\widehat{\mathbf{A} \times \mathbf{B}}$ may be described as tuples $\langle S, f, g, A, B\rangle$, with S a set, A an object in $\mathbf{A}$, B an object in $\mathbf{B}$, and f and g functions that form a so-called *span:*

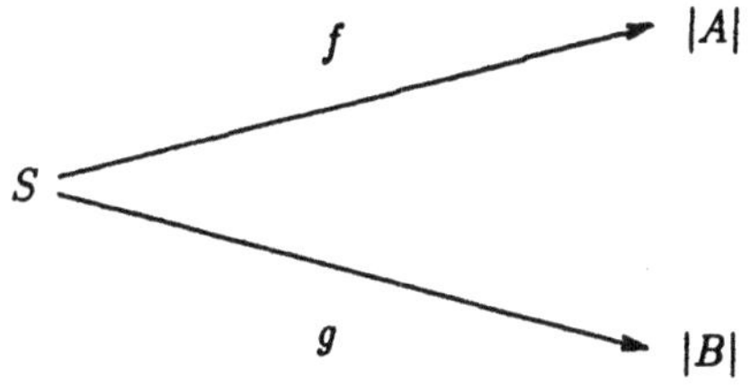

The morphisms $\langle S, f, g, A, B\rangle \to \langle S', f', g', A', B'\rangle$ in $\widehat{\mathbf{A}\times\mathbf{B}}$ may be described as tuples $\langle t, x, y\rangle$, with $x: A \to A'$ a morphism in $\mathbf{A}$, $y: B \to B'$ a morphism in $\mathbf{B}$, and $t: S \to S'$ a function such that:

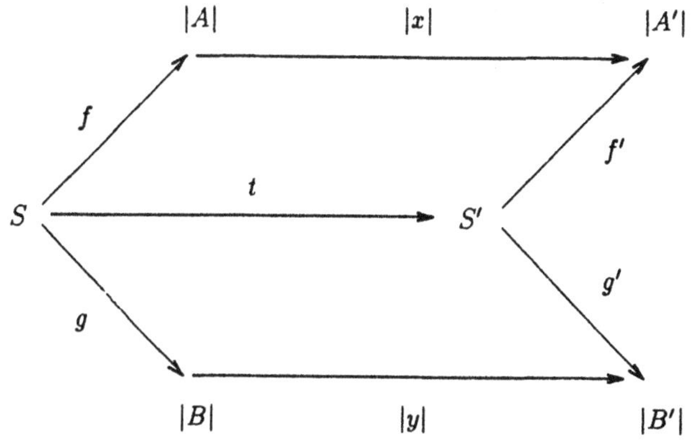

Morphisms in the scone of a product category resemble *transformations of structors,* described in [Fre93].

Let us now concentrate on the subcategory $\widetilde{\mathbf{A}\times\mathbf{B}}$ of $\widehat{\mathbf{A}\times\mathbf{B}}$. Working through the definition of $\widetilde{\mathbf{A}\times\mathbf{B}}$, we may regard the objects of this subcategory as tuples $\langle S, f, g, A, B\rangle$, where f and g are the coordinate functions of an inclusion $S \hookrightarrow |A| \times |B|$. In other words, since S , f and g determine a subset of $|A| \times |B|$, an object $\langle S, f, g, A, B\rangle$ is essentially an ordinary binary relation on $|A| \times |B|$. We will therefore simplify notation and suppress the coordinate functions f and g . We will write $S: |A| \longrightarrow\!\!\!\!\times |B|$ to indicate that S is a binary relation on $|A| \times |B|$. This notation makes it possible to express properties of relations using diagrams instead of formulas. As noted above, for morphisms $\langle t, x, y\rangle$ between objects of $\widetilde{\mathbf{A}\times\mathbf{B}}$, the set map t is uniquely determined by $\mathbf{A}$ and $\mathbf{B}$ maps x and y , and the relation S . Specifically, t is the restriction of $|x| \times |y|$ to S . Omitting the redundant set map, the definition of morphism has an appealing diagrammatic presentation as a pair $\langle x, y\rangle$ satisfying:

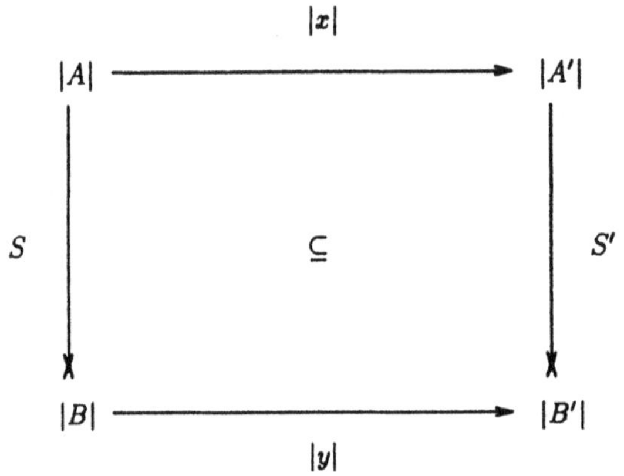

The "$\subseteq$" in this diagram means that the relation $|y| \circ S$ is included in the relation $S' \circ |x|$. It is readily seen that this condition, expressible as a universal Horn clause, states exactly that for all morphisms $a: \mathbf{1} \rightarrow A$ in $\mathbf{A}$ and $b: \mathbf{1} \rightarrow B$ in $\mathbf{B}$:

$$a\, S\, b \quad \text{implies} \quad (x \circ a)\, S'\, (y \circ b),$$

where we use relational notation $a\, S\, b$ to indicate that S relates a to b , and similarly for S' .

It is instructive to write out the exponentials (function spaces) in the cartesian closed category $\widetilde{\mathbf{A}\times\mathbf{B}}$. Suppressing the coordinates of inclusions, $\langle S, A, B\rangle \Rightarrow \langle S', A', B'\rangle$ is given by $\langle R, A \Rightarrow A', B \Rightarrow B'\rangle$, where $R: |A \Rightarrow A'| \longrightarrow\!\!\!\!\times |B \Rightarrow B'|$ is the binary relation such that for all morphisms $e: \mathbf{1} \rightarrow A \Rightarrow A'$ in $\mathbf{A}$ and $d: \mathbf{1} \rightarrow B \Rightarrow B'$ in $\mathbf{B}$:

$$e\, R\, d \text{ iff } a\, S\, b \text{ implies } \mathbf{App} \circ \langle e, a\rangle\, S'\, \mathbf{App} \circ \langle d, b\rangle$$

$$\text{for all } a: \mathbf{1} \rightarrow A \text{ in } \mathbf{A}, b: \mathbf{1} \rightarrow B \text{ in } \mathbf{B}.$$

We may also write out products in a similar fashion. The subscone $\tilde{\mathbf{C}}$ is a ccc and a full subcategory of $\hat{\mathbf{C}}$, and the restriction of the forgetful representation $\hat{\mathbf{C}} \rightarrow \mathbf{C}$ to $\tilde{\mathbf{C}}$ is also a ccc-representation. The main properties of $\widetilde{\mathbf{A}\times\mathbf{B}}$ are summarized in the following proposition, which is easily verified using the ideas presented in the above discussion.

Proposition 3. *Let* $\mathbf{C} = \mathbf{A} \times \mathbf{B}$ *be the cartesian product of two ccc's. Then* $\tilde{\mathbf{C}}$ *is a cartesian closed category, with canonical functor* $\tilde{\mathbf{C}} \longrightarrow \mathbf{C}$ *a representation of cartesian closed categories. Furthermore, products and exponentials are given by*

$$\langle S, A, B\rangle \times \langle S', A', B'\rangle = \langle S \times S', A \times A', B \times B'\rangle,$$

$$\langle S, A, B\rangle \Rightarrow \langle S', A', B'\rangle = \langle S \Rightarrow S', A \Rightarrow A', B \Rightarrow B'\rangle,$$

where relations $S \times S'$ *and* $S \Rightarrow S'$ *have the following characterizations:*

$$a\,(S \times S')\,b \textit{ iff } (\mathbf{Proj}_1 \circ a)\, S\, (\mathbf{Proj}_1 \circ b) \textit{ and } (\mathbf{Proj}_2 \circ a)\, S'\, (\mathbf{Proj}_2 \circ b)$$

$$e\,(S \Rightarrow S')\,d \textit{ iff } \forall a: \mathbf{1} \rightarrow A. \forall b: \mathbf{1} \rightarrow B.\ a\, S\, b \textit{ implies } \mathbf{App} \circ \langle e, a\rangle\, S'\, \mathbf{App} \circ \langle d, b\rangle.$$

The descriptions of products and exponentials in this proposition correspond to the definition of logical relations given *e.g.*, in [Mit90].

Note that although much of the development above uses *Set*, we do not assume that functors $|\cdot|_{\mathbf{A}}: \mathbf{A} \longrightarrow Set$ and $|\cdot|_{\mathbf{B}}: \mathbf{B} \longrightarrow Set$ are one-to-one on morphisms. In other words, we do not assume that categories $\mathbf{A}$ and $\mathbf{B}$ are well-pointed (generated by $\mathbf{1}$). In the argument above, we are simply using the one-to-one correspondence between the morphisms $C \rightarrow C'$ and the morphisms $\mathbf{1} \rightarrow C \Rightarrow C'$, which holds in any cartesian closed category. In the special case of well-pointed cartesian closed categories, we may compare sconing to logical relations for Henkin models of simply typed lambda calculus (see [Sta85, Mit90]). We shall do so in the next section.

We conclude this section with a "binary version" of a general sconing framework given by means of comma categories in Proposition 2. This binary version (as well as the k-ary version) can be obtained from Proposition 2 itself.

Proposition 4. *Let* $\mathbf{A}$, $\mathbf{B}$, *and* $\mathbf{C}'$ *be cartesian closed categories and assume* $\mathbf{C}'$ *has equalizers. Let* $\mathbf{F}: \mathbf{A} \longrightarrow \mathbf{C}'$ *and* $\mathbf{G}: \mathbf{B} \longrightarrow \mathbf{C}'$ *be functors that preserve finite products up to isomorphism. Let* $\mathbf{H}: \mathbf{A} \times \mathbf{B} \longrightarrow \mathbf{C}'$ *be the functor given by* $\mathbf{H}(A, B) = \mathbf{F}(A) \times \mathbf{G}(B)$ *, and likewise on morphisms. Then the comma category* $(\mathbf{C}' \downarrow \mathbf{H})$ *is a cartesian closed category and the canonical functor* $(\mathbf{C}' \downarrow \mathbf{H}) \longrightarrow \mathbf{A} \times \mathbf{B}$ *is a representation of cartesian closed categories.*

5 Sconing and Logical Relations

In this section, we describe the connection between the subscone and the so-called logical relations over Henkin models. We rely on the basic correspondence between well-pointed ccc's and Henkin models described, *e.g.*, in [LS86, MS89]. Logical relations are discussed *e.g.*, in [Sta85, Mit90]. If $\mathbf{C}$ is the cartesian closed category determined by a Henkin model $\mathcal{A}$, then the logical predicates over $\mathcal{A}$ have a direct correspondence with the ccc-representations from a free cartesian closed category into the subscone of $\mathbf{C}$. Since binary relations are often more intuitive than predicates, we will continue to illustrate the main ideas using binary relations. We begin by reviewing the free cartesian closed category over some set of types.

For any set S of type constants, there is a *free* cartesian closed category $\mathcal{F}[S]$ generated by S. The objects of the category $\mathcal{F}[S]$ are types built from the type constants in S and the morphisms $A \rightarrow B$ are equivalence classes of lambda terms of type B with exactly one free variable of type A, where the equivalence relation is given by the equational rules [LS86, MS89]. The freeness of $\mathcal{F}[S]$ means that for any cartesian closed category $\mathbf{D}$ and any assignment $S \rightarrow \mathbf{D}$ mapping the type constants in S to objects of $\mathbf{D}$ there is a unique representation of cartesian closed categories such that

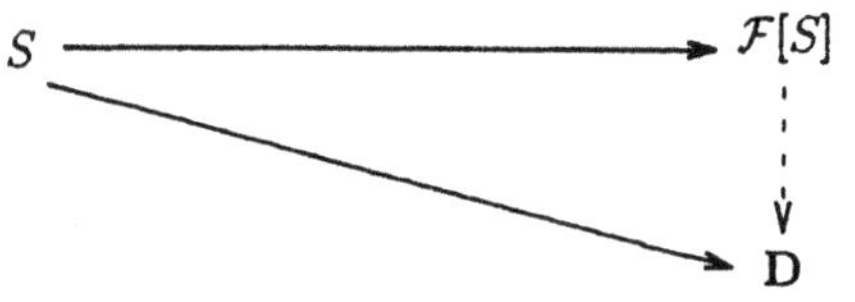

where the map $S \rightarrow \mathcal{F}[S]$ is the inclusion of S into the set of objects of $\mathcal{F}[S]$. The representation of $\mathcal{F}[S]$ into $\mathbf{D}$ is the categorical equivalent of a meaning function $\mathcal{A}[\![\cdot]\!]$ associated with a Henkin model $\mathcal{A}$. Because a representation is a functor that preserves cartesian closed structure, this representation gives both a map from simple types over S to objects of $\mathbf{D}$ and a map from of simply typed lambda terms to morphisms of $\mathbf{D}$. Moreover, these functions together preserve all typing and equational rules of simply typed lambda calculus. One difference between the Henkin model and categorical settings is that we usually consider the mapping from type expressions to sets $\sigma \mapsto A^\sigma$ part of the Henkin model $\mathcal{A}$, whereas the associations of type constants to objects is not part of a cartesian closed category (hence the dependence on a function $S \rightarrow \mathbf{D}$). Another difference is that the meaning function for Henkin models is a map defined on terms, whereas a representation of $\mathcal{F}[S]$ is a map on equivalence classes. However, this is an inessential difference since either form of map may be obtained from the other in an obvious way. Additional discussion and details may be found in [LS86, MS89] and elsewhere.

Because the subscone $\tilde{\mathbf{C}}$ is a cartesian closed category, any assignment of type constants S to objects of $\tilde{\mathbf{C}}$ determines a representation from $\mathcal{F}[S]$ to $\tilde{\mathbf{C}}$. For product categories, if $\mathbf{A}$ and $\mathbf{B}$ are given by Henkin models $\mathcal{A}$ and $\mathcal{B}$ of simply typed lambda calculus, then $\mathbf{C} = \mathbf{A} \times \mathbf{B}$ and $\tilde{\mathbf{C}}$ are again well-pointed cartesian closed categories. A mapping from simple types to objects of $\tilde{\mathbf{C}}$, as mentioned above, is a type-indexed family of relations $R^\sigma \subseteq A^\sigma \times B^\sigma$ between the Henkin models. It is easy to see from Proposition 3 that the family of relations determined by the object map of a representation $\mathcal{F}[S] \rightarrow \tilde{\mathbf{C}}$ is in fact a logical relation.

Proposition 5. *Let* $\mathbf{A}$ *and* $\mathbf{B}$ *be well-pointed cartesian closed categories and let* $\mathcal{A}$ *and* $\mathcal{B}$ *be the corresponding Henkin models of simply typed lambda calculus. Then a logical relation on* $\mathcal{A} \ltimes \mathcal{B}$ *is exactly the object part of a representation of the free cartesian closed category (on a given set of generators) in the cartesian closed category* $\widetilde{\mathbf{A}\times\mathbf{B}}$.

This correspondence and the Basic Lemma for logical relations discussed in [Mit90] are consequences of a straightforward diagram, given in the following proposition.

Proposition 6. *Let* S *be a set of type constants (symbols), let* $\mathbf{C}$ *be a cartesian closed category, and consider any mapping* $S \rightarrow \tilde{\mathbf{C}}$ *from type constants to objects of the subscone of* $\mathbf{C}$. *Composing with the canonical functor* $\tilde{\mathbf{C}} \rightarrow \mathbf{C}$, *we obtain a corresponding map* $S \rightarrow \mathbf{C}$. *Given these two maps from* S, *and the injection of* S *into the types of the free ccc* $\mathcal{F}[S]$ *over* S, *there exist unique ccc representations commuting with the canonical functor* $\tilde{\mathbf{C}} \rightarrow \mathbf{C}$ *as follows:*

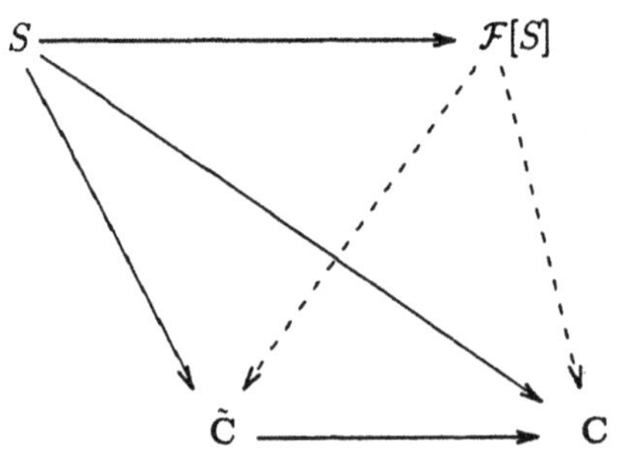

Proof. Let functions from S to objects of the ccc's $\mathbf{C}$, $\tilde{\mathbf{C}}$ and $\mathcal{F}[S]$ be given as in the statement of the proposition. It is a trivial consequence of the definitions that the triangle involving S, $\tilde{\mathbf{C}}$ and $\mathbf{C}$ commutes. By the freeness of $\mathcal{F}[S]$, the representations

$\mathcal{F}[S] \to \mathbf{C}$ and $\mathcal{F}[S] \to \tilde{\mathbf{C}}$ are uniquely determined. This implies that the two triangles with vertices S, $\mathcal{F}[S]$ and either $\tilde{\mathbf{C}}$ or $\mathbf{C}$ commute. Finally, since the canonical functor $\tilde{\mathbf{C}} \to \mathbf{C}$ is a ccc-representation (see Proposition 3), the two uniquely determined representations of $\mathcal{F}[S]$ must commute with this functor.

The Basic Lemma follows from this diagram. Indeed, let us consider the special case that $\mathbf{A}$ and $\mathbf{B}$ are given by Henkin models $\mathcal{A}$ and $\mathcal{B}$ of simply typed lambda calculus. Then both $\mathbf{C} = \mathbf{A} \times \mathbf{B}$ and $\tilde{\mathbf{C}}$ are well-pointed cartesian closed categories. The representation $\mathcal{F}[S] \to \mathbf{C}$ maps a typed lambda term $x{:}\sigma \triangleright M{:}\tau$ to the pair of morphisms giving the meaning of M in $\mathcal{A}$ as a function of x and, respectively, the meaning of M in $\mathcal{B}$. The representation $\mathcal{F}[S] \to \tilde{\mathbf{C}}$ maps a typed lambda term $x{:}\sigma \triangleright M{:}\tau$ to a morphism from some subset $S \subseteq \sigma_{\mathbf{A}} \times \sigma_{\mathbf{B}}$ to a subset $S' \subseteq \tau_{\mathbf{A}} \times \tau_{\mathbf{B}}$, where $\sigma_{\mathbf{A}}$ is the object of $\mathbf{A}$ given by type expression σ over type constants S, according to the chosen map $S \to \mathbf{C}\,(= \mathbf{A} \times \mathbf{B})$, and similarly for the other subscripted type expressions. As noted in Section 4.1, such a morphism in the subscone is a product map $f \times g$ with $f{:}\sigma_{\mathbf{A}} \to \tau_{\mathbf{A}}$ and $g{:}\sigma_{\mathbf{B}} \to \tau_{\mathbf{B}}$ which maps a pair $\langle u, v\rangle$ with $u\ S\ v$ to a pair $\langle f(u), g(v)\rangle$ with $f(u)\ S'\ g(v)$. Because the two representations of $\mathcal{F}[S]$ commute with the forgetful functor $\tilde{\mathbf{C}} \to \mathbf{C}$, f must be the meaning $f = \mathcal{A}[\![x{:}\sigma \triangleright M{:}\tau]\!]$ of the term M in $\mathcal{A}$, regarded as a function of the free variable x, and similarly $g = \mathcal{B}[\![x{:}\sigma \triangleright M{:}\tau]\!]$. Because the term $x{:}\sigma \triangleright M{:}\tau$ was arbitrary, and σ may be a cartesian product of any types, we have shown that given logically related interpretations of the free variables, the meaning of any term in $\mathcal{A}$ is logically related to the meaning of this term in $\mathcal{B}$. This is precisely the Basic Lemma for logical relations over Henkin models.

6 Generalizations of sconing

The proof of Proposition 6 in fact establishes the general version of Proposition 6 where instead of $\tilde{\mathbf{C}}$ one considers any ccc $\mathbf{D}$ that has a ccc-representation in $\mathbf{C}$. For instance, $\mathbf{D}$ might be a subcategory of the comma category $(\mathbf{C}' \downarrow \mathbf{F})$ given in Proposition 2, so that the restriction of the canonical ccc-representation $(\mathbf{C}' \downarrow \mathbf{F}) \longrightarrow \mathbf{C}$ to $\mathbf{D}$ is a ccc-representation. We also remind the reader of the binary version given in Proposition 4. In the general setting that would yield the above diagram with $\mathbf{A} \times \mathbf{B}$ instead of $\mathbf{C}$ and with $\mathbf{D}$ instead of $\tilde{\mathbf{C}}$.

In this section we compare this categorical generalization with three forms of logical relations that have already appeared in the literature, Kripke logical relations [Plo80, MM91], cpo logical relations [MS76, Rey74, CP92], and the relational setting over PER models discussed in Section 4 of [BFS90]. Kripke logical relations over ordinary Henkin models were first used in [Plo80] in a characterization of lambda definability. Kripke logical relations were then adapted to Kripke lambda models in [MM91]. Inductive relations (as well as strict inductive relations) are widely used in relating denotational semantics and in proofs by fixed-point induction. Relations over PER models may be used to derive parametricity properties of lambda definable functions, see Section 4 of [BFS90].

These cases of generalized logical relations are still "concrete" enough so that it makes sense to consider an analogue of a subscone, a certain subcategory of a comma category consisting of those objects for which relevant mappings are given by inclusions. Hence in these cases we will be able to assume that the objects of the subcategory $\mathbf{D}$ of the comma category also have this form.

6.1 Sconing with Kripke models

Let $\mathcal{P}$ be a partially ordered set. $\mathcal{P}$ may be considered as a category whose objects are the elements of $\mathcal{P}$, and where for each p, p' in $\mathcal{P}$ there is at most one morphism $p \to p'$, which exists iff $p \le p'$ in $\mathcal{P}$. It is shown in Section 4 of [MM91] that each Kripke lambda model with $\mathcal{P}$ as the set of "possible worlds" determines a cartesian closed category $\mathbf{C}$ and a finite product preserving functor $\mathbf{F}$ from $\mathbf{C}$ to the functor category $Set^{\mathcal{P}}$. (The objects of $\mathbf{C}$ are types, not interpretations of types given by the Kripke lambda model, but the morphisms $\sigma \to \tau$ in $\mathbf{C}$ are the natural transformations $\Phi_\sigma \to \Phi_\tau$ induced by the global elements of the Kripke model of type $\sigma \Rightarrow \tau$, where Φ_σ is the interpretation of type σ. The functor $\mathbf{F}$, given by Φ, need not be a ccc-representation. Furthermore, any ccc is equivalent as a category to a ccc arising this way, see [MM91]. This framework covers the case when $\mathbf{C}$ arises from an ordinary Henkin model, see our Section 5. Indeed, we let $\mathbf{F}(C)$ for each object C of $\mathbf{C}$ be the constant functor $\mathcal{P} \to Set$ whose value is the domain set in the given Henkin model that corresponds to C.)

Let $\mathbf{D}$ be the full subcategory of the comma category $(Set^{\mathcal{P}} \downarrow \mathbf{F})$ that consists of objects $\langle S, f, A\rangle$ with $f_p: S(p) \hookrightarrow \mathbf{F}(A)(p)$ an inclusion for each p in $\mathcal{P}$. We may conveniently omit f when referring to objects of $\mathbf{D}$. $\mathbf{D}$ is a ccc. In describing exponentials (*i.e.*function spaces) in $\mathbf{D}$ we may use natural transformations $\mathbf{F}(app_{A,A'}): \mathbf{F}(A \Rightarrow A') \times \mathbf{F}(A) \longrightarrow \mathbf{F}(A')$ because the functor $\mathbf{F}$ preserves binary products. For each object C in $\mathbf{C}$ and each $p \le p'$ in $\mathcal{P}$, we also use the transition function $\mathbf{F}(C)(p \le p'): \mathbf{F}(C)(p) \to \mathbf{F}(C)(p')$ given by the action of the functor $\mathbf{F}(C): \mathcal{P} \to Set$ on the order of $\mathcal{P}$. Specifically, $\langle S, A\rangle \Rightarrow \langle S', A'\rangle$ may be given as $\langle P, A \Rightarrow A'\rangle$, where $A \Rightarrow A'$ is taken in $\mathbf{C}$ and $P: \mathcal{P} \to Set$ is the functor such that the set $P(p)$ consists of all t in $\mathbf{F}(A \Rightarrow A')(p)$ for which for all $p' \ge p$ and all a in $\mathbf{F}(A)(p')$, if a belongs to $S(p')$, then the element $\mathbf{F}(app_{A,A'})_{p'}(\mathbf{F}(A \Rightarrow A')(p \le p')(t), a)$ in $\mathbf{F}(A')(p')$ belongs to $S'(p')$. In other words, the object parts of the free ccc-representations in $\mathbf{D}$ in this case amount to, in the terminology of [MM91], the Kripke logical predicates on the Kripke lambda model that corresponds to $\mathbf{F}$.

In the binary case, let $\mathbf{A}$ and $\mathbf{B}$ be cartesian closed categories and $\mathbf{F}: \mathbf{A} \to Set^{\mathcal{P}}$ and $\mathbf{G}: \mathbf{A} \to Set^{\mathcal{P}}$ finite product preserving functors given by Kripke lambda models $\mathcal{A}$ and $\mathcal{B}$, each with $\mathcal{P}$ as the poset of possible worlds. Let $\mathbf{H}: \mathbf{A} \times \mathbf{B} \to Set^{\mathcal{P}}$ be the functor given by $\mathbf{H}(A, B) = \mathbf{F}(A) \times \mathbf{G}(B)$. One now continues as above, with $\mathbf{C} = \mathbf{A} \times \mathbf{B}$ and with $\mathbf{H}$ instead of $\mathbf{F}$. In this case the object parts of the free ccc-representations in $\mathbf{D}$ are exactly the Kripke logical relations on Kripke lambda models $\mathcal{A}$ and $\mathcal{B}$.

6.2 Sconing with cpo domains

Let Cpo be the category whose objects are posets with suprema of countably infinite chains, *i.e.*cpos, and whose morphisms are monotone maps that preserve suprema of countably infinite chains, *i.e.*continuous maps. Cpo is a cartesian closed category with equalizers. Finite products and equalizers are given as in Set, and $P \Rightarrow Q$ consists of all continuous maps $P \to Q$, with pointwise order. Let $Pcpo$ be the full subcategory of Cpo whose objects are pointed cpos, *i.e.*cpos with the least element, usually denoted $\perp$. Because $P \Rightarrow Q$ as given in Cpo is pointed if Q is pointed, with $\perp$ in $P \Rightarrow Q$ being the constant function with value $\perp$ in Q, $Pcpo$ is a sub-ccc of Cpo. We shall denote the inclusion functor by $\mathbf{I}$.

While in $Pcpo$ one loses equalizers, one is compensated by gaining, for each object P, a canonical morphism $fix_P: (P \Rightarrow P) \to P$ that yields fixed-points. If f is an element of $(P \Rightarrow P)$ given by a continuous map $P \to P$, then $fix_P(f)$ is the least fixed-point of f, given by $fix_P(f) = \bigvee\{f^n(\bot)\}$. Thus the ccc $Pcpo$ has the property that for each object P, there exists a canonical morphism $fix_P: (P \Rightarrow P) \to P$ such that for every morphism f targeted at $P \Rightarrow P$, the morphism $p = fix_P \circ f$ targeted at P satisfies $app \circ \langle f, fix_P \circ f\rangle = fix_P \circ f$.

A ccc with this property may be called a *ccc with fixed-point operators*. For any such ccc $\mathbf{C}$, the free ccc-representation $\mathcal{F}[S] \longrightarrow \mathbf{C}$ (induced, as always, by a mapping $S \longrightarrow \mathbf{C}$ that interprets type constants in $\mathbf{C}$) factors through the free ccc with fixed-points on the given collection S of type constants. This latter category $\mathcal{F}_{fix}[S]$ is the term category of simply typed lambda calculus (on the given collection S of type constants) with a constant fix_σ of type $(\sigma \Rightarrow \sigma) \Rightarrow \sigma$ for each type σ, for which we impose

$$\Gamma \rhd M(fix_\sigma M) = fix_\sigma M : \sigma.$$

In addition, the ccc-representation $\mathcal{F}_{fix}[S] \longrightarrow \mathbf{C}$ involved in the factorization of $\mathcal{F}[S] \longrightarrow \mathbf{C}$ also preserves the canonical fixed-point operators.

Let us consider the full subcategory of the comma category $(Cpo \downarrow \mathbf{I})$ whose objects are of the form $\langle P, f, A\rangle$, where f is the inclusion $P \hookrightarrow A$. In other words, P is a sub-cpo of pcpo A. Yet another equivalent description is that P is an *inductive predicate* on A in the sense that P is monotone in A and for any chain Q in A, if $Q \subseteq P$, then $\bigvee Q \in P$. We omit f when referring to these objects. In this ccc the exponentials, *i.e.*function spaces $\langle P, A\rangle \Rightarrow \langle P', A'\rangle$ may be described as $\langle R, A \Rightarrow A'\rangle$, where $A \Rightarrow A'$ is taken in Cpo, *i.e.*it is the cpo of continuous functions $g: A \to A'$ ordered pointwise, and where R consists of all such g for which $g(a) \in P'$ whenever $a \in P$. The object parts of free ccc-representations in this ccc are exactly inductive logical predicates.

It is also useful to consider a further subcategory $\mathbf{D}$ whose objects are of the form $\langle P, A\rangle$, where $P \hookrightarrow A$ is a sub-pcpo, *i.e.*in addition to being inductive, P is also *strict* in the sense that $\bot_A \in P$. The ccc structure is still the same, but in $\mathbf{D}$ we also get canonical fixed-point operators, and the forgetful ccc-representation $\mathbf{D} \to Pcpo$ preserves them. The object parts of free ccc-representations in $\mathbf{D}$ are exactly strict, inductive logical predicates. As we observed above, each such ccc-representation can be lifted to a free ccc-representation that preserves fixed-point operators. In the binary case, so we let $\mathbf{H}(A, B) = A \times B$ and we consider $\mathbf{D}$ be the full subcategory of $(Cpo \times Cpo \downarrow \mathbf{H})$ whose objects are of the form $\langle S, A, B\rangle$, where S is a sub-pcpo of pcpo $A \times B$, *i.e.*S is a strict, inductive relation from A to B.

It is immediate to generalize this entire discussion from chains to directed families. Furthermore, it would be instructive to work out a treatment of sconing with cpos from the point of view of computational monads [Mgg89]. In particular, it seems useful to see objects of the form $\langle P, A\rangle$ with $P \hookrightarrow A$ a sub-pcpo, as the result of lifting in $(Cpo \downarrow \mathbf{I})$.

6.3 Sconing with partial equivalence relations

Let Per be the category whose objects are partial equivalence relations (*i.e.*symmetric, transitive relations) on natural numbers, and whose morphisms $R \to S$ are equivalence classes of partial recursive functions f that are defined on the domain of R and that preserve R in the sense that whenever $n\ R\ k$, then $f(n)\ S\ f(k)$. Two such partial recursive functions f, f' are considered equivalent iff whenever $n\ R\ k$, then $f(n)\ S\ f'(k)$. It is folklore that Per is a ccc. The "indiscriminate" relation that relates any two numbers serves as a terminal object. Given a computable pairing function, the product

$R \times S$ may be described as relating $\langle n,k\rangle$ and $\langle n',k'\rangle$ iff $n\ R\ n'$ and $k\ S\ k'$. Given a gödelnumbering of partial recursive functions, $R \Rightarrow S$ may be described as relating e and e' iff e and e' are gödelnumbers of partial recursive functions that represent per morphisms $R \to S$ and are equivalent as such.

Let $\mathbf{I}$ be the identity functor on *Per*. Let $\mathbf{D}$ be the full subcategory of the comma category $(Per \downarrow \mathbf{I})$ of objects $\langle S,f,A\rangle$, where f is named by the identity function on the natural numbers, and where S is saturated in A in the sense that $S = A \circ S \circ A$ as binary relations on natural numbers (*i.e.* S is a restriction of A to a subset of the domain of A that is a union of A-equivalence classes). We omit f when referring to objects of $\mathbf{D}$. It is readily verified that $\langle S,A\rangle \Rightarrow \langle S',A'\rangle$ in $\mathbf{D}$ is given by $\langle R, A \Rightarrow A'\rangle$, where the right component is as in *Per*, and where $e\ R\ e'$ iff $e\ A \Rightarrow A'\ e'$ and the partial recursive functions with gödelnumbers e and e' also represent per morphisms $S \to S'$ and are equal as such.

In the binary case the relevant comma category is $(Per \times Per \downarrow \mathbf{H})$, where $\mathbf{H}$ is the binary product functor in *Per*. The objects of the relevant full subcategory may be presented in the form $\langle S,A,B\rangle$, where S is an ordinary relation from the domain of A to the domain of B subject to the saturation condition $S = A \circ S \circ B$, see Section 4 in [BFS90]. The ccc structure may be described as above.

7 Implicit polymorphism

In this section we turn our attention to a slight extension of simply typed lambda calculus. Syntactically, the language $\lambda^{\to,t}$ we consider is still very close to simply typed lambda calculus. The only difference is that types may contain type variables as well as basic type symbols. However, our interest in $\lambda^{\to,t}$ lies in the fact that the terms of the calculus depend on type variables, and therefore we have a simple form of polymorphism. We will refer to this as "implicit" polymorphism since the quantification over type variables is implicit, rather than explicit.

Given a cartesian closed category $\mathbf{C}$, let $Obj\,\mathbf{C}$ be the set (or class) of objects of $\mathbf{C}$. As with any set (or class), $Obj\,\mathbf{C}$ may be considered as a discrete category, the only morphisms of which are the identities. We may interpret the types of $\lambda^{\to,t}$ as functors from $Obj\,\mathbf{C}$ to $\mathbf{C}$, and terms as natural transformations. Note that to interpret types with n type variables, we need functors $(Obj\,\mathbf{C})^n \to \mathbf{C}$, and so the interpretation naturally involves a different category for each number of type variables. The reader will note that the functors $(Obj\,\mathbf{C})^n \to \mathbf{C}$ are basically just n-ary functions from the set (or class) $Obj\,\mathbf{C}$ to $Obj\,\mathbf{C}$. In order to account for type substitution, these functor categories must "fit together" as n varies. A proper general categorical framework for this motivating example is based on indexed or fibred categories, see [See87, Pit87, Mog91]. This framework, discussed below, generalizes readily to explicit polymorphism. In addition, a special case of this categorical view is a construction on Henkin models described, *e.g.*, in [Mit90].

7.1 Categorical models of implicit polymorphism

If we consider the type expressions apart from lambda terms, we have an "algebraic" language with type constants (basic types), type variables, and the binary operations $\times$ and $\Rightarrow$. We may regard any such language as an "algebraic theory," in Lawvere's terminology, or a category with a terminator, binary products, and a distinguished object representing the carrier of the algebra. For definiteness, we will assume a "base" category $\mathbf{B}$ with distinguished object V which *generates* $\mathbf{B}$, in the sense that each object is a finite power V^n for some natural number n. This gives us the terminator as V^0. Intuitively, the arrows $V^i \to V^j$ in $\mathbf{B}$ may be regarded as j-tuples of type expressions, each having all type variables among $t_1, \ldots, t_i$. Composition in $\mathbf{B}$ corresponds to substitution of types for type variables. This relatively standard interpretation of an algebraic language is explained in [KR77], for example. It is helpful to note that we may regard an arrow $V^i \to V^j$ as either a j-tuple of types, or a context $\Gamma = \{x_1 : \tau_1, \ldots, x_j : \tau_j\}$ of length j, with all type expressions over the same set of i type variables.

We now consider the interpretation of terms. For each finite set of type variables, $\lambda^{\to,t}$ includes all simply typed terms over these variables, regarded as type constants. Therefore, we expect a categorical interpretation to provide a ccc for each finite number of type variables. More specifically, if we regard a morphism $g: V^i \to V$ of $\mathbf{B}$ as a context $\{x : \tau\}$, we would expect to have some kind of assignment "from" g for each term $\{x : \tau\} \triangleright M : \sigma$. We may use the type variable structure of $\mathbf{B}$ to index categories of types and terms. We assume that for each object V^i in $\mathbf{B}$, there is a category $\mathbf{F}_i$ whose objects correspond to type expressions with i type variables and morphisms represent terms of these types. This assignment is assumed to be functorial in i, *i.e.* $\mathbf{F}$ is a (contravariant) functor to Cat, the category of categories. The structure of $\mathbf{B}$ is linked to $\mathbf{F}_i$ by assuming that objects of $\mathbf{F}_i$ are the same as morphims $V^i \to V$ in $\mathbf{B}$. Similar indexed categories may be found in [See87, Pit87, Mog91].

Definition 7. An *iml-category* is given by:

- a base category $\mathbf{B}$ with finite products, generated by some distinguished object V,
- a functor $\mathbf{F}: \mathbf{B}^{op} \longrightarrow Cat$ such that:
  - for each object I in $\mathbf{B}$, the category $\mathbf{F}(I)$ is cartesian closed,
  - for each morphism $f: I \to J$ in $\mathbf{B}$, the functor $\mathbf{F}(f): \mathbf{F}(J) \to \mathbf{F}(I)$ is a ccc representation,
  - for each object I in $\mathbf{B}$, the class of objects $Obj(\mathbf{F}(I))$ is the same as the class of morphisms $I \to V$ in $\mathbf{B}$,
  - for each morphism $f: I \to J$ in $\mathbf{B}$, $\mathbf{F}(f)$ acts on objects by composition.

This definition may be rephrased by means of the category $Class$ of classes and mappings and the category Ccc of cartesian closed categories and representations thereof. In particular, an *iml*-category may be seen as a base category $\mathbf{B}$ with finite products generated by some distinguished object V, together with a functor $\mathbf{F}: \mathbf{B}^{op} \longrightarrow Ccc$ such that the functor $Obj \circ \mathbf{F}: \mathbf{B}^{op} \longrightarrow Class$ is the representable functor $\mathbf{B}(\cdot, V)$. Here $Obj: Ccc \longrightarrow Class$ is the forgetful functor that takes a ccc to its underlying set (or class) of objects and a ccc representation to the associated object map. (Probably the most expedient way of dealing with the question of a set vs. a class of objects is to regard "smallness" as relative to each example.) Let us also observe that following [Pit87, Mog91], we use a tighter definition of indexed categories than usual in category theory; we are asking for data up to equality rather than just up to canonical isomorphisms.

Because the objects of $\mathbf{B}$ have the form V^n for some natural number n, we may identify the objects of $\mathbf{B}$ with the natural numbers. This simplifies notation. The category $\mathbf{F}(n)$, which we also write as $\mathbf{F}_n$, is often called the *fiber over* n. Again, the intuition behind this definition is that n stands for a set of n distinct type variables, the objects of $\mathbf{F}_n$ for type expressions on these type variables, the morphisms of $\mathbf{F}_n$ for typed terms over these type variables, and $\mathbf{F}(f)$ for type substitution. This intuition may be made precise by defining a *term iml-category* for each $\lambda^{\rightarrow,t}$ language and theory, analogously to [See87, Pit87].

Example 1. Given any ccc $\mathbf{C}$, we may construct an *iml*-category $\mathbf{C}_{iml}$ as follows. We let the base category $\mathbf{B}$ have natural numbers as objects, and let the morphisms $i \rightarrow j$ be functors $(Obj\,\mathbf{C})^i \rightarrow (Obj\,\mathbf{C})^j$. Fiber $\mathbf{F}_n$ over n is the functor category $(Obj\,\mathbf{C})^n \rightarrow \mathbf{C}$. We emphasize that these functors are basically just functions from n-tuples of objects to objects, and therefore the ccc structure of $\mathbf{F}_n$ is in this example given pointwise (*i.e.*, objectwise) by the ccc structure of $\mathbf{C}$. For instance, $(F \Rightarrow G)(A) = F(A) \Rightarrow G(A)$. Finally, if $H: m \rightarrow n$ is a morphism of the base category, and therefore a functor $(Obj\,\mathbf{C})^m \rightarrow (Obj\,\mathbf{C})^n$, we define the functor $\mathbf{F}(H): \mathbf{F}_n \rightarrow \mathbf{F}_m$ by composition with H. More precisely, $\mathbf{F}(H): \mathbf{F}_n \rightarrow \mathbf{F}_m$ takes a morphism $t: F \rightarrow G$ of the fiber $\mathbf{F}_n$ to the morphism $(t \circ H): F \circ H \rightarrow G \circ H$ of the fiber $\mathbf{F}_m$, where for any m-tuple A of objects of $\mathbf{C}$ $(t \circ H)_A = t_{H(A)}$. It is easy to check that $\mathbf{F}(H): \mathbf{F}_n \rightarrow \mathbf{F}_m$ is a ccc representation. We thus see that $\mathbf{C}_{iml}$ is an *iml*-category for any cartesian closed category $\mathbf{C}$. □

Example 2. A more subtle example of an *iml*-category may be found in the setting used in [Gir86]. Consider the category Qd_{emb} of qualitative domains and embedding-projection pairs. This is *not* a cartesian closed category. Let the morphisms $k \rightarrow n$ of the base category $\mathbf{B}$ be n-tuples of functors $(Qd_{emb})^k \longrightarrow Qd_{emb}$ that preserve pullbacks and directed colimits. The fiber $\mathbf{F}_n$ is the ccc whose objects are functors $(Qd_{emb})^n \longrightarrow Qd_{emb}$ that preserve pullbacks and directed colimits, *i.e.*variable types, in the terminology of [Gir86]. The morphisms $t: F \rightarrow G$ of the fiber $\mathbf{F}_n$ are certain "stable" families of stable maps $t_P: F(P) \rightarrow G(P)$, where P ranges over n-tuples of qualitative domains, *i.e.*in the terminology of [Gir86], objects of variable type $F \Rightarrow G$. Here $(F \Rightarrow G)(P) = F(P) \Rightarrow G(P)$, the latter in the sense of the ccc Qd of qualitative domains and stable maps. The action of $F \Rightarrow G$ on embeddings is explained in [Gir86]. If $H: k \rightarrow n$ is a morphism in the base category $\mathbf{B}$, then the ccc-representation $\mathbf{F}(H): \mathbf{F}_n \rightarrow \mathbf{F}_k$ takes a morphism $t: F \rightarrow G$ of the fiber $\mathbf{F}_n$ to the morphism $(t \circ H): F \circ H \rightarrow G \circ H$ of the fiber $\mathbf{F}_k$, where for any k-tuple Q of qualitative domains $(t \circ H)_Q = t_{H(Q)}$. □

Example 3. Let us now describe the *iml*-category PER. We recall the category Per of pers and per morphisms discussed in Section 6.3. Let the morphisms $k \rightarrow n$ of the base category be n-tuples of maps from $(ObjPer)^k$ to $ObjPer$. The fiber $\mathbf{F}_n$ is a ccc whose objects are maps from $(ObjPer)^n$ to $ObjPer$, and whose morphisms $t: F \rightarrow G$ are "realizable" families of per morphisms $t_P: F(P) \rightarrow G(P)$, where P ranges over n-tuples of pers, given by a single partial recursive function that names each per morphism t_P. Two partial recursive functions name the same realizable family of per morphisms $t_P: F(P) \rightarrow G(P)$ iff they name the same per morphism t_P for each n-tuple P of pers. If $H: k \rightarrow n$ is a morphism in the base category, then the ccc-representation $\mathbf{F}(H): \mathbf{F}_n \rightarrow \mathbf{F}_k$ takes a morphism $t: F \rightarrow G$ of the fiber $\mathbf{F}_n$ to the morphism $(t \circ H): F \circ H \rightarrow G \circ H$ of the fiber $\mathbf{F}_k$, where for any k-tuple Q of pers $(t \circ H)_Q = t_{H(Q)}$. □

So far we have defined *iml*-categories and discussed several examples. Recalling that the objects of base categories are identified with the natural numbers, let us now define *iml*-representations:

Definition 8. Let $(\mathbf{B},\mathbf{F})$, $(\mathbf{B}',\mathbf{F}')$ be *iml*-categories. An *iml*-representation $(\mathbf{B},\mathbf{F}) \longrightarrow (\mathbf{B}',\mathbf{F}')$ is given by:

- a functor $T: \mathbf{B} \to \mathbf{B}'$ that preserves finite products so that $T(n) = n$, together with
- a natural transformation $t: \mathbf{F} \longrightarrow (\mathbf{F}' \circ T)$ such that for each n, the functor $t_n: \mathbf{F}_n \to \mathbf{F}'_n$ is a ccc-representation,

such that for any morphism $H: n \to \mathbf{1}$ in $\mathbf{B}$, the morphism $T(H): n \to \mathbf{1}$ in $\mathbf{B}'$ is the object $t_n(H)$ in the fiber $\mathbf{F}'_n$ when H is considered as an object in the fiber $\mathbf{F}_n$.

A special case of this definition is when an *iml*-category $(\mathbf{B},\mathbf{F})$ is a ***sub-iml-category*** of an *iml*-category $(\mathbf{B}',\mathbf{F}')$ in the sense that base category $\mathbf{B}$ is a subcategory of base category $\mathbf{B}'$, (the objects of either are natural numbers), for each n the fiber $\mathbf{F}_n$ is a sub-ccc of the fiber $\mathbf{F}'_n$, and for each morphism $H: k \to n$ the ccc-representation $\mathbf{F}(H): \mathbf{F}_n \to \mathbf{F}_k$ is a restriction of the ccc-representation $\mathbf{F}'(H): \mathbf{F}'_n \to \mathbf{F}'_k$. For instance, the *iml*-category PER described in Example 3 is a sub-*iml*-category of Per_{iml}, defined in Example 1.

Let $(\mathbf{B},\mathbf{F})$ be any *iml*-category. The intuitive motivation, mentioned above, that relates morphisms of $\mathbf{B}$ to types of $\lambda^{\to,t}$ and morphisms of fibers to terms of $\lambda^{\to,t}$ can now be restated in a precise way. For any *iml*-category $(\mathbf{B},\mathbf{F})$ there is a unique *iml*-representation from the term *iml*-category (mentioned just before Example 1) to $(\mathbf{B},\mathbf{F})$ that extends a given assignment of objects of the fiber $\mathbf{F}_0$ to basic type symbols. In other words, the term *iml*-category on the given set S of basic type symbols is the free *iml*-category on S. Indeed, for each n, types and terms with n type variables are mapped to the objects and morphisms of the fiber $\mathbf{F}_n$, respectively, by the unique ccc-representation of the free ccc in the ccc $\mathbf{F}_n$. These ccc-representations commute with type substitution.

On the other hand, while it may seem at first that for any ccc $\mathbf{C}$ the canonical ccc-representation $\tilde{\mathbf{C}} \to \mathbf{C}$ naturally induces an *iml*-representation $(\tilde{\mathbf{C}})_{iml} \to \mathbf{C}_{iml}$, a second thought reveals there is a problem constructing a functor $t_n: (\tilde{\mathbf{C}}_{iml})_n \to (\mathbf{C}_{iml})_n$ for each n. Although the canonical ccc-representation $\tilde{\mathbf{C}} \to \mathbf{C}$ serves for the simplest case, $n = 0$, a problem already arises for $n = 1$. A functor t_1 from $(Obj\,\tilde{\mathbf{C}}) \to \tilde{\mathbf{C}}$ to $(Obj\,\mathbf{C}) \to \mathbf{C}$ is required. This only seems possible, for arbitrary $\mathbf{C}$, if, for each map $(Obj\,\tilde{\mathbf{C}}) \to (Obj\,\tilde{\mathbf{C}})$ sending $\langle S, A\rangle$ to $\langle S', A'\rangle$, the object A' is chosen as a function of A, independent of S. This problem is addressed by the concept of *relator*.

7.2 Relators

Two of the main concepts discussed in this paper are *relators* and *relator transformations*. These concepts and their basic properties will be derived in Section 7.3 by a combination of sconing and the construction described in Example 1. Nevertheless, let us first spell out a concrete presentation of relators and relator transformations, not necessarily in most general terms. As with logical relations and sconing for simply typed lambda calculus, binary relators seem to give the most intuitive and visual picture of the general case.

Definition 9. Let $\mathbf{C},\mathbf{D}$ be categories with terminators. A (binary) *relator* from $\mathbf{C}$ to $\mathbf{D}$ consists of:

- an object map $F: Obj\,\mathbf{C} \longrightarrow Obj\,\mathbf{D}$, together with

- a mapping that to any binary relation $S: |A| \rightarrow\!\!\!\!\times |B|$ associates a binary relation $S': |F(A)| \rightarrow\!\!\!\!\times |F(B)|$.

The more general case of a binary relator of n arguments is defined similarly, using $\mathbf{C}^n$ in place of $\mathbf{C}$ as the domain. In other words, a binary relator of n arguments from $\mathbf{C}$ to $\mathbf{D}$ consists of an object map $F: (Obj\,\mathbf{C})^n \longrightarrow Obj\,\mathbf{D}$, together with a mapping that to any n-tuple of binary relations $S_i: |A_i| \rightarrow\!\!\!\!\times |B_i|$, where $i = 1, \ldots, n$, associates a binary relation
$S': |F(A_1, \ldots, A_n)| \rightarrow\!\!\!\!\times |F(B_1, \ldots, B_n)|$. Even more generally, k-ary relators take k-ary relations to k-ary relations. Further generalizations will be considered in Section 7.3. When $\mathbf{C} = \mathbf{D}$ we shall speak of relators *on* $\mathbf{C}$.

Note. In [AbJ91] a relator is also required to map identity relations to identity relations. This stricter notion also fits in our framework, see Example 4.

Definition 10. Let F, G be relators from $\mathbf{C}$ to $\mathbf{D}$. A *relator transformation* from F to G is a family of morphisms $t_A: F(A) \to G(A)$ in $\mathbf{D}$, with A ranging over the objects of $\mathbf{C}$, such that for each binary relation $S: |A| \rightarrow\!\!\!\!\times |B|$

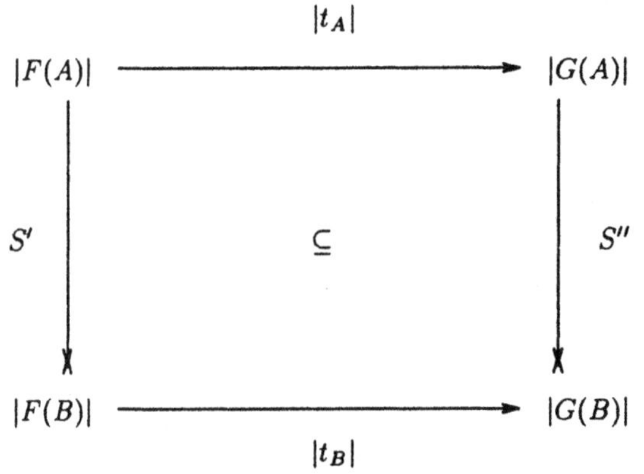

where S' and S'' are the relations assigned to S by F and G, respectively.

We remind the reader that, as in Section 4.1, the "$\subseteq$" in this diagram means that the relation $|t_B| \circ S'$ is included in the relation $S'' \circ |t_A|$. It is readily seen that this condition, expressible as a universal Horn clause, states exactly that for all morphisms $a: \mathbf{1} \to F(A)$ and $b: \mathbf{1} \to F(B)$ in $\mathbf{D}$:

$$a\, S'\, b \quad \text{implies} \quad (t_A \circ a)\, S''\, (t_B \circ b),$$

where we use relational notation $a\ S'\ b$ to indicate that S' relates a to b, and similarly for S''.

These concepts were motivated by work of Wadler [Wad89]. In a modified form (see Example 4), they have been applied to the so-called strictness analysis [AbJ91]. We also note that although relators are slightly more general than *tabular structors* proposed in [Fre93], the Horn clause condition involved in the definition of relator transformations is basically the same as the condition defining transformations of tabular structors given in [Fre93]. Tabular structors map morphisms to relations rather than relations to relations, so the change from tabular structors to relators is required to account for type substitution.

There are three natural notions of composition involving relators and relator transformations. The first one is composition of relators, which corresponds to substitution of types for type variables in types. It is transparent that composing two relators yields a relator. (Tabular structors, on the other hand, do not compose.) Secondly, one has composition of relator transformations, which corresponds to substitution of terms for variables in terms. If $t: F \rightarrow G$ and $u: G \rightarrow H$ are relator transformations between relators from $\mathbf{C}$ to $\mathbf{D}$, let $(u \circ t)$ be the family $(u \circ t)_A = u_A \circ t_A$ of morphisms in $\mathbf{D}$, with A ranging over the objects of $\mathbf{C}$. It is straightforward to check that this family is a relator transformation from F to H. The third notion of composition corresponds to substitution of types for type variables in terms. If $t: F \rightarrow G$ is a relator transformation between relators from $\mathbf{C}$ to $\mathbf{D}$ and if H is a relator from $\mathbf{B}$ to $\mathbf{C}$, then consider the family $(t \circ H)_A = t_{H(A)}$ of morphisms in $\mathbf{D}$, where A ranges over the objects of $\mathbf{B}$. It is again readily checked that this is a relator transformation, this time from $F \circ H$ to $G \circ H$.

7.3 Sconing and *iml*-categories

Let us now return to the problem mentioned at the end of Section 7.1. An interesting application of the *iml*-category construction described in Example 1 arises when the ccc in question is a scone $\hat{\mathbf{C}}$, or the subscone $\tilde{\mathbf{C}}$, see Section 4. By Example 1, both $(\hat{\mathbf{C}})_{iml}$, and $(\tilde{\mathbf{C}})_{iml}$ are iml-categories. We may regard $(\hat{\mathbf{C}})_{iml}$ as a kind of "scone" of $\mathbf{C}_{iml}$, and $(\tilde{\mathbf{C}})_{iml}$ as a kind of "subscone" of $\mathbf{C}_{iml}$. However, both of these categories are in a sense too generous in the collection of morphisms of the base category. For instance, without restricting our attention to a smaller sub-*iml*-category than $(\tilde{\mathbf{C}})_{iml}$, we would not have a straightforward forgetful *iml*-representation of this new *iml*-category in $\mathbf{C}_{iml}$ itself. The reader will recall that the existence of a ccc-representation of $\tilde{\mathbf{C}}$ in $\mathbf{C}$ was essential for the sconing method, and particularly in Propositions 5 and 6.

The problem may be stated more precisely: since even $(\tilde{\mathbf{C}})_{iml}$ is constructed using *all* functions from $Obj\tilde{\mathbf{C}}$ to $Obj\tilde{\mathbf{C}}$, there may be morphisms in the new base category which do not correspond to functions from $Obj\,\mathbf{C}$ to $Obj\,\mathbf{C}$. Therefore, given an arbitrary function $\Phi: Obj\tilde{\mathbf{C}} \longrightarrow Obj\tilde{\mathbf{C}}$, we would like to ensure the existence of a function $F: Obj\,\mathbf{C} \longrightarrow Obj\,\mathbf{C}$ with

$$\Phi(S, A) = (S', F(A))$$

for every object $\langle S, A\rangle$ of $\tilde{\mathbf{C}}$. Notice that if there is such a function on $Obj\,\mathbf{C}$, it is unique. The equation above can be restated as

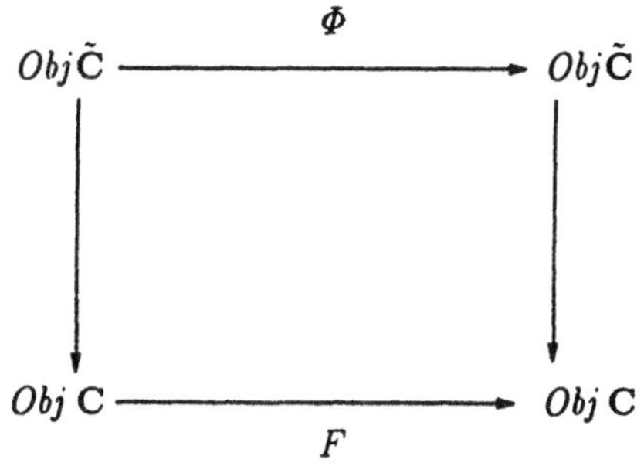

where vertical maps are given by the canonical functor that forgets S in $\langle S, A\rangle$. When there is a function F with this property, we say that Φ is an *extension* of F.

A function $\Phi\colon Obj\tilde{\mathbf{C}} \longrightarrow Obj\tilde{\mathbf{C}}$ that extends some function $F\colon Obj\,\mathbf{C} \longrightarrow Obj\,\mathbf{C}$ is just a (unary) relator on $\mathbf{C}$. In other words, the unary relators on $\mathbf{C}$ are object maps on $\tilde{\mathbf{C}}$ that extend object maps on the "underlying" category $\mathbf{C}$. Equivalently, a (unary) relator on $\mathbf{C}$ consists of an object map $F\colon Obj\,\mathbf{C} \longrightarrow Obj\,\mathbf{C}$, together with a mapping that associates a subset $S' \hookrightarrow |F(A)|$ to any subset $S \hookrightarrow |A|$. More generally, it is possible to define relators that map n-tuples of k-ary relations to k-ary relations, for each k and n. Let $\mathbf{E} = \mathbf{C}^k$. A relator given by a map from $(Obj\tilde{\mathbf{E}})^n$ to $Obj\tilde{\mathbf{E}}$ will be called a k-ary relator of n arguments. The definition of unary relator of n arguments (so $k = 1$) is therefore derived as in the diagram above, using $\mathbf{C}^n$ in place of $\mathbf{C}$ in the domain of F and $(\tilde{\mathbf{C}})^n$ in place of $\tilde{\mathbf{C}}$ in the domain of Φ. We thus obtain the unary case of Definition 9.

Rather than consider arbitrary natural transformations between (unary) relators on $\mathbf{C}$, we will be interested in natural transformations that "extend" natural transformations of $\mathbf{C}_{iml}$. More precisely, let $\Phi, \Psi\colon Obj\tilde{\mathbf{C}} \longrightarrow \tilde{\mathbf{C}}$ be functors from the discrete category $Obj\tilde{\mathbf{C}}$ to $\tilde{\mathbf{C}}$. While these functors are essentially object maps $Obj\tilde{\mathbf{C}} \longrightarrow Obj\tilde{\mathbf{C}}$, the natural transformations between them are families of *arbitrary* morphisms $\tau_{(S,A)}\colon \Phi(S,A) \rightarrow \Psi(S,A)$ in $\tilde{\mathbf{C}}$, with (S,A) ranging over arbitrary objects of $\tilde{\mathbf{C}}$. (These are *not* the same as natural transformations between functors $Obj\tilde{\mathbf{C}} \longrightarrow Obj\tilde{\mathbf{C}}$ into the discrete category.) If Φ, Ψ are relators that extend object maps $F, G\colon Obj\,\mathbf{C} \longrightarrow Obj\,\mathbf{C}$, a natural transformation $\tau\colon \Phi \longrightarrow \Psi$ that extends a (necessarily unique) natural transformation t between the associated functors $F, G\colon Obj\,\mathbf{C} \longrightarrow \mathbf{C}$ is precisely a relator transformation from Φ to Ψ, as in Definition 10. It is worth observing that if τ is a relator transformation extending t, then in fact t also determines τ uniquely. (We observed a similar fact in regard to the morphisms in subscones in Section 4.) Therefore, we often say that t is a relator transformation In other words, if Φ, Ψ are relators on $\mathbf{C}$ with object maps $F, G\colon Obj\,\mathbf{C} \longrightarrow Obj\,\mathbf{C}$, a relator transformation t from Φ to Ψ is a family of morphisms $t_A\colon F(A) \rightarrow G(A)$ in $\mathbf{C}$, with A ranging over arbitrary objects of $\mathbf{C}$, such that for any subset $S \hookrightarrow |A|$, t_A induces a (necessarily unique) morphism $\Phi(S,A) \rightarrow \Psi(S,A)$ in $\tilde{\mathbf{C}}$.

It can now be easily verified that relators and relator transformations form an *iml*-category.

Proposition 11. *Let* $\mathbf{C}$ *be a ccc and let* $\mathbf{D} = \tilde{\mathbf{C}}$. *The category of (unary) relators of* n *arguments and relator transformations on a ccc* $\mathbf{C}$ *is a sub-ccc of the fiber* $F_n = \mathbf{D}^{(Obj\,\mathbf{D})^n}$ *of the iml-category* $\mathbf{D}_{iml}$. *Furthermore, let the morphisms* $k \rightarrow n$ *of a new base category* $\mathbf{B}'$ *be* n*-tuples of relators of* k *arguments over* $\mathbf{C}$. *Then the indexed category* $(\mathbf{B}', Rel\mathbf{C})$ *of relators and relator transformations on* $\mathbf{C}$ *is a sub-iml-category of* $\mathbf{D}_{iml}$ *and the canonical forgetful functor* $(\mathbf{B}', Rel\mathbf{C}) \longrightarrow \mathbf{C}_{iml}$ *is an iml-representation.*

We saw in Section 4.1 that the binary case of sconing is an instance of sconing. The notion of binary relator on a ccc $\mathbf{C}$ given in Definition 9 may be similarly derived from the notion of unary relator just discussed. Notice that it is not sufficient to simply let $\mathbf{D}$ be the subscone of $\mathbf{C} \times \mathbf{C}$ and then proceed as in Proposition 11 because we would get too many object maps. Indeed, we would get all maps $(Obj\,\mathbf{C})^2 \longrightarrow (Obj\,\mathbf{C})^2$, whereas for binary relators we need only those maps induced by pairs of maps $F', F''\colon Obj\,\mathbf{C} \longrightarrow Obj\,\mathbf{C}$, *i.e.* those mapping a pair of objects (A, B) to $(F'(A), F''(B))$. (In addition, we should restrict further $F' = F''$, but this further restriction can be dealt with naturally, see below.) Here the relevant *iml*-category is the product *iml*-category $\mathbf{C}_{iml} \times \mathbf{C}_{iml}$. In general, the product of two *iml*-categories is the *iml*-category whose base category morphisms $k \to n$ are pairs of such morphisms from the two given base categories, and whose fiber over n is the ordinary product of the two given fibers over n. The two projections are *iml*-representations.

With this in mind, we can revisit the commutative square above but with $\tilde{\mathbf{C}}$ replaced by $\mathbf{D}$, where $\mathbf{D}$ is the subscone of $\mathbf{C} \times \mathbf{C}$, and with $\mathbf{C}$ replaced by $\mathbf{C} \times \mathbf{C}$. Moreover, F ranges over objects of the fiber over $\mathbf{1}$ in the product *iml*-category $\mathbf{C}_{iml} \times \mathbf{C}_{iml}$. Such maps $\Phi: \mathbf{D} \longrightarrow \mathbf{D}$ are given by two object maps $F', F'': Obj\,\mathbf{C} \longrightarrow Obj\,\mathbf{C}$ by the requirement $\Phi(S, A, B) = (S', F'(A), F''(B))$. The restriction on transformations follows the unary case: we consider only the families of morphisms $\Phi(S, A, B) \to \Psi(S, A, B)$ in $\mathbf{D}$, with (S, A, B) ranging over the objects of $\mathbf{D}$, that extend the families of morphisms in $\mathbf{C} \times \mathbf{C}$ given by pairs of morphisms $F'(A) \to G'(A)$, $F''(B) \to G''(B)$ in $\mathbf{C}$. In this way we obtain a sub-ccc of the fiber over $\mathbf{1}$ of $\mathbf{D}_{iml}$. Binary relators on $\mathbf{C}$ and relator transformations are clearly a further subcategory of this ccc, and as such they inherit the ccc structure.

For n arguments, we consider the analogous commutative square condition, with $\mathbf{D}^n$ in place of $\mathbf{D}$ in the domain of Φ, and with $(\mathbf{C} \times \mathbf{C})^n$ in place of $\mathbf{C} \times \mathbf{C}$ in the domain of F, obtaining a sub-ccc of the fiber over n of $\mathbf{D}_{iml}$. We have constructed an *iml*-category (with base morphisms given in the obvious way so as to have an *iml*-category) for which the canonical forgetful functor to $\mathbf{C}_{iml} \times \mathbf{C}_{iml}$ is an *iml*-representation. Furthermore, binary relators and their transformations form a sub-*iml*-category of the *iml*-category we have just constructed. Thus the image of the free *iml*-representation of the term *iml*-category in the *iml*-category we have constructed contains only binary relators, *i.e.* the free *iml*-representation factors through the *iml*-category of binary relators and relator transformations.

Having observed a specialization of Proposition 11 to the binary case (k-ary case is analogous), let us also observe that a similar reasoning allows us to generalize Proposition 11 along the lines of Proposition 2 and Section 6. For this purpose let $\mathbf{C}$ and $\mathbf{C}'$ be cartesian closed categories, with $\mathbf{C}'$ having equalizers, and let $\mathbf{F}: \mathbf{C} \longrightarrow \mathbf{C}'$ be a functor that preserves finite products up to isomorphism. Let ccc $\mathbf{D}$ be a full subcategory of the comma category $(\mathbf{C}' \downarrow \mathbf{F})$, so that the restriction of the canonical ccc-representation $(\mathbf{C}' \downarrow \mathbf{F}) \longrightarrow \mathbf{C}$ to $\mathbf{D}$ is a ccc-representation. For now let us also assume that this ccc-representation $\mathbf{D} \to \mathbf{C}$ is surjective on objects. (This additional surjectivity condition does hold in interesting cases for Kripke models, cpos, and pers.) Let us reconsider the commutative square given above with this new choice of $\mathbf{D}$ in place of $\tilde{\mathbf{C}}$. We say that Φ is a (unary) $\mathbf{D}$-*relator* on $\mathbf{C}$ that extends a (necessarily unique) object map F. If Φ, Ψ are $\mathbf{D}$-relators that extend object maps $F, G: Obj\,\mathbf{C} \longrightarrow Obj\,\mathbf{C}$, a natural transformation $\tau: \Phi \longrightarrow \Psi$ that extends a (necessarily unique) natural transformation t between the associated functors $F, G: Obj\,\mathbf{C} \longrightarrow \mathbf{C}$ is said to be a $\mathbf{D}$-relator transformation from Φ *to* Ψ. (While in general t does not determines τ, we have already remarked in Section 6 that Kripke models, cpos, and pers are "concrete" enough so that we may assume that in the objects of $\mathbf{D}$ the mapping components are given by inclusions, and hence that t determines τ uniquely.) We obtain:

Proposition 12. *Let* $\mathbf{C}$ *and* $\mathbf{C}'$ *be cartesian closed categories,* $\mathbf{C}'$ *with equalizers. Let* $\mathbf{G}: \mathbf{C} \longrightarrow \mathbf{C}'$ *be a functor that preserves finite products up to isomorphism. Let ccc* $\mathbf{D}$ *be a full subcategory of the comma category* $(\mathbf{C}' \downarrow \mathbf{G})$, *so that the restriction of the canonical ccc-representation* $(\mathbf{C}' \downarrow \mathbf{G}) \longrightarrow \mathbf{C}$ *to* $\mathbf{D}$ *is a ccc-representation. Furthermore, let this ccc-representation* $\mathbf{D} \to \mathbf{C}$ *be surjective on objects. Then the category of* *(unary)* $\mathbf{D}$ *-relators of* n *arguments and* $\mathbf{D}$ *-relator transformations on a ccc* $\mathbf{C}$ *is a sub-ccc of the fiber* $F_n = \mathbf{D}^{(Obj\,\mathbf{D})^n}$ *of the iml-category* $\mathbf{D}_{iml}$*. Furthermore, let the morphisms* $k \to n$ *of a new base category* $\mathbf{B}'$ *be* n*-tuples of* $\mathbf{D}$ *-relators of* k *arguments over* $\mathbf{C}$*. Then the indexed category* $(\mathbf{B}', Rel_{\mathbf{D}}\mathbf{C})$ *of* $\mathbf{D}$ *-relators and relator transformations on* $\mathbf{C}$ *is a sub-iml-category of* $\mathbf{D}_{iml}$ *and the canonical forgetful functor* $(\mathbf{B}_{\mathbf{D}}, Rel_{\mathbf{D}}\mathbf{C}) \longrightarrow \mathbf{C}_{iml}$ *is an iml-representation.*

Example 4. Let us illustrate the notions involved in Proposition 12 in the setting used in [AbJ91] for the purposes of the so-called strictness analysis. In this example $\mathbf{C}'$ is the category Cpo of cpos and continuous maps, $\mathbf{C}$ is the full subcategory $Pcpo$ of pointed cpos, and $\mathbf{G}$ is the inclusion functor $\mathbf{I}$. We concentrate on the binary case, so we let $\mathbf{H}(A,B) = A \times B$ and we let $\mathbf{D}$ be the full subcategory of $(Cpo \times Cpo \downarrow \mathbf{H})$ whose objects are of the form $\langle S, A, B\rangle$, where S is a sub-pcpo of pcpo $A \times B$, *i.e.*S is a strict, inductive relation from A to B, see Section 6.2. $\mathbf{D}$ is a sub-ccc of the comma category and furthermore $\mathbf{D}$ is a ccc with fixed-point operators. A binary pcpo-relator on $Pcpo$ consists of a mapping F from pcpos to pcpos, together with a mapping that to any strict, inductive relation $S\colon A \longrightarrow\!\!\!\!\times B$ associates a strict, inductive relation $S'\colon F(A) \longrightarrow\!\!\!\!\times F(B)$. If F, G are pcpo-relators on pcpos, a relator transformation from F to G is a family of continuous maps $t_A\colon F(A) \to G(A)$, with A ranging over pcpos, such that for each strict, inductive relation $S\colon A \longrightarrow\!\!\!\!\times B$ and for the two strict, inductive relations $S'\colon F(A) \longrightarrow\!\!\!\!\times F(B)$ and $S''\colon G(A) \longrightarrow\!\!\!\!\times G(B)$ assigned to S by F and by G, respectively, it is the case that the composite inductive relation $t_B \circ S'$ is included in the inductive relation $S'' \circ t_A$. This condition states exactly that for all elements a in pcpo $F(A)$ and b in pcpo $F(B)$, $a\,S'\,b$ implies $(t_A(a))\,S''\,(t_B(b))$. We have described the fiber over $\mathbf{1}$ in the *iml*-category of pcpo-relators over $Pcpo$ and their transformations, a special case of Proposition 12. This *iml*-category has fixed-point operators in the sense that each fiber is a ccc with fixed-point operators and they are preserved by the ccc-representations $\mathbf{F}_n \to \mathbf{F}_k$ induced by morphisms $k \to n$ in the base category. Pcpo-relators over $Pcpo$ that take identity relations to identity relations form a sub-*iml*-category with fixed-point operators. Thus the free *iml*-representation of the $\lambda^{\to,t}$ term *iml*-category in the *iml*-category of pcpo-relators over $Pcpo$ (determined by an interpretation of type constants in $\mathbf{D}$) factors through the sub-*iml*-category of pcpo-relators that take identities to identities. Furthermore, since this sub-*iml*-category also has fixed-point operators, the latter *iml*-representation factors through the term *iml*-category of $\lambda^{\to,t}$ with fixed-point operators, as in Section 6.2. Thus the main soundness theorem in [AbJ91] follows. □

7.4 Relators over *iml*-categories

Proposition 11 may also be generalized in another way. The reader will notice that in our definition of (unary) relators on a ccc $\mathbf{C}$, the required object maps $F\colon Obj\,\mathbf{C} \longrightarrow Obj\,\mathbf{C}$ are exactly the objects of the fiber over $\mathbf{1}$ in the *iml*-category $\mathbf{C}_{iml}$, where the fiber over $\mathbf{0}$ is the ccc $\mathbf{C}$. In other words, we have constructed an *iml*-category of relators and relator transformations, starting from a particular kind of *iml*-category, $\mathbf{C}_{iml}$.

However, the construction easily generalizes to arbitrary *iml*-categories. Let us recall that in any *iml*-category $(\mathbf{B}, \mathbf{F})$, the objects of the fiber $\mathbf{F}_n$ over n are exactly the morphisms $n \to \mathbf{1}$ in the given base category $\mathbf{B}$. In particular, an object of the fiber $\mathbf{F}_1$ is a morphism $\mathbf{1} \to \mathbf{1}$ in $\mathbf{B}$, and hence it composes in $\mathbf{B}$ with any morphism $\mathbf{0} \to \mathbf{1}$ to produce again a morphism $\mathbf{0} \to \mathbf{1}$. In this way an object of the fiber $\mathbf{F}_1$

induces an object map $F: Obj\mathbf{F}_0 \longrightarrow Obj\mathbf{F}_0$. Hence in the more general definition of a relator over an *iml*-category $\mathbf{F}$ it makes sense to consider only those object maps $F: Obj\mathbf{F}_0 \longrightarrow Obj\mathbf{F}_0$ induced by the objects of the fiber $\mathbf{F}_1$ by composition. An object of the fiber $\mathbf{F}_n$ similarly induces an object map of n arguments on $\mathbf{F}_0$ by composition with n-tuples of morphisms $\mathbf{0} \to \mathbf{1}$ (*i.e.*with morphisms $\mathbf{0} \to n$) in $\mathbf{B}$.

Definition 13. A (unary) relator of n arguments over an *iml*-category $(\mathbf{B}, \mathbf{F})$ is given by an object A of the fiber $\mathbf{F}_n$, together with a (unary) relator of n arguments on $\mathbf{F}_0$, whose object map is induced by A by composition.

Relator transformations over an *iml*-category $(\mathbf{B}, \mathbf{F})$ may be defined in a similar manner. Let Φ, Ψ be relators over $(\mathbf{B}, \mathbf{F})$, whose object maps F, G are induced by objects A, B of the fiber $\mathbf{F}_1$. Any morphism $x: A \to B$ in the fiber $\mathbf{F}_1$ induces a family of morphisms $t_C: F(C) \to G(C)$ in the fiber $\mathbf{F}_0$, with C ranging over the objects of the fiber $\mathbf{F}_0$, as follows. Because $\mathbf{F}(C): \mathbf{F}_1 \longrightarrow \mathbf{F}_0$ is a functor , it is the case that $\mathbf{F}(C)(x): \mathbf{F}(C)(A) \to \mathbf{F}(C)(B)$. Furthermore, the functor $\mathbf{F}(C)$ acts on objects by composition, so $\mathbf{F}(C)(A) = A \circ C$ and $\mathbf{F}(C)(B) = B \circ C$, and thus in fact $\mathbf{F}(C)(x): A \circ C \to B \circ C$. But $F(C) = A \circ C$ and $G(C) = B \circ C$. We shall be interested only in those relator transformations $\Phi \longrightarrow \Psi$ for which the associated families t are induced by morphisms of the fiber $\mathbf{F}_1$ by composition, as just described. Similarly, for relators of n arguments over $(\mathbf{B}, \mathbf{F})$, we shall be interested only in those relator transformations for which the associated transformations between object maps are induced by morphisms of the fiber $\mathbf{F}_n$.

Definition 14. Let $(A, \Phi), (B, \Psi)$ be relators of n arguments over an *iml*-category $(\mathbf{B}, \mathbf{F})$. Let $F, G: (Obj\mathbf{F}_0)^n \longrightarrow \mathbf{F}_0$ be functors induced by A and by B , respectively. A relator transformation $(A, \Phi) \longrightarrow (B, \Psi)$ over $(\mathbf{B}, \mathbf{F})$ is given by a morphism $x: A \to B$ of the fiber $\mathbf{F}_n$ together with a relator transformation $\Phi \longrightarrow \Psi$ whose associated transformation $F \longrightarrow G$ is induced by the morphism x by composition.

Proposition 11 thus extends to:

Theorem 15. *Let* $(\mathbf{B}, \mathbf{F})$ *be an iml-category. Let the morphisms* $k \to n$ *of a new base category* $\mathbf{B}'$ *be* n*-tuples of relators of* k *arguments over* $(\mathbf{B}, \mathbf{F})$*. Then the indexed category* $(\mathbf{B}', \tilde{\mathbf{F}})$ *whose fiber over* n *is the category of relators of* n *arguments and relator transformations over* $(\mathbf{B}, \mathbf{F})$ *is an iml-category. Furthermore, the canonical forgetful functor* $(\mathbf{B}', \tilde{\mathbf{F}}) \longrightarrow (\mathbf{B}, \mathbf{F})$ *is an iml-representation.*

Proof. Let us outline the main points. We shall actually show that when dealing with relators and relator transformations over $(\mathbf{B}, \mathbf{F})$, both of which are defined as certain pairs, in the left coordinate one may work in $(\mathbf{B}, \mathbf{F})$ and in the right coordinate one may work in the *iml*-category of relators and relator transformations on ccc $\mathbf{F}_0$ given in Proposition 11.

First, observe that Definition 13 is stable under composition: if (A, Φ) is a relator of n arguments over $(\mathbf{B}, \mathbf{F})$ and $(B_1, \Psi_1), \ldots, (B_n, \Psi_n)$ relators of k arguments over $(\mathbf{B}, \mathbf{F})$, then $(A \circ \langle B_1, \ldots, B_n \rangle, \Phi \circ \langle \Psi_1, \ldots, \Psi_n \rangle)$ is a relator of k arguments over $(\mathbf{B}, \mathbf{F})$. Here the n-tuple $\langle B_1, \ldots, B_n \rangle$ is the morphism $k \to n$ in the given base category $\mathbf{B}$ with products. In this way each morphism $k \to n$ of the new base category $\mathbf{B}'$ induces a mapping from relators of n arguments to relators of k arguments, *i.e.*from the objects of the new fiber $(\tilde{\mathbf{F}})_n$ over n to the objects of the new fiber $(\tilde{\mathbf{F}})_k$ over k. This mapping will be the object part of the required functor $(\tilde{\mathbf{F}})_n \to (\tilde{\mathbf{F}})_k$.

Similarly, let $(x,\tau):(A,\Phi) \longrightarrow (A',\Phi')$ be a relator transformation between relators of n arguments over $(\mathbf{B},\mathbf{F})$ and let $k \to n$ be a morphism in the new base category $\mathbf{B}'$ given by an n-tuple $\langle(B_1,\Psi_1),\ldots,(B_n,\Psi_n)\rangle$ of relators of k arguments over $(\mathbf{B},\mathbf{F})$. We shall define the associated relator transformation between relators of k arguments over $(\mathbf{B},\mathbf{F})$. Let $a:k \to n$ be the morphism in the starting base category $\mathbf{B}$ given by $\langle B_1,\ldots,B_n\rangle$. Let $D = \mathbf{F}(a)(A) = A \circ a$, $D' = \mathbf{F}(a)(A') = A' \circ a$, and $y = \mathbf{F}(a)(x)$. Let $\Theta = \Phi \circ \langle\Psi_1,\ldots,\Psi_n\rangle$, $\Theta' = \Phi' \circ \langle\Psi_1,\ldots,\Psi_n\rangle$, and $\sigma = \tau \circ \langle\Psi_1,\ldots,\Psi_n\rangle$, as defined in the *iml*-category of relators on $\mathbf{F}_0$. Then $(y,\sigma):(D,\Theta) \longrightarrow (D',\Theta')$ is the required relator transformation between relators of k arguments over $(\mathbf{B},\mathbf{F})$. The reader will readily check that this assignment is functorial, *i.e.*it preserves composition of relator transformations.

We also verify that each new fiber $(\tilde{\mathbf{F}})_n$ is a ccc in such way that the ccc structure is preserved under composition. Finite products in $(\tilde{\mathbf{F}})_n$ are clearly given coordinatewise and hence preserved under composition. Let us calculate $(A,\Phi) \Rightarrow (B,\Psi)$ in $(\tilde{\mathbf{F}})_n$. Here A, B are objects in $\mathbf{F}_n$ and Φ,Ψ are relators of n arguments on the ccc $\mathbf{F}_0$, whose object maps are induced by composition by A, B, respectively. Note that for any morphism $C: 0 \to n$ in the given base category $\mathbf{B}$, $(A \Rightarrow B)\circ C = (A\circ C) \Rightarrow (B \circ C)$ because $\mathbf{F}(C):\mathbf{F}_n \longrightarrow \mathbf{F}_0$ is a ccc representation such that $\mathbf{F}(C)(D) = D \circ C$ for any object D in the ccc $\mathbf{F}_n$. But this means exactly that $A \Rightarrow B$ induces the object map of $\Phi \Rightarrow \Psi$ as determined in the ccc of relators of n arguments on ccc $\mathbf{F}_0$, see Proposition 11.

Finally, erasing the right coordinate yields both a functor $L:\mathbf{B}' \longrightarrow \mathbf{B}$ that preserves finite products so that $L(n) = n$, as well as a natural transformation $l:\tilde{\mathbf{F}} \longrightarrow (\mathbf{F} \circ L)$ such that for each n, $l_n:(\tilde{\mathbf{F}})_n \to \mathbf{F}_n$ is a ccc representation.

Example 5. The discussion of the so-called total objects in the Appendix D of [Gir86] provides an interesting special case of Theorem 15. The starting *iml*-category $(\mathbf{B},\mathbf{F})$ is given in Example 2. Object maps of the relevant relators on the ccc Qd of qualitative domains and stable maps are precisely the object maps of functors from the category Qd_{emb} of qualitative domains and embeddings to Qd_{emb} that preserve pullbacks and directed colimits. For any type σ in n type variables, the free *iml*-representation in $(\mathbf{B},\mathbf{F})$ assigns to σ a functor $A:(Qd_{emb})^n \longrightarrow Qd_{emb}$ that preserves pullbacks and directed colimits. By Theorem 15, the canonical forgetful functor from the *iml*-category of relators over $(\mathbf{B},\mathbf{F})$ to $(\mathbf{B},\mathbf{F})$ is an *iml*-representation. Furthermore, because free *iml*-representation (of the term *iml*-category) in any given *iml*-category is unique, the unique free *iml*-representation in $(\mathbf{B},\mathbf{F})$ must factor through the free *iml*-representation to the *iml*-category of relators over $(\mathbf{B},\mathbf{F})$. Therefore this latter representation assigns to type σ a relator of n arguments over $(\mathbf{B},\mathbf{F})$ that must be of the form (A,Φ). In this particular case, this means simply that the object map of A is the object map of the relator Φ of n arguments on Qd. The total elements of A, defined in the Appendix D of [Gir86], are in our terminology those morphisms $1 \to A$ in $\mathbf{F}_n$ that induce relator transformations $1 \to \Phi$. Thus the first order part of Theorem D.1 in [Gir86] follows from our Theorem 15. □

Theorem 15 also specializes to the m-ary case, by working over the m-fold power of a given *iml*-category (defined below Proposition 11). For instance, a binary relator over an *iml*-category $(\mathbf{B},\mathbf{F})$ is given by an object A of the fiber $\mathbf{F}_1$, together with a unary relator on $\mathbf{F}_0 \times \mathbf{F}_0$, whose object map is induced by the object $\langle A, A\rangle$ of the fiber $(\mathbf{F} \times \mathbf{F})_1$ by composition.

Let us also mention that Theorem 15 may be generalized along the lines of Proposition 12. The reader will note that the additional condition in Proposition 12 that the canonical ccc-representation $\mathbf{D} \to \mathbf{C}$ is surjective on objects may now be omitted because the object map of a $\mathbf{D}$-relator over a given *iml*-category $(\mathbf{B}, \mathbf{F})$ is given explicitly by an object of the fiber $\mathbf{F}_1$.

Theorem 16. *Let* $(\mathbf{B}, \mathbf{F})$ *be an iml-category. Let* $\mathbf{C}'$ *be a cartesian closed categories with equalizers. Let* $\mathbf{G}: \mathbf{F}_0 \longrightarrow \mathbf{C}'$ *be a functor that preserves finite products up to isomorphism. Let ccc* $\mathbf{D}$ *be a full subcategory of the comma category* $(\mathbf{C}' \downarrow \mathbf{G})$, *so that the restriction of the canonical ccc-representation* $(\mathbf{C}' \downarrow \mathbf{G}) \longrightarrow \mathbf{C}$ *to* $\mathbf{D}$ *is a ccc-representation. Let the morphisms* $k \to n$ *of the new base category* $\mathbf{B}'$ *be* n*-tuples of* $\mathbf{D}$*-relators of* k *arguments over* $(\mathbf{B}, \mathbf{F})$. *Then the indexed category* $(\mathbf{B}', \tilde{\mathbf{F}})$ *whose fiber over* n *is the category of* $\mathbf{D}$*-relators of* n *arguments and* $\mathbf{D}$*-relator transformations over* $(\mathbf{B}, \mathbf{F})$ *is an iml-category. Furthermore, the canonical forgetful functor* $(\mathbf{B}', \tilde{\mathbf{F}}) \longrightarrow (\mathbf{B}, \mathbf{F})$ *is an iml-representation.*

Example 6. Let us give an example of the binary case of Theorem 16. Recalling Section 6.3, let $\mathbf{D}$ be the full sub-ccc of the comma category $(Per \times Per \downarrow \mathbf{H})$ whose objects are saturated binary relations on pers, *i.e.*objects of the form $\langle S, P, Q \rangle$, where S is a saturated binary relation from the domain of per P to the domain of per Q. Let the starting *iml*-category $(\mathbf{B}, \mathbf{F})$ be the *iml*-category PER discussed in Example 3, a sub-*iml*-category of Per_{iml}. In particular, the fiber over 0 in PER is the ccc Per of pers and per morphisms. A (binary) $\mathbf{D}$-relator over PER is simply a (binary) $\mathbf{D}$-relator on Per, *i.e.*it consists of a mapping F from pers to pers and a mapping Φ: $Obj\,\mathbf{D} \longrightarrow Obj\,\mathbf{D}$ such that $\Phi(S, P, Q) = (S', F(P), F(Q))$. However, a $\mathbf{D}$-relator transformation over PER is not simply a $\mathbf{D}$-relator transformation on Per, because it must be given by a realizable family of per morphisms. Indeed, let $\langle F, \Phi \rangle$ and G, Ψ be $\mathbf{D}$-relators over PER. A $\mathbf{D}$-relator transformation $t: \langle F, \Phi \rangle \longrightarrow \langle G, \Psi \rangle$ over PER is given by a partial recursive function f that names each per morphism $t_P: F(P) \to G(P)$, with P ranging over pers, and such that for every saturated relation S from the domain of P to the domain of Q and the associated saturated relations S' and S'' given by Φ and Ψ, respectively, it is the case that the composite relation $t_Q \circ S'$ is included in the relation $S'' \circ t_P$. (Each per morphism induces a saturated relation.) In other words, for all pers P and Q and all numbers n in the domain of $F(P)$ and k in the domain of $F(Q)$, $n\ S'\ k$ implies $f(n)\ S''\ f(k)$. Thus the first order part of the Soundness Theorem in Section 4.7 of [BFS90] is a special case of our Theorem 16. □

Example 7. Another example of this more general situation is the *iml*-category of pcpo-relators whose associated object maps are not just arbitrary maps from pcpos to pcpos, but functors from the category $Pcpo_{emb}$ of pcpos and embedding-projection pairs to $Pcpo_{emb}$. Here the fiber $\mathbf{F}_n$ of the starting *iml*-category is the ccc whose objects are functors $(Pcpo_{emb})^n \longrightarrow Pcpo_{emb}$ and the morphisms $t: F \to G$ of the fiber $\mathbf{F}_n$ are certain families of continuous maps $t_P: F(P) \to G(P)$, where P ranges over n-tuples of pcpos, see [CGW89]. □

References

[ACC93] M. Abadi, L. Cardelli, and P.-L. Curien. Formal parametric polymorphism. In *Proc. 20-th ACM Symposium on Principles of Programming Languages*, 1993.

[AbJ91] S. Abramsky and T.P. Jensen. A relational approach to strictness analysis for higher-order polymorphic functions. In *Proc. 18-th ACM Symposium on Principles of Programming Languages*, 1991.

[BFS90] E.S. Bainbridge, P.J. Freyd, A. Scedrov, and P.J. Scott. Functorial Polymorphism. *Theoretical Computer Science*, 70:35–64, 1990. Corrigendum *ibid.*, 71:431, 1990.

[BTC88] V. Breazu-Tannen and T. Coquand. Extensional models for polymorphism. *Theoretical Computer Science*, 59:85–114, 1988.

[BMM90] K. B. Bruce, A. R. Meyer, and J. C. Mitchell. The semantics of second-order lambda calculus. *Information and Computation*, 85(1):76–134, 1990. Reprinted in *Logical Foundations of Functional Programming*, ed. G. Huet, Addison-Wesley (1990) 213–273.

[CMS91] L. Cardelli, J.C. Mitchell, S. Martini, and A. Scedrov. An extension of system F with Subtyping. To appear in *Information and Computation*. Extended abstract in T.Ito and A.R. Meyer (eds.), *Theoretical Aspects of Computer Software*, pages 750–770. Springer-Verlag LNCS 526, 1991.

[CGW89] T. Coquand, C.A. Gunter, and G. Winskel. Domain theoretic models of polymorphism. *Information and Computation*, 81:123–167, 1989.

[CP92] R.L. Crole and A.M. Pitts. New foundations for fixpoint computations: FIX-hyperdoctrines and the FIX-logic. *Information and Computation*, 98:171–210, 1992.

[Fre93] P.J. Freyd. Structural polymorphism. *Theoretical Computer Science*, to appear.

[FrS90] P.J. Freyd and A. Scedrov. *Categories, Allegories*. Mathematical Library, North-Holland, 1990.

[FRR92] P.J. Freyd, E.P. Robinson, and G. Rosolini. Functorial parametricity. In *Proc. 7-th Annual IEEE Symposium on Logic in Computer Science*, pages 444–452, 1992.

[FRS92] P.J. Freyd, E.P. Robinson, and G. Rosolini. Dinaturality for free. In M.P. Fourman, P.T. Johnstone, and A.M. Pitts, eds., *Applications of Categories in Computer Science*, pages 107–118. London Math. Soc. Lecture Note Series, vol. 177, Cambridge Univ. Press, 1992.

[Gir86] J.-Y. Girard. The system F of variable types, fifteen years later. *Theoretical Computer Science*, 45:159–192, 1986. Reprinted in *Logical Foundations of Functional Programming*, ed. G. Huet, Addison-Wesley (1990) 87–126.

[GLT89] J.-Y. Girard, Y. Lafont, and P. Taylor. *Proofs and Types*. Cambridge Tracts in Theoretical Computer Science, Cambridge University Press, 1989.

[GSS91] J.-Y. Girard, A. Scedrov, and P.J. Scott. Normal forms and cut-free proofs as natural transformations. In: Y.N. Moschovakis, editor, *Logic from Computer Science, Proc. M.S.R.I. Workshop, Berkeley, 1989*. M.S.R.I. Series, Springer-Verlag, 1991.

[Has90] R. Hasegawa. Categorical data types in parametric polymorphims. Manuscript, 1990.

[Has91] R. Hasegawa. Parametricity of extensionally collapsed term models of polymorphism and their categorical properties. In T.Ito and A.R. Meyer (eds.), *Theoretical Aspects of Computer Software*, pages 495–512. Springer-Verlag LNCS 526, 1991.

[KR77] A. Kock, G.E. Reyes. Doctrines in categorical logic. In J. Barwise, editor, *Handbook of Mathematical Logic*, pages 283–313. North-Holland, 1977.

[Laf88] Y. Lafont. *Logiques, Categories & Machines*. Thèse de Doctorat, Université Paris VII, 1988.

[LS86] J. Lambek and P.J. Scott. *Introduction to Higher-Order Categorical Logic*. Cambridge studies in advanced mathematics 7, Cambridge University Press, 1986.

[MaR92] Q. Ma and J.C. Reynolds. Types, abstraction, and parametric polymorphism, Part 2. In S. Brookes *et al.*, editors, *Mathematical Foundations of Programming Semantics, Proceedings 1991*, pages 1–40. Springer-Verlag LNCS 598, 1992.

[McL71] S. Mac Lane. *Categories for the Working Mathematician*. Graduate Texts in Mathematics, Springer-Verlag, 1971.

[Mai91] H. Mairson. Outline of a proof theory of parametricity. In *Proc. 5-th Intern. Symp. on Functional Programming and Computer Architecture*, 1991.

[MS76] R.E. Milne and C. Strachey. *A theory of programming language semantics*. Chapman and Hall, London, and Wiley, New York, 1976.

[Mit86] J.C. Mitchell. A type-inference approach to reduction properties and semantics of polymorphic expressions. In *ACM Conference on LISP and Functional Programming*, pages 308–319, August 1986. Revised version in *Logical Foundations of Functional Programming*, ed. G. Huet, Addison-Wesley (1990) 195–212.

[Mit88] J.C. Mitchell. Polymorphic type inference and containment. *Information and Computation*, 76(2/3):211–249, 1988. Reprinted in *Logical Foundations of Functional Programming*, ed. G. Huet, Addison-Wesley (1990) 153–194.

[Mit90] J.C. Mitchell. Type systems for programming languages. In J. van Leeuwen, editor, *Handbook of Theoretical Computer Science*, pages 365–458. North-Holland, 1990.

[MM85] J.C. Mitchell and A.R. Meyer. Second-order logical relations. In *Logics of Programs*, pages 225–236. Springer-Verlag LNCS 193, June 1985.

[MM91] J.C. Mitchell and E. Moggi. Kripke-style models for typed lambda calculus. *Annals Pure Appl. Logic*, 51:99–124, 1991.

[MS89] J.C. Mitchell and P.J. Scott. Typed lambda calculus and cartesian closed categories. In J. W. Gray and A. Scedrov, editors, *Categories in Computer Science and Logic, Contemporary Math.*, vol. 92, pages 301-316. Amer. Math. Society, 1989.

[Mgg89] E. Moggi. Computational lambda calculus and monads. In *Proc. 4th IEEE Symposium on Logic in Computer Science*, pages 14–23, IEEE Computer Society Press, 1989.

[Mog91] E. Moggi. A category-theoretic account of program modules. *Math. Structures in Computer Science*, 1(1):103–139, 1991.

[OHT93] P.W. O'Hearn and R.D. Tennent. Relational parametricity and local variables. In *Proc. 20-th ACM Symposium on Principles of Programming Languages*, 1993.

[Pit87] A.M. Pitts. Polymorphism is set-theoretic, constructively. In *Category Theory and Computer Science, Proceedings Edinburgh, 1987*, pages 12–39. Springer-Verlag LNCS volume 283, 1987.

[Plo80] G.D. Plotkin. Lambda definability in the full type hierarchy. In *To H.B. Curry: Essays on Combinatory Logic, Lambda Calculus and Formalism*, pages 363–373. Academic Press, 1980.

[Rey74] J.C. Reynolds. On the relation between direct and continuation semantics. In *Second Colloq. Automata, Languages and Programming*, pages 141–156. Springer-Verlag LNCS, 1974.

[Rey83] J.C. Reynolds. Types, abstraction, and parametric polymorphism. In R. E. A. Mason, editor, *Information Processing '83*, pages 513–523. North-Holland, 1983.

[ScS82] A. Scedrov and P.J. Scott. A note on the Friedman slash and Freyd covers. In A.S. Troelstra and D. van Dalen, editors, *The L. E. J. Brouwer Symposium*, pages 443–452. North-Holland, 1982.

[See87] R.A.G. Seely. Categorical semantics for higher order polymorphic lambda calculus. *Journal of Symbolic Logic*, 52:969–989, 1987.

[Sta85] R. Statman. Logical relations and the typed lambda calculus. *Information and Control*, 65:85–97, 1985.

[Str67] C. Strachey. Fundamental concepts in programming languages. Unpublished lecture notes, International Summer School in Computer Programming, Copenhagen, August, 1967.

[Wad89] P. Wadler. Theorems for free! In *4th Internat. Symp. on Functional Programming Languages and Computer Architecture, London*, pages 347–359, Assoc. for Comp. Machinery, 1989.

SOLVING 3-SATISFIABILITY IN LESS THAN $1,579^n$ STEPS

Ingo Schiermeyer
Lehrstuhl C für Mathematik
Technische Hochschule Aachen
W-5100 Aachen, Germany

Abstract

In this paper we describe and analyse an improved algorithm for solving the 3-Satisfiability problem. If F is a boolean formula in conjunctive normal form with n variables and r clauses, then we will show that this algorithm solves the Satisfiability problem for formulas with at most three literals per clause in time less than $O(1,579^n)$.

1 INTRODUCTION

Let $V = \{v_1, v_2, \ldots, v_n\}$ be a set of boolean variables. For each variable v_i there is a *positive literal*, denoted by v_i, and a *negative literal*, denoted by $\overline{v}_i$. The literal v_i has value *true* if and only if the variable v_i has value *true*, and the literal $\overline{v}_i$ has value *true* if and only if the variable v_i has value *false*. The literals v_i and $\overline{v}_i$ will be called *complemented* and $L = \{v_1, \overline{v}_1, v_2, \overline{v}_2, \ldots, v_n, \overline{v}_n\}$ is the set of literals corresponding to V. For $L' \subseteq L$ we set $Lit(L') := \{v, \overline{v} \mid v \in L' \text{ or } \overline{v} \in L'\}$. A *k-clause* $(k \geq 1)$ is a subset of k different literals of L. A *truth assignment* (short: *assignment*) is a mapping t which assigns each variable of V the value *true* or the value *false*. For each variable v we define: $t(v)$ has value *true* if and only if $t(\overline{v})$ has value *false*. An assignment t *satisfies* a clause K if and only if $t(x) = true$ for at least one literal x of K; an assignment t *satisfies* a *formula* F of clauses if and only if it satisfies each clause of F. A formula F is called *satisfiable* if there is at least one satisfying truth assignment for F.

For some k with $1 \leq k \leq n$ we consider formulas F with r clauses $K_1, K_2, ..., K_r$ of lengths $k_1, k_2, ..., k_r$ in conjunktive normal form (CNF). The *Satisfiability Problem (Sat)* is the decision problem whether F is satisfiable. If all clauses have length k for some $k \geq 1$ then we speak of the *k-Satisfiability problem (k-Sat)*. For the purpose of this paper we assume that no clause contains complemented literals and that all clauses are pairwise different.

It is well-known that k-Sat is solvable in polynomial time for $k = 1, 2$ and NP-complete for $k \geq 3$ (cf. [Co],[GJ]). The best known algorithm with respect to its running time is due to Monien and Speckenmeyer [MS]. Especially, it solves the 3-Satisfiability problem in less than $1,619^n$ steps. Branching at a 2-clause leads to the constant $1,619$.

A basic concept in this algorithm is the notion of an autark truth assignment which is defined as follows. Let $L' \subseteq L$ and $t : Lit(L') \rightarrow \{true, false\}$ be an assignment. Then t is called *autark* in F if and only if

$$\forall K \in F : K \cap Lit(L') \neq \emptyset \Rightarrow \exists x \in K \cap Lit(L') : t(x) = true.$$

The importance for their approach is based upon the following consideration. If $t : Lit(L') \rightarrow \{true, false\}$ is autark in F, then all clauses $K \in F$ with $K \cap Lit(L') \neq \emptyset$ are simultaneously satisfied by t without having fixed any of the literals not in $Lit(L')$. If t is not autark in F, then there is a clause $K \in F$ such that $K \cap Lit(L') \neq \emptyset$ and $\forall x \in K \cap Lit(L') : t(x) = false$. Thus the formula F', which is computed by evaluating F according to t contains the clause $K' = K - Lit(L')$ with $k' < k$.

The basic idea of our 3-Sat algorithm may now be described as follows: Instead of branching at a single clause K this 3-Sat algorithm searches for a set of two, three or four clauses having some variables in common. It then examines whether there is an autark truth assignment or branches at this set of clauses. If none of these situations occurs then it branches (as usual) at a 2-clause. A precise calculation of the number of 2-clauses then leads to the improvement of the constant to $1,579$.

2 THE 3-SAT ALGORITHM

Let $K_3(n,r)$ denote the set of all formulas F which contain at most n variables and r clauses for which each clause $K \in F$ has at most three literals.

3-Sat Algorithm

Input: $F \in K_3(n,r)$
Output: *satisfiable*, if F is satisfiable
unsatisfiable, else

Prozedur $Sat(F)$;

<u>begin</u>

1. <u>if</u> $F = \emptyset$ <u>then</u> F is *satisfiable*;

2. <u>if</u> $\Box \in F$ <u>then</u> F is *unsatisfiable*;

3. $M := \{y \in L \mid \overline{y} \notin L\}$; (* pure literals *)

   <u>if</u> $M \neq \emptyset$ <u>then</u>
   <u>begin</u> $F' := \{K \in F \mid K \cap M = \emptyset\}; Sat(F')$
   <u>if</u> F' satisfiable <u>then</u> F is *satisfiable*
   <u>else</u> F is *unsatisfiable*
   <u>end</u>

4. determine the length l of a shortest clause in F;

5. if $l = 1$ then
begin (5.)
determine the set M of all 1-clauses;
if there exists some i, $1 \leq i \leq n$, such that $x_i, \overline{x}_i \in M$
then F is *unsatisfiable*
else begin (A1)
$F' := \{K - \overline{y} \mid K \in F, K \cap y = \emptyset\}$ for all $y \in M$; $Sat(F')$;
if F' *satisfiable* then F is *satisfiable*
else F is *unsatisfiable*
end (A1);
end (5.);

6. if $l = 2$ then
begin (B1-F)
if there are two clauses $K_1 = x_1 \vee x_2$ and $K_2 = x_1 \vee \overline{x}_2$
then begin (B1)
$F' = \{K - \overline{x}_1 \mid K \in F, K \cap x_1 = \emptyset\}$; $Sat(F')$;
if F' satisfiable then F is *satisfiable*
else F is *unsatisfiable*
end (B1);
if there are two clauses $K_1 = x_1 \vee x_2$ and $K_2 = \overline{x}_1 \vee \overline{x}_2$
then begin (B2)
$F' = \{(K \cup \overline{x}_1) - x_2 \mid K \in F, x_2 \in K, K \cap \overline{x}_2 = \emptyset, K \cap x_1 = \emptyset\}$;
$\cup\{(K \cup x_1) - \overline{x}_2 \mid K \in F, \overline{x}_2 \in K, K \cap x_2 = \emptyset, K \cap \overline{x}_1 = \emptyset\}$; $Sat(F')$;
if F' satisfiable then F is *satisfiable*
else F is *unsatisfiable*
end (B2);
branch := *true*;
while *branch* do
begin (B3-F)
$p := 2$;
if there are two clauses $K_1 = x_1 \vee x_2$ and $K_2 = \overline{x}_1 \vee \overline{x}_3$ then
begin (B3)
$T_1 := \{x_1, x_3\}; T_2 := \{\overline{x}_1, x_2\}$; *branch* := *false*;
end (B3);
(* in the following a case like (B3) will be abbreviated as
follows: (B3) $K_1 = x_1 \vee x_2, K_2 = \overline{x}_1 \vee \overline{x}_3$
$T_1 := \{x_1, x_3\}; T_2 := \{\overline{x}_1, x_2\}$; *)
(B4) $K_1 = x_1 \vee x_2$ and $K_2 = x_1 \vee x_3$
$T_1 := \{x_1\}; T_2 := \{\overline{x}_1, x_2, x_3\}$;
eliminate all 3-clauses $K = x_1 \vee x_2 \vee x_3$ for which one
of the 2-clauses $x_1 \vee x_2, x_1 \vee x_3$ or $x_2 \vee x_3$ belongs to F;
(C1) $K_1 = x_1 \vee x_2, K_2 = x_3 \vee x_4$ and $K_3 = x_1 \vee \overline{x}_2 \vee \overline{x}_3$
$T_1 := \{x_1, x_3\}; T_2 := \{\overline{x}_3, x_4\}$;
(C2) $K_1 = x_1 \vee x_2$ and $K_2 = x_1 \vee \overline{x}_2 \vee x_3$

$T_1 := \{x_1\}; T_2 := \{\overline{x}_1, x_2, x_3\};$
if there are three clauses $K_1 = x_1 \vee x_2$, $K_2 = x_3 \vee x_4$ and
$K_3 = \overline{x}_1 \vee \overline{x}_2 \vee \overline{x}_3$ then
begin (C3)
if no other clause contains $\overline{x}_3$ then
begin
$T_1 := \{x_1\}; T_2 := \{\overline{x}_1, x_2, x_3\};$
end
else
begin
$T_1 := \{x_3\}; E_1 := \{x_2\}; D_1 := \{\overline{x}_1\}; T_2 := \{\overline{x}_3, x_4\};$
end;
branch := *false*;
end (C3);
if there are two clauses $K_1 = x_1 \vee x_2$ and $K_2 = \overline{x}_1 \vee \overline{x}_2 \vee x_3$ then
begin (C4)
if no other clause contains x_3 then
begin
$T_1 := \{\overline{x}_3\}; E_1 := \{x_2\}; D_1 := \{\overline{x}_1\}; T_2 := \{x_1, x_2, x_3\};$
end
else
begin
$T_1 := \{x_3\}; T_2 := \{\overline{x}_3\}; E_2 := \{x_2\}; D_2 := \{\overline{x}_1\};$
end;
branch := *false*;
end (C4);
(C5) $K_1 = x_1 \vee x_2$, $K_2 = x_3 \vee x_4$ and $K_3 = \overline{x}_1 \vee \overline{x}_2 \vee x_3$
$T_1 := \emptyset; E_1 := \{x_2\}; D_1 := \{\overline{x}_1\}; T_2 := \{x_1, x_2, x_3\};$
$p := 3;$
(D1) $K_1 = x_1 \vee x_2$, $K_2 = x_3 \vee x_4$, $K_3 = x_5 \vee x_6$
and $K_4 = \overline{x}_1 \vee \overline{x}_3 \vee \overline{x}_5$
$T_1 := \{x_1, x_3, \overline{x}_5, x_6\}; T_2 := \{x_1, \overline{x}_3, x_4\}; T_3 := \{\overline{x}_1, x_2\};$
(D2) $K_1 = x_1 \vee x_2$, $K_2 = x_3 \vee x_4$ and $K_3 = \overline{x}_1 \vee \overline{x}_3 \vee x_5$
$T_1 := \{x_1, x_3, x_5\}; T_2 := \{x_1, \overline{x}_3, x_4\}; T_3 := \{\overline{x}_1, x_2\};$
if there are four clauses $K_1 = x_1 \vee x_2$, $K_2 = x_3 \vee x_4$,
$K_3 = x_5 \vee x_6$ and $K_4 = x_1 \vee \overline{x}_3 \vee x_5$ then
begin (D3)
if $\overline{x}_2 \vee y_1 \vee y_2$ belongs to F
for each clause $\overline{x}_3 \vee y_1 \vee y_2 \in F \setminus K_4$ then
begin
$p := 2; T_1 := \{x_1\}; T_2 := \{\overline{x}_1, x_2, x_3\};$
end
else
begin
$p := 3; T_1 := \{x_1\}; T_2 := \{\overline{x}_1, x_2, x_3, x_5\}; T_3 := \{\overline{x}_1, x_2, \overline{x}_3, x_4\};$
end;

    branch := *false*;
    end (D3);
(D4) $K_1 = x_1 \vee x_2, K_2 = x_3 \vee x_4$ and $K_3 = x_1 \vee \overline{x}_3 \vee x_5$
    $T_1 := \{x_1, x_3\}; T_2 := \{\overline{x}_1, x_2, x_3, x_5\}; T_3 := \{\overline{x}_3, x_4\};$
if there are three clauses $K_1 = x_1 \vee x_2, K_2 = x_3 \vee x_4$ and
$K_3 = x_1 \vee x_3 \vee x_5$ then
    begin (D5)
    if $\overline{x}_2 \vee y_1 \vee y_2$ belongs to F
    for each clause $\overline{x}_3 \vee y_1 \vee y_2$ then
      begin
      $p := 2; T_1 := \{x_1\}; T_2 := \{\overline{x}_1, x_2, x_3\};$
      end
    else
      begin
      $p := 3; T_1 := \{x_1\}; T_2 := \{\overline{x}_1, x_2, x_3\}; T_3 := \{\overline{x}_1, x_2, \overline{x}_3, x_4, x_5\};$
      end;
    branch := *false*;
    end (D5);
if there are four clauses $K_1 = x_1 \vee x_2, K_2 = x_3 \vee x_4,$
$K_3 = x_5 \vee x_6$ and $K_4 = x_1 \vee x_3 \vee x_5$ then
    begin (D6)
    if $\overline{x}_2 \vee y_1 \vee y_2$ belongs to F
    for each clause $\overline{x}_3 \vee y_1 \vee y_2$ then
      begin
      $p := 2; T_1 := \{x_1\}; T_2 := \{\overline{x}_1, x_2, x_3\};$
      end
    else
      begin
      if $\overline{x}_2 \vee y_5 \vee y_6$ belongs to F
      for each clause $\overline{x}_5 \vee y_5 \vee y_6$ then
        begin
        $T_1 := \{x_1\}; T_2 := \{\overline{x}_1, x_2, x_5\};$
        end
      else
        begin
        $p := 3; T_1 := \{x_1\}; T_2 := \{\overline{x}_1, x_2, x_3\}; T_3 := \{\overline{x}_1, x_2, \overline{x}_3, x_4, x_5\};$
        end;
      end;
    branch := *false*;
    end (D6);
$p := 2;$
(E1) $K_1 = x_1 \vee x_2, K_2 = x_1 \vee x_3 \vee x_4$ and $K_3 = \overline{x}_2 \vee x_3 \vee \overline{x}_4$
    $T_1 := \{x_1\}; T_2 := \{\overline{x}_1, x_2, x_3\};$
$p := 3;$
(E2) $K_1 = x_1 \vee x_2, K_2 = x_1 \vee x_3 \vee x_4$ and $K_3 = \overline{x}_2 \vee \overline{x}_3 \vee \overline{x}_4$
    $T_1 := \{x_1\}; T_2 := \{\overline{x}_1, x_2, x_3, \overline{x}_4\}; T_3 := \{\overline{x}_1, x_2, x_3, \overline{x}_4\};$

(E3) $K_1 = x_1 \vee x_2, K_2 = x_1 \vee x_3 \vee x_4$ and $K_3 = \overline{x}_2 \vee \overline{x}_3 \vee x_5$
    $T_1 := \{x_1\}; T_2 := \{\overline{x}_1, x_2, x_3, x_5\}; T_3 := \{\overline{x}_1, x_2, \overline{x}_3, x_4\};$
(E4) $K_1 = x_1 \vee x_2, K_2 = x_1 \vee x_3 \vee x_4$ and $K_3 = \overline{x}_2 \vee x_3 \vee x_5$
    $T_1 := \{x_1\}; T_2 := \{\overline{x}_1, x_2, x_3\}; T_3 := \{\overline{x}_1, x_2, \overline{x}_3, x_4, x_5\};$
$p := 2$; determine a clause $K_1 = x_1 \vee x_2$;
if $((K \cup \overline{x}_1) - x_2) \in F$ for all $K \in F \setminus K_1$ with $K \cap x_2 \neq \emptyset$ then
    begin (F)
    $T_1 := \{x_1, \overline{x}_2\}; T_2 := \{\overline{x}_1, x_2\};$ $branch := false;$
    end;
else begin
    $T_1 := \{x_1, \overline{x}_2\}; T_2 := \{x_2\};$ $branch := false;$
    end (F);
end (B3-F);

7. $naut := true; i := 1;$
while $i \leq p$ and naut do
    begin
    if $y = true$ for all $y \in T_i$ induces an autark
    truth assignment on $Lit(T_i)$ then
        begin
        $F_i := \{K \in F \mid K \cap Lit(T_i) = \emptyset\}; naut := false; Sat(F_i);$
        if F_i satisfiable then F is $satisfiable$;
            else F is $unsatisfiable$;
        end
    else $i := i + 1;$
    end

8. if $naut$ then
  begin
  $unsat := true; i := 1;$
  while $i \leq p$ and $unsat$ do
    begin
    $F_i := \{K - \overline{T}_i \mid K \in F, K \cap T_i = \emptyset\};$ if $D_j \neq \emptyset$ then begin
    $F_j := \{(K \cup D_j) - E_j \mid K \in F_j, E_j \in K,$
        $K \cap \overline{E}_j = \emptyset, K \cap \overline{D}_j = \emptyset\}$
        $\cup\{(K \cup \overline{D}_j) - \overline{E}_j \mid K \in F_j, \overline{E}_j \in K,$
        $K \cap E_j = \emptyset, K \cap D_j = \emptyset\};$
    $Sat(F_i);$
    end;
    if F_i satisfiable
    then begin
        $unsat := false;$ F is $satisfiable$;
        end
    else $i := i + 1;$
    end

```
            if i = p + 1 then F is unsatisfiable;
        end;
      end (B1-F)


  9. begin (A2)
     determine a clause K = x1 ∨ x2 ∨ x3; naut := true; i := 1;
     while i ≤ 3 and naut do
           begin
           if x1 = ... x_{i-1} = false, x_i = true induces an
           autark truth assignment on Lit({x1,...,x_i}) then
             begin
             F' := {K ∈ F | K ∩ Lit({x1,...,x_i}) = ∅};
             naut := false; Sat(F');
             if F' satisfiable then F is satisfiable;
                               else F is unsatisfiable;
             end
           else i := i + 1;
           end


 10. if naut then
       begin (10.)
       unsat := true; i := 1;
       while i ≤ 3 and unsat do
             begin
             F_i := {K − {x1,...,x_{i-1}, x̄_i} | K ∈ F,
             K ∩ {x̄1,...,x̄_{i-1}, x_i} = ∅}; Sat(F_i);
             if F_i satisfiable then
               begin
               unsat := false; F is satisfiable;
               end
             else i := i + 1;
             end;
       if i = 4 then F is unsatisfiable
       end (10.)
     end (A2)

end;
```

Correctness of the algorithm: At first the algorithm tests whether $F = \emptyset$ or $\square \in F$. Then F is satisfiable or unsatisfiable, respectively, and the algorithm stops. Next the algorithm computes the set M of all pure literals and assigns value *true* to all literals of M. Otherwise the length l of a shortest clause in F is determined. In the case $l = 1$ the set M of all 1-clauses is computed. If M contains two complemented literals x_i and $\overline{x}_i$, then F is unsatisfiable and the algorithm stops. Otherwise all variables y of

M are assigned value *true*. Moreover, *Sat* is recalled only once instead of $|M|$-times by considering simultaneously all 1-clauses. In the case $l = 2$ the existence of two clauses $K_1 = x_1 \vee x_2$ and $K_2 = x_1 \vee \overline{x}_2$ (B1) requires that x_1 is assigned value *true*. In (B2) the algorithm sets $x_2 := \overline{x}_1$ and in (B3) and (B4) all other possible configurations are tested, in which two 2-clauses have at least one variable in common. If none of the situations (B1) - (B4) occurs, then the set of 2-clauses forms a partition of pairwise variables-disjoint 2-clauses. After that all 3-clauses $K = x_1 \vee x_2 \vee x_3$ are eliminated, for which one of the 2-clauses $K_1 = x_1 \vee x_2$, $K_2 = x_1 \vee x_3$ or $K_3 = x_2 \vee x_3$ belongs to F, since K is always satisfied if K_1, K_2 or K_3 is satisfied. In (C1) - (C5) all other possible configurations are tested, in which a 3-clause has two variables in common with a 2-clause. In (D1) - (D6) all possible configurations are tested, in which a 3-clause has exactly one variable in common with each of two or three 2-clauses. If there are a 2-clause $K_1 = x_1 \vee x_2$ and two 3-clauses $K_2 = x_1 \vee y_1 \vee y_2$ and $K_3 = \overline{x}_2 \vee y_3 \vee y_4$, then in (E1) - (E4) the algorithm tests, whether K_2 and K_3 have one or two variables in common. Finally, after the test of (E4), either $\{y_1, y_2\} = \{y_3, y_4\}$ or K_2 and K_3 have no variables in common and the algorithm branches as indicated in (F). If one of the branching situations (B3) - (F) occurs then the autarkness of all corresponding partial assignments is tested. If none of these partial assignments is autark then the algorithm branches. Likewise in the case $l = 3$ the autarkness of possible partial assignments is tested before subproblems are generated.

3 COMPLEXITY-ANALYSIS

Lemma 3.1 *Let $F \in K_3(n,r)$. Then for all subproblems F_i generated in the while-loop of statement 8. or 10. there is a clause $K_j \in F_i$ with $k_j < k$.*

Proof: If statement 8. will be executed then the following must have happened during the preceeding execution of statement 7.. For each possible branching (B3) - (F) the assignments t_i are not autark, which are given by $x_j = true$ for all x_j of T_i. Hence there is a clause $K \in F$ such that $K \cap Lit(T_i) \neq \emptyset$ and $\forall x \in (K \cap Lit(T_i)) : t_i = false$. Thus $K \cap \overline{T_i} = \emptyset$ and therefore $K_j := (K - T_i) \in F_i$, where $k_j < k$.

If 10. will be executed then during the execution of statement 9. for all i with $1 \leq i \leq l$ the assignment $t_i : x_1 = \ldots = x_{i-1} = false, x_i = true$ was not autark in F. Hence there is a clause $K \in F$, such that $K \cap Lit(\{x_1, \ldots, x_i\}) \neq \emptyset$ and $\forall x \in K \cap Lit(\{x_1, \ldots, x_i\} : t_i(x) = false$. Thus $K \cap \{\overline{x}_1, \ldots, \overline{x}_{i-1}, x_i\} = \emptyset$ and therefore $K_j := (K - \{x_1, \ldots, x_{i-1}, \overline{x}_i\} \in F_i$, where $k_j < k$.

□

Lemma 3.2 *The execution time of Sat between two successive calls is bounded by $O(n \cdot r^4)$ when applied to $F \in K_3(n,r)$.*

Proof: For statements 1., 2. and 3. the bound certainly holds. Considering all clauses of F and counting which variables occur both positive and negative, the execution time of statement 4. can also be bounded by $O(n \cdot r)$. For statement 5. we determine the set M of all 1-clauses and count, whether a variable $x_i \in M$ occurs both positive and negative in M. This can be done within $O(n \cdot r)$ steps. In statement 6. for each possible branching (B3) - (F) the structure of a subset of two, three or four clauses is examined. This requires $O(n \cdot r^4)$ steps. If 3-clauses are eliminated (between branchings (B4) and (C1)), then the 2-clauses form already a partition. Again this can be done within $O(n \cdot r)$ steps. In statement 7. the autarkness of two, three or four assignments t_i is tested, which is possible within $O(n \cdot r)$ steps. Analogously we obtain $O(n \cdot r)$ for statement 8.. Finally, in statement 9. the algorithm branches at a 3-clause, the autarkness of the three corresponding assignments is tested and, if none of them is autark, then three subproblems are generated. Again this can be done within $O(n \cdot r)$ steps (cf. [MS]). Altogether the execution time can be bounded by $O(n \cdot r^4)$.

□

In order to compute the running time of the algorithm 3-Sat it remains to compute the number $T(F)$ of recursive calls *Sat* has to perform to evaluate F. Define

$$\begin{aligned} T_3(n) &:= max\{T(F) \mid F \in K_3(n,r)\} \text{ and} \\ T_3'(n) &:= max\{T(F) \mid F \in K_3(n,r), \exists K_i \in F : k_i < 3\}. \end{aligned}$$

Lemma 3.3

$$\begin{aligned} T_3(0) &= T_3'(0) = 1 \\ T_3(n) &\leq max\Big(\{T_3(n-i)+1 \mid 1 \leq i \leq 3\} \cup \{1 + \sum_{i=1}^{3} T_3'(n-i)\} \cup \{1\}\Big) \\ T_3'(n) &\leq max\Big(\{T_3(n-|T_i|)+1 \mid 1 \leq i \leq p\} \cup \{1 + \sum_{i=1}^{p} T_3'(n-|T_i|)\} \cup \{1\}\Big) \end{aligned}$$

Proof: Let $F \in K_3(n,r)$. If $F = \emptyset$ or $\Box \in F$, then $T(F) = 1$. Likewise $T(F) = 1$ if there is a variable x_i such that $x_i, \overline{x}_i \in M$.

First of all let $l = 1$ or $l = 3$ and K be a shortest clause of length l. If the assignment $x_1 = \ldots = x_{i-1} = false, x_i = true$, $1 \leq i \leq 3$, is autark in F during the execution of statement 9., then all clauses $K \in F$ with $K \cap Lit(K) \neq \emptyset$ are satisfied. The resulting formula F', which will be called recursively by $3 - Sat$, then belongs to $K_3(n-i, r-1)$ and thus $T(F) = T(F') + 1 \leq T_3(n-i) + 1$. If none of these assignments is autark then each subproblem F_i generated during the execution of statement 10. belongs to $K_3(n-i, r-1)$ and contains a clause K_j with $k_j < 3$ by lemma 3.1. Therefore

$$T(F) \leq 1 + \sum_{i=1}^{l} T(F_i) \leq 1 + \sum_{i=1}^{l} T_3'(n-i).$$

If $l = 2$ then the proof looks similarly. This time the autarkness of p assignments, $2 \leq p \leq 4$, with $|T_i|$ variables is tested. If one of them is autark in F then

$$T(F) = T(F') + 1 \leq T_3(n - |T_i|) + 1.$$

If none of them is autark then by lemma 3.1

$$T(F) \leq 1 + \sum_{i=1}^{p} T_3'(n - |T_i|).$$

□

For a formula $F \in K_3(n,r)$ let m be the maximal number of pairwise variables-disjoint 1-clauses and 2-clauses. In order to estimate $T_3(n)$ and $T_3'(n)$ we choose two functions $F_3(n)$ and $F_3'(n)$ as follows:

$$\begin{aligned} F_3(n) &= c_1\beta^n \\ F_3'(n) &= c_2\alpha^m\beta^{n-m},\ m \geq 1, \end{aligned}$$

where $\beta = 1,5782, \alpha = 1,4401 < \beta^3 - \beta^2$ and c_1, c_2 are two sufficiently large constants with $c_1 = 0,8171 \cdot \alpha \cdot c_2$.

Lemma 3.4

$$T_3(n) \leq F_3(n)$$

Proof: The proof will be performed by induction on n. The proposition holds for $n = 0$ and $n = 1$.

Now suppose that the proposition holds for all $k < n$ with $n \geq 2$. Without restriction let $T_3(n) > 1$. If in statement 9. an assignment t_i is autark, then by lemma 3.3

$$T_3(n) \leq T_3(n-i) + 1 = 1 + c_1\beta^{n-i} < c_1\beta^n = F_3(n)$$

for $1 \leq i \leq 3$. If all assignments t_i for $1 \leq i \leq 3$ are not autark in F, then by 3.3

$$T_3(n) \leq 1 + \sum_{i=1}^{3} T_3'(n-i) \leq 1 + c_2\alpha\beta^{n-2} + c_2\alpha\beta^{n-3} + c_2\alpha\beta^{n-4},$$

since $m \geq 1$ by lemma 3.1 for each subproblem $F_i, 1 \leq i \leq 3$. Now

$$\begin{aligned} 1 + c_2\alpha\beta^{n-2} + c_2\alpha\beta^{n-3} + c_2\alpha\beta^{n-4} &= 1 + \frac{c_1}{0,8171} \cdot (\beta^{n-2} + \beta^{n-3} + \beta^{n-4}) \\ &= 1 + c_1\beta^n \cdot \frac{1}{0,8171} \cdot (\frac{1}{\beta^2} + \frac{1}{\beta^3} + \frac{1}{\beta^4}) \\ &< c_1\beta^n. \end{aligned}$$

Therefore, $T_3(n) \leq F_3(n)$.

□

In order to prove an analog estimation $T_3'(n) \leq F_3'(n)$ for $T_3'(n)$ we need an intensification of the corresponding estimation for $T_3'(n)$ in lemma 3.3.

Lemma 3.5 *Let $F \in K_3(n,r)$ be a formula generated by one of the branchings (B3) - (F). Let P be the variables-disjoint partition of the 2-clauses with $|P| = m$, which are induced by the initial (maximal) partition. If $T(F) \leq c_2\alpha^m\beta^{n-m}f(\alpha,\beta)$, then for $\epsilon = 1 - (\frac{1}{\alpha} + \frac{1}{\alpha\beta^2}) > 0$ and all $t \geq 1$*

$$T(F) \leq max\Big\{\{1\} \cup \{(2^t - 1 + c_2\alpha^m\beta^{n-m}f(\alpha,\beta) \cdot max(\frac{\alpha}{\beta}, \frac{c_1}{\alpha \cdot c_2}, (1-\epsilon)^t)\}\Big\}.$$

Proof: Since F has been generated by a branching there is at least a 1-clause or a 2-clause not contained in the partition P. We distinguish three cases and perform the proof by induction on t.

$t = 1$

1. If the partition P of the 1-clauses and 2-clauses in F is not maximal then at least one 1-clause or 2-clause can be added. Then the estimation can be improved by a factor of at least $\frac{\alpha}{\beta}$. Therefore,

$$T(F) \leq 1 + c_2\alpha^{m+1}\beta^{n-(m+1)}.$$

2. There is no branching in the evaluation of F. If $\Box \in F$ or there is a variable x_i with $x_i, \overline{x}_i \in F$, then $T(F) = 1$. Otherwise F contains 1-clauses or there is an autark assignment t_i in F with respect to a possible branching. Suppose that during the evaluation variables of q of the 1-clauses and 2-clauses are assigned truth values, $1 \leq q \leq m$. Then

$$\begin{aligned} T(F) &\leq 1 + c_1\alpha^{m-q}\beta^{n-m-(|T_i|-q)} \\ &\leq 1 + c_1\alpha^{m-1}\beta^{n-m}. \end{aligned}$$

3. There are two 2-clauses which have at least one variable in common. Then $3-Sat$ tests (B1), (B2) and the branchings (B3) and (B4). If (B1) or (B2) occurs or there is an autark assignment $t_i, 1 \leq i \leq 2$, then we obtain again $T(F) \leq 1 + c_1\alpha^{m-1}\beta^{n-m}$. Otherwise it will be branched at some $(B_j), 3 \leq j \leq 4$, and we obtain

$$\begin{aligned} T(F) &\leq 1 + T(F_1) + T(F_2) \\ &\leq 1 + c_2\alpha^m\beta^{n-m} f_{B_j}(\alpha,\beta) \\ &\leq 1 + c_2\alpha^m\beta^{n-m} \cdot \max_{3\leq j\leq 4} f_{B_j}(\alpha,\beta) \\ &= 1 + c_2\alpha^m\beta^{n-m} \cdot (\frac{1}{\alpha} + \frac{1}{\alpha^2\beta}) \\ &= 1 + c_2\alpha^m\beta^{n-m}(1-\epsilon). \end{aligned}$$

$t \to t+1$ In cases 1. and 2. the obtained estimations hold and thus

$$\begin{aligned} T(F) &\leq max\Big\{\{1\} \cup \{1 + c_2\alpha^m\beta^{n-m} f(\alpha,\beta) \cdot max(\frac{\alpha}{\beta}, \frac{c_1}{\alpha \cdot c_2})\}\Big\} \\ &\leq max\Big\{\{1\} \cup \{2^{t+1} - 1 + c_2\alpha^m\beta^{n-m} f(\alpha,\beta) \cdot max(\frac{\alpha}{\beta}, \frac{c_1}{\alpha \cdot c_2})\}\Big\}. \end{aligned}$$

Likewise in case 3. we have $T(F) \leq 1 + c_1\alpha^{m-1}\beta^{n-m}$ for (B1) or (B2) or an autark assignment $t_i, 1 \leq i \leq 2$, in the branchings (B3) and (B4). Otherwise it will be branched at some $B_j, 3 \leq j \leq 4$, and we obtain

$$\begin{aligned} &T(F) \leq 1 + T(F_1) + T(F_2) \\ &\leq 1 + c_2\alpha^m\beta^{n-m} f(\alpha,\beta) f_{B_j}(\alpha,\beta) \\ &= 1 + c_2\alpha^m\beta^{n-m} f(\alpha,\beta)\Big(f^1_{B_j}(\alpha,\beta) + f^2_{B_j}(\alpha,\beta)\Big) \\ &\leq 1 + max\Big\{\{1\} \cup \{(2^t - 1) + c_2\alpha^m\beta^{n-m} f(\alpha,\beta) f^1_{B_j}(\alpha,\beta) \cdot max(\tfrac{\alpha}{\beta}, \tfrac{c_1}{\alpha\cdot c_2}, (1-\epsilon)^t)\}\Big\} \\ &\quad + max\Big\{\{1\} \cup \{(2^t - 1) + c_2\alpha^m\beta^{n-m} f(\alpha,\beta) f^2_{B_j}(\alpha,\beta) \cdot max(\tfrac{\alpha}{\beta}, \tfrac{c_1}{\alpha\cdot c_2}, (1-\epsilon)^t)\}\Big\} \\ &\leq 1 + max\Big\{\{(2^{t+1} - 2) + c_2\alpha^m\beta^{n-m} f(\alpha,\beta)(1-\epsilon) \cdot max(\tfrac{\alpha}{\beta}, \tfrac{c_1}{\alpha\cdot c_2}, (1-\epsilon)^t)\}\Big\} \\ &\leq max\Big\{\{1\} \cup \{(2^{t+1} - 1) + c_2\alpha^m\beta^{n-m} f(\alpha,\beta) \cdot max(\tfrac{\alpha}{\beta}, \tfrac{c_1}{\alpha\cdot c_2}, (1-\epsilon)^{t+1})\}\Big\}. \end{aligned}$$

□

We now show an intensification of lemma 3.5.

Lemma 3.6 *Let $F \in K_3(n,r)$ be a formula generated by one of the branchings (B3) - (F). Let P be the variables-disjoint partition of the 2-clauses with $|P| = m$, which are induced by the initial (maximal) partition. Moreover, F contains two more (distinct) 2-clausels which are different from the 2-clauses in P. If $T(F) \leq c_2\alpha^m\beta^{n-m}f(\alpha,\beta)$ then for $\epsilon = 1 - (\frac{1}{\alpha} + \frac{1}{\alpha\beta^2}) > 0$ and all $t \geq 1$*

$$\begin{aligned} T(F) &\leq max\Big\{\{1\} \cup \{(2^t - 1 + c_2\alpha^m\beta^{n-m}f(\alpha,\beta) \\ &\quad \cdot max(\frac{\alpha^2}{\beta^2}, \frac{\alpha}{\beta}(1-\epsilon)^t, \frac{c_1}{\alpha \cdot c_2}, (1-\epsilon)^t)\}\Big\}. \end{aligned}$$

Proof: We perform the proof as for lemma 3.5. If the partition P can be extended by the two 2-clauses, then the estimation can be improved by a factor of at least $\frac{\alpha^2}{\beta^2}$. If P can be extended by exactly one of the two 2-clauses, then we apply lemma 3.5 to this clause and P as well as to the remaining 2-clause and obtain $\frac{\alpha}{\beta}(1-\epsilon)^t$ as an estimation. The estimations $\frac{c_1}{\alpha \cdot c_2}$ and $(1-\epsilon)^t$ result as in lemma 3.5.

□

Lemma 3.7

$$T_3'(n) \leq F_3'(n)$$

Proof: In the following we consider (B1) - (F). For (B3), (B4) and (E1) - (E3) we have $f(\alpha,\beta) < 1$ and thus

$$T(F) \leq 1 + c_2\alpha^m\beta^{n-m}f(\alpha,\beta) < c_2\alpha^m\beta^{n-m}.$$

For (D3), (D5) and (D6) there exists (in the second case) by lemma 3.6 some t_2 such that

$$max(\frac{\alpha^2}{\beta^2}, \frac{\alpha}{\beta}(1-\epsilon)^{t_2}, \frac{c_1}{\alpha \cdot c_2}, (1-\epsilon)^t) < max(\frac{\alpha^2}{\beta^2}, 0,8171) = \frac{\alpha^2}{\beta^2}.$$

We then show with lemma 3.5 and lemma 3.6 that $T(F) < c_2\alpha^m\beta^{n-m}$. For all remaining branchings of (C1) - (E4) there exists by lemma 3.5 some t_1 such that

$$max(\frac{\alpha}{\beta}, \frac{c_1}{\alpha \cdot c_2}, (1-\epsilon)^{t_1}) < max(\frac{\alpha}{\beta}, 0,8171) = \frac{\alpha}{\beta}.$$

We show that $f(\alpha,\beta) < \frac{\beta}{\alpha}$ and thus by lemma 3.5

$$T(F) \leq 1 + c_2\alpha^m\beta^{n-m}f(\alpha,\beta) \cdot \frac{\alpha}{\beta} < c_2\alpha^m\beta^{n-m}.$$

The branching (F) will be treated seperately. Altogether we have $T_3'(n) \leq F_3'(n)$.

□

B1 $K_1 = x_1 \vee x_2, K_2 = x_1 \vee \overline{x}_2$
Then x_1 has to be assigned value *true*.

$$T(F) \leq 1 + c_2\alpha^{m-1}\beta^{n-m} < c_2\alpha^m\beta^{n-m}$$

B2 $K_1 = x_1 \vee x_2, K_2 = \overline{x}_1 \vee \overline{x}_2$
Then we necessarily have $x_2 := \overline{x}_1$.

$$\begin{aligned} T(F) &\leq 1 + c_1 \cdot \alpha^{m-1}\beta^{n-m} \\ &= 1 + 0,8171 \cdot c_2\alpha^m\beta^{n-m} < c_2\alpha^m\beta^{n-m} \end{aligned}$$

In the following the clauses of the partition P are underlined.

B3 $\underline{K_1} = x_1 \vee x_2, K_2 = \overline{x}_1 \vee x_3$

$$\begin{aligned} T(F) &\leq 1 + c_2(\alpha^{m-2}\beta^{n-m} + \alpha^{m-1}\beta^{n-m-1}) \\ &= 1 + c_2\alpha^m\beta^{n-m}(\frac{1}{\alpha^2} + \frac{1}{\alpha\beta}) \Rightarrow f_{B3}(\alpha,\beta) \; < \; 0,923 \end{aligned}$$

B4 $\underline{K_1} = x_1 \vee x_2, K_2 = x_1 \vee x_3$

$$\begin{aligned} T(F) &\leq 1 + c_2(\alpha^{m-1}\beta^{n-m} + \alpha^{m-2}\beta^{n-m-1}) \\ &= 1 + c_2\alpha^m\beta^{n-m}(\frac{1}{\alpha} + \frac{1}{\alpha^2\beta}) \Rightarrow f_{B4}(\alpha,\beta) \; < \; 0,99993 \end{aligned}$$

C1 $\underline{K_1} = x_1 \vee x_2, \underline{K_2} = x_3 \vee x_4, K_3 = x_1 \vee \overline{x}_2 \vee \overline{x}_3$

$$\begin{aligned} T(F) &\leq 1 + c_2(\alpha^{m-2}\beta^{n-m} + \alpha^{m-1}\beta^{n-m-1}) \\ &= 1 + c_2\alpha^m\beta^{n-m}(\frac{1}{\alpha^2} + \frac{1}{\alpha\beta}) \Rightarrow f_{C1}(\alpha,\beta) \; < \; 0,923 \end{aligned}$$

C2 $\underline{K_1} = x_1 \vee x_2, K_2 = x_1 \vee \overline{x}_2 \vee x_3$

$$\begin{aligned} T(F) &\leq 1 + c_2(\alpha^{m-1}\beta^{n-m} + \alpha^{m-2}\beta^{n-m-1}) \\ &= 1 + c_2\alpha^m\beta^{n-m}(\frac{1}{\alpha} + \frac{1}{\alpha^2\beta}) \Rightarrow f_{C2}(\alpha,\beta) \; < \; 0,99993 \end{aligned}$$

C3 $\underline{K_1} = x_1 \vee x_2, \underline{K_2} = x_3 \vee x_4, K_3 = \overline{x}_1 \vee \overline{x}_2 \vee \overline{x}_3$

1. There is no other 3-clause containing $\overline{x}_3$

$$\begin{aligned} T(F) &\leq 1 + c_2(\alpha^{m-1}\beta^{n-m} + \alpha^{m-2}\beta^{n-m-1}) \\ &= 1 + c_2\alpha^m\beta^{n-m}(\frac{1}{\alpha} + \frac{1}{\alpha^2\beta}) \Rightarrow f_{C3}(\alpha,\beta) \; < \; 0,99993 \end{aligned}$$

2. There is another 3-clause containing $\overline{x}_3$. Then F_1 has a 1-clause or a 2-clause.

$$\begin{aligned} T(F) &\leq 1 + c_2(\alpha^{m-2}\beta^{n-m} + \alpha^{m-1}\beta^{n-m-1}) \\ &= 1 + c_2\alpha^m\beta^{n-m}(\frac{1}{\alpha^2} + \frac{1}{\alpha\beta}) \Rightarrow f_{C3}(\alpha,\beta) \; < \; 0,923 \end{aligned}$$

C4 $\underline{K_1} = x_1 \vee x_2, K_2 = \overline{x}_1 \vee \overline{x}_2 \vee x_3$

1. There is no other 3-clause containing x_3.

$$\begin{aligned} T(F) &\leq 1 + c_2(\alpha^{m-1}\beta^{n-m-1} + \alpha^{m-1}\beta^{n-m-2}) \\ &= 1 + c_2\alpha^m\beta^{n-m}(\frac{1}{\alpha\beta} + \frac{1}{\alpha\beta^2}) \Rightarrow f_{C4}(\alpha,\beta) \; < \; 0,719 \end{aligned}$$

2. There is another 3-clause containing x_3. Then F_2 has a 1-clause or a 2-clause.

$$\begin{aligned} T(F) &\leq 1 + c_2(\alpha^{m}\beta^{n-m-1} + \alpha^{m-1}\beta^{n-m-1}) \\ &= 1 + c_2\alpha^m\beta^{n-m}(\frac{1}{\beta} + \frac{1}{\alpha\beta}) \Rightarrow f_{C4}(\alpha,\beta) \; < \; 1,074 \end{aligned}$$

C5 $\underline{K_1} = x_1 \vee x_2, K_2 = x_3 \vee x_4, K_3 = \overline{x}_1 \vee \overline{x}_2 \vee x_3$

$$\begin{aligned} T(F) &\leq 1 + c_2(\alpha^{m-1}\beta^{n-m} + \alpha^{m-2}\beta^{n-m-1}) \\ &= 1 + c_2\alpha^m\beta^{n-m}(\frac{1}{\alpha} + \frac{1}{\alpha^2\beta}) \Rightarrow f_{C5}(\alpha,\beta) \; < \; 0,99993 \end{aligned}$$

D1 $\underline{K_1} = x_1 \vee x_2, \underline{K_2} = x_3 \vee x_4, \underline{K_3} = x_5 \vee x_6, K_4 = \overline{x}_1 \vee \overline{x}_3 \vee \overline{x}_5$

$$\begin{aligned} T(F) &\leq 1 + c_2(\alpha^{m-3}\beta^{n-m-1} + \alpha^{m-2}\beta^{n-m-1} + \alpha^{m-1}\beta^{n-m-1}) \\ &= 1 + c_2\alpha^m\beta^{n-m}(\frac{1}{\alpha^3\beta} + \frac{1}{\alpha^2\beta} + \frac{1}{\alpha\beta}) \Rightarrow f_{D1}(\alpha,\beta) \; < \; 0,958 \end{aligned}$$

D2 $\underline{K_1} = x_1 \vee x_2, \underline{K_2} = x_3 \vee x_4, K_3 = \overline{x}_1 \vee \overline{x}_3 \vee x_5$

$$\begin{aligned} T(F) &\leq 1 + c_2(\alpha^{m-3}\beta^{n-m} + \alpha^{m-2}\beta^{n-m-1} + \alpha^{m-1}\beta^{n-m-1}) \\ &= 1 + c_2\alpha^m\beta^{n-m}(\frac{1}{\alpha^3} + \frac{1}{\alpha^2\beta} + \frac{1}{\alpha\beta}) \Rightarrow f_{D2}(\alpha,\beta) \; < \; 1,081 \end{aligned}$$

D3 $\underline{K_1} = x_1 \vee x_2, \underline{K_2} = x_3 \vee x_4, \underline{K_3} = x_5 \vee x_6, K_4 = x_1 \vee \overline{x}_3 \vee x_5$

1. For each further clause $\overline{x}_3 \vee y_1 \vee y_2$ the clause $\overline{x}_2 \vee y_1 \vee y_2$ belongs to F.

$$\begin{aligned} T(F) &\leq 1 + c_2(\alpha^{m-1}\beta^{n-m} + \alpha^{m-2}\beta^{n-m-1}) \\ &= 1 + c_2\alpha^m\beta^{n-m}(\frac{1}{\alpha} + \frac{1}{\alpha^2\beta}) \Rightarrow f_{D3}(\alpha,\beta) \; < \; 0,99993 \end{aligned}$$

2. There are two clauses $\overline{x}_2 \vee y_1 \vee y_2$ and $\overline{x}_3 \vee y_3 \vee y_4$ with $y_1 \vee y_2 \not\equiv y_3 \vee y_4$.

$$\begin{aligned} T(F) &\leq 1 + c_2(\alpha^{m-1}\beta^{n-m} \cdot \frac{\alpha}{\beta} + \alpha^{m-3}\beta^{n-m-1} \cdot \frac{\alpha^2}{\beta^2} + \alpha^{m-2}\beta^{n-m-2} \cdot \frac{\alpha}{\beta}) \\ &= 1 + c_2\alpha^m\beta^{n-m}(\frac{1}{\beta} + \frac{2}{\alpha\beta^3}) < c_2\alpha^m\beta^{n-m} \end{aligned}$$

D4 $\underline{K_1} = x_1 \vee x_2, \underline{K_2} = x_3 \vee x_4, K_3 = x_1 \vee \overline{x}_3 \vee x_5$

$$\begin{aligned} T(F) &\leq 1 + c_2(\alpha^{m-1}\beta^{n-m} + \alpha^{m-2}\beta^{n-m-2} + \alpha^{m-2}\beta^{n-m-2}) \\ &= 1 + c_2\alpha^m\beta^{n-m}(\frac{1}{\alpha} + \frac{2}{\alpha^2\beta^2}) \Rightarrow f_{D4}(\alpha,\beta) \; < \; 1,082 \end{aligned}$$

D5 $\underline{K_1} = x_1 \vee x_2, \underline{K_2} = x_3 \vee x_4, K_3 = x_1 \vee x_3 \vee x_5$

1. For each further clause $\overline{x}_3 \vee y_1 \vee y_2$ the clause $\overline{x}_2 \vee y_1 \vee y_2$ belongs to F.

$$\begin{aligned} T(F) &\leq 1 + c_2(\alpha^{m-1}\beta^{n-m} + \alpha^{m-2}\beta^{n-m-1}) \\ &= 1 + c_2\alpha^m\beta^{n-m}(\frac{1}{\alpha} + \frac{1}{\alpha^2\beta}) \Rightarrow f_{D5}(\alpha,\beta) \; < \; 0,99993 \end{aligned}$$

2. There are two clauses $\overline{x}_2 \vee y_1 \vee y_2$ and $\overline{x}_3 \vee y_3 \vee y_4$ with $y_1 \vee y_2 \not\equiv y_3 \vee y_4$.

$$\begin{aligned} T(F) &\leq 1 + c_2(\alpha^{m-1}\beta^{n-m} \cdot \frac{\alpha}{\beta} + \alpha^{m-2}\beta^{n-m-1} \cdot \frac{\alpha^2}{\beta^2} + \alpha^{m-2}\beta^{n-m-3} \cdot \frac{\alpha}{\beta}) \\ &= 1 + c_2\alpha^m\beta^{n-m}(\frac{1}{\beta} + \frac{1}{\beta^3} + \frac{1}{\alpha\beta^4}) < c_2\alpha^m\beta^{n-m} \end{aligned}$$

D6 $\underline{K_1} = x_1 \vee x_2, \underline{K_2} = x_3 \vee x_4, \underline{K_3} = x_5 \vee x_6, K_4 = x_1 \vee x_3 \vee x_5$
If $\overline{x}_2 \vee y_1 \vee y_2$ belongs to F for each clause $\overline{x}_3 \vee y_1 \vee y_2$ then we obtain (as in the first case of (D5)) $f_{D6}(\alpha, \beta) < 0,99993$. If there are two clauses $\overline{x}_2 \vee y_1 \vee y_2$ and $\overline{x}_3 \vee y_3 \vee y_4$ with $y_1 \vee y_2 \not\equiv y_3 \vee y_4$ then we distinguish two cases.

1. For each clause $\overline{x}_5 \vee y_5 \vee y_6$ the clause $\overline{x}_2 \vee y_5 \vee y_6$ belongs to F.

$$\begin{aligned} T(F) &\leq 1 + c_2(\alpha^{m-1}\beta^{n-m} + \alpha^{m-2}\beta^{n-m-1}) \\ &= 1 + c_2\alpha^m\beta^{n-m}(\frac{1}{\alpha} + \frac{1}{\alpha^2\beta}) \Rightarrow f_{D6}(\alpha, \beta) < 0,99993 \end{aligned}$$

2. There are two clauses $\overline{x}_2 \vee y_5 \vee y_6$ and $\overline{x}_5 \vee y_7 \vee y_8$ with $y_5 \vee y_6 \not\equiv y_7 \vee y_8$.

$$\begin{aligned} T(F) &\leq 1 + c_2(\alpha^{m-1}\beta^{n-m} \cdot \frac{\alpha}{\beta} + \alpha^{m-2}\beta^{n-m-1} \cdot \frac{\alpha^2}{\beta^2} + \alpha^{m-3}\beta^{n-m-2} \cdot \frac{\alpha^2}{\beta^2}) \\ &= 1 + c_2\alpha^m\beta^{n-m}(\frac{1}{\beta} + \frac{1}{\beta^3} + \frac{1}{\alpha\beta^4}) < c_2\alpha^m\beta^{n-m} \end{aligned}$$

E1 $\underline{K_1} = x_1 \vee x_2, K_2 = x_1 \vee x_3 \vee x_4, K_3 = \overline{x}_2 \vee x_3 \vee \overline{x}_4, \exists K_4 = \overline{x}_1 \vee x_5 \vee x_6$

$$\begin{aligned} T(F) &\leq 1 + c_2(\alpha^{m-1+1}\beta^{n-m-1} + \alpha^{m-1}\beta^{n-m-2}) \\ &= 1 + c_2\alpha^m\beta^{n-m}(\frac{1}{\beta} + \frac{1}{\alpha\beta^2}) \Rightarrow f_{E1}(\alpha, \beta) < 0,913 \end{aligned}$$

E2 $\underline{K_1} = x_1 \vee x_2, K_2 = x_1 \vee x_3 \vee x_4, K_3 = \overline{x}_2 \vee \overline{x}_3 \vee \overline{x}_4, \exists K_4 = \overline{x}_1 \vee x_5 \vee x_6$

$$\begin{aligned} T(F) &\leq 1 + c_2(\alpha^{m-1+1}\beta^{n-m-1} + \alpha^{m-1}\beta^{n-m-3} + \alpha^{m-1}\beta^{n-m-3}) \\ &= 1 + c_2\alpha^m\beta^{n-m}(\frac{1}{\beta} + \frac{2}{\alpha\beta^3}) \Rightarrow f_{E2}(\alpha, \beta) < 0,987 \end{aligned}$$

E3 $\underline{K_1} = x_1 \vee x_2, K_2 = x_1 \vee x_3 \vee x_4, K_3 = \overline{x}_2 \vee \overline{x}_3 \vee x_5, \exists K_4 = \overline{x}_1 \vee x_6 \vee x_7$

$$\begin{aligned} T(F) &\leq 1 + c_2(\alpha^{m-1+1}\beta^{n-m-1} + \alpha^{m-1}\beta^{n-m-3} + \alpha^{m-1}\beta^{n-m-3}) \\ &= 1 + c_2\alpha^m\beta^{n-m}(\frac{1}{\beta} + \frac{2}{\alpha\beta^3}) \Rightarrow f_{E3}(\alpha, \beta) < 0,987 \end{aligned}$$

E4 $\underline{K_1} = x_1 \vee x_2, K_2 = x_1 \vee x_3 \vee x_4, K_3 = \overline{x}_2 \vee x_3 \vee x_5$

$$\begin{aligned} T(F) &\leq 1 + c_2(\alpha^{m-1}\beta^{n-m} + \alpha^{m-1}\beta^{n-m-2} + \alpha^{m-1}\beta^{n-m-4}) \\ &= 1 + c_2\alpha^m\beta^{n-m}(\frac{1}{\alpha} + \frac{1}{\alpha\beta^2} + \frac{1}{\alpha\beta^4}) \Rightarrow f_{E4}(\alpha, \beta) < 1,086 \end{aligned}$$

F $K = x_1 \vee x_2$ We distinguish two cases.

1. For each clause $K = x_2 \vee y_1 \vee y_2 \in F$ the clause $\overline{x}_1 \vee y_1 \vee y_2$ belongs also to F. For $x_1 = true$ we then can assign $x_2 = false$, since

$$(\overline{x}_1 \vee y_1 \vee y_2) \wedge (x_2 \vee y_1 \vee y_2) = (y_1 \vee y_2) \wedge (x_2 \vee y_1 \vee y_2) \equiv y_1 \vee y_2$$

and K is satisfied. Moreover, there exists a clause $\overline{x}_2 \vee x_3 \vee x_4$.

$$\begin{aligned} T(F) &\leq 1 + c_2(\alpha^{m-1+1}\beta^{n-m-2} + \alpha^{m-1+1}\beta^{n-m-2}) \\ &= 1 + c_2\alpha^m\beta^{n-m} \cdot \frac{2}{\beta^2} \Rightarrow f_F(\alpha,\beta) < 0,803 \end{aligned}$$

2. If 1. does not hold then there are two clauses $\overline{x}_1 \vee y_1 \vee y_2$ and $x_2 \vee y_3 \vee y_4$.

$$\begin{aligned} T(F) &\leq 1 + c_2(\alpha^{m-1+2}\beta^{n-m-3} + \alpha^{m-1+1}\beta^{n-m-1}) \\ &= 1 + c_2\alpha^m\beta^{n-m}(\frac{\alpha}{\beta^3} + \frac{1}{\beta}) < c_2\alpha^m\beta^{n-m} \text{ for } \alpha < \beta^3 - \beta^2 \end{aligned}$$

With lemmata 3.2, 3.4 and 3.7 we obtain the main result of this chapter.

Theorem 3.8 *The algorithm $3-Sat$ computes every formula $F \in K_3(n,r)$ in $O(1,5782^n \cdot nr^4)$ steps.*

Corollary 3.9 *The algorithm $3-Sat$ computes every formula $F \in K_3(n,r)$ in less than $O(1,579^n)$ steps.*

References

[Co] S. A. Cook, *the Complexity of Theorem-Proving Procedures*, Proc. 3rd Ann. ACM Symp. on Theory of Computing, New York, 1971, 151 - 158.

[GJ] M. R. Garey and D. S. Johnson, *Computers and Intractability, A Guide to the Theory of NP-Completeness*, W. H. Freeman and Company, New York, 1979.

[MS] B. Monien and E. Speckenmeyer, *Solving Satisfiability in less than 2^n Steps*, Discrete Appl. Math. 10 (1985) 287 - 295.

Kleene's Slash and Existence of Values of Open Terms in Type Theory

Jan M. Smith

Department of Computer Science, University of Göteborg/Chalmers
S-412 96 Göteborg, Sweden
smith@cs.chalmers.se

1 Introduction

For most typed languages, a closed object of a type can be computed to a value, that is, to a term beginning with a constructor. An object depending on free variables, however, cannot in general be computed to a value: the normal form of such an object may begin with a selector or a variable.

In this paper we will give a necessary and sufficient condition on the type of a free variable for an open object containing that variable to be computable to a value. The condition is formulated in terms of the type theoretic interpretation of Kleene's slash [8] given in Smith [13], and corresponds to the condition in [8] satisfied by formulas which slashes themselves, that is, to formulas satisfying $C \mid C$. The existence of values of open objects is of interest, for instance, in partial evaluation [7] and pattern matching [2, 4].

When optimizing programs extracted from proofs, an important role is played by sets corresponding to Harrop formulas [6] since they are "without computational content" [12, 16]. We will define what it means for a set to be without computational content and then show that a set is without computational content if and only if it slashes itself; the sets satisfying this condition strictly contain the Harrop sets introduced in [13].

In the formulation of Martin-Löf's type theory that we are considering, the judgemental equality $a = b \in A$ is understood as definitional equality; it is by having an intensional equality that it becomes possible to obtain results concerning computations when interpreting Kleene's slash. Lambek and Scott [9] used Kleene's slash in a type theory, formulated within category theory and with an extensional equality; in that approach, results of the kind we have in this paper cannot be obtained.

In the interpretation in [13] of Kleene's slash for arithmetic, Martin-Löf's type theory without universes was used. Since the only way to obtain dependent sets when not having a universe is by the propositional equality, no really, from the computational point of view, interesting dependent sets can be constructed in the type theory considered in [13]. In this paper, which is focused on computations, we leave out the propositional equality but add a universe and extend the results in [13] to this theory. The main point of a universe is, in this context, that it can be used to define dependent sets by recursion. However, the definitions and results below are of interest already in the case of no dependent sets. Besides the universe, the only basic sets we consider are the natural numbers N, the one element set $\top$ and the empty set $\bot$, but the definitions and results below can easily be extended to a theory which include other sets defined by strictly positive inductive definitions.

We follow the notation in [11] and, by the terminology of Martin-Löf's type theory, I will use the word "set" instead of "type," which commonly is used in connection with programming languages.

2 Definition of slash in type theory

In arithmetic, slash is a relation $\Gamma \mid A$ between a list Γ of closed formulas and a closed formula A; this is in type theory translated to a relation $\Gamma \mid t \in A$ where Γ is a context and $t \in A$ a judgement in the context Γ.

$\Gamma \mid t \in A$ is defined by an inductive definition of the same kind as the definition of the computability predicate $Comp_A(t)$ when using Tait's method [15] to prove that closed typable terms are normalizable. In fact, the computability predicate can be obtained from the slash by letting Γ be the empty context, that is, $Comp_A(t) \;=\; \mid t \in A$. The possibility of introducing a parameter similar to the context Γ in the definition of the computability predicate has also been realized by Hallnäs [5]. For a more detailed discussion of the type theoretic interpretation of slash, see [13], and for normalization proofs of Martin-Löf's type theory, I refer to [10, 14, 3].

The definition of $\Gamma \mid t \in A$ is by the following clauses.

1. $\Gamma \mid t \in A{+}B$ if $\Gamma \vdash t \in A{+}B$, $\Gamma \vdash t = inl(a) \in A{+}B$ for some term a such that $\Gamma \mid a \in A$, or $\Gamma \vdash t = inr(b) \in A{+}B$ for some term b such that $\Gamma \mid b \in B$.

2. $\Gamma \mid t \in \Pi(A, B)$ if $\Gamma \vdash t \in \Pi(A, B)$, there exists a term $b(x)$ such that $\Gamma,\, x \in A \vdash b(x) \in B(x)$, for all terms a, $\Gamma \mid a \in A$ implies $\Gamma \mid b(a) \in B(a)$, and $\Gamma \vdash t = \lambda x.b(x) \in \Pi(A, B)$.

3. $\Gamma \mid t \in \Sigma(A, B)$ if $\Gamma \vdash t \in \Sigma(A, B)$, there exist terms a and b such that $\Gamma \mid a \in A$, $\Gamma \mid b \in B(a)$, and $\Gamma \vdash t = \langle a, b \rangle \in \Sigma(A, B)$.

4. $\Gamma \mid t \in N$ if $\Gamma \vdash t \in N$ and $\Gamma \vdash t = \overline{n} \in N$ for some numeral $\overline{n}$.

5. $\Gamma \mid t \in \top$ if $\Gamma \vdash t \in \top$ and $\Gamma \vdash t = tt \in \top$.

6. $\Gamma \mid t \in \bot$ does not hold for any t.

When including a universe U in type theory, the definition of slash must be extended to the set U and the decoding set $Set(a)$ of an element a in U. The definition of slash for these sets is made simultaneously, reflecting that the definitions of U and $Set(a)$ depend on each other.

7. $\Gamma \mid t \in U$ and $\Gamma \mid s \in Set(t)$ if $\Gamma \vdash t \in U$, $\Gamma \vdash s \in Set(t)$ and one of the following clauses holds.

   (a) $\Gamma \vdash t = a \widehat{+} b \in U$ for some terms a and b, $\Gamma \vdash a \in U$ and $\Gamma \vdash b \in U$, such that $\Gamma \mid a \in U$ and $\Gamma \mid b \in U$. $\Gamma \mid s \in Set(t)$ then means that $\Gamma \mid s \in Set(a) + Set(b)$.

   (b) $\Gamma \vdash t = \widehat{\Pi}(a, b) \in U$ for some terms a and b, $\Gamma \vdash a \in U$ and $\Gamma,\ x \in Set(a) \vdash b(x) \in U$, such that $\Gamma \mid a \in U$ and, for all terms u, $\Gamma \mid u \in Set(a)$ implies $\Gamma \mid b(u) \in U$. $\Gamma \mid s \in Set(t)$ then means that $\Gamma \mid s \in \Pi(Set(a), (x)Set(b(x)))$.

   (c) $\Gamma \vdash t = \widehat{\Sigma}(a, b) \in U$ for some terms a and b, $\Gamma \vdash a \in U$ and $\Gamma,\ x \in Set(a) \vdash b(x) \in U$, such that $\Gamma \mid a \in U$ and, for all terms u, $\Gamma \mid u \in Set(a)$ implies $\Gamma \mid b(u) \in U$. $\Gamma \mid s \in Set(t)$ then means that $\Gamma \mid s \in \Sigma(Set(a), (x)Set(b(x)))$.

   (d) $\Gamma \vdash t = \widehat{N} \in U$. $\Gamma \mid s \in Set(t)$ means that $\Gamma \mid s \in N$.

   (e) $\Gamma \vdash t = \widehat{\top} \in U$. $\Gamma \mid s \in Set(t)$ means that $\Gamma \mid s \in \top$.

   (f) $\Gamma \vdash t = \widehat{\bot} \in U$. $\Gamma \mid s \in Set(t)$ means that $\Gamma \mid s \in \bot$.

Note that when not having a universe, there are no dependent sets and the definition of $\Gamma \mid t \in A$ is then an inductive definition on the structure of the set A, which simply is built up from the basic sets by $+$, $\rightarrow$ and $\times$.

The main result for the interpretation of slash in type theory is the following theorem.

Theorem 1. *Let Δ be a context $x_1 \in D_1, \ldots, x_m \in D_m(x_1, \ldots, x_{m-1})$ and $d_1, \ldots, d_m$ terms such that $\Gamma \mid d_i \in D_i(d_1, \ldots, d_{i-1})$, $0 < i \leq m$. Then $\Delta \vdash a(x_1, \ldots, x_m) \in A(x_1, \ldots, x_m)$ implies $\Gamma \mid a(d_1, \ldots, d_m) \in A(d_1, \ldots, d_m)$.*

The proof is by induction on the length of the derivation of $\Delta \vdash a(x_1, \ldots, x_m) \in A(x_1, \ldots, x_m)$ and follows closely a normalization proof for type theory based on Tait's method; in fact, the context Γ will act just as a parameter in the proof and the usual normalization proof is obtained when Γ is the empty context. The proof for type theory without a universe is given in [13]; the proof with a universe is a straightforward extension of that, given the following lemma.

Lemma 1. *Let Δ be a context and $d_1, \ldots, d_m$ terms as in theorem 1. Then $\Delta \vdash A(x_1, \ldots, x_m) = B(x_1, \ldots, x_m)$, $\Delta \vdash a(x_1, \ldots, x_m) \in A(x_1, \ldots, x_m)$ and $\Gamma \mid a(d_1, \ldots, d_m) \in A(d_1, \ldots, d_m)$ implies $\Gamma \mid a(d_1, \ldots, d_m) \in B(d_1, \ldots, d_m)$.*

This lemma is needed for the induction step in the proof concerning the rule

$$\frac{a \in A \qquad A = B}{a \in B}$$

For the theory without a universe, this lemma is easy to prove by structural induction on the set A. When including a universe, the lemma is no longer trivial, but a proof of it can be obtained from the proof of proposition 1 in Coquand [3], which is a corresponding result for the computability predicate. Coquand's proof is by a computability argument and, again, Γ can be introduced as a parameter in that proof without any changes in the proof.

An alternative way of proving the lemma is to use a crucial result in [3] on the uniqueness of values together with the predicativity of Martin-Löf's set theory. Predicativity means here that if a set is introduced then its parts must already have been defined; for instance, if at a certain stage the set $\Pi(A, B)$ is introduced then we must at earlier stages know that A is a set and that $B(a)$ is a set for all $a \in A$. The predicativity can be seen directly from the rules of forming sets, but there is also a metamathematical proof of this in Aczel [1] for set theory with one universe, that is, the theory we are considering. Predicativity gives that we obtain a well-ordering from the transitive closure of the relation $<$ defined by

(i) $D < D + E$,

(ii) $E < D + E$,

(iii) $D < \Pi(D, E)$,

(iv) $E(d) < \Pi(D, E)$ for all $d \in D$,

(v) $D < \Sigma(D, E)$ and

(vi) $E(d) < \Sigma(D, E)$ for all $d \in D$.

In [3] it is shown that if two sets on constructor form are definitionally equal then the constructors are the same and the parts are definitionally equal; for instance, if $\Pi(A, B) = C(D, E)$ then C is Π, $A = D$ and $x \in A \vdash B(x) = E(x)$. Using this result and the fact that $\Gamma \mid a \in A$ implies that A is definitionally equal to a set on constructor form, we can use the above well-ordering and straightforwardly prove the lemma by transfinite induction.

From theorem 1 we get

Corollary 1 *Let* $z \in C \mid c \in C$. *Then* $z \in C \vdash a(z) \in A$ *implies* $z \in C \mid a(c) \in A$.

3 Sets without computational content

Kleene [8] showed that if a formula C satisfies the extended disjunction and existence properties, that is, for all formulas A, B and $A(x)$,

$$C \vdash A \vee B \text{ implies } C \vdash A \text{ or } C \vdash B \tag{ED}$$

$$C \vdash \exists x A(x) \text{ implies } C \vdash A(t) \text{ for some term } t \tag{EE}$$

then $C \mid C$. We will show a corresponding result for type theory; the extended disjunction and existence properties for a formula will then correspond to that a set is without computational content. Intuitively, a set C is without computational content if $z \in C \vdash t \in A$ implies that t has a value even if A is a set with several constructors.

For the remaining of this section we will assume that every element of a set with only one constructor can be expanded to constructor form; for function sets this corresponds η-conversion.

η-rule for $\top$

$$\frac{c \in \top}{c = tt \in \top}$$

η-rule for Π

$$\frac{c \in \Pi(A,B)}{c = \lambda x.apply(c,x) \in \Pi(A,B)}$$

η-rule for Σ

$$\frac{c \in \Sigma(A,B)}{c = \langle fst(c), snd(c)\rangle \in \Sigma(A,B)}$$

In the theory we are considering, the sets with more than one constructor are disjoint unions, the set of natural numbers and the universe. Hence, we say that a set C is *without computational content* if it satisfies following clauses.

1. If $z \in C \vdash t \in A+B$ then $z \in C \vdash t = inl(a) \in A{+}B$ for some term a such that $z \in C \vdash a \in A$, or $z \in C \vdash t = inr(b) \in A{+}B$ for some term b such that $z \in C \vdash b \in B$.

2. If $z \in C \vdash t \in N$ then $z \in C \vdash t = \overline{n} \in N$ for some numeral $\overline{n}$.

3. If $z \in C \vdash t \in U$ then one of the following holds.

   (a) $z \in C \vdash t = a \widehat{+} b \in U$ for some terms a and b such that $z \in C \vdash a \in U$ and $z \in C \vdash b \in U$.

   (b) $z \in C \vdash t = \widehat{\Pi}(a,b) \in U$ for some terms a and b such that $z \in C \vdash a \in U$ and $z \in C, x \in Set(a) \vdash b(x) \in U$.

   (c) $z \in C \vdash t = \widehat{\Sigma}(a,b) \in U$ for some terms a and b such that $z \in C \vdash a \in U$ and $z \in C, x \in Set(a) \vdash b(x) \in U$.

   (d) $z \in C \vdash t = \widehat{N} \in U$.

   (e) $z \in C \vdash t = \widehat{\top} \in U$.

   (f) $z \in C \vdash t = \widehat{\bot} \in U$.

The definition of slash and corollary 1 give

Corollary 2 *If $z \in C \mid z \in C$ then C is without computational content.*

Examples of sets without computational content are sets corresponding to Harrop formulas. The Harrop sets H_Γ in a context Γ are inductively defined by

(i) $\top$ is in H_Γ,

(ii) if A is in H_Γ and $B(x)$ in $H_{\Gamma, x\in A}$, then $\Sigma(A, B)$ is in H_Γ,

(iii) if A is a set in Γ and $B(x)$ in $H_{\Gamma, x\in A}$, then $\Pi(A, B)$ is in H_Γ.

If H is a Harrop set in the empty context we simply say that H is a Harrop set.

Theorem 2 *If H is a Harrop set then $z \in H \mid z \in H$.*

Proof. By induction on the definition of a Harrop set in a context Γ it is straightforward to show that if H is a Harrop set in the context Γ then $\Gamma,\ z \in H \mid z \in H$. Alternatively, the theorem can be obtained from theorem 3 in [13], which is a corresponding result in the more general situation of no η-rules.

Corollary 2 and theorem 2 give

Corollary 3 *Harrop sets are without computational content.*

There are more sets than the Harrop sets which satisfy $z \in C \mid z \in C$ and, hence, are without computational content; two examples are $\bot \rightarrow N$ and $\Pi(N, (n)F(n))$ where

$$\begin{cases} F(0) & = \top \\ F(succ(n)) & = N \rightarrow F(n). \end{cases}$$

Formally, we have to use the universe to introduce the family F:

$$F(n) \;=\; Set(natrec(n, \widehat{\top}, (x, y)\widehat{N} \widehat{\rightarrow} y)).$$

The next theorem gives, as a special case, the converse of corollary 2.

Theorem 3 *Let C be without computational content. Then $z \in C \vdash a \in A$ implies that $z \in C \mid a \in A$.*

Proof. We first discuss the simpler case of type theory without a universe; we can in this case use structural induction on the formation of the set $A(x_1, \ldots, x_{m-1})$ in the context $x_1 \in G_1, \ldots, x_m \in G_m(x_1, \ldots, x_{m-1})$ to prove that

if $z \in C,\ x_1 \in G_1, \ldots,\ x_m \in G_m(x_1, \ldots, x_{m-1}) \vdash a(x_1, \ldots, x_{m-1}) \in A(x_1, \ldots, x_{m-1})$ and $g_1, \ldots,\ g_m$ are terms such that $z \in C \mid g_i \in G_i(g_1, \ldots, g_{i-1})$, $0 < i \leq m$, then $z \in C \mid a(g_1, \ldots, g_m) \in A(g_1, \ldots, g_m)$.

I exemplify the proof for A equal to $\Pi(D, E)$ and $D + E$ and, to simplify notation, we leave out the context $x_1 \in G_1, \ldots,\ x_m \in G_m(x_1, \ldots, x_{m-1})$ and the corresponding substitutions of slashable terms $g_1, \ldots,\ g_m$.

Let $z \in C \vdash a \in \Pi(D, E)$. By the η-rule for Π, $z \in C \vdash a = \lambda x.apply(a, x) \in \Pi(D, E)$. Since $z \in C,\ x \in D \vdash apply(a, x) \in E(x)$, the induction hypothesis gives that $z \in C \mid d \in D$ implies $z \in C \mid apply(a, d) \in E(d)$. Hence, by the definition of slash, $z \in C \mid a \in \Pi(D, E)$.

For the +-case, let $z \in C \vdash a \in D + E$. Since C is without computational content, $z \in C \vdash a = inl(d) \in D + E$ for some term d such that $z \in C \vdash d \in D$ or $z \in C \vdash a = inl(e) \in D + E$ for some term e such that $z \in C \vdash e \in E$. Assume that the first case holds. By the induction hypothesis, $z \in C \mid d \in D$. Hence, by the definition of slash, $z \in C \mid a \in D + E$. The second case is handled in the same way.

For type theory with a universe, structural induction cannot be used since sets may then also be of the form $Set(a)$ where $a \in U$. Since C is without computational content, $z \in C \vdash A\ set$ implies that A is definitionally equal to a basic set, a set on +-form, a set on Π-form, or a set on Σ-form; hence, we can use the well-ordering introduced in the proof of lemma 1. The proof of the theorem for set theory with a universe can now proceed as above for the case without a universe, using transfinite induction instead of structural induction.

Since $z \in C \vdash z \in C$, we obtain from theorem 3

Corollary 4 *If C is without computational content then $z \in C \mid z \in C$.*

Corollaries 2 and 4 give a characterization of sets without computational content in terms of slash: a set C is without computational content if and only if $z \in C \mid z \in C$.

Acknowledgement. Normalization proofs for Martin-Löf's type theory with a universe contain subtle details and I would like to thank Thierry Coquand for many discussions on the topic and also for suggesting the alternative proof of lemma 1.

References

[1] Peter Aczel. The strength of Martin-Löf's type theory with one universe. In *Proceedings of the Symposium on Mathematical Logic, Oulu, 1974*, pages 1–32. Report No 2, Department of Philosophy, University of Helsinki, 1977.

[2] Rod Burstall. Proving Properties of Programs by Structural Induction. *Computer Journal*, 12(1):41–48, 1969.

[3] Thierry Coquand. An algorithm for testing conversion in type theory. In *Logical Frameworks.* Cambridge University Press, 1991.

[4] Thierry Coquand. Pattern matching with dependent types. In *In the informal proceeding from the logical framework workshop at Båstad*, June 1992.

[5] Lars Hallnäs. Partial Inductive Definitions. *Theoretical Computer Science*, (87), 1991.

[6] R. Harrop. Concerning formulas of the types $A \rightarrow B \vee C$, $A \rightarrow (\exists x)B(x)$ in intuitionistic formal systems. *Journal of Symbolic Logic*, 25:27–32, 1960.

[7] N. D. Jones, P. Sestoft, and H. Søndergaard. Mix: A self-applicable partial evaluator for experiments in compiler generation. *Lisp and Symbolic Computation*, (2):9–50, 1989.

[8] S. C. Kleene. Disjunction and existence under implication in elementary intuitionistic formalisms. *Journal of Symbolic Logic*, 27:11–18, 1962.

[9] J. Lambek and P. J. Scott. New proofs of some intuitionistic principles. *Zeit. f Math. Logik und Grundlagen d. Math.*, 29:493–504, 1983.

[10] Per Martin-Löf. An Intuitionistic Theory of Types. Technical report, University of Stockholm, 1972.

[11] Bengt Nordström, Kent Petersson, and Jan M. Smith. *Programming in Martin-Löf's Type Theory. An Introduction.* Oxford University Press, 1990.

[12] Christine Paulin-Mohring. *Extraction de Programmes dans le Calcul des Constructions.* PhD thesis, L'Universite Paris VII, 1989.

[13] Jan M. Smith. An interpretation of kleene's slash in type theory. In G. Huet, G. Plotkin, and C. Jones, editors, *Informal Proceedings of the Second Workshop on Logical Frameworks*, pages 337–342. Esprit Basic Research Action, May 1991. To appear in G. Huet and G. Plotkin, editors, Logical Frameworks, Cambridge University Press.

[14] Catarina Svensson. A normalization proof for Martin-Löf's type theory. Licentiate Thesis, Chalmers University of Technology and University of Göteborg, Sweden, March 1990.

[15] W. W. Tait. Intensional interpretation of functionals of finite type I. *Journal of Symbolic Logic*, 32(2):198–212, 1967.

[16] Yukihide Takayama. Extracting redundancy-free programs from proofs in first order arithmetic. *To appear in Journal of Symbolic Computation.*

Negation-Complete Logic Programs

Robert F. Stärk

CIS, Universität München, Leopoldstraße 139, D–8000 München 40
⟨staerk@cis.uni-muenchen.de⟩

Abstract. We give a short, direct proof that a logic program is negation-complete if, and only if, it has the cut-property. The property *negation-complete* refers to three-valued models, the *cut-property* is defined in terms of ESLDNF-computations only.

1 Introduction

Let P be a general logic program and $\mathbf{F}$ be the set of goals which have a finitely failed SLDNF-tree. We will use the following two definitions. The first definition uses three-valued models, the second not.

Definition (Shepherdson, [6], p. 58). A program P is *negation-complete* iff for every goal $?- L_1, \ldots, L_n$
if $comp(P) \models_3 \neg\exists(L_1 \wedge \ldots \wedge L_n)$ then $(L_1, \ldots, L_n) \in \mathbf{F}$.

Of course, the original definition due to Shepherdson is for classical logic and not for three-valued logic.

Definition. A program P has the *cut-property* iff for every goal Γ, for every clause $A :- L_1, \ldots, L_n$ of P and every substitution σ,
if for all $1 \leq i \leq n$ $(\Gamma, \neg L_i\sigma) \in \mathbf{F}$ then $(\Gamma, \neg A\sigma) \in \mathbf{F}$.

In this definition we assume that $\neg\neg R(\vec{t})$ is replaced by $R(\vec{t})$. The main theorem will be that the two properties are equivalent. Therefore, in order to show that a program is negation-complete it is sufficient and necessary to prove that it has the cut-property. In most cases, this is easier to do than to show that it is negation-complete. For instance, in [9] it is shown that *decomposable* programs have the cut-property. This class contains the definite programs and the allowed programs.

The equivalence of negation-complete and cut-property was first proved in [9] using proof-theoretic methods. We give here a direct, more traditional proof. We will show that if a program P has the cut-property and if a goal $?- L_1, \ldots, L_n$ does not finitely fail then there exists a maximal consistent and P-saturated (or in terms of logic programming: P-supported) set from which one can extract a three-valued structure in which the formula $\neg\exists(L_1 \wedge \ldots \wedge L_n)$ is not true.

The cut-property is related to the well-known proof-theoretic notions in the following way. Consider a program clause of the form $A :- L_1, \ldots, L_n$. In a sequent calculus a natural rule for this clause is the following.

$$\frac{\Gamma \supset \Delta, L_1\sigma \quad \cdots \quad \Gamma \supset \Delta, L_n\sigma}{\Gamma \supset \Delta, A\sigma}$$

In combination with the sequent $L_1, \ldots, L_n \supset A$ this is a special case of the cut rule in LK. We say that a sequent $\Gamma \supset \Delta$ of literals is true if, and only if, $(\Gamma, \neg\Delta) \in \mathbf{F}$. Now, the above rule is sound under this interpretation if, and only if, the program has the cut-property.

2 ESLDNF-Resolution and the Completion

We summarize the definition of the *completion* of a logic program. The completion of a logic program was introduced by Clark in [1]. Let $\mathcal{L}$ be a first order language with a set of function symbols *Fun* and a set of relation symbols *Rel*. A program P is a finite set of clauses of the form $A :- L_1, \ldots, L_n$, where A is a positive literal, $0 \leq n$, and $L_1, \ldots, L_n$ are positive or negative literals. A goal is an expression of the form $?- L_1, \ldots, L_n$. We assume that programs and goals do not contain equality. Capital greek letters Γ, Δ, Λ, Π, ... denote finite lists of literals. Thus clauses will be denoted by $A :- \Pi$ and goals by $?- \Gamma$ or simply by Γ. The empty goal is $\emptyset$. We write $\Gamma \wedge A$ for a goal of the form (Δ_0, A, Δ_1).

Let P be a program and R be a n-ary relation symbol of $\mathcal{L}$. We assume that there are m clauses in P whose head is of the form $R(\ldots)$ and that the i-th clause is of the form

$$R(t_{i,1}[\vec{y}], \ldots, t_{i,n}[\vec{y}]) :- L_{i,1}[\vec{y}], \ldots, L_{i,k(i)}[\vec{y}].$$

This clause has $k(i)$ literals in its body. The *defining formula* for R is defined by

$$D_R[x_1, \ldots, x_n] := \bigvee_{i=1}^{m} \exists \vec{y} \left(\bigwedge_{j=1}^{n} x_j \approx t_{i,j}[\vec{y}] \wedge \bigwedge_{j=1}^{k(i)} L_{i,j}[\vec{y}] \right)$$

and the *completed definition* of R in P is the formula

$$\forall x_1 \ldots \forall x_n \Big(R(x_1, \ldots, x_n) \leftrightarrow D_R[x_1, \ldots, x_n] \Big).$$

The completion $comp(P)$ is obtained from P by taking all completed definitions of all relations of *Rel* and the following equality and freeness axioms for $\mathcal{L}$, the so called theory *CET* (Clark's equational theory).

(1) $x \approx x$
(2) $x \approx y \rightarrow y \approx x$
(3) $x \approx y \wedge y \approx z \rightarrow x \approx z$
(4) $x_1 \approx y_1 \wedge \ldots \wedge x_n \approx y_n \rightarrow f(x_1, \ldots, x_n) \approx f(y_1, \ldots, y_n)$ [if f is n-ary]
(5) $f(x_1, \ldots, x_n) \approx f(y_1, \ldots, y_n) \rightarrow x_i \approx y_i$ [if f is n-ary and $1 \leq i \leq n$]
(6) $f(x_1, \ldots, x_n) \not\approx g(y_1, \ldots, y_m)$ [if f is n-ary, g is m-ary, and $f \neq g$]
(7) $t(x) \not\approx x$ [if $t(x)$ is a term, $t(x) \neq x$, and x occurs in $t(x)$]

As in [7] we use an extension of SLDNF-resolution, where it is allowed to select non-ground negative literals. We define a set $\mathbf{R}$ of pairs $\langle \Gamma, \theta \rangle$ and a set $\mathbf{F}$ of goals such that $\langle \Gamma, \theta \rangle \in \mathbf{R}$ or $\Gamma \; \mathbf{R} \; \theta$ means 'the goal Γ returns answer θ' or 'the goal Γ succeeds with answer θ' and $\Gamma \in \mathbf{F}$ means 'the goal Γ finitely fails'. The sets $\mathbf{R}$ and $\mathbf{F}$ are the least sets satisfying the following closure conditions.

(R1) $\emptyset$ **R** ε. (The empty goal succeeds with answer the identity substitution.)
(R2) If $\Gamma = \Delta \wedge A$, $B :- \Pi$ is a variant of a clause of P such that var(Γ) is disjoint from var($B :- \Pi$), φ is a most general unifier of A and B, and $(\Delta \wedge \Pi)\varphi \mathbf{R} \chi$ then $\Gamma \mathbf{R} (\varphi\chi) \upharpoonright \text{var}(\Gamma)$.
(R3) If $\Gamma = \Delta \wedge \neg A$, $A \in \mathbf{F}$, and $\Delta \mathbf{R} \theta$ then $\Gamma \mathbf{R} \theta$.
(F1) If $\Gamma = \Delta \wedge A$ and, for each variant $B :- \Pi$ of a clause of P and most general unifier φ of A and B, $(\Delta \wedge \Pi)\varphi \in \mathbf{F}$ then $\Gamma \in \mathbf{F}$.
(F2) If $\Gamma = \Delta \wedge \neg A$, $A \mathbf{R} \theta$, and θ is a renaming substitution for A then $\Gamma \in \mathbf{F}$.

We do not require in (R3) and (F2) that the negative literal $\neg A$ must be ground. If we did then we would obtain exactly the definition of Kunen in [4] which is equivalent to the definition of SLDNF-resolution in Lloyd [5]. Our extension of SLDNF-resolution is called ESLDNF-resolution and it is also sound for the completion. This means that for any program P

(1) if $(L_1, \ldots, L_n) \mathbf{R} \theta$ then $comp(P) \models \forall(L_1\theta \wedge \ldots \wedge L_n\theta)$,
(2) if $(L_1, \ldots, L_n) \in \mathbf{F}$ then $comp(P) \models \neg\exists(L_1 \wedge \ldots \wedge L_n)$.

We will use the fact that the set $\mathbf{F}$ is closed under the following structural rules. Properties (4) and (5) are not trivial. Full proofs of (1)–(5) using the inductive definition of ESLDNF-resolution with $\mathbf{R}$ and $\mathbf{F}$ can be found in [8].

(1) Substitution: If $\Gamma \in \mathbf{F}$ then $\Gamma\sigma \in \mathbf{F}$.
(2) Permutation: If $\Gamma \in \mathbf{F}$ and Δ is a permutation of Γ then $\Delta \in \mathbf{F}$.
(3) Weakening: If $\Gamma \in \mathbf{F}$ then $\Gamma, L \in \mathbf{F}$.
(4) Contraction: If $\Gamma, L, L \in \mathbf{F}$ then $\Gamma, L \in \mathbf{F}$.
(5) Cut: If $\Gamma, A \in \mathbf{F}$ and $\Gamma, \neg A \in \mathbf{F}$ then $\Gamma \in \mathbf{F}$.

Note, that we do not fix any selection rule. The contraction rule, for example, is read as follows. If the goal Γ, L, L fails under some selection rule then the goal Γ, L fails under some selection rule.

Following [2] and [3] we use the strong three-valued logic of Kleene. A three-valued structure $\mathfrak{M}$ is a 4-tuple $\left\langle M, \sim, (f^{\mathfrak{M}})_{f \in Fun}, (R^{\mathfrak{M}})_{R \in Rel} \right\rangle$ such that

(1) M is a non empty set: the universe of $\mathfrak{M}$,
(2) $\sim$ is a binary relation on M: the interpretation of equality,
(3) $f^{\mathfrak{M}}$ is a function from M^n into M for every n-ary $f \in Fun$,
(4) $R^{\mathfrak{M}}$ is a function from M^n into $\{\mathbf{t}, \mathbf{f}, \mathbf{u}\}$ for every n-ary $R \in Rel$,

The universe M of $\mathfrak{M}$ is denoted by $|\mathfrak{M}|$ and the relation $\sim$ is denoted by $\sim_{\mathfrak{M}}$. For every term $t[\vec{x}]$ and elements $\vec{a} \in M$ the value $\mathfrak{M}(t[\vec{a}]) \in M$ is defined as usual. For a formula $A[\vec{x}]$ and elements $\vec{a} \in M$ the truth value $\mathfrak{M}(A[\vec{a}]) \in \{\mathbf{t}, \mathbf{f}, \mathbf{u}\}$ is defined according to Kleene's truth table. Equivalence means 'same truth-value'.

A monotonic operator Φ_P on three-valued structures has been defined in [2]. If $\mathfrak{M}$ is a three-valued structure then $\Phi_P(\mathfrak{M})$ differs from $\mathfrak{M}$ only in the interpretation of the relation symbols. If the completed definition of R in P is $\forall\vec{x}\,(R(\vec{x}) \leftrightarrow D_R[\vec{x}])$ then for all $\vec{a} \in |\mathfrak{M}|$

$$R^{\Phi_P(\mathfrak{M})}(\vec{a}) = \mathfrak{M}(D_R[\vec{a}]).$$

The relation $\mathfrak{M} \leq \mathfrak{N}$ means that $\mathfrak{M}$ differs from $\mathfrak{N}$ only in the interpretation of the relation symbols and that, for all relation symbols $R \in Rel$ and $\vec{a} \in |\mathfrak{M}|$, (1) if $R^{\mathfrak{M}}(\vec{a}) = \mathbf{t}$ then $R^{\mathfrak{N}}(\vec{a}) = \mathbf{t}$ and (2) if $R^{\mathfrak{M}}(\vec{a}) = \mathbf{f}$ then $R^{\mathfrak{N}}(\vec{a}) = \mathbf{f}$. ESLDNF-resolution is also sound for three-valued logic.

(1) If $(L_1, \ldots, L_n)\ \mathbf{R}\ \theta$ then $comp(P) \models_3 \forall(L_1\theta \wedge \ldots \wedge L_n\theta)$.
(2) If $(L_1, \ldots, L_n) \in \mathbf{F}$ then $comp(P) \models_3 \neg\exists(L_1 \wedge \ldots \wedge L_n)$.

3 The Main Theorem

A substitution σ is called a *solution* of a set of equations E iff for all $s \approx t$ in E we have $s\sigma = t\sigma$. We say that a goal Γ fails modulo a set of equations E (written $\Gamma \in \mathbf{F}/E$) iff, for every solution σ of E, $\Gamma\sigma \in \mathbf{F}$.

Lemma 1. *Let the defining formula for the relation symbol R in P be*

$$D_R[x_1, \ldots, x_n] = \bigvee_{i=1}^{m} \exists\vec{y} \left(\bigwedge_{j=1}^{n} x_j \approx t_{i,j}[\vec{y}] \wedge \bigwedge_{j=1}^{k(i)} L_{i,j}[\vec{y}] \right).$$

If for every $1 \leq i \leq m$

$$L_{i,1}[\vec{u}], \ldots, L_{i,k(i)}[\vec{u}], R(s_1, \ldots, s_n), \Gamma \in \mathbf{F}/E \cup \{s_1 \approx t_{i,1}[\vec{u}], \ldots, s_n \approx t_{i,n}[\vec{u}]\}$$

and $\mathrm{var}(R(s_1, \ldots, s_n), \Gamma, E) \cap \{\vec{u}\} = \emptyset$ *then* $R(s_1, \ldots, s_n), \Gamma \in \mathbf{F}/E$.

Proof. Assume that σ is a solution of E. Without loss of generality we may assume that σ does not act on the variables $\vec{u}$. Assume that $1 \leq i \leq m$ and that φ is a most general unifier of $R(s_1\sigma, \ldots, s_n\sigma)$ and $R(t_{i,1}[\vec{u}], \ldots, t_{i,n}[\vec{u}])$. The composition $\sigma\varphi$ is a solution of

$$E \cup \{s_1 \approx t_{i,1}[\vec{u}], \ldots, s_n \approx t_{i,n}[\vec{u}]\}.$$

Hence, by assumption, $L_{i,1}[\vec{u}]\varphi, \ldots, L_{i,k(i)}[\vec{u}]\varphi, R(s_1, \ldots, s_n)\sigma\varphi, \Gamma\sigma\varphi \in \mathbf{F}$. Since i was chosen arbitrarily, $R(s_1, \ldots, s_n)\sigma, R(s_1, \ldots, s_n)\sigma, \Gamma\sigma \in \mathbf{F}$. By contraction, we obtain that $R(s_1, \ldots, s_n)\sigma, \Gamma\sigma \in \mathbf{F}$. □

Let P be a program, E an infinite set of equations, and Γ an infinite set of literals.

(1) We write $\Gamma \in \mathbf{F}/E$ iff there exists a finite set $E_0 \subseteq E$ and a goal $\Gamma_0 \subseteq \Gamma$ such that $\Gamma_0 \in \mathbf{F}/E_0$.
(2) $\langle E, \Gamma \rangle$ is called *P-supported* iff for every positive literal $R(s_1, \ldots, s_n)$ in Γ there exists an instance $R(t_1, \ldots, t_n) :- L_1, \ldots, L_k$ of a clause of P such that for all $1 \leq j \leq n$ the equation $(s_j \approx t_j) \in E$ and for all $1 \leq j \leq k$ the literal $L_j \in \Gamma$.
(3) $\langle E, \Gamma \rangle$ is called *maximal consistent for P* iff
   (a) $\Gamma \notin \mathbf{F}/E$ and,
   (b) for all positive literals A, if $\Gamma, A \notin \mathbf{F}/E$ then $A \in \Gamma$.

Lemma 2. *Let P be a program. If $(L_1,\ldots,L_r) \notin \mathbf{F}$ then there exist a set of equations E and a set of literals Γ such that $\langle E,\Gamma\rangle$ is P-supported and maximal consistent and $\{L_1,\ldots,L_r\} \subseteq \Gamma$.*

Proof. Assume that $(L_1,\ldots,L_r) \notin \mathbf{F}$. We construct two increasing sequences $(E_k)_{k\in\omega}$ and $(\Gamma_k)_{k\in\omega}$ such that $\Gamma_k \notin \mathbf{F}/E_k$ for every $k \in \omega$. For $k = 0$ we set $E_0 := \emptyset$ and $\Gamma_0 := (L_1,\ldots,L_r)$. Let $A_0, A_1,\ldots$ be an enumeration of all positive literals of $\mathcal{L}$. Assume that E_k and Γ_k have been constructed and that $\Gamma_k \notin \mathbf{F}/E_k$.

Case $k = 2j$: We assume that $\Gamma_k = \Delta_0, R(s_1,\ldots,s_n), \Delta_1$ and $R(s_1,\ldots,s_n)$ is the leftmost positive literal in Γ_k, and that the defining formula of R in P is

$$D_R[x_1,\ldots,x_n] = \bigvee_{i=1}^{m} \exists\vec{y}\left(\bigwedge_{j=1}^{n} x_j \approx t_{i,j}[\vec{y}] \wedge \bigwedge_{j=1}^{k(i)} L_{i,j}[\vec{y}]\right).$$

Let $\vec{u}$ be a string of variables which do not occur in E_k or Γ_k. By Lemma 1, there exists an $1 \le i \le m$ such that $\Delta_0, L_{i,1}[\vec{u}],\ldots,L_{i,k(i)}[\vec{u}], R(s_1,\ldots,s_n), \Delta_1$ is not in $F/E_k \cup \{s_1 \approx t_{i,1}[\vec{u}],\ldots,s_n \approx t_{i,n}[\vec{u}]\}$. We set

$$E_{k+1} := E_k \cup \{s_1 \approx t_{i,1}[\vec{u}],\ldots,s_n \approx t_{i,n}[\vec{u}]\},$$
$$\Gamma_{k+1} := \Delta_0, \Delta_1, L_{i,1}[\vec{u}],\ldots,L_{i,k(i)}[\vec{u}], R(s_1,\ldots,s_n).$$

Case $k = 2j+1$: If $(\Gamma_k, A_j) \notin \mathbf{F}/E_k$ then $\Gamma_{k+1} := (\Gamma_k, A_j)$ else $\Gamma_{k+1} := \Gamma_k$. In both cases $E_{k+1} := E_k$. Finally we set

$$E := \bigcup_{k\in\omega} E_k \quad \text{and} \quad \Gamma := \bigcup_{k\in\omega} \Gamma_k.$$

It is easy to see that $\langle E,\Gamma\rangle$ is P-supported and maximal consistent. For the maximality, assume that B is a positive literal and $\Gamma, B \notin \mathbf{F}/E$. Since $A_0, A_1,\ldots$ is an enumeration of all literals of $\mathcal{L}$, there exists a $j < \omega$ such that $A_j = B$. Let $k := 2j+1$. Since $E_k \subseteq E$ and $(\Gamma_k, A_j) \subseteq \Gamma, B$, we have $\Gamma_k, A_j \notin \mathbf{F}/E_k$. Hence A_j is in Γ_{k+1} and therefore $B \in \Gamma$. □

Lemma 3. *If P has the cut-property and if $\langle E,\Gamma\rangle$ is P-supported and maximal consistent then there exists a three-valued term model $\mathfrak{M}$ of comp(P) such that $\mathfrak{M}(L) \neq \mathbf{f}$ for every literal $L \in \Gamma$.*

Proof. Here $\mathfrak{M}(L)$ denotes the truth value of L in $\mathfrak{M}$ under the canonical variable assignment which assigns to a variable x the element x of the free term structure for $\mathcal{L}$. Let $\langle E,\Gamma\rangle$ be P-supported and maximal consistent. We will prove a weaker statement. We will show that there exists a three-valued term structure $\mathfrak{M}$ which satisfies *CET* such that $\Phi_P(\mathfrak{M}) \le \mathfrak{M}$ and $\mathfrak{M}(L) \neq \mathbf{f}$ for every literal $L \in \Gamma$. By Lemma 1 of [7], it then follows that there exists a three-valued structure $\mathfrak{N} \le \mathfrak{M}$ such that $\Phi_P(\mathfrak{N}) = \mathfrak{N}$. Hence $\mathfrak{N}$ is a model of *comp*(P) and $\mathfrak{N}(L) \neq \mathbf{f}$ for all $L \in \Gamma$. The structure $\mathfrak{M}$ is defined as follows.

(1) $|\mathfrak{M}|$ is the set of all terms of $\mathcal{L}$ (with variables).
(2) If $f \in \mathit{Fun}$ is n-ary and $t_1,\ldots,t_n \in |\mathfrak{M}|$ then $f^{\mathfrak{M}}(t_1,\ldots,t_n) := f(t_1,\ldots,t_n)$.

(3) For $s, t \in |\mathfrak{M}|$ we define $s \sim_{\mathfrak{M}} t$ if there exists a finite sequence $E_0 \subseteq E$ such that, for every solution σ of E_0, $s\sigma = t\sigma$.

(4) For a relation symbol $R \in Rel$ and $\vec{a} \in |\mathfrak{M}|$ we define

$$R^{\mathfrak{M}}(\vec{a}) := \begin{cases} \mathbf{t}, & \text{if } \Gamma, \neg R(\vec{a}) \in \mathbf{F}/E; \\ \mathbf{f}, & \text{if } \Gamma, R(\vec{a}) \in \mathbf{F}/E; \\ \mathbf{u}, & \text{otherwise.} \end{cases}$$

This definition is possible, since if $\Gamma, \neg R(\vec{a}) \in \mathbf{F}/E$ and $\Gamma, R(\vec{a}) \in \mathbf{F}/E$ then, by the cut rule, $\Gamma \in \mathbf{F}/E$. But, by assumption, $\langle E, \Gamma \rangle$ is consistent and therefore $\Gamma \notin \mathbf{F}/E$.

It is easy to see that $\sim_{\mathfrak{M}}$ satisfies the equational theory *CET*, since every finite subset $E_0 \subseteq E$ has a solution. In a next step we show that $\mathfrak{M}(L) \neq \mathbf{f}$ for every literal $L \in \Gamma$. Suppose that $R(\vec{a}) \in \Gamma$ and $\mathfrak{M}(R(\vec{a})) = \mathbf{f}$. By definition, this means that $\Gamma, R(\vec{a}) \in \mathbf{F}/E$. Therefore $\Gamma \in \mathbf{F}/E$. Contradiction. Suppose that $\neg R(\vec{a}) \in \Gamma$ and $\mathfrak{M}(\neg R(\vec{a})) = \mathbf{f}$. Then we have $\mathfrak{M}(R(\vec{a})) = \mathbf{t}$ and, by definition, this means that $\Gamma, \neg R(\vec{a}) \in \mathbf{F}/E$. Hence $\Gamma \in \mathbf{F}/E$. Contradiction.

Now we show that $\Phi_P(\mathfrak{M}) \leq \mathfrak{M}$. Assume that the defining formula for the n-ary relation symbol R of $\mathcal{L}$ in P is

$$D_R[x_1, \ldots, x_n] = \bigvee_{i=1}^{m} \exists \vec{y} \left(\bigwedge_{j=1}^{n} x_j \approx t_{i,j}[\vec{y}] \wedge \bigwedge_{j=1}^{k(i)} L_{i,j}[\vec{y}] \right).$$

First we show that $\mathfrak{M}(D_R[a_1, \ldots, a_n]) = \mathbf{t}$ implies that $\mathfrak{M}(R(a_1, \ldots, a_n)) = \mathbf{t}$.

Assume that $\mathfrak{M}(D_R[a_1, \ldots, a_n]) = \mathbf{t}$. Then there exists an $1 \leq i \leq m$ and elements $\vec{b} \in |\mathfrak{M}|$ such that $a_j \sim_{\mathfrak{M}} t_{i,j}[\vec{b}]$ for $1 \leq j \leq n$ and $\Gamma, \neg L_{i,j}[\vec{b}] \in \mathbf{F}/E$ for $1 \leq j \leq k(i)$. By definition, there exists a finite subset $E_0 \subseteq E$ and a goal $\Gamma_0 \subseteq \Gamma$ such that for every solution σ of E_0, for $1 \leq j \leq n$, $a_j\sigma = t_{i,j}[\vec{b}]\sigma$ and, for $1 \leq j \leq k(i)$, $\Gamma_0\sigma, \neg L_{i,j}[\vec{b}]\sigma \in \mathbf{F}$. Since P has the cut-property, it follows that, for every solution σ of E_0, we have $\Gamma_0\sigma, \neg R(a_1, \ldots, a_n)\sigma \in \mathbf{F}$. Hence $\Gamma, \neg R(a_1, \ldots, a_n) \in \mathbf{F}/E$ and, by definition, $\mathfrak{M}(R(a_1, \ldots, a_n)) = \mathbf{t}$.

Finally we show that $\mathfrak{M}(R(a_1, \ldots, a_n)) \neq \mathbf{f}$ implies $\mathfrak{M}(D_R[a_1, \ldots, a_n]) \neq \mathbf{f}$. Assume that $\mathfrak{M}(R(a_1, \ldots, a_n)) \neq \mathbf{f}$. By definition, $\Gamma, R(a_1, \ldots, a_n) \notin \mathbf{F}/E$. Since $\langle E, \Gamma \rangle$ is maximal consistent, it follows that $R(a_1, \ldots, a_n) \in \Gamma$. Since $\langle E, \Gamma \rangle$ is P-supported, there exists an $1 \leq i \leq m$ and terms $\vec{b} \in |\mathfrak{M}|$ such that for $1 \leq j \leq n$ the equation $(a_j \approx t_{i,j}[\vec{b}]) \in E$ and for $1 \leq j \leq k(i)$ the literal $L_{i,j}[\vec{b}] \in \Gamma$. For $1 \leq j \leq n$ we have $\mathfrak{M}(a_j \approx t_{i,j}[\vec{b}]) = \mathbf{t}$. For $1 \leq j \leq k(i)$ we have $\mathfrak{M}(L_{i,j}[\vec{b}]) \neq \mathbf{f}$. Hence $\mathfrak{M}(D_R[a_1, \ldots, a_n]) \neq \mathbf{f}$. □

Theorem. *A program P is negation-complete if, and only if, it has the cut-property.*

Proof. Assume that P is negation-complete, $A :- L_1, \ldots, L_n$ is a clause of P, and σ is a substitution such that, for $1 \leq i \leq n$, $(M_1, \ldots, M_k, \neg L_i\sigma) \in \mathbf{F}$. By the soundness of ESLDNF-resolution, $comp(P) \models_3 \neg\exists(M_1 \wedge \ldots \wedge M_k \wedge \neg L_i\sigma)$ for $1 \leq i \leq n$. From this it follows that $comp(P) \models_3 \neg\exists(M_1 \wedge \ldots \wedge M_k \wedge \neg A\sigma)$.

Since P is negation-complete, we obtain that $(M_1, \ldots, M_k, \neg A\sigma) \in \mathbf{F}$. Hence P has the cut-property.

Assume that P has the cut-property and $(L_1, \ldots, L_n) \notin \mathbf{F}$. By Lemma 2, there exist a set of equations E and a set of literals Γ such that $\langle E, \Gamma \rangle$ is P-supported and maximal consistent and such that $\{L_1, \ldots, L_n\} \subseteq \Gamma$. Since P has the cut-property, by Lemma 3, there exists a three-valued term model $\mathfrak{M}$ of $comp(P)$ such that $\mathfrak{M}(L) \neq \mathbf{f}$ for every literal $L \in \Gamma$. Therefore we have $comp(P) \not\models_3 \neg\exists(L_1 \wedge \ldots \wedge L_n)$. □

The reader may ask whether this theorem also holds for ordinary SLDNF-resolution, where the negative selected literals are ground. The answer is yes. But for ordinary SLDNF-resolution definite programs no longer have the cut-property. Consider for example the second clause of the definition of the `member` relation.

```
member(X, [X|L]).
member(X, [Y|L]) :- member(X, L).
```

Let $\sigma := \{\mathtt{X}/\mathtt{a}, \mathtt{L}/[\mathtt{a}, \mathtt{b}]\}$. We apply σ to the second clause and obtain

```
member(a, [Y, a, b]) :- member(a, [a, b]).
```

The negation of the body, the goal ?− ¬`member(a, [a, b])`, fails in SLDNF-resolution, but the negation of the head, the goal ?− ¬`member(a, [Y, a, b])`, does not, it flounders. However, in ESLDNF-resolution ¬`member(a, [Y, a, b])` $\in \mathbf{F}$. This follows with (F2), since `member(a, [Y, a, b])` $\mathbf{R}\ \varepsilon$ and ε is a renaming substitution for `member(a, [Y, a, b])`.

References

1. K. L. Clark. Negation as failure. In H. Gallaire and J. Minker, editors, *Logic and Data Bases*, pages 293–322. Plenum Press, New York, 1978.
2. M. Fitting. A Kripke-Kleene semantics for logic programs. *Journal of Logic Programming*, 2:295–312, 1985.
3. K. Kunen. Negation in logic programming. *Journal of Logic Programming*, 4(4):289–308, 1987.
4. K. Kunen. Signed data dependencies in logic programs. *Journal of Logic Programming*, 7(3):231–245, 1989.
5. J. W. Lloyd. *Foundations of Logic Programming*. Springer, Berlin, second edition, 1987.
6. J. C. Shepherdson. Negation in logic programming. In J. Minker, editor, *Foundations of Deductive Databases and Logic Programming*, pages 19–88. Morgan Kaufmann, Los Altos, 1987.
7. R. F. Stärk. A complete axiomatization of the three-valued completion of logic programs. *J. of Logic and Computation*, 1(6):811–834, 1991.
8. R. F. Stärk. Cut-property and negation as failure. Technical report, University of Berne, Switzerland, 1992.
9. R. F. Stärk. *The Proof Theory of Logic Programs with Negation*. PhD thesis, University of Berne, 1992.

Logical characterization of bounded query classes II: Polynomial-time oracle machines

Iain A. Stewart

Computing Laboratory, Univ. Newcastle upon Tyne, Claremont Tower, Claremont Road, Newcastle upon Tyne, NE1 7RU, England.

1. INTRODUCTION

This paper is a companion to [Ste92b] and continues the investigation of the logic $(\pm HP)^*[FO_s]$, introduced in [Ste92a]. In [Ste92b] the logics $\text{Bool}[HP^1[FO_s]]$, $(\pm HP)^1[FO_s]$, and $(\pm DTC)^*[HP^*[FO_s]]$ were studied with the latter two being shown to capture the complexity class $\mathbf{L}^{\mathbf{NP}}$: the former was shown to capture the complexity class $\mathbf{L}^{\mathbf{NP}}[O(1)]$, with the outcome being that it is extremely unlikely that $\text{Bool}[HP^1[FO_s]]$ is as expressible as either of the logics $(\pm HP)^1[FO_s]$ or $(\pm DTC)^*[HP^*[FO_s]]$ ($\text{Bool}[HP^1[FO_s]]$ is the Boolean closure of $HP^1[FO_s]$).

In this paper we show that the logics $(\pm HP)^*[FO_s]$ and $(\pm HP)^1[FO_s]$ actually have the same expressibility: that is, all nested applications of the operator *HP* in a sentence of $(\pm HP)^*[FO_s]$ can be replaced by a single application. Consequently, the hierarchy of logics

$$(\mathbf{NP} = HP^1[FO_s] \subseteq)\ (\pm HP)^1[FO_s] \subseteq (\pm HP)^2[FO_s] \subseteq \ldots \subseteq (\pm HP)^*[FO_s]$$

collapses to $(\pm HP)^1[FO_s]$ (a logic is equated with the class of problems described by the sentences of that logic). This result should be compared with the collapse of similar hierarchies of logics such as

$$(\mathbf{NL} =)\ TC^1[FO_s] = (\pm TC)^1[FO_s] = (\pm TC)^2[FO_s] = \ldots = (\pm TC)^*[FO_s],$$

as established in [Imm87, Imm88].

We achieve our results by providing logical characterizations of some bounded-query complexity classes contained within $\mathbf{P}^{\mathbf{NP}}$ and by applying an existing result due to Buss and Hay. Bounded-query complexity classes are defined by restricting the access of an oracle Turing machine to its oracle by, for instance, limiting the number of oracle queries or the number of batches of parallel oracle queries made in any computation.

2. BASIC DEFINITIONS

A *vocabulary* $\tau = <\underline{R}_1, ..., \underline{R}_k, \underline{C}_1, ..., \underline{C}_m>$ is a tuple of *relation symbols* $\{\underline{R}_i : i = 1, ..., k\}$ and *constant symbols* $\{\underline{C}_i : i = 1, ..., m\}$. A *(finite) structure of size n* over the vocabulary τ is a tuple $S = <\{0, 1, ..., n-1\}, R_1, ..., R_k, C_1, ..., C_m>$ consisting of a *universe* $|S| = \{0, 1, ..., n-1\}$, *relations* $R_1, ..., R_k$, and *constants* $C_1, ..., C_m$. The size of some structure S is also denoted by $|S|$. We denote the set of all structures over τ by STRUCT(τ) (henceforth, we do not distinguish between relations (resp. constants) and relation (resp. constant) symbols, and we assume that all structures have size at least 2). A *problem* of arity t (≥ 0) over τ is a subset of $\text{STRUCT}_t(\tau) = \{(S, \mathbf{u}) : S \in \text{STRUCT}(\tau), \mathbf{u} \in |S|^t\}$.

The language of the *first-order logic* $FO_s(\tau)$ over the vocabulary τ is as expected except there is also a binary relation symbol s and two constant symbols 0 and *max*. A formula $\phi \in FO_s(\tau)$, with free variables those of the t-tuple $\mathbf{x}$, is interpreted in the set $\text{STRUCT}_t(\tau)$, and for each $S \in \text{STRUCT}(\tau)$ of size n and $\mathbf{u} \in |S|^t$, $(S, \mathbf{u}) \models \phi(\mathbf{x})$ iff $\phi^S(\mathbf{u})$ holds, where $\phi^S(\mathbf{u})$ denotes the obvious interpretation of ϕ in S, except that the binary relation symbol s is interpreted as the successor relation on $|S|$, the constant symbol 0 is interpreted as $0 \in |S|$, the constant symbol *max* is interpreted as $n-1 \in |S|$, and each variable of $\mathbf{x}$ is given the corresponding value from $\mathbf{u}$. Let ϕ be a formula, over the t-tuple of variables $\mathbf{x}$, of $FO_s(\tau)$. Then ϕ *represents* (or *specifies* or *describes*) the problem $\{(S, \mathbf{u}) : (S, \mathbf{u}) \in \text{STRUCT}_t(\tau), (S, \mathbf{u}) \models \phi(\mathbf{x})\}$ of arity t.

We add new operators (corresponding to problems) to first-order logic to form new logics. For example, if *HP* is the problem of arity 2 over the vocabulary $\tau_2 = <E>$, where E is a relation symbol of arity 2, defined as

$HP = \{(S, u, v) \in \text{STRUCT}_2(\tau)$: there is an Hamiltonian path in the digraph S from u to $v\}$

then $(\pm HP)^*[FO_s]$ denotes the logic formed from FO_s by allowing an unlimited number of nested applications of the operator *HP*, where $HP[\lambda \mathbf{x}, \mathbf{y} \phi^S(\mathbf{x}, \mathbf{y})]$, for some formula $\phi \in (\pm HP)^*[FO_s]$, some k-tuples of distinct variables $\mathbf{x}$ and $\mathbf{y}$, and some relevant structure S, denotes the digraph with vertices indexed by the tuples of $|S|^k$, and where there is an edge from $\mathbf{u}$ to $\mathbf{v}$ iff there is an Hamiltonian path in the digraph described by $\phi^S(\mathbf{x}, \mathbf{y})$ from $\mathbf{u}$ to $\mathbf{v}$ (this is the logic (*FO+HP*) of [Ste92a]). We write $(\pm HP)^i[FO_s]$ (resp. $HP^*[FO_s]$) to denote the sub-logic of $(\pm HP)^*[FO_s]$ where all formulae have at most i nested applications of the operator HP (resp. where no operator HP appears within the scope of a negation sign): the sub-logic $HP^i[FO_s]$ is defined similarly. We can extend these logics using new operators in exactly the same way that we extended FO_s.

In order to compare problems, we use the notion of a logical translation. Let $\tau' = <R_1, ..., R_k, C_1, ..., C_m>$ be some vocabulary where each R_i is a relation symbol of arity a_i and each C_j is a constant symbol, and let $L(\tau)$ be some logic over the vocabulary τ (being an extension of FO_s as described above). Then the formulae of $\Sigma = \{\phi_i(\mathbf{x}_i), \psi_j(\mathbf{y}_j) : i = 1, ..., k, j = 1, ..., m\} \subseteq L(\tau)$, where

(i) each formula ϕ_i (resp. ψ_j) is over the qa_i (resp. q) distinct variables $\mathbf{x}_i$ (resp. $\mathbf{y}_j$), for some fixed positive integer q;

(ii) for each $j = 1, ..., m$ and for each $S \in$ STRUCT(τ)

$$S \models \exists x_1 ... \exists x_q [\psi_j(x_1, ..., x_q) \wedge \forall y_1 ... \forall y_q[\psi_j(y_1, ..., y_q) \Leftrightarrow (x_1 = y_1 \wedge ... \wedge x_q = y_q)]],$$

are called *τ'-descriptive*. For each $S \in$ STRUCT(τ), the *τ'-translation of S w.r.t. Σ* is the structure $S' \in$ STRUCT(τ') with universe $|S|^q$, defined as follows: for all $i = 1, ..., k$ and for any tuples $\{\mathbf{u}_1, ..., \mathbf{u}_{a_i}\} \subseteq |S'| = |S|^q$:

$$R_i^{S'}(\mathbf{u}_1, ..., \mathbf{u}_{a_i}) \text{ holds iff } (S, (\mathbf{u}_1, ..., \mathbf{u}_{a_i})) \models \phi_i(\mathbf{x}_i),$$

and for all $j = 1, ..., m$ and for any tuple $\mathbf{u} \in |S'| = |S|^q$:

$$C_j^{S'} = \mathbf{u} \text{ iff } (S, \mathbf{u}) \models \psi_j(\mathbf{y}_j)$$

(tuples are ordered lexicographically with $(0, 0, ...0) < (0, 0, ..., 1) < ...$, etc.). Let Ω and Ω' be problems over the vocabularies τ and τ', respectively. Let Σ be a set of τ'-descriptive formulae from some logic $L(\tau)$, and for each $S \in$ STRUCT(τ), let $\sigma(S) \in$ STRUCT(τ') denote the τ'-translation of S w.r.t. Σ. Then Ω' is an *L-translation* of Ω iff for each $S \in$ STRUCT(τ), $S \in \Omega$ iff $\sigma(S) \in \Omega'$.

Let $\phi \in FO_s(\tau)$, for some vocabulary τ, be of the form

$$(\alpha_1 \wedge \beta_1) \vee ... \vee (\alpha_l \wedge \beta_l)$$

where each α_i is a conjunction of the logical atomic relations, s, =, and their negations; each β_i is atomic or negated atomic; and if $i \neq j$ then α_i and α_j are mutually exclusive. Then ϕ is a *projective formula*. Consequently, we have the notion of one problem being a *projection translation* of another.

Define

$DTC = \{(S, u, v) \in \text{STRUCT}_2(\tau_2)$: there is a path in the digraph S from u to v such that each vertex on the path, except for possibly v, has out-degree 1 in S, i.e., the path is *deterministic*$\}$,

and let

$DTC(0, max) = \{S \in \mathrm{STRUCT}(\tau_2)$: there is a deterministic path in the digraph S from 0 to $max\}$,

with $HP(0, max)$ defined similarly. Henceforth, we identify a logic with the class of problems of (arity 0) described by the sentences of that logic.

THEOREM 2.1. *(a)* $FO_S \neq \mathbf{L} = DTC^1[FO_S] = (\pm DTC)^*[FO_S]$ *([Imm87]).*

(b) $HP^*[FO_S] = HP^1[FO_S] = \mathbf{NP}$ *([Ste92a]).*

(c) DTC(0, max) and HP(0, max) are complete for **L** *and* **NP**, *respectively, via projection translations ([Imm87, Ste90]).*

We mention that extensions of first-order logic (capturing **NP**) using other operators are considered in [Ste91a, Ste91b, Ste91c], and we remind the reader of the results given in the Introduction.

Due to space limitations, we omit the definitions concerning bounded-query complexity classes and simply refer the reader to [Wag90] for definitions of all concepts mentioned in this paper. However, we state the following theorem in order that the reader can see the fundamental relationships between some important bounded-query classes (for (i) see [Wag90]; for (ii) see [Wag90, Köb87]; and for (iii) see [Bei88, Wag90]).

THEOREM 2.2. *(i)* $\mathbf{L}^{\mathbf{NP}} = \mathbf{L}_{||}^{\mathbf{NP}} = \mathbf{L}^{\mathbf{NP}}[O(\log n)] = \mathbf{P}_{||}^{\mathbf{NP}} = \mathbf{P}^{\mathbf{NP}}[O(\log n)]$;

(ii) $\mathbf{L}_{||}^{\mathbf{NP}}[O(\log n)] = \mathbf{P}_{||}^{\mathbf{NP}}[O(\log n)] = \leq_{ftt}^{log}(\mathbf{NP}) = \leq_{ftt}^{P}(\mathbf{NP})$;

(iii) $\mathbf{L}_{||}^{\mathbf{NP}}[O(1)] = \mathbf{L}^{\mathbf{NP}}[O(1)] = \mathbf{P}_{||}^{\mathbf{NP}}[O(1)] = \mathbf{P}^{\mathbf{NP}}[O(1)]$.

We end with a result, due to Buss and Hay, which will be important later on.

THEOREM 2.3. ([BH88]) *Let M be a polynomial-time DOTM with an oracle in* **NP** *which makes k batches of parallel oracle queries (a batch of oracle queries is made in parallel if a complete list of them is produced before any of them is made). Then there is a polynomial-time DOTM with an oracle in* **NP** *which accepts the same set of strings as M and makes all its oracle queries in parallel.*

3. THE MAIN RESULTS

We begin with a general-purpose result, the proof of which introduces a technique that is used throughout. Let L be some logic and suppose that ϕ is some projective formula. Consider replacing some occurence $R(\mathbf{x})$ in ϕ, where R is a relation symbol, with a formula $\psi(\mathbf{x})$ of L. If the formula ϕ' is obtained from ϕ by making some replacements of this form then we say that ϕ' is *projective in formulae of L*.

PROPOSITION 3.1. *Suppose that ϕ is a sentence of $HP^1[L]$, where L is some logic. Then ϕ is equivalent to a sentence of the form*

$$HP[\lambda x, y, \psi(x, y)](\mathbf{0}, \mathbf{max}),$$

where ψ is projective in formulae of L.

PROOF. By definition, ϕ is built from formulae $\phi_1, \phi_2, ..., \phi_t$ of L using first-order constructs and positive applications of the operator HP so that no nesting of this operator occurs (we may assume that each ϕ_i has at least one free variable and is used exactly once in ϕ). For $i = 1, ..., t$, let R_i be a new relation symbol of arity a_i, where a_i is the length of the tuple $\mathbf{z}_i$ of free variables of ϕ_i. Let ϕ' be the sentence formed by replacing the occurrence of each formula $\phi_i(\mathbf{z}_i)$ in ϕ by $R_i(\mathbf{z}_i)$, and let the vocabulary τ' be formed from $\tau(\phi)$ (that is, the vocabulary of ϕ) by also including the relation symbols $R_1, ..., R_t$. In particular, $\phi' \in HP^1[FO_s]$ and so, by Theorem 2.1, ϕ' is equivalent to a sentence of the form $HP[\lambda \mathbf{x}, \mathbf{y}, \psi'](\mathbf{0}, \mathbf{max})$ where ψ' is projective. Let ψ be this sentence except that each occurrence of $R_i(\mathbf{w}_i)$ in ψ', where $\mathbf{w}_i$ is some tuple of variables and constant symbols of length a_i, is replaced by $\phi_i(\mathbf{w}_i)$: hence, ψ is a sentence of HP^1[L].

For each structure $S \in$ STRUCT($\tau(\phi)$), let $S' \in$ STRUCT(τ') be the expansion of S such that for each i = 1, ..., t and for any tuple $\mathbf{u} \in |S|^{a_i}$

$$R_i^{S'}(\mathbf{u}) \text{ holds iff } (S, \mathbf{u}) \models \phi_i(\mathbf{z}_i).$$

Clearly, $S \models \phi$ iff $S' \models \phi'$ iff $S' \models HP[\lambda \mathbf{x}, \mathbf{y}, \psi'(\mathbf{x}, \mathbf{y})](\mathbf{0}, \mathbf{max})$ iff $S \models \psi$, and so the result follows. □

Even though, in the statement of Proposition 3.1, ϕ is a sentence we can clearly apply this result (and, similarly, other results in this paper) to formulae with free variables by replacing each free variable in ϕ with a new constant symbol and considering this new sentence to be over the vocabulary formed from $\tau(\phi)$ by also including these new constant symbols.

The technique of replacing sub-formulae of some formula by atomic formulae involving new relation symbols, as in the proof of Proposition 3.1, can regularly be used to good effect, so much so that we choose to give it the name of *atomic substitution*. Henceforth, instead of giving proofs of certain results in detail, we shall merely state that the results can be proved by atomic substitution and refer the reader back to the proof of Proposition 3.1.

A formula $\phi \in (\pm HP)^*[FO_s]$ is a *d.p.n.(1)-formula* of $(\pm HP)^*[FO_s]$ (where d.p.n. stands for "disjunction of a positive and a negative") if it is of the form

$$HP[\lambda \mathbf{x}_p, \mathbf{y}_p, \phi_p](\mathbf{0}, \mathbf{max}) \vee \neg HP[\lambda \mathbf{x}_n, \mathbf{y}_n, \phi_n](\mathbf{0}, \mathbf{max}),$$

where ϕ_p and ϕ_n are projective formulae. The formula ϕ is a *d.p.n.(i)-formula* of $(\pm HP)^*[FO_s]$, for some $i > 1$, if it is of the above form where ϕ_p and ϕ_n are d.p.n.$(i-1)$-formulae. There are analogous definitions for similarly defined logics.

THEOREM 3.2. *Let $k > 0$. Then every sentence of $HP^2[(\pm HP)^k[FO_s]]$ of the form*

$$HP[\lambda \mathbf{x}, \mathbf{y}, HP[\lambda \mathbf{u}, \mathbf{v}, \phi](\mathbf{0}, \mathbf{max})](\mathbf{0}, \mathbf{max})$$

where $\phi \in (\pm HP)^k[FO_s]$, is equivalent to one of the form $HP[\lambda \mathbf{x}, \mathbf{y}, \psi](\mathbf{0}, \mathbf{max})$, where ψ is a d.p.n.(k)-formula of $(\pm HP)^*[FO_s]$.

PROOF. We proceed by induction on k. Suppose $k = 1$. Let the sentence $\phi \in HP^2[(\pm HP)^1[FO_s]]$ be of the form

$$HP[\lambda \mathbf{x}, \mathbf{y}, HP[\lambda \mathbf{u}, \mathbf{v}, \psi](\mathbf{0}, \mathbf{max})](\mathbf{0}, \mathbf{max}),$$

where $\psi \in (\pm HP)^1[FO_s]$. By atomic substitution and Theorem 2.1, ϕ is equivalent to a sentence of the form $HP[\lambda \mathbf{x}, \mathbf{y}, \theta](\mathbf{0}, \mathbf{max})$, where θ is projective in formulae of $HP^1[FO_s]$. By Theorem 2.1, θ is equivalent to a d.p.n.(1)-formula of $(\pm HP)^*[FO_s]$.

Suppose, as our induction hypothesis, that the result holds for all m such that $0 < m < k$. Let the sentence $\phi \in HP^2[(\pm HP)^k[FO_s]]$ be of the form

$$HP[\lambda \mathbf{x}_1, \mathbf{y}_1, HP[\lambda \mathbf{x}_2, \mathbf{y}_2, \psi](\mathbf{0}, \mathbf{max})](\mathbf{0}, \mathbf{max}),$$

where $\psi \in (\pm HP)^k[FO_s]$. By atomic substitution and Theorem 2.1, ϕ is equivalent to a sentence of the form $HP[\lambda \mathbf{x}, \mathbf{y}, \theta](\mathbf{0}, \mathbf{max})$, where θ is projective in formulae of $HP^1[(\pm HP)^{k-1}[FO_s]]$. Clearly, θ is equivalent to a formula of the form $\theta_p \vee \neg\theta_n$, where θ_p and θ_n are formulae of $HP^1[(\pm HP)^{k-1}[FO_s]]$. By Proposition 3.1, ϕ is equivalent to a sentence of the form

$$HP[\lambda \mathbf{x}, \mathbf{y}, (HP[\lambda \mathbf{x}_1, \mathbf{y}_1, \psi_1](\mathbf{0}, \mathbf{max}) \vee \neg HP[\lambda \mathbf{x}_2, \mathbf{y}_2, \psi_2](\mathbf{0}, \mathbf{max}))](\mathbf{0}, \mathbf{max}),$$

where ψ_1 and ψ_2 are formulae of $(\pm HP)^{k-1}[FO_s]$. The result follows after an application of the induction hypothesis (we actually apply the induction hypothesis to $HP[\lambda u, v, HP[\lambda \mathbf{x}_i, \mathbf{y}_i, \psi_i](\mathbf{0}, \mathbf{max})](0, max)$, where u and v are new variables, which is equivalent to $HP[\lambda \mathbf{x}_i, \mathbf{y}_i, \psi_i](\mathbf{0}, \mathbf{max})$, for $i = 1, 2$). □

COROLLARY 3.3. *Any sentence of the logic $ATC^1[(\pm HP)^k[FO_s]]$ is equivalent to one of the form $ATC^1[\lambda x, y, \psi_1, x\ \psi_2](0, max)$, where ψ_1 and ψ_2 are d.p.n.(k)-formulae of $(\pm HP)^k[FO_s]$ if $k > 0$ and projective otherwise. An analogous result holds for the logics $STC^1[(\pm HP)^k[FO_s]]$, $TC^1[(\pm HP)^k[FO_s]]$, and $DTC^1[(\pm HP)^k[FO_s]]$ (see [Imm87] for definitions of STC and TC).*

PROOF. Let ϕ be a sentence of $ATC^1[(\pm HP)^k[FO_s]]$. Then, by atomic substitution and Theorem 4.7 of [Imm87], ϕ is equivalent to a sentence of the form $ATC^1[\lambda \mathbf{x}, \mathbf{y}, \psi_1, \mathbf{x}, \psi_2](\mathbf{0}, \mathbf{max})$, where ψ_1 and ψ_2 are projective in formulae of $HP^1[(\pm HP)^{k-1}[FO_s]]$. The result follows by Theorem 3.2. For the other logics, the results follow similarly except we use Theorem 4.1, 3.3, or 4.2 of [Imm87] in place of Theorem 4.7 of [Imm87], above, as appropriate. □

We now use the normal form theorems above to establish a connection between sub-logics of $(\pm HP)^*[FO_s]$ and complexity classes contained in $\mathbf{P}^{\mathbf{NP}}$. Let τ be some vocabulary. Then for any $S \in \text{STRUCT}(\tau)$ of size n, $e_\tau(S)$ denotes the encoding of S as a string of 0's and 1's where a relation R of arity a is encoded as a string of n^a 0's and 1's (denoting whether $R^S(0, 0, ..., 0)$, $R^S(0, 0, ..., 1)$, etc., hold) and a constant is encoded as its binary representation.

THEOREM 3.4. *If ϕ is a sentence of $(\pm HP)^k[FO_s]$, for some $k > 0$, then there is a polynomial-time DOTM M with oracle $e_{\tau_2}(HP(0, max))$ which accepts the encoding of the problem represented by ϕ and makes exactly k batches of parallel oracle queries. In particular, $(\pm HP)^*[FO_s] \subseteq P^{NP}$.*

PROOF. Let ϕ be a sentence of $(\pm HP)^k[FO_s]$, for some $k > 0$. By Corollary 3.3, ϕ is equivalent to a sentence of the form $DTC[\lambda \mathbf{x}, \mathbf{y}, \psi](\mathbf{0}, \mathbf{max})$, where ψ is a d.p.n.(k)-formula of $(\pm HP)^*[FO_s]$. That is, ψ is equivalent to a formula of the form

$$HP[\lambda \mathbf{x}_1, \mathbf{y}_1, \psi_1](\mathbf{0}, \mathbf{max}) \vee \neg HP[\lambda \mathbf{x}_2, \mathbf{y}_2, \psi_2](\mathbf{0}, \mathbf{max}),$$

where ψ_1 and ψ_2 are d.p.n.(k−1)-formulae if $k > 1$ or projective formulae otherwise. Let $|\mathbf{x}| = |\mathbf{y}| = p$, $|\mathbf{x}_1| = |\mathbf{y}_1| = p_1$, and $|\mathbf{x}_2| = |\mathbf{y}_2| = p_2$.

Suppose that whether the formula $\psi_1(\mathbf{x}, \mathbf{y}, \mathbf{x}_1, \mathbf{y}_1)$ (resp. $\psi_2(\mathbf{x}, \mathbf{y}, \mathbf{x}_2, \mathbf{y}_2)$) holds or does not hold in some relevant structure, given some values for $\mathbf{x}$, $\mathbf{y}$, $\mathbf{x}_1$, and $\mathbf{y}_1$ (resp. $\mathbf{x}$, $\mathbf{y}$, $\mathbf{x}_2$, and $\mathbf{y}_2$), can be determined by a polynomial-time DOTM M_1 (resp. M_2) with oracle $e_{\tau_2}(HP(0, max))$ making exactly k−1 batches of parallel oracle queries (this is certainly the case when $k = 1$ as ψ_1 and ψ_2 are projective). Consider the following informal description of a polynomial-time DOTM M:

IF the input string ω is not the encoding of some relevant structure THEN
reject (after making some dummy oracle queries): otherwise, ω must be the encoding of some relevant structure S;

FOR each pair $(\mathbf{x}, \mathbf{y}) \in |S|^p \times |S|^p$ in lexicographic order DO
FOR each pair $(\mathbf{x}_1, \mathbf{y}_1) \in |S|^{p1} \times |S|^{p1}$ in lexicographic order DO
simulate M_1 on input $(\mathbf{x}, \mathbf{y}, \mathbf{x}_1, \mathbf{y}_1)$, calling this simulation $M_1(\mathbf{x}, \mathbf{y}, \mathbf{x}_1, \mathbf{y}_1)$, until the oracle consultation state is entered, with the simulation $M_1(\mathbf{x}, \mathbf{y}, \mathbf{x}_1, \mathbf{y}_1)$ in ID $ID_1(1, \mathbf{x}, \mathbf{y}, \mathbf{x}_1, \mathbf{y}_1)$ (using the oracle tape of M);
store the ID $ID_1(1, \mathbf{x}, \mathbf{y}, \mathbf{x}_1, \mathbf{y}_1)$;
FOR each pair $(\mathbf{x}_2, \mathbf{y}_2) \in |S|^{p2} \times |S|^{p2}$ in lexicographic order DO
simulate M_2 on input $(\mathbf{x}, \mathbf{y}, \mathbf{x}_2, \mathbf{y}_2)$, calling this simulation $M_2(\mathbf{x}, \mathbf{y}, \mathbf{x}_2, \mathbf{y}_2)$, until the oracle consultation state is entered, with the simulation $M_2(\mathbf{x}, \mathbf{y}, \mathbf{x}_2, \mathbf{y}_2)$ in ID $ID_2(1, \mathbf{x}, \mathbf{y}, \mathbf{x}_2, \mathbf{y}_2)$;
store the ID $ID_2(1, \mathbf{x}, \mathbf{y}, \mathbf{x}_2, \mathbf{y}_2)$;

query the oracle for the batch of oracle strings on M's oracle tape and store the answers to the queries;

FOR each pair $(\mathbf{x}, \mathbf{y}) \in |S|^p \times |S|^p$ in lexicographic order DO
FOR each pair $(\mathbf{x}_1, \mathbf{y}_1) \in |S|^{p1} \times |S|^{p1}$ in lexicographic order DO
resume the simulation $M_1(\mathbf{x}, \mathbf{y}, \mathbf{x}_1, \mathbf{y}_1)$ from $ID_1(1, \mathbf{x}, \mathbf{y}, \mathbf{x}_1, \mathbf{y}_1)$ using the relevant oracle answer, and proceed until the oracle consultation state is entered, with the simulation $M_1(\mathbf{x}, \mathbf{y}, \mathbf{x}_1, \mathbf{y}_1)$ in ID $ID_1(2, \mathbf{x}, \mathbf{y}, \mathbf{x}_1, \mathbf{y}_1)$;
store the ID $ID_1(2, \mathbf{x}, \mathbf{y}, \mathbf{x}_1, \mathbf{y}_1)$;
FOR each pair $(\mathbf{x}_1, \mathbf{y}_1) \in |S|^{p2} \times |S|^{p2}$ in lexicographic order DO
resume the simulation $M_2(\mathbf{x}, \mathbf{y}, \mathbf{x}_2, \mathbf{y}_2)$ from $ID_2(1, \mathbf{x}, \mathbf{y}, \mathbf{x}_2, \mathbf{y}_2)$ using the relevant oracle answer, and proceed until the oracle consultation state is entered, with the simulation $M_2(\mathbf{x}, \mathbf{y}, \mathbf{x}_2, \mathbf{y}_2)$ in ID $ID_2(2, \mathbf{x}, \mathbf{y}, \mathbf{x}_2, \mathbf{y}_2)$;
store the ID $ID_2(2, \mathbf{x}, \mathbf{y}, \mathbf{x}_2, \mathbf{y}_2)$;

query the oracle for the batch of oracle strings on M's oracle tape and store the answers to the queries;

...

FOR each pair $(\mathbf{x}, \mathbf{y}) \in |S|^p \times |S|^p$ in lexicographic order DO
FOR each pair $(\mathbf{x}_1, \mathbf{y}_1) \in |S|^{p1} \times |S|^{p1}$ in lexicographic order DO

resume the simulation $M_1(\mathbf{x}, \mathbf{y}, \mathbf{x}_1, \mathbf{y}_1)$ from $\mathrm{ID}_1(k, \mathbf{x}, \mathbf{y}, \mathbf{x}_1, \mathbf{y}_1)$ using the relevant oracle answer until $M_1(\mathbf{x}, \mathbf{y}, \mathbf{x}_1, \mathbf{y}_1)$ halts;

IF $M_1(\mathbf{x}, \mathbf{y}, \mathbf{x}_1, \mathbf{y}_1)$ has accepted THEN write a 1 on the oracle tape, otherwise write a 0;

FOR each pair $(\mathbf{x}_2, \mathbf{y}_2) \in |S|^{p2} \times |S|^{p2}$, in lexicographic order DO

resume the simulation $M_2(\mathbf{x}, \mathbf{y}, \mathbf{x}_2, \mathbf{y}_2)$ from $\mathrm{ID}_2(k, \mathbf{x}, \mathbf{y}, \mathbf{x}_2, \mathbf{y}_2)$ using the relevant oracle answer until $M_2(\mathbf{x}, \mathbf{y}, \mathbf{x}_2, \mathbf{y}_2)$ halts;

IF $M_2(\mathbf{x}, \mathbf{y}, \mathbf{x}_2, \mathbf{y}_2)$ has accepted THEN write a 1 on the oracle tape, otherwise write a 0;

** note that we have now computed whether $\psi_1{}^S(\mathbf{x}, \mathbf{y}, \mathbf{x}_1, \mathbf{y}_1)$ holds and whether $\psi_2{}^S(\mathbf{x}, \mathbf{y}, \mathbf{x}_2, \mathbf{y}_2)$ holds, for all possible $\mathbf{x}, \mathbf{y}, \mathbf{x}_1, \mathbf{y}_1, \mathbf{x}_2, \mathbf{y}_2$ **

query the oracle for the batch of $2|S|^{2p}$ oracle strings (of length $|S|^{2p1}$ or $|S|^{2p2}$) on M's oracle tape and store the answers to the queries;

** note that we have now computed whether $HP[\lambda \mathbf{x}_1, \mathbf{y}_1, \psi_1{}^S(\mathbf{x}, \mathbf{y}, \mathbf{x}_1, \mathbf{y}_1)](\mathbf{0}, \mathbf{max})$ holds and whether $HP[\lambda \mathbf{x}_2, \mathbf{y}_2, \psi_2{}^S(\mathbf{x}, \mathbf{y}, \mathbf{x}_2, \mathbf{y}_2)](\mathbf{0}, \mathbf{max})$ holds, for all possible $\mathbf{x}, \mathbf{y}$ **

using the answers to the most recent oracle queries, decide whether the input string ω is to be accepted (this can be done in logspace).

It should be clear that M makes k batches of parallel oracle queries and that the DOTM M with oracle $e_{\tau 2}(HP(0, max))$ accepts the set of strings represented by the sentence ϕ. The result follows by induction on k. □

Theorem 3.4 enables us to present a refined view of the structure of $\mathbf{P}^{\mathbf{NP}}$; that is, refined with respect to that presented in [CGH88] and [Wag90]. This view is given in Fig. 1, where the notation used is as follows:

for even j, $HP^i(j)$ is the class of problems represented by a sentence of the form

$$((\ldots((\psi_1 \vee \neg\psi_2) \wedge \psi_3) \vee \ldots) \vee \neg\psi_j),$$

where each $\psi_k \in HP^1[(\pm HP)^{i-1}[FO_s]]$;

for odd j, $HP^i(j)$ is the class of problems represented by a sentence of the form

$$((\ldots((\psi_1 \vee \neg\psi_2) \wedge \psi_3) \vee \ldots) \wedge \psi_j),$$

where each $\psi_k \in HP^1[(\pm HP)^{i-1}[FO_s]]$.

By Theorem 2.1.2 of [CGH88], it is easy to see that $HP^1(j) = \mathbf{NP}(j)$ when j is odd, and $HP^1(j) = \mathbf{co\text{-}NP}(j)$ when j is even ($\mathbf{NP}(j)$ is a level of the Boolean hierarchy; see [CGH88, CGH89]). Hence, the Boolean Hierarchy **BH** can be written as

$$\mathbf{P} \subseteq HP^1(1) \subseteq co\text{-}HP^1(2) \subseteq HP^1(3) \subseteq \ldots \subseteq HP^1(2j-1) \subseteq co\text{-}HP^1(2j) \subseteq \ldots,$$

and so

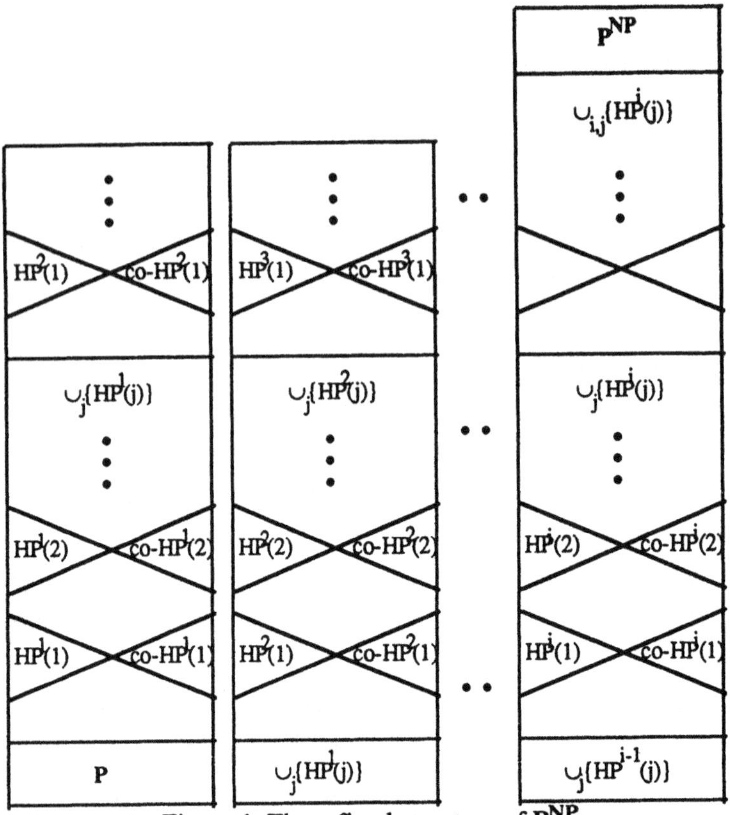

Figure 1 The refined structure of $\mathbf{P^{NP}}$

$$\mathbf{BH} = \cup\{HP^1(2j-1), co\text{-}HP^1(2j) : j = 1, 2, ...\} = \cup\{HP^1(j) : j = 1, 2, ...\}$$

(this is Theorem 2.1.4 of [CGH88]). Also, by Theorem 2.2, the above, and Corollary 3.6 of [Ste92b], we have that

$$\mathbf{BH} = \mathbf{L^{NP}}[O(1)] = \mathbf{P^{NP}}[O(1)] = \text{Bool}[HP^1[FO_s]] \subseteq HP^2(1) \cap co\text{-}HP^2(1).$$

As we can see from Fig. 1, there is a naturally defined hierarchy within $\mathbf{P^{NP}}$ extending the Boolean Hierarchy. Note that this hierarchy, with its elements ordered by inclusion, has ordinal type ω^2 (here ω is the ordinal type of the set of natural numbers and not a string over $\{0, 1\}$, as it usually is) whereas the ordinal type of **BH** (and **PH**, for that matter) is ω. We now relate this hierarchy with classes of polynomial-time DOTM's with oracles in **NP**.

PROPOSITION 3.5. *Let M be a polynomial-time DOTM whose oracle queries are made in parallel. Suppose also that the set of strings accepted by M with oracle X is the encoding of some problem over the vocabulary τ. Then there are formulae ϕ_0, ϕ_1, $\psi \in ATC^1[FO_s(\tau)]$ with free variables those of z and **q**, z and **q**, and **q**,* *respectively, where z is a k-tuple and q is an m-tuple, such that for any structure $S \in STRUCT(\tau)$ and tuples $u \in |S|^k$ and $v \in |S|^m$*

$(S, u, v) \models \phi_0(z, q)$ *iff the* u^{th} *symbol of the* v^{th} *oracle string is a 0 when* $e_\tau(S)$ *is input to* M^X;

$(S, u, v) \models \phi_1(z, q)$ *iff the* u^{th} *symbol of the* v^{th} *oracle string is a 1 when* $e_\tau(S)$ *is input to* M^X;

$(S, u) \models \psi(q)$ *iff when* $e_\tau(S)$ *is input to* M^X, *M makes exactly v oracle calls.*

PROOF. Augment τ with m new constant symbols to obtain the vocabulary τ_a, and augment τ_a with k new constant symbols to obtain the vocabulary τ_b. Clearly, a structure S together with tuples **u** and **v** (resp. a tuple **v**), as above, can be considered as a structure over τ_b (resp. τ_a). Each of the three predicates in the statement of Proposition 3.5 can clearly be verified by a deterministic polynomial-time Turing machine (simulating M^X), and so the existence of the formulae ϕ_0, ϕ_1, and ψ is implied by Theorem 4.7 of [Imm87]. □

THEOREM 3.6. *Let M be a polynomial-time DOTM such that for any oracle X,* M^X *makes exactly k batches of parallel oracle queries, for some* $k \geq 0$. *Let X be a particular oracle in NP and suppose that the set of strings accepted by* M^X *is the encoding of a problem over the vocabulary* τ. *Then there is a sentence* Φ *of* $ATC^1[(\pm HP)^k[FO_s(\tau)]]$ *of the form*

$$ATC[\lambda x, y, \Psi_1, x, \Psi_2](0, max),$$

where Ψ_1 *and* Ψ_2 *are d.p.n.(k)-formulae if* $k > 0$ *and projective formulae otherwise, such that for any structure* $S \in STRUCT(\tau)$

$$S \models \Phi \text{ iff } e_\tau(S) \text{ is accepted by } M^X.$$

PROOF. We may clearly assume that M^X initially checks to see whether the input string is the encoding of some structure over τ and rejects immediately if it isn't (with some dummy oracle queries). Hence, if oracle calls are made in some computation of M^X on some input string ω then we can be sure that $\omega = e_\tau(S)$, for some structure S over τ. By the proof of Theorem 5.3 of [Ste92a], we may clearly assume that the oracle X is $e_{\tau_2}(HP(0, max))$ and that every string for which the oracle is consulted is of length n^{2m}, for some fixed m, where n is the size of the structure S, above. Also, by inserting dummy queries if necessary, we may assume that each batch of queries consists of n^m queries.

We proceed by induction on k. When $k = 0$ the result follows by Theorem 4.7 of [Imm87]. Suppose, as our induction hypothesis (IH), that the result holds for all i such that $0 \le i < k$, and let M be as in the statement of the theorem. By Proposition 3.5, there clearly exists a formula $\phi \in ATC^1[FO_s]$, with free variables those of the m-tuples **x**, **y**, and **z**, such that for all $S \in$ STRUCT(τ) and tuples **u**, **v**, $\mathbf{w} \in |S|^m$, $(S, \mathbf{u}, \mathbf{v}, \mathbf{w}) \models \phi(\mathbf{x}, \mathbf{y}, \mathbf{z})$ iff when $e_\tau(S)$ is input to M with oracle $e_{\tau 2}(HP(0, max))$, the $(\mathbf{u}, \mathbf{v})^{\text{th}}$ symbol of the $\mathbf{w}^{\text{th}}$ query of the first batch of parallel oracle queries is a 1. Let R be a new relation symbol of arity m and let τ' be the vocabulary formed by augmenting τ with this relation symbol R. For any structure $S \in$ STRUCT(τ), let $S_0 \in$ STRUCT(τ') be the expansion of S such that for each tuple $\mathbf{w} \in |S_0|^m = |S|^m$

$$R^{S_0}(\mathbf{w}) \text{ holds iff } (S, \mathbf{w}) \models HP[\lambda \mathbf{x}, \mathbf{y}, \phi(\mathbf{x}, \mathbf{y}; \mathbf{z})](\mathbf{0}, \mathbf{max}).$$

Hence, $R^{S_0}(\mathbf{w})$ holds iff the oracle $e_{\tau 2}(HP(0, max))$ answers "yes" to the $\mathbf{w}^{\text{th}}$ query of the first batch of parallel oracle queries in the computation of M on input $e_\tau(S)$.

Let the polynomial-time DOTM M' with oracle $e_{\tau 2}(HP(0, max))$ be defined as follows:

(a) the set of strings accepted is the encoding of a problem over τ', and an initial check is made to see whether the input string is indeed the encoding of a structure over τ', with immediate rejection if not (after making some dummy oracle queries);

(b) M' on input $e_{\tau'}(S')$, for some $S' \in$ STRUCT(τ'), simulates M (with oracle $e_{\tau 2}(HP(0, max))$) on input $e_\tau(S)$, where S is S' restricted to τ, except that instead of making the first batch of parallel oracle queries, M' uses the answers held in the relation $R^{S'}$ (and so in this stage, M' proceeds as M does except that nothing is written on the oracle tape).

It should be clear that for all $S \in$ STRUCT(τ), M with oracle $e_{\tau 2}(HP(0, max))$ accepts $e_\tau(S)$ iff M' with oracle $e_{\tau 2}(HP(0, max))$ accepts $e_{\tau'}(S_0)$ (where S_0 is as above). We may also assume that M' makes exactly k–1 batches of parallel oracle queries for any oracle.

By (IH) there is a sentence Φ' of $ATC^1[(\pm HP)^{k-1}[FO_s(\tau')]]$ of the form $ATC[\lambda \mathbf{x}_1, \mathbf{y}_1, \psi_1, \mathbf{x}_1, \psi_2](\mathbf{0}, \mathbf{max})$, where ψ_1 and ψ_2 are d.p.n.(k–1)-formulae of $(\pm HP)^{k-1}[FO_s(\tau')]$ if $k > 1$ and projective formulae otherwise, such that for any structure $S' \in$ STRUCT(τ')

$$S' \models \Phi' \text{ iff } e_{\tau'}(S') \text{ is accepted by } M' \text{ with oracle } e_{\tau 2}(HP(0, max)).$$

Let the formulae Ψ_1 and Ψ_2 be the formulae ψ_1 and ψ_2, respectively, with any occurence $R(\mathbf{t})$, for some tuple of variables and constant symbols **t**, replaced by the formula $HP[\lambda \mathbf{x}, \mathbf{y}, \phi(\mathbf{x}, \mathbf{y}; \mathbf{t})](\mathbf{0}, \mathbf{max})$. Define Φ as

$$ATC[\lambda \mathbf{x}_1, \mathbf{y}_1, \Psi_1, \mathbf{x}_1, \Psi_2](\mathbf{0}, \mathbf{max}).$$

Then for any $S \in$ STRUCT(τ), $e_\tau(S)$ is accepted by M with oracle $e_{\tau 2}(HP(0, max))$ iff $e_{\tau'}(S_0)$ is accepted by M' with oracle $e_{\tau 2}(HP(0, max))$ iff $S_0 \models \Phi'$ iff $S \models \Phi$. The sentence Φ is in $ATC^1[(\pm HP)^k[FO_s]]$, and so the result follows by Corollary 3.3. □

Denote the class of (encodings of) problems accepted by polynomial-time DOTM's with an oracle from **NP** and which make at most $f(n)$ batches of parallel oracle queries where each batch has size at most $g(n)$ by $\mathbf{P}^{\mathbf{NP}}[||f(n), \#g(n)]$ (the class of problems $\mathbf{P}^{\mathbf{NP}}[||O(f), \#O(g)]$, $\mathbf{P}^{\mathbf{NP}}[||f(n), \#\text{poly}]$, and $\mathbf{P}^{\mathbf{NP}}[||O(f), \#\text{poly}]$ should be obvious).

COROLLARY 3.7. *$ATC^1[(\pm HP)^k[FO_s]] = P^{NP}[\,||k, \#poly]$.*

PROOF. The proof that $ATC^1[(\pm HP)^k[FO_s]] \subseteq \mathbf{P}^{\mathbf{NP}}[||k, \#\text{poly}]$ follows similarly to the proof of Theorem 3.4, and the result follows by Theorem 3.6. □

We can now apply the result due to Buss and Hay, given earlier as Theorem 2.3, to show that the extension of the Boolean hierarchy as depicted in Fig. 1 actually collapses.

COROLLARY 3.8. *The following are all logical characterizations of the complexity class L^{NP}:*

(i) $HP^i(j)$, for $i > 1$ and $j \geq 1$;

(ii) co-$HP^i(j)$, for $i > 1$ and $j \geq 1$;

(iii) $(\pm HP)^*[FO_s]$;

(iv) $(\pm HP)^1[FO_s]$,

and every sentence of the logic $(\pm HP)^1[FO_s]$ is equivalent to one of the form

$$\exists x_1 \exists x_2 \ldots \exists x_m [HP[\lambda x, y, \psi_1(x, y)](\mathbf{0}, \mathbf{max}) \wedge \neg HP[\lambda x, y, \psi_2(x, y)](\mathbf{0}, \mathbf{max})],$$

where ψ_1 and ψ_2 are projective formulae.

PROOF. By Theorem 2.3, any set of strings accepted by a polynomial-time DOTM with an oracle from **NP** which makes k batches of parallel oracle queries, for any input string, can be accepted by one which makes 1 batch of parallel oracle queries, for any input string. Hence, the result follows by Theorems 2.2, 2.3, an 3.4, and Corollary 3.9 of [Ste92b]. □

COROLLARY 3.9. *If the logics $(\pm HP)^*[FO_s]$ and $Bool[HP^1[FO_s]]$ have the same expressibility then the Polynomial Hierarchy collapses.*

PROOF. Immediate from Corollary 3.10 of [Ste92b]. □

4. CONCLUSION

We have shown that the logics $(\pm HP)^*[FO_s]$ and $(\pm HP)^1[FO_s]$ are of the same expressibility, and both capture $\mathbf{P}_{||}^{\mathbf{NP}}$. This result gives us the weakest possible hint that it might be wiser to try and show that $(\pm STC)^*[FO_s]$ collapses to $(\pm STC)^1[FO_s]$ as opposed to trying to show that $STC^1[FO_s]$ is closed under complementation: an attempt to use the methods of [Imm88] to achieve this latter result has failed (see [BCD89]).

As to achieving the former result, one approach might be to consider logspace DOTM's with oracles in **NSYMLOG**. In particular, one could try to show that such oracle machines are equivalent to logspace DOTM's which make all their oracle queries in parallel (as is the case when the oracle is in **NP**) and then to code such computations as formulae of $DTC^1[STC^1[FO_s]]$. This is just a suggestion, but it should be clear that a consideration of oracle machines with restricted access to oracles not necessarily in **NP** should be undertaken.

REFERENCES

[BCD89] A. BORODIN, S. A. COOK, P. W. DYMOND, W. L. RUZZO, AND M. TOMPA, Two applications of inductive counting for complementation problems, *SIAM J. Comput.*, **18**, 3 (1989), 559-578.

[Bei88] R. J. BEIGEL, Bounded queries to SAT and the Boolean Hierarchy, to appear, *Theoret. Comput. Sci.*.

[BH88] S. R. BUSS AND L. HAY, On truth-table reducibility to SAT and the Difference Hirarchy over NP, *Proc. 3rd Symp. on Structure in Complexity Theory*, IEEE Press (1988), 224-233.

[CGH88] J. CAI, T. GUNDERMANN, J. HARTMANIS, L. A. HEMACHANDRA, V. SEWELSON, K. W. WAGNER, AND G. WECHSUNG, The Boolean hierarchy I: Structural properties, *SIAM J. Comput.*, **17**, 6 (1988), 1232-1252.

[CGH89] J. CAI, T. GUNDERMANN, J. HARTMANIS, L. A. HEMACHANDRA, V. SEWELSON, K. W. WAGNER, AND G. WECHSUNG, The Boolean hierarchy II: Applications, *SIAM J. Comput.*, **18**, 1 (1989), 95-111.

[Imm87] N. IMMERMAN, Languages that capture complexity classes, *SIAM J. Comput.*, **16**, 4 (1987), 760-778.

[Imm88] N. IMMERMAN, Nondeterministic space is closed under complementation, *SIAM J. Comput.*, **17**, 5 (1988), 935-938.

[Köb87] J. KÖBLER, personal communication as cited in [Wag90].

[KSW87] J. KÖBLER, U. SCHÖNING, AND K. W. WAGNER, The difference and the truth-table hierarchies for NP, *RAIRO Inform. Theory*, **21** (1987), 419-435.

[Ste91a] I. A. STEWART, Comparing the expressibility of languages formed using NP-complete operators, *J. Logic Computat.*, **1**, 3 (1991), 305-330.

[Ste91b] I. A. STEWART, On completeness for NP via projection translations, *Math. Systems Theory*, to appear.

[Ste91c] I. A. STEWART, Complete problems involving Boolean labelled structures and projection translations, *J. Logic Computat.*, **1**, 6 (1991), 861-882.

[Ste92a] I. A. STEWART, Using the Hamiltonian operator to capture NP, *J. Comput. System Sci.*, **45**, 1 (1992), 127-151.

[Ste92b] I. A. STEWART, Logical characterizations of bounded query classes I: logspace oracle machines, *L.N.C.S.* 620 (1992), 470-479.

[Wag90] K. W. WAGNER, Bounded query classes, *SIAM J. Comput.*, **19**, 5 (1990), 833-846.

On Asymptotic Probabilities of Monadic Second Order Properties

Jerzy Tyszkiewicz*

Institute of Informatics, University of Warsaw,
ul. Banacha 2, 02-097 Warszawa, Poland.
jurekty@mimuw.edu.pl

Abstract. We propose a new, general and easy method for proving nonexistence of asymptotic probabilities of monadic second-order sentences in classes of finite structures where first-order extension axioms hold almost surely.

1 Introduction

1.1 The problem

In this paper we discuss logical problems of *random structure theory.*

Let us consider a class of finite structures over some fixed signature, equipped with a probability space structure. This probability is usually assumed to be only *finitely* additive. Then we draw one structure at random and ask:

- how does the drawn structure look like?
- does the drawn structure have some particular property?

Those questions are typical in random structure theory. To turn to the logical part of it, look at the drawn structure through the logical glasses: we can only notice properties definable in some particular logic. Then new questions become natural:

- does every property we can observe have a probability (is it measurable)?
- if so, what is the value of this probability?
- is it possible to compute this probability, and what is the complexity of the computation?

It becomes clear from the above that the random structure theory is closely connected to combinatorics, finite model theory, mathematical logic and computer science. An exposition of the logical part of the random structure theory may be found in a nice survey by Compton [1].

One of the problems we would like to pursue in this paper is the following: It was empirically observed that first-order 0–1 laws sometimes extend to fixpoint 0–1 laws, sometimes to monadic second-order 0–1 laws, but almost never to both of these logics at the same time. There was no theoretical explanation of this fact so far. One of our main intentions, except presenting some new results and new proofs of some already known results, is to make a first step towards such explanation. It will follow from uniform proof method we propose for nonconvergence results in monadic logic.

* Research partially supported by KBN grant 2 1192 91 01.

1.2 Definitions

Throughout the paper we are dealing with first–order and monadic second–order (monadic, for short) logics over some fixed, finite signature σ (with equality). We assume that σ contains exclusively relation symbols, and therefore functions are represented as restricted relations.

Formulas of first–order logic are built from atomic formulas of the form $R(\mathbf{x})$, where R is a symbol in σ and $\mathbf{x}$ is a vector of variables of length equal to arity of R, and using usual connectives: $\wedge$, $\vee$, $\neg, \rightarrow$, and quantifiers $\forall$, $\exists$.

In formulas of monadic logic set variables are allowed to occur in atomic formulas of the form $x \in X$. Also set quantification, denoted $\forall$ and $\exists$, is allowed.

We use uppercase letters, or two–letter uppercase abbreviations, to denote set variables, and lowercase ones to denote first–order variables. Therefore it will always be clear what kind of quantification we have in mind.

Let $\mathcal{A}$ be a set of finite structures $\mathbf{A}$ over the signature σ, such that the universe $|\mathbf{A}|$ of $\mathbf{A}$ is some initial segment of natural numbers. Let $\mathcal{A}(n)$ be a subset of $\mathcal{A}$ containing all structures $\mathbf{A}$ with carrier set (of cardinality) $|\mathbf{A}| = n = \{0, \ldots, n-1\}$. To avoid pathological cases we assume that for each positive $n \in \mathbb{N}$ the set $\mathcal{A}(n)$ is nonempty.

Let for $n = 0, 1, \ldots$ μ_n be a probability distribution on $\mathcal{A}(n)$. We write μ for $\{\mu_n\}_{n \in \mathbb{N}}$, and call μ, somehow loosely, also a distribution. The pair $\langle \mathcal{A}, \mu \rangle$ is an object of our study in this paper.

For any subset $D \subseteq \mathcal{A}$ we define

$$\mu_n(D) = \mu_n(D \cap \mathcal{A}(n)).$$

If $D = \{\mathbf{A} \in \mathcal{A} \mid \mathbf{A} \models \varphi\}$ for some sentence φ of the logic under consideration, then we write $\mu_n(\varphi)$, instead of $\mu_n(D)$. We are interested in asymptotic properties of $\mu_n(\varphi)$, and especially whether the limit $\mu(\varphi) = \lim_{n \to \infty} \mu_n(\varphi)$ exists, for sentences φ of the logic under consideration. If it exists, we call it an *asymptotic probability of* φ. If this is the case for every sentence of the logic L, we say that the *convergence law* holds (for L and μ). If, in addition, every sentence has probability either 0 or 1, we say that the *0–1 law* holds. Sometimes, instead of writing $\mu(\varphi) = 1$, we say that φ *holds almost surely (a.s. in short)*.

The following family of examples of probability distributions on the class $\mathcal{G}$ of all finite graphs $\mathbf{G} = \langle |\mathbf{G}|, E \rangle$ was first introduced and studied by Erdős and Rényi in [3]. Let $p = p(n)$ be a function from $\mathbb{N}$ into the real interval [0,1]. Then we define the probability model $\mathcal{G}(n,p) = \langle \mathcal{G}, \mu^p \rangle$ with $\mathcal{G}$ being the class of all undirected finite graphs, and $\mu_n^p(\{\mathbf{G}\}) = p^e(1-p)^{\binom{n}{2}-e}$, where e is the number of edges in $\mathbf{G}$. Equivalently, one obtains a random graph $\mathbf{G} \in \mathcal{G}(n,p)$ as a result of the following experiment: For every pair of vertices in $\{0, \ldots, n-1\}$, independently of other pairs, one tosses a coin with outcomes 1 (edge) with probability $p(n)$ and 0 (non-edge) with probability $1 - p(n)$. After $\binom{n}{2}$ tosses the random graph is constructed.

Another example is the following:

Let $\mathcal{A}$ be arbitrary class of finite structures, satisfying the conditions we formulated at the beginning. We then define the *uniform labelled probability distribution on* $\mathcal{A}$. Namely, we set

$$\mu_{|\mathbf{A}|}(\{\mathbf{A}\}) = \begin{cases} 1/|\mathcal{A}(n)| & \text{if } \mathbf{A} \in \mathcal{A}(n), \\ 0 & \text{if } \mathbf{A} \notin \mathcal{A}(n). \end{cases}$$

E.g., if $p = const. = 1/2$ then the μ^p and the uniform labelled distributions on $\mathcal{G}$ coincide.

Unlabelled distribution is the one in which equivalence classes of the isomorphism relation are equiprobable rather than structures themselves.

For a finite structure $\mathbf{A} \in \mathcal{A}(n)$ we define a first-order quantifier-free formula $[\{x_0,\ldots,x_{n-1}\} \simeq \mathbf{A}]$, called the *diagram of* $\mathbf{A}$. Let v be a valuation such that $v(x_i) = i$ for $i = 0,\ldots,n-1$. Then we define $[\{x_0,\ldots,x_{n-1}\} \simeq \mathbf{A}]$ to be the conjunction of all formulas in the set

$$\{R(x_{i_1},\ldots,x_{i_k}) \mid R \in \sigma,\ 0 \le i_1,\ldots,i_k < n,\ \mathbf{A}, v \models R(x_{i_1},\ldots,x_{i_k})\} \cup$$

$$\{\neg R(x_{i_1},\ldots,x_{i_k}) \mid R \in \sigma,\ 0 \le i_1,\ldots,i_k < n,\ \mathbf{A}, v \models \neg R(x_{i_1},\ldots,x_{i_k})\}.$$

1.3 Organization of the paper

The paper is organized as follows: In the second section we give a new proof of a classical result of Kaufmann and Shelah [6], stating that there are monadic second-order properties without asymptotic probability with respect to uniform, labelled probability on the class of finite graphs. In the third section we show how the proof can be improved to work for so called *sparse* random graphs $\mathcal{G}(n,p)$ with $p = p(n)$ bounded away from 1 and satisfying $p(n) \ge n^{-\alpha}$ for some $\alpha < 1$. The fourth section is devoted to adaptation of the same proof to the case of random (uniform, labelled) partial orders, and random K_{m+1}–free graphs. In the fifth section we discuss inter-relations of monadic second-order and fixpoint logics, which can be deduced from our results. In the last, sixth section, we present some final remarks.

2 The Kaufmann and Shelah Theorem

In this section we give a new proof of the following, classical result of Kaufmann and Shelah.

Theorem 1 (Kaufmann and Shelah [6]). *Then there are monadic second-order sentences without asymptotic uniform labelled probability in the class of finite graphs* $\mathcal{G}$.

Proof. First short description of the main idea: consider a pair $(\mathbf{G},\mathbf{H})$ of graphs $\mathbf{G} \subseteq \mathbf{H}$ with $|\mathbf{G}| = \{0,\ldots,k-1\}$ and $|\mathbf{H}| \setminus |\mathbf{G}| = \{k,\ldots,\ell-1\}$. From now on writing $(\mathbf{G},\mathbf{H})$ we tacitly make an assumption that $\mathbf{G} \subseteq \mathbf{H}$.

Denote by $Ext(\mathbf{G},\mathbf{H})$ the following sentence, usually called an *extension axiom*:

$$\forall x_0,\ldots,x_{k-1}\ ([\{x_0,\ldots,x_{k-1}\} \simeq \mathbf{G}] \to (\exists x_k,\ldots,x_{\ell-1}\ [\{x_0,\ldots,x_{\ell-1}\} \simeq \mathbf{H}])),$$

i.e., the one expressing *"every copy of* $\mathbf{G}$ *extends to a copy of* $\mathbf{H}$*"*.

It is well known since the paper of Fagin [4] that for arbitrary two graphs $\mathbf{G} \subsetneq \mathbf{H}$ the equality $\mu(Ext(\mathbf{G},\mathbf{H})) = 1$ holds. In particular $\mu(Ext(\emptyset,\mathbf{G})) = 1$ and $\mu(Ext(\mathbf{G},\mathbf{H})) = 1$. Then, assuming additionally that $|\mathbf{H}|$ is of much greater cardinality than $|\mathbf{G}|$, the graph of the function $n \mapsto \mu_n(Ext(\mathbf{G},\mathbf{H}))$ looks like the one on figure 1, at the end of the paper.

It is important that $\mu_n(Ext(\mathbf{G},\mathbf{H})) = 1$ for $n < |\mathbf{G}|$, and $\mu_n(Ext(\mathbf{G},\mathbf{H})) = 1 - \mu_n(Ext(\emptyset,\mathbf{G}))$ for $n < |\mathbf{H}|$.

Then it is natural to expect that a sentence expressing $\bigwedge_{i\in\mathbb{N}} Ext(\mathbf{G}_i,\mathbf{H}_i)$, with cardinalities of $(\mathbf{G}_i,\mathbf{H}_i)$ growing very fast with i, should have no asymptotic probability, as we may naturally expect

$$\mu_n(\bigwedge_{i\in\mathbb{N}} Ext(\mathbf{G}_i,\mathbf{H}_i)) \approx \min_{i\in\mathbb{N}} \mu_n(Ext(\mathbf{G}_i,\mathbf{H}_i)),$$

as in the figure 2 at the end of the paper.

Now let us be more precise.

Let M be a deterministic, one tape Turing Machine that accepts numbers in unary expansion as its arguments, and always halts.

Define

$$size_M(m) = m + space_M(m) \cdot time_M(m),$$

where $time_M(m)$ denotes the number of steps of computation of M on input m, and similarly for $space_M$.

We construct a monadic formula $\varphi(X)$ with the property that whenever $\mathbf{G} \models \varphi(X)$ is true, the cardinality of X is equal to $size_M(m)$ for some m.

Indeed, take $\varphi(X)$ to be:

$$\exists M,U\ \exists EC,OC,ER,OR\ \varphi(M,U,EC,OC,ER,OR),$$

where φ is the conjunction of the following conditions:

1. $M,U \subseteq X$
2. $EC,OC,ER,OR \subseteq U$
3. $M \cap U = \emptyset,\ M \cup U = X,$
4. $EC \cap OC = \emptyset,\ EC \cup OC = U,$
5. $ER \cap OR = \emptyset,\ ER \cup OR = U,$
6. $\langle U,E\rangle$ is a square grid, such that EC and OC (ER and OR, resp.) are unions of disjoint chains, which are even and odd columns (rows, resp.) of the grid,
7. E is a bijection from M onto the first $|M|$ elements in the first row of the grid,
8. there are no other edges, except possible edges between vertices of M.

X extends M – we call extensions of that shape *grid extensions*. A figure showing grid extension is to be found at the end of the paper. Now it is routine to express, adding more set variables that correspond to letters in tape cells, control states and head positions that the grid with these sets represents successful computation of M on input $1\ldots1$ of length $|M|$.

Now similarly, for another Turing machine N we can write a formula $\gamma(X,Y)$ with the property that whenever $\mathbf{G} \models \gamma(X,Y)$, then $|Y \setminus X|$ is equal to $size_N(|X|)$. It is to be done by treating X exactly as M in φ above, but without quantifying it.

Now let the function $g : \mathbb{N} \to \mathbb{N}$ be defined as follows:

$$g(m) = 1 + \begin{pmatrix} \text{least } n > g(m-1) \text{ such that } \mu_n(\bigwedge Ext(\mathbf{G},\mathbf{H})) \geq \\ 1 - 1/m, \text{ where we conjunct over all grid exten-} \\ \text{sions } (\mathbf{G},\mathbf{H}) \text{ with } |\mathbf{H}| \leq m \end{pmatrix} \qquad (1)$$

Clearly g is recursive and strictly increasing.

Let a machine N compute some space constructible function $h > g$ in the way that $h(m) = space_N(m)$, taking unary strings as inputs and producing unary strings as outputs.

Now let M be a one tape deterministic Turing Machine that takes unary input strings and outputs also unary strings. Moreover, let M compute a total function f that for $m > 0$ satisfies:

$$f(m) > g(size_N(size_M(m-1))).$$

Let us see how to make a monadic sentence without asymptotic probability from N and M.

Let $\varphi(X)$ and $\gamma(X,Y)$ be formulas constructed for M and N, respectively. Consider sentence

$$\mathbf{Ext} \equiv \forall X \ (\varphi(X) \to (\exists Y \ \gamma(X,Y))).$$

Observe that indeed **Ext** is equivalent to the infinitary conjunction of the form $\bigwedge_{m\in\mathbb{N}} Ext(\mathbf{G}_m, \mathbf{H}_m)$, where $\mathbf{G}_m$ is a grid extension of $\emptyset$ with $size_M(m)$ elements, and $\mathbf{H}_m$ is a grid extension of $\mathbf{G}_m$ with $size_N(size_M(m))$ elements.

We claim that Ext has no limiting probability.

First let $n = g(size_M(m)) - 1$ for some m. Then, by construction of g,

$$\mu_n(Ext(\emptyset, \mathbf{H})) \geq 1 - 1/m$$

for all grid extensions $(\emptyset, \mathbf{H})$ with $|\mathbf{H}| \leq size_M(m)$. Therefore in random $\mathbf{G}$ with n elements there is a choice of X to satisfy $\varphi(X)$ of cardinality $|X| = size_M(m)$, with probability $\geq 1-1/m$. But in this case there is no choice of Y to satisfy $\gamma(X,Y)$ since then it would be $|Y \setminus X| \geq g(size_M(m)) > n$, which is impossible. Therefore $\mu_n(\mathbf{Ext}) \leq 1/m$.

In this part of the computation we essentially check that the function $n \mapsto \mu_n(Ext(\mathbf{G}_m, \mathbf{H}_m))$ *will have* a minimum with value close to 0 in n.

Secondly, let $n = g(size_N(size_M(m-1)))$ for some m. Then, as $size_M(m) > n$, all X's that may satisfy $\varphi(X)$ are of cardinalities $size_M(0), \ldots, size_M(m-1)$. But $\mu_n(\bigwedge Ext(\mathbf{G},\mathbf{H})) \geq 1 - 1/m$, where we conjunct over all grid $(\mathbf{G},\mathbf{H})$ with $|\mathbf{H}| \leq size_N(size_M(m-1))$, hence for every X that satisfies $\varphi(X)$ there is a choice of Y to satisfy $\gamma(X,Y)$, with probability no less than $1-1/m$. Therefore $\mu_n(\mathbf{Ext}) \geq 1-1/m$.

In this part of the computation we essentially check that $\mu_n(Ext(\mathbf{G}_m, \mathbf{H}_m))$ *will not have* a minimum with value close to 0 before $\mu_{n'}(\bigwedge_{i=0}^{m-1} Ext(\mathbf{G}_i, \mathbf{H}_i))$ becomes close to 1 for some $n' > |\mathbf{H}_{m-1}|$.

We immediately infer that **Ext** has no asymptotic probability. □

3 Sparse random graphs

We next show how to improve our proof of Kaufmann and Shelah Theorem to obtain the following, much more general result:

Theorem 2. *Let* $p : \mathbb{N} \to [0,1]$ *be a function such that* $n^{-\alpha} \leq \beta$ *for some* $0 < \alpha, \beta < 1$, *and all large* n. *Then there exist monadic second-order sentences without asymptotic probability with respect to random graphs* $\mathcal{G}(n,p)$.

It should be stressed that Spencer and Shelah [10] proved a first-order 0–1 law for random graphs $\mathcal{G}(n, n^{-\alpha})$ with α *irrational.* Therefore our nonconvergence result is almost optimal.

Proof. We will describe step by step the necessary changes of our proof of Kaufmann and Shelah Theorem to make it work for sparse random graphs. First we give informal description of the changes. Precise formulations follow it, and are stated as separate definitions and lemmas.

1. It is not true that $\mu^p(Ext(\mathbf{G},\mathbf{H})) = 1$ holds for arbitrary grid extension $(\mathbf{G},\mathbf{H})$. Therefore we find an improved notion of extensions, called *k–improved grid extensions,* for which the equality holds. It is still possible to encode computations of Turing machines in such extensions.
2. We change the function *size* suitably.
3. We define the function g by equality similar to (1):

$$g(m) = 1 + \begin{pmatrix} \text{least } n > g(m-1) \text{ such that } \mu_n(\bigwedge Ext(\mathbf{G},\mathbf{H})) \geq \\ 1-1/m, \text{ where we conjunct over all } k\text{–improved grid} \\ \text{extensions } (\mathbf{G},\mathbf{H}) \text{ with } |\mathbf{H}| \leq m \end{pmatrix}$$

   As g defined above need not be recursive, we show that there exists a recursive function $\widetilde{g}$ such that $\widetilde{g} \geq g$.

Precise formulations of the points above follow:

1. Extension axioms Let for a graph $\mathbf{G}$ the symbol $e(\mathbf{G})$ denote the number of edges in $\mathbf{G}$, and $v(\mathbf{G})$ the number of vertices of $\mathbf{G}$. We say that the pair $(\mathbf{G},\mathbf{H})$ is *safe for exponent* α iff for every S with $\mathbf{G} \subset S \subseteq \mathbf{H}$:

$$(v(S) - v(\mathbf{G})) - \alpha \cdot (e(S) - e(\mathbf{G})) > 0.$$

Now we cite the theorem:

Theorem 3 (Ruciński and Vince [9]). *If* $(\mathbf{G},\mathbf{H})$ *is safe for exponent* $\alpha < 1$ *and* $0 < \beta < p(n) \geq n^{-\alpha}$ *for some constant* β, *then* $\mu^p(Ext(\mathbf{G},\mathbf{H})) = 1$. □

According to the above result we construct k–improved grid extensions, safe for arbitrary given exponent $\alpha > 0$. They are obtained from standard grid extensions by adding many new vertices: we add k intermediate vertices on edges between vertices of $X \setminus M$, and $2k$ intermediate vertices on edges connecting elements of M with elements of $X \setminus M$. The added vertices we call *new,* in contrary to the *old* ones. (See figure 4 at the end of the paper.)

The only thing we need is to show that choosing k large enough we can make such extension safe for any given $0 < \alpha < 1$. Of course all necessary relations are still monadic definable.

Lemma 4. *If positive natural k is chosen so that $\alpha < 2k/2k{+}1$, then the k-improved grid extension is safe for exponent α.*

Proof. Let (M, X) be an improved grid extension, and let $M \subset S \subseteq X$. Let us call a subset H of $S \setminus M$ a *hole* if it induces a cycle in S inside of which there is no other vertex of S ("inside" should be understood to refer to the situation on figures 3 and 4). Let us call a *root* an old vertex in S that is connected by some path made of new vertices in S to some vertex in M. Now let v denote the number of vertices in $S \setminus M$, e the number of edges in S, excluding edges between vertices inside M, h the number of holes in S, and r the number of roots in S.

Then, as it is easy to see,

$$e - v \leq r + h. \tag{2}$$

Observe that each hole in S has at least $4k$ new vertices on its boundaries, and each new vertex in S cannot be counted in this way more than twice. Each root is connected to an element in M by a path of $2k$ new vertices, which do not lie on a boundary of any hole, and, moreover, are different for different roots. Therefore we get

$$v \geq 2kr + 2kh. \tag{3}$$

Taking (2) and (3) together we get

$$\frac{v}{e} = \frac{v}{v + (e - v)} \geq \frac{2k(r + h)}{2k(r + h) + (r + h)} = \frac{2k}{2k + 1} > \alpha,$$

so indeed $v - \alpha \cdot e > 0$, as desired. □

2. The function size. Once we have the notion of k–grid extension, the definition of the function *size* is modified in obvious way.

3. The functions g and $\tilde{g}$. We show that g defined in the proof of theorem 2 is *majorized* by a recursive function. Our estimates generally come from the fact that there exist rational numbers and $0 < \alpha, \beta < 1$ such that $n^{-\alpha} \leq p(n) \leq \beta$ holds for all sufficiently large n.

First of all, we reduce our attention to k–improved (we will omit this word in the sequel) grid extensions, which are safe for exponent α.

Now we modify our random graph structure. We allow three valued $\{0, \frac{1}{2}, 1\}$ logic for edge existence. Now between two given vertices there may exist: non-edge, half-edge and edge. Moreover, we suitably change the probability space. For every pair of vertices from $\{0, \ldots, n-1\}$, independently of other pairs, we toss a three-sided die, which gives outcomes: 0 with probability β, 1 with probability $n^{-\alpha}$ and $\frac{1}{2}$ with probability $1 - \beta - n^{-\alpha}$. After $\binom{n}{2}$ tosses we obtain random three valued graph. This defines a probability space $\langle \mathcal{G}^3, \mu^{\alpha,\beta} \rangle$, where $\mathcal{G}^3$ stands for the class of finite three valued graphs.

The informal idea is as follows: the half-edge should be understood as "not yet decided: there is an edge or not". Usually a $\mu^{\alpha,\beta}$-random graph in $\mathcal{G}^3$ contains many such undecided places. But it is possible that the places that are already decided allow one to verify that whatever will be decided about half-edges later, extension axioms are satisfied. We want to show that this possibility holds with large probability.

The class $\mathcal{G}$ of standard, two valued graphs we denote $\mathcal{G}^2$, to avoid confusion. It is natural that one may assume $\mathcal{G}^2 \subseteq \mathcal{G}^3$, and therefore it is clear what it means that two graphs $\mathbf{G} \in \mathcal{G}^3$ and $\mathbf{H} \in \mathcal{G}^2$ are isomorphic, etc. In particular, μ^p is also a probability distribution in $\mathcal{G}^3$.

Now it is routine, following the proof of theorem 3 presented in [9], to prove also the following:

Lemma 5. *Let pair* $(\mathbf{G}, \mathbf{H})$ *with* $\mathbf{G}, \mathbf{H} \in \mathcal{G}^3$, *but without half-edges in* $\mathbf{H}$, *except between vertices of* $\mathbf{G}$, *be safe for exponent* α. *Definition of safety ignores edges inside* $\mathbf{G}$, *so the above makes sense. Then with asymptotic probability one every subgraph of a random three-valued graph* $\mathbf{K}$ *that is isomorphic to* $\mathbf{G}$ *can be extended to a subgraph isomorphic to* $\mathbf{H}$. □

Let $(\mathbf{G}, \mathbf{H})$ be as in the above lemma. The property that every isomorphic copy of $\mathbf{G}$ extends to a copy of $\mathbf{H}$ we denote by $Ext(\mathbf{G}, \mathbf{H})$, and call it also an extension axiom. Note that formally $Ext(\mathbf{G}, \mathbf{H})$ is not a first-order sentence, unless $\mathbf{G}, \mathbf{H} \in \mathcal{G}^2$.

Now we define the function $\tilde{g}$ by

$$\tilde{g}(m) = 1 + \left(\begin{array}{l}\text{least } n > \tilde{g}(m-1) \text{ such that } \mu_n^{\alpha,\beta}(\bigwedge Ext(\mathbf{G},\mathbf{H})) \geq \\ 1 - 1/m, \text{ where we conjunct over all } k\text{-improved grid} \\ \text{extensions } (\mathbf{G},\mathbf{H}) \text{ with } |\mathbf{H}| \leq m, \text{ with } \mathbf{G},\mathbf{H} \in \mathcal{G}^3 \text{ like} \\ \text{in the last lemma}\end{array}\right)$$

It can be easily observed that $\tilde{g}$ is recursive.

Lemma 6. *For every natural number* n

$$\tilde{g}(n) \geq g(n).$$

Proof. It suffices to prove the following:

For all n such that $n^{-\alpha} \leq p(n) \leq \beta$ the inequality

$$\mu_n^{\alpha,\beta}(\bigwedge Ext(\mathbf{G},\mathbf{H})) \leq \mu_n^p(\bigwedge Ext(\mathbf{G},\mathbf{H}))$$

holds, where in both sides we conjunct over all k-improved grid extensions $(\mathbf{G}, \mathbf{H})$ with $|\mathbf{H}| \leq m$, possibly with $\mathbf{G} \in \mathcal{G}^3$.

We fix n satisfying $n^{-\alpha} \leq p(n) \leq \beta$.

For $\mathbf{K} \in \mathcal{G}^3(n)$ and $i = 2, 3$ let $Cl^i(\mathbf{K})$ denote the set of all graphs $\mathbf{G} \in \mathcal{G}^i(n)$ such that if there is edge (non-edge, resp.) between u and v in $\mathbf{K}$, then there is edge (non-edge, resp.) between u and v in $\mathbf{G}$. In particular, always $Cl^2(\mathbf{K}) = Cl^3(\mathbf{K}) \cap \mathcal{G}^2(n) \subseteq Cl^3(\mathbf{K})$.

Let $\mathbf{K}' \in \mathcal{G}^3$ be any graph obtained from $\mathbf{K} \in \mathcal{G}^3$ by replacing some of its half-edges by non-edges or edges, in arbitrary way. Then it is not difficult to observe that if $\bigwedge Ext(\mathbf{G}, \mathbf{H})$ is true in $\mathbf{K}$, then so is in $\mathbf{K}'$.

By the above observation it becomes clear that if $\mathbf{K} \models \bigwedge Ext(\mathbf{G}, \mathbf{H})$, then $\mathbf{K}' \models \bigwedge Ext(\mathbf{G}, \mathbf{H})$ for every $\mathbf{K}' \in Cl^i(\mathbf{K})$.

Moreover, it is easy to observe that for two graphs $\mathbf{K}, \mathbf{K}' \in \mathcal{G}^i$, the sets $Cl^i(\mathbf{K})$ and $Cl^i(\mathbf{K}')$ are either disjoint, or one is included in the other.

Let $\mathcal{K} = \{\mathbf{K}_0, \ldots, \mathbf{K}_m\}$ be the set of all graphs in $\mathcal{G}^3(n)$ in which the sentence $\bigwedge Ext(\mathbf{G}, \mathbf{H})$ is true. Then maximal sets among $Cl^3(\mathbf{K}_j)$, $j = 0, \ldots, m$ partition $\mathcal{K}$ into disjoint subsets.

Now, in order to finish the proof, it suffices to check that always

$$\mu_n^p(Cl^2(\mathbf{K}_j)) \geq \mu_n^{\alpha,\beta}(Cl^3(\mathbf{K}_j)).$$

Indeed, if we look at a graph as a sequence of outcomes in $\binom{n}{2}$ die tosses, $Cl^i(\mathbf{K}_j)$ is the set of such sequences that have 0's and 1's in some specified places, and "anything" in other places. Now it is easier to get 1 in a specified place according to μ_n^p than according to $\mu_n^{\alpha,\beta}$, and similarly for 0. Obtaining "anything" is equally easy.

Now the claim easily follows. □

It is left for the reader to verify that the changes described above guarantee that our proof of Kaufmann and Shelah Theorem still works for $\mathcal{G}(n,p)$, and therefore that the proof of theorem 2 is finished. □

4 Partial orders and K_{m+1}–free graphs

First we improve our proof of Kaufmann and Shelah Theorem to make it work for random partial orders. The theorem we obtain is not new – it has been proved by Compton, as reported in [2].

Theorem 7 (Compton [2]). *There are monadic second-order sentences without asymptotic probability with respect to uniform labelled probability distribution on the class of all finite partial orders.*

Proof. The only problem in our case is to find another representation of computations of Turing machines, as the one found for graphs cannot be embedded in random partial orders. To do so we present now part of the first–order description of a random partial order (i.e., a part of the complete axiomatization of the almost sure theory), after cf. [2]. We assume $\leq$ to be the symbol of the ordering relation, subject to random choice.

With labelled asymptotic probability 1 a partial order will have no chains of length greater than 3. Thus, almost every partial order can be partitioned into 3 levels: L_0, the set of minimal elements, L_1, the set of elements immediately succeeding elements in L_0, and L_2, the set of elements immediately succeeding elements in L_1.

First we add three unary relations to the signature for the levels L_0, L_1, and L_2. Now we formulate extension axioms for random partial orders.

Let $x_1, \ldots, x_m$, $y_1, \ldots, z_k$ and $y_1, \ldots, y_\ell$ be variables, all of them different. Let $S \subseteq \{1, \ldots, m\} \times \{1, \ldots, \ell\}$ be arbitrary. The following formula is an extension axiom, and holds a.s. in a random partial order:

$$(\forall x_1,\ldots,x_m \in L_1)\ (\forall z_1,\ldots,z_k \in L_0)$$
$$[(\bigwedge_{1\le i<j\le m} x_i \ne x_j \wedge \bigwedge_{1\le i<j\le k} z_i \ne z_j) \rightarrow$$
$$((\exists y_1,\ldots,y_\ell \in L_0)\ \bigwedge_{1\le i<j\le \ell} y_i \ne y_j \wedge \bigwedge_{1\le i\le k,\ 1\le j\le \ell} z_i \ne y_j \wedge$$
$$\bigwedge_{(i,j)\in S} x_i \ge y_j \wedge \bigwedge_{(i,j)\notin S} x_i \not\ge y_j)].$$

A formula resulting from the above by interchanging places of L_0 and L_1, and changing $\ge$ into $\le$, is also an axiom.

Now we would like to represent, in a way definable in first–order logic, graphs in random partial orders so that all extension axioms for graphs are a.s. true about these representations.

Namely, a graph **G** on vertex set $U_0 \subseteq L_0$ is represented by a 4–tuple of subsets of a random partial order $\langle U_0, U_1, U_2, U_3\rangle$, such that:

1. $U_0, U_2 \subseteq L_0$; $U_1, U_3 \subseteq L_1$;
2. $\le$ is a bijection from U_0 into U_1 (denoted φ_0);
3. $\ge$ is a bijection from U_1 into U_2 (denoted φ_1);
4. $\le$ is a bijection from U_2 into U_3 (denoted φ_2);
5. for $u, v \in U_0$ let $\varphi_2(\varphi_1(\varphi_0(u))) \ge v$ iff there is an edge from u to v in **G**.

We leave for the reader easy verification that the graph extension axioms of the form $Ext(\mathbf{G},\mathbf{H})$ hold a.s., when we look at their representations in random partial orders. Now our proof of Kaufmann and Shelah result applies here, and the thesis follows immediately. □

In contrast to the previous result, the following is, to the best of author's knowledge, a new one. A $K_{m+1}-$ free graph is a one having no subgraph isomorphic to K_{m+1}, the complete graph on $m+1$ vertices.

Theorem 8. *There are monadic second–order sentences without asymptotic probability with respect to uniform labelled probability distribution on the class of all finite $K_{m+1}-$ free graphs, for $m \ge 2$.*

Proof. Once more we need a part of the first–order description of a random $K_{m+1}-$ free graph, this time provided by Kolaitis, Prömel, and Rothschild [7, 8].

A *spindle* connecting two vertices x and y in a $K_{m+1}-$ free graph is a subgraph isomorphic to K_{m-1} such that all its vertices are adjacent to both x and y. It appears that with labelled asymptotic probability 1 the relation of being connected by a spindle is an equivalence relation of index m, and no two vertices in an equivalence class of this relation are adjacent.

Suppose that the edge relation is denoted by E and that $L_0,\ldots,L_{m-1}$ are the equivalence classes for the spindle connection relation in a random $K_{m+1}-$ free graph.

Let $x_1,\ldots,x_m$, $y_1,\ldots,z_k$ and $y_1,\ldots,y_\ell$ be variables, all of them different. Let $S \subseteq \{1,\ldots,m\} \times \{1,\ldots,\ell\}$ be arbitrary. The following formula is an extension axiom, and holds a.s. in a random $K_{m+1}-$ free graph:

$$(\forall x_1,\ldots,x_m \in L_1)(\forall z_1,\ldots,z_k \in L_0)$$
$$\big[(\textstyle\bigwedge_{1\le i<j\le m} x_i \neq x_j \wedge \bigwedge_{1\le i<j\le k} z_i \neq z_j) \rightarrow$$
$$((\exists y_1,\ldots,y_\ell \in L_0) \textstyle\bigwedge_{1\le i<j\le \ell} y_i \neq y_j \wedge \bigwedge_{1\le i\le k,\ 1\le j\le \ell} z_i \neq y_j$$
$$\textstyle\bigwedge_{(i,j)\in S} E(x_i,y_j) \wedge \bigwedge_{(i,j)\notin S} \neg E(x_i,y_j))\big].$$

A formula resulting from the above by interchanging places of L_0 and L_1 is also an axiom.

Now we can easily observe, that almost identical representation as in random partial orders is appropriate here. The equivalence classes L_2 to L_{m-1} are ignored in our construction. □

5 Monadic second–order logic vs. fixpoint logic in the theory of asymptotic probabilities

In this section we shall investigate connections between behaviours of asymptotic probabilities in fixpoint logic and in monadic logic. It was empirically observed that first–order 0–1 laws sometimes extend to fixpoint 0–1 laws, sometimes to monadic second–order 0–1 laws, but almost never to both of these logics at the same time.

First of all, in order to support the above claim, we recall some results concerning behaviour of asymptotic probabilities in fixpoint and monadic second–order logics. Results for monadic second–order logic can be found, together with references, in [1, 2], while their fixpoint counterparts are to be found in [11, 12]. In all the cases below the first–order 0–1 law holds.

1. Graphs and uniform, labelled or unlabelled probabilities: the fixpoint 0–1 law holds, nonconvergence in monadic logic;
2. Partial orders and uniform, labelled probabilities: the fixpoint 0–1 law holds, nonconvergence in monadic logic;
3. Graphs that are unions of cycles and uniform, unlabelled probabilities: the monadic 0–1 law holds, nonconvergence in fixpoint logic;
4. Random graphs $\mathcal{G}(n,p)$ with recursive $p = p(n)$ such that either for all $\varepsilon > 0$ $n^{-1-\varepsilon} \ll p(n) \ll n^{-1}$ or $n^{-1} \ll p(n) \ll n^{-1}\log n$: the monadic 0–1 law holds, nonconvergence in fixpoint logic;
5. Equivalence relations and uniform, labelled probabilities: both monadic and fixpoint 0–1 laws hold.

The last entry above is essentially degenerated: it is known that both fixpoint and monadic logics collapse to first–order logic on equivalence relations. We will return to this observation later.

As we saw in previous sections, the fact that appropriate extension axioms hold with asymptotic probability one is a strong premise against the convergence law for monadic logic. At the same time, almost all known proofs of 0–1 laws for fixpoint logic are based on extension axioms. Therefore it seems that the 0–1 law for fixpoint logic is a strong premise against the convergence law for monadic logic. As far as we know, this is the first step towards explanation of the observation we mentioned at the beginning of this section.

Of course it would be nice to have some theorem of the form:

"If both fixpoint and monadic 0–1 laws hold for $\langle \mathcal{A}, \mu \rangle$, then ...".

At first glance it seems promising that *"both fixpoint and monadic logics almost surely collapse to first-order logic in $\langle \mathcal{A}, \mu \rangle$"* could be placed as a thesis in the above. Unfortunately, this is not true. Appropriate counterexample is constructed in [12], and cited also in [13]. Only the fixpoint logic is mentioned there, but the 0–1 law for monadic logic can be proved in a way similar to that for the fixpoint logic. So the hypothetic theorem–explanation has to be modified. We suggest to add restriction to *recursive distributions* only, the latter being defined in [13]. The above mentioned counterexample is nonrecursive. To the best of author's knowledge, no known recursive distribution violates the suggested formulation, which follows:

Conjecture 9. *Let $\langle \mathcal{A}, \mu \rangle$ be such that the relation $I_\mu = \{\langle \mathbf{A}, q \rangle \mid \mu_{|\mathbf{A}|}(\{\mathbf{A}\} \geq q\} \subseteq \mathcal{A} \times \mathbb{Q}$ is recursive.*

If both fixpoint and monadic 0–1 laws hold for $\langle \mathcal{A}, \mu \rangle$, then both fixpoint logic and monadic second-order logic a.s. collapse to first-order logic.

For recursive distributions we have a tool which may provide a handle for attacking the conjecture. Namely, it is proved in [13], that if fixpoint 0–1 law holds for a recursive distribution, then fixpoint logic is a.s. bounded, and hence a.s. collapses to first-order logic. Similar property, even for recursive distributions, is not true for monadic logic.

6 Final remarks

6.1 Further results

It can be shown that all our results for labelled distributions are true for unlabelled ones, as well. In fact, in all cases we considered, except the one of sparse random graphs, the extension axioms that hold a.s. for labelled distribution, are also known to hold a.s. for the unlabelled one (see [1]).

Our proof method also applies to: arbitrary classes of relational structures over arbitrary similarity type with at least one at least binary relation symbol (but without constants or functions) with either uniform labelled or uniform unlabelled probabilities, uniform labelled or unlabelled d– complexes for $d > 0$, and most of classes given by parametric conditions with uniform labelled or unlabelled probabilities. Descriptions of these classes, together with references, are to be found in [1].

6.2 Existential monadic logic

Kaufmann in [5] showed that 0–1 law for uniform labelled random graphs fails even for existential monadic logic. Our proof method can also be applied in this case. However, the suitable changes are greater than in previously presented cases.

Their informal description is as follows: replace the subformula $\exists Y\ \gamma(X,Y)$ in the sentence **Ext** by

$$\exists y\ \forall x\ (x \in X \rightarrow E(x,y)).$$

The modified **Ext** is then universal monadic second-order sentence. The existential sentence we then obtain negating the modified **Ext**.

To apply our proof method we have to show the following:

> for every pair $(\mathbf{G},\mathbf{H})$ such that $|\mathbf{H}|-|\mathbf{G}|=1$ and the only vertex of $|\mathbf{H}|\setminus|\mathbf{G}|$ is incident to all vertices of $\mathbf{G}$, the graph of the function $n \mapsto \mu_n(Ext(\mathbf{G},\mathbf{H}))$ has the shape presented on figure 1 at the end of the paper.

The proof can be based on well known combinatorial estimates for random graphs. Then our proof method can be used.

Acknowledgment This paper benefited very much from sugestions of anonymous referees.

References

1. Compton, K.J., 0–1 laws in logic and combinatorics, *Proc. NATO Advanced Study Institute on Algorithms and Order* (I. Rival, ed.), Reidel, Dordrecht (1988).
2. Compton, K.J., The Computational Complexity of Asymptotic Problems I: Partial Orders, *Information and Computation* **78**(1988), pp. 108–123.
3. Erdős, P., and Rényi, A., On the evolution of random graphs, *Magyar Tud. Akad. Mat. Kutató Int. Kőzl.* **5**(1960), pp. 17–61.
4. Fagin, R., Probabilities on finite models, *J. Symbolic Logic* **41**(1976), pp. 50–58.
5. Kaufmann, M. A counterexample to the 0–1 law for existential monadic second-order logic, *CLI Internal Note 32, Computational Logic Inc., Dec. 1987.*
6. Kaufmann, M., Shelah, S., On random models of finite power and monadic logic, *Discrete Math.* **54**(1985), pp. 285–293.
7. Kolaitis, Ph.G., Prõmel, H., and Rothschild, B., Asymptotic enumeration and a 0–1 law for $m-$ clique free graphs, *Bull. Amer. Math. Soc. (N.S.)* **13**(1985), pp. 160–162.
8. Kolaitis, Ph.G., Prõmel, H., and Rothschild, B., $K_{l+1}-$ free graphs: asymptotic structure and a 0–1 law, preprint.
9. Ruciński, A., and Vince, A., Strongly balanced graphs and random graphs, *J. Graph Th.* **10**(1986), pp. 251–264.
10. Shelah, S., Spencer, J., Zero–one law for sparse random graphs, *Journal of the AMS,* **1**(1988), pp. 97–115.
11. Tyszkiewicz, J., Infinitary queries and their asymptotic probabilities I: Properties definable in Transitive Closure Logic *in: E. Börger et al. (eds.), Proc. Computer Science Logic '91,* LNCS 626, Springer Verlag, pp. 396–410.
12. Tyszkiewicz, J., Infinitary queries and their asymptotic probabilities II: Properties definable in Least Fixed Point Logic *to appear in: A. Frieze et al., eds., Proc. Random Graphs '91, Wiley (?).*
13. Tyszkiewicz, J., On asymptotic probabilities in logics that capture DSPACE(log n) in presence of ordering, *to appear in Proc. CAAP '93.*

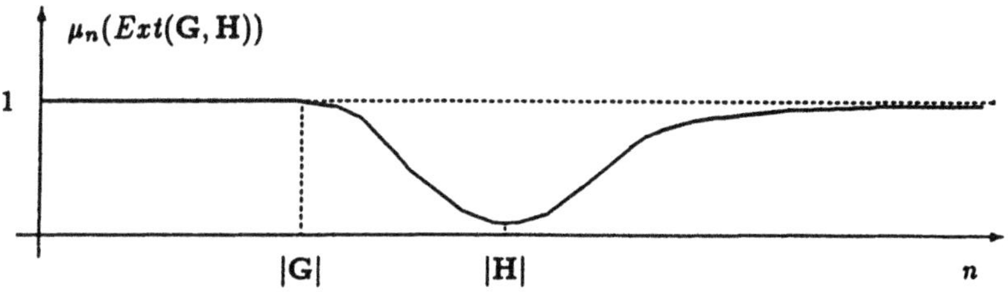

Fig. 1. Graph of the function $n \mapsto \mu_n(Ext(\mathbf{G},\mathbf{H}))$.

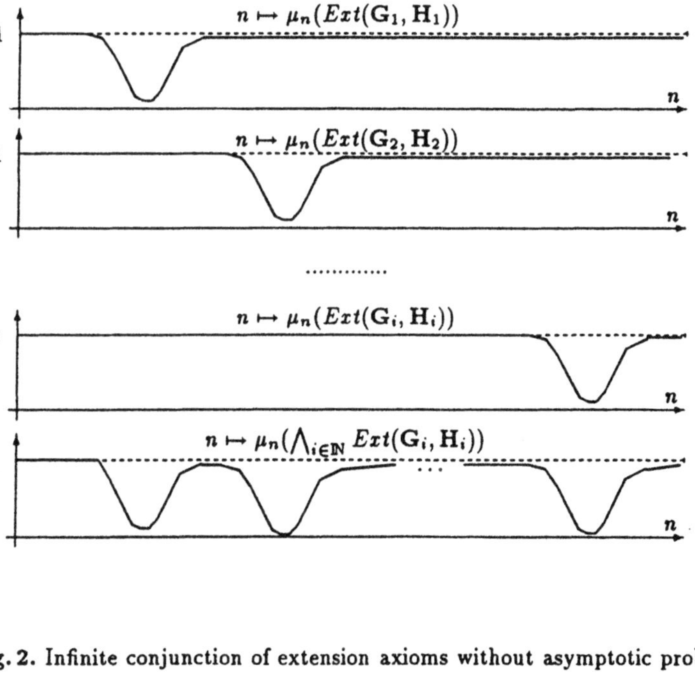

Fig. 2. Infinite conjunction of extension axioms without asymptotic probability

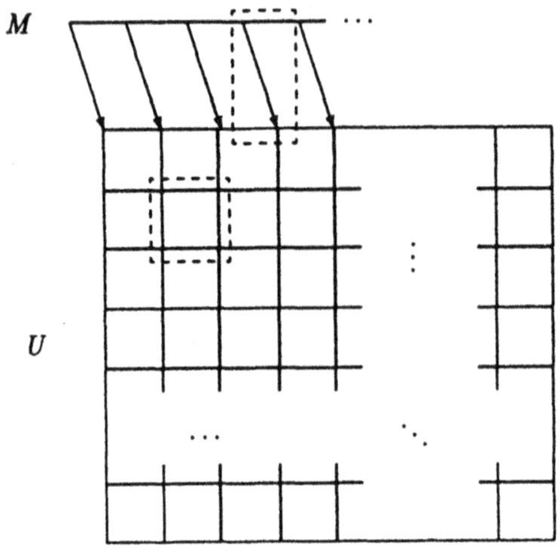

Fig. 3. A grid extension.

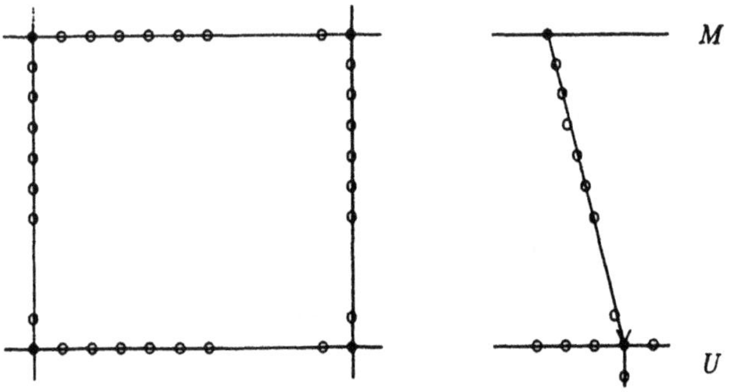

Fig. 4. Details of Fig. 3 *after* improvement. New vertices are denoted by circles, and old ones by filled circles. There are k new vertices on each side of the square. There are $2k$ new vertices on each path from a vertex in M to a root.

Springer-Verlag and the Environment

We at Springer-Verlag firmly believe that an international science publisher has a special obligation to the environment, and our corporate policies consistently reflect this conviction.

We also expect our business partners – paper mills, printers, packaging manufacturers, etc. – to commit themselves to using environmentally friendly materials and production processes.

The paper in this book is made from low- or no-chlorine pulp and is acid free, in conformance with international standards for paper permanency.

Lecture Notes in Computer Science

For information about Vols. 1–629
please contact your bookseller or Springer-Verlag

Vol. 630: W. R. Cleaveland (Ed.), CONCUR '92. Proceedings. X, 580 pages. 1992.

Vol. 631: M. Bruynooghe, M. Wirsing (Eds.), Programming Language Implementation and Logic Programming. Proceedings, 1992. XI, 492 pages. 1992.

Vol. 632: H. Kirchner, G. Levi (Eds.), Algebraic and Logic Programming. Proceedings, 1992. IX, 457 pages. 1992.

Vol. 633: D. Pearce, G. Wagner (Eds.), Logics in AI. Proceedings. VIII, 410 pages. 1992. (Subseries LNAI).

Vol. 634: L. Bougé, M. Cosnard, Y. Robert, D. Trystram (Eds.), Parallel Processing: CONPAR 92 – VAPP V. Proceedings. XVII, 853 pages. 1992.

Vol. 635: J. C. Derniame (Ed.), Software Process Technology. Proceedings, 1992. VIII, 253 pages. 1992.

Vol. 636: G. Comyn, N. E. Fuchs, M. J. Ratcliffe (Eds.), Logic Programming in Action. Proceedings, 1992. X, 324 pages. 1992. (Subseries LNAI).

Vol. 637: Y. Bekkers, J. Cohen (Eds.), Memory Management. Proceedings, 1992. XI, 525 pages. 1992.

Vol. 639: A. U. Frank, I. Campari, U. Formentini (Eds.), Theories and Methods of Spatio-Temporal Reasoning in Geographic Space. Proceedings, 1992. XI, 431 pages. 1992.

Vol. 640: C. Sledge (Ed.), Software Engineering Education. Proceedings, 1992. X, 451 pages. 1992.

Vol. 641: U. Kastens, P. Pfahler (Eds.), Compiler Construction. Proceedings, 1992. VIII, 320 pages. 1992.

Vol. 642: K. P. Jantke (Ed.), Analogical and Inductive Inference. Proceedings, 1992. VIII, 319 pages. 1992. (Subseries LNAI).

Vol. 643: A. Habel, Hyperedge Replacement: Grammars and Languages. X, 214 pages. 1992.

Vol. 644: A. Apostolico, M. Crochemore, Z. Galil, U. Manber (Eds.), Combinatorial Pattern Matching. Proceedings, 1992. X, 287 pages. 1992.

Vol. 645: G. Pernul, A M. Tjoa (Eds.), Entity-Relationship Approach – ER '92. Proceedings, 1992. XI, 439 pages, 1992.

Vol. 646: J. Biskup, R. Hull (Eds.), Database Theory – ICDT '92. Proceedings, 1992. IX, 449 pages. 1992.

Vol. 647: A. Segall, S. Zaks (Eds.), Distributed Algorithms. X, 380 pages. 1992.

Vol. 648: Y. Deswarte, G. Eizenberg, J.-J. Quisquater (Eds.), Computer Security – ESORICS 92. Proceedings. XI, 451 pages. 1992.

Vol. 649: A. Pettorossi (Ed.), Meta-Programming in Logic. Proceedings, 1992. XII, 535 pages. 1992.

Vol. 650: T. Ibaraki, Y. Inagaki, K. Iwama, T. Nishizeki, M. Yamashita (Eds.), Algorithms and Computation. Proceedings, 1992. XI, 510 pages. 1992.

Vol. 651: R. Koymans, Specifying Message Passing and Time-Critical Systems with Temporal Logic. IX, 164 pages. 1992.

Vol. 652: R. Shyamasundar (Ed.), Foundations of Software Technology and Theoretical Computer Science. Proceedings, 1992. XIII, 405 pages. 1992.

Vol. 653: A. Bensoussan, J.-P. Verjus (Eds.), Future Tendencies in Computer Science, Control and Applied Mathematics. Proceedings, 1992. XV, 371 pages. 1992.

Vol. 654: A. Nakamura, M. Nivat, A. Saoudi, P. S. P. Wang, K. Inoue (Eds.), Parallel Image Analysis. Proceedings, 1992. VIII, 312 pages. 1992.

Vol. 655: M. Bidoit, C. Choppy (Eds.), Recent Trends in Data Type Specification. X, 344 pages. 1993.

Vol. 656: M. Rusinowitch, J. L. Rémy (Eds.), Conditional Term Rewriting Systems. Proceedings, 1992. XI, 501 pages. 1993.

Vol. 657: E. W. Mayr (Ed.), Graph-Theoretic Concepts in Computer Science. Proceedings, 1992. VIII, 350 pages. 1993.

Vol. 658: R. A. Rueppel (Ed.), Advances in Cryptology – EUROCRYPT '92. Proceedings, 1992. X, 493 pages. 1993.

Vol. 659: G. Brewka, K. P. Jantke, P. H. Schmitt (Eds.), Nonmonotonic and Inductive Logic. Proceedings, 1991. VIII, 332 pages. 1993. (Subseries LNAI).

Vol. 660: E. Lamma, P. Mello (Eds.), Extensions of Logic Programming. Proceedings, 1992. VIII, 417 pages. 1993. (Subseries LNAI).

Vol. 661: S. J. Hanson, W. Remmele, R. L. Rivest (Eds.), Machine Learning: From Theory to Applications. VIII, 271 pages. 1993.

Vol. 662: M. Nitzberg, D. Mumford, T. Shiota, Filtering, Segmentation and Depth. VIII, 143 pages. 1993.

Vol. 663: G. v. Bochmann, D. K. Probst (Eds.), Computer Aided Verification. Proceedings, 1992. IX, 422 pages. 1993.

Vol. 664: M. Bezem, J. F. Groote (Eds.), Typed Lambda Calculi and Applications. Proceedings, 1993. VIII, 433 pages. 1993.

Vol. 665: P. Enjalbert, A. Finkel, K. W. Wagner (Eds.), STACS 93. Proceedings, 1993. XIV, 724 pages. 1993.

Vol. 666: J. W. de Bakker, W.-P. de Roever, G. Rozenberg (Eds.), Semantics: Foundations and Applications. Proceedings, 1992. VIII, 659 pages. 1993.

Vol. 667: P. B. Brazdil (Ed.), Machine Learning: ECML – 93. Proceedings, 1993. XII, 471 pages. 1993. (Subseries LNAI).

Vol. 668: M.-C. Gaudel, J.-P. Jouannaud (Eds.), TAPSOFT '93: Theory and Practice of Software Development. Proceedings, 1993. XII, 762 pages. 1993.

Vol. 669: R. S. Bird, C. C. Morgan, J. C. P. Woodcock (Eds.), Mathematics of Program Construction. Proceedings, 1992. VIII, 378 pages. 1993.

Vol. 670: J. C. P. Woodcock, P. G. Larsen (Eds.), FME '93: Industrial-Strength Formal Methods. Proceedings, 1993. XI, 689 pages. 1993.

Vol. 671: H. J. Ohlbach (Ed.), GWAI-92: Advances in Artificial Intelligence. Proceedings, 1992. XI, 397 pages. 1993. (Subseries LNAI).

Vol. 672: A. Barak, S. Guday, R. G. Wheeler, The MOSIX Distributed Operating System. X, 221 pages. 1993.

Vol. 673: G. Cohen, T. Mora, O. Moreno (Eds.), Applied Algebra, Algebraic Algorithms and Error-Correcting Codes. Proceedings, 1993. X, 355 pages 1993.

Vol. 674: G. Rozenberg (Ed.), Advances in Petri Nets 1993. VII, 457 pages. 1993.

Vol. 675: A. Mulkers, Live Data Structures in Logic Programs. VIII, 220 pages. 1993.

Vol. 676: Th. H. Reiss, Recognizing Planar Objects Using Invariant Image Features. X, 180 pages. 1993.

Vol. 677: H. Abdulrab, J.-P. Pécuchet (Eds.), Word Equations and Related Topics. Proceedings, 1991. VII, 214 pages. 1993.

Vol. 678: F. Meyer auf der Heide, B. Monien, A. L. Rosenberg (Eds.), Parallel Architectures and Their Efficient Use. Proceedings, 1992. XII, 227 pages. 1993.

Vol. 679: C. Fermüller, A. Leitsch, T. Tammet, N. Zamov, Resolution Methods for the Decision Problem. VIII, 205 pages. 1993. (Subseries LNAI).

Vol. 680: B. Hoffmann, B. Krieg-Brückner (Eds.), Program Development by Specification and Transformation. XV, 623 pages. 1993.

Vol. 681: H. Wansing, The Logic of Information Structures. IX, 163 pages. 1993. (Subseries LNAI).

Vol. 682: B. Bouchon-Meunier, L. Valverde, R. R. Yager (Eds.), IPMU '92 – Advanced Methods in Artificial Intelligence. Proceedings, 1992. IX, 367 pages. 1993.

Vol. 683: G.J. Milne, L. Pierre (Eds.), Correct Hardware Design and Verification Methods. Proceedings, 1993. VIII, 270 Pages. 1993.

Vol. 684: A. Apostolico, M. Crochemore, Z. Galil, U. Manber (Eds.), Combinatorial Pattern Matching. Proceedings, 1993. VIII, 265 pages. 1993.

Vol. 685: C. Rolland, F. Bodart, C. Cauvet (Eds.), Advanced Information Systems Engineering. Proceedings, 1993. XI, 650 pages. 1993.

Vol. 686: J. Mira, J. Cabestany, A. Prieto (Eds.), New Trends in Neural Computation. Proceedings, 1993. XVII, 746 pages. 1993.

Vol. 687: H. H. Barrett, A. F. Gmitro (Eds.), Information Processing in Medical Imaging. Proceedings, 1993. XVI, 567 pages. 1993.

Vol. 688: M. Gauthier (Ed.), Ada-Europe '93. Proceedings, 1993. VIII, 353 pages. 1993.

Vol. 689: J. Komorowski, Z. W. Ras (Eds.), Methodologies for Intelligent Systems. Proceedings, 1993. XI, 653 pages. 1993. (Subseries LNAI).

Vol. 690: C. Kirchner (Ed.), Rewriting Techniques and Applications. Proceedings, 1993. XI, 488 pages. 1993.

Vol. 691: M. Ajmone Marsan (Ed.), Application and Theory of Petri Nets 1993. Proceedings, 1993. IX, 591 pages. 1993.

Vol. 692: D. Abel, B.C. Ooi (Eds.), Advances in Spatial Databases. Proceedings, 1993. XIII, 529 pages. 1993.

Vol. 693: P. E. Lauer (Ed.), Functional Programming, Concurrency, Simulation and Automated Reasoning. Proceedings, 1991/1992. XI, 398 pages. 1993.

Vol. 694: A. Bode, M. Reeve, G. Wolf (Eds.), PARLE '93. Parallel Architectures and Languages Europe. Proceedings, 1993. XVII, 770 pages. 1993.

Vol. 695: E. P. Klement, W. Slany (Eds.), Fuzzy Logic in Artificial Intelligence. Proceedings, 1993. VIII, 192 pages. 1993. (Subseries LNAI).

Vol. 696: M. Worboys, A. F. Grundy (Eds.), Advances in Databases. Proceedings, 1993. X, 276 pages. 1993.

Vol. 697: C. Courcoubetis (Ed.), Computer Aided Verification. Proceedings, 1993. IX, 504 pages. 1993.

Vol. 698: A. Voronkov (Ed.), Logic Programming and Automated Reasoning. Proceedings, 1993. XIII, 386 pages. 1993. (Subseries LNAI).

Vol. 699: G. W. Mineau, B. Moulin, J. F. Sowa (Eds.), Conceptual Graphs for Knowledge Representation. Proceedings, 1993. IX, 451 pages. 1993. (Subseries LNAI).

Vol. 700: A. Lingas, R. Karlsson, S. Carlsson (Eds.), Automata, Languages and Programming. Proceedings, 1993. XII, 697 pages. 1993.

Vol. 701: P. Atzeni (Ed.), LOGIDATA+: Deductive Databases with Complex Objects. VIII, 273 pages. 1993.

Vol. 702: E. Börger, G. Jäger, H. Kleine Büning, S. Martini, M. M. Richter (Eds.), Computer Science Logic. Proceedings, 1992. VIII, 439 pages. 1993.

Vol. 703: M. de Berg, Ray Shooting, Depth Orders and Hidden Surface Removal. X, 201 pages. 1993.

Vol. 704: F. N. Paulisch, The Design of an Extendible Graph Editor. XV, 184 pages. 1993.

Vol. 705: H. Grünbacher, R. W. Hartenstein (Eds.), Field-Programmable Gate Arrays. Proceedings, 1992. VIII, 218 pages. 1993.

Vol. 706: H. D. Rombach, V. R. Basili, R. W. Selby (Eds.), Experimental Software Engineering Issues. Proceedings, 1992. XVIII, 261 pages. 1993.

Vol. 707: O. M. Nierstrasz (Ed.), ECOOP '93 – Object-Oriented Programming. Proceedings, 1993. XI, 531 pages. 1993.